T0189579

# Lecture Notes in Computer Science 13803

More information about this series at https://link.springer.com/bookseries/558

Leonid Karlinsky · Tomer Michaeli ·
Ko Nishino (Eds.)

# Computer Vision – ECCV 2022 Workshops

Tel Aviv, Israel, October 23–27, 2022
Proceedings, Part III

*Editors*
Leonid Karlinsky
IBM Research - MIT-IBM Watson AI Lab
Massachusetts, USA

Tomer Michaeli 🆔
Technion – Israel Institute of Technology
Haifa, Israel

Ko Nishino 🆔
Kyoto University
Kyoto, Japan

ISSN 0302-9743 ISSN 1611-3349 (electronic)
Lecture Notes in Computer Science
ISBN 978-3-031-25065-1 ISBN 978-3-031-25066-8 (eBook)
https://doi.org/10.1007/978-3-031-25066-8

This Springer imprint is published by the registered company Springer Nature Switzerland AG
The registered company address is: Gewerbestrasse 11, 6330 Cham, Switzerland

# Foreword

Organizing the European Conference on Computer Vision (ECCV 2022) in Tel-Aviv during a global pandemic was no easy feat. The uncertainty level was extremely high, and decisions had to be postponed to the last minute. Still, we managed to plan things just in time for ECCV 2022 to be held in person. Participation in physical events is crucial to stimulating collaborations and nurturing the culture of the Computer Vision community.

There were many people who worked hard to ensure attendees enjoyed the best science at the 17th edition of ECCV. We are grateful to the Program Chairs Gabriel Brostow and Tal Hassner, who went above and beyond to ensure the ECCV reviewing process ran smoothly. The scientific program included dozens of workshops and tutorials in addition to the main conference and we would like to thank Leonid Karlinsky and Tomer Michaeli for their hard work. Finally, special thanks to the web chairs Lorenzo Baraldi and Kosta Derpanis, who put in extra hours to transfer information fast and efficiently to the ECCV community.

We would like to express gratitude to our generous sponsors and the Industry Chairs Dimosthenis Karatzas and Chen Sagiv, who oversaw industry relations and proposed new ways for academia-industry collaboration and technology transfer. It's great to see so much industrial interest in what we're doing!

Authors' draft versions of the papers appeared online with open access on both the Computer Vision Foundation (CVF) and the European Computer Vision Association (ECVA) websites as with previous ECCVs. Springer, the publisher of the proceedings, has arranged for archival publication. The final version of the papers is hosted by SpringerLink, with active references and supplementary materials. It benefits all potential readers that we offer both a free and citeable version for all researchers, as well as an authoritative, citeable version for SpringerLink readers. Our thanks go to Ronan Nugent from Springer, who helped us negotiate this agreement. Last but not least, we wish to thank Eric Mortensen, our publication chair, whose expertise made the process smooth.

October 2022

Rita Cucchiara
Jiří Matas
Amnon Shashua
Lihi Zelnik-Manor

# Preface

Welcome to the workshop proceedings of the 17th European Conference on Computer Vision (ECCV 2022). This year, the main ECCV event was accompanied by 60 workshops, scheduled between October 23–24, 2022. We received 103 workshop proposals on diverse computer vision topics and unfortunately had to decline many valuable proposals because of space limitations. We strove to achieve a balance between topics, as well as between established and new series. Due to the uncertainty associated with the COVID-19 pandemic around the proposal submission deadline, we allowed two workshop formats: hybrid and purely online. Some proposers switched their preferred format as we drew near the conference dates. The final program included 30 hybrid workshops and 30 purely online workshops. Not all workshops published their papers in the ECCV workshop proceedings, or had papers at all. These volumes collect the edited papers from 38 out of the 60 workshops. We sincerely thank the ECCV general chairs for trusting us with the responsibility for the workshops, the workshop organizers for their hard work in putting together exciting programs, and the workshop presenters and authors for contributing to ECCV.

October 2022

Tomer Michaeli
Leonid Karlinsky
Ko Nishino

# Organization

## General Chairs

| | |
|---|---|
| Rita Cucchiara | University of Modena and Reggio Emilia, Italy |
| Jiří Matas | Czech Technical University in Prague, Czech Republic |
| Amnon Shashua | Hebrew University of Jerusalem, Israel |
| Lihi Zelnik-Manor | Technion – Israel Institute of Technology, Israel |

## Program Chairs

| | |
|---|---|
| Shai Avidan | Tel-Aviv University, Israel |
| Gabriel Brostow | University College London, UK |
| Giovanni Maria Farinella | University of Catania, Italy |
| Tal Hassner | Facebook AI, USA |

## Program Technical Chair

| | |
|---|---|
| Pavel Lifshits | Technion – Israel Institute of Technology, Israel |

## Workshops Chairs

| | |
|---|---|
| Leonid Karlinsky | IBM Research - MIT-IBM Watson AI Lab, USA |
| Tomer Michaeli | Technion – Israel Institute of Technology, Israel |
| Ko Nishino | Kyoto University, Japan |

## Tutorial Chairs

| | |
|---|---|
| Thomas Pock | Graz University of Technology, Austria |
| Natalia Neverova | Facebook AI Research, UK |

## Demo Chair

| | |
|---|---|
| Bohyung Han | Seoul National University, South Korea |

## Social and Student Activities Chairs

| | |
|---|---|
| Tatiana Tommasi | Italian Institute of Technology, Italy |
| Sagie Benaim | University of Copenhagen, Denmark |

## Diversity and Inclusion Chairs

Xi Yin                          Facebook AI Research, USA
Bryan Russell                   Adobe, USA

## Communications Chairs

Lorenzo Baraldi                 University of Modena and Reggio Emilia, Italy
Kosta Derpanis                  York University and Samsung AI Centre Toronto,
                                Canada

## Industrial Liaison Chairs

Dimosthenis Karatzas            Universitat Autònoma de Barcelona, Spain
Chen Sagiv                      SagivTech, Israel

## Finance Chair

Gerard Medioni                  University of Southern California and Amazon,
                                USA

## Publication Chair

Eric Mortensen                  MiCROTEC, USA

## Workshops Organizers

**W01 - AI for Space**

Tat-Jun Chin                    The University of Adelaide, Australia
Luca Carlone                    Massachusetts Institute of Technology, USA
Djamila Aouada                  University of Luxembourg, Luxembourg
Binfeng Pan                     Northwestern Polytechnical University, China
Viorela Ila                     The University of Sydney, Australia
Benjamin Morrell                NASA Jet Propulsion Lab, USA
Grzegorz Kakareko               Spire Global, USA

**W02 - Vision for Art**

Alessio Del Bue                 Istituto Italiano di Tecnologia, Italy
Peter Bell                      Philipps-Universität Marburg, Germany
Leonardo L. Impett              École Polytechnique Fédérale de Lausanne
                                (EPFL), Switzerland
Noa Garcia                      Osaka University, Japan
Stuart James                    Istituto Italiano di Tecnologia, Italy

## W03 - Adversarial Robustness in the Real World

| | |
|---|---|
| Angtian Wang | Johns Hopkins University, USA |
| Yutong Bai | Johns Hopkins University, USA |
| Adam Kortylewski | Max Planck Institute for Informatics, Germany |
| Cihang Xie | University of California, Santa Cruz, USA |
| Alan Yuille | Johns Hopkins University, USA |
| Xinyun Chen | University of California, Berkeley, USA |
| Judy Hoffman | Georgia Institute of Technology, USA |
| Wieland Brendel | University of Tübingen, Germany |
| Matthias Hein | University of Tübingen, Germany |
| Hang Su | Tsinghua University, China |
| Dawn Song | University of California, Berkeley, USA |
| Jun Zhu | Tsinghua University, China |
| Philippe Burlina | Johns Hopkins University, USA |
| Rama Chellappa | Johns Hopkins University, USA |
| Yinpeng Dong | Tsinghua University, China |
| Yingwei Li | Johns Hopkins University, USA |
| Ju He | Johns Hopkins University, USA |
| Alexander Robey | University of Pennsylvania, USA |

## W04 - Autonomous Vehicle Vision

| | |
|---|---|
| Rui Fan | Tongji University, China |
| Nemanja Djuric | Aurora Innovation, USA |
| Wenshuo Wang | McGill University, Canada |
| Peter Ondruska | Toyota Woven Planet, UK |
| Jie Li | Toyota Research Institute, USA |

## W05 - Learning With Limited and Imperfect Data

| | |
|---|---|
| Noel C. F. Codella | Microsoft, USA |
| Zsolt Kira | Georgia Institute of Technology, USA |
| Shuai Zheng | Cruise LLC, USA |
| Judy Hoffman | Georgia Institute of Technology, USA |
| Tatiana Tommasi | Politecnico di Torino, Italy |
| Xiaojuan Qi | The University of Hong Kong, China |
| Sadeep Jayasumana | University of Oxford, UK |
| Viraj Prabhu | Georgia Institute of Technology, USA |
| Yunhui Guo | University of Texas at Dallas, USA |
| Ming-Ming Cheng | Nankai University, China |

## W06 - Advances in Image Manipulation

| | |
|---|---|
| Radu Timofte | University of Würzburg, Germany, and ETH Zurich, Switzerland |
| Andrey Ignatov | AI Benchmark and ETH Zurich, Switzerland |
| Ren Yang | ETH Zurich, Switzerland |
| Marcos V. Conde | University of Würzburg, Germany |
| Furkan Kınlı | Özyeğin University, Turkey |

## W07 - Medical Computer Vision

| | |
|---|---|
| Tal Arbel | McGill University, Canada |
| Ayelet Akselrod-Ballin | Reichman University, Israel |
| Vasileios Belagiannis | Otto von Guericke University, Germany |
| Qi Dou | The Chinese University of Hong Kong, China |
| Moti Freiman | Technion, Israel |
| Nicolas Padoy | University of Strasbourg, France |
| Tammy Riklin Raviv | Ben Gurion University, Israel |
| Mathias Unberath | Johns Hopkins University, USA |
| Yuyin Zhou | University of California, Santa Cruz, USA |

## W08 - Computer Vision for Metaverse

| | |
|---|---|
| Bichen Wu | Meta Reality Labs, USA |
| Peizhao Zhang | Facebook, USA |
| Xiaoliang Dai | Facebook, USA |
| Tao Xu | Facebook, USA |
| Hang Zhang | Meta, USA |
| Péter Vajda | Facebook, USA |
| Fernando de la Torre | Carnegie Mellon University, USA |
| Angela Dai | Technical University of Munich, Germany |
| Bryan Catanzaro | NVIDIA, USA |

## W09 - Self-Supervised Learning: What Is Next?

| | |
|---|---|
| Yuki M. Asano | University of Amsterdam, The Netherlands |
| Christian Rupprecht | University of Oxford, UK |
| Diane Larlus | Naver Labs Europe, France |
| Andrew Zisserman | University of Oxford, UK |

## W10 - Self-Supervised Learning for Next-Generation Industry-Level Autonomous Driving

| | |
|---|---|
| Xiaodan Liang | Sun Yat-sen University, China |
| Hang Xu | Huawei Noah's Ark Lab, China |

Fisher Yu                          ETH Zürich, Switzerland
Wei Zhang                          Huawei Noah's Ark Lab, China
Michael C. Kampffmeyer             UiT The Arctic University of Norway, Norway
Ping Luo                           The University of Hong Kong, China

## W11 - ISIC Skin Image Analysis

M. Emre Celebi                     University of Central Arkansas, USA
Catarina Barata                    Instituto Superior Técnico, Portugal
Allan Halpern                      Memorial Sloan Kettering Cancer Center, USA
Philipp Tschandl                   Medical University of Vienna, Austria
Marc Combalia                      Hospital Clínic of Barcelona, Spain
Yuan Liu                           Google Health, USA

## W12 - Cross-Modal Human-Robot Interaction

Fengda Zhu                         Monash University, Australia
Yi Zhu                             Huawei Noah's Ark Lab, China
Xiaodan Liang                      Sun Yat-sen University, China
Liwei Wang                         The Chinese University of Hong Kong, China
Xiaojun Chang                      University of Technology Sydney, Australia
Nicu Sebe                          University of Trento, Italy

## W13 - Text in Everything

Ron Litman                         Amazon AI Labs, Israel
Aviad Aberdam                      Amazon AI Labs, Israel
Shai Mazor                         Amazon AI Labs, Israel
Hadar Averbuch-Elor                Cornell University, USA
Dimosthenis Karatzas               Universitat Autònoma de Barcelona, Spain
R. Manmatha                        Amazon AI Labs, USA

## W14 - BioImage Computing

Jan Funke                          HHMI Janelia Research Campus, USA
Alexander Krull                    University of Birmingham, UK
Dagmar Kainmueller                 Max Delbrück Center, Germany
Florian Jug                        Human Technopole, Italy
Anna Kreshuk                       EMBL-European Bioinformatics Institute,
                                   Germany
Martin Weigert                     École Polytechnique Fédérale de Lausanne
                                   (EPFL), Switzerland
Virginie Uhlmann                   EMBL-European Bioinformatics Institute, UK

| | |
|---|---|
| Peter Bajcsy | National Institute of Standards and Technology, USA |
| Erik Meijering | University of New South Wales, Australia |

## W15 - Visual Object-Oriented Learning Meets Interaction: Discovery, Representations, and Applications

| | |
|---|---|
| Kaichun Mo | Stanford University, USA |
| Yanchao Yang | Stanford University, USA |
| Jiayuan Gu | University of California, San Diego, USA |
| Shubham Tulsiani | Carnegie Mellon University, USA |
| Hongjing Lu | University of California, Los Angeles, USA |
| Leonidas Guibas | Stanford University, USA |

## W16 - AI for Creative Video Editing and Understanding

| | |
|---|---|
| Fabian Caba | Adobe Research, USA |
| Anyi Rao | The Chinese University of Hong Kong, China |
| Alejandro Pardo | King Abdullah University of Science and Technology, Saudi Arabia |
| Linning Xu | The Chinese University of Hong Kong, China |
| Yu Xiong | The Chinese University of Hong Kong, China |
| Victor A. Escorcia | Samsung AI Center, UK |
| Ali Thabet | Reality Labs at Meta, USA |
| Dong Liu | Netflix Research, USA |
| Dahua Lin | The Chinese University of Hong Kong, China |
| Bernard Ghanem | King Abdullah University of Science and Technology, Saudi Arabia |

## W17 - Visual Inductive Priors for Data-Efficient Deep Learning

| | |
|---|---|
| Jan C. van Gemert | Delft University of Technology, The Netherlands |
| Nergis Tömen | Delft University of Technology, The Netherlands |
| Ekin Dogus Cubuk | Google Brain, USA |
| Robert-Jan Bruintjes | Delft University of Technology, The Netherlands |
| Attila Lengyel | Delft University of Technology, The Netherlands |
| Osman Semih Kayhan | Bosch Security Systems, The Netherlands |
| Marcos Baptista Ríos | Alice Biometrics, Spain |
| Lorenzo Brigato | Sapienza University of Rome, Italy |

## W18 - Mobile Intelligent Photography and Imaging

| | |
|---|---|
| Chongyi Li | Nanyang Technological University, Singapore |
| Shangchen Zhou | Nanyang Technological University, Singapore |

| Ruicheng Feng | Nanyang Technological University, Singapore |
| Jun Jiang | SenseBrain Research, USA |
| Wenxiu Sun | SenseTime Group Limited, China |
| Chen Change Loy | Nanyang Technological University, Singapore |
| Jinwei Gu | SenseBrain Research, USA |

## W19 - People Analysis: From Face, Body and Fashion to 3D Virtual Avatars

| Alberto Del Bimbo | University of Florence, Italy |
| Mohamed Daoudi | IMT Nord Europe, France |
| Roberto Vezzani | University of Modena and Reggio Emilia, Italy |
| Xavier Alameda-Pineda | Inria Grenoble, France |
| Marcella Cornia | University of Modena and Reggio Emilia, Italy |
| Guido Borghi | University of Bologna, Italy |
| Claudio Ferrari | University of Parma, Italy |
| Federico Becattini | University of Florence, Italy |
| Andrea Pilzer | NVIDIA AI Technology Center, Italy |
| Zhiwen Chen | Alibaba Group, China |
| Xiangyu Zhu | Chinese Academy of Sciences, China |
| Ye Pan | Shanghai Jiao Tong University, China |
| Xiaoming Liu | Michigan State University, USA |

## W20 - Safe Artificial Intelligence for Automated Driving

| Timo Saemann | Valeo, Germany |
| Oliver Wasenmüller | Hochschule Mannheim, Germany |
| Markus Enzweiler | Esslingen University of Applied Sciences, Germany |
| Peter Schlicht | CARIAD, Germany |
| Joachim Sicking | Fraunhofer IAIS, Germany |
| Stefan Milz | Spleenlab.ai and Technische Universität Ilmenau, Germany |
| Fabian Hüger | Volkswagen Group Research, Germany |
| Seyed Ghobadi | University of Applied Sciences Mittelhessen, Germany |
| Ruby Moritz | Volkswagen Group Research, Germany |
| Oliver Grau | Intel Labs, Germany |
| Frédérik Blank | Bosch, Germany |
| Thomas Stauner | BMW Group, Germany |

## W21 - Real-World Surveillance: Applications and Challenges

| Kamal Nasrollahi | Aalborg University, Denmark |
| Sergio Escalera | Universitat Autònoma de Barcelona, Spain |

Radu Tudor Ionescu            University of Bucharest, Romania
Fahad Shahbaz Khan            Mohamed bin Zayed University of Artificial
                              Intelligence, United Arab Emirates
Thomas B. Moeslund            Aalborg University, Denmark
Anthony Hoogs                 Kitware, USA
Shmuel Peleg                  The Hebrew University, Israel
Mubarak Shah                  University of Central Florida, USA

## W22 - Affective Behavior Analysis In-the-Wild

Dimitrios Kollias             Queen Mary University of London, UK
Stefanos Zafeiriou            Imperial College London, UK
Elnar Hajiyev                 Realeyes, UK
Viktoriia Sharmanska          University of Sussex, UK

## W23 - Visual Perception for Navigation in Human Environments: The JackRabbot Human Body Pose Dataset and Benchmark

Hamid Rezatofighi             Monash University, Australia
Edward Vendrow                Stanford University, USA
Ian Reid                      University of Adelaide, Australia
Silvio Savarese               Stanford University, USA

## W24 - Distributed Smart Cameras

Niki Martinel                 University of Udine, Italy
Ehsan Adeli                   Stanford University, USA
Rita Pucci                    University of Udine, Italy
Animashree Anandkumar         Caltech and NVIDIA, USA
Caifeng Shan                  Shandong University of Science and Technology,
                              China
Yue Gao                       Tsinghua University, China
Christian Micheloni           University of Udine, Italy
Hamid Aghajan                 Ghent University, Belgium
Li Fei-Fei                    Stanford University, USA

## W25 - Causality in Vision

Yulei Niu                     Columbia University, USA
Hanwang Zhang                 Nanyang Technological University, Singapore
Peng Cui                      Tsinghua University, China
Song-Chun Zhu                 University of California, Los Angeles, USA
Qianru Sun                    Singapore Management University, Singapore
Mike Zheng Shou               National University of Singapore, Singapore
Kaihua Tang                   Nanyang Technological University, Singapore

## W26 - In-Vehicle Sensing and Monitorization

| | |
|---|---|
| Jaime S. Cardoso | INESC TEC and Universidade do Porto, Portugal |
| Pedro M. Carvalho | INESC TEC and Polytechnic of Porto, Portugal |
| João Ribeiro Pinto | Bosch Car Multimedia and Universidade do Porto, Portugal |
| Paula Viana | INESC TEC and Polytechnic of Porto, Portugal |
| Christer Ahlström | Swedish National Road and Transport Research Institute, Sweden |
| Carolina Pinto | Bosch Car Multimedia, Portugal |

## W27 - Assistive Computer Vision and Robotics

| | |
|---|---|
| Marco Leo | National Research Council of Italy, Italy |
| Giovanni Maria Farinella | University of Catania, Italy |
| Antonino Furnari | University of Catania, Italy |
| Mohan Trivedi | University of California, San Diego, USA |
| Gérard Medioni | Amazon, USA |

## W28 - Computational Aspects of Deep Learning

| | |
|---|---|
| Iuri Frosio | NVIDIA, Italy |
| Sophia Shao | University of California, Berkeley, USA |
| Lorenzo Baraldi | University of Modena and Reggio Emilia, Italy |
| Claudio Baecchi | University of Florence, Italy |
| Frederic Pariente | NVIDIA, France |
| Giuseppe Fiameni | NVIDIA, Italy |

## W29 - Computer Vision for Civil and Infrastructure Engineering

| | |
|---|---|
| Joakim Bruslund Haurum | Aalborg University, Denmark |
| Mingzhu Wang | Loughborough University, UK |
| Ajmal Mian | University of Western Australia, Australia |
| Thomas B. Moeslund | Aalborg University, Denmark |

## W30 - AI-Enabled Medical Image Analysis: Digital Pathology and Radiology/COVID-19

| | |
|---|---|
| Jaime S. Cardoso | INESC TEC and Universidade do Porto, Portugal |
| Stefanos Kollias | National Technical University of Athens, Greece |
| Sara P. Oliveira | INESC TEC, Portugal |
| Mattias Rantalainen | Karolinska Institutet, Sweden |
| Jeroen van der Laak | Radboud University Medical Center, The Netherlands |
| Cameron Po-Hsuan Chen | Google Health, USA |

| | |
|---|---|
| Diana Felizardo | IMP Diagnostics, Portugal |
| Ana Monteiro | IMP Diagnostics, Portugal |
| Isabel M. Pinto | IMP Diagnostics, Portugal |
| Pedro C. Neto | INESC TEC, Portugal |
| Xujiong Ye | University of Lincoln, UK |
| Luc Bidaut | University of Lincoln, UK |
| Francesco Rundo | STMicroelectronics, Italy |
| Dimitrios Kollias | Queen Mary University of London, UK |
| Giuseppe Banna | Portsmouth Hospitals University, UK |

### W31 - Compositional and Multimodal Perception

| | |
|---|---|
| Kazuki Kozuka | Panasonic Corporation, Japan |
| Zelun Luo | Stanford University, USA |
| Ehsan Adeli | Stanford University, USA |
| Ranjay Krishna | University of Washington, USA |
| Juan Carlos Niebles | Salesforce and Stanford University, USA |
| Li Fei-Fei | Stanford University, USA |

### W32 - Uncertainty Quantification for Computer Vision

| | |
|---|---|
| Andrea Pilzer | NVIDIA, Italy |
| Martin Trapp | Aalto University, Finland |
| Arno Solin | Aalto University, Finland |
| Yingzhen Li | Imperial College London, UK |
| Neill D. F. Campbell | University of Bath, UK |

### W33 - Recovering 6D Object Pose

| | |
|---|---|
| Martin Sundermeyer | DLR German Aerospace Center, Germany |
| Tomáš Hodaň | Reality Labs at Meta, USA |
| Yann Labbé | Inria Paris, France |
| Gu Wang | Tsinghua University, China |
| Lingni Ma | Reality Labs at Meta, USA |
| Eric Brachmann | Niantic, Germany |
| Bertram Drost | MVTec, Germany |
| Sindi Shkodrani | Reality Labs at Meta, USA |
| Rigas Kouskouridas | Scape Technologies, UK |
| Ales Leonardis | University of Birmingham, UK |
| Carsten Steger | Technical University of Munich and MVTec, Germany |
| Vincent Lepetit | École des Ponts ParisTech, France, and TU Graz, Austria |
| Jiří Matas | Czech Technical University in Prague, Czech Republic |

## W34 - Drawings and Abstract Imagery: Representation and Analysis

| | |
|---|---|
| Diane Oyen | Los Alamos National Laboratory, USA |
| Kushal Kafle | Adobe Research, USA |
| Michal Kucer | Los Alamos National Laboratory, USA |
| Pradyumna Reddy | University College London, UK |
| Cory Scott | University of California, Irvine, USA |

## W35 - Sign Language Understanding

| | |
|---|---|
| Liliane Momeni | University of Oxford, UK |
| Gül Varol | École des Ponts ParisTech, France |
| Hannah Bull | University of Paris-Saclay, France |
| Prajwal K. R. | University of Oxford, UK |
| Neil Fox | University College London, UK |
| Ben Saunders | University of Surrey, UK |
| Necati Cihan Camgöz | Meta Reality Labs, Switzerland |
| Richard Bowden | University of Surrey, UK |
| Andrew Zisserman | University of Oxford, UK |
| Bencie Woll | University College London, UK |
| Sergio Escalera | Universitat Autònoma de Barcelona, Spain |
| Jose L. Alba-Castro | Universidade de Vigo, Spain |
| Thomas B. Moeslund | Aalborg University, Denmark |
| Julio C. S. Jacques Junior | Universitat Autònoma de Barcelona, Spain |
| Manuel Vázquez Enríquez | Universidade de Vigo, Spain |

## W36 - A Challenge for Out-of-Distribution Generalization in Computer Vision

| | |
|---|---|
| Adam Kortylewski | Max Planck Institute for Informatics, Germany |
| Bingchen Zhao | University of Edinburgh, UK |
| Jiahao Wang | Max Planck Institute for Informatics, Germany |
| Shaozuo Yu | The Chinese University of Hong Kong, China |
| Siwei Yang | Hong Kong University of Science and Technology, China |
| Dan Hendrycks | University of California, Berkeley, USA |
| Oliver Zendel | Austrian Institute of Technology, Austria |
| Dawn Song | University of California, Berkeley, USA |
| Alan Yuille | Johns Hopkins University, USA |

## W37 - Vision With Biased or Scarce Data

| | |
|---|---|
| Kuan-Chuan Peng | Mitsubishi Electric Research Labs, USA |
| Ziyan Wu | United Imaging Intelligence, USA |

**W38 - Visual Object Tracking Challenge**

Matej Kristan                   University of Ljubljana, Slovenia
Aleš Leonardis                  University of Birmingham, UK
Jiří Matas                      Czech Technical University in Prague,
                                   Czech Republic
Hyung Jin Chang                 University of Birmingham, UK
Joni-Kristian Kämäräinen        Tampere University, Finland
Roman Pflugfelder               Technical University of Munich, Germany,
                                   Technion, Israel, and Austrian Institute of
                                   Technology, Austria
Luka Čehovin Zajc               University of Ljubljana, Slovenia
Alan Lukežič                    University of Ljubljana, Slovenia
Gustavo Fernández               Austrian Institute of Technology, Austria
Michael Felsberg                Linköping University, Sweden
Martin Danelljan                ETH Zurich, Switzerland

# Contents – Part III

# W06 - Advances in Image Manipulation: Reports

# W06 - Advances in Image Manipulation: Reports

Image manipulation is a key computer vision task, aiming at the restoration of degraded image content, the filling in of missing information, or the needed transformation and/or manipulation to achieve a desired target (with respect to perceptual quality, contents, or performance of apps working on such images). Recent years have witnessed an increased interest from the vision and graphics communities in these fundamental topics of research. Not only has there been a constantly growing flow of related papers but also substantial progress has been achieved.

Each step forward eases the use of images by people or computers for the fulfillment of further tasks, as image manipulation serves as an important frontend. Not surprisingly then, there is an ever growing range of applications in fields such as surveillance, the automotive industry, electronics, remote sensing, medical image analysis, etc. The emergence and ubiquitous use of mobile and wearable devices offer another fertile ground for additional applications and faster methods.

This workshop aimed to provide an overview of the new trends and advances in those areas. Moreover, it offered an opportunity for academic and industrial attendees to interact and explore collaborations.

October 2022

Radu Timofte
Andrey Ignatov
Ren Yang
Marcos V. Conde
Furkan Kınlı

# Reversed Image Signal Processing and RAW Reconstruction. AIM 2022 Challenge Report

Marcos V. Conde(✉), Radu Timofte, Yibin Huang, Jingyang Peng,
Chang Chen, Cheng Li, Eduardo Pérez-Pellitero, Fenglong Song, Furui Bai,
Shuai Liu, Chaoyu Feng, Xiaotao Wang, Lei Lei, Yu Zhu, Chenghua Li,
Yingying Jiang, Yong A, Peisong Wang, Cong Leng, Jian Cheng, Xiaoyu Liu,
Zhicun Yin, Zhilu Zhang, Junyi Li, Ming Liu, Wangmeng Zuo, Jun Jiang,
Jinha Kim, Yue Zhang, Beiji Zou, Zhikai Zong, Xiaoxiao Liu,
Juan Marín Vega, Michael Sloth, Peter Schneider-Kamp, Richard Röttger,
Furkan Kınlı, Barış Özcan, Furkan Kıraç, Li Leyi, S. M. Nadim Uddin,
Dipon Kumar Ghosh, and Yong Ju Jung

Computer Vision Lab, CAIDAS, University of Würzburg, Würzburg, Germany
{marcos.conde-osorio,radu.timofte}@uni-wuerzburg.de
https://data.vision.ee.ethz.ch/cvl/aim22/,
https://github.com/mv-lab/AISP/

**Abstract.** Cameras capture sensor RAW images and transform them
into pleasant RGB images, suitable for the human eyes, using their integrated Image Signal Processor (ISP). Numerous low-level vision tasks
operate in the RAW domain (*e.g.* image denoising, white balance) due
to its linear relationship with the scene irradiance, wide-range of information at 12bits, and sensor designs. Despite this, RAW image datasets
are scarce and more expensive to collect than the already large and public RGB datasets. This paper introduces the AIM 2022 Challenge on
Reversed Image Signal Processing and RAW Reconstruction. We aim to
recover raw sensor images from the corresponding RGBs without metadata and, by doing this, "reverse" the ISP transformation. The proposed
methods and benchmark establish the state-of-the-art for this low-level
vision inverse problem, and generating realistic raw sensor readings can
potentially benefit other tasks such as denoising and super-resolution.

**Keywords:** Computational photography · Image signal processing ·
Image synthesis · Inverse problems · Low-level vision · Raw images

## 1 Introduction

The majority of low-level vision and computational photography tasks use RGB
images obtained from the in-camera Image Signal Processor (ISP) [13] that con-
verts the camera sensor's raw readings into perceptually pleasant RGB images,
suitable for the human visual system. One of the reasons is the accessibility and
amount of RGB datasets. Multiple approaches have been proposed to model the
RAW to RGB transformation (*i.e.* ISP) using deep neural networks. We can
highlight FlexISP [19], DeepISP [40] and PyNET [22, 25, 26].

© The Author(s), under exclusive license to Springer Nature Switzerland AG 2023
L. Karlinsky et al. (Eds.): ECCV 2022 Workshops, LNCS 13803, pp. 3–26, 2023.
https://doi.org/10.1007/978-3-031-25066-8_1

However, the characteristics of raw camera's sensor data (*e.g.* linear relationship with scene irradiance at each in 12–14 bits, unprocessed signal and noise samples) are often better suited for the inverse problems that commonly arise in low-level vision tasks such as denoising, deblurring, super-resolution [2,6,17,38]. Professional photographers also commonly choose to process RAW images by themselves to produce images with better visual effects [28]. Unfortunately, RAW image datasets are not as common and diverse as their RGB counterparts, therefore, CNN-based approaches might not reach their full potential. To bridge this gap, we introduce the AIM Reversed ISP Challenge and review current solutions. We can find metadata-based raw reconstruction methods for de-render or unprocess the RGB image back to its original raw values [34,35,37], these methods usually require specific metadata stored as an overhead (*e.g.* 64 KB). Similarly, UPI [8] proposes a generic camera ISP model composed of five canonical and invertible steps [15,28]. This simple approach can produce realistic raw sensor data for training denoising models [8]. However, it requires specific camera parameters (*e.g.* correction matrices, digital gains) that are generally inaccessible.

Learning-based approaches [3,13,36,45,47] attempt to learn the ISP, and how to reverse it, in an end-to-end manner. CycleISP [47] uses a 2-branch model to learn the RAW2RGB and RGB2RAW transformations. InvISP [45] models the camera ISP as an invertible function using normalizing-flows [31]. These learned methods are essentially non-interpretable black-boxes. MBISPLD [13] proposes a novel learnable, invertible and interpretable model-based ISP, based on classical designs [15,28]. In a more complex domain, Arad *et al.* proposed the recovery of whole-scene hyperspectral (HS) information from a RGB image [5].

Despite other approaches explored how to recover RAW information [8,35–37], to the best of our knowledge, this is the first work to analyze exhaustively this novel (or at least, not so studied) inverse problem.

**Our Contribution.** In this paper, we introduce the first benchmark for learned raw sensor image reconstruction. We use data from 4 different smartphone sensors and analyze 14 different learned methods (*i.e.* no metadata or prior information about the camera is required).

This challenge is one of the AIM 2022 associated challenges: reversed ISP [14], efficient learned ISP [25], super-resolution of compressed image and video [46], efficient image super-resolution [20], efficient video super-resolution [21], efficient Bokeh effect rendering [23], efficient monocular depth estimation [24], Instagram filter removal [32] (Fig. 1).

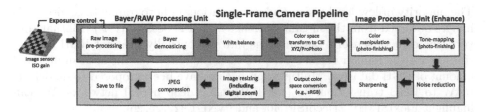

**Fig. 1.** Classical Image Signal Processor (ISP). Image from Delbracio *et al.* [15].

## 2   AIM 2022 Reversed ISP Challenge

The objectives of the AIM 2022 Reversed ISP Challenge are: (i) to propose the first challenge and benchmark for RAW image reconstruction and Reversed Image Signal Processing; (ii) to establish the state-of-the-art in this low-level vision inverse problem; (iii) to compare different contemporary solutions and analyze their drawbacks. The challenge consists on two tracks: Track 1 for the analysis of the *Samsung S7* sensor (Sony IMX260); Track 2 for the analysis of the *Huawei P20 Pro* sensor (Sony IMX380). Both tracks are structured in the same way: participants use the provided training datasets and baseline[1], and submit their results (*i.e.* generated RAW images) through an online server with a public leaderboard[2] where they can compare with other methods. Finalists must submit reproducible code and a detailed factsheet describing their solution. **Overview:** 80 unique participants registered at the challenge, with 11 teams competing in the final testing phase. More than 200 submissions were received.

### 2.1   Datasets

We aim to promote research and reproducibility in this area, for this reason, we select public available datasets for training and testing the methods. Moreover, we select smartphone's datasets, as these are, by design, more complex ISPs than DSLR cameras [15,28] and therefore more challenging to model and "reverse".

- Track 1. Samsung **S7** Dataset from DeepISP [40]. Raw and processed RGB images of the 110 different scenes captured by the phone camera with normal exposure time. The images have resolution 3024×4032 (12MP Sony IMX260). Image pairs are aligned enough to use pixel-wise metrics.
- Track 2. ETH Huawei **P20** Pro Dataset (PyNET) [26]. RAW and RGB images obtained with Huawei's built-in ISP (12.3 MP Sony Exmor IMX380). The photos were captured in automatic mode, and default settings were used throughout the whole collection procedure. Original images have resolution 3840 × 5120. However, the pairs of raw sensor data and RGB have a notable misalignment (over 8px).
- SSID Dataset [1] clean RAW-RGB image pairs of the Samsung Galaxy **S6** Edge phone (16MP Sony IMX240), captured under normal brightness.
- RAW-to-RAW Dataset [4]. Authors provide natural scene RAW and camera-ISP rendered RGB pairs using the Samsung Galaxy **S9** (Sony IMX345).

Datasets are manually filtered. Considering the original full-resolution (FR) images, we unify RAWs to a common RGGB pattern, next, we extract non-overlapping crops (504 × 504 in track 1, 496 × 496 in track 2) and save them as 4-channel 10-bit images in .npy format. Some teams used directly FR images.

---

[1] https://github.com/mv-lab/AISP/.
[2] https://codalab.lisn.upsaclay.fr/competitions/5079.

## 2.2  Evaluation and Results

We use 20% of the S6 dataset [40] and P20 [26] as validation and test sets for each track. Since the provided S6 and P20 training and testing ("Test1") sets are public, we use an additional internal test set ("Test2") not disclosed until the end of the challenge. For the Track 1 (S7), the internal test contains additional images from the previously explained S9 [4] and S6 [1] datasets. For the Track 2 (P20) we use extra noisy images from [26]. This allows us to evaluate the generalization capabilities and robustness of the proposed solutions.

Table 1 represents the challenge benchmark [14]. The best 3 approaches achieve above 30 dB PSNR, which denotes a great reconstruction of the original RAW readings captured by the camera sensor. Attending to "Test2", the proposed methods can generalize and produce realistic RAW data for unseen similar sensors. For instance, models trained using S7 [40] data, can produce realistic RAWs for S9 [4] and S6 [1] (previous and next-generation sensors).

Attending to the PSNR and SSIM values, we can see a clear fidelity-perception tradeoff [7] $i.e.$ at the "Test2" of Track 2, the top solutions scored the same SSIM and notably different PSNR values. To explore this, we provide extensive qualitative comparisons of the proposed methods in the Appendix A.

**Table 1.** AIM Reversed ISP Challenge Benchmark. Teams are ranked based on their performance on Test1 and Test2, an internal test set to evaluate the generalization capabilities and robustness of the proposed solutions. The methods (*) have trained using extra data from [40], and therefore only results on the internal datasets are relevant. CycleISP [47] was reported by multiple participants

| Team | Track 1 (Samsung S7) | | | | Track 2 (Huawei P20) | | | |
| | Test1 | | Test2 | | Test1 | | Test2 | |
| name | PSNR ↑ | SSIM ↑ | PSNR ↑ | SSIM ↑ | PSNR ↑ | SSIM ↑ | PSNR ↑ | SSIM ↑ |
|---|---|---|---|---|---|---|---|---|
| NOAHTCV | 31.86 | 0.83 | 32.69 | 0.88 | 38.38 | 0.93 | 35.77 | 0.92 |
| MiAlgo | 31.39 | 0.82 | 30.73 | 0.80 | 40.06 | 0.93 | 37.09 | 0.92 |
| CASIA LCVG (*) | 30.19 | 0.81 | 31.47 | 0.86 | 37.58 | 0.93 | 33.99 | 0.92 |
| HIT-IIL | 29.12 | 0.80 | 30.22 | 0.87 | 36.53 | 0.91 | 34.25 | 0.90 |
| SenseBrains | 28.36 | 0.80 | 30.08 | 0.86 | 35.47 | 0.92 | 32.63 | 0.91 |
| CS^2U (*) | 29.13 | 0.79 | 29.95 | 0.84 | - | - | - | - |
| HiImage | 27.96 | 0.79 | - | - | 34.40 | 0.94 | 32.13 | 0.90 |
| 0noise | 27.67 | 0.79 | 29.81 | 0.87 | 33.68 | 0.90 | 31.83 | 0.89 |
| OzU VGL | 27.89 | 0.79 | 28.83 | 0.83 | 32.72 | 0.87 | 30.69 | 0.86 |
| PixelJump | 28.15 | 0.80 | - | - | - | - | - | - |
| CVIP | 27.85 | 0.80 | 29.50 | 0.86 | - | - | - | - |
| CycleISP [47] | 26.75 | 0.78 | - | - | 32.70 | 0.85 | - | - |
| UPI [8] | 26.90 | 0.78 | - | - | - | - | - | - |
| U-Net Base | 26.30 | 0.77 | - | - | 30.01 | 0.80 | - | - |

In Sect. 3, we introduce the solutions from each team. All the proposed methods are learned-based and do not require metadata or specific camera parameters. Multiple approaches can serve as a plug-in augmentation.

# 3   Proposed Methods and Teams

The complete information about the teams can be consulted in the Appendix B.

## 3.1   NOAHTCV

*Yibin Huang, Jingyang Peng, Chang Chen, Cheng Li, Eduardo Pérez-Pellitero, and Fenglong Song*

This method is illustrated in Fig. 2, it is based on MBISPLD [13], a model-based interpretable approach. The only remarkable differences are (i) a modified inverse tone mapping block and (ii) LocalNet, a custom CNN added on top to improve performance in exchange of losing invertibility and efficiency. The "inverse tone mapping" is formulated as a convex combination of several monotonic curves, parameterized in a polynomial form. Note that these curves are fixed and not learned together with the model. The "LocalNet" is designed to predict conditions for feature modulation, and can be viewed as a variant of local mapping. The output maintains fine details and inverts only tone and color. Authors emphasize in their ablation studies that this LocalNet contributes significantly to improve performance. The team adopts Half Instance Normalization (HIN) [10] as a basic component in this network. **Ablation studies:** The original MBISPLD [13] achieved 34.02 dB using 0.54M parameters (8 GMACs) and a runtime of 14 ms, after adding the custom inverse tone mapping and LocalNet, the model can achieve 36.20 dB (improvement of 2 dB) in exchange of being ×10 more complex (80 GMACs, 5.7M parameters) and ×2 slower. Authors also compare with other methods such as UPI [8], DRN [50] and ESRGAN [43]; the base method MBISPLD [13] and their proposed modification achieved the best results.

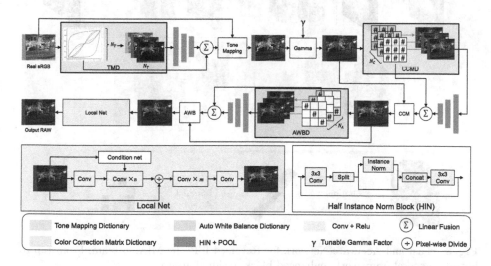

**Fig. 2.** NOAHTCV method overview, inspired in MBISPLD [13].

## 3.2  MiAlgo

*Furui Bai, Shuai Liu, Chaoyu Feng, Xiaotao Wang, Lei Lei*

The proposed method has an end-to-end encoder-decoder structure, illustrated in Fig. 3. The approach is a lightweight structure that exploits the powerful capabilities of information encoding-decoding, and therefore, it is able to handle full-resolution (FR) inputs efficiently. The UNet-like structure consists of many sampling blocks and residual groups to obtain deep features. The main components are the "residual group" [47] and the "enhanced block" [13], these can be visualized Fig. 3. Authors emphasize that since many modules in the ISP are based on full-resolution (*e.g.* global tone mapping, auto white balance, and lens shading correction), FR training and testing improves performance in comparison to patch-based approaches. The method was implemented in PyTorch 1.8, and was trained roughly 13 h on 4× V100 GPU using FR images, L1 loss and SSIM loss, and batch size 1. Authors correct misalignments and black leve of RAW images. We must highlight that the method is able to **process 4K images in 18 ms** in a single V100 GPU. Moreover, it generalizes successfully to noisy inputs and other unseen sensors (see Table 1). The code will be open-sourced.

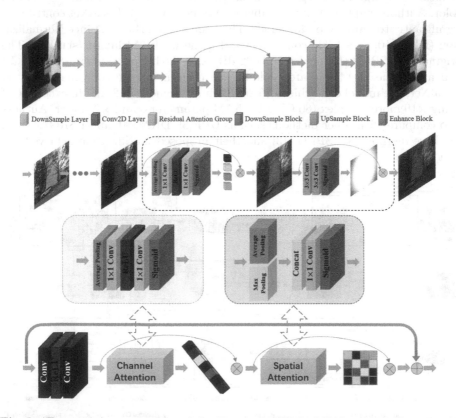

**Fig. 3.** "Fast and Accurate Network for Reversed ISP" by MiAlgo team. From top to bottom: general framework, enhanced block, residual group.

### 3.3    CASIA LCVG

*Yu Zhu, Chenghua Li, Cong Leng, Jian Cheng*

The team proposes a new reversed ISP network (RISPNet) [16] to achieve efficient RGB to RAW mapping. This is a novel encoder-decoder network with a third-order attention (TOA) module. Since attention [42,44] facilitates the complex trade-off between achieving spatial detail and high level contextual information, it can help to recover high dynamic range RAW and thus making the whole recovery process a more manageable step. The architecture of RISPNet is illustrated in Fig. 4, it consists of multi-scale hierarchical design incorporating the basic RISPblock, which is a residual attention block combining layer normalization, depth-wise convolutions and channel-spatial attention [9,44].

We must note that the team uses the DeepISP S7 dataset [40], besides the proposed challenge dataset, this implies that **the results are not accurate** for this method, since they have potentially trained with extra validation data, nevertheless this should not affect the results in the "Test2" sets or "Track2".

The method was implemented in Pytorch 1.10, and was trained roughly 2 days on 8× A100 GPU using cropped images, AdamW, L1 loss and batch size 32. For data augmentation, the team uses horizontal and vertical flips.

The model is relatively efficient, it can process a 504px RGBs in 219 ms using self-ensembles (*i.e.* fuse 8 inputs obtained by flip and rotate operations), which improves their performance by 0.2 dB. Note that this model has 464M parameters, and therefore, it is only suitable for offline RAW synthesis.

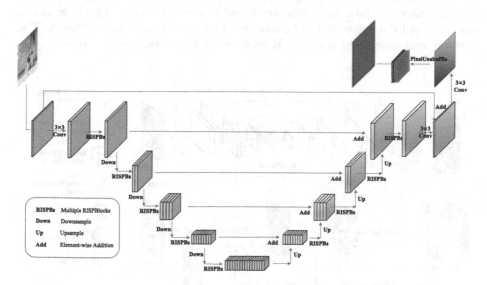

**Fig. 4.** RISPNet by CASIA LCVG.

## 3.4   HIT-IIL

*Xiaoyu Liu, Zhicun Yin, Zhilu Zhang, Junyi Li, Ming Liu, Wangmeng Zuo*

The team's solution is "Learning Reverse ISP with Inaccurately Aligned Supervision", inspired by their previous work [51]. In order to decrease the burden of the Reverse ISP Network, they take advantage of the traditional inverse tone mapping and gamma expansion [8,13]. The pipeline of the method is shown in Fig. 5. Note that the team adapts the solution for each track. At Track 1, since the misalignment is small, they use a 9M parameters model (LiteISP 8.88M, GCM 0.04M). Considering the spatial misalignment of inputs (RGBs) and targets (RAWs), they adopt a pre-trained PWC-Net [41] to warp the target RAW images aligned with RGB inputs. GCM [51] is also applied to diminish the effect of color inconsistency in the image alignment. However, since GCM alone is not enough to handle the issue satisfactorily, they use the color histogram to match the GCM output with sRGB images for better color mapping. Based on this framework, they also adopt Spectral Normalization Layer (SN) to improve model's stability and Test-time Local Converter [12] to alleviate the inconsistency of image sizes between training and testing. By doing this, they can effectively improve performance. At Track 2, due to the lens shading effect and strong misalignment, they use larger patches of 1536px randomly sampled from the FR image to train the model. Moreover they replace LiteISPNet [51] with NAFNet [9] as RevISPNet (see Fig. 5), this increases model's complexity to 116M parameters. The method was implemented in PyTorch and trained using 2× V100 GPU for 5 days using cropped images, AdamW and batch size 6. Flip and rotation ×8 data augmentation is used during training. The inference is done using FR images and self-ensembles to improve color inconsistency. The authors emphasize that general FR inference is better than patch-based.

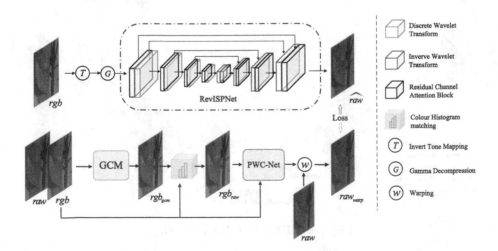

**Fig. 5.** Solution from HIT-IIL team, based on LiteISPNet [51] and NAFNet [9].

## 3.5 CS^2U

*Yue Zhang, Beiji Zou*

The team proposes "Learned Reverse ISP with soft supervision" [53], an efficient network and a new loss function for the reversed ISP learning. They propose an encoder-decoder architecture called **SSDNet** inspired by the Transformer [9,42] architecture and is straightforward. The method is illustrated in Fig. 6. They exploit the unique property of RAW data, i.e. high dynamic range (HDR), by suggesting to relax the supervision to a multivariate Gaussian distribution in order to learn images that are reasonable for a given supervision. Equipped with the above two components, the method achieves an effective RGB to RAW mapping. We must note that the team uses the DeepISP S7 dataset [40], besides the proposed challenge dataset, this implies that **the results are not accurate** for this method, since they have potentially trained with extra validation data, nevertheless this should not affect the results in the "Test2" sets or "Track2".

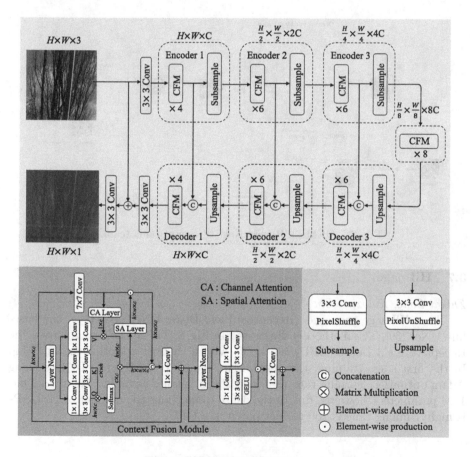

**Fig. 6.** SSDNet by CS^2U team.

### 3.6   SenseBrains

*Jun Jiang, Jinha Kim*

The team proposes "OEMask-ReverseNet" [27], a multi-step refinement process integrating an overexposure mask. The key points are: instead of mapping RGB to RAW (Bayer pattern), the pipeline trains from RGB to demosaiced RAW [17]; the multi-step process [8] of this reverse ISP has greatly enhanced the performance of the baseline U-Net; the refinement pipeline can enhance other "reverse ISPs". The method is illustrated in Fig. 8: (i) unprocess the input RGB [8] and obtain the demosaiced output [17]; (ii) using two independent U-Net [39] networks (`U-Net(OE)`, `U-Net(NOE)`), estimate an overexposure mask; (iii) apply the mask to the previously obtained demosaiced RGB and finally mosaic it in a 4-channel (RGGB) RAW. They also propose a network to refine the predicted RAWs in the YUV space. The team implemented the method in Pytorch and trained using a combination of LPIPS [49], MS-SSIM-L1 and L2 loss (Fig. 7).

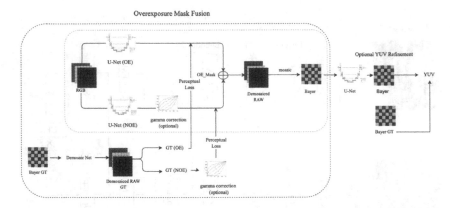

**Fig. 7.** "OEMask ReverseNet" by SenseBrains team.

### 3.7   HiImage

*Zhikai Zong, Xiaoxiao Liu*

Team HiImage proposed "Receptive Field Dense UNet" based on the baseline UNet [39]. The method is illustrated in Fig. 8. The authors augmented the Receptive Field Dense (RFD) block [50] with a channel attention module [44]. In the first stage training, the network was trained to minimize L1 loss using Adam optimizer, learning rate 1e-4, and 256px patches as input. The model has 11M parameters, however, due to the dense connections the number of GFLOPs is high, running time is 200 ms per patch in a GPU.

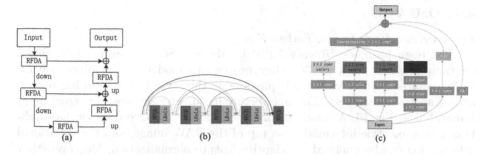

**Fig. 8.** "Receptive Field Dense UNet" by HiImage team.

### 3.8  0noise

*Juan Marín Vega, Michael Sloth, Peter Schneider-Kamp, Richard Röttger*

The team introduces RISP, a Reversed ISP network for RGB to RAW mapping. Authors combine model-based [13] approaches and custom layers for an end-to-end RGB o RAW mapping network. In particular, they utilize Tone Mapping and Lens Correction layers from MBISPLD [13], and combine it with a custom Color Shift and a Mosaicing Layer. The method is illustrated in Fig. 9. The tone mapping, color shift and lens shading are applied by a point-wise multiplication. The mosaicing layer produces an output of $\frac{h}{2}\frac{w}{2}$ with a $2 \times 2$ convolution with stride 2. The method is implemented using Pytorch. The network was trained using Adam for 200 epochs and learning rate $1e - 4$, and $250 \times 250$ patches. Authors use a combination of the L1 and VGG19 losses. We must highlight that this model is the **smallest among the proposed ones**, the model has only 0.17M parameters, it is fast and able to process 4K images under a second. Based on internal experiments, authors emphasize that there is no clear benefit in using a computationally heavy network as CycleISP [47] in comparison to lighter or model-based approaches [13]. The code will be open-sourced.

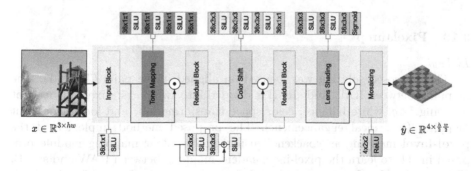

**Fig. 9.** "RISP" approach by 0noise team.

## 3.9   OzU VGL

*Furkan Kınlı, Barış Özcan, Furkan Kıraç*

The team proposes "Reversed ISP by Reverse Style Transfer" [33]. Based on their previous work [29,30], they propose to model the non-linear operations required to reverse the ISP pipeline as the style factor by using reverse style transferring strategy. The architecture has an encoder-decoder structure as illustrated in Fig. 10. Authors assume the final sRGB output image has additional injected style information on top of the RAW image, and this additional information can be removed by adaptive feature normalization. Moreover, they employ a wavelet-based discriminator network, which provides a discriminative regularization to the final RAW output of the main network. This architecture has 86.3M parameters, mostly due to the style projectors for each residual block. This method was implemented in PyTorch and trained for 52 epochs using patches, Adam with the learning rate of $1e - 4$. The code will be open-sourced.

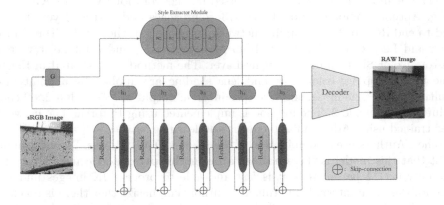

**Fig. 10.** "Reversed ISP by Reverse Style Transfer" approach by OzU VGL.

## 3.10   PixelJump

*Li Leyi*

The proposed solution was focused on aligned image pairs (Track 1). The mapping between RAW and sRGB is divided into two parts: pixel-independent mapping (*e.g.* white balance, color space transformation) and local-dependent mapping (*e.g.* local engancement). The proposed method implemented the **pixel-level** mapping as "backend" using global colour mapping module proposed in [11] to learn the pixel-independent mapping between RAW and sRGB. Then, the **local enhancement** module (encoder-decoder [52] and ResNet structure [18]) is added to deal with the local-dependent mapping. Ultimately, the outputs of these three modules were fused by a set of learned weights from the weight predictor inspired by [48]. The method is implemented in PyTorch and trained using 8× RTX 3090 GPU, and batch size 2; it processes a patch in 40 ms on GPU (Fig. 11).

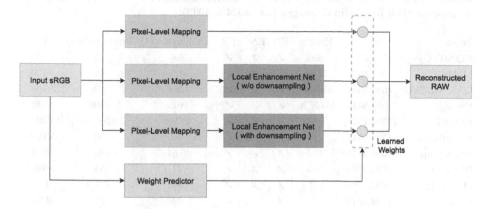

**Fig. 11.** PixelJump method overview.

## 3.11 CVIP

*SM Nadim Uddin, Dipon Kumar Ghosh, Yong Ju Jung*

The team proposes "Reverse ISP Estimation using Recurrent Dense Channel-Spatial Attention". This method is illustrated in Fig. 12. Authors compare their method directly with CycleISP [47] and improve considerably (+5 dB). The method was implemented in PyTorch 1.8 and trained for 2 days using a RTX

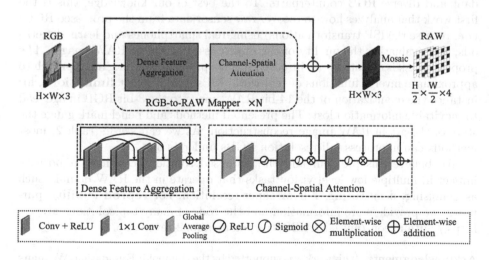

**Fig. 12.** CVIP method overview.

**Table 2.** Team information summary. "Input" refers to the input image size used during training, most teams used the provided patches (*i.e.* 504px or 496px). ED indicates the use of "Extra Datasets" besides the provided challenge datasets. ENS indicates if the solution is an "Ensemble" of multiple models. FR indicates if the model can process "Full-Resolution" images (*i.e.* 3024 × 4032)

| Team | Input | Epochs | ED | ENS | FR | # Params. (M) | Runtime (ms) | GPU |
|------|-------|--------|----|-----|----|----|----|-----|
| NOAHTCV | (504,504) | 500 | ✗ | ✗ | ✓ | 5.6 | 25 | V100 |
| MiAlgo | (3024,4032) | 3000 | ✗ | ✗ | ✓ | 4.5 | 18 | V100 |
| CASIA LCVG | (504,504) | 300K it. | ✓ | ✓ | ✓ | 464 | 219 | A100 |
| CS^2U | (504,504) | 276K it. | ✓ | ✓ | ✓ | 105 | 1300 | 3090 |
| HIT-IIL | (1536,1536) | 1000 | ✗ | ✗ | ✓ | 9/116 | 19818 (cpu) | V100 |
| SenseBrains | (504,504) | 220 | ✗ | ✓ | ✓ | 69 | 50 | V100 |
| PixelJump | (504,504) | 400 | ✗ | ✓ | ✓ | 6.64 | 40 | 3090 |
| HiImage | (256, 256) | 600 | ✗ | ✗ | ✓ | 11 | 200 | 3090 |
| OzU VGL | (496, 496) | 52 | ✗ | ✗ | ✓ | 86 | 6 | 2080 |
| CVIP | (504,504) | 75 | ✗ | ✗ | ✓ | 2.8 | 400 | 3090 |
| 0noise | (504,504) | 200 | ✗ | ✗ | ✓ | 0.17 | 19 | Q6000 |

3090 GPU. MAD and SSIM were used as loss functions. The method has only 2.8M parameters (724 GFLOP) and it can unprocess a 252px RGB in 10 ms (using GPU). This method can generalize to unseen sensors as shown in Table 1, however, authors detected a bias towards indoor images.

# 4   Conclusions

The AIM 2022 Reversed ISP Challenge promotes a novel direction of research aiming at solving the availability of RAW image datasets by leveraging the abundant and diverse RGB counterparts. To the best of our knowledge, this is the first work that analyzes how to recover raw sensor data from the processed RGBs (*i.e.* reverse the ISP transformations) using real smartphones and learned methods, and evaluating the quality of the unprocessed synthetic RAW images. The proposed task is essentially an ill-posed inverse problem, learning an approach to approximate inverse functions of real-world ISPs have implicit **limitations**, for instance the quantization of the 14-bit RAW image to the 8-bit RGB image lead to inevitable information lost. The presented methods and benchmark gauge the state-of-the-art in RAW image reconstruction. As we can see in Table 2, most methods can unprocess full-resolution RGBs under 1s on GPU.

We believe this problem and the proposed solutions can have a positive impact in multiple low-level vision tasks that operate in the RAW domain such as denoising, demosaicing, hdr or super-resolution. For **reproducibility** purposes, we provide more information and the code of many top solutions at: https://github.com/mv-lab/AISP/tree/main/aim22-reverseisp.

**Acknowledgments.** This work was supported by the Humboldt Foundation. We thank the sponsors of the AIM and Mobile AI 2022 workshops and challenges: AI Witchlabs, MediaTek, Huawei, Reality Labs, OPPO, Synaptics, Raspberry Pi, ETH Zürich (Computer Vision Lab) and University of Würzburg (Computer Vision Lab). We also thank Andrey Ignatov and Eli Schwartz for their datasets used in this challenge.

## A    Appendix 1: Qualitative Results

As we pointed out in Sect. 2.2, we find a clear fidelity-perception tradeoff [7] in the solutions, *i.e.* some methods achieve high PSNR values, yet the generated RAW images do no look realistic. In this report, due to the visualization and space constraints, we only include samples from the best 8 ranked teams in Table 1. We provide dditional qualitative comparisons including all the teams, and supplementary material at:
https://github.com/mv-lab/AISP/tree/main/aim22-reverseisp.

The following Figs. 13, 14, 15, 16 show a complete comparison between the most competitive methods in Track 1 and 2. Overall, the RAW reconstruction quality is high, however, we can appreciate clear colour and texture differences between the methods. Note that the RAW images are visualized using a simple

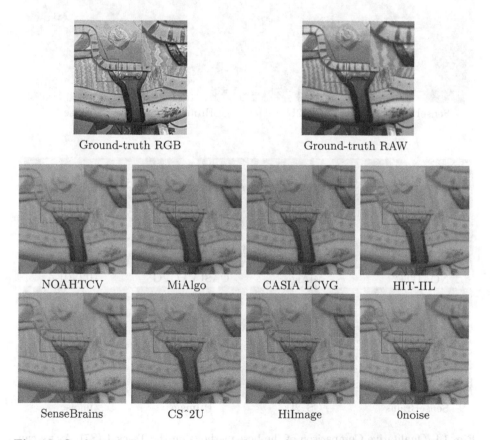

**Fig. 13.** Qualitative Comparison on the Track 1 (S7). We can appreciate that most methods recover accurately colors, yet struggle at recovering light intensity.

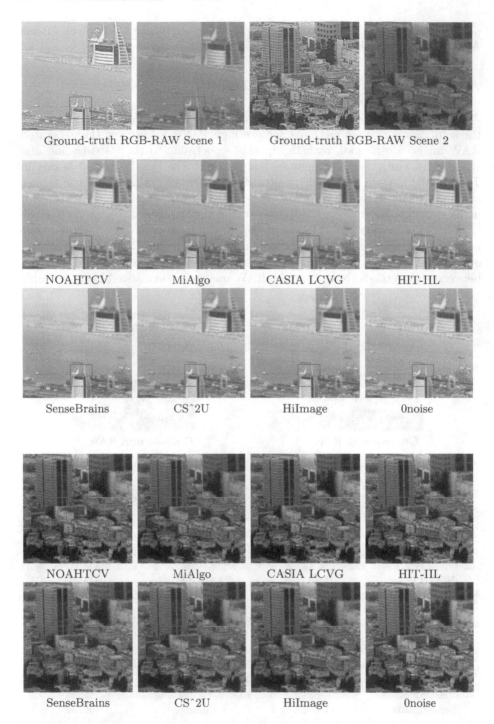

Ground-truth RGB-RAW Scene 1          Ground-truth RGB-RAW Scene 2

NOAHTCV          MiAlgo          CASIA LCVG          HIT-IIL

SenseBrains          CS^2U          HiImage          0noise

NOAHTCV          MiAlgo          CASIA LCVG          HIT-IIL

SenseBrains          CS^2U          HiImage          0noise

**Fig. 14.** Qualitative Comparison of the best methods on the Track 1 (S7). As we can see, these methods can recover detailed RAW images without artifacts.

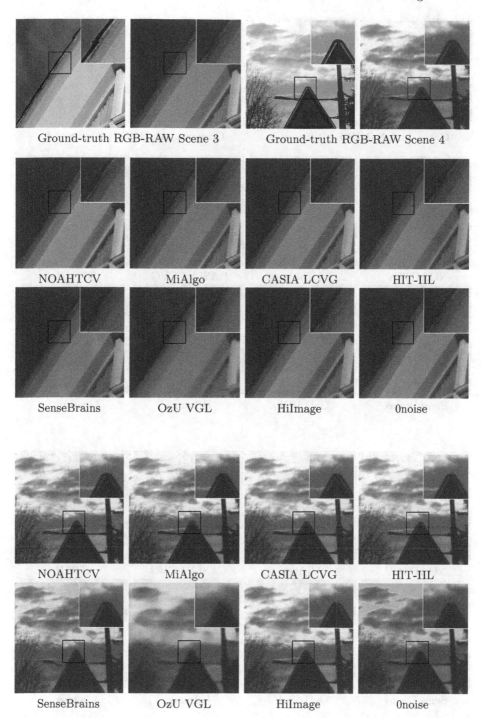

**Fig. 15.** Qualitative Comparison on the Track 2 (P20). Details and intensities are recovered even for these non-aligned images (i.e. slightly overexposed sky).

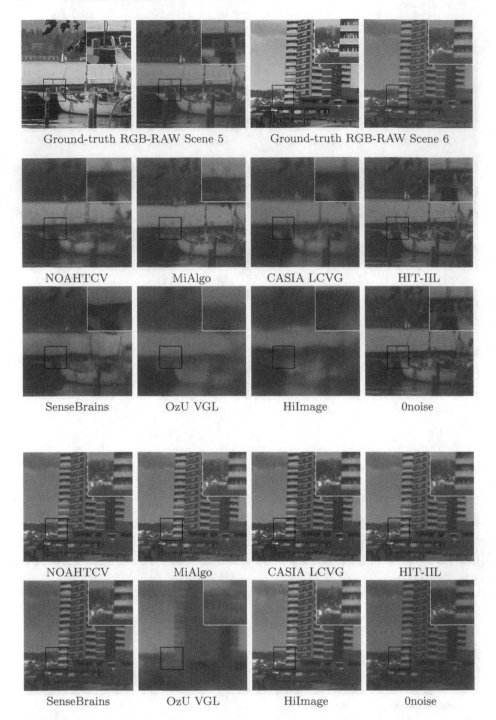

**Fig. 16.** Qualitative Comparison of the top performing methods on Track 2 (P20). In these challenging non-aligned samples we can appreciate clear differences between methods, many cannot recover high-frequencies and produce blur artifacts.

green-average demosaic, tone mapping and gamma operators. Moreover, in the case of the Track 2 (P20) images are strongly misaligned.

# B Appendix 2: Teams and Affiliations

## NOAHTCV

**Members:** Yibin Huang, Jingyang Peng, Chang Chen, Cheng Li, Eduardo Pérez-Pellitero, and Fenglong Song

**Affiliations:** Huawei Noah's Ark Lab

**Contact:** songfenglong@huawei.com

## MiAlgo

**Members:** Furui Bai, Shuai Liu, Chaoyu Feng, Xiaotao Wang, Lei Lei

**Affiliations:** Xiaomi Inc., China

**Contact:** baifurui@xiaomi.com

## CASIA LCVG

**Members:** Yu Zhu, Chenghua Li, Yingying Jiang, Yong A, Peisong Wang, Cong Leng, Jian Cheng

**Affiliations:** Institute of Automation, Chinese Academy of Sciences; MAICRO; AiRiA

**Contact:** zhuyu.cv@gmail.com

## HIT-IIL

**Members:** Xiaoyu Liu, Zhicun Yin, Zhilu Zhang, Junyi Li, Ming Liu, Wangmeng Zuo

**Affiliations:** Harbin Institute of Technology, China

**Contact:** liuxiaoyu1104@gmail.com

## SenseBrains

**Members:** Jun Jiang, Jinha Kim

**Affiliations:** SenseBrain Technology

**Contact:** jinhakim@sensebrain.site, jinhakim@mit.edu

## CS²U

**Members:** Yue Zhang, Beiji Zou

**Affiliations:** School of Computer Science and Engineering, Central South University; Hunan Engineering Research Center of Machine Vision and Intelligent Medicine.

**Contact:** yuezhang@csu.edu.cn

**HiImage**

**Members:** Zhikai Zong, Xiaoxiao Liu

**Affiliations:** Qingdao Hi-image Technologies Co., Ltd. (Hisense Visual Technology Co., Ltd.)

**Contact:** zzksdu@163.com

**0noise**

**Members:** Juan Marín Vega, Michael Sloth, Peter Schneider-Kamp, Richard Röttger

**Affiliations:** University of Southern Denmark, Esoft Systems

**Contact:** marin@imada.sdu.dk

**OzU VGL**

**Members:** Furkan Kınlı, Barış Özcan, Furkan Kıraç

**Affiliations:** Özyeğin University

**Contact:** furkan.kinli@ozyegin.edu.tr

**PixelJump**

**Members:** Li Leyi

**Affiliations:** Zhejiang University

**Contact:** lileyi@zju.edu.cn

**CVIP**

**Members:** SM Nadim Uddin, Dipon Kumar Ghosh, Yong Ju Jung

**Affiliations:** Computer Vision and Image Processing (CVIP) Lab, School of Computing, Gachon University.

**Contact:** smnadimuddin@gmail.com

## References

1. Abdelhamed, A., Lin, S., Brown, M.S.: A high-quality denoising dataset for smartphone cameras. In: Proceedings of the IEEE Conference on Computer Vision and Pattern Recognition, pp. 1692–1700 (2018)
2. Abdelhamed, A., Timofte, R., Brown, M.S.: NTIRE 2019 challenge on real image denoising: methods and results. In: Proceedings of the IEEE/CVF Conference on Computer Vision and Pattern Recognition Workshops (2019)
3. Afifi, M., Abdelhamed, A., Abuolaim, A., Punnappurath, A., Brown, M.S.: CIE XYZ Net: unprocessing images for low-level computer vision tasks. IEEE Trans. Pattern Anal. Mach. Intell. **44**(9), 4688–4700 (2021)
4. Afifi, M., Abuolaim, A.: Semi-supervised raw-to-raw mapping. arXiv preprint arXiv:2106.13883 (2021)
5. Arad, B., Timofte, R., Ben-Shahar, O., Lin, Y.T., Finlayson, G.D.: NTIRE 2020 challenge on spectral reconstruction from an RGB image. In: Proceedings of the IEEE/CVF Conference on Computer Vision and Pattern Recognition Workshops, pp. 446–447 (2020)

6. Bhat, G., Danelljan, M., Timofte, R.: NTIRE 2021 challenge on burst super-resolution: methods and results. In: Proceedings of the IEEE/CVF Conference on Computer Vision and Pattern Recognition, pp. 613–626 (2021)
7. Blau, Y., Michaeli, T.: The perception-distortion tradeoff. In: Proceedings of the IEEE Conference on Computer Vision and Pattern Recognition, pp. 6228–6237 (2018)
8. Brooks, T., Mildenhall, B., Xue, T., Chen, J., Sharlet, D., Barron, J.T.: Unprocessing images for learned raw denoising. In: Proceedings of the IEEE/CVF Conference on Computer Vision and Pattern Recognition, pp. 11036–11045 (2019)
9. Chen, L., Chu, X., Zhang, X., Sun, J.: Simple baselines for image restoration. arXiv preprint arXiv:2204.04676 (2022)
10. Chen, L., Lu, X., Zhang, J., Chu, X., Chen, C.: HINet: half instance normalization network for image restoration. In: Proceedings of the IEEE/CVF Conference on Computer Vision and Pattern Recognition, pp. 182–192 (2021)
11. Chen, X., Zhang, Z., Ren, J.S., Tian, L., Qiao, Y., Dong, C.: A new journey from SDRTV to HDRTV. In: Proceedings of the IEEE/CVF International Conference on Computer Vision, pp. 4500–4509 (2021)
12. Chu, X., Chen, L., Chen, C., Lu, X.: Improving image restoration by revisiting global information aggregation. arXiv preprint arXiv:2112.04491 (2021)
13. Conde, M.V., McDonagh, S., Maggioni, M., Leonardis, A., Pérez-Pellitero, E.: Model-based image signal processors via learnable dictionaries. In: Proceedings of the AAAI Conference on Artificial Intelligence, vol. 36, pp. 481–489 (2022)
14. Conde, M.V., Timofte, R., et al.: Reversed image signal processing and RAW reconstruction. AIM 2022 challenge report. In: Proceedings of the European Conference on Computer Vision (ECCV) Workshops (2022)
15. Delbracio, M., Kelly, D., Brown, M.S., Milanfar, P.: Mobile computational photography: a tour. arXiv preprint arXiv:2102.09000 (2021)
16. Dong, X., Zhu, Y., Li, C., Wang, P., Cheng, J.: RISPNet: a network for reversed image signal processing. In: Proceedings of the European Conference on Computer Vision (ECCV) Workshops (2022)
17. Gharbi, M., Chaurasia, G., Paris, S., Durand, F.: Deep joint demosaicking and denoising. ACM Trans. Graph. (ToG) 35(6), 1–12 (2016)
18. He, K., Zhang, X., Ren, S., Sun, J.: Identity mappings in deep residual networks. In: Leibe, B., Matas, J., Sebe, N., Welling, M. (eds.) ECCV 2016. LNCS, vol. 9908, pp. 630–645. Springer, Cham (2016). https://doi.org/10.1007/978-3-319-46493-0_38
19. Heide, F., et al.: FlexISP: a flexible camera image processing framework. ACM Trans. Graph. (ToG) 33(6), 1–13 (2014)
20. Ignatov, A., Timofte, R., Denna, M., Younes, A., et al.: Efficient and accurate quantized image super-resolution on mobile NPUs, mobile AI & AIM 2022 challenge: report. In: Proceedings of the European Conference on Computer Vision (ECCV) Workshops (2022)
21. Ignatov, A., Timofte, R., Kuo, H.K., Lee, M., Xu, Y.S., et al.: Real-time video super-resolution on mobile NPUs with deep learning, mobile AI & AIM 2022 challenge: report. In: Proceedings of the European Conference on Computer Vision (ECCV) Workshops (2022)

22. Ignatov, A., et al.: AIM 2020 challenge on learned image signal processing pipeline. In: Bartoli, A., Fusiello, A. (eds.) ECCV 2020. LNCS, vol. 12537, pp. 152–170. Springer, Cham (2020). https://doi.org/10.1007/978-3-030-67070-2_9

23. Ignatov, A., Timofte, R., et al.: Efficient bokeh effect rendering on mobile GPUs with deep learning, mobile AI & AIM 2022 challenge: report. In: Proceedings of the European Conference on Computer Vision (ECCV) Workshops (2022)

24. Ignatov, A., Timofte, R., et al.: Efficient single-image depth estimation on mobile devices, mobile AI & AIM 2022 challenge: report. In: Proceedings of the European Conference on Computer Vision (ECCV) Workshops (2022)

25. Ignatov, A., Timofte, R., et al.: Learned smartphone ISP on mobile GPUs with deep learning, mobile AI & AIM 2022 challenge: report. In: Proceedings of the European Conference on Computer Vision (ECCV) Workshops (2022)

26. Ignatov, A., Van Gool, L., Timofte, R.: Replacing mobile camera ISP with a single deep learning model. In: Proceedings of the IEEE/CVF Conference on Computer Vision and Pattern Recognition (CVPR) Workshops (2020)

27. Jiang, J., Kim, J., Gu, J.: Overexposure mask fusion: generalizable reverse ISP multi-step refinement. In: Proceedings of the European Conference on Computer Vision (ECCV) Workshops (2022)

28. Karaimer, H.C., Brown, M.S.: A software platform for manipulating the camera imaging pipeline. In: Leibe, B., Matas, J., Sebe, N., Welling, M. (eds.) ECCV 2016. LNCS, vol. 9905, pp. 429–444. Springer, Cham (2016). https://doi.org/10.1007/978-3-319-46448-0_26

29. Kınlı, F., Özcan, B., Kıraç, F.: Patch-wise contrastive style learning for instagram filter removal. In: Proceedings of the IEEE/CVF Conference on Computer Vision and Pattern Recognition (CVPR) Workshops, pp. 578–588 (2022)

30. Kinli, F., Ozcan, B., Kirac, F.: Instagram filter removal on fashionable images. In: Proceedings of the IEEE/CVF Conference on Computer Vision and Pattern Recognition (CVPR) Workshops, pp. 736–745 (2021)

31. Kobyzev, I., Prince, S.J., Brubaker, M.A.: Normalizing flows: an introduction and review of current methods. IEEE Trans. Pattern Anal. Mach. Intell. $43(11)$, 3964–3979 (2020)

32. Kınlı, F.O., Menteş, S., Özcan, B., Kirac, F., Timofte, R., et al.: AIM 2022 challenge on instagram filter removal: methods and results. In: Proceedings of the European Conference on Computer Vision (ECCV) Workshops (2022)

33. Kınlı, F.O., Özcan, B., Kirac, F.: Reversing image signal processors by reverse style transferring. In: Proceedings of the European Conference on Computer Vision (ECCV) Workshops (2022)

34. Nam, S., Punnappurath, A., Brubaker, M.A., Brown, M.S.: Learning SRGB-to-raw-RGB de-rendering with content-aware metadata. In: Proceedings of the IEEE/CVF Conference on Computer Vision and Pattern Recognition (CVPR), pp. 17704–17713 (2022)

35. Nguyen, R.M.H., Brown, M.S.: Raw image reconstruction using a self-contained SRGB-jpeg image with only 64 kb overhead. In: Proceedings of the IEEE Conference on Computer Vision and Pattern Recognition (CVPR) (2016)

36. Punnappurath, A., Brown, M.S.: Learning raw image reconstruction-aware deep image compressors. IEEE Trans. Pattern Anal. Mach. Intell. **42**(4), 1013–1019 (2019)

37. Punnappurath, A., Brown, M.S.: Spatially aware metadata for raw reconstruction. In: Proceedings of the IEEE/CVF Winter Conference on Applications of Computer Vision, pp. 218–226 (2021)

38. Qian, G., et al.: Rethinking the pipeline of demosaicing, denoising and super-resolution. arXiv preprint arXiv:1905.02538 (2019)

39. Ronneberger, O., Fischer, P., Brox, T.: U-Net: convolutional networks for biomedical image segmentation. In: Navab, N., Hornegger, J., Wells, W.M., Frangi, A.F. (eds.) MICCAI 2015. LNCS, vol. 9351, pp. 234–241. Springer, Cham (2015). https://doi.org/10.1007/978-3-319-24574-4_28

40. Schwartz, E., Giryes, R., Bronstein, A.M.: DeepISP: toward learning an end-to-end image processing pipeline. IEEE Trans. Image Process. **28**(2), 912–923 (2018)

41. Sun, D., Yang, X., Liu, M.Y., Kautz, J.: PWC-net: CNNs for optical flow using pyramid, warping, and cost volume. In: Proceedings of the IEEE Conference on Computer Vision and Pattern Recognition, pp. 8934–8943 (2018)

42. Vaswani, A., et al.: Attention is all you need. In: Advances in Neural Information Processing Systems, vol. 30 (2017)

43. Wang, X., et al.: ESRGAN: enhanced super-resolution generative adversarial networks. In: Proceedings of the European Conference on Computer Vision (ECCV) Workshops (2018)

44. Woo, S., Park, J., Lee, J.Y., Kweon, I.S.: CBAM: convolutional block attention module. In: Proceedings of the European Conference on Computer Vision (ECCV), pp. 3–19 (2018)

45. Xing, Y., Qian, Z., Chen, Q.: Invertible image signal processing. In: Proceedings of the IEEE/CVF Conference on Computer Vision and Pattern Recognition, pp. 6287–6296 (2021)

46. Yang, R., Timofte, R., et al.: AIM 2022 challenge on super-resolution of compressed image and video: dataset, methods and results. In: Proceedings of the European Conference on Computer Vision (ECCV) Workshops (2022)

47. Zamir, S.W., et al.: CycleISP: real image restoration via improved data synthesis. In: Proceedings of the IEEE/CVF Conference on Computer Vision and Pattern Recognition, pp. 2696–2705 (2020)

48. Zeng, H., Cai, J., Li, L., Cao, Z., Zhang, L.: Learning image-adaptive 3D lookup tables for high performance photo enhancement in real-time. IEEE Trans. Pattern Anal. Mach. Intell. **44**(4), 2058–2073 (2020)

49. Zhang, R., Isola, P., Efros, A.A., Shechtman, E., Wang, O.: The unreasonable effectiveness of deep features as a perceptual metric. In: Proceedings of the IEEE Conference on Computer Vision and Pattern Recognition, pp. 586–595 (2018)

50. Zhang, Y., Tian, Y., Kong, Y., Zhong, B., Fu, Y.: Residual dense network for image super-resolution. In: Proceedings of the IEEE Conference on Computer Vision and Pattern Recognition, pp. 2472–2481 (2018)

51. Zhang, Z., Wang, H., Liu, M., Wang, R., Zhang, J., Zuo, W.: Learning raw-to-SRGB mappings with inaccurately aligned supervision. In: Proceedings of the IEEE/CVF International Conference on Computer Vision, pp. 4348–4358 (2021)
52. Zhu, Yu., et al.: EEDNet: enhanced encoder-decoder network for AutoISP. In: Bartoli, A., Fusiello, A. (eds.) ECCV 2020. LNCS, vol. 12537, pp. 171–184. Springer, Cham (2020). https://doi.org/10.1007/978-3-030-67070-2_10
53. Zou, B., Zhang, Y.: Learned reverse ISP with soft supervision. In: Proceedings of the European Conference on Computer Vision (ECCV) Workshops (2022)

# AIM 2022 Challenge on Instagram Filter Removal: Methods and Results

Furkan Kınlı[1]([✉]), Sami Menteş[1], Barış Özcan[1], Furkan Kıraç[2], Radu Timofte[2], Yi Zuo[3], Zitao Wang[3], Xiaowen Zhang[3], Yu Zhu[4], Chenghua Li[4], Cong Leng[4,5,6], Jian Cheng[4,5,6], Shuai Liu[7], Chaoyu Feng[7], Furui Bai[7], Xiaotao Wang[7], Lei Lei[7], Tianzhi Ma[8], Zihan Gao[8], Wenxin He[8], Woon-Ha Yeo[9], Wang-Taek Oh[9], Young-Il Kim[9], Han-Cheol Ryu[9], Gang He[10], Shaoyi Long[10], S. M. A. Sharif[11,12], Rizwan Ali Naqvi[11,12], Sungjun Kim[11,12], Guisik Kim[13], Seohyeon Lee[13], Sabari Nathan[14], and Priya Kansal[14]

[1] Özyeğin University, Istanbul, Turkey
furkan.kinli@ozyegin.edu.tr
[2] University of Würzburg, Würzburg, Germany
[3] IPIU Laboratory, Xidian University, Xi'an, China
[4] Institute of Automation, Chinese Academy of Sciences, Beijing, China
[5] MAICRO, Nanjing, China
[6] AiRiA, Nanjing, China
[7] Xiaomi Inc., Beijing, China
[8] Xidian University, Xi'an, China
[9] Sahmyook University, Seoul, South Korea
[10] Xidian University, Xi'an, Shaanxi, China
[11] FS Solution, Seoul, South Korea
[12] Sejong University, Seoul, South Korea
[13] Chung-Ang University, Seoul, Republic of Korea
[14] Couger Inc., Tokyo, Japan

**Abstract.** This paper introduces the methods and the results of AIM 2022 challenge on Instagram Filter Removal. Social media filters transform the images by consecutive non-linear operations, and the feature maps of the original content may be interpolated into a different domain. This reduces the overall performance of the recent deep learning strategies. The main goal of this challenge is to produce realistic and visually plausible images where the impact of the filters applied is mitigated while preserving the content. The proposed solutions are ranked in terms of the PSNR value with respect to the original images. There are two prior studies on this task as the baseline, and a total of 9 teams have competed in the final phase of the challenge. The comparison of qualitative results of the proposed solutions and the benchmark for the challenge are presented in this report.

**Keywords:** Filter removal · Image restoration · Image-to-image translation

L. Karlinsky et al. (Eds.): ECCV 2022 Workshops, LNCS 13803, pp. 27–43, 2023.
https://doi.org/10.1007/978-3-031-25066-8_2

# 1 Introduction

Understanding the content of social media images is a crucial task for industrial applications in different domains. The images shared on social media platforms generally contain visual filters applied where these filters lead to different consecutive transformations on top of the images or inject varied combinations of distractive factors like noise or blurring. Filtered images are the challenging subjects for the recent deep learning strategies (e.g., CNN-based architectures) since they are not robust to these transformations or distractive factors. These filters basically lead to interpolating the feature maps of the original content, and the final outputs for the original content and its filtered version are not the same due to this interpolation factor. Earlier studies addressing this issue present some prior solutions which focus on the filter classification [4,8,9,34] or learning the inverse of a set of transformations applied [3,28]. Recent studies [21,22] introduce the idea of recovering the original image from its filtered version as pre-processing in order to improve the overall performance of the models employed for visual understanding of the content.

We can define that the main goal of Instagram filter removal is to produce realistic and pleasant images in which the impact of the filters applied is mitigated while preserving the original content within. Jointly with the AIM workshop, we propose an AIM challenge on Instagram Filter Removal: the task of recovering the set of Instagram-filtered images to their original versions with high fidelity. The main motivation in this challenge is to increase attention on this research topic, which may be helpful for the applications that primarily target social media images.

This challenge is a part of the AIM 2022 Challenges: Real-Time Image Super-Resolution [12], Real-Time Video Super-Resolution [13], Single-Image Depth Estimation [15], Learned Smartphone ISP [16], Real-Time Rendering Realistic Bokeh [14], Compressed Input Super-Resolution [35] and Reversed ISP [10]. The results obtained in the other competitions and the description of the proposed solutions can be found in the corresponding challenge reports.

# 2 Challenge

In this challenge, the participants are expected to provide a solution that removes Instagram filters from the given images. Provided solutions are ranked in terms of the Peak Signal-to-Noise Ratio (PSNR) of their outputs with respect to the ground truth. Each entry was required to submit the model file along with a runnable code file and the outputs obtained by the model.

## 2.1 Challenge Data

We have used the IFFI dataset [22] in this challenge, which has 500 and 100 instances of original images and their filtered versions as the training and validation sets, respectively. IFFI dataset[1] are provided for the training and validation

---

[1] https://github.com/birdortyedi/instagram-filter-removal-pytorch.

phases in this challenge. For the final evaluation phase, we have shared another 100 instances of images with 11 filters (*i.e.* available for the annotation) for the private test set. We have followed the same procedure given in [22] during the annotation. The filters are picked among the ones in the training and validation sets, but available on Instagram at the time of annotation for the private test set, which are *Amaro, Clarendon, Gingham, He-Fe, Hudson, Lo-Fi, Mayfair, Nashville, Perpetua, Valencia,* and *X-ProII.* All of the samples are originally collected in high-resolution (*i.e.,* 1080 px), then we have to downsample them to the low-resolution (*i.e.,* 256 px) in order to follow the previous benchmark given in [22]. To avoid anti-aliasing, gaussian blur is applied prior to the downsampling. High-resolution images are also provided to the participants in the case of using their pre-processing strategy. The final evaluation has been done on low-resolution images. The participants were free to use the extra data in their training process.

**Table 1.** The benchmark for the private test of IFFI dataset on Instagram Filter Removal challenge.

| Team | Username | Framework | PSNR | SSIM | Runtime (s) | CPU/GPU | Extra data |
|------|----------|-----------|------|------|-------------|---------|------------|
| Fivewin | zuoyi | PyTorch | **34.70** | **0.97** | 0.91 | GPU | No |
| CASIA LCVG | zhuqingweiyu | PyTorch | 34.48 | 0.96 | 0.43 | GPU | No |
| MiAlgo | mialgo_ls | PyTorch | 34.47 | **0.97** | 0.40 | GPU | No |
| Strawberry | penwaterman | PyTorch | 34.39 | 0.96 | 10.47 | GPU | No |
| SYU-HnVLab | Una | PyTorch | 33.41 | 0.95 | 0.04 | GPU | No |
| XDER | clearlon | PyTorch | 32.19 | 0.95 | 0.05 | GPU | No |
| CVRG | CVRG | PyTorch | 31.78 | 0.95 | 0.06 | GPU | No |
| CVML | sheee7 | PyTorch | 30.93 | 0.94 | **0.02** | GPU | No |
| Couger AI | SabariNathan | TensorFlow | 30.83 | 0.94 | 10.43 | CPU | Yes |
| IFRNet [22] | Baseline 1 | PyTorch | 30.46 | - | 0.60 | GPU | No |
| CIFR [21] | Baseline 2 | PyTorch | 30.02 | - | 0.62 | GPU | No |

## 2.2 Evaluation

Considering Instagram Filter removal is an image-to-image translation problem where the images in a specific filter domain are translated into the output images in the original domain, we have measured the performance of the submissions by the common fidelity metrics in the literature (*i.e.,* Peak Signal-to-Noise Ratio (PSNR) and the Structural Similarity (SSIM)). The participants have uploaded the outputs that they have obtained to CodaLab, and all evaluation process has been done on the CodaLab servers. We have shared the evaluation script on the challenge web page. In final evaluation phase, we ranked the submitted solutions by the average PSNR value on the private test set. Since it may be hard to differentiate the differences among the top solutions, we did not employ a user study in the evaluation process of this challenge.

## 2.3 Submissions

The participants were allowed to submit the unfiltered versions of the given images for 100 different instances in PNG format for validation and final evaluation phases. Each team could submit their outputs without any limits in the validation phase, while it is only allowed at most 3 times for the final evaluation phase. As mentioned before, we have created a private test set for the final evaluation phase and shared it with the participants without the original versions. After the submissions are completed, we have validated the leaderboard by double-checking the scores by the outputs sent in the submission files.

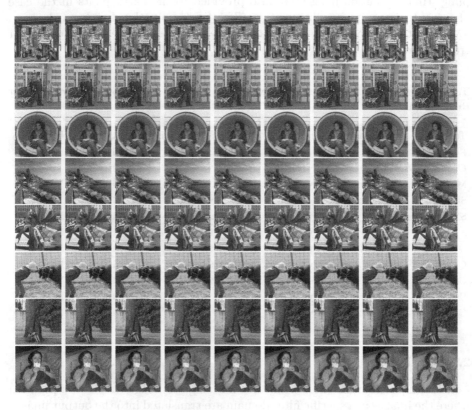

GT    Fivewin   CASIA   MiAlgo   SYU   XDER   CVRG   CVML   Couger

**Fig. 1.** Qualitative results on the private test set, submitted by the participants. From top to bottom, filter types: *X-ProII, He-Fe, Amaro, Perpetua, Mayfair, Gingham, Lo-Fi, Clarendon*; image IDs: 38, 14, 47, 88, 63, 11, 23, 55.

# 3  Results

This section introduces the benchmark results of Instagram Filter Removal challenge. It also presents the qualitative results of all proposed solutions. Lastly, the details of all solutions proposed for this challenge are described.

## 3.1  Overall Results

From 114 registered participants, 9 teams entered the final phase and submitted the valid results, codes, executables, and factsheets. Table 1 summarizes the final challenge results and presents PSNR and SSIM scores of all submitted solutions on the private test set. Figure 1 demonstrates some examples among the qualitative results of all submitted solutions. Top solutions follow different deep learning strategies. The first team in the rankings has employed Transformer-based [33] with a test-time augmentation mechanism. The team in the second rank has proposed a simple encoder-decoder architecture and a specialized defiltering module. They mainly focus on the training settings, instead of the network structure. Next, the MiAlgo team has used a two-stage learning approach where the first stage is responsible for processing the low-resolution input images and the second stage refines the output by using residual groups and different attention modules. Multi-scale color attention, style-based Transformer architecture, gated mechanism, two-branch approach, and a UNet-like architecture represent the other solutions followed by the rest of the participants.

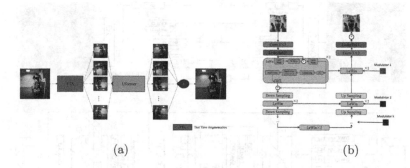

(a)                                    (b)

**Fig. 2.** The proposed solution by Fivewin Team. **(a)** The main structure of UTFR is shown in this figure. **(b)** The main structure of Uformer is shown in this figure.

## 3.2  Solutions

**Fivewin.** The Fivewin team proposed the UTFR method to restore the image after filtering (A U-shaped transformer for filter removal, abbreviated: UTFR). Figure 2 illustrates the main structure of the UTFR. We consider the filter as alternative image noise, so removing the filter is equivalent to denoising the image. We introduce the TTA mechanism (Test-Time Augmentation) based on the Uformer [33] model to expand the amount of data and enhance the generalization of the model by data augmentation during the training of the model. Finally, the same data enhancement method is applied to the test images separately in the test and the changed images are fused to get the final results.

UTFR accepted an image size of $256 \times 256 \times 3$, trained for 250 epochs, with an initial learning rate of 0.0002, using the AdamW optimizer [27] ($\beta_1 = 0.9, \beta_2 = 0.999$, eps $= 1e^{-8}$, weight-decay $= 0.02$). A cosine annealing method was used to perform the learning rate decay to $1e^{-6}$. The learning rate was gradually warmed up (*i.e.*, increased) in the optimizer by warm-up, with the batch size set to 6. We used the Charbonnier loss function that can better handle outliers to approximate the $\mathcal{L}_1$ loss to improve the performance of the model, and finally used the PSNR as the evaluation metric for image restoration.

**CASIA LCVG.** The CASIA LCVG team proposed a simple encoder-decoder network for the Instagram Filter Removal task. As shown in Fig. 4, DefilterNet consists of the basic Unit Defilter Block (DFB). Instead of focusing on designing complex network structures, this work tackles the problem from a different perspective, namely the training setting. They only use the normal end-to-end training method and try to enhance DFilterNet by investigating the training settings. Especially, they found that data augmentation plays an important role in this training method, especially Mixup [39], which effectively improves the DefilterNet by 2 dB.

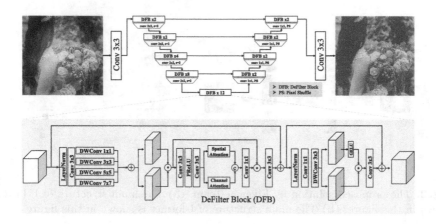

**Fig. 3.** The architecture of DefilterNet proposed by CASIA LCVG.

They train DefilterNet with AdamW optimizer ($\beta_1 = 0.9, \beta_2 = 0.9$) and L1 loss for $300K$ iterations with the initial learning rate $2e^{-4}$ gradually reduced to $1e^{-6}$ with the cosine annealing. The training image size is $256 \times 256$ and batch size is 24. For data augmentation, we use horizontal and vertical flips, especially involves Mixup [39].

**MiAlgo.** The MiAlgo team proposed a Two-stage Network for the Instagram Filter Removal task. In the first stage, we adopt the BigUNet to process the low-resolution input images. The BigUNet is improved based on MWCNN and RCAN. We replaced the convolutional layer in MWCNN with the residual group (RG, with channel attention layer) in RCAN to enhance the reconstruction ability of the network. In the second stage, we downsample the high-resolution input, then concatenate it with the output of the first stage, and feed it into the refine network. The refine network contains 5 residual groups with channel attention and spatial attention, which can get better details, as well as more accurate local color, local brightness, etc.

The BigUNet is first trained for 4000 epochs with a learning rate of $1e-4$. Then freeze the parameters of BigUNet, and train the refine network for 2000 epochs with a learning rate of $1e-4$. Finally, the overall finetune is trained for 2000 epochs with a learning rate of $1e-5$. All stages use Charbonnier loss and use cosine annealing for learning rate decay, the batch size is set to 16, and the whole image is used for training. In the early stage of training, the filter style is randomly selected; in the later stage of training, we increase the weight of the hard styles, that is hard example mining. At test time, we fuse the outputs of the first and second stages.

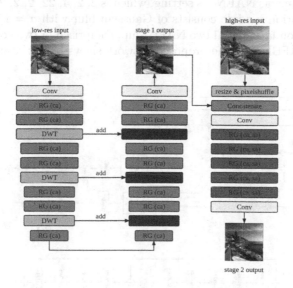

**Fig. 4.** The architecture of Two-stage Network proposed by MiAlgo.

**Strawberry.** The solution of Strawberry is composed of different architectures, such as Uformer [33] Pix2Pix [17], VRT [23], RNN-MBP-Local [5], and Cycle-GAN [40]. The pre-training weights are not used for the experiments. In the early stage of training, the team simply enhanced the data, processed brightness and contrast, and introduced random noise. No additional data is used in the experiments. The method is the fusion of the result after training several main models, and different models need to be trained separately. The test-time augmentation method is preferred for a single model to dynamically fuse the results.

**SYU-HnVLab.** The SYU-HnVLab team proposed Multi-Scale Color Attention Network for Instagram Filter Removal (CAIR) [36]. The proposed CAIR is modified from NAFNet [6] with two improvements: 1) Color attention (CA) mechanism inspired by the color correction scheme of CycleISP [38], and 2) CAIR takes multi-scale input images. The main difference between the filter image and the original image is the *color* of images. Therefore, it is essential to consider color information to remove filters from filter images effectively. Through the CA mechanism, useful color information can be obtained from the multi-scale input images. We downscale an input image ($H \times W$) by a factor of 2 in each level, as described in Fig. 5. The input image of each level and the higher level image pass together through the color attention module, and the color attentive features are obtained. These features are concatenated with the features from higher-level and then passes through NAFBlock. NAFBlock and the modules in NAFBlock are described in Fig. 5. The number of NAFBlocks of each level ($n_1$-$n_7$) is set the same as NAFNet's setting, which is 2, 2, 4, 22, 2, 2, 2, respectively. The color attention module consists of Gaussian blur with $\sigma = 12$ module, two $1 \times 1$ convolution layers, and two NAFGroup, comprised of two convolution layers and two NAFBlocks. The proposed network shows better results with lower

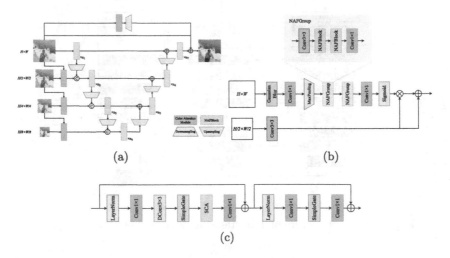

**Fig. 5.** The proposed solution by SYU-HnVLab Team. **(a)** The main architecture of the proposed multi-scale color attention network (CAIR). **(b)** The proposed color attention module. $\otimes$: element-wise multiplication, $\oplus$: element-wise summation. **(c)** NAFBlock [6]

computational and memory complexity when using the color attention mechanism than the original NAFNet under the same configuration.

Before training the model, we resize high-resolution (HR) images (1080 × 1080) into 256 × 256 due to a lack of training dataset. For resizing, an image is interpolated through resampling using pixel area relation (*i.e.*, OpenCV interpolation option INTER_AREA) for image quality. As a result, we use three sets of images including HR images cropped in 256, resized HR images, and the low resolution (256 × 256) images. This data augmentation helps improve the flexibility of the trained model. In addition, images are randomly flipped and rotated with the probability of 0.5 in the training process. We use AdamW [27] optimizer with $\beta_1 = 0.9, \beta_2 = 0.9$, weight decay $1e^{-4}$. The initial learning rate is set to $1e^{-3}$ and gradually reduces to $1e^{-6}$ with the cosine annealing schedule [26]. It is trained for up to 200K iterations. The training mini-batch size is 64, and the patch size is 256 × 256. We use Peak Signal-to-Noise Ratio (PSNR) metric as the loss function, which is PSNR loss. The number of channels in each NAFBlock is set to 32. As in [25, 31], we use the self-ensemble method for test time augmentation (TTA), which produces seven new variants for each input. Therefore, each filter removal image is obtained by taking the average of all outputs processed by the network from augmented inputs. Additionally, we improve the output of CAIR through ensemble learning.

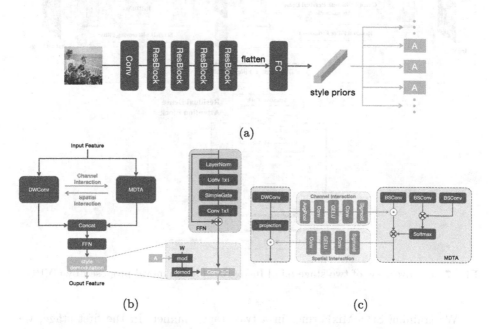

**Fig. 6.** The proposed solution by XDER team. **(a)** The style priors extraction network. **(b)** The interaction mechanism. **(c)** The StyleMixformer Block. ⊙: element-wise multiplication, ⊗: matrix multiplication, ⊕: element-wise addition. $A$ denotes a learned affine transform from style priors.

**XDER.** The XDER team proposed a StyleMixFormer network for the Instagram Filter Removal task. Inspired by Kınlı et al. [22], we consider this problem as a multi-style conversion problem by Assuming that each filter introduces global and local style manipulations to the original image. In order to extract style priors information from the filtered image, we design a style priors extraction module as shown in Fig. 6a. Compared with AdaIN, we introduce a more effective style demodulation [18] to remove style of filtered image. As shown in Fig. 6b, we propose a style demodulation Mixing block in order to fully extract the features of the filtered image and transform the style of the feature map. Inspired by Chen et al. [7], we extract global and local features by attention mechanism and depth-wise convolution, respectively, and interact information between the two features. The details of the information interaction are shown in Fig. 6c. Unlike Chen et al. [7] we replace the local-window self-attention block with multi-Dconv head transposed attention (MDTA) [37] which has been experimentally shown to be effective on low-level vision. For the feed-forward network, we adopt the SimpleGate mechanism [6]. We use the Unet network as our overall architecture following Chen et al. [6].

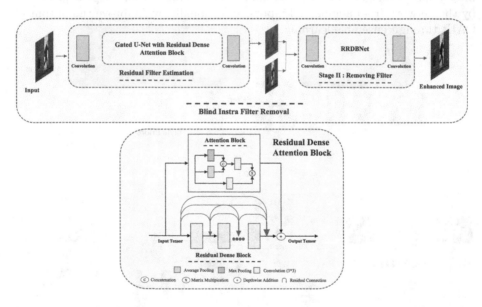

**Fig. 7.** The overview of two-stage blind Instagram filter removal proposed by CVRG.

We training StyleMixFormer in a two stage manner. In the first stage, we use a 4-layer ResNet to extract the style priors of different filtered images with ArcFace loss [11], which can effectively narrow intra-class distances and increase inter-class distances. In the second stage, we train StyleMixFormer with L1 loss. For the second stage, we used the L1 loss and Adam optimizer for training. The initialized learning rate is 5e-4, the batch size is 16, and the total number of iterations is 600k. More detail and the source code can be found at https://github.com/clearlon/StyleMixFormer.

**CVRG.** Figure 7 illustrates the overview of the proposed two-stage solution. CVRG Team has developed the stage-I based on Gated U-Net [29,30] with residual dense attention block [1]. Stage-II of the solution comprises residual-in-residual dense blocks (RRDB) [32]. It leverages the residual filters (from stage-I) and the input images to obtain the final output. The $\mathcal{L}_1$ + Gradient loss was utilized to optimize both stages. Also, both networks were optimized with Adam optimizer [19], where the hyperparameters were tuned as $\beta_1 = 0.9$, $\beta_2 = 0.99$, and learning rate $= 5e-4$. We trained our model for 65 epochs with a constant batch size of 4. It takes around four days to complete the training process. We conducted our experiments on a machine comprised of an AMD Ryzen 3200G central processing unit (CPU) clocked at 3.6 GHz, a random-access memory of 16 GB, and n single Nvidia Geforce GTX 1080 (8 GB) graphical processing unit (GPU).

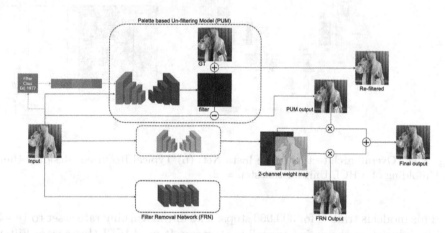

**Fig. 8. Overall architecture of the CVML.** *i.e.*, **two-branch model.** The model removes the filter from the input image in two ways and integrates each result.

**CVML.** The CVML team proposed a two-branch model for the Instagram Filter Removal task. One network learns an intrinsic filter from an input image with a color palette. Another network learns unfiltered images directly from the input images. This approach fuses each unfiltered image to get the final result. The proposed model has the advantage of being lightweight, fast inference time, and fine quality for resulting images.

The model named Palette-based Un-filtering Model (PUM) is a convolutional model with an encoder-decoder structure. The palette and the filtered images are inserted and the model learns an intrinsic Instagram filter. The palette has colors representing the corresponding filter. The intrinsic filter is obtained as the output of the network and subtracted from the filtered image to obtain an unfiltered PUM output image. The intrinsic filter is used once again to recreate the filtered image by adding to the ground truth image. With a re-filtered image, the missing features can be checked in the intrinsic filter. The color palette is generated by extracting five average colors with the K-means algorithm. In this stage, all corresponding images of each filter are added. The Filter Removal Network (FRN), is a model of a simple encoder-decoder structure, which creates an unfiltered image directly from a filtered image. Output images from PUM and FRN are fused with a 2-channel weight map to obtain the final result. The weight map is used to determine which portion of each output will be used for the final output.

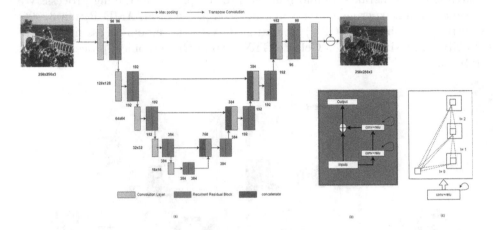

**Fig. 9.** (a) Overall architecture of the Insta Net; (b) Typical Recurrent Residual Unit (c) Unfolding of a RCL Unit for timestep = 2.

This model is trained for 300,000 steps. The initial learning rate is set to $1e-4$ and is reduced by the cosine annealing strategy. It used IFFI (Instagram Filter Removal on Fashionable Images) dataset images with low-resolution ($256 \times 256$) for the experiments and a color palette generated for each filter. This solution has performed data augmentation through 90, 180, and 270 degrees rotation. It takes about 1 day to train our model and about 0.025 s. per image to test.

**CougerAI.** As shown in Fig. 9, the proposed architecture is a U-Net-based structure that includes an encoding path to down-sample features and a decoding path to retrieve the required image. In each step of the encoding path, the input features are passed to a convolutional layer with $3 \times 3$ filter and then two recurrent residual blocks [2,24] with time step 2. The output of the recurrent residual block is down-sampled by the factor of two using max pooling operation. The image is down-sampled 5 times in the encoding path. The same structure is used in the decoding path, where the up-sampled output is concatenated with the same-step output of the encoding path using a skip connection. After that, it is passed to the convolutional block and two subsequent recurrent residue blocks like the encoding path. As we have taken this task as an image restoration task, we subtracted the input image to retrieve the original unfiltered image. For the purpose of training, the images are normalized between 0 to 1. To update the weights during training, we used the Adam optimizer [20]. The learning rate is initialized with 0.001 and reduced by 10 percent after 15 epochs if the validation loss does not improve. The batch size is set to 2. The model is evaluated using the Peak-Signal-to Noise Ratio (PSNR) and Structural Similarity Index (SSIM). The model is trained for 200 epochs on a single 16 GB NVIDIA Tesla K80 GPU in Google Colab pro.

## 4    Teams and Affiliations

### 4.1    Organizers of AIM 2022 Challenge on Instagram Filter Removal

 – Furkan Kınlı[1], Sami Menteş[1], Barış Özcan[1], Furkan Kıraç[1], Radu Timofte[2]

*Affiliations:*

 – [1] Özyeğin University, Türkiye.
 – [2] University of Würzburg, Germany.

### 4.2    Fivewin

 – Yi Zuo, Zitao Wang, Xiaowen Zhang

*Affiliations:*

 – IPIU Laboratory, Xidian University.

### 4.3    CASIA LCVG

 – Yu Zhu[1], Chenghua Li[1], Cong Leng[1,2,3], Jian Cheng[1,2,3]

*Affiliations:*

 – [1]Institute of Automation, Chinese Academy of Sciences, Beijing, China.
 – [2]MAICRO, Nanjing, China.
 – [3]AiRiA, Nanjing, China.

### 4.4    MiAlgo

 – Shuai Liu, Chaoyu Feng, Furui Bai, Xiaotao Wang, Lei Lei

*Affiliations:*

 – Xiaomi Inc., China.

## 4.5  Strawberry

– Tianzhi Ma, Zihan Gao, Wenxin He

*Affiliations:*

– Xidian University.

## 4.6  SYU-HnVLab

– Woon-Ha Yeo, Wang-Taek Oh, Young-Il Kim, Han-Cheol Ryu

*Affiliation:*

– Sahmyook University, Seoul, South Korea.

## 4.7  XDER

– Gang He, Shaoyi Long

*Affiliations:*

– Xidian University, Xi'an Shaanxi, China.

## 4.8  CVRG

– S. M. A. Sharif, Rizwan Ali Naqvi, Sungjun Kim

*Affiliations:*

– FS Solution, South Korea.
– Sejong University, South Korea.

## 4.9  CVML

– Guisik Kim, Seohyeon Lee

*Affiliations:*

– Chung-Ang University, Republic of Korea.

## 4.10    Couger AI

– Sabari Nathan, Priya Kansal

*Affiliations:*

– Couger Inc, Japan.

**Acknowledgements.** We thank the sponsors of the AIM and Mobile AI 2022 workshops and challenges: AI Witchlabs, MediaTek, Huawei, Reality Labs, OPPO, Synaptics, Raspberry Pi, ETH Zürich (Computer Vision Lab) and University of Würzburg (Computer Vision Lab).

# References

1. A Sharif, S., Naqvi, R.A., Biswas, M., Kim, S.: A two-stage deep network for high dynamic range image reconstruction. In: Proceedings of the IEEE/CVF Conference on Computer Vision and Pattern Recognition, pp. 550–559 (2021)
2. Alom, M.Z., Hasan, M., Yakopcic, C., Taha, T.M., Asari, V.K.: Recurrent residual convolutional neural network based on U-Net (R2U-Net) for medical image segmentation. arXiv preprint arXiv:1802.06955 (2018)
3. Bianco, S., Cusano, C., Piccoli, F., Schettini, R.: Artistic photo filter removal using convolutional neural networks. J. Electron. Imaging **27**(1), 011004 (2017). https://doi.org/10.1117/1.JEI.27.1.011004
4. Bianco, S., Cusano, C., Schettini, R.: Artistic photo filtering recognition using CNNs. In: Bianco, S., Schettini, R., Trémeau, A., Tominaga, S. (eds.) CCIW 2017. LNCS, vol. 10213, pp. 249–258. Springer, Cham (2017). https://doi.org/10.1007/978-3-319-56010-6_21
5. Chao, Z., et al.: Deep recurrent neural network with multi-scale bi-directional propagation for video deblurring. In: AAAI (2022)
6. Chen, L., Chu, X., Zhang, X., Sun, J.: Simple baselines for image restoration. arXiv preprint arXiv:2204.04676 (2022)
7. Chenl, Q., et al.: Mixformer: mixing features across windows and dimensions. In: Proceedings of the IEEE/CVF Conference on Computer Vision and Pattern Recognition, pp. 5249 5259 (2022)
8. Chen, Y.H., Chao, T.H., Bai, S.Y., Lin, Y.L., Chen, W.C., Hsu, W.H.: Filter-invariant image classification on social media photos, MM 2015, pp. 855–858. Association for Computing Machinery, New York (2015). https://doi.org/10.1145/2733373.2806348
9. Chu, W.T., Fan, Y.T.: Photo filter classification and filter recommendation without much manual labeling. In: 2019 IEEE 21st International Workshop on Multimedia Signal Processing (MMSP), pp. 1–6. IEEE (2019)
10. Conde, M.V., Timofte, R., et al.: Reversed image signal processing and raw reconstruction. AIM 2022 challenge report. In: Proceedings of the European Conference on Computer Vision Workshops (ECCVW) (2022)
11. Deng, J., Guo, J., Xue, N., Zafeiriou, S.: Arcface: additive angular margin loss for deep face recognition. In: Proceedings of the IEEE/CVF Conference on Computer Vision and Pattern Recognition, pp. 4690–4699 (2019)

12. Ignatov, A., Timofte, R., Denna, M., Younes, A., et al.: Efficient and accurate quantized image super-resolution on mobile NPUs, mobile AI & AIM 2022 challenge: report. In: Proceedings of the European Conference on Computer Vision (ECCV) Workshops (2022)
13. Ignatov, A., Timofte, R., Kuo, H.K., Lee, M., Xu, Y.S., et al.: Real-time video super-resolution on mobile NPUs with deep learning, mobile AI & AIM 2022 challenge: report. In: Proceedings of the European Conference on Computer Vision (ECCV) Workshops (2022)
14. Ignatov, A., Timofte, R., et al.: Efficient bokeh effect rendering on mobile GPUs with deep learning, mobile AI & AIM 2022 challenge: report. In: Proceedings of the European Conference on Computer Vision (ECCV) Workshops (2022)
15. Ignatov, A., Timofte, R., et al.: Efficient single-image depth estimation on mobile devices, mobile AI & AIM 2022 challenge: report. In: Proceedings of the European Conference on Computer Vision (ECCV) Workshops (2022)
16. Ignatov, A., Timofte, R., et al.: Learned smartphone ISP on mobile GPUs with deep learning, mobile AI & AIM 2022 challenge: report. In: Proceedings of the European Conference on Computer Vision (ECCV) Workshops (2022)
17. Isola, P., Zhu, J.Y., Zhou, T., Efros, A.A.: Image-to-image translation with conditional adversarial networks. CVPR (2017)
18. Karras, T., Laine, S., Aittala, M., Hellsten, J., Lehtinen, J., Aila, T.: Analyzing and improving the image quality of StyleGAN. In: Proceedings of CVPR (2020)
19. Kingma, D.P., Ba, J.: Adam: a method for stochastic optimization. arXiv preprint arXiv:1412.6980 (2014)
20. Kingma, D.P., Ba, J.: Adam: a method for stochastic optimization. In: Bengio, Y., LeCun, Y. (eds.) 3rd International Conference on Learning Representations, ICLR 2015, San Diego, CA, USA, 7–9 May 2015, Conference Track Proceedings (2015). http://arxiv.org/abs/1412.6980
21. Kınlı, F., Özcan, B., Kıraç, F.: Patch-wise contrastive style learning for instagram filter removal. In: Proceedings of the IEEE/CVF Conference on Computer Vision and Pattern Recognition (CVPR) Workshops, pp. 578–588 (2022)
22. Kinli, F., Ozcan, B., Kirac, F.: Instagram filter removal on fashionable images. In: Proceedings of the IEEE/CVF Conference on Computer Vision and Pattern Recognition (CVPR) Workshops, pp. 736–745 (2021)
23. Liang, J., et al.: VRT: a video restoration transformer. arXiv preprint arXiv:2201.12288 (2022)
24. Liang, M., Hu, X.: Recurrent convolutional neural network for object recognition. In: Proceedings of the IEEE Conference on Computer Vision and Pattern Recognition, pp. 3367–3375 (2015)
25. Lim, B., Son, S., Kim, H., Nah, S., Mu Lee, K.: Enhanced deep residual networks for single image super-resolution. In: Proceedings of the IEEE Conference on Computer Vision and Pattern Recognition Workshops, pp. 136–144 (2017)
26. Loshchilov, I., Hutter, F.: SGDR: stochastic gradient descent with warm restarts. arXiv preprint arXiv:1608.03983 (2016)
27. Loshchilov, I., Hutter, F.: Decoupled weight decay regularization. arXiv preprint arXiv:1711.05101 (2017)
28. Sen, M., Chakraborty, P.: A deep convolutional neural network based approach to extract and apply photographic transformations. In: Nain, N., Vipparthi, S.K., Raman, B. (eds.) CVIP 2019. CCIS, vol. 1148, pp. 155–162. Springer, Singapore (2020). https://doi.org/10.1007/978-981-15-4018-9_14
29. Sharif, S., Naqvi, R.A., Biswas, M.: SAGAN: adversarial spatial-asymmetric attention for noisy Nona-Bayer reconstruction. arXiv preprint arXiv:2110.08619 (2021)

30. Sharif, S., Naqvi, R.A., Biswas, M., Loh, W.K.: Deep perceptual enhancement for medical image analysis. IEEE J. Biomed. Health Inform. **26**(10), 4826–4836 (2022)
31. Timofte, R., Rothe, R., Van Gool, L.: Seven ways to improve example-based single image super resolution. In: Proceedings of the IEEE Conference on Computer Vision and Pattern Recognition, pp. 1865–1873 (2016)
32. Wang, X., et al.: ESRGAN: enhanced super-resolution generative adversarial networks. In: Proceedings of the European Conference on Computer Vision (ECCV) Workshops (2018)
33. Wang, Z., Cun, X., Bao, J., Zhou, W., Liu, J., Li, H.: Uformer: a general U-shaped transformer for image restoration. In: Proceedings of the IEEE/CVF Conference on Computer Vision and Pattern Recognition, pp. 17683–17693 (2022)
34. Wu, Z., Wu, Z., Singh, B., Davis, L.: Recognizing instagram filtered images with feature de-stylization. In: Proceedings of the AAAI Conference on Artificial Intelligence, vol. 34, no. 07, pp. 12418–12425 (2020). https://doi.org/10.1609/aaai. v34i07.6928. https://ojs.aaai.org/index.php/AAAI/article/view/6928
35. Yang, R., Timofte, R., et al.: AIM 2022 challenge on super-resolution of compressed image and video: dataset, methods and results. In: Proceedings of the European Conference on Computer Vision (ECCV) Workshops (2022)
36. Yeo, W.H., Oh, W.T., Kang, K.S., Kim, Y.I., Ryu, H.C.: CAIR: fast and lightweight multi-scale color attention network for instagram filter removal. In: Proceedings of the European Conference on Computer Vision (ECCV) Workshops (2022)
37. Zamir, S.W., Arora, A., Khan, S., Hayat, M., Khan, F.S., Yang, M.H.: Restormer: efficient transformer for high-resolution image restoration. In: CVPR (2022)
38. Zamir, S.W., et al.: CycleISP: real image restoration via improved data synthesis. In: Proceedings of the IEEE/CVF Conference on Computer Vision and Pattern Recognition, pp. 2696–2705 (2020)
39. Zhang, H., Cisse, M., Dauphin, Y.N., Lopez-Paz, D.: Mixup: beyond empirical risk minimization. arXiv preprint arXiv:1710.09412 (2017)
40. Zhu, J.Y., Park, T., Isola, P., Efros, A.A.: Unpaired image-to-image translation using cycle-consistent adversarial networks. In: 2017 IEEE International Conference on Computer Vision (ICCV) (2017)

# Learned Smartphone ISP on Mobile GPUs with Deep Learning, Mobile AI & AIM 2022 Challenge: Report

Andrey Ignatov[1,2]([✉]), Radu Timofte[1,2,3], Shuai Liu[4], Chaoyu Feng[4],
Furui Bai[4], Xiaotao Wang[4], Lei Lei[4], Ziyao Yi[5], Yan Xiang[5], Zibin Liu[5],
Shaoqing Li[5], Keming Shi[5], Dehui Kong[5], Ke Xu[5], Minsu Kwon[6], Yaqi Wu[7],
Jiesi Zheng[8], Zhihao Fan[9], Xun Wu[10], Feng Zhang[7,8,9,10], Albert No[11],
Minhyeok Cho[11], Zewen Chen[12], Xiaze Zhang[13], Ran Li[14], Juan Wang[12],
Zhiming Wang[10], Marcos V. Conde[3], Ui-Jin Choi[3], Georgy Perevozchikov[15],
Egor Ershov[15], Zheng Hui[16], Mengchuan Dong[17], Xin Lou[1], Wei Zhou[17],
Cong Pang[17], Haina Qin[12], and Mingxuan Cai[12]

[1] Computer Vision Lab, ETH Zurich, Zurich, Switzerland
andrey@vision.ee.ethz.ch
[2] AI Witchlabs, Zurich, Switzerland
[3] University of Wuerzburg, Wuerzburg, Germany
{radu.timofte,marcos.conde-osorio}@uni-wuerzburg.de
[4] Xiaomi Inc., Beijing, China
liushuai21@xiaomi.com
[5] Sanechips Co. Ltd., Shanghai, China
yi.ziyao@sanechips.com.cn
[6] ENERZAi, Seoul, Korea
minsu.kwon@enerzai.com
[7] Harbin Institute of Technology, Harbin, China
[8] Zhejiang University, Hangzhou, China
[9] University of Shanghai for Science and Technology, Shanghai, China
[10] Tsinghua University, Beijing, China
[11] Hongik University, Seoul, Korea
albertno@hongik.ac.kr
[12] Institute of Automation, Chinese Academy of Sciences, Beijing, China
{chenzewen2022,qinhaina2020}@ia.ac.cn
[13] School of Computer Science, Fudan University, Shanghai, China
[14] Washington University in St. Louis, Seattle, USA
[15] Moscow Institute of Physics and Technology, Moscow, Russia
perevozchikov.gp@phystech.edu
[16] Alibaba DAMO Academy, Beijing, China
huizheng.hz@alibaba-inc.com
[17] ShanghaiTech University, Shanghai, China

Andrey Ignatov and Radu Timofte are the main Mobile AI & AIM 2022 challenge organizers. The other authors participated in the challenge.
Mobile AI 2022 Workshop website:
https://ai-benchmark.com/workshops/mai/2022/
Appendix A contains the authors' team names and affiliations.

**Abstract.** The role of mobile cameras increased dramatically over the past few years, leading to more and more research in automatic image quality enhancement and RAW photo processing. In this Mobile AI challenge, the target was to develop an efficient end-to-end AI-based image signal processing (ISP) pipeline replacing the standard mobile ISPs that can run on modern smartphone GPUs using TensorFlow Lite. The participants were provided with a large-scale Fujifilm UltraISP dataset consisting of thousands of paired photos captured with a normal mobile camera sensor and a professional 102MP medium-format FujiFilm GFX100 camera. The runtime of the resulting models was evaluated on the Snapdragon's 8 Gen 1 GPU that provides excellent acceleration results for the majority of common deep learning ops. The proposed solutions are compatible with all recent mobile GPUs, being able to process Full HD photos in less than 20–50 ms while achieving high fidelity results. A detailed description of all models developed in this challenge is provided in this paper.

**Keywords:** Mobile AI Challenge · Learned ISP · Mobile cameras · Photo enhancement · Mobile AI · Deep learning · AI Benchmark

# 1   Introduction

Visualized RAW Image     MediaTek Dimensity 820 ISP Photo     Fujifilm GFX 100 Photo

**Fig. 1.** Example set of full-resolution images (top) and crops (bottom) from the collected Fujifilm UltraISP dataset. From left to right: original RAW visualized image, RGB image obtained with MediaTek's built-in ISP system, and Fujifilm GFX100 target photo.

Nowadays, the cameras are ubiquitous mainly due to the tremendous success and adoption of modern smartphones. Over the years, the quality of the smartphone cameras continuously improved due to advances in both hardware and software. Currently, due to their versatility, the critical improvements are coming from the advanced image processing algorithms employed, *e.g.*, to perform color reconstruction or adjustment, noise removal, super-resolution, high dynamic range processing. The image enhancement task can be effectively solved with deep learning-based approaches. The critical part is the acquisition of appropriate (paired) low and high-quality ground truth images for training. For the first time the end-to-end mobile photo quality enhancement problem was tackled in [19,20]. The authors proposed to directly map the images from a low-quality smartphone camera to the higher-quality images from a high-end DSLR camera. The introduced DPED dataset was later employed in many competitions [28,36] and works [15,16,49,53,61,68] that significantly advanced the research on this problem. The major shortcoming of the proposed methods is that they are working on the images produced by cameras' built-in ISPs and, thus, they are not using a significant part of the original sensor data lost in the ISP pipeline. In [39] the authors proposed to replace the smartphone ISP with a deep neural network learned to map directly the RAW Bayer sensor data to the higher-quality images captured by a DSLR camera. For this, a *Zurich RAW to RGB* dataset containing RAW-RGB image pairs from a mobile camera sensor and a high-end DSLR camera was collected. The proposed learned ISP reached the quality level of commercial ISP system of the Huawei P20 camera phone, and these results were further improved in [7,33,37,43,60]. In this challenge, we use a more advanced FujiFlim UltraISP dataset [23,27] and additional efficiency-related constraints on the developed solutions. We target deep learning solutions capable to run on mobile GPUs. This is the second installment after the challenge conducted in conjunction with the Mobile AI 2021 CVPR workshop [18].

The deployment of AI-based solutions on portable devices usually requires an efficient model design based on a good understanding of the mobile processing units (*e.g.* CPUs, NPUs, GPUs, DSP) and their hardware particularities, including their memory constraints. We refer to [30,34] for an extensive overview of mobile AI acceleration hardware, its particularities and performance. As shown in these works, the latest generations of mobile NPUs are reaching the performance of older-generation mid-range desktop GPUs. Nevertheless, a straightforward deployment of neural networks-based solutions on mobile devices is impeded by (i) a limited memory (*i.e.*, restricted amount of RAM) and (ii) a limited or lacking support of many common deep learning operators and layers. These impeding factors make the processing of high resolution inputs impossible with the standard NN models and require a careful adaptation or re-design to the constraints of mobile AI hardware. Such optimizations can employ a combination of various model techniques such as 16-bit/8-bit [4,41,42,73] and low-bit [3,40,52,67] quantization, network pruning and compression [4,25,47,51,55], device- or NPU-specific adaptations, platform-aware neural architecture search [14,63,69,72], *etc.*

The majority of competitions aimed at efficient deep learning models use standard desktop hardware for evaluating the solutions, thus the obtained models rarely show acceptable results when running on real mobile hardware with many specific constraints. In this *Mobile AI challenge*, we take a radically different approach and propose the participants to develop and evaluate their models directly on mobile devices. The goal of this competition is to design a fast and performant deep learning-based solution for the learned smartphone ISP problem. For this, the participants were provided with the Fujifilm UltraISP dataset consisting of thousands of paired photos captured with a normal mobile camera sensor and a professional 102MP medium-format FujiFilm GFX100 camera. The efficiency of the proposed solutions was evaluated on the Snapdragon 8 Gen 1 mobile platform capable of accelerating floating-point and quantized neural networks. All solutions developed in this challenge are fully compatible with the TensorFlow Lite framework [64], thus can be efficiently executed on various Linux and Android-based IoT platforms, smartphones and edge devices.

This challenge is a part of the *Mobile AI & AIM 2022 Workshops and Challenges* consisting of the following competitions:

- Learned Smartphone ISP on Mobile GPUs
- Power Efficient Video Super-Resolution on Mobile NPUs [29]
- Quantized Image Super-Resolution on Mobile NPUs [32]
- Efficient Single-Image Depth Estimation on Mobile Devices [24]
- Realistic Bokeh Effect Rendering on Mobile GPUs [38]
- Super-Resolution of Compressed Image and Video [74]
- Reversed Image Signal Processing and RAW Reconstruction [6]
- Instagram Filter Removal [45]

The results and solutions obtained in the previous *MAI 2021 Challenges* are described in our last year papers:

- Learned Smartphone ISP on Mobile NPUs [18]
- Real Image Denoising on Mobile GPUs [17]
- Quantized Image Super-Resolution on Mobile NPUs [31]
- Real-Time Video Super-Resolution on Mobile GPUs [57]
- Single-Image Depth Estimation on Mobile Devices [21]
- Quantized Camera Scene Detection on Smartphones [22]

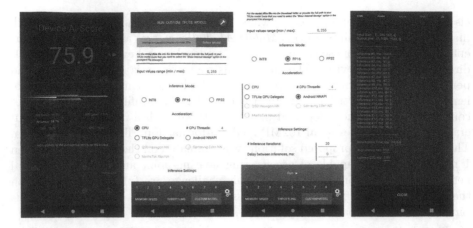

**Fig. 2.** Loading and running custom TensorFlow Lite models with AI Benchmark application. The currently supported acceleration options include Android NNAPI, TFLite GPU, Hexagon NN, Qualcomm QNN, MediaTek Neuron and Samsung ENN delegates as well as CPU inference through TFLite or XNNPACK backends. The latest app version can be downloaded at https://ai-benchmark.com/download.

## 2    Challenge

In order to design an efficient and practical deep learning-based solution for the considered task that runs fast on mobile devices, one needs the following tools:

1. A large-scale high-quality dataset for training and evaluating the models. Real, not synthetically generated data should be used to ensure a high quality of the obtained model;
2. An efficient way to check the runtime and debug the model locally without any constraints as well as the ability to check the runtime on the target evaluation platform.

This challenge addresses all the above issues. Real training data, tools, and runtime evaluation options provided to the challenge participants are described in the next sections.

### 2.1    Dataset

In this challenge, we use the Fujifilm UltraISP dataset collected using the Fujifilm GFX100 medium format 102 MP camera capturing the target high-quality images, and a popular Sony IMX586 Quad Bayer mobile camera sensor that can be found in tens of mid-range and high-end mobile devices released in the past 3 years. The Sony sensor was mounted on the MediaTek Dimensity 820 development board, and was capturing both raw and processed (by its built-in ISP system) 12MP images. The Dimensity board was rigidly attached to the Fujifilm camera, and they were shooting photos synchronously to ensure that the image content is identical. The dataset contains over 6 thousand daytime image pairs captured at a wide variety of places with different illumination and weather conditions. An example set of full-resolution photos from the Fujifilm UltraISP dataset is shown in Fig. 1. As the collected RAW-RGB image pairs were not perfectly aligned, they were initially matched using the state-of-the-art deep learning based dense matching algorithm [66] to extract $256 \times 256$ pixel patches from the original photos. It should be mentioned that all alignment operations were performed on Fujifilm RGB images only, therefore RAW photos from the Sony sensor remained unmodified, exhibiting the same values as read from the sensor.

## 2.2  Local Runtime Evaluation

When developing AI solutions for mobile devices, it is vital to be able to test the designed models and debug all emerging issues locally on available devices. For this, the participants were provided with the *AI Benchmark* application [30, 34] that allows to load any custom TensorFlow Lite model and run it on any Android device with all supported acceleration options. This tool contains the latest versions of *Android NNAPI, TFLite GPU, Hexagon NN, Qualcomm QNN, MediaTek Neuron* and *Samsung ENN* delegates, therefore supporting all current mobile platforms and providing the users with the ability to execute neural networks on smartphone NPUs, APUs, DSPs, GPUs and CPUs.

To load and run a custom TensorFlow Lite model, one needs to follow the next steps:

1. Download AI Benchmark from the official website[1] or from the Google Play[2] and run its standard tests.
2. After the end of the tests, enter the *PRO Mode* and select the *Custom Model* tab there.
3. Rename the exported TFLite model to *model.tflite* and put it into the *Download* folder of the device.
4. Select mode type *(INT8, FP16, or FP32)*, the desired acceleration/inference options and run the model.

These steps are also illustrated in Fig. 2.

---

[1] https://ai-benchmark.com/download.
[2] https://play.google.com/store/apps/details?id=org.benchmark.demo.

**Table 1.** Mobile AI 2022 learned smartphone ISP challenge results and final rankings. The runtime values were obtained on Full HD (1920 × 1088) resolution images on the Snapdragon 8 Gen 1 mobile platform. The results of the MicroISP and PyNET-V2 Mobile models are provided for the reference. * The solution submitted by team *Multimedia* had corrupted weights due to incorrect model conversion, this issue was fixed after the end of the challenge.

| Team | Author | Framework | Model Size, MB | PSNR↑ | SSIM↑ | CPU Runtime, ms ↓ | GPU Runtime, ms ↓ | Final Score |
|------|--------|-----------|----------------|-------|-------|-------------------|-------------------|-------------|
| MiAlgo | mialgo_ls | TensorFlow | 0.014 | 23.33 | 0.8516 | **135** | **6.8** | **14.87** |
| ENERZAi | MinsuKwon | TensorFlow | 0.077 | 23.8 | 0.8652 | 208 | 18.9 | 10.27 |
| HITZST01 | Jaszheng | Keras/TensorFlow | 0.060 | 23.89 | 0.8666 | 712 | 34.3 | 6.41 |
| MINCHO | Minhyeok | TensorFlow | 0.067 | 23.65 | 0.8658 | 886 | 41.5 | 3.8 |
| ENERZAi | MinsuKwon | TensorFlow | 4.5 | 24.08 | 0.8778 | 45956 | 212 | 1.35 |
| HITZST01 | Jaszheng | Keras/TensorFlow | 1.2 | **24.09** | 0.8667 | 4694 | 482 | 0.6 |
| JMU-CVLab | nanashi | Keras/TensorFlow | 0.041 | 23.22 | 0.8281 | 3487 | 182 | 0.48 |
| Rainbow | zheng222 | TensorFlow | 1.0 | 21.66 | 0.8399 | 277 | 28 | 0.36 |
| CASIA 1st | Zevin | PyTorch/TensorFlow | 205 | **24.09** | **0.884** | 14792 | 1044 | 0.28 |
| MiAlgo | mialgo_ls | PyTorch/TensorFlow | 117 | 23.65 | 0.8673 | 15448 | 1164 | 0.14 |
| DANN-ISP | gosha20777 | TensorFlow | 29.4 | 23.1 | 0.8648 | 97333 | 583 | 0.13 |
| Multimedia * | lillythecutie | PyTorch/ OpenVINO | 0.029 | 23.96 | 0.8543 | 293 | 11.4 | 21.24 |
| SKD-VSP | dongdongdong | PyTorch/TensorFlow | 78.9 | 24.08 | 0.8778 | > 10 min | Failed | N.A |
| CHannel Team | sawyerk2212 | PyTorch/TensorFlow | 102.0 | 22.28 | 0.8482 | > 10 min | Failed | N.A. |
| MicroISP 1.0 [27] | Baseline | TensorFlow | 0.152 | 23.87 | 0.8530 | 973 | 23.1 | 9.25 |
| MicroISP 0.5 [27] | Baseline | TensorFlow | 0.077 | 23.60 | 0.8460 | 503 | 15.6 | 9.43 |
| PyNET-V2 Mobile [23] | Baseline | TensorFlow | 3.6 | 24.72 | 0.8783 | 8342 | 194 | 3.58 |

## 2.3 Runtime Evaluation on the Target Platform

In this challenge, we use the the *Qualcomm Snapdragon 8 Gen 1* mobile SoC as our target runtime evaluation platform. The considered chipset demonstrates very decent AI Benchmark scores and can be found in the majority of flagship Android smartphones released in 2022. It can efficiently accelerate floating-point networks on its Adreno 730 GPU with a theoretical FP16 performance of 5 TFLOPS. The models were parsed and accelerated using the TensorFlow Lite GPU delegate [46] demonstrating the best performance on this platform when using general deep learning models. All final solutions were tested using the aforementioned AI Benchmark application.

## 2.4 Challenge Phases

The challenge consisted of the following phases:

I. *Development:* the participants get access to the data and AI Benchmark app, and are able to train the models and evaluate their runtime locally;
II. *Validation:* the participants can upload their models to the remote server to check the fidelity scores on the validation dataset, and to compare their results on the validation leaderboard;
III. *Testing:* the participants submit their final results, codes, TensorFlow Lite models, and factsheets.

## 2.5 Scoring System

All solutions were evaluated using the following metrics:

– Peak Signal-to-Noise Ratio (PSNR) measuring fidelity score,

**Table 2.** Mean Opinion Scores (MOS) of all solutions submitted during the final phase of the MAI 2022 challenge and achieving a PSNR score of at least 23 dB. Visual results were assessed based on the reconstructed 12MP full resolution images.

| Team | Author | Framework | Model Size, MB | PSNR↑ | SSIM↑ | MOS Score |
|------|--------|-----------|----------------|-------|-------|-----------|
| HITZST01 | Jaszheng | Keras/TensorFlow | 1.2 | **24.09** | 0.8667 | 3.1 |
| ENERZAi | MinsuKwon | TensorFlow | 4.5 | 24.08 | 0.8778 | 3.1 |
| CASIA 1st | Zevin | PyTorch/TensorFlow | 205 | **24.09** | **0.884** | 3.0 |
| Multimedia | lillythecutie | PyTorch/ OpenVINO | 0.029 | 23.96 | 0.8543 | 3.0 |
| HITZST01 | Jaszheng | Keras/TensorFlow | 0.060 | 23.89 | 0.8666 | 3.0 |
| MINCHO | Minhyeok | TensorFlow | 0.067 | 23.65 | 0.8658 | 3.0 |
| ENERZAi | MinsuKwon | TensorFlow | 0.077 | 23.8 | 0.8652 | 2.8 |
| DANN-ISP | gosha20777 | TensorFlow | 29.4 | 23.1 | 0.8648 | 2.8 |
| MiAlgo | mialgo_ls | TensorFlow | 0.014 | 23.33 | 0.8516 | 2.5 |
| JMU-CVLab | nanashi | Keras/TensorFlow | 0.041 | 23.22 | 0.8281 | 2.3 |
| MiAlgo | mialgo_ls | PyTorch/TensorFlow | 117 | 23.65 | 0.8673 | 2.2 |

– Structural Similarity Index Measure (SSIM), a proxy for perceptual score,
– The runtime on the target Snapdragon 8 Gen 1 platform.

In this challenge, the participants were able to submit their final models to two tracks. In the first track, the score of each final submission was evaluated based on the next formula ($C$ is a constant normalization factor):

$$\text{Final Score} = \frac{2^{2 \cdot \text{PSNR}}}{C \cdot \text{runtime}},$$

In the second track, all submissions were evaluated only based on their visual results as measured by the corresponding Mean Opinion Scores (MOS). This was done to allow the participants to develop larger and more powerful models for the considered task.

During the final challenge phase, the participants did not have access to the test dataset. Instead, they had to submit their final TensorFlow Lite models that were subsequently used by the challenge organizers to check both the runtime and the fidelity results of each submission under identical conditions. This approach solved all the issues related to model overfitting, reproducibility of the results, and consistency of the obtained runtime/accuracy values.

## 3   Challenge Results

From the above 140 registered participants, 11 teams entered the final phase and submitted valid results, TFLite models, codes, executables, and factsheets. The proposed methods are described in Sect. 4, and the team members and affiliations are listed in Appendix A.

### 3.1   Results and Discussion

Tables 1 and 2 demonstrate the fidelity, runtime and MOS results of all solutions submitted during the final test phase. Models submitted to the 1st and 2nd challenge tracks were evaluated together since the participants had to upload the corresponding TensorFlow Lite models in both cases. In the 1st tack, the overall best results were achieved by team *Multimedia*. The authors proposed a novel *eReopConv* layer that consists of a large number of convolutions that are fused during the final model exporting stage to improve the its runtime while maintaining the fidelity scores. This approach turned out to be very efficient as the model submitted by this team was able to achieve one of the best PSNR and MOS scores as well as a runtime of less than 12 ms on the target Snapdragon platform. Unfortunately, the original TFLite file submitted by this team had corrupted weights caused by incorrect model conversion (the initial PyTorch model was converted to TFLite via ONNX), and this issue was fixed only after the end of the challenge.

The best runtime on the Snapdragon 8 Gen 1 was achieved by team *MiAlgo*, which solution is able to process one Full HD resolution photo on its GPU under 7 ms. This efficiency was achieved due to a very shallow structure of the proposed 3-layer neural network, which architecture was inspired by the last year MAI challenge winner [18]. The visual results obtained by this model are also satisfactory, though significantly fall behind the results of the solution from team *Multimedia*. The second best result in the 1st challenge track was obtained by team *ENERZAi* that proposed a UNet-based model with channel attention blocks. The final structure of this solution was obtained using the neural architecture search modified to take the computational complexity of the model as an additional key penalty parameter.

After evaluating the visual quality of the proposed models, we obtained quite similar MOS scores for half of the submissions. A detailed inspection of the results revealed that their overall quality can be generally considered as relatively comparable, though none of the models was able to perform an ideal image reconstruction: each model had some issues either with color rendition or with noise suppression/texture rendering. Solutions with a MOS score of less than 2.8 were usually having several issues or were exhibiting noticeable image corruptions. These results highlighted again the difficulty of an accurate assessment of the results obtained in learned ISP task as the conventional fidelity metrics are often not indicating the real image quality.

## 4   Challenge Methods

This section describes solutions submitted by all teams participating in the final stage of the MAI 2022 Learned Smartphone ISP challenge.

### 4.1   MiAlgo

For track 1, team MiAlgo proposed a smaller three-convolution structure based on last year's MAI 2021 winning solution [18] (Fig. 3). The authors reduced

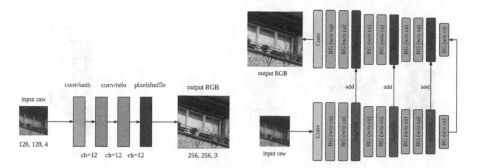

**Fig. 3.** Model architectures proposed by team MiAlgo for the 1st (left) and 2nd (right) challenge tracks.

the number of convolutional channels from 16 to 12, which also decreased the inference time by about 1/4. Besides that, the authors additionally used the distillation technique to remove the misalignment of some raw-RGB pairs in order to make the model converge better, which improved the PSNR score by about 0.3 dB. The model was trained for 10K epochs with L1 loss. The parameters were optimized with the Adam [44] algorithm using a batch size of 32 and a learning rate of 1e−4 that was decreased within the training.

For track 2, the authors proposed a 4-level UNet-based structure (Fig. 3, right). Several convolutional layers in the UNet [58] architecture were replaced with a residual group (RG, without channel attention layer) from RCAN [76] to enhance the reconstruction ability of the network. The authors used average pooling for the down-sampling layer and deconvolution for the up-sampling layer. The number of channels for each model level is 32, 64, 128, and 256 respectively. The model was first trained for 2K epochs with L1 loss, and then fine-tuned for about 2K epochs with the L1 and VGG loss functions. The initial learning rate was set to 1e–4 and was decreased within the training.

### 4.2 Multimedia

Inspired by a series of research on model re-parameterization from Ding [8–10], team Multimedia proposed an *enormous Re-parameter Convolution (eReop-Conv)* layer to replace the standard convolution. eReopConv has a large structure during training to learn superior proprieties but transforms into a less structured block during inference and retains these proprieties. The training and inference structures are shown in Fig. 4: unlike the RepConv in the RepVGG [10] that contain a 3 × 3 and 1 × 1 convolution layers in two branches during the training procedure, this method retains multi-branch convolution layers with different kernel sizes. *E.g.*, eRepConv with 5 × 5 kernel size has ten convolution layers with a kernel size ranging from 1 × 1 to 5 × 5 to collect enough information from different receptive fields. During the inference procedure, the training parameters are re-parameterized by continuous linear transforms.

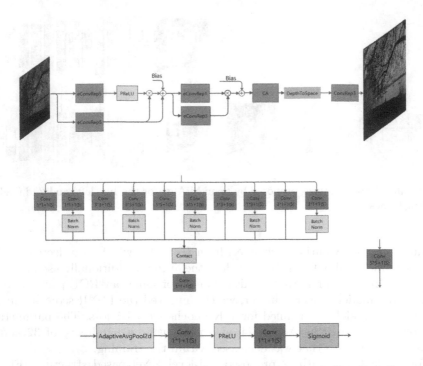

**Fig. 4.** The overall model architecture (top), the structure of the *eReopConv* block during the training and inference stages (middle), and the architecture of the CA module (bottom) proposed by team Multimedia.

As many spatial features are extracted by the eRepConv, a spatial attention mechanism could effectively improve the network performance. However, classical spatial attention blocks like [70] would increase the computational complexity because of the max pooling and average pooling operations. To save the extra costs, the authors utilize another eRepconv to produce a fine-granularity spatial attention block that has a specific attention matrix for each spatial feature in different channels. A learnable parameter is also added as a bias to expand this nonlinear expressivity. Although the fine-granularity spatial attention improves the performance, too many multiple operators and all channels fused by one convolution operator will inevitably make the model hard to train. To fix this, channel attention is added at the end of the network to help the network to converge since it could identify the importance of each channel.

Besides the Charbonnier loss and the cosine similarity loss, the authors propose to use an additional patch loss to enhance the image quality by considering the patch level information. In the patch loss, the ground truth $y$ and the generated $y'$ images are divided into a set of patches $\{y_{p1}, y_{p2}, ..., y_{pn}\}$ and $\{y'_{p1}, y'_{p2}, ...y'_{pn}\}$, then the mean and the variance of the difference between the generated and the ground truth patches is calculated and used as a weight for each pixel in the patch. The exact formulation of this loss function is:

$$L_P = \sum_{i=0}^{n} e^{mean(y_{pi}-y'_{pi})^{-1}+var(y_{pi}-y'_{pi})^{-1}} \times |y - y'|, \tag{1}$$

where *mean* represents how big the difference between the two images is, and *var* reflects whether the two images have similar changing rates.

The model is trained in two stages. During the warming-up training stage, besides the RAW-to-RGB transformation task, the model also learns to perform masked raw data recovery: raw image is divided into a set of $3 \times 3$ patches, 50% of them are randomly masked out, and the network is trained to recover the masked patches. This MAE-liked [13] strategy has two benefits. First, defective pixels often occur in the raw data, and it is quite useful to be able to recover these pixels. Secondly, since it is harder for a network to perform these two tasks simultaneously, the resulting model is usually more robust. The learning rate of the warming-up stage is set to 1e–6, and the model is trained with the L1 loss only. Next, in the second normal train stage the learning rate is set to 1e–3 and decayed with a cosine annealing scheduler, and the model is optimized with the Adam for only 350 epochs.

### 4.3   ENERZAi

For track 1, team ENERZAi first constructed a search space for the target model architecture based on the UNet [58] and Hourglass architectures. Since a right balance between image colors is important when constructing RGB images from raw data, a channel attention module [76] was also included in the model space. Optimal model design (Fig. 5) was obtained using an architecture search algorithm that was based on the evolutionary algorithm [56] and was taking latency into account by penalizing computationally heavy models.

The authors used L1 loss, MS-SIM and ResNet50-based perceptual loss functions to train the model. In addition, the authors developed a differentiable approximate histogram feature inspired by the Histogan [1] paper to consider the color distribution of the constructed RGB image. For each R, G, and B channel, the histogram differences between the constructed image and the target image were calculated and added to the total loss. The Adam optimizer with a learning rate of 0.001 was used for training the model, and the learning rate was halved each several epochs.

For track 2 (Fig. 5, bottom), the authors used the same approach but without penalizing computationally heavy models, which resulted in the increased number of convolutional layers and channel sizes.

### 4.4   HITZST01

Team HITZST01 proposed the RFD-CSA architecture [71] for the considered problem demonstrated in Fig. 6. The model consists of there modules: the Source Features Module, the Enhance Features Module and the Upsample Features Module. The purpose of the Source Features Module is to extract rough features

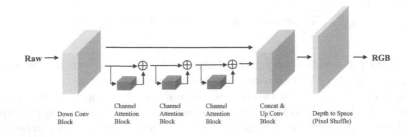

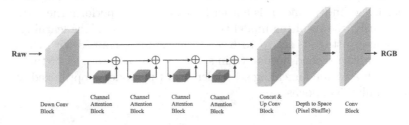

**Fig. 5.** Model designs obtained by team ENERZAi for the 1st (top) and 2nd (bottom) challenge tracks using the neural architecture search.

from the original raw images. This module is based on the ECB/Conv3 architecture proposed in [75] with the re-parameterization technique used to boost the performance while keeping the architecture efficient.

The Enhance Features Module is designed to extract effective features at multiple model levels. This module consists of $n$ lightweight multi-level feature extraction structures (with $n$ equal to 2 and 3 for models submitted to the 1st and 2nd challenge tracks, respectively). These structures are modified Residual feature distillation blocks (RFDB) proposed in [50], where the CCA-layer in RFDB is replaced with the CSA block. The authors additionally added a long-term residual connection in this module to avoid the performance degradation caused by model's depth and to boost its efficiency. The Upsample Features Module generates the final image reconstruction results. To improve the performance, ECB/Conv3 blocks are added at the beginning and at the end of this module. The model was trained with a combination of the Charbonnier and SSIM losses using the Adam optimizer with the initial learning rate set to 1e–4 and halved every 200 epochs.

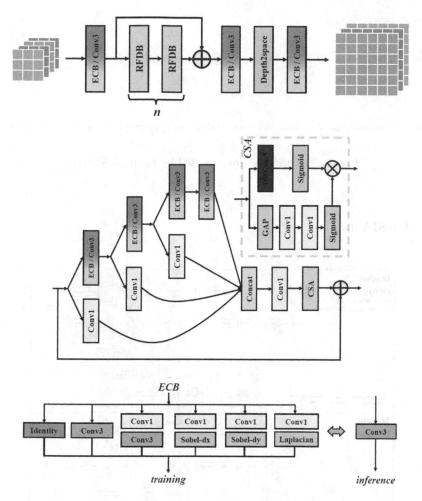

**Fig. 6.** The RFD-CSA architecture proposed by team HITZST01.

## 4.5 MINCHO

The architecture of the model developed by team MINCHO consists of the two main parts (Fig. 7). The first part is used to extract image features and consists of one convolutional blocks with *ReLU* activations, depthwise convolutional blocks, pointwise convolutional blocks with *ReLU* activations, and a skip connection. The second part of the model is the ISP part that is based on the Smallnet architecture [18] with three convolutional and one pixel-shuffle layer. The model was first trained with L1 loss only, and then fine-tuned with a combination of the L2, perceptual-based VGG-19 and SSIM loss functions. Model parameters were optimized using the ADAM algorithm with $\beta_1 = 0.9, \beta_2 = 0.99, \epsilon = 10^{-8}$, a learning rate of $10^{-4}$, and a batch size of 32.

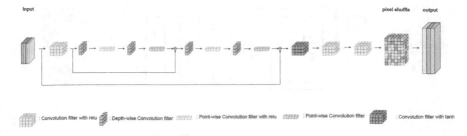

**Fig. 7.** The architecture proposed by team MINCHO.

## 4.6   CASIA 1st

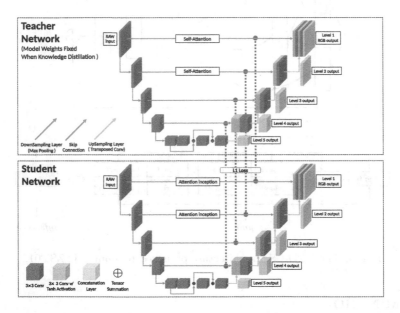

**Fig. 8.** An overview of the model and teacher-guided training strategy proposed by team CASIA 1st.

Team CASIA 1st proposed a two-stage teacher-guided model training strategy for the learned ISP problem (Fig. 8). At the first stage, the authors trained a teacher network (TN) to get good PSNR and SSIM scores. They used the PUNET network [18] as a baseline to develop an attention-aware based Unet (AA-Unet) architecture utilizing a self-attention module. Inspired by [39], the authors applied a multi-loss constraint to features at different model levels.

In the second stage, the authors designed a relatively tiny student network (SN) to inherit knowledge from the teacher network through model distillation.

The student network is very similar to the teacher model. Inspired by [62], the authors connected three attention modules (CBAM, ECA, and SEA) in parallel to form the attention inception module that was used to replace all self-attention modules in the TN to reduce the time complexity. Since the self-attention and attention inception modules have the same input/output shapes, the model distillation strategy was straightforward: the feature maps from each level in the TN model were extracted and used as soft-labels to train SN.

The models were trained with the MSE, L1, SSIM, VGG and edge [59] loss functions. When training the TN, they applied MSE loss to level 2, 3 and 4, SSIM loss to level 2 and 3, and MSE, SSIM, VGG, Edge loss to level 1. When training SN, they fixed the model weight of TN and applied L1 loss to each level with the soft-label. Similar to TN, MSE, SSIM and VGG loss are also applied to level 1. Both TN and SN model parameters were optimized using Adam with a batch size of 12, a learning rate of 5e–5 and a weight decay of 5e–5.

## 4.7 JMU-CVLab

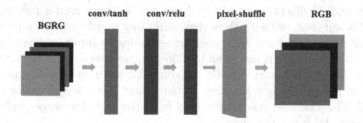

**Fig. 9.** Model architecture used by the JMU-CVLab team.

Team JMU-CVLab based their solution on the Smallnet model developed in the previous MAI 2021 challenge [18] (Fig. 9). The authors replaced the PixelShuffle layer with the Depth2Space op and added two well-known non-linear ISP operations: tone mapping and gamma correction [5]. These operations were applied to the final RGB reconstruction result followed by an additional CBAM attention block [70] applied to allow the model to learn more complex features. The model was trained to minimize the L1 and SSIM losses for 60 epochs with the Adam optimizer. A batch size of 32 with a learning rate of 0.0001 was used, basic augmentations including flips and rotations were applied to the training data.

## 4.8 DANN-ISP

Team DANN-ISP developed a domain adaptation method for the considered problem, which is illustrated in Fig. 10. The authors trained their model to generate RGB images with both source and target domains as inputs. Two

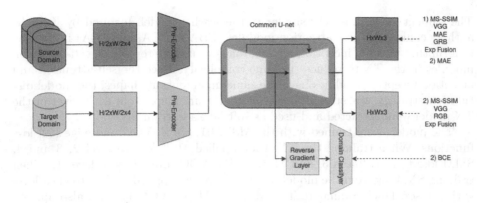

**Fig. 10.** Domain adaptation strategy developed by team DANN-ISP.

pre-encoders were used to reduce the significant domain gap between different cameras by extracting individual and independent features from each one. The architecture of the pre-encoders consisted of three $3 \times 3$ convolutional layers with 8, 16 and 32 filters, respectively. Next, the authors used a lightweight U-Net-like [58] autoencoder with three downsampling and four upsampling blocks. It takes 32-channel features from each pre-encoder as input and produces two outputs: a 3-channel RGB image and a 256-dimensional feature vector from its bottleneck. Finally, a binary domain classifier [12] with an inverse gradient [11] was used to reduce the gap between domains and increase the performance of the model. This classifier was constructed from the global average pooling layer and two dense layers at the end.

First, the model was pre-trained using only source domain data (in this case – using the Zurich-RAW-to-RGB dataset [39]) with the L1, MS-SSIM, VGG-based, color and exposure fusion losses. At the second stage, both source and target domain data were used together, and the model was trained to minimize a combination of the above losses plus the binary domain classifier loss (cross-entropy).

### 4.9  Rainbow

Team rainbow proposed a U-Net based AWBUnet model for this task. In traditional ISP pipelines, demosaicing can be solved by a neural network naturally, and white balance needs to be determined based on the input image. Therefore, the authors propose an RGB gain module (Fig. 11) consisting of a global average pooling (GAP) layer and three fully connected layers to adjust the white balance (RG channels) and brightness (RGB channels). The model was trained using a mini-batch size of 256, random horizontal flips were applied for data augmentation. The model was trained to minimizing the MSE loss function using the Adam optimizer. The initial learning rate was set to 2e–4 and halved every 100K iterations.

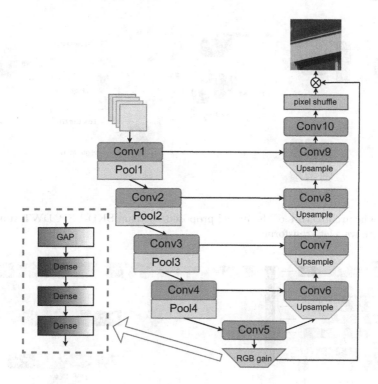

**Fig. 11.** AWBUnet model proposed by team rainbow.

## 4.10  SKD-VSP

Team SKD-VSP applied the discrete wavelet transform to divide the image into high-frequency and low-frequency parts, and used two different networks to process these two parts separately (Fig. 12). After processing, the inverse discrete wavelet transform was used to restore the final image. The architecture of the considered networks is based on the EDSR [48] model. Since the low-frequency part contains most of the global image information, the authors used 16 residual blocks with 256 channels in each block. For the high frequency part, 8 residual blocks with 64 channels were used due to its simple structure. When DWT operates on the image, the high-frequency part is divided into high-frequency information in three directions: horizontal, vertical and diagonal, so the input and output shape of the high-frequency network is $3 \times 3 \times \frac{H}{2} \times \frac{W}{2}$.

The authors used the L1 Loss for training the high frequency part, while for the low frequency part a combination of the L1, perceptual VGG-based and SSIM losses were utilized. Adam optimizer was used to train the model for 100 epochs with the initial learning rate set to 1e–3 and attenuated by a factor of 0.5 every 25 epochs.

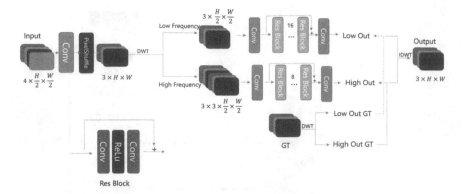

**Fig. 12.** The architecture of the model proposed by team SKD-VSP. DWT stands for the discrete wavelet transform.

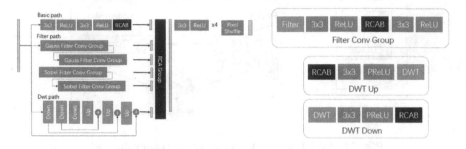

**Fig. 13.** GaUss-DWT model architecture proposed by the CHannel Team.

## 4.11    CHannel Team

CHannel Team proposed a multi-path GaUss-DWT model, which architecture is illustrated in Fig. 13. The model uses three paths to extract different information aspects from raw images: the basic path, the filter path, and the DWT path. The basic path consists of two convolutional layers and a residual attention block. This simple structure provides the model with a basic understanding of the image. The filter path consists of two different filters: Gauss and Sobel that extract chroma and texture information separately. Each convolutional group contains a fixed filter, two convolutional layers, and a residual channel attention block. The DWT path was inspired by the MW-ISPNet [37] model that attains high fidelity results in the AIM 2020 challenge. However, its structure was simplified by replacing its residual channel attention groups with residual channel attention blocks. This modification allowed to reduce the runtime of the model while attaining satisfactory results. Finally, a residual channel attention group integrates the extracted information from all three paths. Its output is then processed by four convolutional layers and a pixel shuffle layer to get the final image result. The model was trained with a combination of the perceptual VGG-based,

MSE, L1, MS-SSIM and edge loss functions. Model parameters were optimized using the Adam algorithm with a learning rate of 1e–4 and a batch size of 8.

## 5   Additional Literature

An overview of the past challenges on mobile-related tasks together with the proposed solutions can be found in the following papers:

- Learned End-to-End ISP: [23,27,33,37]
- Perceptual Image Enhancement: [28,36]
- Bokeh Effect Rendering: [26,35]
- Image Super-Resolution: [2,36,54,65]

**Acknowledgements.** We thank the sponsors of the Mobile AI and AIM 2022 workshops and challenges: AI Witchlabs, MediaTek, Huawei, Reality Labs, OPPO, Synaptics, Raspberry Pi, ETH Zürich (Computer Vision Lab) and University of Würzburg (Computer Vision Lab).

## A   Teams and Affiliations

**Mobile AI 2022 Team**

*Title:*
Mobile AI 2022 Learned Smartphone ISP Challenge
*Members:*
Andrey Ignatov[1,2] *(andrey@vision.ee.ethz.ch)*, Radu Timofte[1,2,3]
*Affiliations:*
[1] Computer Vision Lab, ETH Zurich, Switzerland
[2] AI Witchlabs, Switzerland
[3] University of Wuerzburg, Germany


**MiAlgo**

*Title:*
3Convs and BigUNet for Smartphone ISP
*Members:*
*Shuai Liu (liushuai21@xiaomi.com)*, Chaoyu Feng, Furui Bai, Xiaotao Wang, Lei Lei
*Affiliations:*
Xiaomi Inc., China

## Multimedia

*Title:*
FGARepNet: A real-time end-to-end ISP network based on Fine-Granularity attention and Re-parameter convolution
*Members:*
Ziyao Yi (yi.ziyao@sanechips.com.cn), Yan Xiang, Zibin Liu, Shaoqing Li, Keming Shi, Dehui Kong, Ke Xv
*Affiliations:*
Sanechips Co. Ltd, China


## ENERZAi Research

*Title:*
Latency-Aware NAS and Histogram Feature Loss
*Members:*
Minsu Kwon (minsu.kwon@enerzai.com)
*Affiliations:*
ENERZAi, Seoul, Korea
*enerzai.com*


## HITZST01

*Title:*
Residual Feature Distillation Channel Spatial Attention Network for ISP on Smartphones [71]
*Members:*
Yaqi Wu[1] (titimasta@163.com), Jiesi Zheng[2], Zhihao Fan[3], Xun Wu[4], Feng Zhang
*Affiliations:*
[1] Harbin Institute of Technology, China
[2] Zhejiang University, China
[3] University of Shanghai for Science and Technology, China
[4] Tsinghua University, China


## MINCHO

*Title:*
Mobile-Smallnet: Smallnet with MobileNet blocks for an end-to-end ISP Pipeline
*Members:*
Albert No (albertno@hongik.ac.kr), Minhyeok Cho
*Affiliations:*
Hongik University, Korea

## CASIA 1st

*Title:*
Learned Smartphone ISP Based On Distillation Acceleration
*Members:*
*Zewen Chen[1] (chenzewen2022@ia.ac.cn)*, Xiaze Zhang[2], Ran Li[3], Juan Wang[1], Zhiming Wang[4]
*Affiliations:*
[1] Institute of Automation, Chinese Academy of Sciences, China
[2] School of Computer Science, Fudan University, China
[3] Washington University in St. Louis
[4] Tsinghua University, China

## JMU-CVLab

*Title:*
Shallow Non-linear CNNs as ISP
*Members:*
*Marcos V. Conde (marcos.conde-osorio@uni-wuerzburg.de)*, Ui-Jin Choi
*Affiliations:*
University of Wuerzburg, Germany

## DANN-ISP

*Title:*
Learning End-to-End Deep Learning Based Image Signal Processing Pipeline Using Adversarial Domain Adaptation
*Members:*
*Georgy Perevozchikov (perevozchikov.gp@phystech.edu)*, Egor Ershov
*Affiliations:*
Moscow Institute of Physics and Technology, Russia

## Rainbow

*Title:*
Auto White Balance UNet for Learned Smartphone ISP
*Members:*
*Zheng Hui (huizheng.hz@alibaba-inc.com)*
*Affiliations:*
Alibaba DAMO Academy, China

**SKD-VSP**

*Title:*
IFS Net-Image Frequency Separation Residual Network
*Members:*
*Mengchuan Dong (mengchuan61@gmail.com)*, Wei Zhou, Cong Pang
*Affiliations:*
ShanghaiTech University, China

**CHannel Team**

*Title:*
GaUss-DWT net
*Members:*
*Haina Qin (qinhaina2020@ia.ac.cn)*, Mingxuan Cai
*Affiliations:*
Institute of Automation, Chinese Academy of Sciences, China

# References

1. Afifi, M., Brubaker, M.A., Brown, M.S.: HistoGAN: controlling colors of GAN-generated and real images via color histograms. In: Proceedings of the IEEE/CVF Conference on Computer Vision and Pattern Recognition, pp. 7941–7950 (2021)
2. Cai, J., Gu, S., Timofte, R., Zhang, L.: Ntire 2019 challenge on real image super-resolution: Methods and results. In: Proceedings of the IEEE/CVF Conference on Computer Vision and Pattern Recognition Workshops (2019)
3. Cai, Y., Yao, Z., Dong, Z., Gholami, A., Mahoney, M.W., Keutzer, K.: Zeroq: A novel zero shot quantization framework. In: Proceedings of the IEEE/CVF Conference on Computer Vision and Pattern Recognition, pp. 13169–13178 (2020)
4. Chiang, C.M., et al.: Deploying image deblurring across mobile devices: a perspective of quality and latency. In: Proceedings of the IEEE/CVF Conference on Computer Vision and Pattern Recognition Workshops, pp. 502–503 (2020)
5. Conde, M.V., McDonagh, S., Maggioni, M., Leonardis, A., Pérez-Pellitero, E.: Model-based image signal processors via learnable dictionaries. In: Proceedings of the AAAI Conference on Artificial Intelligence, vol. 36, pp. 481–489 (2022)
6. Conde, M.V., Timofte, R., et al.: Reversed image signal processing and RAW reconstruction. AIM 2022 challenge report. In: Proceedings of the European Conference on Computer Vision (ECCV) Workshops (2022)
7. Dai, L., Liu, X., Li, C., Chen, J.: AWNET: Attentive wavelet network for image ISP. arXiv preprint arXiv:2008.09228 (2020)
8. Ding, X., et al.: ResRep: lossless CNN pruning via decoupling remembering and forgetting. In: Proceedings of the IEEE/CVF International Conference on Computer Vision, pp. 4510–4520 (2021)
9. Ding, X., Xia, C., Zhang, X., Chu, X., Han, J., Ding, G.: RepMLP: Reparameterizing convolutions into fully-connected layers for image recognition. arXiv preprint arXiv:2105.01883 (2021)

10. Ding, X., Zhang, X., Ma, N., Han, J., Ding, G., Sun, J.: RepVGG: making VGG-style convnets great again. In: Proceedings of the IEEE/CVF Conference on Computer Vision and Pattern Recognition, pp. 13733–13742 (2021)

11. Ganin, Y., Lempitsky, V.: Unsupervised domain adaptation by backpropagation. In: International Conference on Machine Learning, pp. 1180–1189. PMLR (2015)

12. Ganin, Y., Ustinova, E., Ajakan, H., Germain, P., Larochelle, H., Laviolette, F., Marchand, M., Lempitsky, V.: Domain-adversarial training of neural networks. J. Mach. Learn. Res. **17**(1), 2030–2096 (2016)

13. He, K., Chen, X., Xie, S., Li, Y., Dollár, P., Girshick, R.: Masked autoencoders are scalable vision learners. In: Proceedings of the IEEE/CVF Conference on Computer Vision and Pattern Recognition, pp. 16000–16009 (2022)

14. Howard, A., et al.: Searching for mobilenetv3. In: Proceedings of the IEEE/CVF International Conference on Computer Vision, pp. 1314–1324 (2019)

15. Huang, J., et al.: Range scaling global U-Net for perceptual image enhancement on mobile devices. In: Leal-Taixé, L., Roth, S. (eds.) ECCV 2018. LNCS, vol. 11133, pp. 230–242. Springer, Cham (2019). https://doi.org/10.1007/978-3-030-11021-5_15

16. Hui, Z., Wang, X., Deng, L., Gao, X.: Perception-preserving convolutional networks for image enhancement on smartphones. In: Proceedings of the European Conference on Computer Vision (ECCV) Workshops (2018)

17. Ignatov, A., Byeoung-su, K., Timofte, R.: Fast camera image denoising on mobile GPUs with deep learning, mobile AI 2021 challenge: Report. In: Proceedings of the IEEE/CVF Conference on Computer Vision and Pattern Recognition Workshops (2021)

18. Ignatov, A., Chiang, J., Kuo, H.K., Sycheva, A., Timofte, R.: Learned smartphone isp on mobile NPUs with deep learning, mobile AI 2021 challenge: Report. In: Proceedings of the IEEE/CVF Conference on Computer Vision and Pattern Recognition Workshops (2021)

19. Ignatov, A., Kobyshev, N., Timofte, R., Vanhoey, K., Van Gool, L.: DSLR-quality photos on mobile devices with deep convolutional networks. In: Proceedings of the IEEE International Conference on Computer Vision, pp. 3277–3285 (2017)

20. Ignatov, A., Kobyshev, N., Timofte, R., Vanhoey, K., Van Gool, L.: WESPE: weakly supervised photo enhancer for digital cameras. In: Proceedings of the IEEE Conference on Computer Vision and Pattern Recognition Workshops, pp. 691–700 (2018)

21. Ignatov, A., Malivenko, G., Plowman, D., Shukla, S., Timofte, R.: Fast and accurate single-image depth estimation on mobile devices, mobile AI 2021 challenge: Report. In: Proceedings of the IEEE/CVF Conference on Computer Vision and Pattern Recognition Workshops (2021)

22. Ignatov, A., Malivenko, G., Timofte, R.: Fast and accurate quantized camera scene detection on smartphones, mobile AI 2021 challenge: report. In: Proceedings of the IEEE/CVF Conference on Computer Vision and Pattern Recognition Workshops (2021)

23. Ignatov, A., et al.: PyNet-V2 mobile: efficient on-device photo processing with neural networks. In: 2021 26th International Conference on Pattern Recognition (ICPR). IEEE (2022)

24. Ignatov, A., Malivenko, G., Timofte, R., et al.: Efficient single-image depth estimation on mobile devices, mobile AI & AIM 2022 challenge: report. In: European Conference on Computer Vision (2022)

25. Ignatov, A., Patel, J., Timofte, R.: Rendering natural camera bokeh effect with deep learning. In: Proceedings of the IEEE/CVF Conference on Computer Vision and Pattern Recognition Workshops, pp. 418–419 (2020)
26. Ignatov, A., et al.: Aim 2019 challenge on bokeh effect synthesis: methods and results. In: 2019 IEEE/CVF International Conference on Computer Vision Workshop (ICCVW), pp. 3591–3598. IEEE (2019)
27. Ignatov, A., et al.: MicroISP: processing 32mp photos on mobile devices with deep learning. In: European Conference on Computer Vision (2022)
28. Ignatov, A., Timofte, R.: Ntire 2019 challenge on image enhancement: methods and results. In: Proceedings of the IEEE/CVF Conference on Computer Vision and Pattern Recognition Workshops (2019)
29. Ignatov, A., et al.: Power efficient video super-resolution on mobile NPUs with deep learning, mobile AI & AIM 2022 challenge: report. In: European Conference on Computer Vision (2022)
30. Ignatov, A., et al.: AI benchmark: running deep neural networks on android smartphones. In: Leal-Taixé, L., Roth, S. (eds.) ECCV 2018. LNCS, vol. 11133, pp. 288–314. Springer, Cham (2019). https://doi.org/10.1007/978-3-030-11021-5_19
31. Ignatov, A., Timofte, R., Denna, M., Younes, A.: Real-time quantized image super-resolution on mobile NPUs, mobile AI 2021 challenge: report. In: Proceedings of the IEEE/CVF Conference on Computer Vision and Pattern Recognition Workshops (2021)
32. Ignatov, A., Timofte, R., Denna, M., Younes, A., et al.: Efficient and accurate quantized image super-resolution on mobile NPUs, mobile AI & AIM 2022 challenge: report. In: European Conference on Computer Vision (2022)
33. Ignatov, A., et al.: Aim 2019 challenge on raw to RGB mapping: methods and results. In: 2019 IEEE/CVF International Conference on Computer Vision Workshop (ICCVW), pp. 3584–3590. IEEE (2019)
34. Ignatov, A., et al.: AI benchmark: all about deep learning on smartphones in 2019. In: 2019 IEEE/CVF International Conference on Computer Vision Workshop (ICCVW), pp. 3617–3635. IEEE (2019)
35. Ignatov, A., et al.: AIM 2020 challenge on rendering realistic Bokeh. In: Bartoli, A., Fusiello, A. (eds.) ECCV 2020. LNCS, vol. 12537, pp. 213–228. Springer, Cham (2020). https://doi.org/10.1007/978-3-030-67070-2_13
36. Ignatov, A., et al.: PIRM challenge on perceptual image enhancement on smartphones: report. In: Proceedings of the European Conference on Computer Vision (ECCV) Workshops (2018)
37. Ignatov, A., et al.: AIM 2020 challenge on learned image signal processing pipeline. arXiv preprint arXiv:2011.04994 (2020)
38. Ignatov, A., Timofte, R., et al.: Realistic bokeh effect rendering on mobile GPUs, mobile AI & AIM 2022 challenge: report (2022)
39. Ignatov, A., Van Gool, L., Timofte, R.: Replacing mobile camera ISP with a single deep learning model. In: Proceedings of the IEEE/CVF Conference on Computer Vision and Pattern Recognition Workshops, pp. 536–537 (2020)
40. Ignatov, D., Ignatov, A.: Controlling information capacity of binary neural network. Pattern Recogn. Lett. **138**, 276–281 (2020)
41. Jacob, B., et al.: Quantization and training of neural networks for efficient integer-arithmetic-only inference. In: Proceedings of the IEEE Conference on Computer Vision and Pattern Recognition, pp. 2704–2713 (2018)
42. Jain, S.R., Gural, A., Wu, M., Dick, C.H.: Trained quantization thresholds for accurate and efficient fixed-point inference of deep neural networks. arXiv preprint arXiv:1903.08066 (2019)

43. Kim, B.-H., Song, J., Ye, J.C., Baek, J.H.: PyNET-CA: enhanced PyNET with channel attention for end-to-end mobile image signal processing. In: Bartoli, A., Fusiello, A. (eds.) ECCV 2020. LNCS, vol. 12537, pp. 202–212. Springer, Cham (2020). https://doi.org/10.1007/978-3-030-67070-2_12

44. Kingma, D.P., Ba, J.: Adam: a method for stochastic optimization. arXiv preprint arXiv:1412.6980 (2014)

45. Kınlı, F.O., Menteş, S., Özcan, B., Kirac, F., Timofte, R., et al.: AIM 2022 challenge on Instagram filter removal: Methods and results. In: Proceedings of the European Conference on Computer Vision (ECCV) Workshops (2022)

46. Lee, J.,et al.: On-device neural net inference with mobile GPUs. arXiv preprint arXiv:1907.01989 (2019)

47. Li, Y., Gu, S., Gool, L.V., Timofte, R.: Learning filter basis for convolutional neural network compression. In: Proceedings of the IEEE/CVF International Conference on Computer Vision, pp. 5623–5632 (2019)

48. Lim, B., Son, S., Kim, H., Nah, S., Mu Lee, K.: Enhanced deep residual networks for single image super-resolution. In: Proceedings of the IEEE Conference on Computer Vision and Pattern Recognition Workshops, pp. 136–144 (2017)

49. Liu, H., Navarrete Michelini, P., Zhu, D.: Deep networks for image-to-image translation with Mux and Demux layers. In: Leal-Taixé, L., Roth, S. (eds.) ECCV 2018. LNCS, vol. 11133, pp. 150–165. Springer, Cham (2019). https://doi.org/10.1007/978-3-030-11021-5_10

50. Liu, J., Tang, J., Wu, G.: Residual feature distillation network for lightweight image super-resolution. In: Bartoli, A., Fusiello, A. (eds.) ECCV 2020. LNCS, vol. 12537, pp. 41–55. Springer, Cham (2020). https://doi.org/10.1007/978-3-030-67070-2_2

51. Liu, Z., et al.: Metapruning: Meta learning for automatic neural network channel pruning. In: Proceedings of the IEEE/CVF International Conference on Computer Vision, pp. 3296–3305 (2019)

52. Liu, Z., Wu, B., Luo, W., Yang, X., Liu, W., Cheng, K.T.: Bi-Real net: enhancing the performance of 1-bit CNNs with improved representational capability and advanced training algorithm. In: Proceedings of the European Conference on Computer Vision (ECCV), pp. 722–737 (2018)

53. Lugmayr, A., Danelljan, M., Timofte, R.: Unsupervised learning for real-world super-resolution. In: 2019 IEEE/CVF International Conference on Computer Vision Workshop (ICCVW), pp. 3408–3416. IEEE (2019)

54. Lugmayr, A., Danelljan, M., Timofte, R.: Ntire 2020 challenge on real-world image super-resolution: methods and results. In: Proceedings of the IEEE/CVF Conference on Computer Vision and Pattern Recognition Workshops, pp. 494–495 (2020)

55. Obukhov, A., Rakhuba, M., Georgoulis, S., Kanakis, M., Dai, D., Van Gool, L.: T-basis: a compact representation for neural networks. In: International Conference on Machine Learning, pp. 7392–7404. PMLR (2020)

56. Real, E., Aggarwal, A., Huang, Y., Le, Q.V.: Regularized evolution for image classifier architecture search. In: Proceedings of the AAAI Conference on Artificial Intelligence, vol. 33, pp. 4780–4789 (2019)

57. Romero, A., Ignatov, A., Kim, H., Timofte, R.: Real-time video super-resolution on smartphones with deep learning, mobile AI 2021 challenge: report. In: Proceedings of the IEEE/CVF Conference on Computer Vision and Pattern Recognition Workshops (2021)

58. Ronneberger, O., Fischer, P., Brox, T.: U-Net: convolutional networks for biomedical image segmentation. In: Navab, N., Hornegger, J., Wells, W.M., Frangi, A.F. (eds.) MICCAI 2015. LNCS, vol. 9351, pp. 234–241. Springer, Cham (2015). https://doi.org/10.1007/978-3-319-24574-4_28

59. Seif, G., Androutsos, D.: Edge-based loss function for single image super-resolution. In: 2018 IEEE International Conference on Acoustics, Speech and Signal Processing (ICASSP), pp. 1468–1472. IEEE (2018)
60. Silva, J.I.S., et al.: A deep learning approach to mobile camera image signal processing. In: Anais Estendidos do XXXIII Conference on Graphics, Patterns and Images, pp. 225–231. SBC (2020)
61. de Stoutz, E., Ignatov, A., Kobyshev, N., Timofte, R., Van Gool, L.: Fast perceptual image enhancement. In: Proceedings of the European Conference on Computer Vision (ECCV) Workshops (2018)
62. Szegedy, C., Ioffe, S., Vanhoucke, V., Alemi, A.A.: Inception-v4, inception-resnet and the impact of residual connections on learning. In: Thirty-first AAAI Conference on Artificial Intelligence (2017)
63. Tan, M., et al.: MNASNet: platform-aware neural architecture search for mobile. In: Proceedings of the IEEE/CVF Conference on Computer Vision and Pattern Recognition, pp. 2820–2828 (2019)
64. TensorFlow-Lite. https://www.tensorflow.org/lite
65. Timofte, R., Gu, S., Wu, J., Van Gool, L.: Ntire 2018 challenge on single image super-resolution: methods and results. In: Proceedings of the IEEE Conference on Computer Vision and Pattern Recognition Workshops, pp. 852–863 (2018)
66. Truong, P., Danelljan, M., Van Gool, L., Timofte, R.: Learning accurate dense correspondences and when to trust them. arXiv preprint arXiv:2101.01710 (2021)
67. Uhlich, S., et al.: Mixed precision DNNs: All you need is a good parametrization. arXiv preprint arXiv:1905.11452 (2019)
68. Vu, T., Nguyen, C.V., Pham, T.X., Luu, T.M., Yoo, C.D.: Fast and efficient image quality enhancement via DesubPixel convolutional neural networks. In: Leal-Taixé, L., Roth, S. (eds.) ECCV 2018. LNCS, vol. 11133, pp. 243–259. Springer, Cham (2019). https://doi.org/10.1007/978-3-030-11021-5_16
69. Wan, A., et al.: FBNetV2: differentiable neural architecture search for spatial and channel dimensions. In: Proceedings of the IEEE/CVF Conference on Computer Vision and Pattern Recognition, pp. 12965–12974 (2020)
70. Woo, S., Park, J., Lee, J.-Y., Kweon, I.S.: CBAM: convolutional block attention module. In: Ferrari, V., Hebert, M., Sminchisescu, C., Weiss, Y. (eds.) ECCV 2018. LNCS, vol. 11211, pp. 3–19. Springer, Cham (2018). https://doi.org/10.1007/978-3-030-01234-2_1
71. Wu, Y., Zheng, J., Fan, Z., Wu, X., Zhang, F.: Residual feature distillation channel spatial attention network for ISP on smartphone. In: Proceedings of the European Conference on Computer Vision (ECCV) Workshops (2022)
72. Wu, B., et al.: FBNet: hardware-aware efficient convnet design via differentiable neural architecture search. In: Proceedings of the IEEE/CVF Conference on Computer Vision and Pattern Recognition, pp. 10734–10742 (2019)
73. Yang, J., et al.: Quantization networks. In: Proceedings of the IEEE/CVF Conference on Computer Vision and Pattern Recognition, pp. 7308–7316 (2019)
74. Yang, R., Timofte, R., et al.: AIM 2022 challenge on super-resolution of compressed image and video: dataset, methods and results. In: Proceedings of the European Conference on Computer Vision (ECCV) Workshops (2022)
75. Zhang, X., Zeng, H., Zhang, L.: Edge-oriented convolution block for real-time super resolution on mobile devices. In: Proceedings of the 29th ACM International Conference on Multimedia, pp. 4034–4043 (2021)
76. Zhang, Y., Li, K., Li, K., Wang, L., Zhong, B., Fu, Y.: Image super-resolution using very deep residual channel attention networks. In: Proceedings of the European Conference on Computer Vision (ECCV), pp. 286–301 (2018)

# Efficient Single-Image Depth Estimation on Mobile Devices, Mobile AI & AIM 2022 Challenge: Report

Andrey Ignatov[1,2(✉)], Grigory Malivenko[1,2,3], Radu Timofte[1,2,3],
Lukasz Treszczotko[4], Xin Chang[4], Piotr Ksiazek[4], Michal Lopuszynski[4],
Maciej Pioro[4], Rafal Rudnicki[4], Maciej Smyl[4], Yujie Ma[4], Zhenyu Li[5],
Zehui Chen[5], Jialei Xu[5], Xianming Liu[5], Junjun Jiang[5], XueChao Shi[6],
Difan Xu[6], Yanan Li[6], Xiaotao Wang[6], Lei Lei[6], Ziyu Zhang[7], Yicheng Wang[7],
Zilong Huang[7], Guozhong Luo[7], Gang Yu[7], Bin Fu[7], Jiaqi Li[8], Yiran Wang[8],
Zihao Huang[8], Zhiguo Cao[8], Marcos V. Conde[3], Denis Sapozhnikov[3],
Byeong Hyun Lee[9], Dongwon Park[9], Seongmin Hong[9], Joonhee Lee[9],
Seunggyu Lee[9], and Se Young Chun[9]

[1] Computer Vision Lab, ETH Zürich, Zürich, Switzerland
{andrey,radu.timofte}@vision.ee.ethz.ch
[2] AI Witchlabs, Zollikerberg, Switzerland
[3] University of Wuerzburg, Wuerzburg, Germany
{radu.timofte,marcos.conde-osorio}@uni-wuerzburg.de
[4] TCL Research Europe, Warsaw, Poland
lukasz.treszczotko@tcl.com
[5] Harbin Institute of Technology, Harbin, China
zhenyuli17@hit.edu.cn
[6] Xiaomi Inc., Beijing, China
shixuechao@xiaomi.com
[7] Tencent GY-Lab, Shenzhen, China
parkzyzhang@tencent.com
[8] National Key Laboratory of Science and Technology on Multi-Spectral
Information Processing, School of Artificial Intelligence and Automation,
Huazhong University of Science and Technology, Wuhan, China
lijiaqi_mail@hust.edu.cn
[9] Department of Electrical and Computer Engineering, Seoul National University,
Seoul, South Korea
ldlqudgus756@snu.ac.kr

**Abstract.** Various depth estimation models are now widely used on many mobile and IoT devices for image segmentation, bokeh effect rendering, object tracking and many other mobile tasks. Thus, it is very crucial to have efficient and accurate depth estimation models that can run fast on low-power mobile chipsets. In this Mobile AI challenge, the

A. Ignatov, G. Malivenko, and R. Timofte are the Mobile AI & AIM 2022 challenge organizers. The other authors participated in the challenge. Appendix A contains the authors' team names and affiliations.
Mobile AI 2022 Workshop website:
https://ai-benchmark.com/workshops/mai/2022/.

target was to develop deep learning-based single image depth estimation solutions that can show a real-time performance on IoT platforms and smartphones. For this, the participants used a large-scale RGB-to-depth dataset that was collected with the ZED stereo camera capable to generated depth maps for objects located at up to 50 m. The runtime of all models was evaluated on the Raspberry Pi 4 platform, where the developed solutions were able to generate VGA resolution depth maps at up to 27 FPS while achieving high fidelity results. All models developed in the challenge are also compatible with any Android or Linux-based mobile devices, their detailed description is provided in this paper.

**Keywords:** Mobile ai challenge · Depth estimation · Raspberry pi · Mobile AI · Deep learning · AI Benchmark

## 1 Introduction

Single-image depth estimation is an important vision task. There is an ever-increasing demand for fast and efficient single-image depth estimation solutions that can run on mobile devices equipped with low-power hardware. This demand is driven by a wide range of depth-guided problems arising in mixed and augmented reality, autonomous driving, human-computer interaction, image segmentation, bokeh effect synthesis tasks. The depth estimation research from the past decade produced multiple accurate deep learning-based solutions [9,13,15,16,49,53,55,56]. However, these solutions are generally focusing on high fidelity results and are lacking the computational efficiency, thus not meeting various mobile-related constraints, which are key for image processing tasks [22,23,43] on mobile devices. Therefore, most of the current state-of-the-art solutions have high computational and memory requirements even for processing low-resolution input images and are incompatible with the scarcity of resources found on the mobile hardware. In this challenge, we are using a large-scale depth estimation dataset and target the development of efficient solutions capable to meet hardware constraints like the ones

**Fig. 1.** The original RGB image and the corresponding depth map obtained with the ZED 3D camera.

found on the Rasberry Pi 4 platform. This is the second installment of this challenge. The previous edition was in conjunction with Mobile AI 2021 CVPR workshop [24].

The deployment of AI-based solutions on portable devices usually requires an efficient model design based on a good understanding of the mobile processing units (*e.g.*CPUs, NPUs, GPUs, DSP) and their hardware particularities, including their memory constraints. We refer to [33,37] for an extensive overview of mobile AI acceleration hardware, its particularities and performance. As shown in these works, the latest generations of mobile NPUs are reaching the performance of older-generation mid-range desktop GPUs. Nevertheless, a straightforward deployment of neural networks-based solutions on mobile devices is impeded by (i) a limited memory (*i.e.*, restricted amount of RAM) and (ii) a limited or lacking support of many common deep learning operators and layers. These impeding factors make the processing of high resolution inputs impossible with the standard NN models and require a careful adaptation or re-design to the constraints of mobile AI hardware. Such optimizations can employ a combination of various model techniques such as 16-bit/8-bit [10,45,46,77] and low-bit [7,44,60,71] quantization, network pruning and compression [10,27,52,59,63], device- or NPU-specific adaptations, platform-aware neural architecture search [19,67,73,76], *etc.*

The majority of competitions aimed at efficient deep learning models use standard desktop hardware for evaluating the solutions, thus the obtained models rarely show acceptable results when running on real mobile hardware with many specific constraints. In this *Mobile AI challenge*, we take a radically different approach and propose the participants to develop and evaluate their models directly on mobile devices. The goal of this competition is to design a fast and performant deep learning-based solution for a single-image depth estimation problem. For this, the participants were provided with a large-scale training dataset containing over 8K RGB-depth image pairs captured using the ZED stereo camera. The efficiency of the proposed solutions was evaluated on the popular Raspberry Pi 4 ARM-based single-board computer widely used for many machine learning IoT projects. The overall score of each submission was computed based on its fidelity (si-RMSE) and runtime results, thus balancing between the depth map reconstruction quality and the computational efficiency of the model. All solutions developed in this challenge are fully compatible with the TensorFlow Lite framework [68], thus can be efficiently executed on various Linux and Android-based IoT platforms, smartphones and edge devices.

This challenge is a part of the *Mobile AI & AIM 2022 Workshops and Challenges* consisting of the following competitions:

– Efficient Single-Image Depth Estimation on Mobile Devices
– Learned Smartphone ISP on Mobile GPUs [41]
– Power Efficient Video Super-Resolution on Mobile NPUs [32]
– Quantized Image Super-Resolution on Mobile NPUs [35]
– Realistic Bokeh Effect Rendering on Mobile GPUs [42]
– Super-Resolution of Compressed Image and Video [78]

- Reversed Image Signal Processing and RAW Reconstruction [11]
- Instagram Filter Removal [48].

The results and solutions obtained in the previous *MAI 2021 Challenges* are described in our last year papers:

- Single-Image Depth Estimation on Mobile Devices [24]
- Learned Smartphone ISP on Mobile NPUs [21]
- Real Image Denoising on Mobile GPUs [20]
- Quantized Image Super-Resolution on Mobile NPUs [34]
- Real-Time Video Super-Resolution on Mobile GPUs [29]
- Quantized Camera Scene Detection on Smartphones [25].

## 2   Challenge

In order to design an efficient and practical deep learning-based solution for the considered task that runs fast on mobile devices, one needs the following tools:

**Fig. 2.** Loading and running custom TensorFlow Lite models with AI Benchmark application. The currently supported acceleration options include Android NNAPI, TFLite GPU, Hexagon NN, Qualcomm QNN, MediaTek Neuron and Samsung ENN delegates as well as CPU inference through TFLite or XNNPACK backends. The latest app version can be downloaded at https://ai-benchmark.com/download

1. A large-scale high-quality dataset for training and evaluating the models. Real, not synthetically generated data should be used to ensure a high quality of the obtained model;
2. An efficient way to check the runtime and debug the model locally without any constraints as well as the ability to check the runtime on the target evaluation platform.

This challenge addresses all the above issues. Real training data, tools, and runtime evaluation options provided to the challenge participants are described in the next sections.

## 2.1   Dataset

To get real and diverse data for the considered challenge, a novel dataset consisting of RGB-depth image pairs was collected using the ZED stereo camera[1] capable of shooting 2K images. It demonstrates an average depth estimation error of less than 0.2 m for objects located closer than 8 m [64], while more coarse predictions are also available for distances of up to 50 m. Around 8.3K image pairs were collected in the wild over several weeks in a variety of places. For this challenge, the obtained images were downscaled to VGA resolution (640 × 480 pixels) that is typically used on mobile devices for different depth-related tasks. The original RGB images were then considered as inputs, and the corresponding 16-bit depth maps — as targets. A sample RGB-depth image pair from the collected dataset is demonstrated in Fig. 1.

## 2.2   Local Runtime Evaluation

When developing AI solutions for mobile devices, it is vital to be able to test the designed models and debug all emerging issues locally on available devices. For this, the participants were provided with the *AI Benchmark* application [33, 37] that allows to load any custom TensorFlow Lite model and run it on any Android device with all supported acceleration options. This tool contains the latest versions of *Android NNAPI, TFLite GPU, Hexagon NN, Qualcomm QNN, MediaTek Neuron* and *Samsung ENN* delegates, therefore supporting all current mobile platforms and providing the users with the ability to execute neural networks on smartphone NPUs, APUs, DSPs, GPUs and CPUs.

To load and run a custom TensorFlow Lite model, one needs to follow the next steps:

1. Download AI Benchmark from the official website[2] or from the Google Play[3] and run its standard tests.
2. After the end of the tests, enter the *PRO Mode* and select the *Custom Model* tab there.
3. Rename the exported TFLite model to *model.tflite* and put it into the *Download* folder of the device.
4. Select mode type *(INT8, FP16, or FP32)*, the desired acceleration/inference options and run the model.

These steps are also illustrated in Fig. 2.

## 2.3   Runtime Evaluation on the Target Platform

In this challenge, we use the *Raspberry Pi 4* single-board computer as our target runtime evaluation platform. It is based on the *Broadcom BCM2711* chipset

---

[1] https://www.stereolabs.com/zed/.

[2] https://ai-benchmark.com/download.

[3] https://play.google.com/store/apps/details?id=org.benchmark.demo.

containing four Cortex-A72 ARM cores clocked at 1.5 GHz and demonstrates AI Benchmark scores comparable to entry-level Android smartphone SoCs [3]. The Raspberry Pi 4 supports the majority of Linux distributions, Windows 10 IoT build as well as Android operating system. In this competition, the runtime of all solutions was tested using the official TensorFlow Lite 2.5.0 Linux build [69] containing many important performance optimizations for the above chipset, the default *Raspberry Pi OS* was installed on the device. Within the challenge, the participants were able to upload their TFLite models to our dedicated competition platform[4] connected to a real Raspberry Pi 4 board and get instantaneous feedback: the runtime of their solution or an error log if the model contains some incompatible operations. The same setup was also used for the final runtime evaluation.

**Table 1.** MAI 2022 Monocular Depth Estimation challenge results and final rankings. The runtime values were obtained on $640 \times 480$ px images on the Raspberry Pi 4 device. Team *TCL* is the challenge winner. * This model was the challenge winner in the previous MAI 2021 depth estimation challenge.

| Team | Author | Framework | Model Size, MB | si-RMSE↓ | RMSE↓ | LOG10↓ | REL↓ | Runtime, ms ↓ | Final Score |
|------|--------|-----------|----------------|----------|-------|--------|------|---------------|-------------|
| TCL | TCL | TensorFlow | 2.9 | **0.2773** | **3.47** | **0.1103** | 0.2997 | 46 | **298** |
| AIIA HIT | Zhenyu Li | PyTorch/TensorFlow | 1.5 | 0.311 | 3.79 | 0.1241 | 0.3427 | **37** | 232 |
| MiAlgo | ChaoMI | PyTorch/TensorFlow | 1.0 | 0.299 | 3.89 | 0.1349 | 0.3807 | 54 | 188 |
| Tencent GY-Lab | Parkzyzhang | PyTorch/TensorFlow | 3.4 | 0.303 | 3.8 | 0.1899 | 0.3014 | 68 | 141 |
| Tencent GY-Lab* | Parkzyzhang | PyTorch/TensorFlow | 3.4 | 0.2836 | 3.56 | 0.1121 | **0.2690** | 103 | 122 |
| SmartLab | RocheL | TensorFlow | 7.1 | 0.3296 | 4.06 | 0.1378 | 0.3662 | 65 | 102 |
| JMU-CVLab | mvc | PyTorch/TensorFlow | 3.5 | 0.3498 | 4.46 | 0.1402 | 0.3404 | 139 | 36 |
| ICL | Byung Hyun Lee | PyTorch/TensorFlow | 5.9 | 0.338 | 6.73 | 0.3323 | 0.5070 | 142 | 42 |

## 2.4 Challenge Phases

The challenge consisted of the following phases:

I. *Development:* the participants get access to the data and AI Benchmark app, and are able to train the models and evaluate their runtime locally;
II. *Validation:* the participants can upload their models to the remote server to check the fidelity scores on the validation dataset, to get the runtime on the target platform, and to compare their results on the validation leaderboard;
III. *Testing:* the participants submit their final results, codes, TensorFlow Lite models, and factsheets.

## 2.5 Scoring System

All solutions were evaluated using the following metrics:

- Root Mean Squared Error (RMSE) measuring the absolute depth estimation accuracy,

---

[4] https://ml-competitions.com.

- Scale Invariant Root Mean Squared Error (si-RMSE) measuring the quality of relative depth estimation (relative position of the objects),
- Average $\log_{10}$ and Relative (REL) errors [56],
- The runtime on the target Raspberry Pi 4 device.

The score of each final submission was evaluated based on the next formula ($C$ is a constant normalization factor):

$$\text{Final Score} = \frac{2^{-20 \cdot \text{si-RMSE}}}{C \cdot \text{runtime}},$$

During the final challenge phase, the participants did not have access to the test dataset. Instead, they had to submit their final TensorFlow Lite models that were subsequently used by the challenge organizers to check both the runtime and the fidelity results of each submission under identical conditions. This approach solved all the issues related to model overfitting, reproducibility of the results, and consistency of the obtained runtime/accuracy values.

## 3 Challenge Results

From above 70 registered participants, 7 teams entered the final phase and submitted valid results, TFLite models, codes, executables and factsheets. Table 1 summarizes the final challenge results and reports si-RMSE, RMSE, LOG10 and REL measures and runtime numbers for each submitted solution on the final test dataset and on the target evaluation platform. The proposed methods are described in Sect. 4, and the team members and affiliations are listed in Appendix A.

### 3.1 Results and Discussion

All solutions proposed in this challenge demonstrated a very high efficiency, being able to produce depth maps under 150 ms on the Raspberry 4 platform. As expected, all proposed architectures use a U-Net based structure, where MobileNets or EffientNets are generally used in the encoder part to extract relevant features from the input image. Team *TCL* is the winner of the challenge: their model demonstrated the best accuracy results and a runtime of less than 50 ms on the target device. Compared to the last year's winning challenge solution [80], this model is slightly more accurate while more than 2 times faster. The second best result was achieved by team *AIIA HIT*. This solution is able to run at more than 27 FPS on the Raspberry Pi 4, thus demonstrating a nearly real-time performance, which is critical for many depth estimation applications. Overall, we can see a noticeable improvement in the efficiency of the proposed solutions compared to the models produced in the previous Mobile AI depth estimation challenge [24], which allows for faster and more accurate depth estimation models on mobile devices.

## 4    Challenge Methods

This section describes solutions submitted by all teams participating in the final stage of the MAI 2022 Monocular Depth Estimation challenge.

### 4.1    TCL

Team TCL proposed a UNet-like [66] architecture presented in Fig. 3. The input image is first resized from $640 \times 480$ to $160 \times 128$ pixels and then passed to a simplified MobileNetV3-based [19] encoder. The lowest resolution blocks were completely remove from the encoder module, which had a little impact on the overall quality, but led to a noticeable latency drop. The output of the encoder is then passed to the decoder module that contains only Collapsible Linear Blocks (CLB) [5]. Finally, the output of the decoder is resized with a scale factor of 10 from $48 \times 64$ to $480 \times 640$ pixels.

The objective function described in [50] was used as it provided better results than the standard RMSE loss. Model parameters were optimized using the Adam [47] algorithm with the cosine learning rate scheduler.

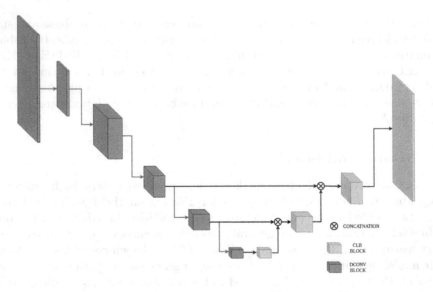

**Fig. 3.** Model architecture proposed by team TCL.

## 4.2 AIIA HIT

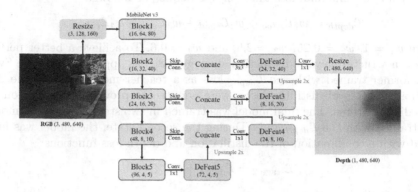

**Fig. 4.** An overview of the model proposed by team AIIA HIT.

Team AIIA HIT followed a similar approach of using a UNet-like architecture with a MobileNetV3-based encoder and an extremely lightweight decoder part that only contains a single $3 \times 3$ convolution and four $1 \times 1$ convolution modules. The architecture of this solution [54] is demonstrated in Fig. 4. The training process included a custom data augmentation strategy called the R$^2$ crop (Random locations and Random size changes of patches). Four different loss functions were used for training. The first SILog [13] loss function was defined as:

$$\mathcal{L}_{silog} = \alpha \sqrt{\frac{1}{N} \sum_i^N e_i^2 - \frac{\lambda}{N^2} (\sum_i^N e_i)^2}, \tag{1}$$

where $e_i = \log \hat{d}_i - \log d_i$ is the log difference between the ground truth $d_i$ and the predicted $\hat{d}_i$ depth maps. $N$ here denotes the number of pixels having valid ground truth values. The other three losses are the gradient loss $\mathcal{L}_{grad}$, the virtual norm loss $\mathcal{L}_{vnl}$ [79], and the robust loss $\mathcal{L}_{robust}$ [4] specified as follows:

$$\mathcal{L}_{grad} = \frac{1}{T} \sum_i \left( \left\| \nabla_x \hat{d}_i - \nabla_x d_i \right\|_1 + \left\| \nabla_y \hat{d}_i - \nabla_y d_i \right\|_1 \right) \tag{2}$$

$$\mathcal{L}_{vnl} = \sum_i^N \left\| \hat{n}_i - n_i \right\|_1, \tag{3}$$

$$\mathcal{L}_{robust} = \frac{|\alpha - 2|}{\alpha} \left( \left( \frac{(x/c)^2}{|\alpha - 2|} \right)^{\alpha/2} - 1 \right), \tag{4}$$

where $\nabla$ is the gradient operator, $\alpha = 1$ and $c = 2$. It should be noted that the $\mathcal{L}_{vnl}$ loos slightly differs from the original implementation as the points here are sampled from the reconstructed point clouds (instead of the ground truth maps) to filter invalid samples as this helped the model to converge at the beginning of the training. The final loss function was defined as follows:

$$\mathcal{L}_{depth} = w_1\mathcal{L}_{silog} + w_2\mathcal{L}_{grad} + w_3\mathcal{L}_{vnl} + w_4\mathcal{L}_{robust}. \tag{5}$$

where $w_1 = 1$, $w_2 = 0.25$, $w_3 = 2.5$, and $w_4 = 0.6$. To achieve a better performance, a structure-aware distillation strategy [58] was applied, where the Swin Transformer trained with the $\mathcal{L}_{depth}$ loss as a teacher model. During the distillation, multi-level distilling losses were adopted to provide a supervision on immediate features. The final model was trained in two stages. First, the model was trained with the $\mathcal{L}_{depth}$ only. During the second state, the model was fine-tuned with a combination of the distillation and depth loss functions:

$$\mathcal{L}_{\text{stage-2}} = \mathcal{L}_{depth} + w\mathcal{L}_{distill} \tag{6}$$

where $w = 10$.

### 4.3  MiAIgo

Team MiAIgo used a modified GhostNet [17] architecture with a reduced number of channels for feature extraction. The overall architecture of the proposed

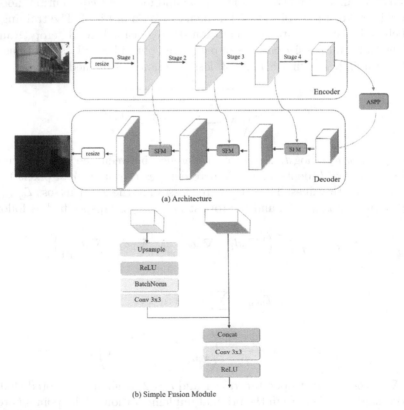

(a) Architecture

(b) Simple Fusion Module

**Fig. 5.** The architecture of the model and the structure of the Simple Fusion Module (SFM) proposed by team MiAIgo.

model is demonstrated in Fig. 5. The input image is first resized from $640 \times 480$ to $128 \times 96$ pixels and then passed to the encoder module consisting of four blocks. The Atrous Spatial Pyramid Pooling (ASPP) [8] module was placed on the top of the encoder to process multi-scale contextual information and increase the receptive field. According to the experiments, ASPP provides a better accuracy with minor inference time expenses. Besides that, the Simple Fusion Module (SFM) was designed and used in the decoder part of the model. The SFM concatenates the outputs of different stages with the decoder feature maps to achieve better fidelity results. An additional nearest neighbor resizing layer was placed after the decoder part to upscale the model output to the target resolution. The model was trained to minimize the SSIM loss function, its parameters were optimized using the Adam algorithm with a learning rate of 8e–4 and a batch size of 8.

### 4.4 Tencent GY-Lab

Team Tencent GY-Lab proposed a U-Net like architecture presented in Fig. 6, where a MobileNet-V3 [19] based encoder is used for dense feature extraction. To reduce the amount of computations, the input image is first resized from $640 \times 480$ to $128 \times 96$ pixels and then passed to the encoder module consisting of five blocks. The outputs of each block are processed by the Feature Fusion Module (FFM) that concatenates them with the decoder feature maps to get better fidelity results. The authors use one additional *nearest neighbor* resizing layer on top of the model to upscale the output to the target resolution. Knowledge distillation [18] is further used to improve the quality of the reconstructed depth maps: a bigger ViT-Large [12] was first trained on the same dataset and then its features obtained before the last activation function were used to guide the smaller network. This process allowed to decrease the si-RMSE score from 0.3304 to 0.3141. The proposed model was therefore trained to minimize a combination of the distillation loss (computed as $L_2$ norm between its features from the last convolutional layer and the above mentioned features from the larger model), and the depth estimation loss proposed in [51]. The network parameters were optimized for 500 epochs using Adam [47] with a learning rate of $8e-3$ and a polynomial decay with a power of 0.9. The model was implemented and trained

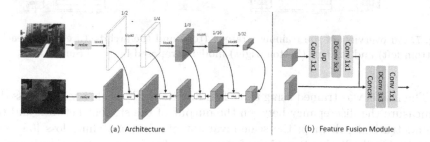

**Fig. 6.** The model architecture and the structure of the Feature Fusion Module (FFM) proposed by team Tencent GY-Lab.

with PyTorch and then converted to TensorFlow Lite using ONNX as an intermediate representation. A more detailed description of the proposed solution is
provided in [80].

## 4.5 SmartLab

The solution proposed by team SmartLab largely relies on transfer learning and
consists of two networks as shown in Fig. 7. The representation ability of the
pretrained teacher network is transferred to the student network via the pairwise distillation [57]. Scale-invariant loss [14] and gradient matching loss [65]
were used to train the student network, guiding the model to predict more accurate depth maps. Both teacher and student networks are based on the encoder-
decoder architecture with the EfficientNet used as a backbone. The network
structure design mainly refers to the MiDaS [65] paper, the encoder uses the
EfficientNet-B1 architecture, and the basic block of the decoder is the Feature
Fusion Module (FFM).

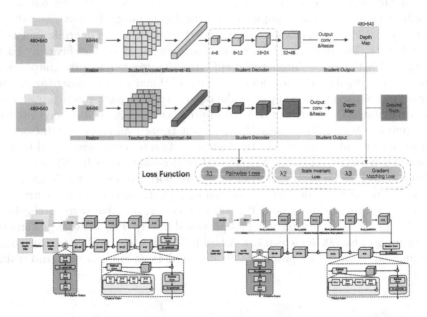

**Fig. 7.** An overview of the training strategy (top) and the architecture of the teacher
(bottom left) and student (bottom right) models proposed by team SmartLab.

The model was trained using three loss functions: the scale invariant loss [14]
to measure the discrepancy between the output of the student network and the
ground truth depth map; the scale-invariant gradient matching loss [65]; and
the pairwise distillation loss [57] to force the student network to produce similar
feature maps as the outputs of the corresponding layers of the teacher network.
The latter loss is defined as:

$$\mathcal{L}_{pa}(S,T) = \frac{1}{w \times h} \sum_i \sum_j (a_{ij}^s - a_{ij}^t)^2, \tag{7}$$

$$a_{ij} = f_i^T f_j / (\|f_i\|_2 \times \|f_j\|_2), \tag{8}$$

where $F_t \in \mathbb{R}^{h \times w \times c_1}$ and $F_s \in \mathbb{R}^{h \times w \times c_2}$ are the feature maps from the teacher and student networks, respectively, and $f$ denotes one row of a feature map ($F_t$ or $F_s$). Model parameters were optimized using the Adam algorithm for 100 epochs with a batch size of 2, input images were cropped to patches of size $64 \times 96$ and flipped randomly horizontally for data augmentation.

## 4.6 JMU-CVLab

The solution developed by team JMU-CVLab was inspired by the FastDepth [75] paper and the last year's winning challenge solution [81], its architecture is shown in Fig. 8. The input image is first resized to $160 \times 128$ pixels and then passed to a MobileNet-V3 backbone pre-trained on the ImageNet dataset that is used for feature extraction. The obtained features are then processed with a series of decoder FFM blocks [74] that perform an upsampling operation and reduce the number of channels by projecting them to a lower space using separable and pointwise convolutions. The model was trained with a combination of the L2, SSIM and si-RMSE loss functions. Model parameters were optimized with the Adam algorithm for 100 epochs using a learning rate 0.0001 and batch size 64. Basic data augmentations were used during the training including horizontal flips, random rotations limited by a 15° angle, color and brightness changes.

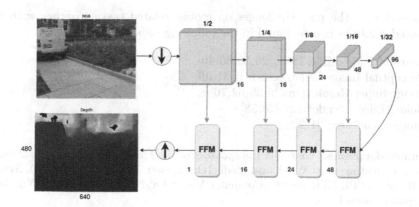

**Fig. 8.** Model architecture proposed by team JMU-CVLab.

## 4.7   ICL

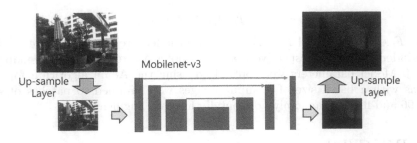

**Fig. 9.** ICL.

Team ICL proposed a solution that also uses the MobileNet-V3 network as a feature extractor (Fig. 9). The authors first fine-tuned the ViT model pre-trained on the KITTI depth estimation dataset [72], and then used it as a teacher network. Next, the proposed student model was first pre-trained for the semantic segmentation task and then fine-tuned on depth data. The student network was trained using a combination of the si-RMSE, gradient and knowledge distillation losses. Model parameters were optimized using the AdamW [61] optimizer with a batch size of 32 and the cosine learning rate scheduler.

## 5   Additional Literature

An overview of the past challenges on mobile-related tasks together with the proposed solutions can be found in the following papers:

- Learned End-to-End ISP: [26, 30, 36, 40]
- Perceptual Image Enhancement: [31, 39]
- Image Super-Resolution: [6, 39, 62, 70]
- Bokeh Effect Rendering: [28, 38]
- Image Denoising: [1, 2].

**Acknowledgements.** We thank the sponsors of the Mobile AI and AIM 2022 workshops and challenges: AI Witchlabs, MediaTek, Huawei, Reality Labs, OPPO, Synaptics, Raspberry Pi, ETH Zürich (Computer Vision Lab) and University of Würzburg (Computer Vision Lab).

## A   Teams and Affiliations

### Mobile AI 2022 Team

*Title:*
Mobile AI 2022 Challenge on Single-Image Depth Estimation on Mobile Devices

*Members:*
Andrey Ignatov[1,2] *(andrey@vision.ee.ethz.ch)*, Grigory Malivenko *(grigory.malivenko @gmail.com)*, Radu Timofte[1,2,3] *(radu.timofte@vision.ee.ethz.ch)*
*Affiliations:*
[1] Computer Vision Lab, ETH Zurich, Switzerland
[2] AI Witchlabs, Switzerland
[3] University of Wuerzburg, Germany.

## TCL

*Title:*
Simplified UNET Architecture For Fast Depth Estimation on Edge Devices
*Members:*
*Lukasz Treszczotko (lukasz.treszczotko@tcl.com)*, Xin Chang, Piotr Ksiazek, Michal Lopuszynski, Maciej Pioro, Rafal Rudnicki, Maciej Smyl, Yujie Ma
*Affiliations:*
TCL Research Europe, Warsaw, Poland.

## AIIA HIT

*Title:*
Towards Fast and Accurate Depth Estimation on Mobile Devices [54]
*Members:*
*Zhenyu Li (zhenyuli17@hit.edu.cn)*, Zehui Chen, Jialei Xu, Xianming Liu, Junjun Jiang
*Affiliations:*
Harbin Institute of Technology, China.

## MiAIgo

*Title:*
SL-Depth : A Superior Lightweight Model for Monocular Depth Estimation on Mobile Devices
*Members:*
*XueChao Shi (shixuechao@xiaomi.com)*, Difan Xu, Yanan Li, Xiaotao Wang, Lei Lei
*Affiliations:*
Xiaomi Inc., China.

## Tencent GY-Lab

*Title:*
A Simple Baseline for Fast and Accurate Depth Estimation on Mobile Devices [80]

*Members:*
*Ziyu Zhang (parkzyzhang@tencent.com)*, Yicheng Wang, Zilong Huang, Guozhong Luo, Gang Yu, Bin Fu
*Affiliations:*
Tencent GY-Lab, China.

## SmartLab

*Title:*
Fast and Accurate Monocular Depth Estimation via Knowledge Distillation
*Members:*
*Jiaqi Li (lijiaqi_mail@hust.edu.cn)*, Yiran Wang, Zihao Huang, Zhiguo Cao
*Affiliations:*
National Key Laboratory of Science and Technology on Multi-Spectral Information Processing, School of Artificial Intelligence and Automation, Huazhong University of Science and Technology, China.

## JMU-CVLab

*Title:*
MobileNetV3 FastDepth
*Members:*
*Marcos V. Conde (marcos.conde-osorio@uni-wuerzburg.de)*, Denis Sapozhnikov
*Affiliations:*
University of Wuerzburg, Germany.

## ICL

*Title:*
Monocular Depth Estimation Using a Simple Network
*Members:*
*Byeong Hyun Lee (ldlqudgus756@snu.ac.kr)*, Dongwon Park, Seongmin Hong, Joonhee Lee, Seunggyu Lee, Se Young Chun
*Affiliations:*
Department of Electrical and Computer Engineering, Seoul National University, South Korea.

# References

1. Abdelhamed, A., Afifi, M., Timofte, R., Brown, M.S.: Ntire 2020 challenge on real image denoising: Dataset, methods and results. In: Proceedings of the IEEE/CVF Conference on Computer Vision and Pattern Recognition Workshops, pp. 496–497 (2020)

2. Abdelhamed, A., Timofte, R., Brown, M.S.: Ntire 2019 challenge on real image denoising: Methods and results. In: Proceedings of the IEEE/CVF Conference on Computer Vision and Pattern Recognition Workshops, pp. 0–0 (2019)
3. Archive, A.B.: http://web.archive.org/web/20210425131428/https://ai-benchmark.com/ranking_processors.html
4. Barron, J.T.: A general and adaptive robust loss function. In: Proceedings of the IEEE/CVF Conference on Computer Vision and Pattern Recognition, pp. 4331–4339 (2019)
5. Bhardwaj, K., et al.: Collapsible linear blocks for super-efficient super resolution. In: Marculescu, D., Chi, Y., Wu, C. (eds.) Proceedings of Machine Learning and Systems. vol. 4, pp. 529–547 (2022). https://proceedings.mlsys.org/paper/2022/file/ac627ab1ccbdb62ec96e702f07f6425b-Paper.pdf
6. Cai, J., Gu, S., Timofte, R., Zhang, L.: Ntire 2019 challenge on real image super-resolution: Methods and results. In: Proceedings of the IEEE/CVF Conference on Computer Vision and Pattern Recognition Workshops, pp. 0–0 (2019)
7. Cai, Y., Yao, Z., Dong, Z., Gholami, A., Mahoney, M.W., Keutzer, K.: Zeroq: a novel zero shot quantization framework. In: Proceedings of the IEEE/CVF Conference on Computer Vision and Pattern Recognition, pp. 13169–13178 (2020)
8. Chen, L.C., Papandreou, G., Schroff, F., Adam, H.: Rethinking atrous convolution for semantic image segmentation (2017). https://doi.org/10.48550/ARXIV.1706.05587, https://arxiv.org/abs/1706.05587
9. Chen, W., Fu, Z., Yang, D., Deng, J.: Single-image depth perception in the wild. arXiv preprint arXiv:1604.03901 (2016)
10. Chiang, C.M., et al.: Deploying image deblurring across mobile devices: a perspective of quality and latency. In: Proceedings of the IEEE/CVF Conference on Computer Vision and Pattern Recognition Workshops, pp. 502–503 (2020)
11. Conde, M.V., Timofte, R., et al.: Reversed image signal processing and RAW reconstruction. AIM 2022 challenge report. In: Proceedings of the European Conference on Computer Vision (ECCV) Workshops (2022). https://doi.org/10.1007/978-3-030-66415-2
12. Dosovitskiy, A., et al.: An image is worth 16x16 words: transformers for image recognition at scale. arXiv preprint arXiv:2010.11929 (2020)
13. Eigen, D., Puhrsch, C., Fergus, R.: Depth map prediction from a single image using a multi-scale deep network. arXiv preprint arXiv:1406.2283 (2014)
14. Eigen, D., Puhrsch, C., Fergus, R.: Depth map prediction from a single image using a multi-scale deep network. In: 27th Proceedings on Advances in Neural Information Processing Systems (2014)
15. Garg, R., B.G., V.K., Carneiro, G., Reid, I.: Unsupervised CNN for single view depth estimation: geometry to the rescue. In: Leibe, B., Matas, J., Sebe, N., Welling, M. (eds.) ECCV 2016. LNCS, vol. 9912, pp. 740–756. Springer, Cham (2016). https://doi.org/10.1007/978-3-319-46484-8_45
16. Godard, C., Mac Aodha, O., Firman, M., Brostow, G.J.: Digging into self-supervised monocular depth estimation. In: Proceedings of the IEEE/CVF International Conference on Computer Vision, pp. 3828–3838 (2019)
17. Han, K., Wang, Y., Tian, Q., Guo, J., Xu, C., Xu, C.: Ghostnet: more features from cheap operations (2019). https://doi.org/10.48550/ARXIV.1911.11907, https://arxiv.org/abs/1911.11907
18. Hinton, G., Vinyals, O., Dean, J.: Distilling the knowledge in a neural network. arXiv preprint arXiv:1503.02531 (2015)
19. Howard, A., et al.: Searching for mobilenetv3. In: Proceedings of the IEEE/CVF International Conference on Computer Vision, pp. 1314–1324 (2019)

20. Ignatov, A., Byeoung-su, K., Timofte, R.: Fast camera image denoising on mobile GPUs with deep learning, mobile AI 2021 challenge: Report. In: Proceedings of the IEEE/CVF Conference on Computer Vision and Pattern Recognition Workshops, pp. 0–0 (2021)
21. Ignatov, A., Chiang, J., Kuo, H.K., Sycheva, A., Timofte, R.: Learned smartphone ISP on mobile NPUs with deep learning, mobile AI 2021 challenge: Report. In: Proceedings of the IEEE/CVF Conference on Computer Vision and Pattern Recognition Workshops, pp. 0–0 (2021)
22. Ignatov, A., Kobyshev, N., Timofte, R., Vanhoey, K., Van Gool, L.: Dslr-quality photos on mobile devices with deep convolutional networks. In: Proceedings of the IEEE International Conference on Computer Vision, pp. 3277–3285 (2017)
23. Ignatov, A., Kobyshev, N., Timofte, R., Vanhoey, K., Van Gool, L.: Wespe: weakly supervised photo enhancer for digital cameras. In: Proceedings of the IEEE Conference on Computer Vision and Pattern Recognition Workshops, pp. 691–700 (2018)
24. Ignatov, A., Malivenko, G., Plowman, D., Shukla, S., Timofte, R.: Fast and accurate single-image depth estimation on mobile devices, mobile AI 2021 challenge: Report. In: Proceedings of the IEEE/CVF Conference on Computer Vision and Pattern Recognition Workshops, pp. 0–0 (2021)
25. Ignatov, A., Malivenko, G., Timofte, R.: Fast and accurate quantized camera scene detection on smartphones, mobile AI 2021 challenge: Report. In: Proceedings of the IEEE/CVF Conference on Computer Vision and Pattern Recognition Workshop,. pp. 0–0 (2021)
26. Ignatov, A., et al.: Pynet-v2 mobile: Efficient on-device photo processing with neural networks. In: 2021 26th International Conference on Pattern Recognition (ICPR), IEEE (2022)
27. Ignatov, A., Patel, J., Timofte, R.: Rendering natural camera bokeh effect with deep learning. In: Proceedings of the IEEE/CVF Conference on Computer Vision and Pattern Recognition Workshops, pp. 418–419 (2020)
28. Ignatov, A., et al.: Aim 2019 challenge on bokeh effect synthesis: methods and results. In: 2019 IEEE/CVF International Conference on Computer Vision Workshop (ICCVW), pp. 3591–3598. IEEE (2019)
29. Ignatov, A., Romero, A., Kim, H., Timofte, R.: Real-time video super-resolution on smartphones with deep learning, mobile AI 2021 challenge: Report. In: Proceedings of the IEEE/CVF Conference on Computer Vision and Pattern Recognition Workshops, pp. 0–0 (2021)
30. Ignatov, A., et al.: MicroISP: processing 32mp photos on mobile devices with deep learning. In: European Conference on Computer Vision (2022)
31. Ignatov, A., Timofte, R.: Ntire 2019 challenge on image enhancement: Methods and results. In: Proceedings of the IEEE/CVF Conference on Computer Vision and Pattern Recognition Workshops, pp. 0–0 (2019)
32. Ignatov, A., et al.: Power efficient super-resolution on mobile NPUs with deep learning, mobile AI & aim 2022 challenge: Report. In: European Conference on Computer Vision (2022)
33. Ignatov, A., et al.: AI benchmark: running deep neural networks on android smartphones. In: Leal-Taixé, L., Roth, S. (eds.) ECCV 2018. LNCS, vol. 11133, pp. 288–314. Springer, Cham (2019). https://doi.org/10.1007/978-3-030-11021-5_19
34. Ignatov, A., Timofte, R., Denna, M., Younes, A.: Real-time quantized image super-resolution on mobile NPUs, mobile AI 2021 challenge: Report. In: Proceedings of the IEEE/CVF Conference on Computer Vision and Pattern Recognition Workshop,. pp. 0–0 (2021)

35. Ignatov, A., Timofte, R., Denna, M., Younes, A., et al.: Efficient and accurate quantized image super-resolution on mobile NPUs, mobile AI & aim 2022 challenge: Report. In: European Conference on Computer Vision (2022)
36. Ignatov, A., et al.: Aim 2019 challenge on raw to RGB mapping: methods and results. In: 2019 IEEE/CVF International Conference on Computer Vision Workshop (ICCVW)., pp. 3584–3590. IEEE (2019)
37. Ignatov, A., et al.: AI benchmark: All about deep learning on smartphones in 2019. In: 2019 IEEE/CVF International Conference on Computer Vision Workshop (ICCVW), pp. 3617–3635. IEEE (2019)
38. Ignatov, A., et al.: AIM 2020 challenge on rendering realistic bokeh. In: Bartoli, A., Fusiello, A. (eds.) ECCV 2020. LNCS, vol. 12537, pp. 213–228. Springer, Cham (2020). https://doi.org/10.1007/978-3-030-67070-2_13
39. Ignatov, A., et al.: PIRM challenge on perceptual image enhancement on smartphones: report. In: Leal-Taixé, L., Roth, S. (eds.) ECCV 2018. LNCS, vol. 11133, pp. 315–333. Springer, Cham (2019). https://doi.org/10.1007/978-3-030-11021-5_20
40. Ignatov, A., et al.: Aim 2020 challenge on learned image signal processing pipeline. arXiv preprint arXiv:2011.04994 (2020)
41. Ignatov, A., Timofte, R., et al.: Learned smartphone ISP on mobile GPUs with deep learning, mobile AI & aim 2022 challenge: Report. In: European Conference on Computer Vision (2022)
42. Ignatov, A., Timofte, R., et al.: Realistic bokeh effect rendering on mobile GPUs, mobile AI & aim 2022 challenge: Report (2022)
43. Ignatov, A., Van Gool, L., Timofte, R.: Replacing mobile camera ISP with a single deep learning model. In: Proceedings of the IEEE/CVF Conference on Computer Vision and Pattern Recognition Workshops, pp. 536–537 (2020)
44. Ignatov, D., Ignatov, A.: Controlling information capacity of binary neural network. Pattern Recogn. Lett. **138**, 276–281 (2020)
45. Jacob, B., et al.: Quantization and training of neural networks for efficient integer-arithmetic-only inference. In: Proceedings of the IEEE Conference on Computer Vision and Pattern Recognition, pp. 2704–2713 (2018)
46. Jain, S.R., Gural, A., Wu, M., Dick, C.H.: Trained quantization thresholds for accurate and efficient fixed-point inference of deep neural networks. arXiv preprint arXiv:1903.08066 (2019)
47. Kingma, D.P., Ba, J.: Adam: a method for stochastic optimization. arXiv preprint arXiv:1412.6980 (2014)
48. Kınlı, F.O., Menteş, S., Özcan, B., Kirac, F., Timofte, R., et al.: Aim 2022 challenge on Instagram filter removal: Methods and results. In: Proceedings of the European Conference on Computer Vision (ECCV) Workshops (2022)
49. Laina, I., Rupprecht, C., Belagiannis, V., Tombari, F., Navab, N.: Deeper depth prediction with fully convolutional residual networks. In: 2016 Fourth International Conference on 3D Vision (3DV). pp. 239–248. IEEE (2016)
50. Lee, J.H., Han, M.K., Ko, D.W., Suh, I.H.: From big to small: Multi-scale local planar guidance for monocular depth estimation (2019). https://doi.org/10.48550/ARXIV.1907.10326, https://arxiv.org/abs/1907.10326
51. Lee, J.H., Han, M.K., Ko, D.W., Suh, I.H.: From big to small: Multi-scale local planar guidance for monocular depth estimation. arXiv preprint arXiv:1907.10326 (2019)
52. Li, Y., Gu, S., Gool, L.V., Timofte, R.: Learning filter basis for convolutional neural network compression. In: Proceedings of the IEEE/CVF International Conference on Computer Vision, pp. 5623–5632 (2019)

53. Li, Z., Snavely, N.: Megadepth: learning single-view depth prediction from internet photos. In: Proceedings of the IEEE Conference on Computer Vision and Pattern Recognition, pp. 2041–2050 (2018)
54. Li, Z., Chen, Z., Xu, J., Liu, X., Jiang, J.: Litedepth: digging into fast and accurate depth estimation on mobile devices. In: Proceedings of the European Conference on Computer Vision (ECCV) Workshops (2022)
55. Liu, F., Shen, C., Lin, G.: Deep convolutional neural fields for depth estimation from a single image. In: Proceedings of the IEEE Conference on Computer Vision and Pattern Recognition, pp. 5162–5170 (2015)
56. Liu, F., Shen, C., Lin, G., Reid, I.: Learning depth from single monocular images using deep convolutional neural fields. IEEE Trans. Pattern Anal. Mach. Intell. **38**(10), 2024–2039 (2015)
57. Liu, Y., Chen, K., Liu, C., Qin, Z., Luo, Z., Wang, J.: Structured knowledge distillation for semantic segmentation. In: Proceedings of the IEEE/CVF Conference on Computer Vision and Pattern Recognition, pp. 2604–2613 (2019)
58. Liu, Y., Shu, C., Wang, J., Shen, C.: Structured knowledge distillation for dense prediction. IEEE Trans. Pattern Anal. Mach. Intell. (99), 1-1 (2020)
59. Liu, Z., et al.: Metapruning: meta learning for automatic neural network channel pruning. In: Proceedings of the IEEE/CVF International Conference on Computer Vision, pp. 3296–3305 (2019)
60. Liu, Z., Wu, B., Luo, W., Yang, X., Liu, W., Cheng, K.-T.: Bi-Real Net: enhancing the performance of 1-Bit CNNs with improved representational capability and advanced training algorithm. In: Ferrari, V., Hebert, M., Sminchisescu, C., Weiss, Y. (eds.) ECCV 2018. LNCS, vol. 11219, pp. 747–763. Springer, Cham (2018). https://doi.org/10.1007/978-3-030-01267-0_44
61. Loshchilov, I., Hutter, F.: Decoupled weight decay regularization. arXiv preprint arXiv:1711.05101 (2017)
62. Lugmayr, A., Danelljan, M., Timofte, R.: Ntire 2020 challenge on real-world image super-resolution: Methods and results. In: Proceedings of the IEEE/CVF Conference on Computer Vision and Pattern Recognition Workshops, pp. 494–495 (2020)
63. Obukhov, A., Rakhuba, M., Georgoulis, S., Kanakis, M., Dai, D., Van Gool, L.: T-basis: a compact representation for neural networks. In: International Conference on Machine Learning, pp. 7392–7404. PMLR (2020)
64. Ortiz, L.E., Cabrera, E.V., Gonçalves, L.M.: Depth data error modeling of the zed 3d vision sensor from stereolabs. ELCVIA: Electr. Lett. Compu. Visi. Image Anal. **17**(1), 0001–15 (2018)
65. Ranftl, R., Lasinger, K., Hafner, D., Schindler, K., Koltun, V.: Towards robust monocular depth estimation: Mixing datasets for zero-shot cross-dataset transfer. IEEE Trans. Pattern Anal. Mach. Intell. (2020)
66. Ronneberger, O., Fischer, P., Brox, T.: U-Net: convolutional networks for biomedical image segmentation. In: Navab, N., Hornegger, J., Wells, W.M., Frangi, A.F. (eds.) MICCAI 2015. LNCS, vol. 9351, pp. 234–241. Springer, Cham (2015). https://doi.org/10.1007/978-3-319-24574-4_28
67. Tan, M., Chen, B., Pang, R., Vasudevan, V., Sandler, M., Howard, A., Le, Q.V.: Mnasnet: Platform-aware neural architecture search for mobile. In: Proceedings of the IEEE/CVF Conference on Computer Vision and Pattern Recognition, pp. 2820–2828 (2019)
68. TensorFlow-Lite: https://www.tensorflow.org/lite
69. TensorFlow-Lite: https://www.tensorflow.org/lite/guide/python

70. Timofte, R., Gu, S., Wu, J., Van Gool, L.: Ntire 2018 challenge on single image super-resolution: Methods and results. In: Proceedings of the IEEE Conference on Computer Vision and Pattern Recognition Workshops,. pp. 852–863 (2018)
71. Uhlich, S., et al.: Mixed precision DNNs: All you need is a good parametrization. arXiv preprint arXiv:1905.11452 (2019)
72. Uhrig, J., Schneider, N., Schneider, L., Franke, U., Brox, T., Geiger, A.: Sparsity invariant CNNS. In: International Conference on 3D Vision (3DV) (2017)
73. Wan, A., et al.: Fbnetv2: differentiable neural architecture search for spatial and channel dimensions. In: Proceedings of the IEEE/CVF Conference on Computer Vision and Pattern Recognition, pp. 12965–12974 (2020)
74. Wang, Y., Li, X., Shi, M., Xian, K., Cao, Z.: Knowledge distillation for fast and accurate monocular depth estimation on mobile devices. In: 2021 IEEE/CVF Conference on Computer Vision and Pattern Recognition Workshops (CVPRW), pp. 2457–2465 (2021). https://doi.org/10.1109/CVPRW53098.2021.00278
75. Wofk, D., Ma, F., Yang, T.J., Karaman, S., Sze, V.: Fastdepth: Fast monocular depth estimation on embedded systems (2019). https://doi.org/10.48550/ARXIV.1903.03273, https://arxiv.org/abs/1903.03273
76. Wu, B., et al.: Fbnet: hardware-aware efficient convnet design via differentiable neural architecture search. In: Proceedings of the IEEE/CVF Conference on Computer Vision and Pattern Recognition, pp. 10734–10742 (2019)
77. Yang, J., et al.: Quantization networks. In: Proceedings of the IEEE/CVF Conference on Computer Vision and Pattern Recognition, pp. 7308–7316 (2019)
78. Yang, R., Timofte, R., et al.: Aim 2022 challenge on super-resolution of compressed image and video: Dataset, methods and results. In: Proceedings of the European Conference on Computer Vision (ECCV) Workshops (2022)
79. Yin, W., Liu, Y., Shen, C., Yan, Y.: Enforcing geometric constraints of virtual normal for depth prediction. In: The IEEE International Conference on Computer Vision (ICCV) (2019)
80. Zhang, Z., Wang, Y., Huang, Z., Luo, G., Yu, G., Fu, B.: A simple baseline for fast and accurate depth estimation on mobile devices. In: Proceedings of the IEEE/CVF Conference on Computer Vision and Pattern Recognition Workshops. pp. 0–0 (2021)
81. Zhang, Z., Wang, Y., Huang, Z., Luo, G., Yu, G., Fu, B.: A simple baseline for fast and accurate depth estimation on mobile devices. In: 2021 IEEE/CVF Conference on Computer Vision and Pattern Recognition Workshops (CVPRW), pp. 2466–2471 (2021). https://doi.org/10.1109/CVPRW53098.2021.00279

# Efficient and Accurate Quantized Image Super-Resolution on Mobile NPUs, Mobile AI & AIM 2022 Challenge: Report

Andrey Ignatov[1,2(✉)], Radu Timofte[1,2,3], Maurizio Denna[4], Abdel Younes[4],
Ganzorig Gankhuyag[5], Jingang Huh[5], Myeong Kyun Kim[5], Kihwan Yoon[5],
Hyeon-Cheol Moon[5], Seungho Lee[5], Yoonsik Choe[6], Jinwoo Jeong[5],
Sungjei Kim[5], Maciej Smyl[7], Tomasz Latkowski[7], Pawel Kubik[7],
Michal Sokolski[7], Yujie Ma[7], Jiahao Chao[8], Zhou Zhou[8], Hongfan Gao[8],
Zhengfeng Yang[8], Zhenbing Zeng[8], Zhengyang Zhuge[9], Chenghua Li[9],
Dan Zhu[10], Mengdi Sun[10], Ran Duan[10], Yan Gao[10], Lingshun Kong[11],
Long Sun[11], Xiang Li[11], Xingdong Zhang[11], Jiawei Zhang[11], Yaqi Wu[11],
Jinshan Pan[11], Gaocheng Yu[12], Jin Zhang[12], Feng Zhang[12], Zhe Ma[12],
Hongbin Wang[12], Hojin Cho[13], Steve Kim[13], Huaen Li[14], Yanbo Ma[14],
Ziwei Luo[15,16], Youwei Li[15,16], Lei Yu[15,16], Zhihong Wen[15,16], Qi Wu[15,16],
Haoqiang Fan[15,16], Shuaicheng Liu[15,16], Lize Zhang[17,18], Zhikai Zong[17,18],
Jeremy Kwon[19], Junxi Zhang[20], Mengyuan Li[20], Nianxiang Fu[20],
Guanchen Ding[20], Han Zhu[20], Zhenzhong Chen[20], Gen Li[21], Yuanfan Zhang[21],
Lei Sun[21], Dafeng Zhang[22], Neo Yang[21], Fitz Liu[21], Jerry Zhao[21],
Mustafa Ayazoglu[23], Bahri Batuhan Bilecen[23], Shota Hirose[24],
Kasidis Arunruangsirilert[24], Luo Ao[24], Ho Chun Leung[25], Andrew Wei[25],
Jie Liu[25], Qiang Liu[25], Dahai Yu[25], Ao Li[16], Lei Luo[16], Ce Zhu[16],
Seongmin Hong[26], Dongwon Park[26], Joonhee Lee[26], Byeong Hyun Lee[26],
Seunggyu Lee[26], Se Young Chun[26], Ruiyuan He[21], Xuhao Jiang[21],
Haihang Ruan[27], Xinjian Zhang[27], Jing Liu[27], Garas Gendy[28], Nabil Sabor[29],
Jingchao Hou[28], and Guanghui He[28]

[1] Computer Vision Lab, ETH Zürich, Zürich, Switzerland
{andrey,radu.timofte}@vision.ee.ethz.ch
[2] AI Witchlabs, Zürich, Switzerland
[3] University of Würzburg, Würzburg, Germany
radu.timofte@uni-wuerzburg.de
[4] Synaptics Europe, Lausanne, Switzerland
{maurizio.denna,abdel.younes}@synaptics.com
[5] Korea Electronics Technology Institute (KETI), Seongnam, South Korea
[6] Yonsei University, Seoul, South Korea
[7] TCL Research Europe, Warsaw, Poland

A. Ignatov, R. Timofte, M. Denna and A. Younes—Are the Mobile AI & AIM 2022
challenge organizers. The other authors participated in the challenge.
Appendix A contains the authors' team names and affiliations.
Mobile AI 2022 Workshop website:
https://ai-benchmark.com/workshops/mai/2022/.

[8] East China Normal University, Shanghai, China
51215902006@stu.ecnu.edu.cn
[9] Institute of Automation, Chinese Academy of Sciences, Beijing, China
[10] BOE Technology Group Co., Ltd., Beijing, China
zhudan@boe.com.cn
[11] Nanjing University of Science and Technology, Nanjing, China
konglingshun@njust.edu.cn
[12] Ant Group, Hangzhou, China
yugaocheng.ygc@antgroup.com
[13] GenGenAI, Seoul, South Korea
jin@gengen.ai
[14] Hefei University of Technology, Hefei, China
huaenli@mail.hfut.edu.cn
[15] Megvii Technology, Beijing, China
[16] University of Electronic Science and Technology of China (UESTC),
Chengdu, China
liao@cqu.edu.cn
[17] Xidian University, Xi'an, China
lzzhang_98@stu.xidian.edu.cn
[18] Qingdao Hi-image Technologies Co., Ltd., Qingdao, China
[19] Seoul, South Korea
[20] Wuhan University, Wuhan, China
sissie_zhang@whu.edu.cn
[21] Beijing, China
[22] Samsung Research, Beijing, China
dfeng.zhang@samsung.com
[23] Aselsan Corporation, Ankara, Turkey
mayazoglu@aselsan.com.tr
[24] Waseda University, Tokyo, Japan
syouta.hrs@akane.waseda.jp
[25] TCL Corporate Research, Hong kong, China
hcleung@tcl.com
[26] Intelligent Computational Imaging Lab, Seoul National University,
Seoul, South Korea
smhongok@snu.ac.kr
[27] Bilibili Inc., Shanghai, China
hhruan@mail.sim.ac.cn
[28] Shanghai Jiao Tong University, Shanghai, China
[29] Assiut University, Asyut, Egypt

**Abstract.** Image super-resolution is a common task on mobile and IoT devices, where one often needs to upscale and enhance low-resolution images and video frames. While numerous solutions have been proposed for this problem in the past, they are usually not compatible with low-power mobile NPUs having many computational and memory constraints. In this Mobile AI challenge, we address this problem and propose the participants to design an efficient quantized image super-resolution solution that can demonstrate a real-time performance on mobile NPUs. The participants were provided with the DIV2K dataset and trained INT8 models to do a high-quality 3X image upscaling. The runtime of all models was evaluated on the Synaptics VS680 Smart Home board with a dedicated edge NPU capable of accelerating quantized neural networks. All proposed solutions are fully compatible with the above NPU, demonstrating an up to 60 FPS rate when reconstructing Full HD resolution images. A detailed description of all models developed in the challenge is provided in this paper.

**Keywords:** Mobile AI challenge · Super-resolution · Mobile NPUs · Mobile AI · Deep learning · Synaptics · AI Benchmark

**Fig. 1.** Sample crop from a 3X bicubically upscaled image and the target DIV2K [3] photo.

# 1   Introduction

Single image super-resolution is a longstanding computer vision problem. Its goal is to restore the original image contents from its downsampled version by recovering lost details. This task draws ever-increasing interest due to the pervasive media contents, cameras and displays and its direct application to real-world problems such as image processing in smartphone cameras (*e.g.*telephoto), enhancement of low-resolution media data and upscaling images and videos to match the high resolution of the displays. There is a rich literature and a multitude of classical (hand-crafted) [20,23,50,66,71,72,74,80–82] and deep learning-based [9,15,16,43,53,59,63,70,73,86] methods have been suggested in the past decades. One of the biggest drawbacks of the existing solutions is that the were not optimized for computational efficiency or mobile-related constraints, but for sheer accuracy in terms of high fidelity scores. Meeting the hardware-constraints from low-power devices is essential for the methods developed for real-world applications of image super-resolution and other tasks related to image processing and enhancement [26,27,47] on mobile devices. In this challenge, we address this drawback and by using the popular DIV2K [3] image super-resolution dataset, we add efficiency-related constraints from mobile NPUs on the images super-resolution solutions.

The deployment of AI-based solutions on portable devices usually requires an efficient model design based on a good understanding of the mobile processing units (*e.g.* CPUs, NPUs, GPUs, DSP) and their hardware particularities, including their memory constraints. We refer to [38,41] for an extensive overview of mobile AI acceleration hardware, its particularities and performance. As shown in these works, the latest generations of mobile NPUs are reaching the performance of older-generation mid-range desktop GPUs. Nevertheless, a straightforward deployment of neural networks-based solutions on mobile devices is impeded by (i) a limited memory (*i.e.*, restricted amount of RAM) and (ii) a limited or lacking support of many common deep learning operators and layers. These impeding factors make the processing of high resolution inputs impossible with the standard NN models and require a careful adaptation or re-design to the constraints of mobile AI hardware. Such optimizations can employ a combination of various model techniques such as 16-bit / 8-bit [11,51,52,83] and low-bit [10,48,61,75] quantization, network pruning and compression [11,32,57,60,65], device- or NPU-specific adaptations, platform-aware neural architecture search [22,68,76,79], *etc.*

The majority of competitions aimed at efficient deep learning models use standard desktop hardware for evaluating the solutions, thus the obtained models rarely show acceptable results when running on real mobile hardware with many specific constraints. In this *Mobile AI challenge*, we take a radically different approach and propose the participants to develop and evaluate their models directly on mobile devices. The goal of this competition is to design a fast and performant quantized deep learning-based solution for image super-resolution problem. For this, the participants were provided with the large-scale DIV2K [3] dataset containing diverse 2K resolution RGB images used to train their models using a downscaling factor of 3. The efficiency of the proposed solutions was evaluated on the Synaptics Dolphin platform featuring a dedicated NPU that can efficiently accelerate INT8 neural networks. The overall score of each submission was computed based on its fidelity and runtime results, thus balancing between the image reconstruction quality and the efficiency of the model. All solutions developed in this challenge are fully compatible with the TensorFlow Lite framework [69], thus can be executed on various Linux and Android-based IoT platforms, smartphones and edge devices.

This challenge is a part of the *Mobile AI & AIM 2022 Workshops and Challenges* consisting of the following competitions:

- Quantized Image Super-Resolution on Mobile NPUs
- Power Efficient Video Super-Resolution on Mobile NPUs [37]
- Learned Smartphone ISP on Mobile GPUs [45]
- Efficient Single-Image Depth Estimation on Mobile Devices [31]
- Realistic Bokeh Effect Rendering on Mobile GPUs [46]
- Super-Resolution of Compressed Image and Video [84]
- Reversed Image Signal Processing and RAW Reconstruction [12]
- Instagram Filter Removal [54]

The results and solutions obtained in the previous *MAI 2021 Challenges* are described in our last year papers:

– Learned Smartphone ISP on Mobile NPUs [25]
– Real Image Denoising on Mobile GPUs [24]
– Quantized Image Super-Resolution on Mobile NPUs [39]

**Fig. 2.** Loading and running custom TensorFlow Lite models with AI Benchmark application. The currently supported acceleration options include Android NNAPI, TFLite GPU, Hexagon NN, Qualcomm QNN, MediaTek Neuron and Samsung ENN delegates as well as CPU inference through TFLite or XNNPACK backends. The latest app version can be downloaded at https://ai-benchmark.com/download

– Real-Time Video Super-Resolution on Mobile GPUs [34]
– Single-Image Depth Estimation on Mobile Devices [28]
– Quantized Camera Scene Detection on Smartphones [29]

## 2    Challenge

In order to design an efficient and practical deep learning-based solution for the considered task that runs fast on mobile devices, one needs the following tools:

1. A large-scale high-quality dataset for training and evaluating the models. Real, not synthetically generated data should be used to ensure a high quality of the obtained model;
2. An efficient way to check the runtime and debug the model locally without any constraints as well as the ability to check the runtime on the target evaluation platform.

This challenge addresses all the above issues. Real training data, tools, and runtime evaluation options provided to the challenge participants are described in the next sections.

## 2.1  Dataset

In this challenge, the participants were proposed to work with the popular DIV2K [3] dataset. It consists from 1000 divers 2K resolution RGB images: 800 are used for training, 100 for validation and 100 for testing purposes. The images are of high quality both aesthetically and in the terms of small amounts of noise and other corruptions (like blur and color shifts). All images were manually collected and have 2K pixels on at least one of the axes (vertical or horizontal). DIV2K covers a large diversity of contents, from people, handmade objects and environments (cities), to flora and fauna and natural sceneries, including underwater. An example set of images is demonstrated in Fig. 1.

## 2.2  Local Runtime Evaluation

When developing AI solutions for mobile devices, it is vital to be able to test the designed models and debug all emerging issues locally on available devices. For this, the participants were provided with the *AI Benchmark* application [38, 41] that allows to load any custom TensorFlow Lite model and run it on any Android device with all supported acceleration options. This tool contains the latest versions of *Android NNAPI, TFLite GPU, Hexagon NN, Qualcomm QNN, MediaTek Neuron* and *Samsung ENN* delegates, therefore supporting all current mobile platforms and providing the users with the ability to execute neural networks on smartphone NPUs, APUs, DSPs, GPUs and CPUs.

To load and run a custom TensorFlow Lite model, one needs to follow the next steps:

1. Download AI Benchmark from the official website[1] or from the Google Play[2] and run its standard tests.
2. After the end of the tests, enter the *PRO Mode* and select the *Custom Model* tab there.
3. Rename the exported TFLite model to *model.tflite* and put it into the *Download* folder of the device.
4. Select mode type *(INT8, FP16, or FP32)*, the desired acceleration/inference options and run the model.

These steps are also illustrated in Fig. 2.

## 2.3  Runtime Evaluation on the Target Platform

In this challenge, we use the *Synaptics VS680 Edge AI SoC* [49] Evaluation Kit as our target runtime evaluation platform. The VS680 Edge AI SoC is integrated into Smart Home solution and it features a powerful NPU designed by *VeriSilicon* and capable of accelerating quantized models (up to 7 TOPS). It supports

---

[1] https://ai-benchmark.com/download.
[2] https://play.google.com/store/apps/details?id=org.benchmark.demo.

Android and can perform NN inference through NNAPI, demonstrating INT8 AI Benchmark scores that are close to the ones of mid-range smartphone chipsets. Within the challenge, the participants were able to upload their TFLite models to an external server and get feedback regarding the speed of their model: the inference time of their solution on the above mentioned NPU or an error log if the network contained incompatible operations and/or improper quantization. Participants' models were first compiled and optimized using Synaptics' SyNAP Toolkit and then executed on the the VS680's NPU using the SyNAP C++ API to achieve the best possible efficiency. The same setup was also used for the final runtime evaluation. The participants were additionally provided with a list of ops supported by this board and model optimization guidance in order to fully utilize the NPU's convolution and tensor processing resources. Besides that, a layer-by-layer timing information was provided for each submitted TFLite model to help the participants to optimize the architecture of their networks.

### 2.4 Challenge Phases

The challenge consisted of the following phases:

I. *Development:* the participants get access to the data and AI Benchmark app, and are able to train the models and evaluate their runtime locally;

II. *Validation:* the participants can upload their models to the remote server to check the fidelity scores on the validation dataset, to get the runtime on the target platform, and to compare their results on the validation leaderboard;

III. *Testing:* the participants submit their final results, codes, TensorFlow Lite models, and factsheets.

### 2.5 Scoring System

All solutions were evaluated using the following metrics:

- Peak Signal-to-Noise Ratio (PSNR) measuring fidelity score,
- Structural Similarity Index Measure (SSIM), a proxy for perceptual score,
- The runtime on the target Synaptics VS680 board.

The score of each final submission was evaluated based on the next formula ($C$ is a constant normalization factor):

$$\text{Final Score} = \frac{2^{2 \cdot \text{PSNR}}}{C \cdot \text{runtime}},$$

During the final challenge phase, the participants did not have access to the test dataset. Instead, they had to submit their final TensorFlow Lite models that were subsequently used by the challenge organizers to check both the runtime and the fidelity results of each submission under identical conditions. This approach solved all the issues related to model overfitting, reproducibility of the results, and consistency of the obtained runtime/accuracy values.

**Table 1.** Mobile AI 2022 Real-Time Image Super-Resolution challenge results and final rankings. During the runtime measurements, the models were performing image upscaling from $640 \times 360$ to $1920 \times 1080$ pixels. $\Delta$ PSNR values correspond to accuracy loss measured in comparison to the original floating-point network. Team *Z6* is the challenge winner.

| Team | Author | Framework | Model Size, KB | PSNR↑ INT8 Model | SSIM↑ | Δ PSNR Drop, FP32 → INT8 | Runtime, ms ↓ CPU | NPU | Speed-Up | Final Score |
|---|---|---|---|---|---|---|---|---|---|---|
| Z6 | Ganzoo | Keras/TensorFlow | 67 | 30.03 | 0.8738 | 0.06 | 809 | 19.2 | 42.1 | 22.22 |
| TCLResearchEurope | Maciejos_s | Keras/TensorFlow | 53 | 29.88 | 0.8705 | 0.13 | 824 | 15.9 | 51.8 | 21.84 |
| ECNUSR | CCjiahao | Keras /TensorFlow | 50 | 29.82 | 0.8697 | 0.10 | 553 | 15.1 | 36.6 | 21.08 |
| LCVG | Zion | Keras/TensorFlow | 48 | 29.76 | 0.8675 | 0.11 | 787 | **15.0** | 52.5 | 19.59 |
| BOE-IOT-AIBD | NBCS | Keras/TensorFlow | 57 | 29.80 | 0.8675 | 0.19 | 877 | 16.1 | 54.5 | 19.27 |
| NJUST | kkkls | TensorFlow | 57 | 29.76 | 0.8676 | 0.16 | 767 | 15.8 | 48.5 | 18.56 |
| Antins_cv | fz | Keras/TensorFlow | 38 | 29.58 | 0.8609 | n.a | **543** | 15.2 | 35.7 | 15.02 |
| GenMedia Group | Stevek | Keras/TensorFlow | 56 | 29.90 | 0.8704 | 0.11 | 855 | 25.6 | 33.4 | 13.91 |
| Vccip | Huaen | PyTorch/TensorFlow | 67 | 29.98 | 0.8729 | n.a | 1042 | 30.5 | 34.2 | 13.07 |
| MegSR | balabala | Keras/TensorFlow | 65 | 29.94 | 0.8704 | n.a | 965 | 29.8 | 32.4 | 12.65 |
| DoubleZ | gideon | Keras/TensorFlow | 63 | 29.94 | 0.8712 | 0.1 | 990 | 30.1 | 32.9 | 12.54 |
| Jeremy Kwon | alan_jaeger | TensorFlow | 48 | 29.80 | 0.8691 | 0.09 | 782 | 25.7 | 30.4 | 12.09 |
| Lab216 | sissie | TensorFlow | 78 | 29.94 | 0.8728 | 0 | 1110 | 31.8 | 34.9 | 11.85 |
| TOVB | jklovezhang | Keras/TensorFlow | 56 | 30.01 | 0.8740 | 0.04 | 950 | 43.3 | 21.9 | 9.60 |
| ABPN [19] | baseline | - | 53 | 29.87 | 0.8686 | n.a | 998 | 36.9 | 27 | 9.27 |
| Samsung Research | xiaozhazha | TensorFlow | 57 | 29.95 | 0.8728 | 0.1 | 941 | 43.2 | 21.8 | 8.84 |
| Rtsisr2022 | rtsisr | Keras/TensorFlow | 68 | 30.00 | 0.8729 | 0.09 | 977 | 46.4 | 21.1 | 8.83 |
| Aselsan Research | deepernewbie | Keras/TensorFlow | 30 | 29.67 | 0.8651 | n.a | 598 | 30.2 | 19.8 | 8.59 |
| Klab_SR | FykAikawa | TensorFlow | 39 | 29.88 | 0.8700 | n.a | 850 | 43.1 | 19.7 | 8.05 |
| TCL Research HK | mrblue | Keras/TensorFlow | 121 | **30.10** | **0.8751** | 0.04 | 1772 | 60.2 | 29.4 | 7.81 |
| RepGSR | yilitiaotiaotang | Keras/TensorFlow | 90 | 30.06 | 0.8739 | 0.02 | 1679 | 61.3 | 27.4 | 7.26 |
| ICL | smhong | Keras/TensorFlow | 55 | 29.76 | 0.8641 | 0.23 | 949 | 43.2 | 22.0 | 6.79 |
| Just A try | kia350 | Keras/TensorFlow | 39 | 29.75 | 0.8671 | n.a | 766 | 42.9 | 17.9 | 6.76 |
| Bilibili AI | Sail | Keras/TensorFlow | 75 | 29.99 | 0.8729 | n.a | 1075 | 68.2 | 15.8 | 5.92 |
| MobileSR | garasgaras | TensorFlow | 82 | 30.02 | 0.8735 | 0.07 | 1057 | 72.4 | 14.6 | 5.82 |
| A+ regression [72] | Baseline | | | 29.32 | 0.8520 | - | - | - | - | - |
| Bicubic Upscaling | Baseline | | | 28.26 | 0.8277 | - | - | - | - | - |

# 3 Challenge Results

From above 250 registered participants, 28 teams entered the final phase and submitted their results, TFLite models, codes, executables and factsheets. Table 1 summarizes the final challenge results and reports PSNR, SSIM and runtime numbers for the valid solutions on the final test dataset and on the target evaluation platform. The proposed methods are described in Sect. 4, and the team members and affiliations are listed in Appendix A.

## 3.1 Results and Discussion

All solutions proposed in this challenge demonstrated an extremely high efficiency, being able to reconstruct a Full HD image on the target Synaptics VS680 board under 15–75 ms, significantly surpassing the basic bicubic image upsampling baseline in terms of the resulting image quality. Moreover, 14 teams managed to beat the last year's top solution [19], demonstrating a better runtime and/or accuracy results. This year, all teams performed INT8 model quantization correctly, and the accuracy drop caused by FP32 → INT8 conversion is less than 0.1 dB in the majority of cases. The general recipe of how to get a fast quantized image super-resolution model compatible with the latest mobile NPUs is as follows:

1. Use a shallow network architecture with or without skip connections;
2. Limit the maximum convolution filter size to $3 \times 3$;
3. Use the depth-to-space layer for the final image upsampling instead of the transposed convolution;
4. Use re-parameterized convolution blocks that are fused into one single convolution layer during the inference for better fidelity results;
5. Replace the nearest neighbor upsampling op with an equivalent $1 \times 1$ convolution layer for a better latency;
6. Clip the model's output values to avoid incorrect output normalization during INT8 quantization; place the clip op before the depth-to-space layer;
7. Fine-tune the network with the quantization aware training algorithm before performing the final INT8 model conversion;
8. Transfer learning might also be helpful in some cases.

Team *Z6* is the challenge winner—the proposed solution demonstrated over 50 FPS on the Synaptics platform while also achieved one of the best fidelity results. Despite the simplicity of the presented SCSRN model, the authors used an advanced training strategy with weights clipping and normalization to get good image reconstruction results. The second best solution was obtained by team *TCLResearchEurope*, which model achieved more than 62 FPS on the Synaptics VS680 board, thus allowing for a 1080P@60FPS video upscaling. The authors used a combination of the convolution re-parametrization, channel pruning and model distillation techniques to get a small but powerful image super-resolution network. The best numerical results in this competition were achieved by team *TCL Research HK* due to a deep structure, convolution re-parametrization, and a large number of skip connections in the proposed model architecture.

When it comes to the Synaptics VS680 smart TV board itself, we can also see a noticeable improvement in its performance compared to the last year's results, which is caused by a number of optimizations in its NPU and NNAPI drivers. In particular, this year it was able to efficiently accelerate all models submitted during the final phase of the competition, showing a speed up of up to 20–55 times compared to CPU-based model execution. This demonstrates that an accurate deep learning-based high-resolution video upsampling on IoT platforms is a reality rather than just a pure concept, and the solutions proposed in this challenge are already capable to reconstruct Full HD videos on this SoC. With additional optimizations, it should also be possible to perform HD to 4K video upsampling, which is nowadays a critical task for many smart TV boards.

## 4　Challenge Methods

This section describes solutions submitted by all teams participating in the final stage of the MAI 2022 Real-Time Image Super-Resolution challenge.

### 4.1　Z6

Team Z6 proposed a compact Skip-Concatenated Image Super Resolution Network (SCSRN) for the considered task (Fig. 3). This model consists of two-seven $3 \times 3$ convolution layers, five Re-Parameterizable blocks (Rep_Block) and

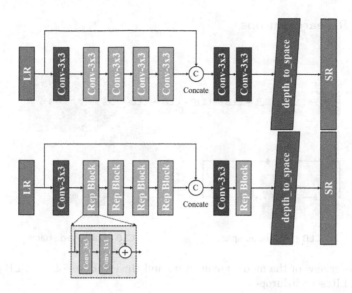

**Fig. 3.** SCSRN model architecture proposed by team Z6 (Top) inference network (Bottom) train network.

a skip connection (concatenation of input image and intermediate feature map directly). The number of channels in the network is set to 32, the depth-to-space op is used to produce the final image output. The authors applied the weight clipping method during the training stage in order to mitigate the performance degradation after INT8 quantization: the weight distribution analysis revealed that the distribution is heavily skewed (asymmetric), especially in the first convolutional layer. In TFLite, the weights quantization is only performed in a symmetric way, thus the quantization error is significantly accumulated from the first layer. Additionally, the inference time was reduced and the correct model output was ensured by clipping the output values before the depth-to-space layer. The network was trained in three stages:

1. In the first stage, the model was trained from scratch using the L1 loss function on patches of size $128 \times 128$ pixels with a mini-batch size of 16. Network parameters were optimized for 800 epochs using the Adam optimizeralgorithm with a learning rate of 1e−3 decreased with the cosine warm-up scheduler [62] with a 0.1% warm-up ratio.
2. Next, the obtained model was further trained to minimize the L2 loss for 200 epochs. The initial learning rate was set to 2e−5 and halved every 60 epochs. At this stage, we also apply a channel shuffle augmentation.
3. In the third stage, the model was fine-tuned with the quantization aware training algorithm for 300 epochs. The initial learning rate was set to 1e−54 and halved every 60 epochs. The authors used the DCT (Discrete Cosine Transform)-domain L1 loss function as the target metric.

## 4.2   TCLResearchEurope

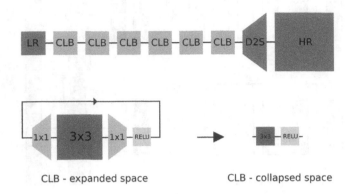

CLB - expanded space            CLB - collapsed space

**Fig. 4.** An overview of the model architecture and the structure of the CLB proposed by team TCLResearchEurope.

The solution proposed by team TCLResearchEurope is based on two core ideas: Structural Reparametrization [13,14,85] and Channel Pruning [4,5]. To accelerate the training process, to limit the memory usage, and to make the quantization aware training in the expanded space possible, an online version of the Reparametrization was adopted [8]. During the training phase, every convolution is represented as a Collapsible Linear Block (CLB) with skip connection (if convolution input channel number is equal to output channel number) with an initial expand ratio of 4. CLB employs similar ideas to the Inverted Residual Block [67], but the depth-wise convolution is replaced with the standard one in this block. The overall model architecture and the structure of the CLB block in both expanded and collapsed states are shown in Fig. 4. The initial number of input and output channels for intermediate convolution layers is set to 48. During the inference state, the model is collapsed into a plain convolutional structure with the depth-to-space layer preceded by the clip-by-value op. No residual connections were used in the model to keep its latency as small as possible, and ReLU was used as an activation function.

Channel Pruning was adopted to further improve the model efficiency. After each trained expanded convolution layer, a small two-layers Channel Ranker network was inserted. Such model is responsible for detecting the least important convolution channels within each layer. Later, during the distillation phase, such channels are, along with the Channel Ranker networks, removed and the model is further fine-tuned. Finally, the authors used the quantization aware training: since the proposed network has a simple feed-forward structure, post-training quantization leads to a significant accuracy drop. Quantizers were inserted right after each collapsed convolution block.

The model was trained on patches of size $128 \times 128$ pixels with a batch size of 16. The training process was done in three stages: first, network parameters were optimized for 1000 epochs using the Adam algorithm with a learning rate of 1e–4 decreased with the cosine scheduler. Next, the model was pruned and trained for another 1000 epochs. Finally, quantization aware training was performed for 100 epochs with a learning rate of 1e–7. During this stage, the authors used an additional loss function called the Self-supervised Quantization-aware Calibration Loss (SQCL) [77]. This loss function is responsible for calibrating the values of parts of the network after quantization to have an approximate distribution as before quantization. SQCL allows to make the training process more stable and decrease the gap in PSNR before and after quantizing the model.

## 4.3    ECNUSR

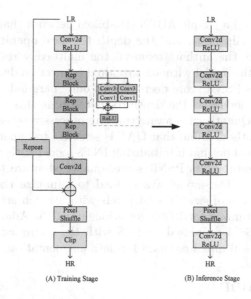

(A) Training Stage          (B) Inference Stage

**Fig. 5.** PureConvSR model proposed by team ECNUSR.

Team ECNUSR proposed a lightweight convolutional network with equivalent transformation (PureConvSR) shown in Fig. 5. Equivalent transformation here refers to the replacement of time-consuming modules or operations such as *clip*, *add* and *concatenate* with convolution modules. In the training stage, the network is composed of $N$ convolution layers, a global skip connection, a pixel shuffle module and a clip module. In the inference stage, the model is only composed of $N + 1$ convolution layers and a pixel shuffle module. These two models are completely equivalent in logic due to the used equivalent transformation and reparameterization. The authors additionally used the EMA training strategy to stabilize the training process and improve the performance of the network. The model was trained using the MSE loss function on patches of size $64 \times 64$ pixels with a batch size of 16, network parameters were optimized using the Adam algorithm.

## 4.4  LCVG

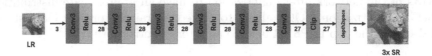

**Fig. 6.** HOPN architecture developed by team LCVG.

Team LCVG proposed a simple ABPN [19]-based network that consists of only $3 \times 3$ convolutions, clip node, and the depth to space operations (Fig. 6). To improve its runtime, the authors removed the multi-copy residual connection proposed in [19] as the model without a residual connection demonstrated lower latency, although its PSNR score decrease was only marginal. The authors also exchanged the clip node and the depth-to-space op as this resulted in a better model runtime. Quantization aware training process was used when training the network. The authors found that QAT is sensitive to hyper-parameters and can easy fall into local optima if initializing INT8 model with pre-trained FP32 weights. Thus, to improve its PSNR score, quantized-aware training was performed from scratch. The model was trained to minimize the mean absolute loss (MAE) on patches of size $64 \times 64$ pixels with a batch size of 16. Network parameters were optimized for 1000K iterations using the Adam algorithm with a learning rate of 6e–4 decreased to 1e–8 with the cosine annealing strategy. Horizontal and vertical flips were used for data augmentation.

## 4.5  BOE-IOT-AIBD

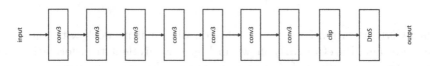

**Fig. 7.** Model architecture proposed by team BOE.

Team BOE designed a small CNN model that takes a low-resolution input image and passes it to a $3 \times 3$ convolutional layer performing feature extraction, followed by six $3 \times 3$ convolutions, one clip node and one depth-to-space layer. The authors found that placing the clip node before the depth-to-space layer is much faster than placing it after the depth-to-space layer. The model was trained to minimize the L1 loss on patches of size $128 \times 128$ pixels with a batch size of 8. Network parameters were optimized for 1000 epochs using the Adam algorithm with a learning rate of 1e–3 decreased by 0.6 every 100 epochs. Quantized-aware training was then applied to improve the accuracy of the resulting INT8 model. During this stage, the model was trained for another 220 epochs with the initial learning rate of 1e–5 decreased by 0.6 every 50 epochs. Random flips and rotations were used for data augmentation (Fig. 7).

## 4.6   NJUST

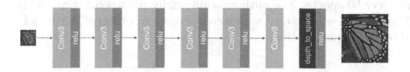

Fig. 8. CNN architecture proposed by team NJUST.

The model architecture developed by team NJUST is very similar to the previous two solutions: the network contains five $3 \times 3$ convolutional layers followed by ReLU activations, and one depth-to-space layer (Fig. 8). No skip connections were used to improve the model efficiency. The network was trained using the L1 loss function on patches of size $64 \times 64$ pixels with a mini-batch size of 16. Model parameters were optimized for 600K iterations with a learning rate of 1e–3 halved every 150K iterations. Quantized-aware training was applied to improve the accuracy of the resulting INT8 model.

## 4.7   Antins_cv

The architecture designed by team Antins_cv is demonstrated in Fig. 9. The model consists of six convolutional layers: after the first convolutional layer, the resulting features are then processed by two network branches. Each branch has two convolutional layers, and the second layer of right right branch is not followed by any activation function. The outputs of these branches are merged pixel-wise using the add operation as it turned to be faster than the concat op. The authors used a pixel shuffle layer to produce the final output image.

The model was first trained to minimize the PSNR loss function on patches of size $128 \times 128$ pixels with a batch size of 32. Network parameters were optimized for 600 epochs using the Adam optimizer with a learning rate of 1e–3 halved every 120 epochs. Then, the authors applied quantization-aware training: MSE loss was used as the target metric, the batch size was set to 8, and the model

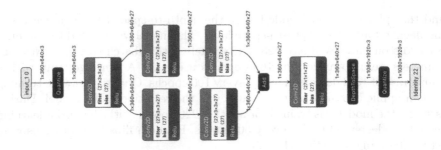

**Fig. 9.** Antins_cv.

was fine-tuned for 220 epochs with the initial learning rate of 1e–4 decreased by half every 50 epochs. The authors additionally proposed a new MovingAvgWithFixedQuantize op to deal with incorrect normalization of the model output caused by INT8 quantization.

## 4.8   GenMedia Group

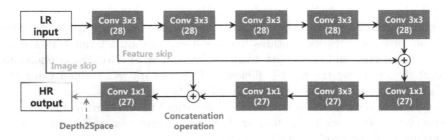

**Fig. 10.** A modified anchor-based plain network proposed by team GenMedia Group.

The model architecture proposed by team GenMedia Group is also inspired by the last year's top solution [19]. The authors added one extra skip connection to the mentioned anchor-based plain net (ABPN) model, and used concatenation followed by a 1×1 convolution instead of a pixel-wise summation at the end of the network (Fig. 10). To avoid potential context switching between CPU and GPU, the order of the clipping function and the depth-to-space operation was rearranged. The authors used the mean absolute error (MAE) as the target objective function. The model was trained on patches of size 96 × 96 pixels with a batch size of 32. Network parameters were optimized for 1000 epochs using the Adam algorithm with a learning rate of 1e–3 decreased by half every

200 epochs. For FP32 training, the authors used SiLU activation function that was later replaced with ReLU when fine-tuning the model with quantization-aware training. Random horizontal and vertical image flips were used for data augmentation.

## 4.9   Vccip

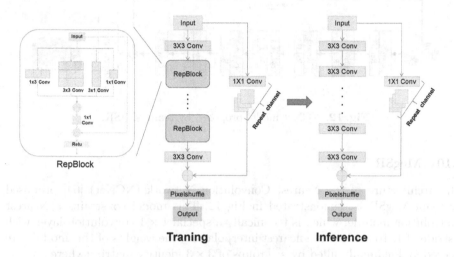

**Fig. 11.** The architecture of the model and the structure of the Re-Parameterizable blocks (RepBlocks) proposed by team Vccip.

Team Vccip derived their solution from the ABPN [19] architecture. The proposed model consists of two $3 \times 3$ convolution layers, six Re-Parameterizable blocks (RepBlocks), and one pixel shuffle layer (Fig. 11). The RepBlock can enhance the representational capacity of a single convolutional layer by combining diverse branches of different scales, including sequences of convolutions, multi-scale convolutions and residual structures. In the inference stage, the RepBlock is re-parameterized into a single convolutional layer for a better latency. As in [19], only a residual part of the SR image is learned. The model was first trained to minimize the Charbonnier and Vgg-based perceptual loss functions on patches of size $64 \times 64$ pixels with a batch size of 16. Network parameters were optimized for 600K iterations using the Adam optimizer with the cyclic learning rate policy. Quantization-aware training was applied to improve the accuracy of the resulting INT8 model.

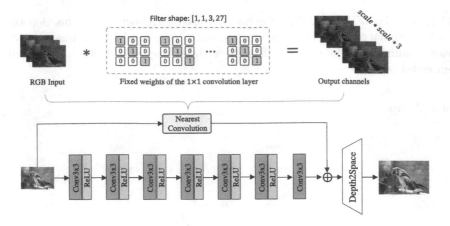

**Fig. 12.** NCNet model proposed by team MegSR.

### 4.10 MegSR

The architecture of the Nearest Convolution Network (NCNet) [64] proposed by team MegSR is demonstrated in Fig. 12. This model constrains a Nearest Convolution module, which is technically a special $1 \times 1$ convolution layer with a stride of 1. To achieve the nearest interpolation, the weights of this module are freezed and manually filled by $s^2$ groups of $3 \times 3$ identity matrix (where $s$ is the upscaling factor), and each group would produce an RGB image, performing a copy operation to repeat the input image. When followed by the depth-to-space operation, this module would reconstruct exactly the same image as in case of the nearest interpolation method, though much faster when running on mobile GPUs or NPUs.

The entire model consists of seven $3 \times 3$ convolution layers with ReLU activations, the number of channels is set to 32. The model was trained to minimize the L1 loss function on patches of size $64 \times 64$ pixels with a batch size of 64. Network parameters were optimized for 500K iterations using the Adam optimizer with the initial learning rate of 1e–3 decreased by half every 200K iterations. Inspired by [58], the authors additionally fine-tuned the model on larger patch size of $128 \times 128$ for another 200K iterations.

### 4.11 DoubleZ

The architecture proposed by team DoubleZ is very similar to the ABPN [19] network: the major difference consists in the residual module, where one convolution layer is used instead of 9 times input image stacking. The authors used ReLU activation to clip the outputs of the model instead of the clip_by_value op as it turned to be faster in the experiments. The model was trained in two stages: first, it was minimizing the L1 loss function on $96 \times 96$ pixel patches using the Adam optimizer with a learning rate of 1e–3 decreased to 1e–5 through the

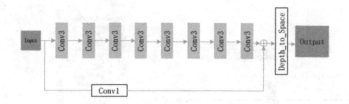

**Fig. 13.** ABPN-based network proposed by team DoubleZ.

training. Then, the model was fine-tuned with the L2 loss. Random flips and rotations were used for data augmentation, quantization-aware training was applied to improve the accuracy of the resulting INT8 model (Fig. 13).

## 4.12   Jeremy Kwon

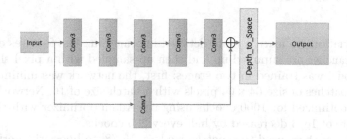

**Fig. 14.** ABPN-based network proposed by team Jeremy Kwon.

Jeremy Kwon used almost the same ABPN-based model design as the previous team (Fig. 14), the main difference between these solutions is the number of convolutional layers. The model was trained to minimize the L1 and MAE losses (the latter one was computed on images resized with the space-to-depth op) on patches of size $96 \times 96$ pixels with a batch size of 16. Network parameters were optimized for 1000 epochs using the Adam optimizer with the initial learning rate of 1e–3 decreased by half every 200 epochs. Quantization-aware training was applied to improve the accuracy of the resulting INT8 model.

## 4.13   Lab216

Team Lab216 proposed a lightweight asymmetric network and applied the contrastive quantization-aware training to improve its performance after quantization (Fig. 15). The model consists of two branches: the first residual branch has only one $3 \times 3$ convolution layer to extract basic features required for image up-sampling, while the second branch contains seven convolution layers with

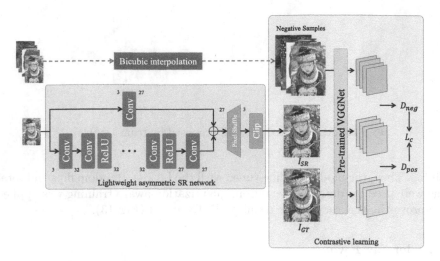

**Fig. 15.** Model architecture trained with the contrastive quantization-aware method proposed by team Lab216.

ReLU activiations to learn a more detailed information. The features extracted by these branches are summed up and then up-sampled with a pixel shuffle op.

The model was trained in two stages: first, the network was minimizing the L1 loss on patches of size $64 \times 64$ pixels with a batch size of 16. Network parameters were optimized for 1000 epochs using the Adam optimizer with the initial learning rate of 1e–3 decreased by half every 200 epochs.

Next, the authors used the contrastive loss [55,78] to boost the performance of quantization-aware training. It works by keeping the anchor close to positive samples while pushing away from negative samples in the considered latent space. In this case, the ground truths images are settled as positive samples, while negative sets are generated with bicubic interpolation. The total loss is defined as:

$$\mathcal{L}_{total} = \mathcal{L}_1 + \lambda \times \mathcal{L}_c,$$

$$\mathcal{L}_c = \frac{d(\phi(I_{SR}), \phi(I_{GT}))}{\sum_{k=1}^{K} d(\phi(I_{SR}), \phi(I_{Neg}^k))},$$

where $K = 16$ is the number of negative samples, $\phi()$ are the intermediate features generated by a pre-trained VGG-19 network. At this stage, the model was trained for 220 epochs with a learning rate initialized at 1e–4 and decreases by half every 50 epochs.

### 4.14    TOVB

Team TOVB proposed an ABPN-based model architecture, where $3 \times 3$ convolutions were replaced with a re-parameterized convolution module (Fig. 16). In total, the network consists of seven RepConv modules. The model was trained

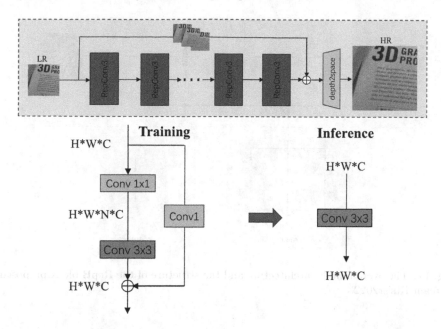

**Fig. 16.** The overall model architecture and the structure of the RepConv block proposed by team TOVB.

in two stages: first, L1 loss was used as the target metric, and model parameters were optimized for 1500 epochs using the Adam algorithm with a batch size of 64 and a learning rate of 2e–4 on patches of size 160 × 160. In the second stage, each RepConv module was fused into a single convolution layer, and the model was fine-tuned for another 30 epochs on patches of size 64 × 64 with a batch size of 16 and a learning rate of 1e–6.

### 4.15  Samsung Research

Team Samsung Research used the ABPN [19] network design. First, the authors trained the model on the ImageNet-2012 data. Then, the network was fine-tuned on the DIV2K and Flickr2K data. Finally, the model was further fine-tuned using the quantization-aware training. The difference of this solution compared from the original ABPN paper is that a smaller learning rate of 1e–5 was used during the quantization-aware training step.

### 4.16  Rtsisr2022

Team Rtsisr2022 adopted the ABPN architecture, but changed the number of channels from 28 to 32 as the model is in general more efficiently executed on NPUs when the number of channels is a multiple of 16. Furthermore, to improve the accuracy results while keeping the same runtime, the authors replaced every

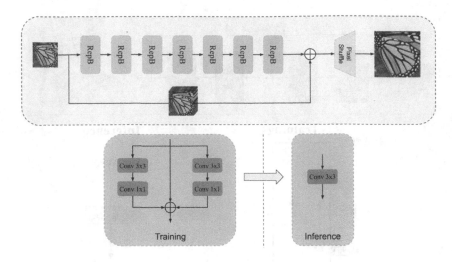

**Fig. 17.** The overall model architecture and the structure of the RepB block proposed by team Rtsisr2022.

$3 \times 3$ convolutional layer with a re-parameterize convolution block (RepB) illustrated in Fig. 17.

The model was trained in three stages: first, L1 loss was used as the target metric, and model parameters were optimized for 2000 epochs using the Adam algorithm with a batch size of 16 and a learning rate of 6e–4 decayed by half every 200 epochs. In the second stage, each RepB module was fused into a single convolution layer, and the model was fine-tuned for another 1000 with a learning rate of 8e–5 decreased by half every 200 epochs. Finally, the authors applied quantization-aware training to improve the accuracy of the resulting INT8 model. At this stage, the model was fine-tuned for 220 epochs with a learning rate of 8e–6 decreased by half every 50 epochs. Patches of size $192 \times 192$ augmented with random flips and rotations were used at all stages.

### 4.17  Aselsan Research

Team Aselsan Research proposed the XCAT model [7] demonstrated in Fig. 18, its distinctive features and their impact on the performance are as follows:

– **Group convolutions with unequal filter groups and dynamic kernels.** Group convolutions [56] include multiple kernels per layer to extract and learn more varying features, compared to a single convolutional layer. Inspired by XLSR [6]'s GConv blocks, XCAT also has repeated group convolution blocks to replace convolutional layers. However, convolutional layers inside the group convolution blocks in XCAT have different layer dimensions

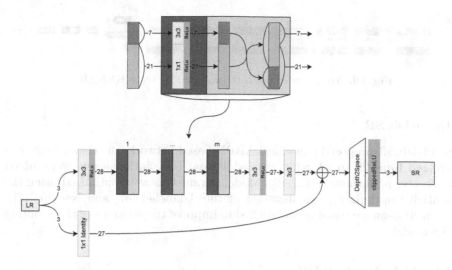

**Fig. 18.** XCAT model architecture proposed by team Aselsan Research.

and dynamic kernel sizes. This allows to pass the same source information between differently specified convolutional layers and allows for a better feature extraction.

- **Cross concatenation.** Instead of using $1 \times 1$ depthwise convolutions for increasing the spatial receptive field of the network as done in XLSR's GConv blocks, the output tensor of each group convolution block is circularly shifted.
- **Intensity-augmented training.** To minimize PSNR difference between the original FP32 and quantized INT8 models, intensity values of the training images are scaled with a randomly chosen constant (1, 0.7, 0.5).
- **Nearest neighborhood up sampling with fixed kernel convolutions.** The authors observed that providing a low resolution input image to the depth-to-space op with accompanying feature tensors increases the PSNR score as opposed to only providing the extracted feature tensors. With this motivation and the inspiration from [19], XCAT adds a repeated input image tensor (where each channel repeated 9 times) to feature tensors, and provides it to the depth-to-space op. For a better performance, a convolutional layer with 3 input channels and 27 output channels is used instead of the tensor copy op, with a $1 \times 1$ untrainable kernel set to serve the same purpose.
- **Clipped ReLU.** Using clipped ReLU at the end of the model allows to deal with incorrect output normalization when performing model quantization.

The model was trained twice. First, the authors used the Charbonnier loss function and optimized model parameters using the Adam algorithm with the initial learning rate of 1e–3 and the warm-up scheduler. During the second stage, the MSE loss function was used, and the initial learning rate was set to 1e–4.

**Fig. 19.** ABPN-based network proposed by team Klab_SR.

## 4.18    Klab_SR

Team Klab_SR proposed a modified ABPN-based network: the channel size was decreased from 28 to 20, while two additional convolution layers were added for a better performance (Fig. 19). Model parameters were optimized using the NadaBelief optimizer (a combination of the Adabelief [87] and Nadam [17]), quantization-aware training was applied to improve the accuracy of the resulting INT8 model.

## 4.19    TCL Research HK

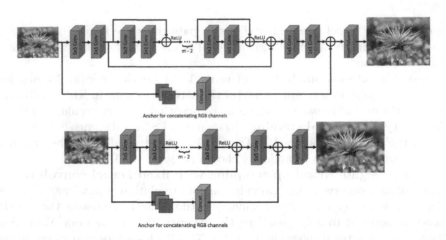

**Fig. 20.** ACSR model architecture proposed by TCL Research HK during the training (top) and inference (bottom) stages.

Team TCL Research HK proposed an anchor-collapsed super resolution (ACSR) network. The model contains the anchor, the feature extraction and the reconstruction modules as shown in Fig. 20. The anchor is a concatenation skip connection layer that stacks the images 9 times in order to learn the image residuals [19]. The feature extraction module consists of several residual blocks with $3 \times 3$ followed by $1 \times 1$ convolutional layers in addition to a ReLU activation layer. The reconstruction module is a $5 \times 5$ followed by $1 \times 1$ convolutional layer with the addition of a pixel shuffle layer. After training, all blocks are collapsed into single convolutional layers [8] to obtain a faster model.

ACSR model was trained to minimize the L1 loss function on patches of size $200 \times 200$ pixels with a batch size of 4. Network parameters were optimized for 1000 epochs using the Adam optimizer with the initial learning rate of 1e–4. Random flips and rotations were used for data augmentation, quantization-aware training was applied to improve the accuracy of the resulting INT8 model.

## 4.20   RepGSR

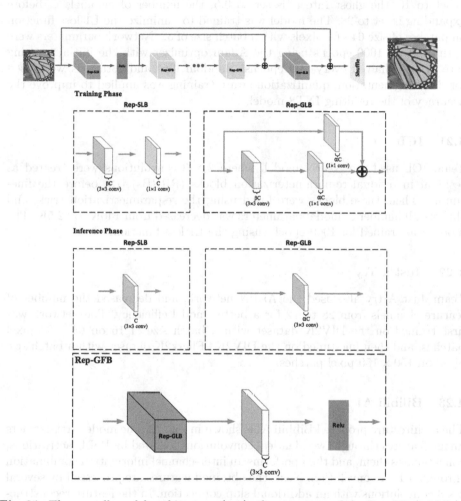

**Fig. 21.** The overall model architecture (top), the structure of the Rep-LB (middle) and Rep-GFB (bottom) blocks proposed by team RepGSR.

To obtain a lightweight super-resolution model, team RepGSR designed a network based on ghost features with re-parametarization. The model is comprised of two module types [Rep-Linear Blocks (Rep-LBs) and Rep-Ghost Features

Blocks (Rep-GFBs)] and two block types [Rep-Simple Linear Blocks (Rep-SLBs) and Rep-Ghost Linear Blocks (Rep-GLBs)], which structure is demonstrated in Fig. 21. The input image is first passed to the rep-SLB block followed by ReLU activation that extracts shallow features. These features are then fed and processed by the following $m$ rep-GFB modules and one rep-SLB block. Finally, a pixel shuffle layer is used to get the final output SR image. During the inference, Rep-SLB and Rep-GLBs blocks are compressed using layer re-parameterization.

The final model has $m = 8$ rep-GFBs blocks, the expansion coefficient $\beta$ is set to 16, the ghost rate $\alpha$ is set to 0.5, the number of channels $C$ before expanding is set to 28. The model was trained to minimize the L1 loss function on patches of size $64 \times 64$ pixels with a batch size of 32. Network parameters were optimized for 1000 epochs using the Adam optimizer with the initial learning rate of 1e–4 halved every 200 epochs. Random flips and rotations were used for data augmentation, quantization-aware training was applied to improve the accuracy of the resulting INT8 model.

### 4.21   ICL

Team ICL used the ABPN model, where $3 \times 3$ convolutions were treated as residual in residual re-parameterization blocks (RRRBs) [18] before the fine-tuning. Then, these blocks were fused using the re-parametrization trick, and the overall number of model parameters was decreased from 140K to 42.5K. The model was trained for 1000 epochs using the L1 loss function.

### 4.22   Just a Try

Team Just A try also used the ABPN network, and decreased the number of feature channels from 28 to 12 for a better model efficiency. The network was first trained on the DIV2K dataset with a batch size of 16 on $64 \times 64$ pixel patches, and then fine-tuned on the DIV2K+Flicrk2K dataset with a batch size of 32 on $150 \times 150$ pixel patches.

### 4.23   Bilibili AI

The architecture proposed Bilibili AI is shown in Fig. 22. The model first extracts image features through two identical convolutions followed by ReLU activations, concatenates them, and then performs an inter-channel information combination through a $1 \times 1$ convolution. Next, the obtained features are processed by several $3 \times 3$ convolutions with an additional skip connection. In the feature reconstruction layer, the authors obtain the final image through a pixel shuffle op and a skip connection adding the input image resized via pixel multiplication. The model was trained to minimize the L1 loss function on patches of size $48 \times 48$ pixels with a batch size of 16. Network parameters were optimized for 1080 epochs using the Adam optimizer with the initial learning rate of 1e–3 decreased by half every 120 epochs.

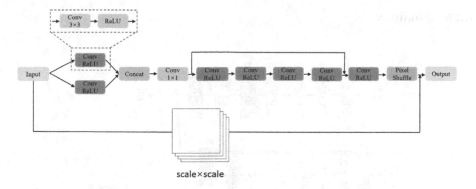

**Fig. 22.** Model architecture proposed by team Bilibili AI.

### 4.24 MobileSR

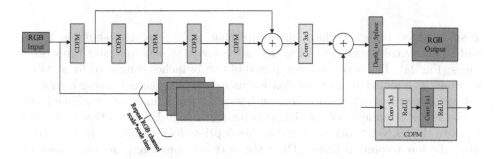

**Fig. 23.** Model architecture proposed by team MobileSR.

The model [21] proposed by team MobileSR contains three blocks: the shallow feature extraction, the deep feature extraction, and the high-resolution image reconstruction blocks shown in Fig. 23. The shallow feature extraction block contains only one channel-and-deep-features-mixer (CDFM, $3 \times 3$ conv followed by $3 \times 3$ conv with ReLU activations after each one). After that, the deep feature extraction module contains 4 stacked CDFM blocks and one skip connection. The final module performing the image reconstruction is using one $3 \times 3$ convolution, an anchor to transfer low-frequency information, and a depth-to-space layer used to produce the final HR image. The model was trained to minimize the L1 loss function on patches of size $64 \times 64$ pixels with a batch size of 32. Network parameters were optimized for 600 epochs using the Adam optimizer with the initial learning rate of 1e–3 halved every 120 epochs. Quantization-aware training was applied to improve the accuracy of the resulting INT8 model.

## 4.25   Noahtcv

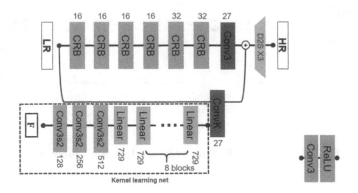

**Fig. 24.** The overall model architecture and the structure of the CRB block (right) proposed by team Noahtcv.

Team noahtcv proposed the architecture that utilizes a pixel shuffling block and seven convolutions on the low resolution space to achieve faster processing times (Fig. 24). The network is comprised of two branches connected by a global residual. The global skip connection learns upsampling kernels using a smaller network comprised of 3 convolutions with stride 2 and 9 linear blocks. Each layer of the kernel learning network (KLN) is passed to a ReLU activation unit. The input to the KLN network is image patches (optionally spatially frequencies too) from the low resolution image. Once the network converges, the top tensor of KLN is fed to the ConvK block generating 27 channels as weight variable to convert it to the inference network.

The model was trained to minimize the L1 loss function on patches of size $64 \times 64$ pixels. Network parameters were optimized using the Adam optimizer with the initial learning rate of 25e-4 decreased by half every 150K iterations. The kernel learning subnet is trained once with the whole network. When the performance converges, the subnet is pruned and the output tensor is fed to the global skip ConvK. The global skip connection becomes a single convolutional layer at inference time.

## 5   Additional Literature

An overview of the past challenges on mobile-related tasks together with the proposed solutions can be found in the following papers:

- Image Super-Resolution: [9,43,63,73]
- Learned End-to-End ISP: [30,35,40,44]
- Perceptual Image Enhancement: [36,43]

- Bokeh Effect Rendering: [33, 42]
- Image Denoising: [1, 2]

**Acknowledgements.** We thank the sponsors of the Mobile AI and AIM 2022 workshops and challenges: Synaptics Inc., AI Witchlabs, MediaTek, Huawei, Reality Labs, OPPO, Raspberry Pi, ETH Zürich (Computer Vision Lab) and University of Würzburg (Computer Vision Lab).

# A    Teams and Affiliations

## Mobile AI 2022 Team

*Title:*
Mobile AI 2022 Image Super-Resolution Challenge
*Members:*
Andrey Ignatov[1,2] *(andrey@vision.ee.ethz.ch)*, Radu Timofte[1,2,3] *(radu.timofte @vision.ee.ethz.ch)*, Maurizio Denna[4] *(maurizio.denna@synaptics.com)*, Abdel Younes[4] *(abdel.younes@synaptics.com)*
*Affiliations:*
[1] Computer Vision Lab, ETH Zürich, Switzerland
[2] AI Witchlabs, Switzerland
[3] University of Würzburg, Germany
[4] Synaptics Europe, Switzerland

## Z6

*Title:*
Skip-Concatenated Image Super Resolution Network (SCSRN) for Mobile Devices
*Members:*
*Ganzorig Gankhuyag[1] (gnzrg25@gmail.com), Jingang Huh[1], Myeong Kyun Kim[1], Kihwan Yoon[1], Hyeon-Cheol Moon[1], Seungho Lee[1], Yoonsik Choe[2], Jinwoo Jeong[1], Sungjei Kim[1]*
*Affiliations:*
[1] Korea Electronics Technology Institute (KETI), South Korea
[2] Yonsei University, South Korea

## TCLResearchEurope

*Title:*
RPQ—Extremely Efficient Super-Resolution Network Via Reparametrization, Pruning and Quantization
*Members:*
*Maciej Smyl (maciej.smyl@gmail.com)*, Tomasz Latkowski, Pawel Kubik, Michal Sokolski, Yujie Ma
*Affiliations:*
TCL Research Europe, Poland

## ECNUSR

*Title:*
PureConvSR: lightweight pure convolutional neural network with equivalent transformation
*Members:*
Jiahao Chao (51215902006@stu.ecnu.edu.cn), Zhou Zhou, Hongfan Gao, Zhengfeng Yang, Zhenbing Zeng
*Affiliations:*
East China Normal University, China

## LCVG

*Title:*
HOPN: A Hardware Optimized Plain Network for Mobile Image Super-Resolution
*Members:*
Zhengyang Zhuge (zyoung2333@gmail.com), Chenghua Li
*Affiliations:*
Institute of Automation, Chinese Academy of Sciences, China

## BOE-IOT-AIBD

*Title:*
Lightweight Quantization CNNNet for Mobile Image Super-Resolution
*Members:*
Dan Zhu (zhudan@boe.com.cn), Mengdi Sun, Ran Duan, Yan Gao
*Affiliations:*
BOE Technology Group Co., Ltd., China

## NJUST

*Title:*
EMSRNet: An Efficient ConvNet for Real-time Image Super-resolution on Mobile Devices
*Members:*
Lingshun Kong (konglingshun@njust.edu.cn), Long Sun, Xiang Li, Xingdong Zhang, Jiawei Zhang, Yaqi Wu, Jinshan Pan
*Affiliations:*
Nanjing University of Science and Technology, China

## Antins_cv

*Title:*
Extremely Light-Weight Dual-Branch Network for Real Time Image Super Resolution

*Members:*
Gaocheng Yu *(yugaocheng.ygc@antgroup.com)*, Jin Zhang, Feng Zhang, Zhe Ma,
Hongbin Wang
*Affiliations:*
Ant Group, China

## GenMedia Group

*Title:*
SkipSkip Video Super-Resolution
*Members:*
Hojin Cho *(jin@gengen.ai)*, Steve Kim
*Affiliations:*
GenGenAI, South Korea

## Vccip

*Title:*
Diverse Branch Re-Parameterizable Net for Mobile Image Super-Resolution
*Members:*
Huaen Li *(huaenli@mail.hfut.edu.cn)*, Yanbo Ma
*Affiliations:*
Hefei University of Technology, China

## MegSR

*Title:*
Fast Nearest Convolution for Real-Time Image Super-Resolution [64]
*Members:*
Ziwei Luo *(ziwei.ro@gmail.com)*, Youwei Li, Lei Yu, Zhihong Wen, Qi Wu, Hao-
qiang Fan, Shuaicheng Liu
*Affiliations:*
Megvii Technology, China
University of Electronic Science and Technology of China (UESTC), China

## DoubleZ

*Title:*
Fast Image Super-Resolution Model
*Members:*
Lize Zhang *(lzzhang_98@stu.xidian.edu.cn)*, Zhikai Zong
*Affiliations:*
Xidian University, China
Qingdao Hi-image Technologies Co., Ltd., China

## Jeremy Kwon

*Title:*
S2R2: Salus Super-Resolution Research
*Members:*
Jeremy Kwon (alan.jaeger0@gmail.com)
*Affiliations:*
None, South Korea

## Lab216

*Title:*
Lightweight Asymmetric Super-Resolution Network with Contrastive Quantized-aware Training
*Members:*
Junxi Zhang (sissie_zhang@whu.edu.cn), Mengyuan Li, Nianxiang Fu, Guanchen Ding, Han Zhu, Zhenzhong Chen
*Affiliations:*
Wuhan University, China

## TOVB

*Title:*
RBPN: Repconv-based Plain Net for Mobile Image Super-Resolution
*Members:*
Gen Li (leegeun@yonsei.ac.kr), Yuanfan Zhang, Lei Sun
*Affiliations:*
None, China

## Samsung Research

*Title:*
ABPN++: Anchor-based Plain Net for Mobile Image Super-Resolution with Pre-training
*Members:*
Dafeng Zhang (dfeng.zhang@samsung.com)
*Affiliations:*
Samsung Research, China

## Rtsisr2022

*Title:*
Re-parameterized Anchor-based Plain Network
*Members:*
Neo Yang (296859095@qq.com), Fitz Liu, Jerry Zhao
*Affiliations:*
None, China

## Aselsan Research

*Title:*
XCAT - Lightweight Single Image Super Resolution Network with Cross Concatenated Heterogeneous Group Convolutions [7]
*Members:*
Mustafa Ayazoglu (mayazoglu@aselsan.com.tr), Bahri Batuhan Bilecen
*Affiliations:*
Aselsan Corporation, Turkey

## Klab_SR

*Title:*
Deeper and narrower SR Model
*Members:*
Shota Hirose (syouta.hrs@akane.waseda.jp), Kasidis Arunruangsirilert, Luo Ao
*Affiliations:*
Waseda University, Japan

## TCL Research HK

*Title:*
Anchor-collapsed Super Resolution for Mobile Devices
*Members:*
Ho Chun Leung (hcleung@tcl.com), Andrew Wei, Jie Liu, Qiang Liu, Dahai Yu
*Affiliations:*
TCL Corporate Research, China

## RepGSR

*Title:*
Super Lightweight Super-resolution Based on Ghost Features with Reparameterization
*Members:*
Ao Li (liao@cqu.edu.cn), Lei Luo, Ce Zhu
*Affiliations:*
University of Electronic Science and Technology, China

## ICL

*Title:*
Neural Network Quantization With Reparametrization And Kernel Range Regularization
*Members:*
Seongmin Hong (smhongok@snu.ac.kr), Dongwon Park, Joonhee Lee, Byeong Hyun Lee, Seunggyu Lee, Se Young Chun
*Affiliations:*
Intelligent Computational Imaging Lab, Seoul National University, South Korea

## Just A try

*Title:*
Plain net for realtime Super-resolution
*Members:*
*Ruiyuan He (164643209@qq.com)*, Xuhao Jiang
*Affiliations:*
None, China

## Bilibili AI

*Title:*
A Robust Anchor-based network for Mobile Super-Resolution
*Members:*
*Haihang Ruan (hhruan@mail.sim.ac.cn)*, Xinjian Zhang, Jing Liu
*Affiliations:*
Bilibili Inc., China

## MobileSR

*Title:*
Real-Time Channel Mixing Net for Mobile Image Super-Resolution [21]
*Members:*
*Garas Gendy[1] (garasgaras@yahoo.com)*, Nabil Sabor[2], Jingchao Hou[1], Guanghui He[1]
*Affiliations:*
[1] Shanghai Jiao Tong University, China
[2] Assiut University, Egypt

# References

1. Abdelhamed, A., Afifi, M., Timofte, R., Brown, M.S.: Ntire 2020 challenge on real image denoising: dataset, methods and results. In: Proceedings of the IEEE/CVF Conference on Computer Vision and Pattern Recognition Workshops, pp. 496–497 (2020)
2. Abdelhamed, A., Timofte, R., Brown, M.S.: Ntire 2019 challenge on real image denoising: Methods and results. In: Proceedings of the IEEE/CVF Conference on Computer Vision and Pattern Recognition Workshops (2019)
3. Agustsson, E., Timofte, R.: Ntire 2017 challenge on single image super-resolution: dataset and study. In: Proceedings of the IEEE Conference on Computer Vision and Pattern Recognition Workshops, pp. 126–135 (2017)
4. Anwar, S., Hwang, K., Sung, W.: Structured pruning of deep convolutional neural networks. ACM J. Emerg. Technol. Comput. Syst. (JETC) 13(3), 1–18 (2017)
5. Anwar, S., Sung, W.: Compact deep convolutional neural networks with coarse pruning. arXiv preprint arXiv:1610.09639 (2016)

6. Ayazoglu, M.: Extremely lightweight quantization robust real-time single-image super resolution for mobile devices. In: Proceedings of the IEEE/CVF Conference on Computer Vision and Pattern Recognition Workshops (2021)
7. Ayazoglu, M., Bilecen, B.B.: XCAT - lightweight quantized single image super-resolution using heterogeneous group convolutions and cross concatenation. In: Proceedings of the European Conference on Computer Vision (ECCV) Workshops (2022)
8. Bhardwaj, K., et al.: Collapsible linear blocks for super-efficient super resolution. Proc. Mach. Learn. Syst. **4**, 529–547 (2022)
9. Cai, J., Gu, S., Timofte, R., Zhang, L.: Ntire 2019 challenge on real image super-resolution: methods and results. In: Proceedings of the IEEE/CVF Conference on Computer Vision and Pattern Recognition Workshops (2019)
10. Cai, Y., Yao, Z., Dong, Z., Gholami, A., Mahoney, M.W., Keutzer, K.: Zeroq: A novel zero shot quantization framework. In: Proceedings of the IEEE/CVF Conference on Computer Vision and Pattern Recognition, pp. 13169–13178 (2020)
11. Chiang, C.M., et al.: Deploying image deblurring across mobile devices: a perspective of quality and latency. In: Proceedings of the IEEE/CVF Conference on Computer Vision and Pattern Recognition Workshops, pp. 502–503 (2020)
12. Conde, M.V., Timofte, R., et al.: Reversed image signal processing and RAW reconstruction. AIM 2022 Challenge Report. In: Proceedings of the European Conference on Computer Vision (ECCV) Workshops (2022)
13. Ding, X., Zhang, X., Han, J., Ding, G.: Diverse branch block: Building a convolution as an inception-like unit. In: Proceedings of the IEEE/CVF Conference on Computer Vision and Pattern Recognition, pp. 10886–10895 (2021)
14. Ding, X., Zhang, X., Ma, N., Han, J., Ding, G., Sun, J.: RepVGG: making VGG-style convnets great again. In: Proceedings of the IEEE/CVF Conference on Computer Vision and Pattern Recognition, pp. 13733–13742 (2021)
15. Dong, C., Loy, C.C., He, K., Tang, X.: Learning a deep convolutional network for image super-resolution. In: Fleet, D., Pajdla, T., Schiele, B., Tuytelaars, T. (eds.) ECCV 2014. LNCS, vol. 8692, pp. 184–199. Springer, Cham (2014). https://doi.org/10.1007/978-3-319-10593-2_13
16. Dong, C., Loy, C.C., He, K., Tang, X.: Image super-resolution using deep convolutional networks. IEEE Trans. Pattern Anal. Mach. Intell. **38**(2), 295–307 (2015)
17. Dozat, T.: Incorporating nesterov momentum into Adam (2016)
18. Du, Z., Liu, D., Liu, J., Tang, J., Wu, G., Fu, L.: Fast and memory-efficient network towards efficient image super-resolution. In: Proceedings of the IEEE/CVF Conference on Computer Vision and Pattern Recognition, pp. 853–862 (2022)
19. Du, Z., Liu, J., Tang, J., Wu, G.: Anchor-based plain net for mobile image super-resolution. In: Proceedings of the IEEE/CVF Conference on Computer Vision and Pattern Recognition Workshops (2021)
20. Freeman, W.T., Jones, T.R., Pasztor, E.C.: Example-based super-resolution. IEEE Comput. Graphics Appl. **22**(2), 56–65 (2002)
21. Gendy, G., nabil sabor, Hou, J., He, G.: Real-time channel mixing net for mobile image super-resolution. In: Proceedings of the European Conference on Computer Vision (ECCV) Workshops (2022)
22. Howard, A., et al.: Searching for mobilenetv3. In: Proceedings of the IEEE/CVF International Conference on Computer Vision, pp. 1314–1324 (2019)
23. Huang, J.B., Singh, A., Ahuja, N.: Single image super-resolution from transformed self-exemplars. In: Proceedings of the IEEE Conference on Computer Vision and Pattern Recognition, pp. 5197–5206 (2015)

24. Ignatov, A., Byeoung-su, K., Timofte, R.: Fast camera image denoising on mobile GPUs with deep learning, mobile AI 2021 challenge: Report. In: Proceedings of the IEEE/CVF Conference on Computer Vision and Pattern Recognition Workshops (2021)

25. Ignatov, A., Chiang, J., Kuo, H.K., Sycheva, A., Timofte, R.: Learned smartphone ISP on mobile NPUs with deep learning, mobile AI 2021 challenge: report. In: Proceedings of the IEEE/CVF Conference on Computer Vision and Pattern Recognition Workshops (2021)

26. Ignatov, A., Kobyshev, N., Timofte, R., Vanhoey, K., Van Gool, L.: DSLR-quality photos on mobile devices with deep convolutional networks. In: Proceedings of the IEEE International Conference on Computer Vision, pp. 3277–3285 (2017)

27. Ignatov, A., Kobyshev, N., Timofte, R., Vanhoey, K., Van Gool, L.: WESPE: weakly supervised photo enhancer for digital cameras. In: Proceedings of the IEEE Conference on Computer Vision and Pattern Recognition Workshop, pp. 691–700 (2018)

28. Ignatov, A., Malivenko, G., Plowman, D., Shukla, S., Timofte, R.: Fast and accurate single-image depth estimation on mobile devices, mobile AI 2021 challenge: report. In: Proceedings of the IEEE/CVF Conference on Computer Vision and Pattern Recognition Workshops (2021)

29. Ignatov, A., Malivenko, G., Timofte, R.: Fast and accurate quantized camera scene detection on smartphones, mobile AI 2021 challenge: report. In: Proceedings of the IEEE/CVF Conference on Computer Vision and Pattern Recognition Workshops (2021)

30. Ignatov, A., et al.: PyNet-V2 mobile: efficient on-device photo processing with neural networks. In: 2021 26th International Conference on Pattern Recognition (ICPR). IEEE (2022)

31. Ignatov, A., Malivenko, G., Timofte, R., et al.: Efficient single-image depth estimation on mobile devices, mobile AI & aim 2022 challenge: Report. In: European Conference on Computer Vision (2022)

32. Ignatov, A., Patel, J., Timofte, R.: Rendering natural camera bokeh effect with deep learning. In: Proceedings of the IEEE/CVF Conference on Computer Vision and Pattern Recognition Workshops, pp. 418–419 (2020)

33. Ignatov, A., et al.: Aim 2019 challenge on bokeh effect synthesis: Methods and results. In: 2019 IEEE/CVF International Conference on Computer Vision Workshop (ICCVW), pp. 3591–3598. IEEE (2019)

34. Ignatov, A., Romero, A., Kim, H., Timofte, R.: Real-time video super-resolution on smartphones with deep learning, mobile AI 2021 challenge: report. In: Proceedings of the IEEE/CVF Conference on Computer Vision and Pattern Recognition Workshops (2021)

35. Ignatov, A., et al.: MicroISP: processing 32mp photos on mobile devices with deep learning. In: European Conference on Computer Vision (2022)

36. Ignatov, A., Timofte, R.: Ntire 2019 challenge on image enhancement: methods and results. In: Proceedings of the IEEE/CVF Conference on Computer Vision and Pattern Recognition Workshops (2019)

37. Ignatov, A., et al.: Power efficient video super-resolution on mobile NPUs with deep learning, mobile AI & aim 2022 challenge: report. In: European Conference on Computer Vision (2022)

38. Ignatov, A., et al.: AI benchmark: running deep neural networks on android smartphones. In: Leal-Taixé, L., Roth, S. (eds.) ECCV 2018. LNCS, vol. 11133, pp. 288–314. Springer, Cham (2019). https://doi.org/10.1007/978-3-030-11021-5_19

39. Ignatov, A., Timofte, R., Denna, M., Younes, A.: Real-time quantized image super-resolution on mobile NPUs, mobile AI 2021 challenge: report. In: Proceedings of the IEEE/CVF Conference on Computer Vision and Pattern Recognition Workshops (2021)

40. Ignatov, A., et al.: Aim 2019 challenge on raw to RGB mapping: methods and results. In: 2019 IEEE/CVF International Conference on Computer Vision Workshop (ICCVW), pp. 3584–3590. IEEE (2019)

41. Ignatov, A., et al.: Ai benchmark: all about deep learning on smartphones in 2019. In: 2019 IEEE/CVF International Conference on Computer Vision Workshop (ICCVW), pp. 3617–3635. IEEE (2019)

42. Ignatov, A., et al.: AIM 2020 challenge on rendering realistic Bokeh. In: Bartoli, A., Fusiello, A. (eds.) ECCV 2020. LNCS, vol. 12537, pp. 213–228. Springer, Cham (2020). https://doi.org/10.1007/978-3-030-67070-2_13

43. Ignatov, A., et al.: Pirm challenge on perceptual image enhancement on smartphones: Report. In: Proceedings of the European Conference on Computer Vision (ECCV) Workshops (2018)

44. Ignatov, A., et al.: Aim 2020 challenge on learned image signal processing pipeline. arXiv preprint arXiv:2011.04994 (2020)

45. Ignatov, A., Timofte, R., et al.: Learned smartphone ISP on mobile GPUs with deep learning, mobile AI & aim 2022 challenge: report. In: European Conference on Computer Vision (2022)

46. Ignatov, A., Timofte, R., et al.: Realistic bokeh effect rendering on mobile GPUs, mobile AI & aim 2022 challenge: report (2022)

47. Ignatov, A., Van Gool, L., Timofte, R.: Replacing mobile camera ISP with a single deep learning model. In: Proceedings of the IEEE/CVF Conference on Computer Vision and Pattern Recognition Workshops, pp. 536–537 (2020)

48. Ignatov, D., Ignatov, A.: Controlling information capacity of binary neural network. Pattern Recogn. Lett. **138**, 276–281 (2020)

49. Inc., S.: https://www.synaptics.com/technology/edge-computing

50. Irani, M., Peleg, S.: Improving resolution by image registration. CVGIP: Graph. Models Image Process. **53**(3), 231–239 (1991)

51. Jacob, B., et al.: Quantization and training of neural networks for efficient integer-arithmetic-only inference. In: Proceedings of the IEEE Conference on Computer Vision and Pattern Recognition, pp. 2704–2713 (2018)

52. Jain, S.R., Gural, A., Wu, M., Dick, C.H.: Trained quantization thresholds for accurate and efficient fixed-point inference of deep neural networks. arXiv preprint arXiv:1903.08066 (2019)

53. Kim, J., Lee, J.K., Lee, K.M.: Accurate image super-resolution using very deep convolutional networks. In: Proceedings of the IEEE Conference on Computer Vision and Pattern Recognition, pp. 1646–1654 (2016)

54. Kinli, F.O., Mentes, S., Ozcan, B., Kirac, F., Timofte, R., et al.: Aim 2022 challenge on Instagram filter removal: methods and results. In: Proceedings of the European Conference on Computer Vision (ECCV) Workshops (2022)

55. Kong, F., et al.: Residual local feature network for efficient super-resolution. In: Proceedings of the IEEE/CVF Conference on Computer Vision and Pattern Recognition, pp. 766–776 (2022)

56. Krizhevsky, A., Sutskever, I., Hinton, G.E.: ImageNet classification with deep convolutional neural networks. In: Advances in Neural Information Processing Systems, vol. 25 (2012)

57. Li, Y., Gu, S., Gool, L.V., Timofte, R.: Learning filter basis for convolutional neural network compression. In: Proceedings of the IEEE/CVF International Conference on Computer Vision, pp. 5623–5632 (2019)
58. Li, Y., et al.: Ntire 2022 challenge on efficient super-resolution: methods and results. In: Proceedings of the IEEE/CVF Conference on Computer Vision and Pattern Recognition, pp. 1062–1102 (2022)
59. Lim, B., Son, S., Kim, H., Nah, S., Mu Lee, K.: Enhanced deep residual networks for single image super-resolution. In: Proceedings of the IEEE Conference on Computer Vision and Pattern Recognition Workshops, pp. 136–144 (2017)
60. Liu, Z., et al.: Metapruning: meta learning for automatic neural network channel pruning. In: Proceedings of the IEEE/CVF International Conference on Computer Vision, pp. 3296–3305 (2019)
61. Liu, Z., Wu, B., Luo, W., Yang, X., Liu, W., Cheng, K.-T.: Bi-real net: enhancing the performance of 1-Bit CNNs with improved representational capability and advanced training algorithm. In: Ferrari, V., Hebert, M., Sminchisescu, C., Weiss, Y. (eds.) ECCV 2018. LNCS, vol. 11219, pp. 747–763. Springer, Cham (2018). https://doi.org/10.1007/978-3-030-01267-0_44
62. Loshchilov, I., Hutter, F.: SGDR: Stochastic gradient descent with warm restarts. arXiv preprint arXiv:1608.03983 (2016)
63. Lugmayr, A., Danelljan, M., Timofte, R.: Ntire 2020 challenge on real-world image super-resolution: Methods and results. In: Proceedings of the IEEE/CVF Conference on Computer Vision and Pattern Recognition Workshops, pp. 494–495 (2020)
64. Luo, Z., et al.: Fast nearest convolution for real-time efficient image super-resolution. In: Proceedings of the European Conference on Computer Vision (ECCV) Workshops (2022)
65. Obukhov, A., Rakhuba, M., Georgoulis, S., Kanakis, M., Dai, D., Van Gool, L.: T-basis: a compact representation for neural networks. In: International Conference on Machine Learning, pp. 7392–7404. PMLR (2020)
66. Park, S.C., Park, M.K., Kang, M.G.: Super-resolution image reconstruction: a technical overview. IEEE Signal Process. Mag. **20**(3), 21–36 (2003)
67. Sandler, M., Howard, A., Zhu, M., Zhmoginov, A., Chen, L.C.: Mobilenetv 2: inverted residuals and linear bottlenecks. In: Proceedings of the IEEE Conference on Computer Vision and Pattern Recognition, pp. 4510–4520 (2018)
68. Tan, M., et al.: MNASnet: platform-aware neural architecture search for mobile. In: Proceedings of the IEEE/CVF Conference on Computer Vision and Pattern Recognition, pp. 2820–2828 (2019)
69. TensorFlow-Lite: https://www.tensorflow.org/lite
70. Timofte, R., Agustsson, E., Van Gool, L., Yang, M.H., Zhang, L.: Ntire 2017 challenge on single image super-resolution: methods and results. In: Proceedings of the IEEE Conference on Computer Vision and Pattern Recognition Workshops, pp. 114–125 (2017)
71. Timofte, R., De Smet, V., Van Gool, L.: Anchored neighborhood regression for fast example-based super-resolution. In: Proceedings of the IEEE International Conference on Computer Vision, pp. 1920–1927 (2013)
72. Timofte, R., De Smet, V., Van Gool, L.: A+: adjusted anchored neighborhood regression for fast super-resolution. In: Cremers, D., Reid, I., Saito, H., Yang, M.-H. (eds.) ACCV 2014. LNCS, vol. 9006, pp. 111–126. Springer, Cham (2015). https://doi.org/10.1007/978-3-319-16817-3_8
73. Timofte, R., Gu, S., Wu, J., Van Gool, L.: Ntire 2018 challenge on single image super-resolution: methods and results. In: Proceedings of the IEEE Conference on Computer Vision and Pattern Recognition Workshops, pp. 852–863 (2018)

74. Timofte, R., Rothe, R., Van Gool, L.: Seven ways to improve example-based single image super resolution. In: Proceedings of the IEEE Conference on Computer Vision and Pattern Recognition, pp. 1865–1873 (2016)
75. Uhlich, S., et al.: Mixed precision DNNs: All you need is a good parametrization. arXiv preprint arXiv:1905.11452 (2019)
76. Wan, A., et al.: Fbnetv2: differentiable neural architecture search for spatial and channel dimensions. In: Proceedings of the IEEE/CVF Conference on Computer Vision and Pattern Recognition, pp. 12965–12974 (2020)
77. Wang, H., Chen, P., Zhuang, B., Shen, C.: Fully quantized image super-resolution networks. In: Proceedings of the 29th ACM International Conference on Multimedia, pp. 639–647 (2021)
78. Wang, Y., et al.: Towards compact single image super-resolution via contrastive self-distillation. arXiv preprint arXiv:2105.11683 (2021)
79. Wu, B., et al.: FBNet: hardware-aware efficient convnet design via differentiable neural architecture search. In: Proceedings of the IEEE/CVF Conference on Computer Vision and Pattern Recognition, pp. 10734–10742 (2019)
80. Yang, C.Y., Yang, M.H.: Fast direct super-resolution by simple functions. In: Proceedings of the IEEE International Conference on Computer Vision, pp. 561–568 (2013)
81. Yang, J., Wright, J., Huang, T., Ma, Y.: Image super-resolution as sparse representation of raw image patches. In: 2008 IEEE Conference on Computer Vision and Pattern Recognition, pp. 1–8. IEEE (2008)
82. Yang, J., Wright, J., Huang, T.S., Ma, Y.: Image super-resolution via sparse representation. IEEE Trans. Image Process. **19**(11), 2861–2873 (2010)
83. Yang, J., et al.: Quantization networks. In: Proceedings of the IEEE/CVF Conference on Computer Vision and Pattern Recognition, pp. 7308–7316 (2019)
84. Yang, R., Timofte, R., et al.: Aim 2022 challenge on super-resolution of compressed image and video: Dataset, methods and results. In: Proceedings of the European Conference on Computer Vision (ECCV) Workshops (2022)
85. Zagoruyko, S., Komodakis, N.: DiracNets: Training very deep neural networks without skip-connections. arXiv preprint arXiv:1706.00388 (2017)
86. Zhang, K., Gu, S., Timofte, R.: Ntire 2020 challenge on perceptual extreme super-resolution: methods and results. In: Proceedings of the IEEE/CVF Conference on Computer Vision and Pattern Recognition Workshops, pp. 492–493 (2020)
87. Zhuang, J., et al.: Adabelief optimizer: adapting stepsizes by the belief in observed gradients. Adv. Neural. Inf. Process. Syst. **33**, 18795–18806 (2020)

# Power Efficient Video Super-Resolution on Mobile NPUs with Deep Learning, Mobile AI & AIM 2022 Challenge: Report

Andrey Ignatov[1,2](✉), Radu Timofte[1,2,3], Cheng-Ming Chiang[4],
Hsien-Kai Kuo[4], Yu-Syuan Xu[4], Man-Yu Lee[4], Allen Lu[4], Chia-Ming Cheng[4],
Chih-Cheng Chen[4], Jia-Ying Yong[4], Hong-Han Shuai[5], Wen-Huang Cheng[5],
Zhuang Jia[6], Tianyu Xu[6], Yijian Zhang[6], Long Bao[6], Heng Sun[6],
Diankai Zhang[7], Si Gao[7], Shaoli Liu[7], Biao Wu[7], Xiaofeng Zhang[7],
Chengjian Zheng[7], Kaidi Lu[7], Ning Wang[7], Xiao Sun[8], HaoDong Wu[8],
Xuncheng Liu[9,10], Weizhan Zhang[9,10], Caixia Yan[9,10], Haipeng Du[9,10],
Qinghua Zheng[9,10], Qi Wang[9,10], Wangdu Chen[9,10], Ran Duan[11],
Mengdi Sun[11], Dan Zhu[11], Guannan Chen[11], Hojin Cho[12], Steve Kim[12],
Shijie Yue[13,14], Chenghua Li[13,14], Zhengyang Zhuge[13,14], Wei Chen[15],
Wenxu Wang[15], Yufeng Zhou[15], Xiaochen Cai[16], Hengxing Cai[16], Kele Xu[17],
Li Liu[17], Zehua Cheng[18], Wenyi Lian[19,20], and Wenjing Lian[19,20]

[1] Computer Vision Lab, ETH Zurich, Zürich, Switzerland
andrey@vision.ee.ethz.ch
[2] AI Witchlabs, Zürich, Switzerland
[3] University of Wuerzburg, Würzburg, Germany
radu.timofte@uni-wuerzburg.de
[4] MediaTek Inc., Hsinchu, Taiwan
{jimmy.chiang,hsienkai.kuo,yu-syuan.xu,my.lee,
allen-cl.lu,cm.cheng,ryan.chen}@mediatek.com, jiaying.ee10@nycu.edu.tw
[5] National Yang Ming Chiao Tung University, Hsinchu, Taiwan
{hhshuai,whcheng}@nycu.edu.tw
[6] Video Algorithm Group, Camera Department, Xiaomi Inc., Beijing, China
{jiazhuang,xutianyu,zhangyijian,baolong,sunheng3}@xiaomi.com
[7] Audio & Video Technology Platform Department, ZTE Corporation, Shenzhen,
China
zhang.diankai@zte.com.cn
[8] Beijing, China
[9] School of Computer Science and Technology, Xi'an Jiaotong University,
Xi'an, China
liuxuncheng123@stu.xjtu.edu.cn
[10] MIGU Video Co., Ltd., Beijing, China

Andrey Ignatov, Radu Timofte, Cheng-Ming Chiang, Hsien-Kai Kuo, Hong-Han Shuai and Wen-Huang Cheng are the main Mobile AI & AIM 2022 challenge organizers. The other authors participated in the challenge.
Appendix A contains the authors' team names and affiliations.
Mobile AI 2022 Workshop website:
https://ai-benchmark.com/workshops/mai/2022/.

[11] BOE Technology Group Co., Ltd., Beijing, China
duanr@boe.com.cn
[12] GenGenAI, Seoul, South Korea
jin@gengen.ai
[13] North China University of Technology, Beijing, China
[14] Institute of Automation, Chinese Academy of Sciences, Beijing, China
[15] State Key Laboratory of Computer Architecture, Institute of Computing
Technology, Beijing, China
chenwei21s@ict.ac.cn
[16] 4Paradigm Inc., Beijing, China
caixc@lamda.nju.edu.cn
[17] National University of Defense Technology, Changsha, China
[18] University of Oxford, Oxford, UK
[19] Uppsala University, Uppsala, Sweden
[20] Northeastern University, Shenyang, China

**Abstract.** Video super-resolution is one of the most popular tasks on mobile devices, being widely used for an automatic improvement of low-bitrate and low-resolution video streams. While numerous solutions have been proposed for this problem, they are usually quite computationally demanding, demonstrating low FPS rates and power efficiency on mobile devices. In this Mobile AI challenge, we address this problem and propose the participants to design an end-to-end real-time video super-resolution solution for mobile NPUs optimized for low energy consumption. The participants were provided with the REDS training dataset containing video sequences for a 4X video upscaling task. The runtime and power efficiency of all models was evaluated on the powerful MediaTek Dimensity 9000 platform with a dedicated AI processing unit capable of accelerating floating-point and quantized neural networks. All proposed solutions are fully compatible with the above NPU, demonstrating an up to 500 FPS rate and 0.2 [Watt/30 FPS] power consumption. A detailed description of all models developed in the challenge is provided in this paper.

**Keywords:** Mobile AI challenge · Video super-resolution · Mobile NPUs · Mobile AI · Deep learning · MediaTek · AI Benchmark

# 1 Introduction

The widespread of mobile devices and the increased demand for various video streaming services from the end users led to a pressing need for video super-resolution solutions that are both efficient and low power device-friendly. A large number of accurate deep learning-based solutions have been introduced for video super-resolution [8,40,49,50,57,59,62,63,69] in the past. Unfortunately, most of these solutions while achieving very good fidelity scores are lacking in terms of computational efficiency and complexity as they are not optimized to meet the specific hardware constraints of mobile devices. Such an optimization is key for processing tasks related to image [15,16,35] and video [58] enhancement on mobile devices. In this challenge, we add efficiency-related constraints on the developed video super-resolution solutions that are validated on mobile NPUs. We employ REDS [57], a popular video super-resolution dataset.

The deployment of AI-based solutions on portable devices usually requires an efficient model design based on a good understanding of the mobile processing units (*e.g.* CPUs, NPUs, GPUs, DSP) and their hardware particularities, including their memory constraints. We refer to [25,29] for an extensive overview of mobile AI acceleration hardware, its particularities and performance. As shown in these works, the latest generations of mobile NPUs are reaching the performance of older-generation mid-range desktop GPUs. Nevertheless, a straightforward deployment of neural networks-based solutions on mobile devices is impeded by (i) a limited memory (*i.e.*, restricted amount of RAM) and (ii) a limited or lacking support of many common deep learning operators and layers. These impeding factors make the processing of high resolution inputs impossible with the standard NN models and require a careful adaptation or re-design to the constraints of mobile AI hardware. Such optimizations can employ a combination of various model techniques such as 16-bit/8-bit [5,38,39,71] and low-bit [4,36,53,67] quantization, network pruning and compression [5,21,45,52,60], device- or NPU-specific adaptations, platform-aware neural architecture search [11,64,68,70], *etc.*

The majority of competitions aimed at efficient deep learning models use standard desktop hardware for evaluating the solutions, thus the obtained models rarely show acceptable results when running on real mobile hardware with many specific constraints. In this *Mobile AI challenge*, we take a radically different approach and propose the participants to develop and evaluate their models

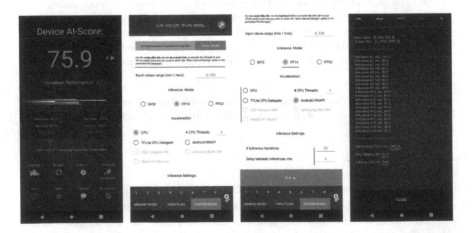

**Fig. 1.** Loading and running custom TensorFlow Lite models with AI Benchmark application. The currently supported acceleration options include Android NNAPI, TFLite GPU, Hexagon NN, Qualcomm QNN, MediaTek Neuron and Samsung ENN delegates as well as CPU inference through TFLite or XNNPACK backends. The latest app version can be downloaded at https://ai-benchmark.com/download

directly on mobile devices. The goal of this competition is to design a fast and power efficient deep learning-based solution for video super-resolution problem. For this, the participants were provided with a large-scale REDS [57] dataset containing the original high-quality and downscaled by a factor of 4 videos. The efficiency of the proposed solutions was evaluated on the MediaTek Dimensity 9000 mobile SoC with a dedicated AI Processing Unit (APU) capable of running floating-point and quantized models. The overall score of each submission was computed based on its fidelity, power consumption and runtime results, thus balancing between the image reconstruction quality and the efficiency of the model. All solutions developed in this challenge are fully compatible with the TensorFlow Lite framework [65], thus can be executed on various Linux and Android-based IoT platforms, smartphones and edge devices.

This challenge is a part of the *Mobile AI & AIM 2022 Workshops and Challenges* consisting of the following competitions:

- Power Efficient Video Super-Resolution on Mobile NPUs
- Quantized Image Super-Resolution on Mobile NPUs [27]
- Learned Smartphone ISP on Mobile GPUs [33]
- Efficient Single-Image Depth Estimation on Mobile Devices [20]
- Realistic Bokeh Effect Rendering on Mobile GPUs [34]
- Super-Resolution of Compressed Image and Video [72]
- Reversed Image Signal Processing and RAW Reconstruction [6]
- Instagram Filter Removal [43]

The results and solutions obtained in the previous *MAI 2021 Challenges* are described in our last year papers:

- Single-Image Depth Estimation on Mobile Devices [17]
- Learned Smartphone ISP on Mobile NPUs [14]
- Real Image Denoising on Mobile GPUs [13]
- Quantized Image Super-Resolution on Mobile NPUs [26]
- Real-Time Video Super-Resolution on Mobile GPUs [61]
- Quantized Camera Scene Detection on Smartphones [18]

## 2 Challenge

In order to design an efficient and practical deep learning-based solution for the considered task that runs fast on mobile devices, one needs the following tools:

1. A large-scale high-quality dataset for training and evaluating the models. Real, not synthetically generated data should be used to ensure a high quality of the obtained model;
2. An easy way to check the runtime and debug the model locally without any constraints as well as the ability to get the runtime and energy consumption on the target evaluation platform.

This challenge addresses all the above issues. Real training data, tools, runtime and power consumption evaluation options provided to the challenge participants are described in the next sections.

## 2.1  Dataset

In this challenge, we use the REDS [57] dataset that serves as a benchmark for traditional video super-resolution task as it contains a large diversity of content and dynamic scenes. Following the standard procedure, we use 240 videos for training, 30 videos for validation, and 30 videos for testing. Each video has sequences of length 100, where every sequence contains video frames of $1280 \times 720$ resolution at 24 fps. To generate low-resolution data, the videos were bicubically downsampled with a factor of 4. The low-resolution video data is then considered as input, and the high-resolution—are the target.

## 2.2  Local Runtime Evaluation

When developing AI solutions for mobile devices, it is vital to be able to test the designed models and debug all emerging issues locally on available devices. For this, the participants were provided with the *AI Benchmark* application [25, 29] that allows to load any custom TensorFlow Lite model and run it on any Android device with all supported acceleration options. This tool contains the latest versions of *Android NNAPI, TFLite GPU, MediaTek Neuron, Hexagon NN, Qualcomm QNN* and *Samsung ENN* delegates, therefore supporting all current mobile platforms and providing the users with the ability to execute neural networks on smartphone NPUs, APUs, DSPs, GPUs and CPUs.

To load and run a custom TensorFlow Lite model, one needs to follow the next steps:

1. Download AI Benchmark from the official website[1] or from the Google Play[2] and run its standard tests.
2. After the end of the tests, enter the *PRO Mode* and select the *Custom Model* tab there.
3. Rename the exported TFLite model to *model.tflite* and put it into the *Download* folder of the device.
4. Select mode type *(INT8, FP16, or FP32)*, the desired acceleration/inference options and run the model.

These steps are also illustrated in Fig. 1.

## 2.3  Runtime Evaluation on the Target Platform

In this challenge, we use the *MediaTek Dimensity 9000* SoC as our target runtime evaluation platform. This chipset contains a powerful APU [44] capable of accelerating floating point, INT16 and INT8 models, being ranked first by AI Benchmark at the time of its release. Within the challenge, the participants were able to upload their TFLite models to a dedicated validation server connected to a real device and get an instantaneous feedback: the power consumption and the

---

[1] https://ai-benchmark.com/download.
[2] https://play.google.com/store/apps/details?id=org.benchmark.demo.

runtime of their solution on the Dimensity 9000 APU or a detailed error log if the model contains some incompatible operations. The models were parsed and accelerated using MediaTek Neuron delegate[3]. The same setup was also used for the final model evaluation. The participants were additionally provided with a detailed model optimization guideline demonstrating the restrictions and the most efficient setups for each supported TFLite op. A comprehensive tutorial demonstrating how to work with the data and how to train a baseline MobileRNN model on the provided videos was additionally released to the participants: https://github.com/MediaTek-NeuroPilot/mai22-real-time-video-sr.

## 2.4   Challenge Phases

The challenge consisted of the following phases:

I. *Development:* the participants get access to the data and AI Benchmark app, and are able to train the models and evaluate their runtime locally;
II. *Validation:* the participants can upload their models to the remote server to check the fidelity scores on the validation dataset, to get the runtime and energy consumption on the target platform, and to compare their results on the validation leaderboard;
III. *Testing:* the participants submit their final results, codes, TensorFlow Lite models, and factsheets.

## 2.5   Scoring System

All solutions were evaluated using the following metrics:

**Table 1.** Mobile AI 2022 Power Efficient Video Super-Resolution challenge results and final rankings. During the runtime and power consumption measurements, the models were upscaling video frames from $180 \times 320$ to $1280 \times 720$ pixels on the MediaTek Dimensity 9000 chipset. Teams *MVideoSR* and *ZX_VIP* are the challenge winners. *Team *SuperDash* used the provided MobileRNN baseline model.

| Team | Author | Framework | Model size, KB | PSNR ↑ | SSIM ↑ | Runtime, ms ↓ | Power consumption, W@30FPS ↓ | Final score |
|---|---|---|---|---|---|---|---|---|
| MVideoSR | erick | PyTorch/TensorFlow | 17 | 27.34 | 0.7799 | 3.05 | **0.09** | 90.9 |
| ZX_VIP | OptimusPrime | PyTorch/TensorFlow | 20 | 27.52 | 0.7872 | 3.04 | 0.10 | 90.7 |
| Fighter | sx | TensorFlow | 11 | 27.34 | 0.7816 | 3.41 | 0.20 | 85.4 |
| XJTU-MIGU SUPER | liuxunchenglxc | TensorFlow | 50 | 27.77 | 0.7957 | 3.25 | 0.22 | 85.1 |
| BOE-IOT-AIBD | DoctoR | Keras/TensorFlow | 40 | 27.71 | 0.7820 | 1.97 | 0.24 | 84.0 |
| GenMedia Group | stevek | Keras/TensorFlow | 135 | 28.40 | 0.8105 | 3.10 | 0.33 | 80.6 |
| NCUT VGroup | ysj | Keras/TensorFlow | 35 | 27.46 | 0.7822 | 1.39 | 0.40 | 75.6 |
| Mortar ICT | work mai | PyTorch/TensorFlow | 75 | 22.91 | 0.7546 | **1.76** | 0.36 | 70.0 |
| RedCat AutoX | caixc | Keras/TensorFlow | 62 | 27.71 | 0.7945 | 7.26 | 0.53 | 69.5 |
| 221B | shermanlian | Keras/TensorFlow | 186 | 28.19 | 0.8093 | 10.1 | 0.80 | 56.8 |
| SuperDash* | xiaoxuan | TensorFlow | 1810 | **28.45** | **0.8171** | 26.8 | 3.73 | −89.3 |
| Upscaling | Baseline | | | 26.50 | 0.7508 | - | - | - |

---

[3] https://github.com/MediaTek-NeuroPilot/tflite-neuron-delegate.

- Peak Signal-to-Noise Ratio (PSNR) measuring fidelity score,
- Structural Similarity Index Measure (SSIM), a proxy for perceptual score,
- The energy consumption on the target Dimensity 9000 platform,
- The runtime on the Dimensity 9000 platform.

The score of each final submission was evaluated based on the next formula:

$$\text{Final Score} = \alpha \cdot \text{PSNR} + \beta \cdot (1 - \text{power consumption}),$$

where $\alpha = 1.66$ and $\beta = 50$. Besides that, the runtime of the solutions should be less than 33 ms when reconstructing one Full HD video frame on the Dimensity 9000 NPU (thus achieving a real-time performance of 30 FPS), otherwise their final score was set to 0. The energy consumption was computed as the amount of watts consumed when processing 30 subsequent video frames (in 1 s).

During the final challenge phase, the participants did not have access to the test dataset. Instead, they had to submit their final TensorFlow Lite models that were subsequently used by the challenge organizers to check both the runtime and the fidelity results of each submission under identical conditions. This approach solved all the issues related to model overfitting, reproducibility of the results, and consistency of the obtained runtime/accuracy values.

## 3 Challenge Results

From above 160 registered participants, 11 teams entered the final phase and submitted valid results, TFLite models, codes, executables and factsheets. Table 1 summarizes the final challenge results and reports PSNR, SSIM and runtime and power consumption numbers for each submitted solution on the final test dataset and on the target evaluation platform. The proposed methods are described in Sect. 4, and the team members and affiliations are listed in Appendix A.

### 3.1 Results and Discussion

All solutions proposed in this challenge demonstrated a very high efficiency, achieving a real-time performance of more than 30 FPS on the target MediaTek Dimensity 9000 platform. The majority of models followed a simple single-frame restoration approach to improve the runtime and power efficiency. Only one model (from *GenMedia Group* team) used all 10 input video frames as an input, and two teams (*RedCat AutoX* and *221B*) proposed RNN-based solutions that used two and three subsequent input video frames, respectively. All designed models have shallow architectures with small convolutional filters and channel sizes. The majority of networks used the depth-to-space (pixel shuffle) op at the end of the model instead of the transposed convolution to avoid the computations when upsampling the images. The sizes of all designed architectures are less than 200 KB, except for the solution from team *SuperDash* that submitted the baseline MobileRNN network.

Teams *MVideoSR* and *ZX_VIP* are the challenge winners. The speed of their models is over 300 FPS on the MediaTek Dimensity 9000 APU, while the power consumption is less than 0.1 W when processing 30 video frames. Both networks used a single-frame uplsampling structure with one residual block and a pixel shuffle op at the end. In terms of the runtime, one of the best results was achieved by team *BOE-IOT-AIBD*, which solution demonstrated over 500 FPS on MediaTek's APU while also showing quite decent fidelity scores. The best fidelity-runtime balance was achieved by the solution proposed by team *GenMedia Group*: its PSNR and SSIM scores are just slightly behind a considerably larger and slower MobileRNN baseline, though its runtime is just around 3 ms per one video frame, same as for the solutions from the first two teams. One can also notice that RNN-based solutions are considerably slower and more power demanding, though they do not necessarily lead to better perceptual and fidelity results. Therefore, we can conclude that right now the standard single- or multi-frame restoration CNN models are more suitable for on-device video super-resolution in terms of the balance between the runtime, power efficiency and visual video restoration quality.

## 4     Challenge Methods

This section describes solutions submitted by all teams participating in the final stage of the MAI 2022 Real-Time Video Super-Resolution challenge.

### 4.1     MVideoSR

In order to meet the low power consumption requirement, the authors used a simple network structure shown in Fig. 2 that has 6 layers, of which only 5 have learnable parameters (4 conv layers and a PReLU activation layer). All intermediate layers have 6 feature channels, pixel shuffle operation is used to upscale the image to the target resolution without performing any computations. When choosing the activation function, the authors compared the commonly used ReLU, Leaky ReLU and Parametric ReLU (PReLU) activations, and the results implied that the PReLU can boost the performance by about 0.05 dB due to its higher flexibility. At the same time, PReLU op has nearly the same power consumption as ReLU/LeakyReLU, thus this activation layer was selected.

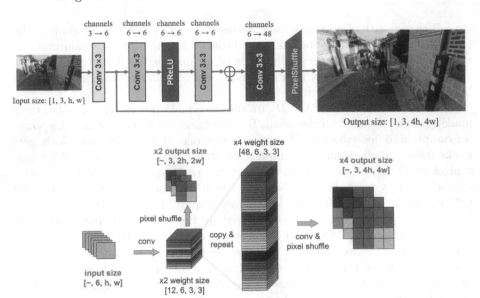

**Fig. 2.** The architecture proposed by team MVideoSR (top), and the used weights repetition strategy in the last convolutional layer when switching from a 2X to a 4X super-resolution problem (bottom).

The authors first trained the proposed network to perform a 2× image super-resolution using the same REDS dataset. For this, the last conv layer of the model was modified to have weights of size $[12, 8, 3, 3]$. After pre-training, these weights were kept and repeated 4 times along the channel dimension (Fig. 2, bottom) to comply with the spatial position of the output obtained after pixel shuffle when performing a 4× super-resolution task. This strategy was used to accelerate the convergence of the model on the final task. Overall, the training of the model was done in five stages:

1. First, the authors trained a 2× super-resolution model from scratch as stated above with a batch size of 64 and the target (high-resolution) patch size of $256 \times 256$. The model was optimized using the L1 loss and the Adam [42] algorithm for 500 K iterations, the initial learning rate was set to 5e−4 and decreases by 0.5 after 200 K and 400 K iterations.
2. Next, a 4× super-resolution model was initialized with weights obtained during the first stage. The same training setup was used except for the learning rate that was initialized with 5e−5 and decreases by 0.5 after 100 K, 300 K and 450 K iterations. After that, the learning rate was set to 2e−4 and the model was trained again for 500 K iterations, where the learning rate was decreased by 0.5 after 200 K iterations.

3. The model was then fine-tuned with the MSE loss function for 1000 K iterations with a learning rate of 2e−4 decreases by 0.5 after 300 K, 600 K and 900 K iterations.
4. During the next 500 K iterations, the model was further fine-tuned with the MSE loss, but the target patch size was changed to 512 × 512. This stage took 500 K iterations, the learning rate was set to 2e−5 and decreases by 0.5 after 100 K, 200 K, 300 K and 400 K iterations.
5. Finally, the obtained model was fine-tuned for another 50 K iterations on patches of size 640 × 640 with a learning rate of 2e−5.

## 4.2   ZX VIP

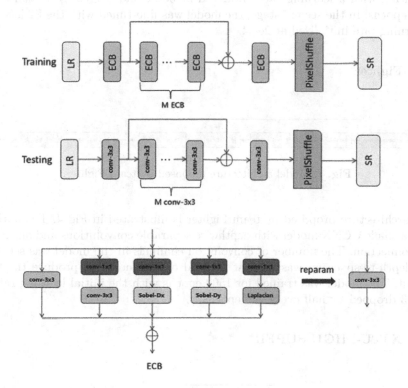

**Fig. 3.** The overall model architecture (top) and the structure of the RCBSR block (bottom) proposed by team ZX VIP.

Inspired by the architecture of the edge-oriented convolution block (ECB) [74], team ZX VIP developed a re-parametrization edge-oriented convolution block

(RCBSR) [9] shown in Fig. 3. During the training stage, the authors took the standard ECB block as a basic backbone unit since it can extract expressive features due to its multi-branch structure. PReLU activations were replaced with ReLU in this block for a greater efficiency. During the inference stage, the multi-branch structure of the ECB was merged into one single 3 × 3 convolution by using a re-parametrization trick. The authors used the depth-to-space op at the end of the model to produce the final output without any computations. The overall model architecture is shown in Fig. 3, the authors used only one ECB block ($M = 1$) in their final model. The network was optimized in two stages: first, is was trained with a batch size of 64 on 512 × 512 pixel images augmented with random flips and rotations. Charbonnier loss was used as the target metric, model parameters were optimized for 4000 epochs using the Adam algorithm with a learning rate initialized at 5e−4 and decreases by half every 1000 epochs. In the second stage, the model was fine-tuned with the L2 loss and a learning rate initialized at 2e−4.

### 4.3    Fighter

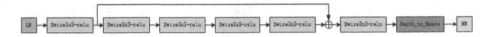

**Fig. 4.** Model architecture proposed by team Fighter.

The architecture proposed by team Fighter is illustrated in Fig. 4. The authors used a shallow CNN model with depthwise separable convolutions and one residual connection. The number of convolution channels in the model was set to 8, the depth-to-space op was used at the end of the model to produce the final output. The model was trained for 1500 epochs with the initial learning rate of 10e−3 dropped by half every 240 epochs.

### 4.4    XJTU-MIGU SUPER

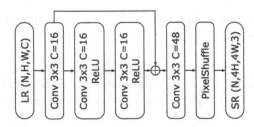

**Fig. 5.** Model architecture developed by team XJTU-MIGU SUPER.

Team XJTU-MIGU SUPER proposed a small CNN-based model originated from genetic algorithm shown in Fig. 5. The network has 16 feature channels in the 1st, 2nd and 3rd layers, and 48 channels in the last convolutional layer. ReLu was used as an activation function in the 2nd and 3rd layers, while the 1st and 4th layers were not followed by any activations. The pixel shuffle operation is used at the end of model to produce the final image output while minimizing the number of computations. First, the models are trained to minimize the L1 loss for 200 epochs using the Adam optimizer with a batch size of 4 and the initial learning rate of 1.6e−2. Then, the learning rate was changed to 0.12 and the model was fine-tuned for another 800 epochs with the batch size of 64. Finally, the batch size was changed again to 4 and the model was trained for another 1600 epochs on the whole dataset.

## 4.5  BOE-IOT-AIBD

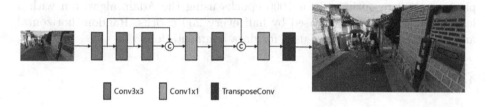

**Fig. 6.** The architecture of the model developed by team BOE-IOT-AIBD.

The solution developed by team BOE-IOT-AIBD is based on the CNN-Net [41] architecture, its structure is illustrated in Fig. 6. This model consists of six convolutional layers and one transposed convolutional layer. ReLU activations are used after the first four and the last convolutional layer, the number of feature channels was set to 25. The authors applied model distillation [10] when training the network, and used the RFDN [51] CNN as a teacher model. In the training phase, the MSE loss function was calculated between the main model's output and the corresponding output from the teacher model, and between the feature maps obtained after the last concat layer of both models having 50 feature channels. The model was trained on patches of size 60 × 80 pixels with a batch size of 4. Network parameters were optimized using the Adam algorithm with a constant learning rate of 1e−4.

## 4.6  GenMedia Group

The model architecture proposed by team GenMedia Group is inspired by the last year's top solution [7] from the MAI image super-resolution challenge [26]. The authors added one extra skip connection to the mentioned anchor-based plain net (ABPN) model, and used concatenation followed by a 1×1 convolution

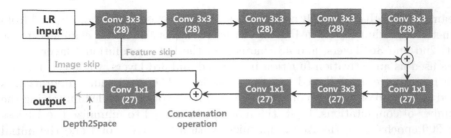

**Fig. 7.** A modified anchor-based plain network proposed by team GenMedia Group.

instead of a pixel-wise summation at the end of the network (Fig. 7). To avoid potential context switching between CPU and GPU, the order of the clipping function and the depth-to-space operation was rearranged. The authors used the mean absolute error (MAE) as the target objective function. The model was trained on patches of size 96 × 96 pixels with a batch size of 32. Network parameters were optimized for 1000 epochs using the Adam algorithm with a learning rate of 1e−3 decreased by half every 200 epochs. Random horizontal and vertical image flips were used for data augmentation.

## 4.7   NCUT VGroup

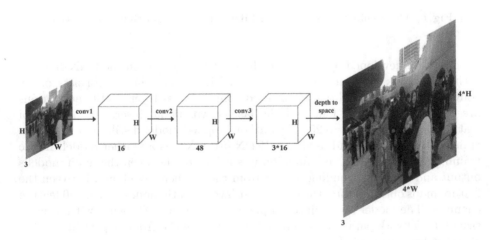

**Fig. 8.** Model architecture proposed by team NCUT.

Team NCUT VGrou also based their solution [73] on the ABPN [7] model (Fig. 8). To improve the runtime, the authors removed four convolution layers and residual connection, thus the resulting network consists of only three convolution layers. The model was trained to minimize the mean error (MAE) loss on patches of size $64 \times 64$ pixels with a batch size of 16. Network parameters were optimized for 168 K iterations using the Adam algorithm with a learning rate of 1e−3 reduced to 1e−8 with the cosine annealing. Random horizontal and vertical image flips were used for data augmentation.

## 4.8   Mortar ICT

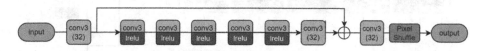

**Fig. 9.** CNN architecture proposed by team Mortar ICT.

Team Mortar ICT used a simple CNN model for the considered video super-resolution task (Fig. 9). The proposed network consists of eight $3 \times 3$ convolutional layers with 32 channels followed by a depth-to-space operation. The authors additionally used a re-parametrization method during the training to expand $3 \times 3$ convolutions to $3 \times 3 + 1 \times 1$ convs that were fused at the inference stage. The model was trained to minimize the L1 loss on patches of size $64 \times 64$ pixels. Network parameters were optimized for 100 epochs using the Adam algorithm with a learning rate of 5e−4 that was decreased by half every 30 epochs.

## 4.9   RedCat AutoX

Team RedCat AutoX used the provided MobileRNN baseline to design their solution (Fig. 10). To improve the model complexity, the authors replaced the residual block in this model with information multi-distillation blocks (IMDBs) [12], and adjusted the number of blocks and base channels to 5 and 8, respectively. The model has a the recurrent residual structure and takes two adjacent video frames $I_t, I_{t+1}$ as an input to see the forward information. To be specific, at time $t$ the model take three parts as an input: the previous hidden state $h_{t-1}$, the current frame at time $t$, and the forward frame at time $t + 1$. A global skip connection, where the input frame is upsampled with a bilinear operation and added to the final output, is used to improve the fidelity and stability of the model. During the training process, L1 loss was used as the target metric. The network was trained for 150 K iterations using the Adam algorithm with an initial learning rate of 3e−3 halved every 15 K steps. The batch size was set to 8, random clipping, horizontal and vertical flips were used for data augmentation.

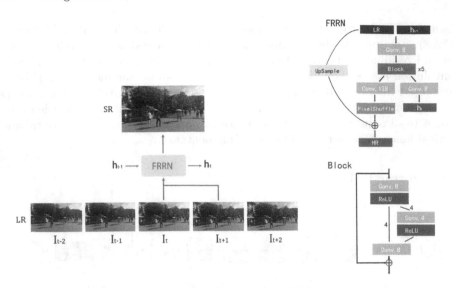

**Fig. 10.** RNN-based model proposed by team RedCat.

## 4.10   221B

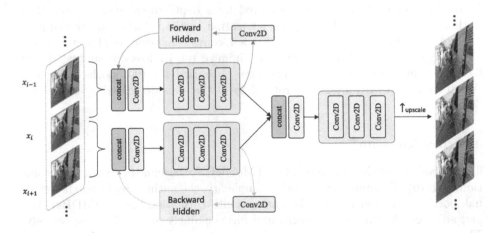

**Fig. 11.** Recurrent network architecture proposed by team 221B.

Team 221B developed an RNN-based model architecture [47] illustrated in Fig. 11. At each time step $t$, the network accepts three subsequent video frames (previous, current and future) and produces the target frame, which is similar to a sliding window multi-frame super-resolution algorithm [48,55,56,69]. Inspired by [37], the authors take advantage of recurrent hidden states to preserve the previous and future information. Specifically, the initial hidden states (forward and backward) are set to 0 and are updated when the window slides to the next frames. Here, the previous frame $x_{t-1}$, the current frame $x_t$ and the forward hidden state are concatenated as a forward group, then the future frame $x_{t+1}$, the current frame $x_t$ and the backward hidden state compose a backward group. Deep features for each group are separately extracted and concatenated to aggregate a multi-frame information to reconstruct the target high-resolution frame. Meanwhile, the extracted features of the forward and backward groups update the corresponding forward and backward hidden states, respectively. The model uses only $3 \times 3$ convolution layers with ReLU activation, a bilinearly upsampled current frame is taken as a residual connection to improve the restoration accuracy [46].

The model was trained to minimize the Charbonnier loss function on patches of size $64 \times 64$ pixels with a batch size of 16. Network parameters were optimized for $150\,K$ iterations using the Adam algorithm with a learning rate of $1e-3$ decreased by half every $50\,K$ iterations.

## 5 Additional Literature

An overview of the past challenges on mobile-related tasks together with the proposed solutions can be found in the following papers:

- Video Super-Resolution: [19,23,58,59]
- Image Super-Resolution: [3,31,54,66]
- Learned End-to-End ISP: [28,32]
- Perceptual Image Enhancement: [24,31]
- Bokeh Effect Rendering: [22,30]
- Image Denoising: [1,2]

**Acknowledgements.** We thank the sponsors of the Mobile AI and AIM 2022 workshops and challenges: AI Witchlabs, MediaTek, Huawei, Reality Labs, OPPO, Synaptics, Raspberry Pi, ETH Zürich (Computer Vision Lab) and University of Würzburg (Computer Vision Lab).

## A    Teams and Affiliations

### Mobile AI & AIM 2022 Team

*Title:*
Mobile AI & AIM 2022 Video Super-Resolution Challenge
*Members:*
Andrey Ignatov[1,2] *(andrey@vision.ee.ethz.ch)*, Radu Timofte[1,2,3] *(radu.timofte@uni-wuerzburg.de)*, Cheng-Ming Chiang[4] *(jimmy.chiang@mediatek.com)*, Hsien-Kai Kuo[4] *(hsienkai.kuo@mediatek.com)*, Yu-Syuan Xu[4] *(yu-syuan.xu@mediatek.com)*, Man-Yu Lee[4] *(my.lee@mediatek.com)*, Allen Lu[4] *(allen-cl.lu@mediatek.com)*, Chia-Ming Cheng[4] *(cm.cheng@mediatek.com)*, Chih-Cheng Chen[4] *(ryan.chen@mediatek.com)*, Jia-Ying Yong[4] *(jiaying.ee10@nycu.edu.tw)*, Hong-Han Shuai[5] *(hhshuai@nycu.edu.tw)*, Wen-Huang Cheng[5] *(whcheng@nycu.edu.tw)*

*Affiliations:*
1 Computer Vision Lab, ETH Zurich, Switzerland
2 AI Witchlabs, Switzerland
3 University of Wuerzburg, Germany
4 MediaTek Inc., Taiwan
5 National Yang Ming Chiao Tung University, Taiwan

## MVideoSR

*Title:*
Extreme Low Power Network for Real-time Video Super Resolution
*Members:*
*Zhuang Jia (jiazhuang@xiaomi.com)*, Tianyu Xu (xutianyu@xiaomi.com), Yijian Zhang (zhangyijian@xiaomi.com), Long Bao (baolong@xiaomi.com), Heng Sun (sunheng3@xiaomi.com)
*Affiliations:*
Video Algorithm Group, Camera Department, Xiaomi Inc., China

## ZX VIP

*Title:*
Real-Time Video Super-Resolution Model [9]
*Members:*
*Diankai Zhang (zhang.diankai@zte.com.cn)*, Si Gao, Shaoli Liu, Biao Wu, Xiaofeng Zhang, Chengjian Zheng, Kaidi Lu, Ning Wang
*Affiliations:*
Audio & Video Technology Platform Department, ZTE Corp., China

## Fighter

*Title:*
Fast Real-Time Video Super-Resolution
*Members:*
*Xiao Sun (2609723059@qq.com)*, HaoDong Wu
*Affiliations:*
None, China

## XJTU-MIGU SUPER

*Title:*
Light and Fast On-Mobile VSR
*Members:*
*Xuncheng Liu (liuxuncheng123@stu.xjtu.edu.cn)*, Weizhan Zhang, Caixia Yan, Haipeng Du, Qinghua Zheng, Qi Wang, Wangdu Chen
*Affiliations:*
School of Computer Science and Technology, Xi'an Jiaotong University, China
MIGU Video Co. Ltd, China

## BOE-IOT-AIBD

*Title:*
Lightweight Quantization CNN-Net for Mobile Video Super-Resolution
*Members:*
Ran Duan (duanr@boe.com.cn), Ran Duan, Mengdi Sun, Dan Zhu, Guannan Chen
*Affiliations:*
BOE Technology Group Co., Ltd., China

## GenMedia Group

*Title:*
SkipSkip Video Super-Resolution
*Members:*
Hojin Cho (jin@gengen.ai), Steve Kim
*Affiliations:*
GenGenAI, South Korea

## NCUT VGroup

*Title:*
EESRNet: A Network for Energy Efficient Super Resolution [73]
*Members:*
Shijie Yue (1161126955@qq.com), Chenghua Li, Zhengyang Zhuge
*Affiliations:*
North China University of Technology, China
Institute of Automation, Chinese Academy of Sciences, China

## Mortar ICT

*Title:*
Real-Time Video Super-Resolution Model
*Members:*
Wei Chen (chenwei21s@ict.ac.cn), Wenxu Wang, Yufeng Zhou
*Affiliations:*
State Key Laboratory of Computer Architecture, Institute of Computing Technology, China

## RedCat AutoX

*Title:*
Forward Recurrent Residual Network
*Members:*
Xiaochen Cai[1] (caixc@lamda.nju.edu.cn), Hengxing Cai[1], Kele Xu[2], Li Liu[2], Zehua Cheng[3]

**Affiliations:**
[1]4Paradigm Inc., Beijing, China
[2]National University of Defense Technology, Changsha, China
[3]University of Oxford, Oxford, United Kingdom


**221B**

*Title:*
Sliding Window Recurrent Network for Efficient Video Super-Resolution [47]
*Members:*
*Wenyi Lian (shermanlian@163.com)*, Wenjing Lian
*Affiliations:*
Uppsala University, Sweden
Northeastern University, China


# References

1. Abdelhamed, A., Afifi, M., Timofte, R., Brown, M.S.: NTIRE 2020 challenge on real image denoising: dataset, methods and results. In: Proceedings of the IEEE/CVF Conference on Computer Vision and Pattern Recognition Workshops, pp. 496–497 (2020)
2. Abdelhamed, A., Timofte, R., Brown, M.S.: NTIRE 2019 challenge on real image denoising: methods and results. In: Proceedings of the IEEE/CVF Conference on Computer Vision and Pattern Recognition Workshops (2019)
3. Cai, J., Gu, S., Timofte, R., Zhang, L.: NTIRE 2019 challenge on real image super-resolution: methods and results. In: Proceedings of the IEEE/CVF Conference on Computer Vision and Pattern Recognition Workshops (2019)
4. Cai, Y., Yao, Z., Dong, Z., Gholami, A., Mahoney, M.W., Keutzer, K.: ZeroQ: a novel zero shot quantization framework. In: Proceedings of the IEEE/CVF Conference on Computer Vision and Pattern Recognition, pp. 13169–13178 (2020)
5. Chiang, C.M., et al.: Deploying image deblurring across mobile devices: a perspective of quality and latency. In: Proceedings of the IEEE/CVF Conference on Computer Vision and Pattern Recognition Workshops, pp. 502–503 (2020)
6. Conde, M.V., Timofte, R., et al.: Reversed image signal processing and RAW reconstruction. AIM 2022 challenge report. In: Karlinsky, L., et al. (eds.) ECCV 2022. LNCS, vol. 13803, pp. 3–26. Springer, Cham (2023)
7. Du, Z., Liu, J., Tang, J., Wu, G.: Anchor-based plain net for mobile image super-resolution. In: Proceedings of the IEEE/CVF Conference on Computer Vision and Pattern Recognition Workshops (2021)
8. Fuoli, D., Gu, S., Timofte, R.: Efficient video super-resolution through recurrent latent space propagation. In: 2019 IEEE/CVF International Conference on Computer Vision Workshop (ICCVW), pp. 3476–3485. IEEE (2019)
9. Gao, S., et al.: RCBSR: re-parameterization convolution block for super-resolution. In: Karlinsky, L., et al. (eds.) ECCV 2022. LNCS, vol. 13802, pp. 540–548. Springer, Cham (2023)
10. Hinton, G., Vinyals, O., Dean, J.: Distilling the knowledge in a neural network. arXiv preprint arXiv:1503.02531 (2015)

11. Howard, A., et al.: Searching for MobileNetV3. In: Proceedings of the IEEE/CVF International Conference on Computer Vision, pp. 1314–1324 (2019)
12. Hui, Z., Gao, X., Yang, Y., Wang, X.: Lightweight image super-resolution with information multi-distillation network. In: Proceedings of the 27th ACM International Conference on Multimedia, pp. 2024–2032 (2019)
13. Ignatov, A., Byeoung-su, K., Timofte, R.: Fast camera image denoising on mobile GPUs with deep learning, mobile AI 2021 challenge: report. In: Proceedings of the IEEE/CVF Conference on Computer Vision and Pattern Recognition Workshops (2021)
14. Ignatov, A., Chiang, J., Kuo, H.K., Sycheva, A., Timofte, R.: Learned smartphone ISP on mobile NPUs with deep learning, mobile AI 2021 challenge: report. In: Proceedings of the IEEE/CVF Conference on Computer Vision and Pattern Recognition Workshops (2021)
15. Ignatov, A., Kobyshev, N., Timofte, R., Vanhoey, K., Van Gool, L.: DSLR-quality photos on mobile devices with deep convolutional networks. In: Proceedings of the IEEE International Conference on Computer Vision, pp. 3277–3285 (2017)
16. Ignatov, A., Kobyshev, N., Timofte, R., Vanhoey, K., Van Gool, L.: WESPE: weakly supervised photo enhancer for digital cameras. In: Proceedings of the IEEE Conference on Computer Vision and Pattern Recognition Workshops, pp. 691–700 (2018)
17. Ignatov, A., Malivenko, G., Plowman, D., Shukla, S., Timofte, R.: Fast and accurate single-image depth estimation on mobile devices, mobile AI 2021 challenge: report. In: Proceedings of the IEEE/CVF Conference on Computer Vision and Pattern Recognition Workshops (2021)
18. Ignatov, A., Malivenko, G., Timofte, R.: Fast and accurate quantized camera scene detection on smartphones, mobile AI 2021 challenge: report. In: Proceedings of the IEEE/CVF Conference on Computer Vision and Pattern Recognition Workshops (2021)
19. Ignatov, A., et al.: PyNet-V2 mobile: efficient on-device photo processing with neural networks. In: 2021 26th International Conference on Pattern Recognition (ICPR). IEEE (2022)
20. Ignatov, A., Malivenko, G., Timofte, R., et al.: Efficient single-image depth estimation on mobile devices, mobile AI & AIM 2022 challenge: report. In: Karlinsky, L., et al. (eds.) ECCV 2022. LNCS, vol. 13803, pp. 71–91. Springer, Cham (2023)
21. Ignatov, A., Patel, J., Timofte, R.: Rendering natural camera bokeh effect with deep learning. In: Proceedings of the IEEE/CVF Conference on Computer Vision and Pattern Recognition Workshops, pp. 418–419 (2020)
22. Ignatov, A., et al.: AIM 2019 challenge on bokeh effect synthesis: methods and results. In: 2019 IEEE/CVF International Conference on Computer Vision Workshop (ICCVW), pp. 3591–3598. IEEE (2019)
23. Ignatov, A., et al.: MicroISP: processing 32MP photos on mobile devices with deep learning. In: Karlinsky, L., et al. (eds.) ECCV 2022. LNCS, vol. 13802, pp. 729–746. Springer, Cham (2023)
24. Ignatov, A., Timofte, R.: NTIRE 2019 challenge on image enhancement: methods and results. In: Proceedings of the IEEE/CVF Conference on Computer Vision and Pattern Recognition Workshops (2019)
25. Ignatov, A., et al.: AI benchmark: running deep neural networks on Android smartphones. In: Leal-Taixé, L., Roth, S. (eds.) ECCV 2018. LNCS, vol. 11133, pp. 288–314. Springer, Cham (2019). https://doi.org/10.1007/978-3-030-11021-5_19

26. Ignatov, A., Timofte, R., Denna, M., Younes, A.: Real-time quantized image super-resolution on mobile NPUs, mobile AI 2021 challenge: report. In: Proceedings of the IEEE/CVF Conference on Computer Vision and Pattern Recognition Workshops (2021)

27. Ignatov, A., Timofte, R., Denna, M., Younes, A., et al.: Efficient and accurate quantized image super-resolution on mobile NPUs, mobile AI & AIM 2022 challenge: report. In: Karlinsky, L., et al. (eds.) ECCV 2022. LNCS, vol. 13803, pp. 92–129. Springer, Cham (2023)

28. Ignatov, A., et al.: AIM 2019 challenge on raw to RGB mapping: methods and results. In: 2019 IEEE/CVF International Conference on Computer Vision Workshop (ICCVW), pp. 3584–3590. IEEE (2019)

29. Ignatov, A., et al.: AI benchmark: all about deep learning on smartphones in 2019. In: 2019 IEEE/CVF International Conference on Computer Vision Workshop (ICCVW), pp. 3617–3635. IEEE (2019)

30. Ignatov, A., et al.: AIM 2020 challenge on rendering realistic bokeh. In: Bartoli, A., Fusiello, A. (eds.) ECCV 2020. LNCS, vol. 12537, pp. 213–228. Springer, Cham (2020). https://doi.org/10.1007/978-3-030-67070-2_13

31. Ignatov, A., et al.: PIRM challenge on perceptual image enhancement on smartphones: report. In: Leal-Taixé, L., Roth, S. (eds.) ECCV 2018. LNCS, vol. 11133, pp. 315–333. Springer, Cham (2019). https://doi.org/10.1007/978-3-030-11021-5_20

32. Ignatov, A., et al.: AIM 2020 challenge on learned image signal processing pipeline. arXiv preprint arXiv:2011.04994 (2020)

33. Ignatov, A., Timofte, R., et al.: Learned smartphone ISP on mobile GPUs with deep learning, mobile AI & AIM 2022 challenge: report. In: Karlinsky, L., et al. (eds.) ECCV 2022. LNCS, vol. 13803, pp. 44–70. Springer, Cham (2023)

34. Ignatov, A., Timofte, R., et al.: Realistic bokeh effect rendering on mobile GPUs, mobile AI & AIM 2022 challenge: report. In: Karlinsky, L., et al. (eds.) ECCV 2022. LNCS, vol. 13803, pp. 153–173. Springer, Cham (2023)

35. Ignatov, A., Van Gool, L., Timofte, R.: Replacing mobile camera ISP with a single deep learning model. In: Proceedings of the IEEE/CVF Conference on Computer Vision and Pattern Recognition Workshops, pp. 536–537 (2020)

36. Ignatov, D., Ignatov, A.: Controlling information capacity of binary neural network. Pattern Recogn. Lett. **138**, 276–281 (2020)

37. Isobe, T., Zhu, F., Jia, X., Wang, S.: Revisiting temporal modeling for video super-resolution. arXiv preprint arXiv:2008.05765 (2020)

38. Jacob, B., et al.: Quantization and training of neural networks for efficient integer-arithmetic-only inference. In: Proceedings of the IEEE Conference on Computer Vision and Pattern Recognition, pp. 2704–2713 (2018)

39. Jain, S.R., Gural, A., Wu, M., Dick, C.H.: Trained quantization thresholds for accurate and efficient fixed-point inference of deep neural networks. arXiv preprint arXiv:1903.08066 (2019)

40. Kappeler, A., Yoo, S., Dai, Q., Katsaggelos, A.K.: Video super-resolution with convolutional neural networks. IEEE Trans. Comput. Imaging **2**(2), 109–122 (2016)

41. Kim, J., Lee, J.K., Lee, K.M.: Accurate image super-resolution using very deep convolutional networks. In: Proceedings of the IEEE Conference on Computer Vision and Pattern Recognition, pp. 1646–1654 (2016)

42. Kingma, D.P., Ba, J.: Adam: a method for stochastic optimization. arXiv preprint arXiv:1412.6980 (2014)

43. Kınlı, F.O., Menteş, S., Özcan, B., Kirac, F., Timofte, R., et al.: AIM 2022 challenge on Instagram filter removal: methods and results. In: Karlinsky, L., et al. (eds.) ECCV 2022. LNCS, vol. 13803, pp. 27–43. Springer, Cham (2023)
44. Lee, Y.L., Tsung, P.K., Wu, M.: Technology trend of edge AI. In: 2018 International Symposium on VLSI Design, Automation and Test (VLSI-DAT), pp. 1–2. IEEE (2018)
45. Li, Y., Gu, S., Gool, L.V., Timofte, R.: Learning filter basis for convolutional neural network compression. In: Proceedings of the IEEE/CVF International Conference on Computer Vision, pp. 5623–5632 (2019)
46. Li, Y., et al.: NTIRE 2022 challenge on efficient super-resolution: methods and results. In: Proceedings of the IEEE/CVF Conference on Computer Vision and Pattern Recognition, pp. 1062–1102 (2022)
47. Lian, W., Lian, W.: Sliding window recurrent network for efficient video super-resolution. In: Karlinsky, L., et al. (eds.) ECCV 2022. LNCS, vol. 13802, pp. 591–601. Springer, Cham (2023)
48. Lian, W., Peng, S.: Kernel-aware raw burst blind super-resolution. arXiv preprint arXiv:2112.07315 (2021)
49. Liang, J., et al.: VRT: a video restoration transformer. arXiv preprint arXiv:2201.12288 (2022)
50. Liu, H., et al.: Video super-resolution based on deep learning: a comprehensive survey. Artif. Intell. Rev. **55**, 5981–6035 (2022). https://doi.org/10.1007/s10462-022-10147-y
51. Liu, J., Tang, J., Wu, G.: Residual feature distillation network for lightweight image super-resolution. In: Bartoli, A., Fusiello, A. (eds.) ECCV 2020. LNCS, vol. 12537, pp. 41–55. Springer, Cham (2020). https://doi.org/10.1007/978-3-030-67070-2_2
52. Liu, Z., et al.: MetaPruning: meta learning for automatic neural network channel pruning. In: Proceedings of the IEEE/CVF International Conference on Computer Vision, pp. 3296–3305 (2019)
53. Liu, Z., Wu, B., Luo, W., Yang, X., Liu, W., Cheng, K.T.: Bi-real net: enhancing the performance of 1-bit CNNs with improved representational capability and advanced training algorithm. In: Proceedings of the European conference on computer vision (ECCV), pp. 722–737 (2018)
54. Lugmayr, A., Danelljan, M., Timofte, R.: NTIRE 2020 challenge on real-world image super-resolution: methods and results. In: Proceedings of the IEEE/CVF Conference on Computer Vision and Pattern Recognition Workshops, pp. 494–495 (2020)
55. Luo, Z., et al.: BSRT: improving burst super-resolution with swin transformer and flow-guided deformable alignment. In: Proceedings of the IEEE/CVF Conference on Computer Vision and Pattern Recognition, pp. 998–1008 (2022)
56. Luo, Z., et al.: EBSR: feature enhanced burst super-resolution with deformable alignment. In: Proceedings of the IEEE/CVF Conference on Computer Vision and Pattern Recognition, pp. 471–478 (2021)
57. Nah, S., et al.: NTIRE 2019 challenge on video deblurring and super-resolution: dataset and study. In: Proceedings of the IEEE/CVF Conference on Computer Vision and Pattern Recognition Workshops (2019)
58. Nah, S., Son, S., Timofte, R., Lee, K.M.: NTIRE 2020 challenge on image and video deblurring. In: Proceedings of the IEEE/CVF Conference on Computer Vision and Pattern Recognition Workshops, pp. 416–417 (2020)
59. Nah, S., et al.: NTIRE 2019 challenge on video super-resolution: methods and results. In: Proceedings of the IEEE/CVF Conference on Computer Vision and Pattern Recognition Workshops (2019)

60. Obukhov, A., Rakhuba, M., Georgoulis, S., Kanakis, M., Dai, D., Van Gool, L.: T-basis: a compact representation for neural networks. In: International Conference on Machine Learning, pp. 7392–7404. PMLR (2020)
61. Romero, A., Ignatov, A., Kim, H., Timofte, R.: Real-time video super-resolution on smartphones with deep learning, mobile AI 2021 challenge: report. In: Proceedings of the IEEE/CVF Conference on Computer Vision and Pattern Recognition Workshops (2021)
62. Sajjadi, M.S., Vemulapalli, R., Brown, M.: Frame-recurrent video super-resolution. In: Proceedings of the IEEE Conference on Computer Vision and Pattern Recognition, pp. 6626–6634 (2018)
63. Shi, W., et al.: Real-time single image and video super-resolution using an efficient sub-pixel convolutional neural network. In: Proceedings of the IEEE Conference on Computer Vision and Pattern Recognition, pp. 1874–1883 (2016)
64. Tan, M., et al.: MnasNet: platform-aware neural architecture search for mobile. In: Proceedings of the IEEE/CVF Conference on Computer Vision and Pattern Recognition, pp. 2820–2828 (2019)
65. TensorFlow-Lite. https://www.tensorflow.org/lite
66. Timofte, R., Gu, S., Wu, J., Van Gool, L.: NTIRE 2018 challenge on single image super-resolution: methods and results. In: Proceedings of the IEEE Conference on Computer Vision and Pattern Recognition Workshops, pp. 852–863 (2018)
67. Uhlich, S., et al.: Mixed precision DNNs: all you need is a good parametrization. arXiv preprint arXiv:1905.11452 (2019)
68. Wan, A., et al.: FBNetV2: differentiable neural architecture search for spatial and channel dimensions. In: Proceedings of the IEEE/CVF Conference on Computer Vision and Pattern Recognition, pp. 12965–12974 (2020)
69. Wang, X., Chan, K.C., Yu, K., Dong, C., Change Loy, C.: EDVR: video restoration with enhanced deformable convolutional networks. In: Proceedings of the IEEE/CVF Conference on Computer Vision and Pattern Recognition Workshops (2019)
70. Wu, B., et al.: FBNet: hardware-aware efficient convnet design via differentiable neural architecture search. In: Proceedings of the IEEE/CVF Conference on Computer Vision and Pattern Recognition, pp. 10734–10742 (2019)
71. Yang, J., et al.: Quantization networks. In: Proceedings of the IEEE/CVF Conference on Computer Vision and Pattern Recognition, pp. 7308–7316 (2019)
72. Yang, R., Timofte, R., et al.: AIM 2022 challenge on super-resolution of compressed image and video: dataset, methods and results. In: Karlinsky, L., et al. (eds.) ECCV 2022. LNCS, vol. 13803, pp. 174–202. Springer, Cham (2023)
73. Yue, S., Li, C., Zhuge, Z., Song, R.: EESRNet: a network for energy efficient super-resolution. In: Karlinsky, L., et al. (eds.) ECCV 2022. LNCS, vol. 13802, pp. xx–yy. Springer, Cham (2023)
74. Zhang, X., Zeng, H., Zhang, L.: Edge-oriented convolution block for real-time super resolution on mobile devices. In: Proceedings of the 29th ACM International Conference on Multimedia, pp. 4034–4043 (2021)

# Realistic Bokeh Effect Rendering on Mobile GPUs, Mobile AI & AIM 2022 Challenge: Report

Andrey Ignatov[1,2]([envelope]), Radu Timofte[1,2,3], Jin Zhang[4], Feng Zhang[4], Gaocheng Yu[4], Zhe Ma[4], Hongbin Wang[4], Minsu Kwon[5], Haotian Qian[6], Wentao Tong[6], Pan Mu[6], Ziping Wang[6], Guangjing Yan[6], Brian Lee[7], Lei Fei[8], Huaijin Chen[7], Hyebin Cho[9], Byeongjun Kwon[9], Munchurl Kim[9], Mingyang Qian[10], Huixin Ma[9], Yanan Li[10], Xiaotao Wang[9], and Lei Lei[10]

[1] Computer Vision Lab, ETH Zürich, Zürich, Switzerland
andrey@vision.ee.ethz.ch
[2] AI Witchlabs, Zollikerberg, Switzerland
[3] University of Wuerzburg, Wuerzburg, Germany
radu.timofte@uni-wuerzburg.de
[4] Ant Group, Hangzhou, China
zj346862@antgroup.com
[5] ENERZAi, Seoul, South Korea
minsu.kwon@enerzai.com
[6] Zhejiang University of Technology, Hangzhou, China
[7] Sensebrain Technology, San Jose, USA
brianlee@sensebrain.site
[8] Tetras.AI, Seoul, China
[9] Korea Advanced Institute of Science and Technology (KAIST),
Yuseong-gu, South Korea
jhb0316@kaist.ac.kr
[10] Xiaomi Inc., Beijing, China
qianmingyang@xiaomi.com

**Abstract.** As mobile cameras with compact optics are unable to produce a strong bokeh effect, lots of interest is now devoted to deep learning-based solutions for this task. In this Mobile AI challenge, the target was to develop an efficient end-to-end AI-based bokeh effect rendering approach that can run on modern smartphone GPUs using TensorFlow Lite. The participants were provided with a large-scale EBB! bokeh dataset consisting of 5K shallow/wide depth-of-field image pairs captured using the Canon 7D DSLR camera. The runtime of the resulting models was evaluated on the Kirin 9000's Mali GPU that provides excellent acceleration results for the majority of common deep learning ops. A detailed description of all models developed in this challenge is provided in this paper.

A. Ignatov and R. Timofte are the main Mobile AI & AIM 2022 challenge organizers. The other authors participated in the challenge.
Appendix A contains the authors' team names and affiliations.
Mobile AI 2022 Workshop website:
https://ai-benchmark.com/workshops/mai/2022/.

L. Karlinsky et al. (Eds.): ECCV 2022 Workshops, LNCS 13803, pp. 153–173, 2023.
https://doi.org/10.1007/978-3-031-25066-8_7

**Keywords:** Mobile AI challenge · Bokeh · Portrait photos · Mobile cameras · Shallow depth-of-field · Mobile AI · Deep learning · AI Benchmark

# 1    Introduction

Bokeh is common technique used to highlight the main object on the photo by blurring all out-of-focus regions. Due to physical limitations of mobile cameras, they cannot produce a strong bokeh effect naturally, thus computer vision-based approaches are often used for this on mobile devices. This topic became very popular over the past few years, and one of the most common solution for this problem consists in segmenting out the main object of interest on the photo and then blurring the background [56,57,70]. Another approach proposed for this task is to blur the image based on the predicted depth map [11,12,67] that can be obtained either using the parallax effect or with stereo vision [2,3]. Finally, an end-to-end deep learning-based solution and the corresponding *EBB!* bokeh effect rendering dataset was proposed in [21], where the authors developed a neural network capable of transforming wide to shallow depth-of-field images automatically. Many other alternative solutions using this dataset were presented in [9,10,22,31,50,51,63,63,69].

**Fig. 1.** Sample wide and shallow depth-of-field image pairs from the EBB! dataset.

The deployment of AI-based solutions on portable devices usually requires an efficient model design based on a good understanding of the mobile processing units (*e.g.* CPUs, NPUs, GPUs, DSP) and their hardware particularities, including their memory constraints. We refer to [26,30] for an extensive overview of mobile AI acceleration hardware, its particularities and performance. As shown in these works, the latest generations of mobile NPUs are reaching the performance of older-generation mid-range desktop GPUs. Nevertheless, a straightforward deployment of neural networks-based solutions on mobile devices is impeded by (i) a limited memory (*i.e.*, restricted amount of RAM) and (ii) a limited or lacking support of many common deep learning operators and layers. These impeding factors make the processing of high resolution inputs impossible with the standard NN models and require a careful adaptation or re-design to the constraints of mobile AI hardware. Such optimizations can employ a combination of various model techniques such as 16-bit/8-bit [7,37,38,65] and low-bit [5,35,46,61] quantization, network pruning and compression [7,21,44,45,49], device- or NPU-specific adaptations, platform-aware neural architecture search [13,58,62,64],etc.

The majority of competitions aimed at efficient deep learning models use standard desktop hardware for evaluating the solutions, thus the obtained models rarely show acceptable results when running on real mobile hardware with many specific constraints. In this *Mobile AI challenge*, we take a radically different approach and propose the participants to develop and evaluate their models directly on mobile devices. The goal of this competition is to design a efficient deep learning-based solution for the realistic bokeh effect rendering problem. For this, the participants were provided with a large-scale *EBB!* dataset containing 5K shallow/wide depth-of-field image pairs collected in the wild with the Canon 7D DSLR camera and 50 mm f/1.8 fast lens. The efficiency of the proposed solutions was evaluated on the Kirin 9000 mobile platform capable of accelerating floating-point and quantized neural networks. The majority of solutions developed in this challenge are fully compatible with the TensorFlow Lite framework [59], thus can be efficiently executed on various Linux and Android-based IoT platforms, smartphones and edge devices.

This challenge is a part of the *Mobile AI & AIM 2022 Workshops and Challenges* consisting of the following competitions:

- Realistic Bokeh Effect Rendering on Mobile GPUs
- Learned Smartphone ISP on Mobile GPUs [34]
- Power Efficient Video Super-Resolution on Mobile NPUs [25]
- Quantized Image Super-Resolution on Mobile NPUs [28]
- Efficient Single-Image Depth Estimation on Mobile Devices [20]
- Super-Resolution of Compressed Image and Video [66]
- Reversed Image Signal Processing and RAW Reconstruction [8]
- Instagram Filter Removal [41]

The results and solutions obtained in the previous *MAI 2021 Challenges* are described in our last year papers:

- Learned Smartphone ISP on Mobile NPUs [15]
- Real Image Denoising on Mobile GPUs [14]
- Quantized Image Super-Resolution on Mobile NPUs [27]

- Real-Time Video Super-Resolution on Mobile GPUs [54]
- Single-Image Depth Estimation on Mobile Devices [17]
- Quantized Camera Scene Detection on Smartphones [18]

## 2　Challenge

In order to design an efficient and practical deep learning-based solution for the considered task that runs fast on mobile devices, one needs the following tools:

1. A large-scale high-quality dataset for training and evaluating the models. Real, not synthetically generated data should be used to ensure a high quality of the obtained model;
2. An efficient way to check the runtime and debug the model locally without any constraints as well as the ability to check the runtime on the target evaluation platform.

This challenge addresses all the above issues. Real training data, tools, and runtime evaluation options provided to the challenge participants are described in the next sections.

**Fig. 2.** Loading and running custom TensorFlow Lite models with AI Benchmark application. The currently supported acceleration options include Android NNAPI, TFLite GPU, Hexagon NN, Qualcomm QNN, MediaTek Neuron and Samsung ENN delegates as well as CPU inference through TFLite or XNNPACK backends. The latest app version can be downloaded at https://ai-benchmark.com/download.

### 2.1　Dataset

One of the biggest challenges in the bokeh rendering task is to get high-quality real data that can be used for training deep models. To tackle this problem, in this challenge we use a popular large-scale *Everything is Better with Bokeh!*

(EBB!) dataset containing more than 10 thousand images was collected in the wild during several months. By controlling the aperture size of the lens, images with shallow and wide depth-of-field were taken. In each photo pair, the first image was captured with a narrow aperture (f/16) that results in a normal sharp photo, whereas the second one was shot using the highest aperture (f/1.8) leading to a strong bokeh effect. The photos were taken during the daytime in a wide variety of places and in various illumination and weather conditions. The photos were captured in automatic mode, the default settings were used throughout the entire collection procedure. An example set of collected images is presented in Fig. 1.

The captured image pairs are not aligned exactly, therefore they were first matched using SIFT keypoints and RANSAC method same as in [16]. The resulting images were then cropped to their intersection part and downscaled so that their final height is equal to 1024 pixels. From the resulting 10 thousand images, 200 image pairs were reserved for testing, while the other 4.8 thousand photo pairs can be used for training and validation.

## 2.2 Local Runtime Evaluation

When developing AI solutions for mobile devices, it is vital to be able to test the designed models and debug all emerging issues locally on available devices. For this, the participants were provided with the *AI Benchmark* application [26, 30] that allows to load any custom TensorFlow Lite model and run it on any Android device with all supported acceleration options. This tool contains the latest versions of *Android NNAPI, TFLite GPU, Hexagon NN, Qualcomm QNN, MediaTek Neuron* and *Samsung ENN* delegates, therefore supporting all current mobile platforms and providing the users with the ability to execute neural networks on smartphone NPUs, APUs, DSPs, GPUs and CPUs.

To load and run a custom TensorFlow Lite model, one needs to follow the next steps:

1. Download AI Benchmark from the official website[1] or from the Google Play[2] and run its standard tests.
2. After the end of the tests, enter the *PRO Mode* and select the *Custom Model* tab there.
3. Rename the exported TFLite model to *model.tflite* and put it into the *Download* folder of the device.
4. Select mode type *(INT8, FP16, or FP32)*, the desired acceleration/inference options and run the model.

These steps are also illustrated in Fig. 2.

---

[1] https://ai-benchmark.com/download.

[2] https://play.google.com/store/apps/details?id=org.benchmark.demo.

## 2.3   Runtime Evaluation on the Target Platform

In this challenge, we use the *Kirin 9000* mobile SoC as our target runtime evaluation platform. The considered chipset demonstrates very decent AI Benchmark scores and can be found in the Huawei Mate 40 Pro/X2 smartphone series. It can efficiently accelerate floating-point networks on its Mali-G78 MP24 GPU with a theoretical FP16 performance of 4.6 TFLOPS. The models were parsed and accelerated using the TensorFlow Lite GPU delegate [43] demonstrating the best performance on this platform when using general deep learning models. All final solutions were tested using the aforementioned AI Benchmark application.

## 2.4   Challenge Phases

The challenge consisted of the following phases:

I.  *Development:* the participants get access to the data and AI Benchmark app, and are able to train the models and evaluate their runtime locally;
II. *Validation:* the participants can upload their models to the remote server to check the fidelity scores on the validation dataset, and to compare their results on the validation leaderboard;
III. *Testing:* the participants submit their final results, codes, TensorFlow Lite models, and factsheets.

**Table 1.** Mobile AI 2022 bokeh effect rendering challenge results and final rankings. The runtime values were obtained on 1024 × 1024 pixel images on the Kirin 9000 mobile platform. The results of the PyNET-V2 model are provided for the reference.

| Team | Author | Framework | Model size, MB | PSNR↑ | SSIM↑ | LPIPS↓ | CPU Runtime, ms ↓ | GPU Runtime, ms ↓ | Final score |
|------|--------|-----------|---------------|-------|-------|--------|------------------|------------------|-------------|
| Antins_cv | xiaokaoji | Keras/TensorFlow | 0.06 | 22.76 | 0.8652 | 0.2693 | **125** | **28.1** | 74 |
| ENERZAi | MinsuKwon | TensorFlow | 30 | 22.89 | 0.8754 | 0.2464 | 1637 | 89.3 | 28 |
| MiAlgo | hxin | TensorFlow | 1.5 | 20.08 | 0.7209 | 0.4349 | 1346 | 112 | 0.5 |
| Sensebrain | brianjsl | PyTorch/TensorFlow | 402 | 22.81 | 0.8653 | 0.2207 | 12879 | Error | N.A. |
| ZJUT-Vision | HaotianQian | TensorFlow | 13 | **23.53** | **0.8796** | **0.1907** | Error | Error | N.A. |
| VIC | hyessol | PyTorch | 127 | 22.77 | 0.8713 | 0.2657 | Error | Error | N.A. |
| PyNET [21] | Baseline | TensorFlow | 181 | 23.28 | 0.8780 | 0.2438 | N.A. | 3512 | 1.2 |

## 2.5   Scoring System

All solutions were evaluated using the following metrics:

- Peak Signal-to-Noise Ratio (PSNR) measuring fidelity score,
- Structural Similarity Index Measure (SSIM) and LPIPS [68] (proxy for perceptual score),
- The runtime on the target Snapdragon 8 Gen 1 platform.

In this challenge, the participants were able to submit their final models to two tracks. In the first track, the score of each final submission was evaluated based on the next formula ($C$ is a constant normalization factor):

$$\text{Final Score} = \frac{2^{2 \cdot \text{PSNR}}}{C \cdot \text{runtime}},$$

In the second track, all submissions were evaluated only based on their visual results as measured by the corresponding Mean Opinion Scores (MOS). This was done to allow the participants to develop larger and more powerful models for the considered task.

During the final challenge phase, the participants did not have access to the test dataset. Instead, they had to submit their final TensorFlow Lite models that were subsequently used by the challenge organizers to check both the runtime and the fidelity results of each submission under identical conditions. This approach solved all the issues related to model overfitting, reproducibility of the results, and consistency of the obtained runtime/accuracy values.

## 3    Challenge Results

From the above 90 registered participants, 6 teams entered the final phase and submitted valid results, TFLite models, codes, executables, and factsheets. The proposed methods are described in Sect. 4, and the team members and affiliations are listed in Appendix A.

**Table 2.** Mean Opinion Scores (MOS) of all solutions submitted during the final phase of the MAI 2022 bokeh effect rendering challenge. Visual results were assessed based on the reconstructed full resolution images. The results of the PyNET-V2 model are provided for the reference.

| Team | Author | Framework | Model size, MB | PSNR↑ | SSIM↑ | LPIPS↓ | MOS score |
|------|--------|-----------|----------------|-------|-------|--------|-----------|
| Antins_cv | xiaokaoji | Keras/TensorFlow | 0.06 | 22.76 | 0.8652 | 0.2693 | 2.6 |
| ENERZAi | MinsuKwon | TensorFlow | 30 | 22.89 | 0.8754 | 0.2464 | 3.5 |
| MiAlgo | hxin | TensorFlow | 1.5 | 20.08 | 0.7209 | 0.4349 | 2.3 |
| Sensebrain | brianjsl | PyTorch/TensorFlow | 402 | 22.81 | 0.8653 | 0.2207 | 3.4 |
| ZJUT-Vision | HaotianQian | TensorFlow | 13 | **23.53** | **0.8796** | **0.1907** | 3.8 |
| VIC | hyessol | PyTorch | 127 | 22.77 | 0.8713 | 0.2657 | 3.4 |
| PyNET [21] | Baseline | TensorFlow | 181 | 23.28 | 0.8780 | 0.2438 | 4.0 |

### 3.1    Results and Discussion

Tables 1 and 2 demonstrate the fidelity, runtime and MOS results of all solutions submitted during the final test phase. Models submitted to the 1st and 2nd challenge tracks were evaluated together since the participants had to upload the corresponding TensorFlow Lite models in both cases. In the 1st tack, the overall best results were achieved by team *Antins_cv*. The authors proposed a small U-Net based model that demonstrated a very good efficiency on the Kirin 9000 SoC, being able to achieve over 30 FPS on its Mali GPU when processing 1024 × 1024 pixel images. Very good runtime and fidelity results were also demonstrated by model designed by team *ENERZAi*: the visual results of this solution were substantially better while its runtime was still less than 90 ms on the target platform.

After evaluating the visual results of the proposed models, we can see that conventional fidelity scores do not necessarily correlate with the perceptual image reconstruction quality. In particular, all solutions except for the baseline PyNET model and the network proposed by team *ZJUT-Vision* produced images that suffered from different corruptions, *e.g.*, issues with texture rendering in the blurred areas, significant resolution drop on the entire image, or almost no changes compared to the input image, which can be observed in case of model trained by team *MiAIgo*. These results highlighted again the difficulty of an accurate assessment of the results obtained in bokeh effect rendering task as the standard metrics are often not indicating the real image quality.

## 4   Challenge Methods

This section describes solutions submitted by all teams participating in the final stage of the MAI 2022 Bokeh Effect Rendering challenge.

### 4.1   Antins_cv

Team Antins_cv proposed a tiny U-Net based model for the considered task, which architecture is illustrated in Fig. 3. The model is composed of three levels: the first one downsamples the original image using a $3 \times 3$ convolution with a stride of 2, while each following level is used to generate more refined features. The final result is produced using several upsampling layers with additional skip connections introduced for feature fusion. Feature upsampling is performed with a $3 \times 3$ convolution followed by the pixel shuffle layer with a scale factor of 2. The model was trained to maximize the PSNR loss function on $1024 \times 1024$ pixel patches with a batch size of 32. Network parameters were optimized for 600 epochs using the Adam [40] algorithm with a learning rate of 1e–3 decreased by half every 120 epochs.

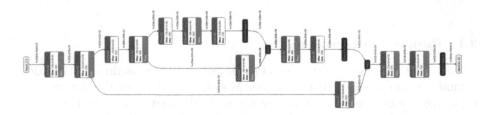

**Fig. 3.** Model architecture proposed by team Antins_cv.

## 4.2  ENERZAi

The model architecture proposed by team ENERZAi is also based on the U-Net design (Fig. 4) that can effectively extract high-level image features. Leaky ReLU with a slope of 0.2 is used as an activation function in all convolution and deconvolution blocks, the depth-to-space op is applied at the end of the model to produce the final image output. The model was trained with a combination of the L1, multiscale-ssim (MS-SSIM) and ResNet-50 perceptual loss functions. In addition, the authors used a differentiable approximate histogram feature loss inspired by the Histogan [1] to consider the color distribution of the constructed RGB image. For each R, G, and B channel, the histogram difference between the constructed and target images was calculated and added to the overall loss function. Network parameters were optimized with a batch size of 1 using the Adam algorithm with a learning rate of 1e–3 halved every 4 epochs.

## 4.3  ZJUT-Vision

The solution proposed by team ZJUT-Vision is based on the BGGAN [51] architecture that uses two consecutive U-Net models (Fig. 6) and two adversarial discriminators. Each U-net generator model contains nine residual blocks, and transposed convolution block is used instead of direct bilinear upsampling to improve the performance of our network. The second U-net model has also two spatial attention and one channel attention block for better visual results. Two discriminators with a different numbers of downsampling conv blocks (5 and 7) are used to extract both local and global information from the image. The authors used the Wasserstein GAN training strategy for optimizing the model.

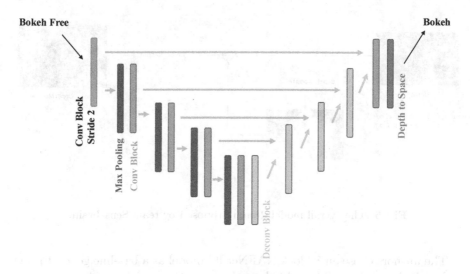

**Fig. 4.** U-Net based model proposed by team ENERZAi.

The total loss consists two components: the generator loss G combining the adversarial, perceptual, L1 and SSIM losses, and the discriminator loss D.

### 4.4 Sensebrain

The solution [42] proposed by team Sensebrain consists of three main parts:

1. A Dense Prediction Transformer (DPT) [52] based depth estimation module that estimates the relative depth of the scene;
2. A Nonlinear Activation Free Network (NAFNet) based generator module [6] that takes the input image together with the predicted depth map, and outputs the depth-aware bokeh image;
3. A dual-receptive field patch-GAN discriminator based on [51].

To generate the bokeh output, the input image is first processed by the depth prediction module to produce a depth map $\mathbf{D}$. The output $\mathbf{D}$ is then concatenated with the original input image $\mathbf{I}$ and then passed to the NAFNet generator $\mathbf{G}$ to produce the final bokeh image $\mathbf{G}(\mathbf{I} \odot \mathbf{D})$.

The depth map of the scenes is estimated from a single monocular image using the dense prediction transformer (DPT) [52] that was pre-trained on a large-scale mixed RGB-to-depth "MIX 5" dataset [53]. The authors found empirically that this network can generate realistic depth maps with smooth boundaries and fine-grain details at object boundaries, which provides extra blurring clues to assist the backbone generator in creating a better bokeh blur. The predicted depth map is also used as a greyscale saliency mask needed for separating the background and foreground when calculating the Bokeh Loss described below.

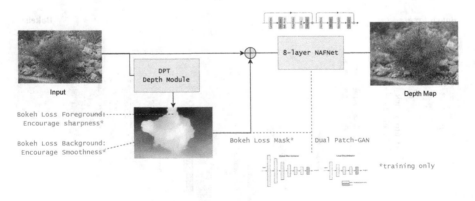

**Fig. 5.** The overall model design proposed by team Sensebrain.

The authors chose an 8-block NAFNet [6] model as a baseline generator network as it demonstrated a good balance between the model complexity and performance. The core idea of the NAFNet is that the traditional high-complexity non-linear operators such as Softmax, GELU, and Sigmoid, can be removed or replaced with multiplications, which is especially useful in reducing the complexity in computationally expensive attention estimation. The encoder size was set to [2,2,2,20], the decoder size – to [2,2,2,2], and two NAF blocks were concatenated in the middle.

The authors used adversarial learning to improve the perceptual quality of the reconstructed images. Similar to [51], a multi-receptive-field Patch-GAN discriminator [36] was used as a part of the final loss function. The model takes two patch-GAN discriminators, one of depth 3 and one of depth 5, and averages the adversarial losses from both discriminators. Besides the conventional L1, SSIM, and discriminator loss functions, the authors also proposed a bokeh-specific loss. The core idea of this loss is that, for a natural bokeh image, the in-focus area should be sharp and the out-of-focus area should be blurred. Using the depth map we generate by the DPT module as a greyscale saliency mask $\mathbf{M}$, the following loss functions are introduced:

- A *Foreground Edge Loss* that encourages sharper edges separating the foreground and the background;
- An *Edge Difference Loss* that encourages similar edge intensities for the input and output image;
- A *Background Blur Loss* that encourages a smoother background with less noise.

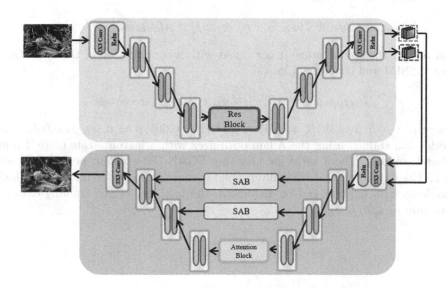

**Fig. 6.** Attention-based double V-Net proposed by team ZJUT.

The *Foreground Edge Loss* $L_{foreedge}$ is used to maximize the intensity of the foreground edges. For this, the input image $\hat{\mathbf{I}}$ is first multiplied with the greyscale saliency mask $\mathbf{M}$, and then the edge map (gradients) is computed using the Sobel filter in the $x, y, xy$, and $yx$ directions. L1-norms of the obtained gradients are then summed up, and the loss is defined as:

$$L_{foreedge}(\hat{\mathbf{I}}, \mathbf{M}) = -\frac{\sum_{z \in \{x,y,xy,yx\}} \left\| S_z(\hat{\mathbf{I}} \cdot \mathbf{M}) \right\|_1}{h_{\hat{\mathbf{I}}} \cdot w_{\hat{\mathbf{I}}}},$$

where $S_z$ is the Sobel convolution operator in the $z$ direction.

The *Edge Difference Loss* is used to minimize the difference in edge strength between the input and the output images:

$$L_{edgediff}(\mathbf{I}, \hat{\mathbf{I}}, \mathbf{M}) = \||L_{foreedge}(\hat{\mathbf{I}} \cdot \mathbf{M})| - |L_{foreedge}(\mathbf{I} \cdot \mathbf{M})|\|.$$

Finally, the *Background Blur Loss* $L_{backblur}$ is used to encourage a smoother blur for the background. The input image $\hat{\mathbf{I}}$ is first multiplied with the inverse of the greyscale mask $\mathbf{M}$, and the loss is defined as a total variation of the scene:

$$L_{backblur}(\hat{\mathbf{I}}, \mathbf{M}) = \frac{1}{h_{\hat{\mathbf{I}}} \cdot w_{\hat{\mathbf{I}}}} TV(\hat{\mathbf{I}} \cdot (1 - \mathbf{M})).$$

The model is trained in two stages. First, it is pre-trained without adversarial losses using the following loss function:

$$L_{pretrain} = \alpha L_1 + \zeta L_{SSIM} + \kappa L_{edgediff} + \mu L_{backblur} + \nu L_{foreedge},$$

and during the second stage it is trained with a combination of the L1, VGG-based, SSIM and adversarial losses:

$$L_{refinement} = \alpha L_1 + \beta L_{VGG} + \zeta L_{SSIM} + \lambda L_{adv},$$

where $\alpha = 0.5, \beta = 0.1, \zeta = 0.05, \lambda = 1, \kappa = 0.005, \mu = 0.1, \nu = 0.005$. The model was trained using the Adam optimizer with a learning rate of 1e–4 and a batch sizeof 2. The $\lambda$ value used for the WGAN-GP gradient penalty was set to 1. The dataset was additionally pruned manually to remove images that had problems with lighting and/or alignment, the final size of the training dataset was around 4225 images.

## 4.5  VIC

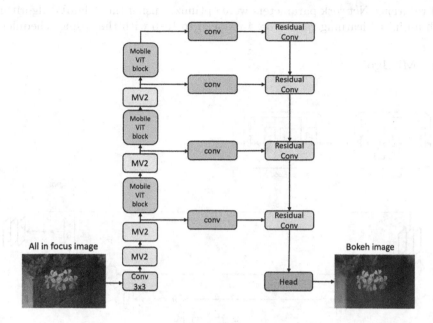

**Fig. 7.** The diagram of the model proposed by team VIC. MV2 represents multiple MovileNetv2 blocks [55].

The overall model architecture proposed by team VIC is similar to the Dense Prediction Transformer (DPT) [52] and is illustrated in Fig. 7. For downsampling, the encoder module uses $3 \times 3$ strided convolutions and MobileNetv2 [55] blocks. To reduce the size of the model and its runtime, the authors used Mobile-ViTv2 [48] blocks in the encoder. Among various versions of the MobileViTv2, the largest one was used to improve the performance. The decoder module reassembles the features corresponding to different resolutions to produce the final bokeh image. Since there are 3 MobileViT blocks in the encoder, the decoder uses 4 input feature maps: an input of the first MobileViT block and the outputs of each MobileViT blocks.

The model was first pre-trained to learn the identity mapping during the first 10 epochs with the L1 loss function. Next, it was trained for another 10 epochs with the L1 loss to perform the considered bokeh effect rendering task. The authors also used the mask made from the DPT [52] disparity map at the end of the training: since this mask can roughly distinguish the foreground and the background, one can apply the L1 loss between the foreground of the predicted image and the foreground of the original all-in-focus image. By doing this, the model can learn how to produce distinct foreground objects. To learn the blur, L1 loss between the background of a predicted image and the background of the ground-truth bokeh image was applied. Feature reconstruction loss [39] was also added without using the mask. The model was trained with a batch size of 2 on images resized to 512 × 768 pixels and augmented with random flips and rotations. Network parameters were optimized using the AdamW algorithm with an initial learning rate of 5e–4 reduced to 1e–6 with the cosine scheduler.

### 4.6  MiAIgo

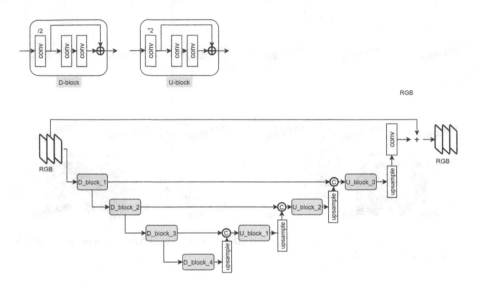

**Fig. 8.** Network architecture developed by team MiAIgo.

Team MiAIgo used an encoder-decoder model design for fast feature extraction (Fig. 8) that produces a high-resolution disparity map used to generate realistic bokeh rendering effect. First, the input image is resized to 1024×1024 px and then processed by seven model blocks that are extracting multi-scale contextual information and generating a high-resolution disparity map. Each block consists of three convolution layers with one additional skip connection. At the end of the decoder module, the obtained disparity map is superimposed with the original image to get the final bokeh output. The model was trained with a combination of the L1, L2 and SSIM loss functions. Network parameters were optimized using the Adam algorithm with a learning rate of 5e–5 and a batch size of 8.

## 5    Additional Literature

An overview of the past challenges on mobile-related tasks together with the proposed solutions can be found in the following papers:

– Learned End-to-End ISP: [19, 23, 29, 33]

- Perceptual Image Enhancement: [24, 32]
- Bokeh Effect Rendering: [22, 31]
- Image Super-Resolution: [4, 32, 47, 60]

**Acknowledgements.** We thank the sponsors of the Mobile AI and AIM 2022 workshops and challenges: AI Witchlabs, MediaTek, Huawei, Reality Labs, OPPO, Synaptics, Raspberry Pi, ETH Zürich (Computer Vision Lab) and University of Würzburg (Computer Vision Lab).

# A    Teams and Affiliations

## Mobile AI 2022 Team

*Title:*
Mobile AI 2022 Learned Smartphone ISP Challenge
*Members:*
Andrey Ignatov[1,2] *(andrey@vision.ee.ethz.ch)*, Radu Timofte[1,2,3]
*Affiliations:*
[1] Computer Vision Lab, ETH Zurich, Switzerland
[2] AI Witchlabs, Switzerland
[3] University of Wuerzburg, Germany

## Antins_cv

*Title:*
A Tiny UNet for Image Bokeh Rendering
*Members:*
*Jin Zhang (zj346862@antgroup.com)*, Feng Zhang, Gaocheng Yu, Zhe Ma, Hongbin Wang
*Affiliations:*
Ant Group, China

## ENERZAi

*Title:*
Bokeh Unet and Histogram Feature Loss
*Members:*
*Minsu Kwon (minsu.kwon@enerzai.com)*
*Affiliations:*
ENERZAi, South Korea

## ZJUT-Vision

*Title:*
Attention-based Double V-Net
*Members:*
Haotian Qian *(1092944263@qq.com)*, Wentao Tong, Pan Mu, Ziping Wang, Guangjing Yan
*Affiliations:*
Zhejiang University of Technology, China

## Sensebrain

*Title:*
Bokeh-Loss GAN: Multi-stage Adversarial Training for Realistic Edge-Aware Bokeh [42]
*Members:*
Brian Lee[1] *(brianlee@sensebrain.site)*, Lei Fei[2], Huaijin Chen[1]
*Affiliations:*
[1] Sensebrain Technology, United States
[2] Tetras.AI, China

## VIC

*Title:*
Mobile-DPT for Bokeh Rendering
*Members:*
Hyebin Cho *(jhb0316@kaist.ac.kr)*, Byeongjun Kwon, Munchurl Kim
*Affiliations:*
Korea Advanced Institute of Science and Technology (KAIST), South Korea

## MiAIgo

*Title:*
Realistic Bokeh Rendering Model on Mobile Devices
*Members:*
Mingyang Qian *(qianmingyang@xiaomi.com)*, Huixin Ma, Yanan Li, Xiaotao Wang, Lei Lei
*Affiliations:*
Xiaomi Inc., China

# References

1. Afifi, M., Brubaker, M.A., Brown, M.S.: Histogan: controlling colors of GAN-generated and real images via color histograms. In: Proceedings of the IEEE/CVF Conference on Computer Vision and Pattern Recognition, pp. 7941–7950 (2021)
2. Barron, J.T., Adams, A., Shih, Y., Hernández, C.: Fast bilateral-space stereo for synthetic defocus. In: Proceedings of the IEEE Conference on Computer Vision and Pattern Recognition, pp. 4466–4474 (2015)
3. Benavides, F.T., Ignatov, A., Timofte, R.: Phonedepth: a dataset for monocular depth estimation on mobile devices. In: Proceedings of the IEEE/CVF Conference on Computer Vision and Pattern Recognition, pp. 0–0 (2022)
4. Cai, J., Gu, S., Timofte, R., Zhang, L.: Ntire 2019 challenge on real image super-resolution: Methods and results. In: Proceedings of the IEEE/CVF Conference on Computer Vision and Pattern Recognition Workshops, pp. 0–0 (2019)
5. Cai, Y., Yao, Z., Dong, Z., Gholami, A., Mahoney, M.W., Keutzer, K.: Zeroq: a novel zero shot quantization framework. In: Proceedings of the IEEE/CVF Conference on Computer Vision and Pattern Recognition, pp. 13169–13178 (2020)
6. Chen, L., Chu, X., Zhang, X., Sun, J.: Simple baselines for image restoration. arXiv preprint arXiv:2204.04676 (2022)
7. Chiang, C.M., et al.: Deploying image deblurring across mobile devices: a perspective of quality and latency. In: Proceedings of the IEEE/CVF Conference on Computer Vision and Pattern Recognition Workshops, pp. 502–503 (2020)
8. Conde, M.V., Timofte, R., et al.: Reversed Image signal Processing and RAW Reconstruction. AIM 2022 Challenge Report. In: Proceedings of the European Conference on Computer Vision (ECCV) Workshops (2022)
9. Dutta, S.: Depth-aware blending of smoothed images for bokeh effect generation. J. Vis. Commun. Image Represent. **77**, 103089 (2021)
10. Dutta, S., Das, S.D., Shah, N.A., Tiwari, A.K.: Stacked deep multi-scale hierarchical network for fast bokeh effect rendering from a single image. In: Proceedings of the IEEE/CVF Conference on Computer Vision and Pattern Recognition, pp. 2398–2407 (2021)
11. in the new Google Camera app, L.B.: https://ai.googleblog.com/2014/04/lens-blur-in-new-google-camera-app.html
12. Ha, H., Im, S., Park, J., Jeon, H.G., So Kweon, I.: High-quality depth from uncalibrated small motion clip. In: Proceedings of the IEEE Conference on Computer Vision and Pattern Recognition, pp. 5413–5421 (2016)
13. Howard, A., et al.: Searching for mobilenetv3. In: Proceedings of the IEEE/CVF International Conference on Computer Vision, pp. 1314–1324 (2019)
14. Ignatov, A., Byeoung-su, K., Timofte, R.: Fast camera image denoising on mobile GPUs with deep learning, mobile AI 2021 challenge: Report. In: Proceedings of the IEEE/CVF Conference on Computer Vision and Pattern Recognition Workshops, pp. 0–0 (2021)
15. Ignatov, A., Chiang, J., Kuo, H.K., Sycheva, A., Timofte, R.: Learned smartphone ISP on mobile NPUs with deep learning, mobile AI 2021 challenge: Report. In: Proceedings of the IEEE/CVF Conference on Computer Vision and Pattern Recognition Workshops, pp. 0–0 (2021)
16. Ignatov, A., Kobyshev, N., Timofte, R., Vanhoey, K., Van Gool, L.: Dslr-quality photos on mobile devices with deep convolutional networks. In: Proceedings of the IEEE International Conference on Computer Vision, pp. 3277–3285 (2017)

17. Ignatov, A., Malivenko, G., Plowman, D., Shukla, S., Timofte, R.: Fast and accurate single-image depth estimation on mobile devices, mobile AI 2021 challenge: Report. In: Proceedings of the IEEE/CVF Conference on Computer Vision and Pattern Recognition Workshops (2021)
18. Ignatov, A., Malivenko, G., Timofte, R.: Fast and accurate quantized camera scene detection on smartphones, mobile AI 2021 challenge: Report. In: Proceedings of the IEEE/CVF Conference on Computer Vision and Pattern Recognition Workshops (2021)
19. Ignatov, A., et al.: Pynet-v2 mobile: efficient on-device photo processing with neural networks. In: 2021 26th International Conference on Pattern Recognition (ICPR). IEEE (2022)
20. Ignatov, A., Malivenko, G., Timofte, R., et al.: Efficient single-image depth estimation on mobile devices, mobile AI & AIM 2022 challenge: Report. In: European Conference on Computer Vision (2022)
21. Ignatov, A., Patel, J., Timofte, R.: Rendering natural camera bokeh effect with deep learning. In: Proceedings of the IEEE/CVF Conference on Computer Vision and Pattern Recognition Workshops, pp. 418–419 (2020)
22. Ignatov, A., et al.: Aim 2019 challenge on bokeh effect synthesis: methods and results. In: 2019 IEEE/CVF International Conference on Computer Vision Workshop (ICCVW), pp. 3591–3598. IEEE (2019)
23. Ignatov, A., et al.: Microisp: Processing 32mp photos on mobile devices with deep learning. In: European Conference on Computer Vision (2022)
24. Ignatov, A., Timofte, R.: Ntire 2019 challenge on image enhancement: Methods and results. In: Proceedings of the IEEE/CVF Conference on Computer Vision and Pattern Recognition Workshops (2019)
25. Ignatov, A., et al.: Power efficient video super-resolution on mobile NPUs with deep learning, mobile AI & aim 2022 challenge: Report. In: European Conference on Computer Vision (2022)
26. Ignatov, A., et al.: AI benchmark: running deep neural networks on android smartphones. In: Leal-Taixé, L., Roth, S. (eds.) ECCV 2018. LNCS, vol. 11133, pp. 288–314. Springer, Cham (2019). https://doi.org/10.1007/978-3-030-11021-5_19
27. Ignatov, A., Timofte, R., Denna, M., Younes, A.: Real-time quantized image super-resolution on mobile NPUs, mobile AI 2021 challenge: Report. In: Proceedings of the IEEE/CVF Conference on Computer Vision and Pattern Recognition Workshops (2021)
28. Ignatov, A., Timofte, R., Denna, M., Younes, A., et al.: Efficient and accurate quantized image super-resolution on mobile NPUs, mobile AI & AIM 2022 challenge: Report. In: European Conference on Computer Vision (2022)
29. Ignatov, A., et al.: Aim 2019 challenge on raw to RGB mapping: methods and results. In: 2019 IEEE/CVF International Conference on Computer Vision Workshop (ICCVW), pp. 3584–3590. IEEE (2019)
30. Ignatov, A., et al.: Ai benchmark: All about deep learning on smartphones in 2019. In: 2019 IEEE/CVF International Conference on Computer Vision Workshop (ICCVW), pp. 3617–3635. IEEE (2019)
31. Ignatov, A., et al.: AIM 2020 challenge on rendering realistic bokeh. In: Bartoli, A., Fusiello, A. (eds.) ECCV 2020. LNCS, vol. 12537, pp. 213–228. Springer, Cham (2020). https://doi.org/10.1007/978-3-030-67070-2_13
32. Ignatov, A., et al.: PIRM challenge on perceptual image enhancement on smartphones: report. In: Leal-Taixé, L., Roth, S. (eds.) ECCV 2018. LNCS, vol. 11133, pp. 315–333. Springer, Cham (2019). https://doi.org/10.1007/978-3-030-11021-5_20

33. Ignatov, A., et al.: Aim 2020 challenge on learned image signal processing pipeline. arXiv preprint arXiv:2011.04994 (2020)
34. Ignatov, A., Timofte, R., et al.: Learned smartphone ISP on mobile GPUs with deep learning, mobile AI & aim 2022 challenge: Report. In: European Conference on Computer Vision (2022)
35. Ignatov, D., Ignatov, A.: Controlling information capacity of binary neural network. Pattern Recogn. Lett. **138**, 276–281 (2020)
36. Isola, P., Zhu, J.Y., Zhou, T., Efros, A.A.: Image-to-image translation with conditional adversarial networks. In: Proceedings of the IEEE Conference on Computer Vision and Pattern Recognition, pp. 1125–1134 (2017)
37. Jacob, B., et al.: Quantization and training of neural networks for efficient integer-arithmetic-only inference. In: Proceedings of the IEEE Conference on Computer Vision and Pattern Recognition, pp. 2704–2713 (2018)
38. Jain, S.R., Gural, A., Wu, M., Dick, C.H.: trained quantization thresholds for accurate and efficient fixed-point inference of deep neural networks. arXiv preprint arXiv:1903.08066 (2019)
39. Johnson, J., Alahi, A., Fei-Fei, L.: Perceptual losses for real-time style transfer and super-resolution. In: Leibe, B., Matas, J., Sebe, N., Welling, M. (eds.) ECCV 2016. LNCS, vol. 9906, pp. 694–711. Springer, Cham (2016). https://doi.org/10.1007/978-3-319-46475-6_43
40. Kingma, D.P., Ba, J.: Adam: a method for stochastic optimization. arXiv preprint arXiv:1412.6980 (2014)
41. Kınlı, F.O., Menteş, S., Özcan, B., Kirac, F., Timofte, R., et al.: Aim 2022 challenge on instagram filter removal: Methods and results. In: Proceedings of the European Conference on Computer Vision (ECCV) Workshops (2022)
42. Lee, B.J., Lei, F., Chen, H., Baudron, A.: Bokeh-loss GAN: multi-stage adversarial training for realistic edge-aware bokeh. In: Proceedings of the European Conference on Computer Vision (ECCV) Workshops (2022)
43. Lee, J., et al.: On-device neural net inference with mobile GPUs. arXiv preprint arXiv:1907.01989 (2019)
44. Li, Y., Gu, S., Gool, L.V., Timofte, R.: Learning filter basis for convolutional neural network compression. In: Proceedings of the IEEE/CVF International Conference on Computer Vision, pp. 5623–5632 (2019)
45. Liu, Z., et al.: Metapruning: Meta learning for automatic neural network channel pruning. In: Proceedings of the IEEE/CVF International Conference on Computer Vision, pp. 3296–3305 (2019)
46. Liu, Z., Wu, B., Luo, W., Yang, X., Liu, W., Cheng, K.T.: Bi-real net: Enhancing the performance of 1-bit cnns with improved representational capability and advanced training algorithm. In: Proceedings of the European conference on computer vision (ECCV), pp. 722–737 (2018)
47. Lugmayr, A., Danelljan, M., Timofte, R.: Ntire 2020 challenge on real-world image super-resolution: Methods and results. In: Proceedings of the IEEE/CVF Conference on Computer Vision and Pattern Recognition Workshops, pp. 494–495 (2020)
48. Mehta, S., Rastegari, M.: Separable self-attention for mobile vision transformers. arXiv preprint arXiv:2206.02680 (2022)
49. Obukhov, A., Rakhuba, M., Georgoulis, S., Kanakis, M., Dai, D., Van Gool, L.: T-basis: a compact representation for neural networks. In: International Conference on Machine Learning, pp. 7392–7404. PMLR (2020)
50. Peng, J., Cao, Z., Luo, X., Lu, H., Xian, K., Zhang, J.: Bokehme: when neural rendering meets classical rendering. In: Proceedings of the IEEE/CVF Conference on Computer Vision and Pattern Recognition, pp. 16283–16292 (2022)

51. Qian, M., et al.: BGGAN: bokeh-glass generative adversarial network for rendering realistic bokeh. In: Bartoli, A., Fusiello, A. (eds.) ECCV 2020. LNCS, vol. 12537, pp. 229–244. Springer, Cham (2020). https://doi.org/10.1007/978-3-030-67070-2_14

52. Ranftl, R., Bochkovskiy, A., Koltun, V.: Vision transformers for dense prediction. In: Proceedings of the IEEE/CVF International Conference on Computer Vision, pp. 12179–12188 (2021)

53. Ranftl, R., Lasinger, K., Hafner, D., Schindler, K., Koltun, V.: Towards robust monocular depth estimation: mixing datasets for zero-shot cross-dataset transfer. IEEE Trans. Pattern Anal. Mach. Intell. **44**, 1623–1637 (2020)

54. Romero, A., Ignatov, A., Kim, H., Timofte, R.: Real-time video super-resolution on smartphones with deep learning, mobile AI 2021 challenge: Report. In: Proceedings of the IEEE/CVF Conference on Computer Vision and Pattern Recognition Workshops, pp. 0–0 (2021)

55. Sandler, M., Howard, A., Zhu, M., Zhmoginov, A., Chen, L.C.: Mobilenetv 2: inverted residuals and linear bottlenecks. In: Proceedings of the IEEE Conference on Computer Vision and Pattern Recognition, pp. 4510–4520 (2018)

56. Shen, X., et al.: Automatic portrait segmentation for image stylization. In: Computer Graphics Forum, vol. 35, pp. 93–102. Wiley Online Library (2016)

57. Shen, X., Tao, X., Gao, H., Zhou, C., Jia, J.: Deep automatic portrait matting. In: Leibe, B., Matas, J., Sebe, N., Welling, M. (eds.) ECCV 2016. LNCS, vol. 9905, pp. 92–107. Springer, Cham (2016). https://doi.org/10.1007/978-3-319-46448-0_6

58. Tan, M., et al.: MnasNet: platform-aware neural architecture search for mobile. In: Proceedings of the IEEE/CVF Conference on Computer Vision and Pattern Recognition, pp. 2820–2828 (2019)

59. TensorFlow-Lite: https://www.tensorflow.org/lite

60. Timofte, R., Gu, S., Wu, J., Van Gool, L.: Ntire 2018 challenge on single image super-resolution: methods and results. In: Proceedings of the IEEE Conference on Computer Vision and Pattern Recognition Workshops, pp. 852–863 (2018)

61. Uhlich, S., et al.: Mixed precision DNNs: all you need is a good parametrization. arXiv preprint arXiv:1905.11452 (2019)

62. Wan, A., et al.: Fbnetv2: differentiable neural architecture search for spatial and channel dimensions. In: Proceedings of the IEEE/CVF Conference on Computer Vision and Pattern Recognition, pp. 12965–12974 (2020)

63. Wang, F., Zhang, Y., Ai, Y., Zhang, W.: Rendering natural bokeh effects based on depth estimation to improve the aesthetic ability of machine vision. Machines **10**(5), 286 (2022)

64. Wu, B., et al.: Fbnet: hardware-aware efficient convnet design via differentiable neural architecture search. In: Proceedings of the IEEE/CVF Conference on Computer Vision and Pattern Recognition, pp. 10734–10742 (2019)

65. Yang, J., et al.: Quantization networks. In: Proceedings of the IEEE/CVF Conference on Computer Vision and Pattern Recognition, pp. 7308–7316 (2019)

66. Yang, R., Timofte, R., et al.: Aim 2022 challenge on super-resolution of compressed image and video: Dataset, methods and results. In: Proceedings of the European Conference on Computer Vision (ECCV) Workshops (2022)

67. Yu, F., Gallup, D.: 3D reconstruction from accidental motion. In: Proceedings of the IEEE Conference on Computer Vision and Pattern Recognition, pp. 3986–3993 (2014)

68. Zhang, R., Isola, P., Efros, A.A., Shechtman, E., Wang, O.: The unreasonable effectiveness of deep features as a perceptual metric. In: Proceedings of the IEEE Conference on Computer Vision and Pattern Recognition, pp. 586–595 (2018)
69. Zheng, B., et al.: Constrained predictive filters for single image bokeh rendering. IEEE Trans. Comput. Imaging **8**, 346–357 (2022)
70. Zhu, B., Chen, Y., Wang, J., Liu, S., Zhang, B., Tang, M.: Fast deep matting for portrait animation on mobile phone. In: Proceedings of the 25th ACM International Conference on Multimedia, pp. 297–305 (2017)

# AIM 2022 Challenge on Super-Resolution of Compressed Image and Video: Dataset, Methods and Results

Ren Yang[1(✉)], Radu Timofte[1,2,19], Xin Li[3], Qi Zhang[3], Lin Zhang[4],
Fanglong Liu[3], Dongliang He[3], Fu Li[3], He Zheng[3], Weihang Yuan[3],
Pavel Ostyakov[5], Dmitry Vyal[5], Magauiya Zhussip[5], Xueyi Zou[5],
Youliang Yan[5], Lei Li[6], Jingzhu Tang[6], Ming Chen[6], Shijie Zhao[6], Yu Zhu[4],
Xiaoran Qin[4], Chenghua Li[4], Cong Leng[4,7,8], Jian Cheng[4,7,8],
Claudio Rota[9], Marco Buzzelli[9], Simone Bianco[9],
Raimondo Schettini[9], Dafeng Zhang[10], Feiyu Huang[10], Shizhuo Liu[10],
Xiaobing Wang[10], Zhezhu Jin[10], Bingchen Li[11], Xin Li[11], Mingxi Li[12],
Ding Liu[12], Wenbin Zou[13,16], Peijie Dong[14], Tian Ye[15],
Yunchen Zhang[17], Ming Tan[16], Xin Niu[14], Mustafa Ayazoglu[18],
Marcos Conde[19], Ui-Jin Choi[20], Zhuang Jia[21], Tianyu Xu[21], Yijian Zhang[21],
Mao Ye[22], Dengyan Luo[22], Xiaofeng Pan[22], and Liuhan Peng[23]

[1] Computer Vision Lab, ETH Zürich, Zürich, Switzerland
ren.yang@vision.ee.ethz.ch
[2] Julius Maximilian University of Würzburg, Würzburg, Germany
radu.timofte@uni-wuerzburg.de
[3] Department of Computer Vision Technology (VIS), Baidu Inc., Beijing, China
lixin41@baidu.com
[4] Institute of Automation, Chinese Academy of Sciences, Beijing, China
[5] Noah's Ark Lab, Huawei, Montreal, Canada
ostyakov.pavel@huawei.com
[6] Multimedia Lab, ByteDance Inc., Beijing, China
lilei.leili@bytedance.com
[7] MAICRO, Nanjing, China
[8] AiRiA, Nanjing, China
[9] University of Milano - Bicocca, Milano, Italy
c.rota30@campus.unimib.it
[10] Samsung Research China - Beijing (SRC-B), Beijing, China
dfeng.zhang@samsung.com
[11] University of Science and Technology of China, Hefei, China
lbc31415926@mail.ustc.edu.cn
[12] ByteDance Inc., Beijing, China
[13] South China University of Technology, Guangzhou, China
[14] National University of Defense Technology, Changsha, China
[15] Jimei University, Xiamen, China
[16] Fujian Normal University, Fuzhou, China
[17] China Design Group Inc., Nanjing, China
[18] Aselsan, Ankara, Turkey
mayazoglu@aselsan.com.tr

© The Author(s), under exclusive license to Springer Nature Switzerland AG 2023
L. Karlinsky et al. (Eds.): ECCV 2022 Workshops, LNCS 13803, pp. 174–202, 2023.
https://doi.org/10.1007/978-3-031-25066-8_8

[19] Computer Vision Lab, Julius Maximilian University of Würzburg,
Würzburg, Germany
marcos.conde-osorio@uni-wuerzburg.de
[20] MegaStudyEdu, Seoul, South Korea
[21] Xiaomi Inc., Beijing, China
jiazhuang@xiaomi.com
[22] University of Electronic Science and Technology of China, Chengdu, China
[23] Xinjiang University, Xinjiang, China
https://www.aselsan.com.tr/

**Abstract.** This paper reviews the Challenge on Super-Resolution of
Compressed Image and Video at AIM 2022. This challenge includes two
tracks. Track 1 aims at the super-resolution of compressed image, and
Track 2 targets the super-resolution of compressed video. In Track 1, we
use the popular dataset DIV2K as the training, validation and test sets.
In Track 2, we propose the LDV 3.0 dataset, which contains 365 videos,
including the LDV 2.0 dataset (335 videos) and 30 additional videos. In
this challenge, there are 12 teams and 2 teams that submitted the final
results to Track 1 and Track 2, respectively. The proposed methods and
solutions gauge the state-of-the-art of super-resolution on compressed
image and video. The proposed LDV 3.0 dataset is available at https://
github.com/RenYang-home/LDV_dataset. The homepage of this challenge is at https://github.com/RenYang-home/AIM22_CompressSR.

**Keywords:** Super-resolution · Image compression · Video compression

# 1  Introduction

Compression plays an important role on the efficient transmission of images and
videos through the band-limited Internet. However, image and video compression unavoidably leads to compression artifacts, which may severely degrade the
visual quality. Therefore, quality enhancement of compressed image and video
has become a popular research topic. However, in the early years, due to the
limitation of devices and band-width, the image and video are usually with low
resolution. Therefore, when we intend to restore them to high resolution and
good quality, we face the challenge to achieve both super-resolution and quality
enhancement of compressed image (Track 1) and video (Track 2).

In the past decade, a great number of works were proposed for single image
super-resolution [21,23,39,47,59,81–83] and there are also plenty of methods
proposed for the reduction of JPEG artifacts [22,25,38,60,81]. Recently, the
blind super-resolution [28,68,80] methods have been proposed. They are able to
use one model to jointly handle the tasks of super-resolution, deblurring, JPEG
artifacts reduction, etc. Meanwhile, video super-solution [8–10,37,46,49,61,64]
and compression artifacts reduction [30,67,69,70,75–77] also has become a popular topic, which aims at adequately exploring the temporal correlation among

frames to facilitate the super-resolution and quality enhancement of videos. NTIRE 2022 [74] is the first challenge we organized on super-resolution of compressed video. The winner method [84] in the NTIRE 2022 challenge successfully outperforms the state-of-the-art method [11].

The AIM 2022 Challenge on Super-Resolution of Compressed Image and Video is one of the AIM 2022 associated challenges: reversed ISP [18], efficient learned ISP [36], super-resolution of compressed image and video [73], efficient image super-resolution [32], efficient video super-resolution [33], efficient Bokeh effect rendering [34], efficient monocular depth estimation [35], Instagram filter removal [41].

The AIM 2022 Challenge on Super-Resolution of Compressed Image and Video steps forward for establishing a benchmark of the super-resolution of JPEG image (Track 1) and HEVC video (Track 2). The methods proposed in this challenge are also have the potential to solve various super-resolution tasks. In this challenge, Track 1 utilizes the DIV2K [1] dataset, and Track 2 uses the proposed LDV 3.0 dataset, which contains 365 videos with diverse content, motion, frame-rate, etc. In the following, we first describe the AIM 2022 Challenge, including the DIV2K [1] dataset and the proposed LDV 3.0 dataset. Then, we introduce the proposed methods and the results.

## 2    AIM 2022 Challenge

The objectives of the AIM 2022 challenge on Super-Resolution of Compressed Image and Video are: (i) to advance the state-of-the-art in super-resolution of compressed inputs; (ii) to compare different solutions; (iii) to promote the proposed LDV 3.0 dataset.

### 2.1    DIV2K [1] Dataset

The DIV2K [1] dataset consists of 1,000 high-resolution images with diverse contents. In Track 1 of AIM 2022 Challenge, we use the training (800 images), validation (100 images) and test (100 images) sets of DIV2K for training, validation and test, respectively.

### 2.2    LDV 3.0 Dataset

The proposed LDV 3.0 dataset is an extension of the LDV 2.0 dataset [71,72,74] with 30 additional videos. Therefore, there are totally 365 videos in the LDV 3.0 dataset. The same as LDV and LDV 2.0, the additional videos in LDV 3.0 are collected from YouTube [27], containing 10 categories of scenes, i.e., *animal, city, close-up, fashion, human, indoor, park, scenery, sports* and *vehicle*, and they are with diverse frame-rates from 24 fps to 60 fps. To ensure the high quality of the groundtruth videos, we only collect the videos with 4K resolution, and without obvious compression artifacts. We downscale the videos to further remove the artifacts, and crop the width and height of each video to the multiples of 8, due to the requirement of the HEVC test model (HM). Besides, we convert videos to the format of YUV 4:2:0, which is the most commonly used format in the existing literature. Note that all source videos in our LDV 3.0 dataset have the licence of *Creative Commons Attribution licence (reuse allowed)*[1], and our LDV 3.0 dataset is used for academic and research proposes.

The Track 2 of AIM 2022 Challenge has the same task as the Track 3 of our NTIRE 2022 Challenge [74]. Therefore, we use the training, validation and test sets of the Track 3 in NTIRE 2022 as the training set (totally 270 videos) for the Track 2 in AIM 2022 Challenge. All videos in the proposed LDV, LDV 2.0 and LDV 3.0 datasets and the splits in NTIRE 2021, NTIRE 2022 and AIM 2022 Challenges are publicly available at https://github.com/RenYang-home/LDV_dataset.

## 2.3  Track 1 – Super-Resolution of Compressed Image

JPEG is the most commonly used image compression standard. Track 1 targets the ×4 super-resolution of the images compressed with JPEG with the quality factor of 10. Specifically, we use the following Python codes to produce the low resolution samples:

```
from PIL import Image
img = Image.open(path_gt + str(i).zfill(4) + '.png')
w, h = img.size
assert w % 4 == 0
assert h % 4 == 0
img = img.resize((int(w/4), int(h/4)), resample=Image.BICUBIC)
img.save(path + str(i).zfill(4) + '.jpg', "JPEG", quality=10)
```

In this challenge, we the version 7.2.0 of the Pillow library.

## 2.4  Track 2 – Super-Resolution of Compressed Video

Track 2 has the same task as the Track 3 in NTIRE 2022 [74], which requires the participants to enhance and meanwhile ×4 super-resolve the HEVC compressed video. In this track, the input videos are first downsampled by the following command:

```
ffmpeg -pix_fmt yuv420p -s WxH -i x.yuv
-vf scale=(W/4)x(H/4):flags=bicubic x_down.yuv
```

where x, W and H indicates the video name, width and height, respectively. Then, the downsampled video is compressed by HM 16.20[2] at QP = 37 with the default Low-Delay P (LDP) setting (*encoder_lowdelay_P_main.cfg*). Note that, we first crop the groundtruth videos to make sure that the downsampled width (W/4) and height (H/4) are integer numbers.

---

[1] https://support.google.com/youtube/answer/2797468?hl=en.

[2] https://hevc.hhi.fraunhofer.de/svn/svn_HEVCSoftware/tags/HM-16.20.

# 3  Challenge Results

## 3.1  Track 1

The PSNR results and the running time of Track 1 are shown in Table 1. In this track, we use the images that are directly upscaled by the bicubic algorithm as the baseline. As we can see from Table 1, all methods proposed in this challenge achieves >1 dB PSNR improvement over the baseline. The PSNR improvement of the top 3 methods are higher than 1.3 dB over the baseline. The VUE Team achieves the best result, that is ~0.1 dB higher the runner-up method. We can also see from Table 1 that the top methods consume high time complexity, while the method of the Giantpandacv Team is the most time-efficient one, whose running time is significantly lower than the methods with higher PSNR. Note that, the data in Table 1 are provided by the participants, so the data may be obtained under different hardware and conditions. Therefore, Table 1 is only for reference. It is hard to guarantee the fairness in comparing time efficiency.

The test and training details are presented in Table 2. As Table 2 shows, most methods use extra training data to improve the performance. In this challenge, Flickr2K [62] is the most popular dataset used in training, in addition to the official training data provided by the organizers. In inference, the self-ensemble strategy [63] is widely utilized. It has been proved to be an effective skill to boost the performance of super-resolution.

## 3.2  Track 2

Table 3 shows the results of Track 2. Similar to Track 1, we use the videos that are directly upscaled by the bicubic algorithm as the baseline performance in this track. The winner team NoahTerminalCV improves the PSNR by more than 2 dB

**Table 1.** Results of Track 1 (×4 super-resolution of JPEG image). The test input is available at https://codalab.lisn.upsaclay.fr/competitions/5076, and the researchers can submit their results to the "testing" phase at the CodaLab server to get the performance of their methods to compare with the numbers in this table.

| Team | PSNR (dB) | Running time (s) | Hardware |
|---|---|---|---|
| VUE | 23.6677 | 120 | Tesla V100 |
| BSR [44] | 23.5731 | 63.96 | Tesla A100 |
| CASIA LCVG [57] | 23.5597 | 78.09 | Tesla A100 |
| SRC-B | 23.5307 | 18.61 | GeForce RTX 3090 |
| USTC-IR [43] | 23.5085 | 19.2 | GeForce 2080ti |
| MSDRSR | 23.4545 | 7.94 | Tesla V100 |
| Giantpandacv | 23.4249 | 0.248 | GeForce RTX 3090 |
| Aselsan Research | 23.4239 | 1.5 | GeForce RTX 2080 |
| SRMUI [17] | 23.4033 | 9.39 | Tesla A100 |
| MVideo | 23.3250 | 1.7 | GeForce RTX 3090 |
| UESTC+XJU CV | 23.2911 | 3.0 | GeForce RTX 3090 |
| cvlab | 23.2828 | 6.0 | GeForce 1080 Ti |
| Bicubic ×4 | 22.2420 | – | – |

**Table 2.** Test and training details of Track 1 (×4 super-resolution of JPEG image).

| Team | Ensemble for test | Extra training data |
|---|---|---|
| VUE | Flip/rotation | ImageNet [19], Flickr2K [62] |
| BSR | Flip/rotation, three models for voting | Flickr2K [62], Flickr2K-L[a] |
| CASIA LCVG | Flip/rotation, three models | ImageNet [19] |
| SRC-B | Flip/rotation | Flickr2K [62] |
| USTC-IR | Flip/rotation | Flickr2K [62] and CLIC datasets |
| MSDRSR | Flip/rotation | Flickr2K [62] and DIV8K [29] |
| Giantpandacv | Flip/rotation, TLC [15] | Flickr2K [62] |
| Aselsan Research | Flip/rotation | Flickr2K [62] |
| SRMUI | Flip/rotation | Flickr2K [62] and MIT 5K [7] |
| MVideo | Flip/rotation | – |
| UESTC+XJU CV | Flip/rotation | – |
| cvlab | – | 4,600 images |
| Bicubic ×4 | – | – |

[a]Flicker2K-L is available as "flickr2k-L.csv" at https://github.com/RenYang-home/AIM22_CompressSR/

**Table 3.** Results of Track 2 (×4 super-resolution of HEVC video). Blue indicates the state-of-the-art method. The test input and groundtruth are available on the homepage (see the abstract) of the challenge.

| Team | PSNR (dB) | Time (s) | Hardware |
|---|---|---|---|
| NoahTerminalCV | 25.1723 | 10 | Tesla V100 |
| NTIRE'22 Winner [84] | 24.1097 | 13.0 | Tesla V100 |
| IVL | 23.0892 | 0.008 | GeForce GTX 1080 |
| Bicubic ×4 | 22.7926 | – | – |

over the baseline, and it successfully beats the winner method [84] in NTIRE 2022, which can be seen as the state-of-the-art method. The IVL method has the very fast running speed, and it is able to achieve the real-time super-resolution on the test videos. In Table 4, we can see that the NoahTerminalCV Team uses a large training set, including 90,000 videos collected from YouTube [27]. This may be obviously beneficial for their test performance. Note that, the data in Table 4 are provided by the participants. It is hard to guarantee the fairness in comparing time efficiency.

**Table 4.** Test and training details of Track 2 (×4 super-resolution of HEVC video). Blue indicates the state-of-the-art method.

| Team | Ensemble for test | Extra training data |
|---|---|---|
| NoahTerminalCV | Flip/rotation | 90K videos[b] from YouTube [27] |
| NTIRE'22 Winner [84] | Flip/rotation, two models | 870 videos from YouTube [27] |
| IVL | – | – |
| Bicubic ×4 | – | – |

[b]Dataset is available as "dataset_Noah.txt" at https://github.com/RenYang-home/AIM22_CompressSR/

## 4   Teams and Methods

### 4.1   VUE Team

The method proposed by the VUE Team is called TCIR: A Transformer and CNN Hybrid Network for Image Restoration. The architecture of TCIR is shown in Fig. 1. Specifically, they decouple this task of Track 1 into two sub-stages. In the first stage, they propose a Transformer and CNN hybrid network (TCIR) to remove JPEG artifacts, and in the second stage they use a finetuned RRDBNet for ×4 super-resolution. The proposed TCIR is based on SwinIR [47] and the main improvements are as follows:

- 1) They conduct ×2 downsampling to the JPEG-compressed input by a convolution with the stride of 2. The main purpose of this downsampling is to save GPU memory and speed up the model. Since the images compressed by JPEG with the quality factor of 10 are very blurry, this does not affect the performance of TCIR.
- 2) Then, they use the new Swinv2 transformer block layer to replace the STL in the original SwinIR to greatly improve the capability of the network.
- 3) In addition, they add several RRDB [83] modules to the basic blocks of TCIR and this combines the advantages of Transformer and CNN.

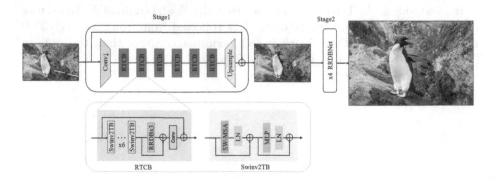

**Fig. 1.** Overview of the TCIR method proposed by the VUE Team.

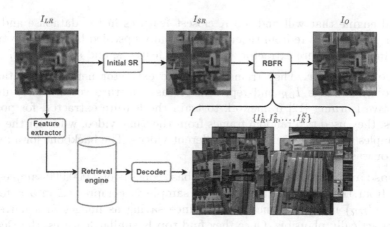

**Fig. 2.** The method proposed by the NoahTerminalCV Team.

## 4.2   NoahTerminalCV Team

The method proposed by the NoahTerminalCV Team is called Enhanced Video Super-Resolution through Reference-Based Frame Refinement. As Fig. 2 shows, the proposed method consists of two subsequent stages. Firstly, they perform an initial super-resolution using a feed-forward multi-frame neural network. Then, the second step is called reference-based frame refinement. They find top K similar images for each low-resolution input frame from the external database. Then, they run a matching correction step for every patch on this input frame to perform a global alignment of reference patches. As a result, the $I_{SR}$, which comes from the first stage, and a set of globally aligned references $\{I_R^1, I_R^2, ..., I_R^K\}$ are obtained. Finally, they are processed with RBFR network ($N_{RBFR}$) to handle residual misalignments and to properly transfer texture, details from reference images to initially super-resolved output $I_{SR}$. The details of training and test are described in the following.

### 4.2.1   Training

**Initial Super-Resolution (Initial SR).** The NoahTerminalCV Team upgraded the BasicVSR++ [10] by increasing channels to 128 and reconstruction blocks to 45. The BasicVSR++ is trained from scratch (except SPyNet) using a pixel-wise $L_1$ objective on the full input images without cropping and fine-tuned using $L_1 + L_2$ objectives. The training phase took about 21 days using 8 NVIDIA Tesla V100 GPUs. They observed a slight performance boost if the model is fine-tuned with a combination of $L_1$ and $L_2$ losses.

**Reference-Based Frame Refinement (RBFR).** The information from subsequent frames is not always enough to produce a high-quality super-resolved image. Therefore, after initial super-resolution using upgraded BasicVSR++, they employed a reference-based refinement strategy. The idea is to design a retrieval engine that will find top K closest features in the database and then transfer details/texture from them to the initially upscaled frame. The retrieval engine includes feature extractor, database of features, and autoencoder.

**Feature Extractor.** They trained a feature extractor network that takes a low-resolution image $I_{LR}$ and represents it as a feature vector. They used a contrastive learning [12] framework to train the feature extractor: for positive samples, they used two random frames from the same video, while for the negative samples we employ frames from different videos. The backbone for a feature extractor was Resnet-34 [31].

**Database and Autoencoder.** The database consists of 2,000,000 samples generated from the training dataset. Each sample is compressed into a feature $z = N_E(I_{HR})$ using the Encoder $N_E$, since saving as images in a naive way is not practically plausible. Once they find top K similar features, the Decoder $N_D$ is used to reconstruct the original input $\hat{I}_{HR} = N_D(z)$.

**Retrieval Engine.** After compressing the database of images using the trained Encoder, obtaining latent representations, and representing all low-resolution versions as a feature vector of size 100 extracted from the trained Feature Extractor, they build an index using the HNSW [53] algorithm from the nmslib [5] library. This algorithm allows searching for the top K nearest neighbors in the database.

**RBFR.** Finally, we train a network $N_{RBFR}$ that takes the result of initial super-resolution $I_{SR}$ and top K similar images from the database $\{I_R^1, I_R^2, ..., I_R^K\}$. The network produces the final prediction $I_O$. We train $N_{RBFR}$ through the L1 objective between $I_O$ and $I_{HR}$. As a $N_{RBFR}$, we use NoahBurstSRNet [3] architecture, since it effectively handles small misalignments and can properly transfer information from reference non-aligned images.

### 4.2.2   Test

**Initial Super-Resolution (Initial SR).** During the inference, in order to upscale the key-frame $I_{LR}^i$, they put it to the initial super-resolution network together with additional frames $I_{LR}^{i-1}, I_{LR}^{i+1}, I_{LR}^{i-2}, I_{LR}^{i+2}$.... The number of additional frames during the inference is set to the full sequence size (up to 600 images).

**Reference-Based Frame Refinement (RBFR).** For RBFR, top K (typically 16) similar images are first obtained using the retrieval engine. Then, the inference is done in a patch-wise manner. They extract a patch from the $I_{SR}$ and use the Template Matching [6] to perform a global alignment and find the most similar patches on the images $\{I_R^1, I_R^2, ..., I_R^K\}$. Then, they put them to the $N_{RBFR}$ to generate the final result.

### 4.3   BSR Team [44]

For most low-level tasks, like image super-resolution, the network is trained on cropped patches rather than the full image. That means the network can only look at the pixels inside the patches during the training phase, even though the network's ability is becoming more and more powerful and the receptive field of the deep neural network could be very large.

The patch size heavily affects the ability of the network. However, with the limitation of the memory and the computing power of GPU, it is not a sanity choose to train the network on the full image. To address the above-mentioned problem, the BSR Team proposes a multi-patches method to greatly increase the receptive field in the training phase while increasing very little memory.

As shown in Fig. 3, they crop low-resolution input patch and its eight surrounding patches as our multi-patches network's input. They use HAT [13] as the backbone, and propose Multi Patches Hybrid Attention Transformer (MPHAT). Compared with HAT [13], MPHAT just simply change the input channel of the network for the multi-patch input. On the validation set of the challenge, the proposed MPHAT achieves the PSNR performance of 23.6266 dB, which is obviously higher than the HAT without multi-patches (23.2854 dB).

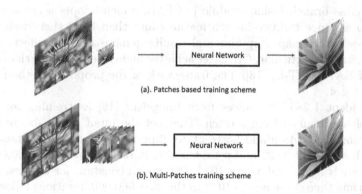

**Fig. 3.** Illustration of the multi patches scheme proposed by the BSR Team. Top images represent general patches based training scheme and bottom image represent multi-patches scheme. Low resolution input patch and its eight surrounding patches are cropped then send to the neural network to reconstruct the super resolution image of the centre patch. The neural network chosen in this competition is HAT [13].

In the training phase, they train the network by using Adam optimizer with $\beta_1 = 0.9$ and $\beta_2 = 0.99$ to minimize the MSE loss. The model is trained for 800,000 iterations with mini-batches of size 32 and patch size 64. The learning rate is initialized as $10^{-4}$ and reduced to half at the 300,000th, 500,000th, 650,000th, 700,000th, 750,000th iterations, respectively.

### 4.4 CASIA LCVG Team [57]

The CASIA LCVG Team proposes a consecutively-interactive dual-branch network (CIDBNet) to take advantage of both convolution and transformer operations, which are good at extracting local features and global interactions, respectively. To better aggregate the two-branch information, they newly introduce an

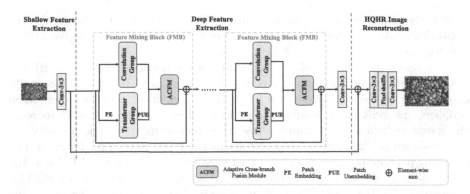

**Fig. 4.** The overall architecture of the Consecutively-Interactive Dual-Branch network (CIDBNet) of the CASIA LCVG Team.

adaptive cross-branch fusion module (ACFM), which adopts a cross-attention scheme to enhance the two-branch features and then fuses them weighted by a content-adaptive map. Experimental results demonstrate the effectiveness of CIDBNet, and in particular, CIDBNet achieves higher performance than a larger variant of HAT (HAT-L) [13]. The framework of the proposed method is illustrated in Fig. 4.

They adopt 1,280,000 images from ImageNet [19] as training set and all the models are trained from scratch. They set the input patch size to $64 \times 64$ and use random rotation and horizontally flipping for data augmentation. The mini-batch size is set to 32 and total training iterations are set to 800,000. The learning rate is initialized as $2 \times 10^{-4}$. It remains constant for the first 270,000 iterations and then decreases to $10^{-6}$ in the next 560,000 iterations following the cosine annealing. They adopt the Adam optimizer ($\beta_1 = 0.9, \beta_2 = 0.9$) to train the model. During test, they first apply self-ensemble trick for each model, which could involves 8 outputs for fusion. Then, they fuse the self-ensembled outputs of the CIDBNet, CIDBNet_NF and CIDBNet_NFE models, respectively.

### 4.5   IVL Team

The architecture proposed by the IVL team for the video challenge track is shown in Fig. 5 and contains three cascaded modules. The first module stacks the input frames (five consecutive frames are used) and extracts deep features from them. The second module aligns the features extracted from the adjacent frames with the features of the target frame. This is achieved by using a Spatio-Temporal Offset Prediction Network (STOPN), which implements a U-Net like architecture to estimate the deformable offsets that are later applied to deform a regular convolution and produce spatially-aligned features. Inspired by [20], STOPN predicts spatio-temporal offsets that are different at each spatial and temporal position. Moreover, as stated by [58], they apply deformable alignment

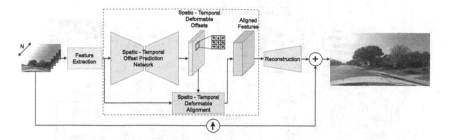

**Fig. 5.** The method proposed by the IVL Team.

at feature level to increase alignment accuracy. The third module contains two groups of standard and transposed convolutions to progressively perform feature fusion and upscaling. The input target frame is then ×4 upscaled using bicubic interpolation and finally added to the network output to produce the final result. They only process the luma channel (Y) of the input frames because it contains the most relevant information on the scene details. The final result is obtained using the restored Y channel and the original chroma channels upscaled using bicubic interpolation, followed by a RGB conversion.

They train the model for 250,000 iterations using a batch size equal to 32. Patches with the size of 96 × 96 pixels are used, and data augmentation with random flip is applied. They set the temporal neighborhood $N$ to five, hence they stacke the target frame with the two previous and the two subsequent frames. The learning rate was initially set to $10^{-3}$ for the first 200,000 iterations, then reduced to $10^{-4}$ for the remaining iterations. They use MSE as the loss function and optimize it using the Adam optimizer.

### 4.6   SRC-B Team

Inspired by SwinIR [47], the SRC-B Team proposes the SwinFIR method with the Swin Transformer [51] and the Fast Fourier Convolution [14]. As shown in Fig. 6, SwinFIR consists of three modules: shallow feature extraction, deep feature extraction and high-quality (HQ) image reconstruction modules. The shallow feature extraction and high-quality (HQ) image reconstruction modules adopt the same configuration as SwinIR. The residual Swin Transformerblock (RSTB) is a residual block with Swin Transformerlayers (STL) and convolutional layers in SwinIR. They all have local receptive fields and cannot extract the global information of the input image. The Fast Fourier Convolution has the ability to extract global features, so they replace the convolution (3 × 3) with Fast Fourier Convolution and a residual module to fuse global and local features, named Spatial-frequency Block (SFB), to improve the representation ability of model.

They use the Adam optimizer with default parameters and the Charbonnier L1 loss [42] to train the model. The initial learning rate is $2 \times 10^{-4}$, and they use the cosine annealing learning rate scheduler [52] with about 500,000 iterations.

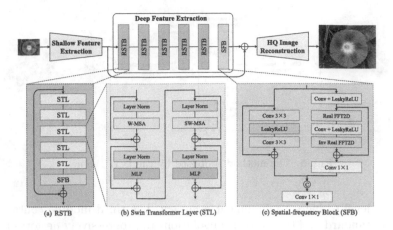

**Fig. 6.** The SwinFIR method proposed by the SRC-B Team.

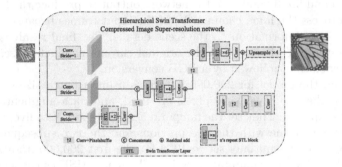

**Fig. 7.** Overview of HST method proposed by the USTC-IR Team. STL block is the Swin Transformer layer from SwinIR [47].

The batch size is 32 and patch size is 64. They use horizontal flip, vertical flip, rotation, RGB perm and mixup [78] for data augmentation.

### 4.7   USTC-IR Team [43]

The USTC-IR Team proposes a Hierarchical Swin Transformer (HST) for compressed image super-resolution, which is inspired by multi-scale-based frameworks [45,54,55] and transformer-based frameworks [46,47]. As shown in Fig. 7, the network is divided into three branches so that it can learn global and local information from different scales. Specifically, the input image is first downsampled to different scales by convolutions. Then, it is fed to different Residual Swin Transformer Blocks (RSTB) from SwinIR [47] to obtain the restored hierarchical features from each scale. To fuse the features from different scales, they super-resolve the low-scale feature and concatenate it with the higher feature. Finally, there is a pixel-shuffle block to implement the 4× super-resolution of features.

The training images are paired-cropped into $64 \times 64$ patches, and augmented by random horizontal flips, vertical flips and rotations. They train the model by the Adam optimizer with the initial learning rate of $2 \times 10^{-4}$. The learning rate is decayed by the factor of 0.5 twice, at the 200,000th and the 300,000 steps, respectively. The network is first trained by Charbonnier loss [42] for about 50,000 steps and finetuned by the MSE loss until convergence.

### 4.8    MSDRSR Team

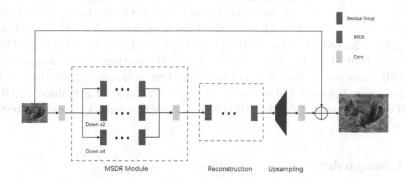

**Fig. 8.** The architecture of the method proposed by the MSDRSR Team. It utilizes a multi-scale degradation removal module, which employs the multi-scale structure to achieve balance between detail enhancement and artifacts removal.

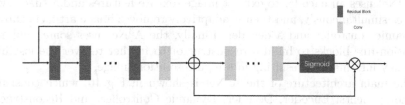

**Fig. 9.** Enhanced residual group.

The architecture of the method proposed by the MSDRSR Team is illustrated in Fig. 8. The details are described in the following.

**MSDR Module.** In the MSDR Module, a Multi-Scale Degradation Removal (MSDR) module is employed after the first convolutional layer. The MSDR module uses several Enhanced Residual Group (ERG) to extract multi-scale features and can achieve a better trade-off between detail enhancement and compression artifacts removal. The architecture of enhanced residual group (ERG) is illustrated in Fig. 9. ERG removes the channel attention module in each residual block, and adds a high-frequency attention block [24] at the end of the residual block. Compared with the original design of residual group, ERG can effectively remove artifacts while reconstructing high-frequency details. Moreover, our proposed ERG is very efficient and does not introduce much runtime overhead.

**Reconstruction Module.** The reconstruction module is build on ESR-GAN [66], which has 23 residual-in-residual dense blocks (RRDB). To further improve the performance [48], they change activation function to SiLU [26].

The MSDRSR employs a two-stage training strategy. In each stage, MSDRSR is first trained with Laplacian pyramid loss with the patch size of 256, and then it is fine-tuned with the MSE loss with the patch size of 640. They augment the training data with random flipping and rotations. In the first stage, MSDRSR is trained on DF2K [1,62] for 100,000 iterations with the batch size of 64. It adopts the Adam optimizer with an initial learning rate of $5 \times 10^{-4}$. The Cosine scheduler is adopted with the minimal learning rate of $5 \times 10^{-5}$. Then, MSDRSR is fine-tuned with learning rate of $10^{-5}$ for 20,000 iterations. In the second stage, MSDRSR loads pre-trained weights from the first stage, and then they add 10 more randomly initialized blocks to the feature extractor. It is trained on DF2K and DIV8K [29] datasets. Then MSDRSR adopts the same training strategy as the first stage.

### 4.9  Giantpandacv

Inspired by previous image restoration and JPEG artifacts removal research [16,38], the Giantpandacv Team proposes the Artifact-aware Attention Network ($A^3$Net) that can use the global semantic information of the image to adaptive control the trade-off between artifacts removal and details restored. Specifically, the $A^3$Net uses an encoder to extract image texture features and artifact-aware features simultaneously, and then it adaptively removes image artifacts through a dynamic controller and a decoder. Finally, the $A^3$Net uses some nonlinear-activation-free blocks to build a reconstructor to further recover the lost high-frequency information, resulting in a high resolution image.

The main architecture of the $A^3$Net is shown in Fig. 10, which consists of four components: Encoder, Decoder, Dynamic Controller, and Reconstructor. The details of these modules of $A^3$Net are described as follows.

**Encoder:** The Encoder aims to extract the deep features and decouple the latent artifact-aware features from the input image. The Encoder contains four scales, each of which has a skip connection to connect to the decoder. In order to improve the computational efficiency of the network, we use 2 Nonlinear Activation Free (NAF) blocks [16] at each scale. The number of output channels in each layer from the first to the fourth scale is set to 64, 128, 256, and 512, respectively. The image features from the encoder are passed into the decoder. At the same time, the global average pooling layer is used to get the artifact-aware features from the image features.

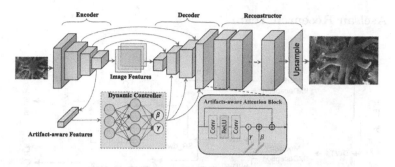

**Fig. 10.** Architecture of the $A^3$Net proposed by the Giantpandacv Team.

**Dynamic Controller:** The dynamic controller is a 3-layer MLP and take as input the artifact-aware features, representing the latent degree of image compression. The main purpose of the dynamic controller is to allow the latent degree of image compression to be flexibly applied to the decoder, thus effectively removing artifacts. Inspired by recent research in spatial feature transform [56,65], we employ dynamic controller to generate modulation parameters pair $(\gamma, \beta)$ which embed on the decoder. Moreover, we used three different final layers of the MLP to accommodate the different scales of features.

**Decoder:** The decoder consists of artifact-aware attention blocks with three different scales. The artifact-aware attention blocks mainly removes artifacts by combining image features and embedded artifact-aware parameters $(\gamma, \beta)$. The number of artifact-aware attention blocks in each scale is set to 4. It can be expressed as follows:

$$F_{out} = \gamma \odot F_{in} + \beta \tag{1}$$

where $F_{in}$ and $F_{out}$ denote the feature maps before and after the affine transformation, and $\odot$ is referred to as element-wise multiplication.

**Reconstructor:** The aim of the reconstructor is to further restore the lost texture details, and then the features are up-sampled to reconstruct a high-resolution image. Specifically, they use a deeper NAF to facilitate the network capturing similar textures over long distances, thus obtaining more texture details.

**Implementation and Training Details:** The numbers of NAF blocks in the each scale of the Encoder and the Reconstructor are flexible and configurable, which are set to 2 and 8, respectively. For the up-scaling module, they use pixel-shuffle to reconstruct a high-resolution image. During training, the $A^3$Net is trained on the crop training dataset with LR and HR pairs. The input pairs are randomly cropped to $512 \times 512$. Random rotation and random horizontal flop are applied for data augmentation. They use the AdamW optimizer with the learning rate of $2 \times 10^{-4}$ learning rate to train the model for 1,000,000 iterations and the learning rate is decayed with the cosine strategy. Weight decay is $10^{-4}$ for all the training periodic.

## 4.10    Aselsan Research Team

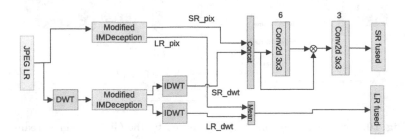

**Fig. 11.** The dual-domain super-resolution network of the Aselsan Research Team.

The Aselsan Research Team proposes a dual-domain super-resolution network, shown in Fig. 11. The network utilizes information in both pixel and wavelet domains. The information of these domains are processed in parallel by a modified IMDeception Network [2] to further increase the receptive field and the capability for processing non-local information. The two branches of the modified IMDeception network generates the super-resolved image and the enhanced low-resolution image, respectively. The super-resolved images are fused through a pixel attention network as used in [4] and enhanced low-resolution images are averaged for fusion. These low-resolution outputs are used during training to further guide the network and add the dual capability to the network. To further boost the performance of the entire network, the structure is encapsulated in a geometric ensembling architecture. We used LR_fused output as well as SR_fused output for training, using LR as guidance through out the optimization. The loss function we used is as follows

$$L = 0.5 \times L2(\text{LR\_fused} - \text{Uncompressed\_LR}) + 0.5 \times L2(\text{SR\_fused} - \text{HR}) \quad (2)$$

Note that almost entire network is shared for this dual purpose, having a secondary and complementary guidance which is able to boost the performance by around 0.05 dB.

They use the Adam optimizer with $\beta_1 = 0.9, \beta_2 = 0.999$ for training. The batch size is set to 8, and the training samples are cropped to $512 \times 512$. The learning rate is initialized to $5^{-4}$ and it is decayed with the factor of 0.75 every 200 epochs (800 iteration in each epoch). The model is totally trained for 2,000 epochs.

## 4.11    SRMUI Team [17]

The method proposed by the SRMUI Team is illustrated in Fig. 12. They propose some modifications of SwinIR [47] (based on Swin Transformer [51]) that enhance the model's capabilities for super-resolution, and in particular, for compressed

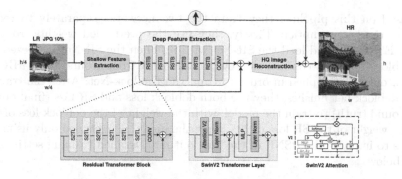

**Fig. 12.** The Swin2SR method of the SRMUI Team.

input SR. They update the original Residual Transformer Block (RSTB) by using the new SwinV2 transformer [50] layers and attention to scale up capacity and resolution. This method has a classical upscaling branch which uses a bicubic interpolation to recover basic structural information. The output of our model is added to the basic upscaled image to enhance it. They also explored different loss functions to make the model more robust to JPEG compression artifacts, being able to recover high-frequency details from the compressed LR image, and therefore it is able to achieve better performance.

### 4.12 MVideo Team

The MVideo Team proposes a two-stage network for compressed image super resolution. The overall architecture of the network is as Fig. 15. The Deblock-Net in first stage takes the compressed low resolution image with JPEG blocky artifacts as input and outputs the enhanced low resolution image. Then the SR-Net in the second stage is applied to the enhanced low resolution image, and generate the final SR image. Both networks use RRDBNet [66] as implementation, while they remove the pixel unshuffle operation from the first Deblock-Net in the beginning and the upsample operation at last. The SR-Net uses the same hyper-parameters as used in ESRGAN [66] (Fig. 13).

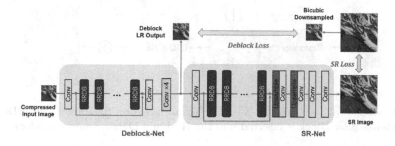

**Fig. 13.** The two-stage network proposed by the MVideo Team.

Based on this pipeline, they train the two networks separately to reduce training time consumption. Firstly, they use the pretrained weights from the official ESRGAN, and load the SR-Net with it. Then the SR-Net is freezed and the deblock loss between bicubic downsampled LR from ground truth HR and deblock output is applied in order to train the Deblock-Net. After the training of the deblock net finishes, they use both deblock loss and SR loss (final output and ground truth) to train the model, with the weight of the deblock loss of 0.01 and the weight of the SR loss of 1.0. Then the model is finetuned only using the SR loss to improve the PSNR of the final output. Detailed training settings are listed below:

- (I) Pre-train the Deblock-Net. Firstly load the pretrained RRDBNet to SR-Net, and then use only deblock loss to train the Deblock-Net. The patch size is 128 and batch size is 32. Training is for 50k iterations using Adam optimizer, and learning rate is initiated with $5 \times 10^{-4}$, which decreases with the factor of 0.5 at the 30,000th and 40,000th iterations, respectively.
- (II) End-to-end training of the two-stage network. Using the pretrained weights from stage (I), they train the full network using both the deblock loss (weight = 0.01) and the SR loss (weight = 1.0), which mainly focuses on the learning of SR-Net. In this stage, patch size, batch size, and Adam optimizer stay the same as above, and they use the CosineAnnealingRestartLR scheduler with all periods of 50,000 for 200,000 iterations.
- (III) Finetuning the last weights using patch size of 512 and batch size of 8. The initial learning rate is set as $2 \times 10^{-5}$ and the learning rate decays by the factor at the 20,000th, 30,000th, and 40,000th iterations, respectively. This stage takes totally 50,000 iterations.
- (IV) The last stage finetunes the model from the previous stage for 50,000 iterations, with the patch size of 256 and the batch size of 8, initial learning rate of $5 \times 10^{-6}$ and multistep scheduler which decreases the learning rate by the factor of 0.5 at the 20,000th, 30,000th and 40,000th iterations, respectively. The final model is used in the inference phase.

### 4.13    UESTC+XJU CV Team

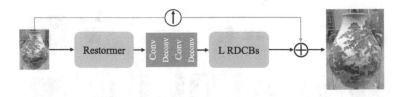

**Fig. 14.** Method of the UESTC+XJU CV Team: based on Restormer [79]. RDCB: A Channel-attention **B**lock is inserted into the **R**esidual **D**ense connection block.

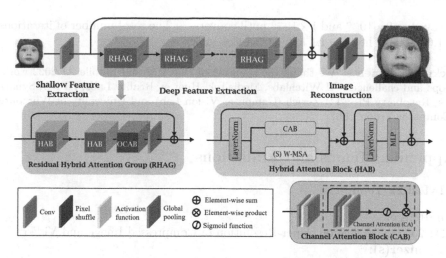

**Fig. 15.** The HAT [13] method used by the cvlab Team in Track 1.

The UESTC+XJU CV Team utilizes Restomer [79] for compressed image enhancement. The overall structure is shown in Fig. 14. First, the compressed image is input into the Restomer (the last layer of the original network is removed), and then upsampled by two convolutional layers and deconvolutional layers. Finally, the feature map is input into the common CNN network, which consists of $L$ (4 in this model) Residual Dense Channel-attention Blocks (RDCBs), and then it is added with the original upsampled image to obtain the reconstructed image. Specially, the channel-attention layer [82] is inserted after the four dense connection layers in the residual dense block [83].

In the training process, the raw image is cropped into patches with the size of $256 \times 256$ as the training samples, and the batch size is set to 8. They also adopt flip and rotation as data augmentation strategies to further expand the dataset. The model is trained by Adam optimizer [40] and cosine annealing learning rate scheduler for $3 \times 10^5$ iterations. The learning rate is initially set to $3 \times 10^{-4}$. They use L2 loss as the loss function.

### 4.14  cvlab Team

The cvlab Team uses HAT [13] as the solution in Track 1. As show in Fig. 15, The number of RHAG blocks is set to 6. The number of HAB blocks in each RHAG block is 6. The number of feature channels is 180. In the training process, the raw and compressed sequences are cropped into $64 \times 64$ patches as the training pairs, and the batch size is set to 4. They also adopt flip and rotation as data augmentation strategies to further expand the dataset. We train all models using Adam [40] optimizer with $\beta_1 = 0.9$, $\beta_2 = 0.999$, $\epsilon = 10^{-6}$, and the learning rate is initially set to $2 \times 10^{-4}$ and decays linearly to $5 \times 10^{-5}$ after 200,000 iterations, which keeps unchanged until 400,000 iterations. Then, the learning rate is further decayed to $2 \times 10^{-5}$ and $1 \times 10^{-5}$ until converged. The total number of iterations is 1,000,000. They use L1 loss as the loss function.

**Acknowledgments.** We thank the sponsors of the AIM and Mobile AI 2022 workshops and challenges: AI Witchlabs, MediaTek, Huawei, Reality Labs, OPPO, Synaptics, Raspberry Pi, ETH Zürich (Computer Vision Lab) and University of Würzburg (Computer Vision Lab).

## Appendix: Teams and Affiliations

### AIM 2022 Team

**Challenge:**
AIM 2022 Challenge on Super-Resolution of Compressed Image and Video
**Organizer(s):**
Ren Yang[1] (ren.yang@vision.ee.ethz.ch),
Radu Timofte[1,2] (radu.timofte@uni-wuerzburg.ch)
**Affiliation(s):**
[1] Computer Vision Lab, ETH Zürich, Switzerland
[2] Julius Maximilian University of Würzburg, Germany

### VUE Team

**Member(s):**
Xin Li[1] (lixin41@baidu.com), Qi Zhang[1], Lin Zhang[2], Fanglong Liu[1], Dongliang He[1], Fu li[1], He Zheng[1], Weihang Yuan[1]
**Affiliation(s):**
[1] Department of Computer Vision Technology (VIS), Baidu Inc.
[2] Institute of Automation, Chinese Academy of Sciences

### NoahTerminalCV Team

**Member(s):**
Pavel Ostyakov (ostyakov.pavel@huawei.com), Dmitry Vyal, Magauiya Zhussip, Xueyi Zou, Youliang Yan
**Affiliation(s):**
Noah's Ark Lab, Huawei

### BSR Team

**Member(s):**
Lei Li (lilei.leili@bytedance.com), Jingzhu Tang, Ming Chen, Shijie Zhao
**Affiliation(s):**
Multimedia Lab, ByteDance Inc.

## CASIA LCVG Team

**Member(s):**
Yu Zhu[1] (zhuyu.cv@gmail.com), Xiaoran Qin[1], Chenghua Li[1], Cong Leng[1,2,3], Jian Cheng[1,2,3]
**Affiliation(s):**
[1] Institute of Automation, Chinese Academy of Sciences, Beijing, China
[2] MAICRO, Nanjing, China
[3] AiRiA, Nanjing, China

## IVL Team

**Member(s):**
Claudio Rota (c.rota30@campus.unimib.it), Marco Buzzelli, Simone Bianco, Raimondo Schettini
**Affiliation(s):**
University of Milano - Bicocca, Italy

## Samsung Research China - Beijing (SRC-B)

**Member(s):**
Dafeng Zhang (dfeng.zhang@samsung.com), Feiyu Huang, Shizhuo Liu, Xiaobing Wang, Zhezhu Jin
**Affiliation(s):**
Samsung Research China - Beijing (SRC-B), China

## USTC-IR

**Member(s):**
Bingchen Li (lbc31415926@mail.ustc.edu.cn), Xin Li
**Affiliation(s):**
University of Science and Technology of China, Hefei, China

## MSDRSR

**Member(s):**
Mingxi Li (li_mx_0318@163.com), Ding Liu[1]
**Affiliation(s):**
[1] ByteDance Inc.

## Giantpandacv Team

**Member(s):**
Wenbin Zou[1,4] (alexzou14@foxmail.com), Peijie Dong[2], Tian Ye[3], Yunchen Zhang[5], Ming Tan[4], Xin Niu[2]
**Affiliation(s):**
[1] South China University of Technology, Guangzhou, China
[2] National University of Defense Technology, Changsha, China
[3] Jimei University, Xiamen, China
[4] Fujian Normal University, Fuzhou, China
[5] China Design Group Inc., Nanjing, China


## Aselsan Research Team

**Member(s):**
Mustafa Ayazoğlu (mayazoglu@aselsan.com.tr)
**Affiliation(s):**
Aselsan (www.aselsan.com.tr), Ankara, Turkey


## SRMUI Team

**Member(s):**
Marcos V. Conde[1] (marcos.conde-osorio@uni-wuerzburg.de), Ui-Jin Choi[2], Radu Timofte[1]
**Affiliation(s):**
[1] Computer Vision Lab, Julius Maximilian University of Würzburg, Germany
[2] MegaStudyEdu, South Korea


## MVideo Team

**Member(s):**
Zhuang Jia (jiazhuang@xiaomi.com), Tianyu Xu, Yijian Zhang
**Affiliation(s):**
Xiaomi Inc.


## UESTC+XJU CV Team

**Member(s):**
Mao Ye (cvlab.uestc@gmail.com), Dengyan Luo, Xiaofeng Pan
**Affiliation(s):**
University of Electronic Science and Technology of China, Chengdu, China

## cvlab Team

**Member(s):**
Liuhan Peng[1] (pengliuhan@gmail.com), Mao Ye[2]
**Affiliation(s):**
[1] Xinjiang University, Xinjiang, China
[2] University of Electronic Science and Technology of China, Chengdu, China

## References

1. Agustsson, E., Timofte, R.: NTIRE 2017 challenge on single image super-resolution: dataset and study. In: Proceedings of the IEEE Conference on Computer Vision and Pattern Recognition Workshops (CVPRW), pp. 126–135 (2017)
2. Ayazoğlu, M.: IMDeception: grouped information distilling super-resolution network. In: Proceedings of the IEEE/CVF Conference on Computer Vision and Pattern Recognition (CVPR) Workshops, pp. 756–765 (2022)
3. Bhat, G., Danelljan, M., Timofte, R.: NTIRE 2021 challenge on burst super-resolution: methods and results. In: Proceedings of the IEEE/CVF Conference on Computer Vision and Pattern Recognition (CVPR) Workshops, pp. 613–626 (2021)
4. Bilecen, B.B., Fişne, A., Ayazoğlu, M.: Efficient multi-purpose cross-attention based image alignment block for edge devices. In: Proceedings of the IEEE/CVF Conference on Computer Vision and Pattern Recognition (CVPR) Workshops, pp. 3639–3648 (2022)
5. Boytsov, L., Naidan, B.: Engineering efficient and effective non-metric space library. In: Brisaboa, N., Pedreira, O., Zezula, P. (eds.) SISAP 2013. LNCS, vol. 8199, pp. 280–293. Springer, Heidelberg (2013). https://doi.org/10.1007/978-3-642-41062-8_28
6. Briechle, K., Hanebeck, U.D.: Template matching using fast normalized cross correlation. In: Optical Pattern Recognition XII, vol. 4387, pp. 95–102. International Society for Optics and Photonics (2001)
7. Bychkovsky, V., Paris, S., Chan, E., Durand, F.: Learning photographic global tonal adjustment with a database of input/output image pairs. In: Proceedings of the IEEE Conference on Computer Vision and Pattern Recognition (CVPR) (2011)
8. Caballero, J., et al.: Real-time video super-resolution with spatio-temporal networks and motion compensation. In: Proceedings of the IEEE Conference on Computer Vision and Pattern Recognition (CVPR), pp. 4778–4787 (2017)
9. Chan, K.C., Wang, X., Yu, K., Dong, C., Loy, C.C.: BasicVSR: the search for essential components in video super-resolution and beyond. In: Proceedings of the IEEE/CVF Conference on Computer Vision and Pattern Recognition (CVPR) (2021)
10. Chan, K.C., Zhou, S., Xu, X., Loy, C.C.: BasicVSR++: improving video super-resolution with enhanced propagation and alignment. arXiv preprint arXiv:2104.13371 (2021)
11. Chen, P., Yang, W., Wang, M., Sun, L., Hu, K., Wang, S.: Compressed domain deep video super-resolution. IEEE Trans. Image Process. 30, 7156–7169 (2021)
12. Chen, T., Kornblith, S., Norouzi, M., Hinton, G.: A simple framework for contrastive learning of visual representations. In: International Conference on Machine Learning (ICML), pp. 1597–1607. PMLR (2020)

13. Chen, X., Wang, X., Zhou, J., Dong, C.: Activating more pixels in image super-resolution transformer. arXiv preprint arXiv:2205.04437 (2022)
14. Chi, L., Jiang, B., Mu, Y.: Fast Fourier convolution. In: Advances in Neural Information Processing Systems (NeurIPS), vol. 33, pp. 4479–4488 (2020)
15. Chu, X., Chen, L., Chen, C., Lu, X.: Improving image restoration by revisiting global information aggregation. arXiv preprint arXiv:2112.04491 (2021)
16. Chu, X., Chen, L., Yu, W.: NAFSSR: stereo image super-resolution using NAFNet. In: Proceedings of the IEEE/CVF Conference on Computer Vision and Pattern Recognition (CVPR), pp. 1239–1248 (2022)
17. Conde, M.V., Choi, U.J., Burchi, M., Timofte, R.: Swin2SR: SwinV2 transformer for compressed image super-resolution and restoration. In: Proceedings of the European Conference on Computer Vision (ECCV) Workshops (2022)
18. Conde, M.V., Timofte, R., et al.: Reversed image signal processing and RAW reconstruction. AIM 2022 challenge report. In: Proceedings of the European Conference on Computer Vision (ECCV) Workshops (2022)
19. Deng, J., Dong, W., Socher, R., Li, L.J., Li, K., Fei-Fei, L.: ImageNet: a large-scale hierarchical image database. In: 2009 IEEE Conference on Computer Vision and Pattern Recognition (CVPR), pp. 248–255 (2009)
20. Deng, J., Wang, L., Pu, S., Zhuo, C.: Spatio-temporal deformable convolution for compressed video quality enhancement. In: Proceedings of the AAAI Conference on Artificial Intelligence, vol. 34, no. 07, pp. 10696–10703 (2020)
21. Deng, X., Yang, R., Xu, M., Dragotti, P.L.: Wavelet domain style transfer for an effective perception-distortion tradeoff in single image super-resolution. In: Proceedings of the IEEE/CVF International Conference on Computer Vision (ICCV), pp. 3076–3085 (2019)
22. Dong, C., Deng, Y., Loy, C.C., Tang, X.: Compression artifacts reduction by a deep convolutional network. In: Proceedings of the IEEE International Conference on Computer Vision (ICCV), pp. 576–584 (2015)
23. Dong, C., Loy, C.C., He, K., Tang, X.: Learning a deep convolutional network for image super-resolution. In: Fleet, D., Pajdla, T., Schiele, B., Tuytelaars, T. (eds.) ECCV 2014. LNCS, vol. 8692, pp. 184–199. Springer, Cham (2014). https://doi.org/10.1007/978-3-319-10593-2_13
24. Du, Z., Liu, D., Liu, J., Tang, J., Wu, G., Fu, L.: Fast and memory-efficient network towards efficient image super-resolution. In: Proceedings of the IEEE/CVF Conference on Computer Vision and Pattern Recognition (CVPR), pp. 853–862 (2022)
25. Ehrlich, M., Davis, L., Lim, S.-N., Shrivastava, A.: Quantization guided JPEG artifact correction. In: Vedaldi, A., Bischof, H., Brox, T., Frahm, J.-M. (eds.) ECCV 2020. LNCS, vol. 12353, pp. 293–309. Springer, Cham (2020). https://doi.org/10.1007/978-3-030-58598-3_18
26. Elfwing, S., Uchibe, E., Doya, K.: Sigmoid-weighted linear units for neural network function approximation in reinforcement learning. Neural Netw. **107**, 3–11 (2018)
27. Google: YouTube. https://www.youtube.com
28. Gu, J., Lu, H., Zuo, W., Dong, C.: Blind super-resolution with iterative kernel correction. In: Proceedings of the IEEE/CVF Conference on Computer Vision and Pattern Recognition (CVPR), pp. 1604–1613 (2019)
29. Gu, S., Lugmayr, A., Danelljan, M., Fritsche, M., Lamour, J., Timofte, R.: DIV8K: DIVerse 8K resolution image dataset. In: 2019 IEEE/CVF International Conference on Computer Vision Workshop (ICCVW), pp. 3512–3516 (2019)

30. Guan, Z., Xing, Q., Xu, M., Yang, R., Liu, T., Wang, Z.: MFQE 2.0: a new approach for multi-frame quality enhancement on compressed video. IEEE Trans. Pattern Anal. Mach. Intell. **43**(3), 949–963 (2019)
31. He, K., Zhang, X., Ren, S., Sun, J.: Deep residual learning for image recognition. In: Proceedings of the IEEE Conference on Computer Vision and Pattern Recognition (CVPR), pp. 770–778 (2016)
32. Ignatov, A., Timofte, R., Denna, M., Younes, A., et al.: Efficient and accurate quantized image super-resolution on mobile NPUs, mobile AI & AIM 2022 challenge: report. In: Proceedings of the European Conference on Computer Vision (ECCV) Workshops (2022)
33. Ignatov, A., Timofte, R., Kuo, H.K., Lee, M., Xu, Y.S., et al.: Real-time video super-resolution on mobile NPUs with deep learning, mobile AI & AIM 2022 challenge: report. In: Proceedings of the European Conference on Computer Vision (ECCV) Workshops (2022)
34. Ignatov, A., Timofte, R., et al.: Efficient bokeh effect rendering on mobile GPUs with deep learning, mobile AI & AIM 2022 challenge: report. In: Proceedings of the European Conference on Computer Vision (ECCV) Workshops (2022)
35. Ignatov, A., Timofte, R., et al.: Efficient single-image depth estimation on mobile devices, mobile AI & AIM challenge: report. In: Proceedings of the European Conference on Computer Vision (ECCV) Workshops (2022)
36. Ignatov, A., Timofte, R., et al.: Learned smartphone ISP on mobile GPUs with deep learning, mobile AI & AIM 2022 challenge: report. In: Proceedings of the European Conference on Computer Vision (ECCV) Workshops (2022)
37. Isobe, T., Jia, X., Gu, S., Li, S., Wang, S., Tian, Q.: Video super-resolution with recurrent structure-detail network. In: Vedaldi, A., Bischof, H., Brox, T., Frahm, J.-M. (eds.) ECCV 2020. LNCS, vol. 12357, pp. 645–660. Springer, Cham (2020). https://doi.org/10.1007/978-3-030-58610-2_38
38. Jiang, J., Zhang, K., Timofte, R.: Towards flexible blind JPEG artifacts removal. In: Proceedings of the IEEE/CVF International Conference on Computer Vision (ICCV), pp. 4997–5006 (2021)
39. Kim, J., Lee, J.K., Lee, K.M.: Accurate image super-resolution using very deep convolutional networks. In: Proceedings of the IEEE Conference on Computer Vision and Pattern Recognition (CVPR), pp. 1646–1654 (2016)
40. Kingma, D.P., Ba, J.: Adam: a method for stochastic optimization. In: Proceedings of the International Conference on Learning Representations (ICLR) (2015)
41. Kınlı, F.O., Menteş, S., Özcan, B., Kirac, F., Timofte, R., et al.: AIM 2022 challenge on Instagram filter removal: methods and results. In: Proceedings of the European Conference on Computer Vision (ECCV) Workshops (2022)
42. Lai, W.S., Huang, J.B., Ahuja, N., Yang, M.H.: Fast and accurate image super-resolution with deep Laplacian pyramid networks. IEEE Trans. Pattern Anal. Mach. Intell. **41**(11), 2599–2613 (2018)
43. Li, B., Li, X., Lu, Y., Liu, S., Feng, R., Chen, Z.: HST: hierarchical swin transformer for compressed image super-resolution. In: Proceedings of the European Conference on Computer Vision (ECCV) Workshops (2022)
44. Li, L., Tang, J., Chen, M., Zhao, S., Li, J., Zhang, L.: Multi-patch learning: looking more pixels in the training phase. In: Proceedings of the European Conference on Computer Vision (ECCV) Workshops (2022)
45. Li, X., Sun, S., Zhang, Z., Chen, Z.: Multi-scale grouped dense network for VVC intra coding. In: Proceedings of the IEEE/CVF Conference on Computer Vision and Pattern Recognition Workshops (CVPRW), pp. 158–159 (2020)

46. Liang, J., et al.: VRT: a video restoration transformer. arXiv preprint arXiv:2201.12288 (2022)
47. Liang, J., Cao, J., Sun, G., Zhang, K., Van Gool, L., Timofte, R.: SwinIR: image restoration using swin transformer. In: Proceedings of the IEEE/CVF International Conference on Computer Vision (ICCV), pp. 1833–1844 (2021)
48. Lin, Z., et al.: Revisiting RCAN: improved training for image super-resolution. arXiv preprint arXiv:2201.11279 (2022)
49. Liu, H., et al.: Video super-resolution based on deep learning: a comprehensive survey. Artif. Intell. Rev. **55**, 5981–6035 (2022). https://doi.org/10.1007/s10462-022-10147-y
50. Liu, Z., et al.: Swin transformer V2: scaling up capacity and resolution. In: Proceedings of the IEEE/CVF Conference on Computer Vision and Pattern Recognition (CVPR), pp. 12009–12019 (2022)
51. Liu, Z., et al.: Swin transformer: hierarchical vision transformer using shifted windows. In: Proceedings of the IEEE/CVF International Conference on Computer Vision (ICCV), pp. 10012–10022 (2021)
52. Loshchilov, I., Hutter, F.: SGDR: stochastic gradient descent with warm restarts. arXiv preprint arXiv:1608.03983 (2016)
53. Malkov, Y.A., Yashunin, D.A.: Efficient and robust approximate nearest neighbor search using hierarchical navigable small world graphs. CoRR abs/1603.09320 (2016). http://arxiv.org/abs/1603.09320
54. Pang, Y., et al.: FAN: frequency aggregation network for real image super-resolution. In: Bartoli, A., Fusiello, A. (eds.) ECCV 2020. LNCS, vol. 12537, pp. 468–483. Springer, Cham (2020). https://doi.org/10.1007/978-3-030-67070-2_28
55. Papyan, V., Elad, M.: Multi-scale patch-based image restoration. IEEE Trans. Image Process. **25**(1), 249–261 (2015)
56. Park, T., Liu, M.Y., Wang, T.C., Zhu, J.Y.: Semantic image synthesis with spatially-adaptive normalization. In: Proceedings of the IEEE/CVF Conference on Computer Vision and Pattern Recognition (CVPR), pp. 2337–2346 (2019)
57. Qin, X., Zhu, Y., Li, C., Wang, P., Cheng, J.: CIDBNet: a consecutively-interactive dual-branch network for JPEG compressed image super-resolution. In: Proceedings of the European Conference on Computer Vision (ECCV) Workshops (2022)
58. Rota, C., Buzzelli, M., Bianco, S., Schettini, R.: Video restoration based on deep learning: a comprehensive survey. Artif. Intell. Rev. (2022). https://doi.org/10.1007/s10462-022-10302-5
59. Sajjadi, M.S., Scholkopf, B., Hirsch, M.: EnhanceNet: single image super-resolution through automated texture synthesis. In: Proceedings of the IEEE International Conference on Computer Vision (ICCV), pp. 4491–4500 (2017)
60. Tai, Y., Yang, J., Liu, X., Xu, C.: MemNet: a persistent memory network for image restoration. In: Proceedings of the IEEE International Conference on Computer Vision (ICCV), pp. 4539–4547 (2017)
61. Tao, X., Gao, H., Liao, R., Wang, J., Jia, J.: Detail-revealing deep video super-resolution. In: Proceedings of the IEEE International Conference on Computer Vision (ICCV), pp. 4472–4480 (2017)
62. Timofte, R., Agustsson, E., Van Gool, L., Yang, M.H., Zhang, L.: NTIRE 2017 challenge on single image super-resolution: methods and results. In: Proceedings of the IEEE Conference on Computer Vision and Pattern Recognition Workshops (CVPRW), pp. 114–125 (2017)
63. Timofte, R., Rothe, R., Van Gool, L.: Seven ways to improve example-based single image super resolution. In: Proceedings of the IEEE Conference on Computer Vision and Pattern Recognition (CVPR), pp. 1865–1873 (2016)

64. Wang, X., Chan, K.C., Yu, K., Dong, C., Change Loy, C.: EDVR: video restoration with enhanced deformable convolutional networks. In: Proceedings of the IEEE/CVF Conference on Computer Vision and Pattern Recognition (CVPR) Workshops (2019)

65. Wang, X., Yu, K., Dong, C., Loy, C.C.: Recovering realistic texture in image super-resolution by deep spatial feature transform. In: Proceedings of the IEEE Conference on Computer Vision and Pattern Recognition (CVPR), pp. 606–615 (2018)

66. Wang, X., et al.: ESRGAN: enhanced super-resolution generative adversarial networks. In: Proceedings of the European Conference on Computer Vision Workshops (ECCVW) (2018)

67. Xu, Y., Gao, L., Tian, K., Zhou, S., Sun, H.: Non-local ConvLSTM for video compression artifact reduction. In: Proceedings of The IEEE International Conference on Computer Vision (ICCV), October 2019

68. Yamac, M., Ataman, B., Nawaz, A.: KernelNet: a blind super-resolution kernel estimation network. In: Proceedings of the IEEE/CVF Conference on Computer Vision and Pattern Recognition Workshops (CVPRW), pp. 453–462 (2021)

69. Yang, R., Mentzer, F., Gool, L.V., Timofte, R.: Learning for video compression with hierarchical quality and recurrent enhancement. In: Proceedings of the IEEE/CVF Conference on Computer Vision and Pattern Recognition (CVPR), pp. 6628–6637 (2020)

70. Yang, R., Sun, X., Xu, M., Zeng, W.: Quality-gated convolutional LSTM for enhancing compressed video. In: Proceedings of the IEEE International Conference on Multimedia and Expo (ICME), pp. 532–537. IEEE (2019)

71. Yang, R., Timofte, R., et al.: NTIRE 2021 challenge on quality enhancement of compressed video: dataset and study. In: Proceedings of the IEEE/CVF Conference on Computer Vision and Pattern Recognition (CVPR) Workshops (2021)

72. Yang, R., Timofte, R., et al.: NTIRE 2021 challenge on quality enhancement of compressed video: methods and results. In: IEEE/CVF Conference on Computer Vision and Pattern Recognition (CVPR) Workshops (2021)

73. Yang, R., Timofte, R., et al.: AIM 2022 challenge on super-resolution of compressed image and video: dataset, methods and results. In: Proceedings of the European Conference on Computer Vision (ECCV) Workshops (2022)

74. Yang, R., Timofte, R., et al.: NTIRE 2022 challenge on super-resolution and quality enhancement of compressed video: dataset, methods and results. In: Proceedings of the IEEE/CVF Conference on Computer Vision and Pattern Recognition (CVPR) Workshops (2022)

75. Yang, R., Xu, M., Liu, T., Wang, Z., Guan, Z.: Enhancing quality for HEVC compressed videos. IEEE Trans. Circ. Syst. Video Technol. **29**(7), 2039–2054 (2018)

76. Yang, R., Xu, M., Wang, Z.: Decoder-side HEVC quality enhancement with scalable convolutional neural network. In: Proceedings of the IEEE International Conference on Multimedia and Expo (ICME), pp. 817–822. IEEE (2017)

77. Yang, R., Xu, M., Wang, Z., Li, T.: Multi-frame quality enhancement for compressed video. In: Proceedings of the IEEE Conference on Computer Vision and Pattern Recognition (CVPR), pp. 6664–6673 (2018)

78. Yoo, J., Ahn, N., Sohn, K.A.: Rethinking data augmentation for image super-resolution: a comprehensive analysis and a new strategy. In: Proceedings of the IEEE/CVF Conference on Computer Vision and Pattern Recognition (CVPR), pp. 8375–8384 (2020)

79. Zamir, S.W., Arora, A., Khan, S., Hayat, M., Khan, F.S., Yang, M.H.: Restormer: efficient transformer for high-resolution image restoration. In: Proceedings of the IEEE/CVF Conference on Computer Vision and Pattern Recognition (CVPR), pp. 5728–5739 (2022)
80. Zhang, K., Liang, J., Van Gool, L., Timofte, R.: Designing a practical degradation model for deep blind image super-resolution. In: Proceedings of the IEEE International Conference on Computer Vision (ICCV), pp. 4791–4800 (2021)
81. Zhang, K., Zuo, W., Chen, Y., Meng, D., Zhang, L.: Beyond a Gaussian denoiser: residual learning of deep CNN for image denoising. IEEE Trans. Image Process. **26**(7), 3142–3155 (2017)
82. Zhang, Y., Li, K., Li, K., Wang, L., Zhong, B., Fu, Y.: Image super-resolution using very deep residual channel attention networks. In: Ferrari, V., Hebert, M., Sminchisescu, C., Weiss, Y. (eds.) ECCV 2018. LNCS, vol. 11211, pp. 294–310. Springer, Cham (2018). https://doi.org/10.1007/978-3-030-01234-2_18
83. Zhang, Y., Tian, Y., Kong, Y., Zhong, B., Fu, Y.: Residual dense network for image super-resolution. In: Proceedings of the IEEE Conference on Computer Vision and Pattern Recognition (CVPR), pp. 2472–2481 (2018)
84. Zheng, M., et al.: Progressive training of a two-stage framework for video restoration. In: Proceedings of the IEEE/CVF Conference on Computer Vision and Pattern Recognition (CVPR) Workshops, pp. 1024–1031 (2022)

# W07 - Medical Computer Vision

# W07 - Medical Computer Vision

The MCV workshop provides an opportunity to students, researchers, and developers in biomedical imaging companies to present, discuss, and learn recent advancements in medical image analysis. The ultimate goal of the workshop is leveraging big data, deep learning, and novel representation to effectively build the next generation of robust quantitative medical imaging parsing tools and products. Prominent applications include large scale cancer screening, computational heart modeling, landmark detection, neural structure, and functional labeling and image-guided intervention. Computer vision advancements and deep learning in particular have rapidly transitioned to the medical imaging community in recent years. Additionally, there is a tremendous growth in startup activity applying medical computer vision algorithms to the healthcare industry. Collecting and accessing radiological patient images is a challenging task. Recent efforts include VISCERAL Challenge and the Alzheimer's Disease Neuroimaging Initiative. The NIH and partners are working on extracting trainable anatomical and pathological semantic labels from radiology reports that are linked to patients' CT/MRI/X-ray images or volumes such as NCI's Cancer Imaging Archive. The MCV workshop aims to encourage the establishment of public medical datasets to be used as unbiased platforms to compare performances on the same set of data for various disease findings.

October 2022

Tal Arbel
Ayelet Akselrod-Balin
Vasileios Belagiannis
Qi Dou
Moti Freiman
Nicolas Padoy
Tammy Riklin Raviv
Mathias Unberath
Yuyin Zhou

# Swin-Unet: Unet-Like Pure Transformer for Medical Image Segmentation

Hu Cao[1], Yueyue Wang[2], Joy Chen[3], Dongsheng Jiang[4(✉)], Xiaopeng Zhang[4], Qi Tian[4(✉)], and Manning Wang[2(✉)]

[1] Technische Universität München, München, Germany
hu.cao@tum.de
[2] Fudan University, Shanghai, China
{yywang17,mnwang}@fudan.edu.cn
[3] Johns Hopkins University, Baltimore, MD, USA
[4] Huawei Technologies, Shanghai, China
dongsheng_jiang@outlook.com, tian.qi1@huawei.com

**Abstract.** In the past few years, convolutional neural networks (CNNs) have achieved milestones in medical image analysis. In particular, deep neural networks based on U-shaped architecture and skip-connections have been widely applied in various medical image tasks. However, although CNN has achieved excellent performance, it cannot learn global semantic information interaction well due to the locality of convolution operation. In this paper, we propose Swin-Unet, which is an Unet-like pure Transformer for medical image segmentation. The tokenized image patches are fed into the Transformer-based U-shaped Encoder-Decoder architecture with skip-connections for local-global semantic feature learning. Specifically, we use a hierarchical Swin Transformer with shifted windows as the encoder to extract context features. And a symmetric Swin Transformer-based decoder with a patch expanding layer is designed to perform the up-sampling operation to restore the spatial resolution of the feature maps. Under the direct down-sampling and up-sampling of the inputs and outputs by $4\times$, experiments on multi-organ and cardiac segmentation tasks demonstrate that the pure Transformer-based U-shaped Encoder-Decoder network outperforms those methods with full-convolution or the combination of transformer and convolution. The codes have been publicly available at the link (https://github.com/HuCaoFighting/Swin-Unet).

**Keywords:** Transformer · Self-attention · Medical image segmentation

## 1 Introduction

Benefiting from the development of deep learning, computer vision technology has been widely used in medical image analysis. Image segmentation is an impor-

---

H. Cao and Y. Wang—Work done as an intern in Huawei Technologies.

L. Karlinsky et al. (Eds.): ECCV 2022 Workshops, LNCS 13803, pp. 205–218, 2023.
https://doi.org/10.1007/978-3-031-25066-8_9

tant part of medical image analysis. Accurate and robust medical image segmentation can play a cornerstone role in computer-aided diagnosis and image-guided clinical surgery [2,11].

Existing medical image segmentation methods mainly rely on fully convolutional neural networks (FCNNs) with U-shaped structure [7,16,21]. The typical U-shaped network, U-Net [21], consists of a symmetric Encoder-Decoder with skip connections. In the encoder, a series of convolutional layers and continuous down-sampling layers are used to extract deep features with large receptive fields. Then, the decoder up-samples the extracted deep features to the input resolution for pixel-level semantic prediction, and the high-resolution features of different scale from the encoder are fused with skip connections to alleviate the loss of spatial information caused by down-sampling. With such an elegant structural design, U-Net has achieved great success in various medical imaging applications. Following this technical route, many algorithms such as 3D U-Net [4], Res-UNet [31], U-Net++ [35] and UNet3+ [15] have been developed for image and volumetric segmentation. The excellent performance of these FCNN-based methods in cardiac segmentation, organ segmentation and lesion segmentation proves that CNN has a strong ability of learning discriminating features.

Currently, although the CNN-based methods have achieved excellent performance in the field of medical image segmentation, there still cannot fully meet the strict requirements of medical applications for segmentation accuracy. Image segmentation is still a challenge task in medical image analysis. Since the intrinsic locality of convolution operation, it is difficult for CNN-based approaches to learn explicit global semantic information interaction [2]. Some studies have tried to address this problem by using atrous convolutional layers [3,9], self-attention mechanisms [22,30], and image pyramids [34]. However, these methods still have limitations in modeling long - range dependencies. Recently, inspired by Transformer's great success in the nature language processing (NLP) domain [27], researchers have tried to bring Transformer into the vision domain [1]. In [6], vision transformer (ViT) is proposed to perform the image recognition task. Taking 2D image patches with positional embeddings as inputs and pre-training on large dataset, ViT achieved comparable performance with the CNN-based methods. Besides, data-efficient image transformer (DeiT) is presented in [23], which indicates that Transformer can be trained on mid-size datasets and that a more robust Transformer can be obtained by combining it with the distillation method. In [18], a hierarchical Swin Transformer is developed. Take Swin Transformer as vision backbone, the authors of [18] achieved state-of-the-art performance on Image classification, object detection and semantic segmentation. The success of ViT, DeiT and Swin Transformer in image recognition task demonstrates the potential for Transformer to be applied in the vision domain.

Motivated by the Swin Transformer's [18] success, we propose Swin-Unet to leverage the power of Transformer for 2D medical image segmentation in this work. To our best knowledge, Swin-Unet is a first pure Transformer-based U-shaped architecture that consists of encoder, bottleneck, decoder, and skip connections. Encoder, bottleneck and decoder are all built based on Swin Transformer block [18]. The input medical images are split into non-overlapping image patches. Each patch is treated as a token and fed into the Transformer-based encoder to learn deep feature representations. The extracted context features are then up-sampled by the decoder with patch expanding layer, and fused with the multi-scale features from the encoder via skip connections, so as to restore the spatial resolution of the feature maps and further perform segmentation prediction. Extensive experiments on multi-organ and cardiac segmentation datasets indicate that the proposed method has excellent segmentation accuracy and robust generalization ability. Concretely, our contributions can be summarized as: (1) Based on Swin Transformer block, we build a symmetric Encoder-Decoder architecture with skip connections. In the encoder, self-attention from local to global is realized; in the decoder, the global features are up-sampled to the input resolution for corresponding pixel-level segmentation prediction. (2) A patch expanding layer is developed to achieve up-sampling and feature dimension increase without using convolution or interpolation operation. (3) It is found in the experiment that skip connection is effective for Transformer, so a pure Transformer-based U-shaped Encoder-Decoder architecture with skip connection is finally constructed, named Swin-Unet.

## 2   Related Work

**CNN-Based Methods:** Early medical image segmentation methods are mainly contour-based and traditional machine learning-based algorithms [12,25]. With the development of deep CNNs, U-Net is proposed in [21] for medical image segmentation. Due to the simplicity and superior performance of the U-shaped structure, various Unet-like methods are constantly emerging, such as Res-UNet [31], Dense-UNet [17], U-Net++ [35] and UNet3+ [15]. And it is also introduced into the field of 3D medical image segmentation, such as 3D-Unet [4] and V-Net [19]. Currently, CNN-based methods have achieved tremendous success in the field of medical image segmentation due to its powerful representation ability.

**Vision Transformers:** Transformer was first proposed for the machine translation task in [27]. In the NLP domain, the Transformer-based methods have achieved the state-of-the-art performance in various tasks [5]. Driven by Transformer's success, the researchers introduced a pioneering vision transformer (ViT) in [6], which achieved the impressive speed-accuracy trade-off on image recognition task. Compared with CNN-based methods, the drawback of ViT is that it requires pre-training on its own large dataset. To alleviate the difficulty in training ViT, Deit [23] describes several training strategies that allow ViT to train well on ImageNet. Recently, several excellent works have been done based on ViT [10,18,28]. It is worth mentioning that an efficient and effective hierarchical vision Transformer, called Swin Transformer, is proposed as a vision backbone in [18]. Based on the shifted windows mechanism, Swin Transformer achieved the state-of-the-art performance on various vision tasks including image classification, object detection and semantic segmentation. In this work, we attempt to use Swin Transformer block as basic unit to build a U-shaped Encoder-Decoder architecture with skip connections for medical image segmentation, thus providing a benchmark comparison for the development of Transformer in the medical image field.

**Self-attention/Transformer to Complement CNNs:** In recent years, researchers have tried to introduce self-attention mechanism into CNN to improve the performance of the network [30]. In [22], the skip connections with additive attention gate are integrated in U-shaped architecture to perform medical image segmentation. However, this is still the CNN-based method. Currently, some efforts are being made to combine CNN and Transformer to break the dominance of CNNs in medical image segmentation [2,11,26]. In [2], the authors combined Transformer with CNN to constitute a strong encoder for 2D medical image segmentation. Similar to [2,26] and [33] use the complementarity of Transformer and CNN to improve the segmentation capability of the model. Currently, various combinations of Transformer with CNN are applied in multi-modal brain tumor segmentation [29] and 3D medical image segmentation [11,32]. Different from the above methods, we try to explore the application potential of pure Transformer in medical image segmentation.

## 3 Method

### 3.1 Architecture Overview

The overall architecture of the proposed Swin-Unet is presented in Fig. 1. Swin-Unet consists of encoder, bottleneck, decoder and skip connections. The basic unit of Swin-Unet is Swin Transformer block [18]. For the encoder, to transform the inputs into sequence embeddings, the medical images are split into non-overlapping patches with patch size of $4 \times 4$. By such partition approach, the feature dimension of each patch becomes to $4 \times 4 \times 3 = 48$. Furthermore, a linear embedding layer is applied to projected feature dimension into arbitrary dimension (represented as C). The transformed patch tokens pass through several Swin Transformer blocks and patch merging layers to generate the hierarchical feature representations. Specifically, patch merging layer is responsible for down-sampling and increasing dimension, and Swin Transformer block is responsible for feature representation learning. Inspired by U-Net [21], we design a symmetric transformer-based decoder. The decoder is composed of Swin Transformer block and patch expanding layer. The extracted context features are fused with multiscale features from encoder via skip connections to complement the loss of spatial information caused by down-sampling. In contrast to patch merging layer, a patch expanding layer is specially designed to perform up-sampling. The patch expanding layer reshapes feature maps of adjacent dimensions into a large feature maps with $2\times$ up-sampling of resolution. In the end, the last patch expanding layer is used to perform $4\times$ up-sampling to restore the resolution of the feature maps to the input resolution ($W \times H$), and then a linear projection layer is applied on these up-sampled features to output the pixel-level segmentation predictions. We would elaborate each block in the following.

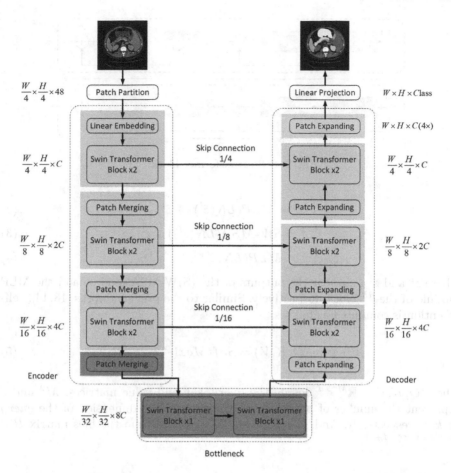

**Fig. 1.** The architecture of Swin-Unet, which is composed of encoder, bottleneck, decoder and skip connections. Encoder, bottleneck and decoder are all constructed based on Swin transformer block.

## 3.2 Swin Transformer Block

Different from the conventional multi-head self attention (MSA) module, Swin transformer block [18] is constructed based on shifted windows. In Fig. 2, two consecutive Swin transformer blocks are presented. Each Swin transformer block is composed of LayerNorm (LN) layer, multi-head self attention module, residual connection and 2-layer MLP with GELU non-linearity. The window-based multi-head self attention (W-MSA) module and the shifted window-based multi-head self attention (SW-MSA) module are applied in the two successive transformer blocks, respectively. Based on such window partitioning mechanism, continuous swin transformer blocks can be formulated as:

$$\hat{z}^l = W\text{-}MSA(LN(z^{l-1})) + z^{l-1}, \tag{1}$$

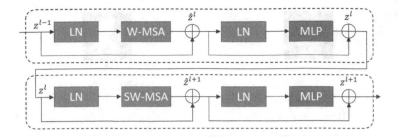

**Fig. 2.** Swin transformer block.

$$z^l = MLP(LN(\hat{z}^l)) + \hat{z}^l, \tag{2}$$

$$\hat{z}^{l+1} = SW\text{-}MSA(LN(z^l)) + z^l, \tag{3}$$

$$z^{l+1} = MLP(LN(\hat{z}^{l+1})) + \hat{z}^{l+1}, \tag{4}$$

where $\hat{z}^l$ and $z^l$ represent the outputs of the (S)W-MSA module and the MLP module of the $l^{th}$ block, respectively. Similar to the previous works [13,14], self-attention is computed as follows:

$$Attention(Q, K, V) = SoftMax(\frac{QK^T}{\sqrt{d}} + B)V, \tag{5}$$

where $Q, K, V \in \mathbb{R}^{M^2 \times d}$ denote the query, key and value matrices. $M^2$ and $d$ represent the number of patches in a window and the dimension of the *query* or *key*, respectively. And, the values in $B$ are taken from the bias matrix $\hat{B} \in \mathbb{R}^{(2M-1) \times (2M+1)}$.

## 3.3   Encoder

In the encoder, the C-dimensional tokenized inputs with the resolution of $\frac{H}{4} \times \frac{W}{4}$ are fed into the two consecutive Swin Transformer blocks to perform representation learning, in which the feature dimension and resolution remain unchanged. Meanwhile, the patch merging layer will reduce the number of tokens (2× down-sampling) and increase the feature dimension to 2× the original dimension. This procedure will be repeated three times in the encoder.

**Patch Merging Layer:** The input patches are divided into 4 parts and concatenated together by the patch merging layer. With such processing, the feature resolution will be down-sampled by 2×. And, since the concatenate operation results the feature dimension increasing by 4×, a linear layer is applied on the concatenated features to unify the feature dimension to the 2× the original dimension.

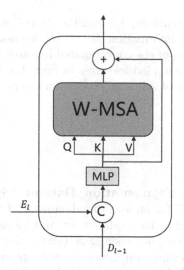

**Fig. 3.** The structure of skip connection.

## 3.4 Bottleneck

Since Transformer is too deep to be converged [24], only two successive Swin Transformer blocks are used to constructed the bottleneck to learn the deep feature representation. In the bottleneck, the feature dimension and resolution are kept unchanged.

## 3.5 Decoder

Corresponding to the encoder, the symmetric decoder is built based on Swin Transformer block. To this end, in contrast to the patch merging layer used in the encoder, we use the patch expanding layer in the decoder to up-sample the extracted deep features. The patch expanding layer reshapes the feature maps of adjacent dimensions into a higher resolution feature map ($2\times$ up-sampling) and reduces the feature dimension to half of the original dimension accordingly.

**Patch Expanding Layer:** Take the first patch expanding layer as an example, before up-sampling, a linear layer is applied on the input features ($\frac{W}{32} \times \frac{H}{32} \times 8C$) to increase the feature dimension to $2\times$ the original dimension ($\frac{W}{32} \times \frac{H}{32} \times 16C$). Then, we use rearrange operation to expand the resolution of the input features to $2\times$ the input resolution and reduce the feature dimension to quarter of the input dimension ($\frac{W}{32} \times \frac{H}{32} \times 16C \rightarrow \frac{W}{16} \times \frac{H}{16} \times 4C$). We will discuss the impact of using patch expanding layer to perform up-sampling in Sect. 4.5.

## 3.6 Skip Connection

Similar to the U-Net [21], the skip connections are used to fuse the multi-scale features from the encoder with the up-sampled features. As shown in Fig. 3, we concatenate the shallow features, $E_l$, and the deep features, $D_{l-1}$, together to reduce the loss of spatial information caused by down-sampling. Followed by a linear layer, the dimension of the concatenated features is remained the same as the dimension of the up-sampled features. In Sect. 4.5, we will detailed discuss the impact of the number of skip connections on the performance of our model.

## 4    Experiments

### 4.1    Datasets

**Synapse Multi-organ Segmentation Dataset (Synapse[1]):** the dataset includes 30 cases with 3779 axial abdominal clinical CT images. Following [2,8], 18 samples are divided into the training set and 12 samples into testing set. And the average Dice-Similarity coefficient (DSC) and average Hausdorff Distance (HD) are used as evaluation metric to evaluate our method on 8 abdominal organs (aorta, gallbladder, spleen, left kidney, right kidney, liver, pancreas, spleen, stomach).

**Automated Cardiac Diagnosis Challenge Dataset (ACDC[2]):** the ACDC dataset is collected from different patients using MRI scanners. For each patient MR image, left ventricle (LV), right ventricle (RV) and myocardium (MYO) are labeled. The dataset is split into 70 training samples, 10 validation samples and 20 testing samples. Similar to [2], only average DSC is used to evaluate our method on this dataset.

### 4.2    Implementation Details

The Swin-Unet is achieved based on Python 3.6 and Pytorch 1.7.0. For all training cases, data augmentations such as flips and rotations are used to increase data diversity. The input image size and patch size are set as $224 \times 224$ and 4, respectively. We train our model on a Nvidia V100 GPU with 32 GB memory. The weights pre-trained on ImageNet are used to initialize the model parameters. During the training period, the batch size is 24 and the popular SGD optimizer with momentum 0.9 and weight decay $1e-4$ is used to optimize our model for back propagation.

### 4.3    Experiment Results on Synapse Dataset

The comparison of the proposed Swin-Unet with previous state-of-the-art methods on the Synapse multi-organ CT dataset is presented in Table 1. Different from TransUnet [2], we add the test results of our own implementations of U-Net [21] and Att-UNet [20] on the Synapse dataset. Experimental results demonstrate that our Unet-like pure transformer method achieves the best performance

---

[1] https://www.synapse.org/!Synapse:syn3193805/wiki/217789.
[2] https://www.creatis.insa-lyon.fr/Challenge/acdc/.

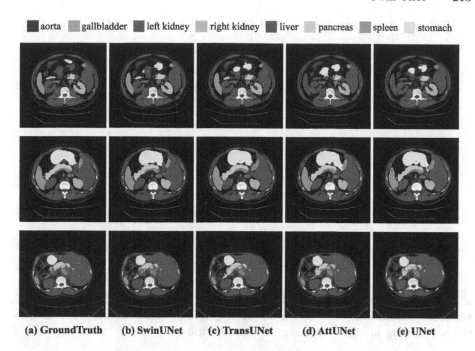

aorta    gallbladder    left kidney    right kidney    liver    pancreas    spleen    stomach

(a) GroundTruth    (b) SwinUNet    (c) TransUNet    (d) AttUNet    (e) UNet

**Fig. 4.** The segmentation results of different methods on the Synapse multi-organ CT dataset.

**Table 1.** Segmentation accuracy of different methods on the Synapse multi-organ CT dataset.

| Methods | DSC↑ | HD↓ | Aorta | Gallbladder | Kidney(L) | Kidney(R) | Liver | Pancreas | Spleen | Stomach |
|---|---|---|---|---|---|---|---|---|---|---|
| V-Net [19] | 68.81 | - | 75.34 | 51.87 | 77.10 | **80.75** | 87.84 | 40.05 | 80.56 | 56.98 |
| DARR [8] | 69.77 | - | 74.74 | 53.77 | 72.31 | 73.24 | 94.08 | 54.18 | 89.90 | 45.96 |
| R50 U-Net [2] | 74.68 | 36.87 | 87.74 | 63.66 | 80.60 | 78.19 | 93.74 | 56.90 | 85.87 | 74.16 |
| U-Net [21] | 76.85 | 39.70 | 89.07 | **69.72** | 77.77 | 68.60 | 93.43 | 53.98 | 86.67 | 75.58 |
| R50 Att-UNet [2] | 75.57 | 36.97 | 55.92 | 63.91 | 79.20 | 72.71 | 93.56 | 49.37 | 87.19 | 74.95 |
| Att-UNet [20] | 77.77 | 36.02 | **89.55** | 68.88 | 77.98 | 71.11 | 93.57 | **58.04** | 87.30 | 75.75 |
| R50 ViT [2] | 71.29 | 32.87 | 73.73 | 55.13 | 75.80 | 72.20 | 91.51 | 45.99 | 81.99 | 73.95 |
| TransUnet [2] | 77.48 | 31.69 | 87.23 | 63.13 | 81.87 | 77.02 | 94.08 | 55.86 | 85.08 | 75.62 |
| SwinUnet | **79.13** | **21.55** | 85.47 | 66.53 | **83.28** | 79.61 | **94.29** | 56.58 | **90.66** | **76.60** |

with segmentation accuracy of 79.13% (DSC↑) and 21.55% (HD↓). Compared with Att-Unet [20] and the recently method TransUnet [2], although our algorithm did not improve much on the DSC evaluation metric, we achieved accuracy improvement of about 4% and 10% on the HD evaluation metric, which indicates that our approach can achieve better edge predictions. The segmentation results of different methods on the Synapse multi-organ CT dataset are shown in Fig. 4. It can be seen from the figure that CNN-based methods tend to have over-segmentation problems, which may be caused by the locality of convolution operation. In this work, we demonstrate that by integrating Transformer with

**Table 2.** Segmentation accuracy of different methods on the ACDC dataset.

| Methods | DSC | RV | Myo | LV |
|---|---|---|---|---|
| R50 U-Net | 87.55 | 87.10 | 80.63 | 94.92 |
| R50 Att-UNet | 86.75 | 87.58 | 79.20 | 93.47 |
| R50 ViT | 87.57 | 86.07 | 81.88 | 94.75 |
| TransUnet | 89.71 | 88.86 | 84.53 | 95.73 |
| SwinUnet | **90.00** | 88.55 | **85.62** | **95.83** |

**Table 3.** Ablation study on the impact of the up-sampling

| Up-sampling | DSC | Aorta | Gallbladder | Kidney (L) | Kidney (R) | Liver | Pancreas | Spleen | Stomach |
|---|---|---|---|---|---|---|---|---|---|
| Bilinear interpolation | 76.15 | 81.84 | 66.33 | 80.12 | 73.91 | 93.64 | 55.04 | 86.10 | 72.20 |
| Transposed convolution | 77.63 | 84.81 | 65.96 | 82.66 | 74.61 | **94.39** | 54.81 | 89.42 | 74.41 |
| Patch expanding | **79.13** | **85.47** | **66.53** | **83.28** | **79.61** | 94.29 | **56.58** | **90.66** | **76.60** |

a U-shaped architecture with skip connections, the pure Transformer approach without convolution can better learn both global semantic information interactions, resulting in better segmentation predictions.

### 4.4 Experiment Results on ACDC Dataset

Similar to the Synapse dataset, the proposed Swin-Unet is trained on ACDC dataset to perform medical image segmentation. The experimental results are summarized in Table 2. By using the image data of MR mode as input, Swin-Unet is still able to achieve excellent performance with an accuracy of 90.00%, which shows that our method has good generalization ability and robustness.

### 4.5 Ablation Study

In order to explore the influence of different factors on the model performance, we conducted ablation studies on Synapse dataset. Specifically, up-sampling, the number of skip connections, input sizes, and model scales are discussed below.

**Effect of Up-Sampling:** Corresponding to the patch merging layer in the encoder, we specially designed a patch expanding layer in the decoder to perform up-sampling and feature dimension increase. To explore the effective of the proposed patch expanding layer, we conducted the experiments of Swin-Unet with bilinear interpolation, transposed convolution and patch expanding layer on Synapse dataset. The experimental results in the Table 3 indicate that the proposed Swin-Unet combined with the patch expanding layer can obtain the better segmentation accuracy.

**Table 4.** Ablation study on the impact of the number of skip connection

| Skip connection | DSC | Aorta | Gallbladder | Kidney (L) | Kidney (R) | Liver | Pancreas | Spleen | Stomach |
|---|---|---|---|---|---|---|---|---|---|
| 0 | 72.46 | 78.71 | 53.24 | 77.46 | 75.90 | 92.60 | 46.07 | 84.57 | 71.13 |
| 1 | 76.43 | 82.53 | 60.44 | 81.36 | 79.27 | 93.64 | 53.36 | 85.95 | 74.90 |
| 2 | 78.93 | **85.82** | 66.27 | **84.70** | **80.32** | 93.94 | 55.32 | 88.35 | **76.71** |
| 3 | **79.13** | 85.47 | **66.53** | 83.28 | 79.61 | **94.29** | **56.58** | **90.66** | 76.60 |

**Table 5.** Ablation study on the impact of the input size

| Input size | DSC | Aorta | Gallbladder | Kidney (L) | Kidney (R) | Liver | Pancreas | Spleen | Stomach |
|---|---|---|---|---|---|---|---|---|---|
| 224 | 79.13 | 85.47 | 66.53 | 83.28 | 79.61 | 94.29 | 56.58 | **90.66** | **76.60** |
| 384 | **81.12** | **87.07** | **70.53** | **84.64** | **82.87** | **94.72** | **63.73** | 90.14 | 75.29 |

**Table 6.** Ablation study on the impact of the model scale

| Model scale | DSC | Aorta | Gallbladder | Kidney (L) | Kidney (R) | Liver | Pancreas | Spleen | Stomach |
|---|---|---|---|---|---|---|---|---|---|
| Tiny | 79.13 | 85.47 | 66.53 | 83.28 | 79.61 | **94.29** | 56.58 | **90.66** | **76.60** |
| Base | **79.25** | **87.16** | **69.19** | **84.61** | **81.99** | 93.86 | **58.10** | 88.44 | 70.65 |

**Effect of the Number of Skip Connections:** The skip connections of our Swin-UNet are added in places of the 1/4, 1/8, and 1/16 resolution scales. By changing the number of skip connections to 0, 1, 2 and 3 respectively, we explored the influence of different skip connections on the segmentation performance of the proposed model. In Table 4, we can see that the segmentation performance of the model increases with the increase of the number of skip connections. Therefore, in order to make the model more robust, the number of skip connections is set as 3 in this work.

**Effect of Input Size:** The testing results of the proposed Swin-Unet with $224 \times 224$, $384 \times 384$ input resolutions as input are presented in Table 5. As the input size increases from $224 \times 224$ to $384 \times 384$ and the patch size remains the same as 4, the input token sequence of Transformer will become larger, thus leading to improve the segmentation performance of the model. However, although the segmentation accuracy of the model has been slightly improved, the computational load of the whole network has increased significantly as well. In order to ensure the running efficiency of the algorithm, the experiments in this paper are based on $224 \times 224$ resolution scale as the input.

**Effect of Model Scale:** Similar to [18], we discuss the effect of network deepening on model performance. It can be seen from Table 6 that the increase of model scale hardly improves the performance of the model, but increases the computational cost of the whole network. Considering the accuracy-speed trade off, we adopt the Tiny-based model to perform medical image segmentation.

## 4.6   Discussion

As we all known, the performance of Transformer-based model is severely affected by model pre-training. In this work, we directly use the training weight of Swin transformer [18] on ImageNet to initialize the network encoder and decoder, which may be a suboptimal scheme. This initialization approach is a simple one, and in the future we will explore the ways to pre-train Transformer end-to-end for medical image segmentation. Moreover, since the input images in this paper are 2D, while most of the medical image data are 3D, we will explore the application of Swin-Unet in 3D medical image segmentation in the following research.

## 5   Conclusion

In this paper, we introduced a novel pure transformer-based U-shaped encoder-decoder for medical image segmentation. In order to leverage the power of Transformer, we take Swin Transformer block as the basic unit for feature representation learning and global semantic information interactive. Extensive experiments on multi-organ and cardiac segmentation tasks demonstrate that the proposed Swin-Unet has excellent performance and generalization ability.

## References

1. Carion, N., Massa, F., Synnaeve, G., Usunier, N., Kirillov, A., Zagoruyko, S.: End-to-end object detection with transformers. In: Vedaldi, A., Bischof, H., Brox, T., Frahm, J.-M. (eds.) ECCV 2020. LNCS, vol. 12346, pp. 213–229. Springer, Cham (2020). https://doi.org/10.1007/978-3-030-58452-8_13
2. Chen, J., et al.: TransUNet: transformers make strong encoders for medical image segmentation. CoRR abs/2102.04306 (2021)
3. Chen, L.C., Papandreou, G., Kokkinos, I., Murphy, K., Yuille, A.L.: DeepLab: semantic image segmentation with deep convolutional nets, atrous convolution, and fully connected CRFs. IEEE Trans. Pattern Anal. Mach. Intell. **40**(4), 834–848 (2018). https://doi.org/10.1109/TPAMI.2017.2699184
4. Çiçek, Ö., Abdulkadir, A., Lienkamp, S.S., Brox, T., Ronneberger, O.: 3D U-Net: learning dense volumetric segmentation from sparse annotation. In: Ourselin, S., Joskowicz, L., Sabuncu, M.R., Unal, G., Wells, W. (eds.) MICCAI 2016. LNCS, vol. 9901, pp. 424–432. Springer, Cham (2016). https://doi.org/10.1007/978-3-319-46723-8_49
5. Devlin, J., Chang, M.W., Lee, K., Toutanova, K.: BERT: pre-training of deep bidirectional transformers for language understanding. In: Proceedings of the 2019 Conference of the North American Chapter of the Association for Computational Linguistics: Human Language Technologies, Volume 1 (Long and Short Papers), pp. 4171–4186. Association for Computational Linguistics, Minneapolis, June 2019. https://doi.org/10.18653/v1/N19-1423. https://www.aclweb.org/anthology/N19-1423
6. Dosovitskiy, A., et al.: An image is worth 16 × 16 words: transformers for image recognition at scale. In: International Conference on Learning Representations (2021)

7. Isensee, F., Jaeger, P.F., Kohl, S.A., Petersen, J., Maier-Hein, K.H.: nnU-Net: a self-configuring method for deep learning-based biomedical image segmentation. Nat. Methods **18**(2), 203–211 (2021)
8. Fu, S., et al.: Domain adaptive relational reasoning for 3D multi-organ segmentation. In: Martel, A.L., et al. (eds.) MICCAI 2020. LNCS, vol. 12261, pp. 656–666. Springer, Cham (2020). https://doi.org/10.1007/978-3-030-59710-8_64
9. Gu, Z., et al.: CE-Net: context encoder network for 2D medical image segmentation. IEEE Trans. Med. Imaging **38**(10), 2281–2292 (2019). https://doi.org/10.1109/TMI.2019.2903562
10. Han, K., Xiao, A., Wu, E., Guo, J., Xu, C., Wang, Y.: Transformer in transformer. In: Ranzato, M., Beygelzimer, A., Dauphin, Y., Liang, P., Vaughan, J.W. (eds.) Advances in Neural Information Processing Systems, vol. 34, pp. 15908–15919. Curran Associates, Inc. (2021). https://proceedings.neurips.cc/paper/2021/file/854d9fca60b4bd07f9bb215d59ef5561-Paper.pdf
11. Hatamizadeh, A., et al.: UNETR: transformers for 3D medical image segmentation. In: 2022 IEEE/CVF Winter Conference on Applications of Computer Vision (WACV), pp. 1748–1758 (2022). https://doi.org/10.1109/WACV51458.2022.00181
12. Held, K., Kops, E., Krause, B., Wells, W., Kikinis, R., Muller-Gartner, H.W.: Markov random field segmentation of brain MR images. IEEE Trans. Med. Imaging **16**(6), 878–886 (1997). https://doi.org/10.1109/42.650883
13. Hu, H., Gu, J., Zhang, Z., Dai, J., Wei, Y.: Relation networks for object detection. In: 2018 IEEE/CVF Conference on Computer Vision and Pattern Recognition, pp. 3588–3597 (2018). https://doi.org/10.1109/CVPR.2018.00378
14. Hu, H., Zhang, Z., Xie, Z., Lin, S.: Local relation networks for image recognition. In: 2019 IEEE/CVF International Conference on Computer Vision (ICCV), pp. 3463–3472 (2019). https://doi.org/10.1109/ICCV.2019.00356
15. Huang, H., et al.: UNet 3+: a full-scale connected UNet for medical image segmentation (2020)
16. Jin, Q., Meng, Z., Sun, C., Cui, H., Su, R.: RA-UNet: a hybrid deep attention-aware network to extract liver and tumor in CT scans. Front. Bioeng. Biotechnol. **8**, 1471 (2020)
17. Li, X., Chen, H., Qi, X., Dou, Q., Fu, C.W., Heng, P.A.: H-DenseUNet: hybrid densely connected UNet for liver and tumor segmentation from CT volumes. IEEE Trans. Med. Imaging **37**(12), 2663–2674 (2018). https://doi.org/10.1109/TMI.2018.2845918
18. Liu, Z., et al.: Swin transformer: hierarchical vision transformer using shifted windows. In: 2021 IEEE/CVF International Conference on Computer Vision (ICCV), pp. 9992–10002 (2021). https://doi.org/10.1109/ICCV48922.2021.00986
19. Milletari, F., Navab, N., Ahmadi, S.A.: V-Net: fully convolutional neural networks for volumetric medical image segmentation. In: 2016 Fourth International Conference on 3D Vision (3DV), pp. 565–571 (2016). https://doi.org/10.1109/3DV.2016.79
20. Oktay, O., et al.: Attention U-Net: learning where to look for the pancreas. In: IMIDL Conference (2018)
21. Ronneberger, O., Fischer, P., Brox, T.: U-Net: convolutional networks for biomedical image segmentation. In: Navab, N., Hornegger, J., Wells, W.M., Frangi, A.F. (eds.) MICCAI 2015. LNCS, vol. 9351, pp. 234–241. Springer, Cham (2015). https://doi.org/10.1007/978-3-319-24574-4_28
22. Schlemper, J., et al.: Attention gated networks: learning to leverage salient regions in medical images. Med. Image Anal. **53**, 197–207 (2019). https://doi.org/10.1016/j.media.2019.01.012

23. Touvron, H., Cord, M., Douze, M., Massa, F., Sablayrolles, A., Jegou, H.: Training data-efficient image transformers distillation through attention. In: Meila, M., Zhang, T. (eds.) Proceedings of the 38th International Conference on Machine Learning. Proceedings of Machine Learning Research, vol. 139, pp. 10347–10357. PMLR, 18–24 July 2021. https://proceedings.mlr.press/v139/touvron21a.html

24. Touvron, H., Cord, M., Sablayrolles, A., Synnaeve, G., Jégou, H.: Going deeper with image transformers. In: 2021 IEEE/CVF International Conference on Computer Vision (ICCV), pp. 32–42 (2021). https://doi.org/10.1109/ICCV48922.2021.00010

25. Tsai, A., et al.: A shape-based approach to the segmentation of medical imagery using level sets. IEEE Trans. Med. Imaging **22**(2), 137–154 (2003). https://doi.org/10.1109/TMI.2002.808355

26. Valanarasu, J.M.J., Oza, P., Hacihaliloglu, I., Patel, V.M.: Medical transformer: gated axial-attention for medical image segmentation. In: de Bruijne, M., et al. (eds.) MICCAI 2021. LNCS, vol. 12901, pp. 36–46. Springer, Cham (2021). https://doi.org/10.1007/978-3-030-87193-2_4

27. Vaswani, A., et al.: Attention is all you need. In: Advances in Neural Information Processing Systems, vol. 30. Curran Associates, Inc. (2017)

28. Wang, W., et al.: Pyramid vision transformer: a versatile backbone for dense prediction without convolutions. In: 2021 IEEE/CVF International Conference on Computer Vision (ICCV), pp. 548–558 (2021). https://doi.org/10.1109/ICCV48922.2021.00061

29. Wang, W., Chen, C., Ding, M., Yu, H., Zha, S., Li, J.: TransBTS: multimodal brain tumor segmentation using transformer. In: de Bruijne, M., et al. (eds.) MICCAI 2021. LNCS, vol. 12901, pp. 109–119. Springer, Cham (2021). https://doi.org/10.1007/978-3-030-87193-2_11

30. Wang, X., Girshick, R., Gupta, A., He, K.: Non-local neural networks. In: 2018 IEEE/CVF Conference on Computer Vision and Pattern Recognition, pp. 7794–7803 (2018). https://doi.org/10.1109/CVPR.2018.00813

31. Xiao, X., Lian, S., Luo, Z., Li, S.: Weighted Res-UNet for high-quality retina vessel segmentation. In: 2018 9th International Conference on Information Technology in Medicine and Education (ITME), pp. 327–331 (2018)

32. Xie, Y., Zhang, J., Shen, C., Xia, Y.: CoTr: efficiently bridging CNN and transformer for 3D medical image segmentation. In: de Bruijne, M., et al. (eds.) MICCAI 2021. LNCS, vol. 12903, pp. 171–180. Springer, Cham (2021). https://doi.org/10.1007/978-3-030-87199-4_16

33. Zhang, Y., Liu, H., Hu, Q.: TransFuse: fusing transformers and CNNs for medical image segmentation. In: de Bruijne, M., et al. (eds.) MICCAI 2021. LNCS, vol. 12901, pp. 14–24. Springer, Cham (2021). https://doi.org/10.1007/978-3-030-87193-2_2

34. Zhao, H., Shi, J., Qi, X., Wang, X., Jia, J.: Pyramid scene parsing network. In: 2017 IEEE Conference on Computer Vision and Pattern Recognition (CVPR), pp. 6230–6239 (2017). https://doi.org/10.1109/CVPR.2017.660

35. Zhou, Z., Rahman Siddiquee, M.M., Tajbakhsh, N., Liang, J.: UNet++: a nested U-Net architecture for medical image segmentation. In: Stoyanov, D., et al. (eds.) DLMIA/ML-CDS -2018. LNCS, vol. 11045, pp. 3–11. Springer, Cham (2018). https://doi.org/10.1007/978-3-030-00889-5_1

# Self-attention Capsule Network for Tissue Classification in Case of Challenging Medical Image Statistics

Assaf Hoogi[1](✉) [iD], Brian Wilcox[2] [iD], Yachee Gupta[2] [iD], and Daniel Rubin[2] [iD]

[1] Ariel University, Ariel, Israel
assafh@ariel.ac.il
[2] Stanford University, Stanford, CA, USA

**Abstract.** We propose the first Self-Attention Capsule Network that was designed to deal with unique core challenges of medical imaging, specifically for tissue classification. These challenges are - significant data heterogeneity with statistics variability across imaging domains, insufficient spatial context and local fine-grained details, and limited training data. Moreover, our proposed method solves limitations of the baseline Capsule Networks (CapsNet) such as handling complicated challenging data and limited computational resources. To cope with these challenges, our method is composed of a self-attention module that simplifies the complexity of the input data such that the CapsNet routing mechanism can be efficiently used, while extracting much richer contextual information, compared with CNNs. To demonstrate the strengths of our method, it was extensively evaluated on three diverse medical datasets and three natural benchmarks. The proposed method outperformed other methods we compared with in classification accuracy but also in robustness, within and across different datasets and domains.

## 1 Background

Tissue classification is a very challenging task, mostly because of the significant variability of intra-modality and inter-modality statistics. Recent advances in computer vision highlight the capabilities of deep learning approaches to solve these challenges, achieving state of the art performances in many classification tasks. The main reason for the success of deep learning is the ability of convolutional neural networks to learn hierarchical representation of the input data. However, these methods do not properly model image data with high level of statistics complexity [14], and as a result they require a large amount of annotated data, and can deal only with a limited data diversity and statistics complexity. Contrary to the natural domain, where huge public benchmarks exist, all these challenges are still typical bottlenecks in the medical domain. Proposed solutions as data augmentation or transfer learning do not supply optimal solutions, mostly because of the high level of heterogeneity in medical data.

Hinton's group recognized the CNNs limitations and the need for better approach of object classification. They proposed the Capsule Network (CapsNet) architecture [29]

B. Wilcox and Y. Gupta—Equal Contributors.

to 1) extract much richer information from each image by supplying a vector of features instead of a single scalar, 2) loose much less information by avoiding any max-pooling step and to 3) learn the spatial correlations between image objects - which is not appropriately done by CNNs. These key ideas are the strengths of CapsNet and are critical for the success of much better image analysis. On the other hand, some of the benefits of capsule networks (e.g. robustness, equivariance) over ordinary CNNs, are still being questioned by several works [9, 23]. In addition, CapsNet requires a significant computational resources due to the dynamic routing procedure. To overcome this limitation, CapsNet uses a relatively shallow architecture but this results a significant difficulty in handling challenging noisy data. **This paper presents few key contributions -**

- A new architecture based on **Self-Attention module that lies within the CapsNet layers (SACN)**, utilizing CapsNet strengths while solving CapsNet weaknesses.
- Our method is the **first work** of self attention and CapsNet **to be applied in the medical imaging domain, tackling its unique challenges** such as high statistics heterogeneity and data complexity, and limited labeled data. It was extensively validated on three medical datasets and three natural benchmarks.
- Our method demonstrates a **significant improvement in accuracy and robustness when analyzing complicated, challenging, high-dimensional heterogeneous data**, comparing with other common-used methods. This helps to minimize false detections.
- Our model shows better **performance-computation balance** than the baseline CapsNet. It **works well under the constraint of limited computational resources**, preventing overkill computation - a well-known limitation of CapsNets.

## 2   Related Work

### 2.1   CapsNet Architecture

Recently developed CapsNet presents a new direction in the field of neural networks. The CapsNet architecture contains three types of layers - the convolutional layer, the primary capsule layer and the classification (digit) layer [29]. Capsule networks are powerful because of two key ideas that distinguish them from the traditional CNNs; 1) *squashing*, where output vectors replace output scalar features of CNNs, and 2) *dynamic routing-by-agreement* that replaces the CNNs' max pooling. Squashing means that instead of having individual neuron sent through non-linearities as is done in CNNs, capsule networks have their output squashed as an entire vector. The squashing function enables a richer representation of the probability that an object is present in the input image. This better representation, in addition to the fact that no max-pooling is applied (i.e. minimizing the information loss), enables CapsNet to handle *limited-size dataset and high complexity of image statistics*, as is typical for medical imaging [14]. In addition, the fact that pooling layers are not part of CapsNet, makes the architecture *equivariant* to changes, considering not only the existence of objects within the image (similar to the CNN's invariance characteristic), but also an additional information about the object's pose. Routing-by-agreement means that it is possible to selectively choose which "parent" in the layer above the current "child" capsule is sent to. For each optional parent, the capsule network can increase or decrease the connection strength, enabling the CapsNet to keep spatial correlations between objects with strong connections only [29]. As a result, the routing-by-agreement mechanism is capable of filtering out noisy predictions of capsules by dynamically adjusting their effects.

However, CapsNet is far from being perfect and recent works start questioning its contribution (especially regarding its routing mechanism). Paik et al. [23] conducted a study, showing that routing mechanisms do not always behave as expected and often produce results that are worse than simple algorithms that assign the connection strengths uniformly or randomly. Attempting to solve the weakness of the routing mechanism, Hahn et al. presented a self-routing mechanism for CapsNet to solve the challenge of complicated input [9]. CapsNet has two main additional drawbacks: (i) high computational load because the feature space is large and the dynamic routing is computationally expensive, and (ii) difficulty in analysis of complicated background/significant input noise. The high computational load constrains the optional choice of network depth (only shallow architectures can be used), leading to a limited performance while analyzing complicated challenging data. Several works present solutions for the CapsNet's weaknesses, however, they are able to show only limited improvement over the baseline CapsNet architecture. Diverse Capsule Networks, introduced in [24], supplied a marginal improvement of 0.31% over the baseline CapsNet accuracy. In [38], the authors explored the effect of a variety of CapsNet modifications, ranging from stacking more capsule layers to trying out different parameters such as increasing the number of primary capsules or customizing an activation function. However, the authors were able to show a 2.57% improvement of the performance accuracy for MNIST dataset, but for more complicated data statistics as CIFAR-10 - their results fell short of the baseline CapsNet.

## 2.2 Self-attention Mechanism

The Self-Attention mechanism is a non-local operation that can help the model to focus on more relevant regions inside the image and to gain better performance for classification tasks with fewer data samples [26] or with more complicated/challenging image backgrounds, exactly as happens in the medical domain. Attention mechanism allows models to learn deeper correlations between objects [19] and helps to discover new interesting patterns within the data [15,22]. Additionally, it helps in modeling long-range, multi-level dependencies across different image regions. Larochelle and Hinton [17] proposed using Boltzmann Machines that choose where to look next to find locations of the most informative intra-class objects, even if they are far away in the image. Reichert et al. proposed a hierarchical model to show that certain aspects of attention can be modeled by Deep Boltzmann Machines [28]. Wang et al. also [37] address the specific problem of CNNs processing information too locally by introducing a Self-Attention mechanism, where the output of each activation is modulated by a subset of other activations. This helps the CNN to consider smaller parts of the image if necessary. Jettley et al. proposed an end-to-end-trainable attention module for convolutional neural network (CNN) architectures built for image classification [13]. The authors were able to show that the learned attention maps neatly highlight the regions of interest while suppressing background clutter. Consequently, their method was able to bootstrap standard CNN architectures for the task of image classification in the natural domain. Attention-based models were also proposed for generative models. In [34], the authors introduce a framework to infer the region of interest for generative models of faces. Their framework is able to pass the relevant information only, through the generative model. Recent technique that focuses on generative adversarial models is called SA-GAN [39]. The authors proposed Self-Attention Generative Adversarial Networks (SA-GAN) that achieve state-of the art generative results on the ImageNet dataset. Recent work that deals specifically with medical data can be found in [21]. This work presents an attention mechanism that is incorporated in the U-Net architecture for tissue/organ segmentation.

## 2.3   Self Attention/Capsule Networks for Tissue Classification

Self Attention has been extensively used for tissue classification in the medical imaging domain, including for brain/lung/liver. Sinha and Dohz introduced multi-scale self-attention for medical image segmentation [31]. Tang et al. proposed a spatial context-aware self-attention model for 3D multi-organ segmentation [33]. Ranjbarzadeh et al. improved the brain tumor segmentation accuracy compared with the state-of-the-art models, by introducing a novel Distance-Wise Attention (DWA) mechanism [27]. Fan et al. developed a Multi-scale Attention Net by introducing two blocks of self attention - Position-wise Attention Block and Multi-scale Fusion Attention Block mechanism to capture rich contextual dependencies based on the attention mechanism [7]. They applied their method for liver and for liver tumor segmentation. Kaul et al. introduced Focusnet - an Attention-Based fully convolutional network for medical image segmentation, and they showed its performance for lung lesion segmentation [16].

Recently, Capsule Networks have been also used for tumor segmentation. Several works show the CapsNet performance for brain [1,2,6], liver [25,35] and lung lesion segmentation [20,32].

## 2.4   Capsule Networks with Self Attention Mechanism

A limited number of works that combine self attention and capsule networks has been published recently [5,8,18,30]. AR CapsNet implements a dynamic routing algorithm where routing between capsules is made through an attention module [4]. The paper of Tsai et al. is another example of attentional routing mechanism between low-level and high-level capsules [36]. DA-CapsNet proposes a dual attention mechanism, where the first one is added after the convolution layer, and the second is added after the primary caps [12]. However, there are a few differences between our presented technique and the previous ones. *First*, all of them were applied for the natural domain (i.e. text, image, audio). None of them was designed for the specific challenges of the Medical Imaging domain. *Second*, previous papers incorporate the self attention mechanism at the same location within the network - between the primary capsules - where the dynamic routing is applied. Dynamic routing and self-attention module share somewhat similar concepts. They both enable to assign higher weights to more dominant aspects. However, incorporating both of them in the same architecture location may cause some degree of role duplication. Contrary to previous works, in our approach each of these key ideas contributes to *different* architecture levels. Our self-attention module is incorporated just after the feature extraction layer (before the capsules layer), while the dynamic routing is applied in its original network location - between low and high level capsules. Means, the self-attention module simplifies the input data and helps to choose the most dominant features that are then considered as an input to the capsules and to the dynamic routing procedure. Both dynamic routing and self-attention are still related but on the other hand - in our framework they also better complete each other. *Third*, the components of our self-attention module is completely different from the previous ones.

*These papers demonstrate the strength of incorporating self-attention mechanisms into capsule networks in the natural domain. However, as far as we know no study has developed such a joint model and evaluated its contribution for medical purposes (in spite of the unique challenges of the medical domain).*

## 3   The Proposed Model

Our proposed model is illustrated in Fig. 1.

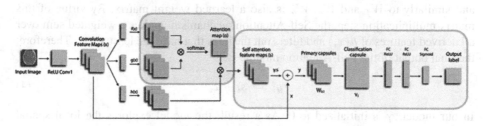

**Fig. 1.** Our proposed SACN architecture. Magenta - self-attention module, Yellow - remaining CapsNet layers (Color figure online)

Let $x \in R^{C \times N}$ be the output feature matrix extracted from the initial convolutional layer of the CapsNet, which is then fed into a Self-Attention module. Let $f(x_1), g(x_2)$ and $h(x)$ be three feature extractors, and $x_1, x_2 \in x$. The features space $x$ contains all input channels $(C)$, while $x_1, x_2$ are subsets of features $(\tilde{C} = C/8)$ within $x$. We chose division by 8 for computational efficiency as we did not see any significant performance decrease, compared with taking all features. $N$ is the number of feature locations from the previous hidden layer. The reduced number of channels $\tilde{C}$ allows us to filter out noisy input channels and care only about the features that are more relevant and dominant for the attention mechanism. $x_1$ and $x_2$ are randomly selected subsets of features that are re-picked every 5 epochs. We found that $C = 512$ supplies the best performance accuracy. $f$ and $g$ are position modules that are used to estimate attention by calculating the correlation between each feature to another. $f(x_i \in x_1)$ and $g(x_j \in x_2)$ represent the $i^{th}$ and $j^{th}$ input feature maps, respectively. Inside the attention module we use 2-D non-strided $1 \times 1$ convolutions and a non-local approach [37]. This helps the CapsNet to model relationships between local regions that are far away from each other and helps to keep the balance between efficiency and long-range dependencies (large receptive fields) by supplying a weighted sum of the features at all image locations. We define the non-local operation as:

$$\eta_{ij}(x) = f(x_i)^T g(x_j) \tag{1}$$

$f(x_i) = W_f x_i$, $g(x_j) = W_g x_j$, where $W_f, W_g \in R^{\tilde{C} \times C}$ are the learned weight matrices. We then compute the softmax of $\eta_{ij}$ to get an output attention map $\alpha_{ij}$,

$$\alpha_{ij} = \frac{exp(\eta_{ij})}{\sum_{i=1}^{N} exp(\eta_{ij})} \tag{2}$$

To obtain the final Self-Attention map, $s \in R^{C \times N}$, which will be the input of the primary CapsNet capsule, we apply matrix multiplication between the attention map $\alpha_{ij}$ and $h(x_i)$,

$$s_j = \sum_{i=1}^{N} \alpha_{ij} h(x_i) \tag{3}$$

where $h(x_i) = W_h x_i$ is the third input feature channel with $C$ channels (see Fig. 1) and similarly to $W_f$ and $W_g$, $W_h$ is also a learned weight matrix. By virtue of this matrix multiplication step, the Self-Attention mechanism applies a weighted sum over all derived features of $h(x_i)$ and filters out the ones that have the least affect. Therefore, the final output of the Self-Attention module is

$$y_i = \gamma s_i + x_i \tag{4}$$

In our model, $\gamma$ is initialized to 0. As a result, the model explores the local spatial information before automatically refining it by the Self-Attention and by considering further regions in the image. Then, the network gradually learns to assign higher weight to the non-local regions. This process allows independence on the user input, contrary to common attention mechanisms.

The final output of the Self-Attention module, $y$, which includes both the self-attention saliency maps and the original feature maps, is then fed into the CapsNet primary layer. Let $v_l$ be the output vector of capsule $l$. The length of the vector, which represents the probability of whether or not a specific object is located in that given location in the image, should be between 0 and 1. To ensure that, we apply a squashing function that keeps the positional information of the object. Short vectors are shrunk to almost 0 length and long vectors are brought to a length slightly below 1. The squashing function is defined as

$$v_l = \frac{|| \sum_k c_{kl} W_{kl} y_k ||^2}{(1 + || \sum_k c_{kl} W_{kl} y_k ||^2)} \frac{\sum_k c_{kl} W_{kl} y_k}{|| \sum_k c_{kl} W_{kl} y_k ||} \tag{5}$$

where $W_{kl}$ is a weight matrix, and $W_{kl} y_k$ is a vector that represents the predicted location of a feature that is part of a lower capsule $k$ in a higher level capsule $l$. $c_{kl}$ are the coupling coefficients between capsule $k$ and all the capsules in the layer above $l$ that are determined by the iterative dynamic routing process

$$c_{kl} = \frac{\exp(b_{kl})}{\sum_n \exp(b_{kn})} \tag{6}$$

$b_{kl}$ are the log prior probabilities that $k^{th}$ capsule should be coupled to $l^{th}$ capsule. $n^{th}$ capsule is a classification capsule that represents a specific class (similar to a digit capsule). To obtain a reconstructed image during training, we use the vector $v_l$ that supplies the coupling coefficient, $c_{kl}$. Then, we feed the correct $v_l$ through two fully connected ReLU layers. The reconstruction loss $L_R(I, \hat{I})$ of the architecture is defined as,

$$L_R(I, \hat{I}) = ||I - \hat{I}||_2^2 \tag{7}$$

where $I$ is the original input image and $\hat{I}$ is the reconstructed image. $L_R(I, \hat{I})$ is used as a regularizer that takes the output of the chosen $v_l$ and learns to reconstruct an image, with the loss function being the sum of squared differences between the outputs of the logistic units and the pixel intensities (L2-Norm). This forces capsules to learn features that are useful for the reconstruction procedure which inherently allows for the model to learn features at near-pixel precision. The reconstruction loss is then added to the following margin loss function, $L_n$,

$$L_n = T_n max(0, \epsilon^+ - ||v_n||)^2 + \lambda(1 - T_n)max(0, ||v_n|| - \epsilon^-)^2 \tag{8}$$

$T_n = 1$ if an instance of class n is present and $\lambda = 0.5$. The margins $\epsilon^+ = 0.9$ and $\epsilon^- = 0.1$ were selected as was suggested in [29]. The end to end SACN architecture is evaluated and its weights are trained by using the total loss function, $L_T$, which is the total of all losses over all classes n,

$$L_T = \sum_n L_n + \xi I_{size} L_R \tag{9}$$

$\xi = 0.0005$ is a regularization factor per channel pixel value - ensuring that the reconstruction loss does not dominate over $L_n$. $I_{size} = H * W * C$ is the number of input values, based on the height, width and number of input channels. Algorithm 1 summarizes the training process of the proposed model.

## 3.1 Implementation Details

Our SACN consists of one convolutional layer with 32 ($5 \times 5$) filter kernels, one primary Capsule layer and one routing iteration. In the Self-Attention module, all convolution layers use $1 \times 1$ kernels and spectral normalization to ensure that gradients are stable during training. Similar to [40], we considered patch-wise CapsNet to improve the localization accuracy because patch-wise approach better reflects local fine-grained details. This is critical in medical imaging, especially when dealing with early subtle changes within the tissue. CapsNet key ideas, which extracts richer information (comparing with CNNs), helps with this task even more. Image patches of $16 \times 16$ pixels

**Data**: $I,G$: Pairs of image $I$ and the ground truth $G$
**Result**: $Y_{out}$: Final instance classification
**while** *not converging* **do**

> **CapsNet Convolutional Layer**: features are extracted and are divided into three output feature vectors $(f(x), g(x), h(x))$.
> **Attention Layer**: Self-Attention map $y_i$ is generated based on the features vectors, attention map $\beta_{ij}$, learned weight matrix $W_h$ and a specific image location $x_i$.
> **Primary and Classification Layers**: the dominant features are then fed into the Primary CapsNet layer and from there to the Classification layer. Output classification $Y_{out}$ is obtained.
> **Calculate the Attention-based CapsNet loss**: $L_T \leftarrow Loss(G, Y_{out})$
> **Back-propagate the loss and compute**: $\frac{\partial L}{\partial W}$
> **Update the weights**: matrices $W$ are updated for both the Self-Attention layer and the CapsNet architecture.

**end**

**Algorithm 1:** Our SACN training process

were extracted, considered as network input. A value of $0.5$ was chosen for the $\lambda$ down-weighting of the loss, together with a weight variance of $0.15$. We chose a batch size of $64$ and a learning rate of $1e^{-3}$. Thirty epochs were sufficient. To ensure the best performance, all hyper-parameters were tuned by using a grid search technique.

## 4   Experiments

### 4.1   Datasets

In this work, we trained and tested our model on three different datasets for the problem of tissue classification. Two datasets that contain images of BrainMR and LungCT were created and annotated at Stanford University. The third dataset is the publicly available Liver Tumor Segmentation dataset (LiTS) [3]. For our purpose, we extracted a random portion of the dataset and created a 2D dataset from the 3D CT volumes. This resulted 1124 annotated liver (CT) cases. Table 1 provides the detailed information such as the number of images, and Global Contrast (GC) Mean and Variance for each dataset. Different values of homogeneity and contrast can substantially affect the performance. Each dataset has its major challenges but the LiverCT is the most difficult one due to high level of heterogeneity of the lesion tissues, while LungCT is the easiest one. The inter- and intra-variability between datasets can be shown in Fig. 2. An external expert annotated two separate regions in each image - normal tissue and cancer lesion. Thirty random patches were extracted from each region, means that each training image supplied 60 random samples to the whole training cohort. A patch that was located on the lesion's boundary was considered as a lesion if more than 70% of its area fell inside the lesion. 80% of each dataset was used for training, 10% percent for validation and 10% for testing.

**Table 1.** Lesion characteristics

| Organ | Modality | Number of images | GC mean | GC variance | GH mean | GH variance | Lesion radius (pxls) |
|---|---|---|---|---|---|---|---|
| Brain | MRI | 369 | 0.056 | 0.029 | 0.907 | 0.003 | $17.42 \pm 9.52$ |
| Lung | CT | 87 | 0.315 | 0.002 | 0.901 | 0.004 | $15.15 \pm 5.77$ |
| Liver (public) | CT | 1124 | 0.289 | 0.011 | 0.819 | 0.005 | $17.48 \pm 8.91$ |

### 4.2  Performance Evaluation

We compared our proposed framework with the *baseline CapsNet* that this work specifically aims to improve, with *Attention-UNet* that was proposed for medical images segmentation (segmentation approach) and with three other methods (classification approach) - *ResNet-50* [10], *DenseNet-40* [11], and the recent *DA-CapsNet* that introduces another framework of self attention module within a CapsNet architecture (for natural domain only) [12]. For the ResNet and DenseNet architectures, we tested their *local and non-local variants*. Patch-wise approach was used. To ensure a fair comparison we chose the optimal set of parameters that supplied the best performance for each method. We evaluated the methods' performance by calculating several statistics such as accuracy, recall, precision and F1 score. Statistical significance between the methods was calculated by using Wilcoxon paired test.

## 5  Results

### 5.1  Qualitative Evaluation

Only for the purpose of *clear visualization*, Fig. 2 presents the classification results of a subset of **randomly** chosen patches. It is clearly seen that **our method deals well with small lesions, highly heterogeneous lesions and low contrast lesions**. Our method can also well distinguish between normal structures within the tissue (e.g. blood vessels in LungCT, normal structures in the BrainMR image) and cancer lesions, despite their substantial similarity. All these challenges, which usually fail common techniques (reflected in lower accuracies in Table 2 for these methods we compared with), are dealt well by our proposed method.

### 5.2  Quantitative Evaluation and Comparison with Other Common Techniques

Table 2 presents the classification accuracy of **all** patches in the testing set, not only those that are colored in Fig. 2. Table 2 shows that our method outperforms all other methods for each dataset that has been analyzed. *First*, the performance **accuracy** within each specific dataset (Liver, Lung, Brain) is consistently higher when using our proposed method. *Second*, the **standard deviation (std)** of the classification accuracy over different images *within* the same dataset is lower when using our proposed method. *Third*, **the robustness and the stability** *across* different datasets are also significantly higher when using our model. Figure 3 presents additional statistics as Precision, Recall

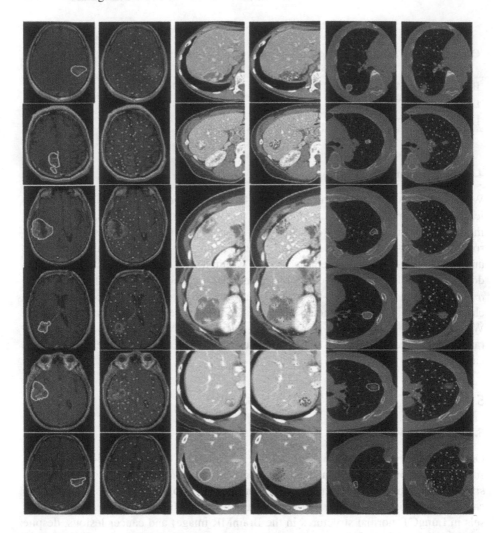

**Fig. 2.** Classification of **RANDOM** patches (only for the purpose of clearer visualization). Each pair of images (in the same row) represents the original image with radiologist's lesion annotation (cyan) and the processed image. Red - classified lesion patches, Green - classified normal patches. 2 left columns - BrainMR, 2 middle columns - LiverCT, 2 right columns - LungCT. (Color figure online)

and F1-score, indicating that our method is able to decrease false classifications (i.e. negative and positive). Our method shows average improvements of 3.17%, 7.75%, 6.25%, and 8.25% for the Accuracy, Precision, Recall and F1 scores - comparing with the DA-CapsNet, baseline CapsNet, and with the non-local versions of ResNet-50 and DenseNet-40. Together with accuracy, these statistics prove the superiority of our method over other methods within a specific dataset, and its robustness across datasets.

For all comparisons that appear in Table 2 and in Fig. 3, Wilcoxon paired test showed **a significant improvement of the tissue classification by using our method, over other methods we compared with** ($p < 0.01$).

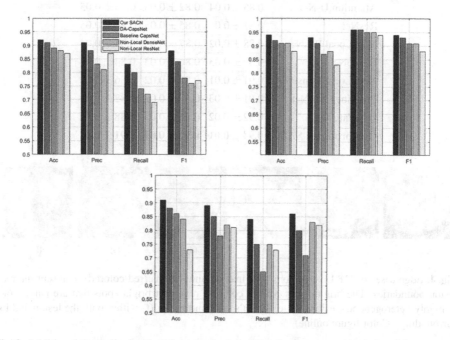

**Fig. 3.** Additional Statistics for the 4 best methods in Table 2 - Accuracy (Acc), Precision (Prec), Recall and F1 score. Upper left - BrainMR, upper right - LungCT. Lower - LiverCT.

Figure 4 shows some examples of edge cases, demonstrating a major challenge of the data - the bright spots. That way, lesions patches can not be properly modeled only by considering the translation invariance of ordinary CNNs. The CapsNet equivariance property has a key contribution. These bright spots that represent heterogeneous lesions/surroundings are the main reason for the larger performance difference (Table 2) and the superiority of our method over the other methods, as none of the others succeeded in classifying these bright regions properly (except for our method).

### 5.3 Ablation Study

**CapsNet Contribution.** Table 2 and Fig. 3 clearly demonstrate that when the image statistics are not too complicated (e.g. LungCT and BrainMR datasets) - the baseline CapsNet outperforms the other vanilla competitors (ResNet and DenseNet), thanks to its unique properties such as equivariance. This property, and the fact that CapsNet extracts richer information from each data sample (using squashing and excluding pooling layers), lead to its superiority. However, for the liver dataset, the classification accuracy obtained by the *baseline* CapsNet drops because of the high heterogeneity of the

**Table 2.** Classification accuracy (mean, std) was calculated over <u>**ALL**</u> test patches (Wilcoxon paired test, $p < 0.01$).

| Dataset | LungCT | BrainMR | LiverCT |
|---|---|---|---|
| Attention U-Net | $0.85 \pm 0.04$ | $0.82 \pm 0.03$ | $0.73 \pm 0.05$ |
| ResNet | $0.87 \pm 0.04$ | $0.85 \pm 0.03$ | $0.83 \pm 0.05$ |
| Non-Local ResNet | $0.88 \pm 0.02$ | $0.87 \pm 0.03$ | $0.84 \pm 0.04$ |
| DenseNet | $0.89 \pm 0.03$ | $0.86 \pm 0.03$ | $0.84 \pm 0.04$ |
| Non-Local DenseNet | $0.91 \pm 0.01$ | $0.88 \pm 0.02$ | $0.86 \pm 0.02$ |
| Baseline CapsNet | $0.91 \pm 0.03$ | $0.89 \pm 0.02$ | $0.82 \pm 0.05$ |
| DA-CapsNet | $0.92 \pm 0.02$ | $0.91 \pm 0.01$ | $0.88 \pm 0.02$ |
| **Our proposed SACN** | $\mathbf{0.94 \pm 0.01}$ | $\mathbf{0.92 \pm 0.01}$ | $\mathbf{0.91 \pm 0.02}$ |

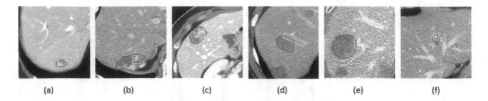

(a)          (b)          (c)          (d)          (e)          (f)

**Fig. 4.** Edge cases of CT Liver lesions. The manual annotations (red colored) represent the real lesion boundaries. The bright spots (yellow colored) represent bright spots that are part of (a–c) highly heterogeneous lesions, (d–e) lesion's surrounding, or (f) within both the lesion and its surrounding. (Color figure online)

patches. As was already shown in the literature - *baseline* CapsNet usually struggles with analysis of challenging data statistics.

When comparing the self-attention/non-local versions of these architectures, our proposed SACN is consistently better than the non-local ResNet/DenseNet. Actually, the superiority of our method becomes even more significant when dealing with complicated and challenging data statistics (LiverCT).

Figure 5 also demonstrates the data complexity with relation to the training/validation losses. It can be clearly seen that for the LiverCT the learning procedure, which is represented by the loss convergence, is noisy and less stable, while for the LungCT the loss curve is much smoother and stable along all epochs. In addition, we can see that the loss of the training and the validation sets are comparable, without a substantial differences between them, thus no over-fitting exists.

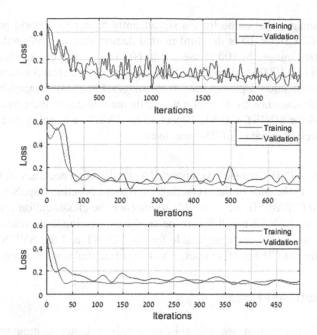

**Fig. 5.** Calculated loss for training and for validation sets. Upper - LiverCT, middle - BrainMR, bottom - Lung CT.

**Self-attention Contribution.** When comparing each vanilla architecture with its non-local variant, the non-local versions always outperform the vanilla ones. For the Cap-sNet architecture, our SACN improves the baseline CapsNet by 5% in average, and the non-local ResNet and non-local DenseNet architectures improve their vanilla versions by 1.67% in average.

**Our Joint Self-attention CapsNet Contribution.** Table 2 and Fig. 3 show that **only by combining** the 1) unique CapsNet capabilities and the 2) self-attention key-idea, we can supply the optimal solution (comparing with all local and non-local versions). None of these aspects alone is able to supply the best performance, robustly enough across datasets and for different complexity levels of data statistics.

### 5.4  Generalizability Across Domains

In order to explore the generalization of the proposed technique to other domains, except for the medical one, we preliminary analyzed several natural benchmarks as well.

**Datasets.** MNIST, SVHN and CIFAR-10 were chosen. The MNIST database includes a training set of 60,000 hand-written digits examples, and a test set of 10,000 examples. The Street View House Numbers (SVHN) is a real-world image dataset. It contains

600,000 digit images that come from a significantly harder real world problem compared to MNIST. CIFAR-10 is the third natural dataset that we analyzed. The dataset consists of 60000 images in 100 classes, with 600 images per class. There are 50000 training images and 10000 test images. On contrary to the medical datasets, we used an image-wise classification approach for natural images. Image-wise analysis was applied because the task was classification of a whole image into a specific class. We chose a batch size of 64 for MNIST, CIFAR-10 and 32 for SVHN, a learning rate of $2e^{-4}$ and 60 epochs. A weight variance of 0.01 was used.

**Results.** For the MNIST dataset, we obtained a classification accuracy of 0.995, which was comparable to the state of the art methods and to the baseline CapsNet architecture. For SVHN, and CIFAR-10, we were able to improve the classification accuracy of the baseline CapsNet, by 3.4% and 3.5%, respectively. Comparing the results with the DA-CapsNet [12], our results are comparable for the MNIST and the SVHN datasets but are 1% better for the CIFAR-10 dataset, which is substantial improvement.

## 6   Computational Load

One of our claims is that we are able to supply a better computational load to performance trade-off. To prove that, we compared our proposed architecture with the baseline Capsule Network and with a deeper Capsule Network that consists of 5-convolutional layers, without Self-Attention module. While our architecture increased the running time by around 10% (but significantly improved the overall performance), compared with the baseline Capsule Network, the deeper network led to two undesired observations; 1) It shutdown around 33% of the processes due to overkill computational load, 2) it increased the running time of the remaining 67% of the processes by around 30%. On contrary, our method is less computational expensive, much more time efficient and never shutdown a process (a well-known CapsNet limitation).

## 7   Discussion and Conclusion

Our Self-Attention Capsule Network framework was proposed to improve classification performance, tackling core challenges of medical imaging such as a wide range of data statistics, significant intra- and inter- domain heterogeneity, limited training and computational resources. Our method was able to extract richer fine-grained spatial context and to prevent overkill computations. **The proposed method outperformed all other methods that we compared with (segmentation or classification approaches), and for all complexity levels of data statistics. Moreover, its superiority becomes even more significant when the image statistics are more complicated and challenging.**

As far as we know, **this is the first work of self attention Capsule networks for medical imaging.** Our work shows an extensive across-domain analysis; We tested our method on medical images but also on some extent of natural images (3 separate datasets each) to demonstrate its robustness and strengths across domains. Our results demonstrate a significant improvement of complex data analysis - better than the limited improvement that other works in literature showed. The proposed method shows significant performance superiority for all data statistics, different datasets and other state of the art methods (Wilcoxon test, $p < 0.01$). This shows the **high generalization, robustness and classification capabilities of our method, proving its remarkable strengths**.

**Acknowledgements.** This work was supported in part by grants from the National Cancer Institute, National Institutes of Health, U01CA142555 and 1U01CA190214.

# References

1. Aminian, M., Khotanlou, H.: Capsnet-based brain tumor segmentation in multimodal MRI images using inhomogeneous voxels in del vector domain. Multimedia Tools Appl. **81**(13), 17793–17815 (2022)
2. Aziz, M.J., et al.: Accurate automatic glioma segmentation in brain MRI images based on capsnet. In: 2021 43rd Annual International Conference of the IEEE Engineering in Medicine and Biology Society (EMBC), pp. 3882–3885. IEEE (2021)
3. Bilic, P., et al.: The liver tumor segmentation benchmark (lits). CoRR abs/1901.04056 (2019). http://arxiv.org/abs/1901.04056
4. Choi, J., Seo, H., Im, S., Kang, M.: Attention routing between capsules. In: Proceedings of the IEEE/CVF International Conference on Computer Vision Workshops (2019)
5. Duarte, K., et al.: Routing with self-attention for multimodal capsule networks (2021)
6. Elmezain, M., Mahmoud, A., Mosa, D.T., Said, W.: Brain tumor segmentation using deep capsule network and latent-dynamic conditional random fields. J. Imaging **8**(7), 190 (2022)
7. Fan, T., Wang, G., Li, Y., Wang, H.: Ma-net: a multi-scale attention network for liver and tumor segmentation. IEEE Access **8**, 179656–179665 (2020)
8. Fu, H., Wang, H., Yang, J.: Video summarization with a dual attention capsule network. In: 2020 25th International Conference on Pattern Recognition (ICPR), pp. 446–451 (2021). https://doi.org/10.1109/ICPR48806.2021.9412057
9. Hahn, T., Pyeon, M., Kim, G.: Self-routing capsule networks. In: NeurIPS (2019)
10. He, K., Zhang, X., Ren, S., Sun, J.: Deep residual learning for image recognition. In: 2016 IEEE Conference on Computer Vision and Pattern Recognition (CVPR) (2016)
11. Huang, G., Liu, Z., Weinberger, K.Q.: Densely connected convolutional networks. CoRR abs/1608.06993 (2016). http://arxiv.org/abs/1608.06993
12. Huang, W., Zhou, F.: Da-capsnet: dual attention mechanism capsule network. Sci. Rep. **10**(1), 1–13 (2020)
13. Jetley, S., Lord, N.A., Lee, N., Torr, P.H.S.: Learn to pay attention. CoRR abs/1804.02391 (2018). http://arxiv.org/abs/1804.02391
14. Jiménez-Sánchez, A., Albarqouni, S., Mateus, D.: Capsule networks against medical imaging data challenges. In: CVII-STENT/LABELS@MICCAI (2018)
15. Jo, Y., et al.: Quantitative phase imaging and artificial intelligence: a review. IEEE J. Sel. Top. Quant. Electron. **25**, 1–14 (2019)
16. Kaul, C., Manandhar, S., Pears, N.: Focusnet: an attention-based fully convolutional network for medical image segmentation. In: 2019 IEEE 16th International Symposium on Biomedical Imaging (ISBI 2019), pp. 455–458 (2019). https://doi.org/10.1109/ISBI.2019.8759477
17. Larochelle, H., Hinton, G.E.: Learning to combine foveal glimpses with a third-order Boltzmann machine. In: Advances in Neural Information Processing Systems, vol. 23. pp. 1243–1251 (2010)

18. Mazzia, V., Salvetti, F., Chiaberge, M.: Efficient-capsnet: capsule network with self-attention routing. Sci. Rep. **11**(1), 1–13 (2021)
19. Mnih, V., Heess, N., Graves, A., Kavukcuoglu, K.: Recurrent models of visual attention. In: Advances in Neural Information Processing Systems, vol. 3 (2014)
20. Mobiny, A., Van Nguyen, H.: Fast CapsNet for lung cancer screening. In: Frangi, A.F., Schnabel, J.A., Davatzikos, C., Alberola-López, C., Fichtinger, G. (eds.) MICCAI 2018. LNCS, vol. 11071, pp. 741–749. Springer, Cham (2018). https://doi.org/10.1007/978-3-030-00934-2_82
21. Oktay, O., et al.: Attention u-net: learning where to look for the pancreas. arXiv preprint arXiv:1804.03999 (2018)
22. Olshausen, B.A., Anderson, C.H., Essen, D.C.V.: A neurobiological model of visual attention and invariant pattern recognition based on dynamic routing of information. J. Neurosci. **13**, 4700–4719 (1993)
23. Paik, I., Kwak, T., Kim, I.: Capsule networks need an improved routing algorithm. In: Lee, W.S., Suzuki, T. (eds.) Proceedings of the Eleventh Asian Conference on Machine Learning. Proceedings of Machine Learning Research, vol. 101, pp. 489–502. PMLR, Nagoya, Japan, 17–19 November 2019
24. Phaye, S.S.R., Sikka, A., Dhall, A., Bathula, D.: Dense and diverse capsule networks: making the capsules learn better (2018)
25. Pino, C., Vecchio, G., Fronda, M., Calandri, M., Aldinucci, M., Spampinato, C.: Twinlivernet: predicting TACE treatment outcome from CT scans for hepatocellular carcinoma using deep capsule networks. In: 2021 43rd Annual International Conference of the IEEE Engineering in Medicine and Biology Society (EMBC), pp. 3039–3043. IEEE (2021)
26. Qian, W., Jiaxing, Z., Sen, S., Zheng, Z.: Attentional neural network: feature selection using cognitive feedback. In: Advances in Neural Information Processing Systems, vol. 27. pp. 2033–2041 (2014)
27. Ranjbarzadeh, R., Bagherian Kasgari, A., Jafarzadeh Ghoushchi, S., Anari, S., Naseri, M., Bendechache, M.: Brain tumor segmentation based on deep learning and an attention mechanism using MRI multi-modalities brain images. Sci. Rep. **11**(1), 1–17 (2021)
28. Reichert, D.P., Seriès, P., Storkey, A.J.: A hierarchical generative model of recurrent object-based attention in the visual cortex. In: ICANN (2011)
29. Sabour, S., Frosst, N., Hinton, G.E.: Dynamic routing between capsules. arXiv preprint arXiv:1710.09829 (2017)
30. Shang, Y., Xu, N., Jin, Z., Yao, X.: Capsule network based on self-attention mechanism. In: 2021 13th International Conference on Wireless Communications and Signal Processing (WCSP), pp. 1–4. IEEE (2021)
31. Sinha, A., Dolz, J.: Multi-scale self-guided attention for medical image segmentation. IEEE J. Biomed. Health Inform. **25**(1), 121–130 (2021). https://doi.org/10.1109/JBHI.2020.2986926
32. Survarachakan, S., Johansen, J.S., Pedersen, M.A., Amani, M., Lindseth, F.: Capsule nets for complex medical image segmentation tasks. In: CVCS (2020)
33. Tang, H., et al.: Spatial context-aware self-attention model for multi-organ segmentation. In: Proceedings of the IEEE/CVF Winter Conference on Applications of Computer Vision, pp. 939–949 (2021)
34. Tang, Y., Srivastava, N., Salakhutdinov, R.: Learning generative models with visual attention. arXiv preprint arXiv:1312.6110 (2013)
35. Tran, M., Ly, L., Hua, B.S., Le, N.: SS-3DCapsNet: self-supervised 3D capsule networks for medical segmentation on less labeled data. In: 2022 IEEE 19th International Symposium on Biomedical Imaging (ISBI), pp. 1–5. IEEE (2022)
36. Tsai, Y.H., Srivastava, N., Goh, H., Salakhutdinov, R.: Capsules with inverted dot-product attention routing. CoRR abs/2002.04764 (2020). https://arxiv.org/abs/2002.04764

37. Wang, X., Girshick, R., Gupta, A., He, K.: Non-local neural networks. arXiv preprint arXiv:1711.07971 (2017)
38. Xi, E., Bing, S., Jin, Y.: Capsule network performance on complex data. CoRR abs/1712.03480 (2017)
39. Zhang, H., Goodfellow, I., Metaxas, D., Odena, A.: Self-attention generative adversarial networks. arXiv preprint arXiv:1805.08318 (2018)
40. Özbulak, G.: Image colorization by capsule networks (2019)

# ReLaX: Retinal Layer Attribution for Guided Explanations of Automated Optical Coherence Tomography Classification

Evan Wen[1]([⊠])([iD]), ReBecca Sorenson[2], and Max Ehrlich[3,4]

[1] The Pingry School, Basking Ridge, NJ, USA
ewen2023@pingry.org
[2] Department of Optometry, Chinle Comprehensive Health Care Facility,
Chinle, AZ, USA
[3] Department of Computer Science, University of Maryland,
College Park, MD, USA
[4] NVIDIA, Santa Clara, CA, USA

**Abstract.** 30 million Optical Coherence Tomography (OCT) imaging tests are issued annually to diagnose various retinal diseases, but accurate diagnosis of OCT scans requires trained eye care professionals who are still prone to making errors. With better systems for diagnosis, many cases of vision loss caused by retinal disease could be entirely avoided. In this work, we present ReLaX, a novel deep learning framework for explainable, accurate classification of retinal pathologies which achieves state-of-the-art accuracy. Furthermore, we emphasize producing both qualitative and quantitative explanations of the model's decisions. While previous works use pixel-level attribution methods for generating model explanations, our work uses a novel retinal layer attribution method for producing rich qualitative and quantitative model explanations. ReLaX determines the importance of each retinal layer by combining heatmaps with an OCT segmentation model. Our work is the first to produce detailed quantitative explanations of a model's predictions in this way. The combination of accuracy and interpretability can be clinically applied for accessible, high-quality patient care.

**Keywords:** Computer-aided detection and diagnosis · Optical imaging/OCT/DOT · Explainability · Neural network · Visualization

## 1 Introduction

Every year there are approximately 30 million Optical Coherence Tomography (OCT) procedures done worldwide [7]. OCT is a non-invasive imaging test that yields cross-sectional slices of a patient's retina [8] which can be used to diagnose a multitude of retinal diseases. It is estimated that up to 11 million people in the United States have some form of macular degeneration [1]. The number of retinal disease patients continues to increase making the need for accurate

L. Karlinsky et al. (Eds.): ECCV 2022 Workshops, LNCS 13803, pp. 236–251, 2023.
https://doi.org/10.1007/978-3-031-25066-8_11

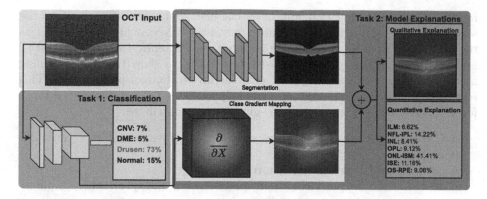

**Fig. 1.** A diagram of ReLaX, a framework for generating accurate, interpretable OCT scan classifications.

diagnosis ever so more important. Accurate OCT diagnosis has traditionally been done by eye care professionals trained to interpret these scans. However, the surplus of patients has been met with a shortage of eye care professionals leading to occurrences of potentially unnecessary cases of vision loss [6].

In order to facilitate a faster diagnosis, many works have used deep learning to analyze retinal imaging. Used correctly, machine learning has the potential to reduce the load on ophthalmic clinics. A common flaw of these models is the lack of interpretability: they are black-boxes. To improve retinal disease patient care, machine learning models must be accurate but also more interpretable. A standalone diagnosis can leave patients skeptical of the model's validity. Another potential benefit of interpretability is when a model's interpretation is statistically suspicious, it can be flagged for review by a doctor for an extra level of validation, this requires **quantitative** as well as **qualitative** explanations. Previous works [10] have attempted to bring a degree of interpretability through attribution based methods such as occlusion testing. However, these approaches are insufficient for a clinical setting. These methods can provide a general idea of the model's interpretation of an image but do not obtain the same meticulousness as a human eye care professional.

ReLaX aims to provide more clear explanations through the use of retinal layer attribution. We present an accurate and interpretable deep learning framework for classifying OCT scans. Retinal layer attribution uses a unique combination of Gradient Weighted Class Activation Mapping (GradCAM) [20] and semantic segmentation to obtain detailed breakdowns of exactly which retinal layers the classification model uses when making its decisions. To our knowledge, we are the first to incorporate the use of segmentation maps for explaining the predictions of a classifier. This allows us to understand the specific behaviors the model learns and analyze why the model makes incorrect decisions. Our method produces qualitative explanations, in the form of heatmaps with highlighted retinal layers and quantitative explanations indicating the percent of model focus on each retinal layer. ReLaX can also go beyond providing explanations by aiding clinical diagnoses through error analysis. Furthermore, our method is completely model agnostic meaning that any CNN classifier and segmentation model are acceptable. The algorithm isn't limited to strictly CNN architectures either as the GradCAM heatmaps can be substituted with attention maps from a ViT architecture [5]. With further analysis, we are able to show that deep learning models interpret medical images in a similar way to human professionals. This proposed approach is evaluated on a publicly available OCT data set. See Fig. 1 for an overview of our algorithm. In summary, we contribute:

- A CNN architecture that achieves state-of-the-art results in OCT classification.
- A novel **retinal layer attribution** concept
- A novel method for producing both **quantitative** and **qualitative** explanations of the machine generated diagnosis.

## 2    Literature Review

### 2.1    OCT Classification

Due to the shortage of experienced eye care professionals and subjectivity in OCT classification, researchers have attempted to apply machine learning algorithms, most notably various types of Convolutional Neural Networks (CNN) [12]. Fauw et al. [3] used a 3D U-Net for segmentation combined with a 3D CNN to classify Normal, Choroidal Neovascularization (CNV), Macular Retinal Edema, Full Macular Hole, Partial Macular Hole, Central Serous Retinopathy (CSR), and Geographic Atrophy. Their models reached or exceeded the performance of human experts. Lu et al. [14] used ResNet-101 to classify Cystoid Macular Edema, Epiretinal Membrane, Macular Hole, and Serous Macular Detachment, their model outperformed two physicians. Yoo et al. [27] used Inception-V3 combined with CycleGAN for data augmentation to classify rarer retinal diseases. Nagasato et al. [15] used a deep convolutional neural network to detect a nonperfusion area in OCTA images. Wang et al. [26] used CliqueNet to classify Age-related Macular Edema (AMD), DME, and Normal. Li et al. [13] used ResNet50 to classify CNV, Diabetic Macular Edema (DME), Drusen, and Normal. Tsuji et al. [25] used a capsule network to classify the same set of diseases. Also on the same data set were the works of Kumar et al. [16] and Asif et al. [2]. Kumar et al. [16] utilized the concept of attention in addition to a deep CNN for OCT classification. Aisf et al. [2] used an architecture similar to ResNet50 with a transfer learning approach to classify the scans. Saleh et al. [19] used an Inception V3 Net for the same task. Of the works listed so far, many have already achieved high enough accuracy for clinical implementation. Most notably Tsuji et al. [25] achieved a classification accuracy of 0.996. The primary drawback of these deep learning solutions is that they operate within a "black box" making it difficult to understand the model's decision-making process. Our work contributes a new CNN architecture which achieves state-of-the-art performance for OCT classification. Furthermore, we achieve a new level of detail in producing model explanations. Previous works were limited to qualitative explanations in heatmaps. However, heatmaps have little to no use when one cannot understand the specific parts of the retina the model is looking at. Due to this constraint, we develop a segmentation-based algorithm for generating quantitative explanations. Our algorithm not only provides novel quantitative explanations but also richer, more in-depth qualitative explanations.

## 2.2 CNN Visualization

Various approaches have been used to visualize the behaviours of CNNs. In Zeiler et al., a deconvolutional network was used to map network activations to the input pixel space and show the input patterns learned by the CNN [28]. Simonyan et al. visualized partial derivatives from predicted class scores and generated saliency maps [21]. Springenberg et al. used a deconvolution approach similar to Zeiler et al. for feature visualization of a CNN [22]. Zhou et al. introduced class activation mapping (CAM), a technique for generating localization maps using the global average pooled convolutional feature maps in the last convolutional layer [29]. Selvaraju et al. generalized the CAM algorithm by making GradCAM, a technique that does not require a specific CNN architecture to develop explanations. CAM, however, requires a CNN with global average pooling followed by a single fully connected layer that outputs the prediction [20]. GradCAM first computes the gradient of the targeted output class with respect to the final convolutional layer's feature map activations. Next, it performs pooling on the gradients to obtain neuron importance weights before performing a weighted combination of the feature maps to compute the GradCAM heatmap. These methods are considered forms of **attribution** as they highlight areas of the input that contribute most to the classification.

## 2.3 Attempts at Explainable OCT Classification

Of the works listed in Sect. 2.1, some have used attribution-based methods to generate heatmaps from model predictions. Tsuji et al. [25] generated heatmaps using an algorithm inspired by CAM [29]. In their work, expert ophthalmologists assessed the heatmap images and confirmed that the activated regions were the correct regions of focus and thus the model was trained accurately. They also performed some error analysis by analyzing heatmaps from their incorrect predictions. However, they note that in wrong predictions the heatmaps still show that the model looked in the correct area. Similarly, Kermany et al. [10] used occlusion testing to generate heatmaps which they also confirmed with human experts. One issue with this form of explanations is that the explanations have little variance between correct and incorrect classifications. How can one estimate the certainty of prediction in a clinical setting when the explanations are always similar? Furthermore, the qualitative explanations are not very robust. The human professionals do seem to agree with the models' areas of focus, however, given that the heatmaps are relatively coarse and human classification can be subjective, the qualitative explanations can still be dubious and difficult to trust. Therefore, the objectives of our proposed algorithm are to

- Demonstrate a stronger connection between human professional examinations and machine learning model interpretations
- More precisely pinpoint the areas being focused on by the model
- Perform meaningful error analysis.

## 3  Methodology

Here we describe ReLaX, our algorithm to produce highly accurate, easily understandable diagnoses. Our algorithm depends on a classification model, GradCAM heatmaps, and a segmentation model which are now discussed in more detail. The motivation of our approach is that OCT scans show each of the retinal layers. Instead of thinking about the attribution-based method in the context of the whole image, we can focus on highlighting the importance of the retinal layers rather than of the individual pixels.

### 3.1  Classification Model

For OCT image classification we build a CNN architecture based off an EfficientNetB2 backbone [24]. We choose the EfficientNet model because of its robust performance on the ImageNet-1k data set [4]. Since EfficientNet is designed for ImageNet classification, it outputs 1000 logits in the final layer. Therefore, we add two fully connected layers of sizes 100 and 4 with a softmax activation on the final layer to perform classification on four retinal diseases.

The full model is trained end-to-end with the Adam [11] Optimizer with a learning rate of 0.001 with the categorical cross-entropy loss function. We train the model for a total of 20 epochs. We use four methods for evaluating our models: accuracy, precision, recall, and F1 score.

### 3.2  Gradient-Weighted Class Activation Mapping

We use Gradient-weighted Class Activation Mapping (GradCAM) [20] to produce heatmaps highlighting the regions the model utilized to make a prediction. The GradCAM algorithm utilizes the inputs to the model, the output of the final convolutional layer, and the output of the model prior to the softmax activation function. We overlay heatmaps on the original images to compare the model's focus during classification of each disease. We also analyze why the model makes incorrect classifications by looking at its resulting heatmap. Standalone GradCAM heatmaps can give a level of interpretability similar to previous works: a visual sense of the regions of focus in the image.

### 3.3  Segmentation Model

To fully implement the idea of retinal layer attribution, we employ a U-Net [17] architecture, pretrained for retinal layer segmentation [18]. Standalone GradCAM heatmaps give a general idea of where the model looked at, but by localizing the heatmap to specific retinal layers, we obtain both a clearer qualitative explanation of the model as well as a thorough quantitative explanation. The U-Net is trained to detect nine retinal layers: region above retina (RaR), inner limiting membrane (ILM), nerve fiber ending to inner plexiform layer (NFL-IPL), inner nuclear layer (INL), outer plexiform layer (OPL), outer nuclear layer to inner segment myeloid (ONL-ISM), inner segment ellipsoid (ISM), outer segment to retinal pigment epithelium (OS-RPE), and region below retina (RbR). The U-Net model is trained on DME scans from the data set presented in Srinivasan et al. [23] for 20 epochs.

### 3.4   Calculating Retinal Layer Attribution

The full proposed ReLaX method consists of applying the segmentation model on our OCT data set before overlaying the GradCAM heatmaps on the segmentation maps. This allows us to obtain a visual breakdown of the model's focus based on the retinal layers. We also calculate percentages of the model's focus on each of the retinal regions excluding the regions above and below the retina. We choose to exclude these regions because they generally are less important for the model, but they have high enough area to potentially alter the percentages of the model's focus. For each of the four retinal disease classifications, we obtain an average retinal layer focus percentage for the seven retinal layers, denoted in Eq. (1), during correct OCT classifications. $R_i$ denotes the model's focus on layer $i$. $S_{i,r,c}$ is a one-hot encoded integer as to whether the pixel at $r, c$ is of retinal layer $i$, and $H_{r,c}$ is the value of the GradCAM heatmap at pixel $r, c$.

$$R_i = 100 * \frac{\sum_{r,c} S_{i,r,c} H_{r,c}}{\sum_{l=1}^{7} \sum_{r,c} S_{l,r,c} H_{r,c}} \tag{1}$$

## 4   Results

### 4.1   Data Set

The OCT scans for this work come from the Kermany labeled OCT data set [10]. The data set contains four classes: Choroidal Neovascularization (CNV), Diabetic Macular Edema (DME), Drusen, and Normal (healthy). The data set utilizes Spectralis OCT machine scans from the Shiley Eye Institute of the University of California San Diego, the California Retinal Research Foundation, the Medical Center Ophthalmology Associates, the Shanghai First People's Hospital, and the Beijing Tongren Eye Center. All images went through an extensive manual classification process by a cohort of retinal specialists to ensure correct labels.

We test class-weighting due to the imbalance in the data set. Drusen had significantly less scans in the training set whereas CNV had significantly more scans. The class weights are evaluated based on the number of scans in the training set. We take the highest number of scans for any label which was 37,205 for CNV and divide by the number of scans for each of the labels to obtain the class weights of 1, 3.279, 4.318, and 1.414 for CNV, DME, Drusen, and Normal respectively.

**Table 1.** Results of our model measured in accuracy, precision, recall, and F1 score. A comparison with previous works is shown; the bold represents the highest scores for each metric. Results shown as reported in previous works.

| Model | Accuracy | Precision | Recall | F1-Score |
|---|---|---|---|---|
| **Ours** | **0.998** | **0.998** | **0.998** | **0.998** |
| Kermany et al. Inception V3 | 0.961 | 0.96125 | 0.961 | 0.963 |
| Tsuji et al. Capsule Network | 0.996 | 0.996 | 0.996 | 0.998 |
| Asif et al. Residual Network | 0.995 | 0.995 | 0.996 | 0.995 |
| Kumar et al. Deep CNN | 0.956 | – | – | – |
| Saleh et al. Inception V3 | 0.984 | – | – | – |
| Li et al. ResNet-50 | 0.973 | 0.963 | 0.985 | 0.975 |

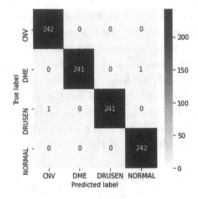

**Fig. 2.** Confusion matrix from our classification model.

## 4.2   Data Preprocessing

The Kermany data set contains a total of 84,484 OCT scans. The data is split into 968 test images and 83,484 training images. Of the 83,484 training images, 37,205 are labeled as CNV, 11,348 are DME, 8616 are Drusen, and 26,315 are Normal. The test data consists of 242 images of each class. Scans are separated by their correct labels into folders containing JPEG image files. This choice of data split was specified by Kermany et al. in the original work and has since been adopted for other recent works. All works in Table 1 are evaluated on the same data split. The dimensions of the original images vary from 384–1536 pixels wide and 496–512 pixels high. All images are rescaled to 260 pixels wide and 260 pixels high using bilinear interpolation. The training data is processed in batches of size 64.

## 4.3   Results of the Classification Models

Table 1 shows the state-of-the-art performance of our classification model. The per-class performance of the model is shown in Fig. 2. As seen in Fig. 2, the model

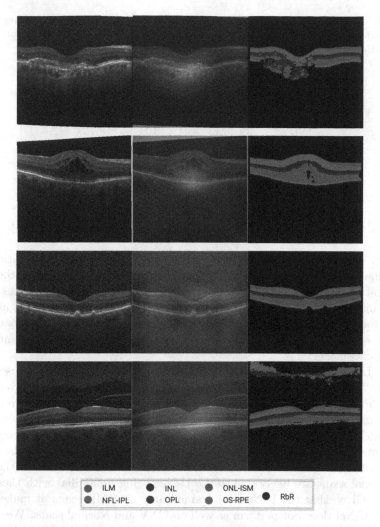

**Fig. 3.** Heatmaps and segmentation maps from the OCT scans. Left: OCT scan, middle: GradCam heatmap, right: segmentation map. The scans are CNV, DME, Drusen, and Normal from top to bottom.

makes only two errors: one mistaking Drusen for CNV and one mistaking DME for Normal. The model performs more accurately than the models presented by Kermany et al. [10], Tsuji et al. [25], and Li et al. [13] who were trained to classify the same four diseases.

## 4.4  Preliminary Qualitative Interpretability Results

Figure 3 shows GradCAM heatmaps and segmented OCT scans of each classification. Heatmaps from the Drusen and Normal scans have less centralized focus,

**Table 2.** The model's mean focus on each of the retinal layers when making correct classifications. Each column provides a breakdown of the model's focus for correct classifications of each disease.

| Layer | CNV | DME | Drusen | Normal |
|---|---|---|---|---|
| ILM | 5.39% | 6.62% | 10.87% | 7.96% |
| NFL-IPL | 14.03% | 14.22% | 23.67% | 20.48% |
| INL | 12.26% | 8.41% | 10.00% | 8.75% |
| OPL | 12.73% | 9.12% | 10.70% | 9.97% |
| ONL-ISM | 35.90% | 41.41% | 27.74% | 30.91% |
| ISE | 9.90% | 11.16% | 9.61% | 13.18% |
| OS-RPE | 9.78% | 9.06% | 7.42% | 8.75% |

but they still emphasize the center of the scan. In the CNV classification, the model has the highest focus on the bottom-most layer of the retina. In the DME classification, the model still focuses on the bottom layer but also focuses on the intraretinal fluid. In the Drusen classification, the model does not have as much central focus. In the Normal classification, the model has more central focus than the Drusen classification. The model looks at the center of the scan to make the Normal classification.

The U-Net model performs well with segmentation on DME, Drusen, and Normal. The model is able to accurately detect each of the retinal layers. The model, however, does have some confusion on the region above the retina on the Normal scan. The scan appears to have some noise in the upper region which is captured in the segmented scan. The CNV scan is the least accurate of the segmented scans. The model has significant difficulty detecting the retinal layers in the irregular subretinal region. The model's difficulty with the CNV and Normal scans are likely due to the U-Net being unfamiliar with these type of scans. Given that the U-Net was trained only on DME scans, it makes sense that the U-Net does not perform as well on CNV and Normal scans. We explain how the inaccuracies in segmentation affect the interpretability algorithm in the next section.

### 4.5  Full ReLaX Interpretability Results

Next, we apply our retinal layer attribution method to produce quantitative explanations. Table 2 shows the $R_i$ (Fig. (1)) values for correct classifications of each type. In all classifications, the ONL-ISM layer is the most focused followed by the NFL-IPL layer. This is likely due to these two regions occupying the most space in the scans. The rest of the layers have some variance between the diseases. CNV has the highest $R_i$ values for INL, OPL, and OS-RPE. DME has the highest $R_i$ for ONL-ISM. Drusen has the highest $R_i$ for ILM and NFL-IPL. Normal has the highest $R_i$ for ISE. In a sense, the interpretability of the model is similar to the reasoning of a human eye care professional. A human

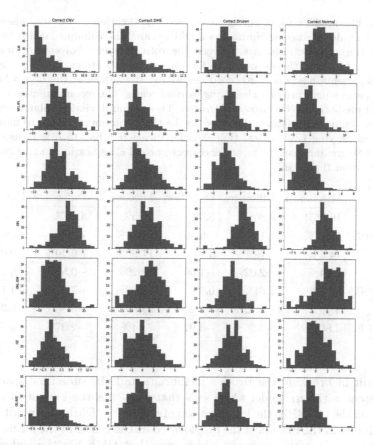

**Fig. 4.** Histograms consisting the numeric deviation from the mean during correct classifications (e.g. a zero value means that the layer involvement for a particular correct classification has the exact same value as the mean layer involvement for that particular classification). Columns: The Pathology Being Classified. Rows: The Retinal Layers

eye care professional looks at certain regions in the scan to make a decision, some regions more important than others. Our model, unlike previous ones, can give the same level of explanation, highlighting which regions are important for the classification. The added interpretability from ReLaX gives newfound quantitative model explanations as well as richer qualitative explanations. The standalone GradCAM heatmaps in Fig. 3 only represent the rough regions where the model focused. ReLaX can name the specific retinal layers of focus and provide a quantitative measure.

## 4.6   Connection to Human Eyecare Professionals

Now we demonstrate that our model is learning the correct areas of the scan to focus on. CNV involves the growth of new blood vessels from the choroid

**Table 3.** The model's focus during its two misclassifications. The first misclassification is the model predicting Normal on a DME scan. The columns labeled Difference represent the numeric difference between the retinal layer involvements during the misclassifications and the mean values for correct classifications. The first deviation column represents how many standard deviations the values in the first column are from the mean when correctly classifying Normal scans. The second misclassification is the model predicting CNV instead of Drusen. The second deviation column represents how many standard deviations the values in the third column are from the mean when correctly classifying CNV scans. (E.g. Misclassification 1 has 16.76% focus on ILM, which is 8.80 greater than the mean for correct normal classification and 5.82 standard deviations from the mean)

| Layer | Misclass. 1 | Difference | Deviation | Misclass. 2 | Difference | Deviation |
|---|---|---|---|---|---|---|
| ILM | 16.76% | 8.80% | 5.82 | 9.95% | 4.56% | 1.74 |
| NFL-IPL | 18.67% | −1.81% | −0.60 | 21.43% | 7.40% | 1.66 |
| INL | 14.54% | 5.79% | 3.01 | 11.96% | −0.30% | −0.07 |
| OPL | 7.94% | −2.02% | −1.66 | 12.20% | −0.53% | −0.14 |
| ONL-ISM | 33.52% | 2.61% | 0.85 | 28.00% | −7.91% | −1.14 |
| ISE | 3.76% | −9.42% | −5.20 | 9.05% | −0.86% | −0.31 |
| OS-RPE | 4.80% | −3.95% | −3.07 | 7.41% | −2.37% | −0.79 |

that result in breaking the Bruch's membrane and the subretinal space. This is consistent with our model which says that CNV has the highest focus of all diseases on the OS-RPE, the lowest region of the retina. DME occurs with excess fluid build up in the macula of the eye, located at the center of the retina. Again, this is consistent with our model which says that DME has the highest focus of all diseases on the ONL-ISM, the layer in the middle of the retina. Drusen is an accumulation of extracellular material which cause elevated RPE. While Drusen has low focus for the ISE and OS-RPE, it has the highest focus on the two regions at the top. This is likely because the model analyzes the peaks of the elevated RPE "humps" which occur closely to the top of the OCT scan. Finally, it makes sense that Normal does not have the most focus on many regions. This is because the model must first rule out the three diseases before making a Normal classification, similar to a human eye care professional.

### 4.7 Error Analysis

When correctly classifying scans, Fig. 4 shows that retinal layer focus largely follow a Gaussian distribution. From a quick visual inspection, a significant majority of the 28 histograms follow a Gaussian distribution. The remaining histograms generally still have peaks close to the mean and thinner ends. This shows that obtaining retinal layer involvement percentages further from the mean (Table 2) implies a lower chance at a correct prediction. The chance of correct prediction is highest when the retinal layer involvements do not significantly deviate from the mean. In addition, Fig. 4 demonstrates that the segmentation errors (Fig. 3) do not significantly impede the quality of explanations. Even though some scans will have segmentation errors, their retinal layer involvements still fall closely to the mean. Using this insight on the distribution of $R_i$ values, we perform error analyses on the model's two misclassifications.

Figure 5 and Table 3 explain the model's incorrect classifications. The Grad-CAM overlayed heatmaps show the model has no specific area of focus in both classifications. The model's focus is nearly evenly distributed across the entire OCT scan instead of a centralized area commonly seen during correct classifications. The deviations from the second column of Table 3 show the ILM, INL, ISE, and OS-RPE are significantly off from the mean $R_i$ for correct Normal classifications. The large deviations show that the model's explanation is too different from the explanation of a correct classification. As shown in Fig. 4, larger deviations directly correlate to lower chances at correct predictions. When a prediction has a statistically suspicious explanation, the scan should be flagged for further review; the model's prediction should not be trusted.

We analyze the probability of correct classification by finding the difference values in the histograms of Fig. 4. An 8.80 difference in ILM involvement for correct normal classification is way off the right end of the histogram. Similarly differences of 5.79, $-9.42$, and $-3.95$ for INL, ISE, and OS-RPE fall very far on the ends of the histogram. This shows the low probability of these classifications being correct.

On the other hand, the other set of deviations from the DME scan mistaken CNV appear to be much closer. This is due to the correct CNV classifications generally having higher variance in their $R_i$. While this example is not as drastic as the other misclassification, there are still some areas of uncertainty. The ILM and NFL-IPL involvements are quite higher than the mean. These difference values correspond to areas of the histogram (Fig. 4 Rows 1 and 2, Column 1) which are much less populated. In a sample size of 242 correct CNV classifications, less than twenty scans have higher ILM deviations. There is a similar story for the NFL-IPL layer involvement. The rest of the layer involvements are relatively close to the mean signaling a somewhat valid interpretation. While the model appears to have a statistically sound quantitative explanation for this classification, the qualitative explanation in the heatmap shows the model is not confident when making this prediction.

## 5   Discussion

Making deep learning suitable for clinical application requires significant progress in explainability. Accurate classification of OCT is a key task for ophthalmic clinics; automation could significantly alleviate the current workload. In previous works for OCT classification, attribution based methods were the best form of model interpretations. However, attribution methods such as occlusion mapping or CAM, fall short of providing the same level of explanation as a human doctor. While deep learning has already shown its potential in classifying OCT, a more in depth form of model explanation could allow full integration of neural networks into the clinical workflow.

Our primary approach to generating greater insight into the classification model is the concept of **retinal layer attribution**. We rethink attribution from a layer-by-layer perspective rather than a pixel-by-pixel perspective. Through our analysis of ReLaX, we find that our classification models have a strong correlation to human judgement. For each disease, the model knows which retinal layer is important for making the correct classification. We verify that human eye care professionals also use the same retinal layers to make diagnoses. This level of interpretability is much stronger than the previous works who simply have human experts verify heatmaps. Having humans verify a heatmap does not fully ensure that the model is behaving correctly. Heatmaps can be very general and humans can be subjective. Our work specifically pinpoints the retinal layers in question and verifies the model's behaviours. Furthermore, ReLaX can also be used for insightful error analysis of the model. Table 3 and Fig. 5 show noticeable indicators that the predictions generated are incorrect. To our knowledge, our work is the first to perform significant error analysis. In a clinical setting, error analysis is very important for maximizing confidence in diagnoses.

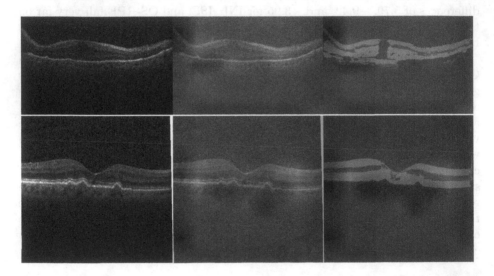

**Fig. 5.** The two misclassifications from our model. Left: OCT scan. Middle: OCT scan with heatmap overlayed, right: Segmentation map with heatmap overlayed. The first row is a DME scan which the model predicted Normal. The second row is a Drusen scan which the model predicted CNV.

We also contribute a state-of-the-art OCT classification model. As shown in Table 1, our model performs more accurately than the previously developed OCT classification models on the Kermany OCT data set. The model we contribute only makes two misclassifications on the entire data set. When paired with a thorough system for error analysis, this model can be clinically implemented with a high degree of confidence. In the cases where the model makes a mistake, it does not go undetected due to the error analysis capabilities of ReLaX.

The main limitations in our work are the lack of publicly-available data for OCT related tasks. We utilize the Kermany OCT data set, but the data set is not sufficiently competitive. While we do contribute a higher accuracy than previous works, several others have developed highly accurate OCT classification models on the data set. The consequence of the noncompetitive data set is that we do not have many incorrect samples to evaluate our proposed ReLaX method. In addition to the lack of available OCT classification data, there is also a lack of ground-truth segmentations for OCT scans. The Kermany data set we utilize does not provide ground truths, thus it is difficult for us to evaluate and improve our segmentation model. While we show that this does not largely impact the efficacy of our algorithm (Fig. 4), accurate segmentation would provide more certainty.

In the future, more publicly-available OCT data sets must be released for the further development of deep learning in ophthalmology. We hope that we can evaluate our approach on a more competitive OCT data set. The Kermany OCT data set has become insufficient for future work as it is possible to achieve nearly perfect accuracy. Furthermore, we wish to improve the accuracy of our segmentation models through employing newer techniques such as the one provided by He et al. [9]. However, our method will require a competitive data set with both ground-truths for OCT classification and OCT segmentation. Another possible direction could be to further strengthen the claims that humans and deep learning models are alike through an eye-tracker study of human professionals analyzing OCT scans.

## 6   Conclusion

In this work, we present ReLaX, a novel retinal layer attribution algorithm for producing rich explanations of OCT classification. We also develop a highly accurate, state-of-the-art model for classifying OCT scans of CNV, DME, Drusen, and Normal. Our classification model performs at an accuracy of 99.8%, higher than all previous works. ReLaX utilizes a novel combination of GradCAM heatmaps and segmented OCT scans to accurately pinpoint the retinal layers important for classification. Our work rethinks attribution-based explanations by focusing on the importance of certain regions in a classification rather than of each pixel. Pixel-based attribution does not accurately demonstrate the important ideas behind a model's decision. Retinal layer attribution can give more-detailed qualitative explanations as well as novel quantitative explanations. ReLaX confirms the similarities between human and computer-aided analysis of retinal imaging. When patients believe that they can get the same level of care from a human and a computer, they are more likely to trust a computer-aided diagnosis. Furthermore, ReLaX gives insightful analytics on the model's likelihood of being correct for a given prediction. Through our investigation, we find that incorrect predictions by the model can often be detected early. In a clinical setting, this prevents false diagnoses which can lead to improper plans of treatment. This new level of interpretability in our model brings deep learning one step closer to real clinical application; patients and doctors can be more confident that they are getting a correct diagnosis. In addition to an accurate diagnosis, patients deserve qualitative and quantitative information as to why the model made such decision rather than a standalone diagnosis. Only then will patients be able to relax.

# References

1. Age-related macular degeneration: facts & figures, January 2020. https://www.brightfocus.org/macular/article/age-related-macular-facts-figures
2. Asif, S., Amjad, K., ul Ain, Q.: Deep residual network for diagnosis of retinal diseases using optical coherence tomography images - interdisciplinary sciences. Comput. Life Sci. SpringerLink (2022). https://link.springer.com/article/10.1007/s12539-022-00533-z#citeas
3. De Fauw, J., et al.: Clinically applicable deep learning for diagnosis and referral in retinal disease, August 2018. https://www.nature.com/articles/s41591-018-0107-6/
4. Deng, J., Dong, W., Socher, R., Li, L.J., Li, K., Fei-Fei, L.: Imagenet: a large-scale hierarchical image database. In: 2009 IEEE Conference on Computer Vision and Pattern Recognition, pp. 248–255. IEEE (2009)
5. Dosovitskiy, A., et al.: An image is worth 16x16 words: transformers for image recognition at scale. CoRR abs/2010.11929 (2020). https://arxiv.org/abs/2010.11929
6. Foot, B., MacEwen, C.: Surveillance of sight loss due to delay in ophthalmic treatment or review: frequency, cause and outcome. Eye (Lond.) 31(5), 771–775 (2017)
7. Fujimoto, J., Swanson, E.: The development, commercialization, and impact of optical coherence tomography. Invest Ophthalmol. Vis. Sci. 57(9), OCT1–OCT13 (2016)
8. Fujimoto, J.G., Pitris, C., Boppart, S.A., Brezinski, M.E.: Optical coherence tomography: an emerging technology for biomedical imaging and optical biopsy. Neoplasia 2(1–2), 9–25 (2000)
9. He, Y., et al.: Fully convolutional boundary regression for retina OCT segmentation. In: Shen, D., et al. (eds.) MICCAI 2019. LNCS, vol. 11764, pp. 120–128. Springer, Cham (2019). https://doi.org/10.1007/978-3-030-32239-7_14
10. Kermany, D.S., et al.: Identifying medical diagnoses and treatable diseases by image-based deep learning. Cell 172(5), 1122–1131 (2018)
11. Kingma, D.P., Ba, J.: Adam: a method for stochastic optimization. CoRR abs/1412.6980 (2015)
12. Krizhevsky, A., Sutskever, I., Hinton, G.E.: Imagenet classification with deep convolutional neural networks. In: Pereira, F., Burges, C.J.C., Bottou, L., Weinberger, K.Q. (eds.) Advances in Neural Information Processing Systems, vol. 25, pp. 1097–1105. Curran Associates, Inc. (2012)
13. Li, F., et al.: Deep learning-based automated detection of retinal diseases using optical coherence tomography images. Biomed. Opt. Express 10(12), 6204–6226 (2019)

14. Lu, W., Tong, Y., Yu, Y., Xing, Y., Chen, C., Shen, Y.: Deep learning-based automated classification of multi-categorical abnormalities from optical coherence tomography images. Transl. Vision Sci. Technol. **7**(6), 41–41 (12 2018). https://doi.org/10.1167/tvst.7.6.41

15. Nagasato, D., et al.: Automated detection of a nonperfusion area caused by retinal vein occlusion in optical coherence tomography angiography images using deep learning. PLoS ONE **14**(11), e0223965 (2019)

16. Puneet, Kumar, R., Gupta, M.: Optical coherence tomography image based eye disease detection using deep convolutional neural network. Health Inf. Sci. Syst. **10**(1), 13 (2022). https://doi.org/10.1007/s13755-022-00182-y

17. Ronneberger, O., Fischer, P., Brox, T.: U-net: convolutional networks for biomedical image segmentation. CoRR abs/1505.04597 (2015). http://arxiv.org/abs/1505.04597

18. Roy, A.G., et al.: Relaynet: retinal layer and fluid segmentation of macular optical coherence tomography using fully convolutional network. CoRR abs/1704.02161 (2017). http://arxiv.org/abs/1704.02161

19. Saleh, N., Abdel Wahed, M., Salaheldin, A.M.: Transfer learning-based platform for detecting multi-classification retinal disorders using optical coherence tomography images. Int. J. Imaging Syst. Technol. **32**(3), 740–752 (2022). https://doi.org/10.1002/ima.22673, https://onlinelibrary.wiley.com/doi/abs/10.1002/ima.22673, _eprint: https://onlinelibrary.wiley.com/doi/pdf/10.1002/ima.22673

20. Selvaraju, R.R., Das, A., Vedantam, R., Cogswell, M., Parikh, D., Batra, D.: Grad-Cam: why did you say that? visual explanations from deep networks via gradient-based localization. CoRR abs/1610.02391 (2016). http://arxiv.org/abs/1610.02391

21. Simonyan, K., Vedaldi, A., Zisserman, A.: Deep inside convolutional networks: visualising image classification models and saliency maps, April 2014. https://arxiv.org/abs/1312.6034

22. Springenberg, J.T., Dosovitskiy, A., Brox, T., Riedmiller, M.: Striving for simplicity: the all convolutional net (2014)

23. Srinivasan, P.P., et al.: Fully automated detection of diabetic macular edema and dry age-related macular degeneration from optical coherence tomography images. Biomed. Opt. Express **5**(10), 3568–3577 (2014). https://doi.org/10.1364/BOE.5.003568, http://www.osapublishing.org/boe/abstract.cfm?URI=boe-5-10-3568

24. Tan, M., Le, Q.V.: Efficientnet: rethinking model scaling for convolutional neural networks. CoRR abs/1905.11946 (2019). http://arxiv.org/abs/1905.11946

25. Tsuji, T., et al.: Classification of optical coherence tomography images using a capsule network. BMC Ophthalmol. **20**(1), 114 (2020)

26. Wang, D., Wang, L.: On oct image classification via deep learning. IEEE Photonics J. **11**(5), 1–14 (2019). https://doi.org/10.1109/JPHOT.2019.2934484

27. Yoo, T.K., Choi, J.Y., Kim, H.K.: Feasibility study to improve deep learning in OCT diagnosis of rare retinal diseases with few-shot classification. Med. Biol. Eng. Comput. **59**(2), 401–415 (2021)

28. Zeiler, M.D., Fergus, R.: Visualizing and understanding convolutional networks. CoRR abs/1311.2901 (2013). http://arxiv.org/abs/1311.2901

29. Zhou, B., Khosla, A., Lapedriza, A., Oliva, A., Torralba, A.: Learning deep features for discriminative localization. In: CVPR (2016)

# Neural Registration and Segmentation of White Matter Tracts in Multi-modal Brain MRI

Noa Barzilay[1], Ilya Nelkenbaum[2], Eli Konen[2], Nahum Kiryati[3],
and Arnaldo Mayer[2](✉)

[1] School of Electrical Engineering, Tel Aviv University, Tel Aviv, Israel
[2] Diagnostic Imaging Department at Sheba Medical Center Affiliated
with the Beverly Sackler School of Medicine, Tel Aviv University, Tel Aviv, Israel
eli.konen@sheba.health.gov.il, arnmayer@gmail.com
[3] Klachky Chair of Image Processing, School of Electrical Engineering,
Tel Aviv University, Tel Aviv, Israel
nk@eng.tau.ac.il

**Abstract.** Pre-surgical mapping of white matter (WM) tracts requires
specific neuroanatomical knowledge and a significant amount of time.
Currently, pre-surgical tractography workflows rely on classical registration tools that prospectively align the multiple brain MRI modalities
required for the task. Brain lesions and patient motion may challenge
the robustness and accuracy of these tool, eventually requiring additional
manual intervention. We present a novel neural workflow for 3-D registration and segmentation of WM tracts in multiple brain MRI sequences.
The method is applied to pairs of T1-weighted (T1w) and directionally encoded color (DEC) maps. Validation is provided on two different
datasets, the Human Connectome Project (HCP) dataset, and a real pre-
surgical dataset. The proposed method outperforms the state-of-the-art
TractSeg and AGYnet algorithms on both datasets, quantitatively and
qualitatively, suggesting its applicability to automatic WM tract mapping in neuro-surgical MRI.

**Keywords:** Convolutional neural networks · Tractography ·
Registration · Multi-modal segmentation · Brain MRI · Neuro-surgical
planning

## 1 Introduction

White matter (WM) tract mapping is of critical importance in neuro-surgical
planning and navigation [6]. The tract reconstruction process, referred to as
tractography, is commonly based on diffusion tensor imaging (DTI) [18]. Tractography requires specific neuro-anatomical knowledge to accurately delineate
fiber seeding ROIs (sROIs) and remove outliers [4]. Until recently, WM tract
mapping had been based on deterministic algorithms, that track the local principal direction of diffusion (PDD), or on probabilistic algorithms, which draw

© The Author(s), under exclusive license to Springer Nature Switzerland AG 2023
L. Karlinsky et al. (Eds.): ECCV 2022 Workshops, LNCS 13803, pp. 252–267, 2023.
https://doi.org/10.1007/978-3-031-25066-8_12

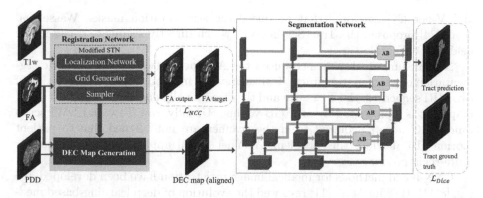

**Fig. 1.** The proposed JTRS framework for joint registration and segmentation of white matter tracts.

the tracts using random walk procedures. Probabilistic tractography or multi-tensor deterministic algorithms [2,14] may better deal with fiber crossings than the single tensor deterministic approach, but require substantial computing time and resources. Tractography is usually computed on diffusion weighted imaging (DWI) acquisitions (diffusion space). Then, the final mapping has to be projected on high resolution anatomical scans (anatomical space), typically T1w. Since the diffusion and anatomical spaces are usually not aligned, accurate spatial registration is typically applied as a pre or post-processing step. This has been accomplished using classical registration algorithms [3,9], often based on mutual information similarity functions [32]. These algorithms have proven robust for neuroscientific research data acquired in optimal conditions. However, in clinical settings they are challenged by patient motion and large brain lesions that appear different than normal brain tissue in diffusion and anatomical images.

In recent years, tractography mapping of streamline trajectories in DWI has been addressed using convolutional neural networks (CNNs) [35,38]. Wasserthal et al. [35] proposed a method that combines three U-Nets to create a tract orientation map, a tract mask, and a start/end region mask for bundle-specific tractography. Zhang et al. [38] introduced DeepWMA, based on demanding whole-brain tractography, for tract classification.

The 3-D curve representation of WM fiber tracts may be useful in neuroscience to sample diffusion signals along tracts. However, neuro-surgeons rather need a volumetric segmentation of the tracts for neurosurgical planning and navigation. Early works on WM tract segmentation adopted various approaches such as markov random field optimization and anatomical priors [5], and 3-D geometric flow [15]. However, the segmentation quality achieved by these techniques is limited. More recently, Pomiecko et al. [26] proposed to use the DeepMedic [16] platform to predict the fiber tract masks. In [36], fiber orientation distribution function (fODF) [30] peaks were passed to a stacked U-net [27] that generated the tract segmentation. In [21], diffusion tensor images were passed to 3-D U-net [7] and V-net [22] networks that produced the segmentation masks. Wasserthal et al. [34] proposed the TractSeg network which directly segments tracts in the field of fODF peaks.

Nelkenbaum et al. [23] introduced the AGYnet network for automatic segmentation of white matter tracts. Unlike previous methods it is based on fusion of two MRI sequences: the T1w scan and the directionally encoded color (DEC) [24] map that is represented by an RGB volume. Usually, raw T1w and DWI scans (and DEC maps) acquired from the scanner are not aligned due to patient motion. To alleviate this problem, AGYnet [4,23] applies a classical tool for a pre-registration.

CNN-based methods for medical image registration have been developed, see e.g. [8,11]. Haskins et al. [11] reviewed the evolution of deep learning-based medical image registration. Fan et al. [8] used hierarchical, dual-supervised learning to estimate the deformation field for 3-D brain registration.

Recently, Jaderberg et al. [12] introduced the spatial transformer network (STN). The STN consists of three main parts: a localization network that computes the transformation parameters, a spatial grid, and a sampling grid. Certain works applied the spatial transformer specifically to the registration task [33].

We introduce an automatic 3-D framework for joint registration and segmentation of WM tracts in multiple brain MRI sequences, referred to as JTRS (Joint Tract Registration and Segmentation). Unaligned anatomical T1w scans and DEC orientation maps are jointly processed to concurrently align the input scans and segment the 3-D tracts.

JTRS avoids separate inter-modality registration by integrating a modified STN module with the segmentation network. By jointly minimizing the registration loss and the segmentation loss, the network will recover the spatial transformation that aligns the scans while segmenting the desired WM tract masks.

JTRS is validated on two different datasets, the Human Connectome Project (HCP) dataset [34], and a real neurosurgical dataset of 75 brain scans of patients referred to brain lesion removal. The clinical dataset is challenging due to significant tissue deformations caused by brain tumors and the common presence of edema. To the best of our knowledge, JTRS is the first 3-D end-to-end neural network for joint registration and segmentation of WM tracts in multiple brain MRI sequences. JTRS is shown to outperform alternative state-of-the-art methods on the challenging neurosurgical dataset.

## 2    Methods

The proposed JTRS, see Fig. 1, combines two main components that are trained simultaneously, one for multi-modal MRI registration and the other for tract segmentation.

JTRS receives as input T1w, PDD (principal direction of diffusion), and FA (fractional anisotropy) 3-D images. The PDD and FA images are not aligned to the T1w scan. However, the FA and PDD images are naturally aligned to each

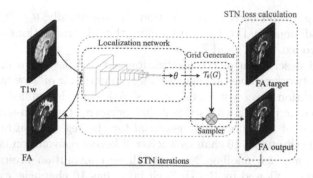

**Fig. 2.** Overview of the modified STN block in the JTRS framework.

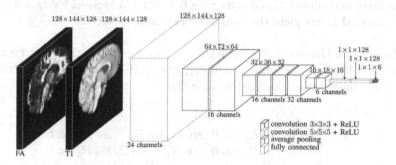

**Fig. 3.** Detailed description of the localization network.

other because they are both derived from the same DWI acquisition. Thus, for simplicity, their registration to the T1w scan can be based on the scalar FA map alone.

In the modified STN block, the parameters of the rigid body transformation between the FA and T1w 3-D images are estimated. The aligned DEC map is generated by resampling, using a transformed sampling grid, of an orientation corrected DEC map. The orientation corrected DEC map is calculated as the product of the input FA image and an orientation corrected PDD image. The orientation correction of the PDD and DEC images adjusts the color to compensate for rotation, see Sect. 2. The T1w image and the aligned DEC map are passed to the segmentation network to produce the tract segmentation. In the following subsections, we focus on the major elements of the JTRS framework.

**Modified STN Block.** Figure 2 shows the modified STN block as implemented in JTRS. It consists of three components: a localization network, a grid generator, and a sampler. Since we need to align MRI scans of the same subject we can limit the algorithm to rigid body transformations.

*Localization Network.* Given two 3-D input images, T1w and FA of height H, width W and depth D, the localization network computes $\theta$, the vector parameterizing a rigid body transformation $\mathcal{T}_\theta$. Specifically, $\theta = [T_x, T_y, T_z, R_x, R_y, R_z]$ is a 6-dimensional vector that specifies the 3-D translations and rotations with respect to axes X, Y, and Z.

A detailed description of the localization network is shown in Fig. 3. The localization network receives as input a concatenated pair of T1w and FA scans. A first convolution layer with 24 $5 \times 5 \times 5$ kernels is applied, followed by ReLU and average pooling with stride 2 that performs the down-sampling while preserving most of the information essential for the registration task [33]. The result proceeds through a 16 channel $3 \times 3 \times 3$ kernel convolution layer followed by ReLU and average pooling. Next are three convolution layers, each with $3 \times 3 \times 3$ kernels followed by ReLU. Each layer has 16 channels, except for the last one, which has 32 channels. The result is forwarded to the last average pooling layer and convolutional layer with 6 $3 \times 3 \times 3$ kernels. Eventually, three fully connected layers yield the final 6-dimensional vector $\theta$.

*Grid Generator.* Given $\theta$ and a 3-D regular sampling grid $G$, the input FA image is resampled by the transformed sampling grid $\mathcal{T}_\theta(G)$ to obtain the output FA image. In homogeneous coordinates, $\mathcal{T}_\theta$ is the $4 \times 4$ matrix

$$\mathcal{T}_\theta = \tag{1}$$

$$\begin{bmatrix} \cos R_z \cos R_y & \cos R_z \sin R_y \sin R_x - \sin R_z \cos R_x \\ \sin R_z \cos R_y & \cos R_z \cos R_x + \sin R_z \sin R_y \sin R_x \\ -\sin R_y & \cos R_y \sin R_x \\ 0 & 0 \end{bmatrix}$$

$$\begin{bmatrix} \sin R_z \sin R_x + \cos R_z \sin R_y \cos R_x & T_x \\ \sin R_z \sin R_y \cos R_x - \cos R_z \sin R_x & T_y \\ \cos R_y \cos R_x & T_z \\ 0 & 1 \end{bmatrix}$$

Adopting the notation of [12], given $G_i = (x_i^t, y_i^t, z_i^t, 1)^T$, a grid point in the output (target) FA image space, the corresponding resampling point in the input (source) FA image space is

$$\begin{pmatrix} x_i^s \\ y_i^s \\ z_i^s \\ 1 \end{pmatrix} = \mathcal{T}_\theta(G_i) = \mathcal{T}_\theta \begin{pmatrix} x_i^t \\ y_i^t \\ z_i^t \\ 1 \end{pmatrix} \tag{2}$$

*Sampler.* Let $\text{FA}^t$ denote the output FA image and let $\text{FA}_i^t$ be its value at a specific point. Let $\text{FA}_{nml}^s$ be the value of the input FA image at a grid point indexed by $n, m, l$. Since the resampling point specified by Eq. 2 is unlikely to coincide with a grid point, the value $\text{FA}_i^s$ of the input FA image at the resampling point must be obtained by interpolation. Specifically,

$$\text{FA}_i^t = \sum_n \sum_m \sum_l \text{FA}_{nml}^s \times k(x_i^s - n; \Phi_x) \times k(y_i^s - m; \Phi_y) \times k(z_i^s - l; \Phi_z) \tag{3}$$

where $\Phi_x, \Phi_y$, and $\Phi_z$ are the parameters of the sampling kernel $k$. Similar to [19], we use a trilinear sampling kernel, that is

$$\text{FA}_i^t = \sum_n \sum_m \sum_l \text{FA}_{nml}^s \max(0, 1 - |x_i^s - n|)$$

$$\max(0, 1 - |y_i^s - m|) \max(0, 1 - |z_i^s - l|) \tag{4}$$

In contrast to [12], the proposed spatial transformer performs a number of iterations (defined by the user) as shown in Fig. 2. In each iteration, the localization network receives as input the T1w image and the FA image as transformed according to the updated transformation. The transformation is progressively refined along the iterative process.

In our application domain, as described in the sequel, we have access to a set of FA images correctly aligned to the corresponding T1w scans. The availability of this ground truth data allows, unlike [12], to define and compute an STN loss function. This loss function, jointly with a segmentation loss function, is used in training as detailed in Subsect. 2.1.

**DEC Orientation Correction.** The alignment of the DEC map amounts to a rigid transformation including rotation and translation components. Since the PDD image (on which the DEC map is based) is a 3-D array of vectors, its rotation requires not only transformation of the tails of these vectors but also rotation of their heads with respect to their tails.

Let $\text{PDD}_i$ denote a vector in the a PDD image, $\text{PDD}_i^{new}$ the corresponding orientation corrected vector and $R$ the $3 \times 3$ matrix representing the rotation component of the rigid transformation. Then,

$$\text{PDD}_i^{new} = \mathcal{R} \cdot \text{PDD}_i \tag{5}$$

**Segmentation Network Architecture.** Inspired by [4,23], the segmentation network is composed of two encoders which operate as two separate analysis paths, and a single decoder which operates as a shared synthesis path (Fig. 1). The number of feature channels at each stage is [20, 40, 80, 160, 320] for the analysis path, and [128, 64, 32, 16] for the synthesis path.

The segmentation network receives as input a 3-D T1w scan and a 3-D DEC map, and processes each of them in a distinct encoding path. The shared synthesis path is fed with the outputs of attention blocks (Fig. 4) each containing two modified [23] attention gates [28]. Each modified attention gate has two high-resolution inputs $X1$, and $X2$, as well as a single low-resolution gate input (Fig. 4).

A modified attention gate uses the low-resolution gate input and the high-resolution input $X2$ to compute a high-resolution feature attention mask $\alpha$. The mask $\alpha$ is used to highlight salient features in the high-resolution input $X1$. The low-level architecture of each modified attention gate is shown in Fig. 5.

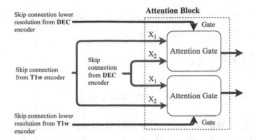

**Fig. 4.** The attention block in the segmentation network.

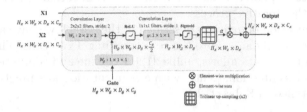

**Fig. 5.** The modified attention gate [23].

The $X1$ and $X2$ inputs receive the skip connections from the two analysis paths. However, unlike [23], the gate input is the skip connection from lower resolution features of the analysis path that correlates with the $X2$ input (Fig. 4), rather than features from the synthesis path.

## 2.1 Loss Function

The registration loss quantifies the disagreement between an output FA image and an FA image correctly aligned to the T1w scan. We use the Normalized Cross Correlation (NCC) [20]

$$\mathcal{L}_{\text{NCC}} = \frac{\sum_i^N p_i g_i}{\sqrt{\sum_i^N p_i^2 \cdot \sum_i^N g_i^2}} \tag{6}$$

where $N$ is the number of voxels, $i$ is a voxel index, $p_i$ is the softmax score and $g_i$ is the binary ground truth.

For segmentation, to accommodate the class imbalance between foreground and background, we use the Dice loss [22]

$$\mathcal{L}_{\text{Dice}} = \frac{2\sum_i^N p_i g_i}{\sum_i^N p_i^2 + \sum_i^N g_i^2} \tag{7}$$

where $N$ is as defined above, $p_i$ is a voxel in the output FA, and $g_i$ is a voxel in the aligned target FA. For joint learning of registration and segmentation, the corresponding networks are trained concurrently, minimizing a joint loss function

$$\mathcal{L}_{\text{total}} = \alpha\mathcal{L}_{\text{NCC}} + \beta\mathcal{L}_{\text{Dice}} \tag{8}$$

where $\alpha$ and $\beta$ are hyper-parameters.

## 2.2 Data Augmentation

In each epoch, a rigid body transformation was applied with probability 0.5 to each brain image. The transformation parameters were sampled from zero mean normally distributed rotations and translations, with standard deviations $(\sigma_x, \sigma_y, \sigma_z) = (2.5, 4, 2.5)$ [degree] for rotations, and $(\sigma_x, \sigma_y, \sigma_z) = (6, 2, 2)$ [pixel] for translations.

Furthermore, to improve the generalization of the STN network, a random rigid body transformation, drawn from the same distribution, was applied with probability 0.5 to each FA and PDD input image pair.

## 2.3 Initialization

Clinical tractography requires time-consuming manual operations by neuroanatomy experts to accurately delineate seeding sROIs and remove tractography outliers. The difficulty of obtaining a sufficient number of annotated (ground truth) masks limits the applicability of segmentation by supervised learning in this task.

In this work, we introduce a transfer learning approach based on the same data but on a different task for which ground truth can be easily produced. Consider the WM segmentation task. High quality WM segmentation can be automatically generated using a proven classical tool such as FSL [13] and regarded as ground truth. Tract segmentation is more specific than WM segmentation. Nevertheless, training for WM segmentation is relevant, to some extent, to tract segmentation as well and leads to better prediction and faster convergence. Thus, JTRS is pre-trained on the WM segmentation task, using aligned T1w and DEC volumes at this stage, and FSL [13] to obtain WM segmentation reference data.

## 3 Experiments and Results

### 3.1 Data

**The Clinical Dataset.** The *Sheba75* dataset consists of pre-operative scans from 75 patients referred for brain lesion removal at the Sheba Medical Center, Tel Hashomer, Israel. These scans were first used by [4]. Sheba75 includes T1w anatomical scans with no contrast injection ($1.25 \times 1.25 \times 1.25\,\text{mm}^3$) and DWI ($1 \times 1 \times 1.26\,\text{mm}^3$ resliced to $1.25 \times 1.25 \times 1.25\,\text{mm}^3$). The T1w and DWI scans are pairwise registered[1] using the SPM12 tool [9]. However, there is no group alignment to any common template, MNI space or atlas. Bilateral ground

---

[1] As described in Subsect. 3.4, we corrupt this given pairwise registration as part of the experimental procedure.

truth (GT) is available for the optic radiation tracts of all cases, and for the motor tracts of 71 cases. The GT was generated using ConTrack software [29] for probabilistic tractography, followed by manual editing. Sheba75 is publicly available in the *Brain Pre-surgical white matter Tractography Mapping* challenge (BrainPTM)[2].

For initialization, 112 T1w and DEC volume pairs from other pre-operative patients were used to generate the WM segmentation GT masks using FSL [13].

**The Neuro-scientific Dataset.** This dataset contains DWI and T1w brain scans from 105 subjects of the Human Connectome Project (HCP) [31]. The dataset includes binary GT segmentation masks for 72 WM tracts from Tract-Seg [34]. Similar to [23,34], the original high-quality DWI data downsampled by resizing to 2.5 mm isotropic resolution. It includes 32 gradient directions with single b0 image, with b-value 1000 s/mm$^2$. The HCP dataset has already undergone distortion correction, motion correction, registration to MNI space and brain extraction. For initialization, WM segmentation GT for the HCP dataset[3] was again generated by FSL [13].

## 3.2  Preprocessing

For the Sheba75 clinical dataset, bias correction in all the T1w scans was performed using SimpleITK [37]. Histogram standardization was implemented using the TorchIO library [25]. All the images in the dataset were cropped to size [128, 144, 128] voxels without removing any brain tissue. The images in the HCP dataset were cropped to size [144, 144, 144], as in [34].

For both datasets, PDD [24] and FA maps were generated from the DWI data using the Dipy python library [10], to produce the DEC map for each subject.

## 3.3  Implementation

JTRS was implemented in Python, using the Tensorflow library [1]. The experiments were performed on a GeForce RTX 3090 GPU with 24 GB memory. A separate network model was trained for each tract type. The Adam optimizer [17] was used with an initial learning rate of 0.0001, reduced after 65 epochs to 0.00001. Performance was validated over five cross-validation folds.

In each cross validation fold, JTRS was trained with a batch size of 1 for 110 epochs. In the first 10 epochs, to facilitate initial training of the registration network, $\alpha$ was set to 1 and $\beta$ was set to 0 in the combined loss function (Eq. 8). In epochs 11–50, joint learning of registration and segmentation was obtained by setting $\alpha$ and $\beta$ to 1. From epoch 51, $\alpha$ was set to 0 to dictate training of the segmentation network alone. The number of STN iterations (see Fig. 2) was set to 4.

---

[2] https://brainptm-2021.grand-challenge.org/.
[3] A cross validation scheme is used in the experimental procedure to prevent contamination of JTRS evaluation by the WM segmentation GT of test cases, see Subsect. 3.3.

As discussed in Subsect. 2.3, the JTRS parameters related to segmentation were initialized by pretraining for WM segmentation, lasting 100 epochs with $\alpha = 0$ and $\beta = 1$.

Certain implementation details are dataset-dependent. *For the clinical dataset*, each fold contained 60 training cases and 15 test cases. Pretraining for WM segmentation was carried out with 112 additional cases. *For the HCP (neuro-scientific) dataset*, each fold contained 63 training cases, 21 validation cases and 21 test cases. For evaluation of the test set, the network parameters were taken from the best epoch on the validation set. The same cross-validation scheme was also used for WM segmentation pretraining. The identical allocation of cases prevents any contamination during pretraining between the tract segmentation training and test sets.

### 3.4   Unregistration of T1w and DEC Data

The need to register the T1w scan to the FA and PDD data (derived from the DWI scan) is an essential motivation for JTRS. However, in the available datasets, both in the clinical Sheba75 dataset and the neuro-scientific HCP dataset, these modalities are already pairwise registered. For performance evaluation of JTRS we artificially unregister these scans, i.e. corrupt the given pairwise registration. Unregistration was performed at random levels taken from two distinct statistical models, one corresponding to typical patient motion, the other representing large patient motion, sometimes encountered in clinical settings.

**Typical Patient Motion.** In this case, unregistration is performed by applying a rigid body transform to the DEC map. The 6 transform parameters were sampled from a zero mean normal distribution, fitted to real transform parameter values observed in typical clinical cases of the Sheba75 dataset[4]. The standard deviations were rounded to $(\sigma_x, \sigma_y, \sigma_z) = (6, 2, 2)$ [pixel] for translations and $(\sigma_x, \sigma_y, \sigma_z) = (2.5, 4, 2.5)$ [degree] for rotations. Note that this unregistration methodology was applied to both the Sheba75 and HCP datasets.

**Large Patient Motion.** This unregistration model accounts for patients changing position or taking a break in the middle of the scan. In this case, we first simulate typical patient motion as described above. We then randomly select one of the 6 transformation parameters. We retain its sign and increase its absolute value by a random value sampled from a uniform distribution: $U(15, 21)$ [pixel] for a translation parameter and $U(9, 15)$ [degree] for a rotation parameter.

### 3.5   Results

We compared JTRS to two state-of-the-art methods for tract segmentation: AGYnet [23] and TractSeg [34], using the implementations provided by the

---

[4] The alignment parameters between the T1w and the DWI scans, as provided by SPM12, were recorded while acquiring the Sheba75 dataset.

262    N. Barzilay et al.

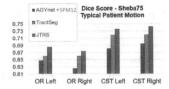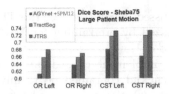

**Fig. 6.** Average Dice on Sheba75 for JTRS, AGYnet, and TractSeg: left-right Corticospinal (CST), and Optic Radiation (OR) tracts. (left) typical and (right) large patient motion.

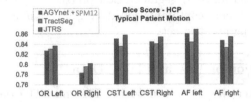

**Fig. 7.** Average Dice score on HCP for JTRS, AGYnet and TractSeg: left-right Corticospinal tract (CST) tracts, Optic Radiation (OR) tracts, Arcuate fascicle (AF).

respective authors. AGYnet calls for pre-registration of the T1w and DEC map [4,23] using SPM12 [9]. TractSeg [34] uses a single imaging modality as input, the peaks of the fODF, hence registration is irrelevant.

For the three networks, performance was evaluated using the Sheba75 and HCP datasets. For JTRS and AGYnet, artificial unregistration of the T1w and DEC map was applied as described in Subsect. 3.4. For TractSeg, since registration is irrelevant, the Sheba75 dataset was used without unregistration. Note that Sheba75 is not aligned to the MNI space, but, left/right, front/back, and up/down orientation of the images were those required for TractSeg. HCP results for TractSeg were taken from the paper [34]. Also, fODF peaks were computed for Sheba75 dataset as required by training and testing with TractSeg in the cross-validation experiments.

The tract segmentation quality was quantified using the Dice score. The Dice scores for both datasets are presented in Figs. 6 and 7 for the optic radiation (OR) and corticospinal (CST) tracts. For the HCP dataset (Fig. 7), results are also provided for the arcuate fasciculus (AF).

The quality of registration achieved between the input FA image and the FA pre-aligned to the T1w scan (referred to as ground truth FA) was assessed using the normalized cross-correlation (NCC). The NCC for both datasets is given in Table 1 (average of left and right hemisphere results).

JTRS outperformed AGYnet and TractSeg in terms of the Dice score for tract segmentation quality. The superiority of the JTRS Dice scores with respect to AGYnet and TractSeg was statistically assessed on the Sheba75 dataset using the Wilcoxon signed-rank test (1 tail) and found to be statistically significant. JTRS outperformed AGYnet and TractSeg with $p < 0.01$ for all tracts except OR right where $p < 0.05$.

**Table 1.** NCC between aligned FA and ground truth FA

| NCC (std) | | | | | | |
|---|---|---|---|---|---|---|
| | Typical patient motion | | | | Large patient motion | |
| | Sheba75 | | HCP | | Sheba75 | |
| | SPM12 | JTRS | SPM12 | JTRS | SPM12 | JTRS |
| OR | 0.936 (0.118) | **0.963** (0.021) | 0.817 (0.014) | **0.959** (0.022) | 0.847 (0.274) | **0.948** (0.029) |
| CST | 0.935 (0.242) | **0.965** (0.219) | 0.817 (0.014) | **0.958** (0.022) | 0.842 (0.330) | **0.954** (0.217) |
| AF | – | – | 0.817 (0.014) | **0.958** (0.021) | – | – |

In Fig. 8, typical tract segmentation results are shown. The top row shows a CST example and the bottom row an OR tract example taken from the Sheba75 dataset. Column (a) is the ground truth segmentation. Columns (b), (c) and (d) display segmentation results using AGYnet, TractSeg and the proposed JTRS respectively. In these examples the 3-D Dice scores are (left/right tract): AGYnet-CST (0.74/0.753), OR (0.756/0.721); TractSeg-CST (0.749/0.767), OR (0.736/0.737); JTRS-CST (0.786/0.774), OR (0.808/0.743). The qualitative and quantitative superiority of JTRS in these examples is evident.

Figure 9 shows the CST segmentation results for a case in the Sheba75 dataset where the left hemisphere's CST (right image half) is close to a large tumor surrounded by edema. We observe that similarly to AGYnet, and better than TractSeg, the proposed JTRS was able to circumnavigate the tumor-edema area at minimal distance, as in the ground truth. The respective 3-D Dice scores are (left/right tract): AGYnet-(0.734/0.67); TractSeg-(0.765/0.711); JTRS-(0.783/0.727). Here again, JTRS clearly outperformed AGYnet and TractSeg in this challenging clinical example.

JTRS provided better registration, in terms of NCC, than the SPM12 procedure called by AGYnet. In addition, comparing the registration results, JTRS appears to be substantially more robust. On the Sheba75 dataset, SPM12 failed to converge in 2.6% of the typical patient motion transformations and in 12% of the large patient motion transformations. The SPM12 failure transformations were not necessarily characterized by large translations or rotations. In contrast, JTRS converged to an accurate registration on all cases, regardless of the patient motion level - typical or large.

On the HCP dataset, registration by SPM12 failed to converge only in a single case with a large patient motion transformation. This may indicate that data quality, in this case high quality neuroscience data (HCP) vs. lower quality clinical neuro-surgical data (Sheba75), is more important for registration robustness than miss-alignment amplitude. The modified STN block, integrated within the proposed JTRS, proves to be especially appropriate for the challenging alignment of DTI and T1w scans of neuro-surgery patients.

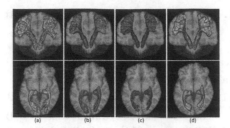

**Fig. 8.** Visualization of tract segmentation for a CST example (top row) and an OR tract example (bottom row), both taken from the Sheba75 dataset: (a) Ground truth, (b) AGYnet, (c) TractSeg, and (d) The proposed JTRS.

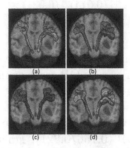

**Fig. 9.** Visualization of CST segmentation for a case in the Sheba75 dataset containing a brain lesion and edema (left hemisphere- right image half): (a) Ground truth, (b) AGYnet, (c) TractSeg, (d) JTRS.

### 3.6    Ablation Studies

The contribution of three JTRS components was evaluated: *joint training* of the registration and segmentation networks, initialization with a pretrained WM segmentation network, and number of STN iterations. We trained five different models and obtained their Dice scores, seen in Fig. 10. The first model is JTRS as proposed in Sect. 4. The second model is JTRS with separate (rather than joint) learning of the registration and segmentation tasks. The registration network has been trained alone for 50 epochs, then the segmentation network was trained alone for 100 epochs. In the third model, we did not use transfer learning for initialization. In the fourth model, the spatial transformer performs a single iteration, as proposed in [12], instead of 4. Finally, in the fifth model, provided for completeness, the registration network is removed from JTRS altogether and replaced by SPM12. The five models were trained on the Sheba75 dataset, unregistered with typical patient motion.

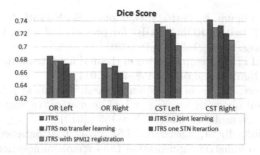

**Fig. 10.** Ablation study results: five scenarios on Sheba75 with typical patient motion.

## 4    Conclusions

In this work, we presented the JTRS framework for automatic segmentation of WM tracts from unaligned multi-modal (T1w and DEC) MRI scans. Registration between the input scans is necessary for tract segmentation. JTRS is a jointly trained neural framework that optimizes registration and segmentation accuracy.

JTRS achieves promising registration and segmentation results, both on the neuro-scientific HCP dataset and the Sheba75 dataset of patients referred to brain lesion removal. Compared to the state-of-the-art AGYnet and TractSeg algorithms, JTRS achieves superior results with statistical significance.

Better robustness of JTRS is clearly seen when applied to the Sheba75 clinical dataset. That dataset contains abnormal signal values, deformations and artifacts due to brain lesions and the resulting mass effect or edema. These phenomena are a hindrance to accurate inter-modality registration. In such cases, the modified STN block in the registration network yielded accurate results while alternative state-of-the art methods often failed.

In an ablation study, we demonstrated the superiority of JTRS joint neural approach over isolated classical pre-registration or separately trained neural registration. Also, performing multiple STN iterations strongly contributed to results accuracy. Transfer learning by pre-training the segmentation network of JTRS for WM segmentation over a larger dataset proved to be a valuable initialization, leading to better, faster and more robust network convergence.

The encouraging results of JTRS suggest it may be further developed into an efficient and fully automatic alternative to the classical presurgical tractography pipeline.

## References

1. Abadi, M., et al.: TensorFlow: a system for large-scale machine learning (2015). https://arxiv.org/abs/1603.04467
2. Anderson, A.W.: Measurement of fiber orientation distributions using high angular resolution diffusion imaging. Magn. Reson. Med. **54**(5), 1194–1206 (2005)
3. Andersson, J.L., Jenkinson, M., Smith, S.: Non-linear registration aka spatial normalisation FMRIB. Technical report TR07JA2, FMRIB Analysis Group, University of Oxford (2007)

4. Avital, I., Nelkenbaum, I., Tsarfaty, G., Konen, E., Kiryati, N., Mayer, A.: Neural segmentation of seeding ROIs (sROIs) for pre-surgical brain tractography. IEEE Trans. Med. Imag. **39**(5), 1655–1667 (2020)

5. Bazin, P.L., et al.: Direct segmentation of the major white matter tracts in diffusion tensor images. Neuroimage **58**(2), 458–468 (2011)

6. Bick, A.S., Mayer, A., Levin, N.: From research to clinical practice: implementation of functional magnetic imaging and white matter tractography in the clinical environment. J. Neurol. Sci. **312**(1), 158–165 (2012)

7. Çiçek, Ö., Abdulkadir, A., Lienkamp, S.S., Brox, T., Ronneberger, O.: 3D U-net: learning dense volumetric segmentation from sparse annotation. In: Ourselin, S., Joskowicz, L., Sabuncu, M.R., Unal, G., Wells, W. (eds.) MICCAI 2016. LNCS, vol. 9901, pp. 424–432. Springer, Cham (2016). https://doi.org/10.1007/978-3-319-46723-8_49

8. Fan, J., Cao, X., Yap, P.T., Shen, D.: BIRNet: brain image registration using dual-supervised fully convolutional networks. Med. Image Anal. **54**, 193–206 (2019)

9. Friston, K., Ashburner, J.: Spm12 [Computer Program]. London, UK, Wellcome Centre for Human Neuroimaging, UCL Queen Square Institute of Neurology (2020)

10. Garyfallidis, E., et al.: Dipy, a library for the analysis of diffusion MRI data. Front. Neuroinform. **8**, 8 (2014)

11. Haskins, G., Kruger, U., Yan, P.: Deep learning in medical image registration: a survey. Mach. Vis. and Appl. **31**(1), 8 (2020)

12. Jaderberg, M., Simonyan, K., Zisserman, A., et al.: Spatial transformer networks. In: Proceedings of Advanced Neural Information Process Systems, pp. 2017–2025 (2015)

13. Jenkinson, M., Beckmann, C.F., Behrens, T.E., Woolrich, M.W., Smith, S.M.: FSL. Neuroimage **62**(2), 782–790 (2012)

14. Jeurissen, B., Tournier, J.D., Dhollander, T., Connelly, A., Sijbers, J.: Multi-tissue constrained spherical deconvolution for improved analysis of multi-shell diffusion MRI data. Neuroimage **103**, 411–426 (2014)

15. Jonasson, L., Bresson, X., Hagmann, P., Cuisenaire, O., Meuli, R., Thiran, J.P.: White matter fiber tract segmentation in DT-MRI using geometric flows. Med. Image Anal. **9**(3), 223–236 (2005)

16. Kamnitsas, K., et al.: Efficient multi-scale 3D CNN with fully connected CRF for accurate brain lesion segmentation. Med. Image Anal. **36**, 61–78 (2017)

17. Kingma, D.P., Ba, J.: Adam: a method for stochastic optimization. In: Proceedings of International Conference on Learning and Representation, pp. 1–41 (2015)

18. Le Bihan, D., et al.: Diffusion tensor imaging: concepts and applications. J. Magn. Reson. Imag. **13**(4), 534–546 (2001)

19. Lee, M.C.H., Petersen, K., Pawlowski, N., Glocker, B., Schaap, M.: TETRIS: template transformer networks for image segmentation with shape priors. IEEE Trans. Med. Imag. **38**(11), 2596–2606 (2019)

20. Lewis, J.P.: Fast normalized cross-correlation. In: Proceedings of Vision Interface, pp. 120–123 (1995)

21. Li, B., de Groot, M., Vernooij, M.W., Ikram, M.A., Niessen, W.J., Bron, E.E.: Reproducible white matter tract segmentation using 3D U-net on a large-scale DTI dataset. In: Shi, Y., Suk, H.-I., Liu, M. (eds.) MLMI 2018. LNCS, vol. 11046, pp. 205–213. Springer, Cham (2018). https://doi.org/10.1007/978-3-030-00919-9_24

22. Milletari, F., Navab, N., Ahmadi, S.A.: V-net: fully convolutional neural networks for volumetric medical image segmentation. In: Proceedings of IEEE 4th International Conference on 3D Vision, pp. 565–571. IEEE, October 2016

23. Nelkenbaum, I., Tsarfaty, G., Kiryati, N., Konen, E., Mayer, A.: Automatic segmentation of white matter tracts using multiple brain MRI sequences. In: Proc. 17th IEEE International Symposium on Biomedical Imaging, pp. 368–371. IEEE, April 2020

24. Pajevic, S., Pierpaoli, C.: Color schemes to represent the orientation of anisotropic tissues from diffusion tensor data: application to white matter fiber tract mapping in the human brain. Magn. Reson. Med. **42**(3), 526–540 (1999)

25. Pérez-García, F., Sparks, R., Ourselin, S.: TorchIO: a Python library for efficient loading, preprocessing, augmentation and patch-based sampling of medical images in deep learning. arXiv:2003.04696v3, January 2021. https://arxiv.org/abs/2003.04696v3

26. Pomiecko, K., Sestili, C., Fissell, K., Pathak, S., Okonkwo, D., Schneider, W.: 3D convolutional neural network segmentation of white matter tract masks from MR diffusion anisotropy maps. In: Proceedings of 15th IEEE International Symposium on Biomedical Imaging, pp. 1–5. IEEE, April 2019

27. Ronneberger, O., Fischer, P., Brox, T.: U-net: convolutional networks for biomedical image segmentation. In: Navab, N., Hornegger, J., Wells, W.M., Frangi, A.F. (eds.) MICCAI 2015. LNCS, vol. 9351, pp. 234–241. Springer, Cham (2015). https://doi.org/10.1007/978-3-319-24574-4_28

28. Schlemper, J., et al.: Attention gated networks: learning to leverage salient regions in medical images. Med. Image Anal. **53**, 197–207 (2019)

29. Sherbondy, A.J., Dougherty, R.F., Ben-Shachar, M., Napel, S., Wandell, B.A.: ConTrack: finding the most likely pathways between brain regions using diffusion tractography. J. Vision **8**(9), 15–15 (2008)

30. Tournier, J.D., Calamante, F., Connelly, A.: Robust determination of the fibre orientation distribution in diffusion MRI: non-negativity constrained super-resolved spherical deconvolution. Neuroimage **35**(4), 1459–1472 (2007)

31. Van Essen, D.C., et al.: The WU-Minn human connectome project: an overview. Neuroimage **80**, 62–79 (2013)

32. Viola, P., Wells, W.M., III.: Alignment by maximization of mutual information. Int. J. Comput. Vision **24**(2), 137–154 (1997)

33. de Vos, B.D., Berendsen, F.F., Viergever, M.A., Staring, M., Išgum, I.: End-to-end unsupervised deformable image registration with a convolutional neural network. In: Cardoso, M.J., et al. (eds.) DLMIA/ML-CDS -2017. LNCS, vol. 10553, pp. 204–212. Springer, Cham (2017). https://doi.org/10.1007/978-3-319-67558-9_24

34. Wasserthal, J., Neher, P., Maier-Hein, K.H.: TractSeg-fast and accurate white matter tract segmentation. Neuroimage **183**, 239–253 (2018)

35. Wasserthal, J., Neher, P.F., Hirjak, D., Maier-Hein, K.H.: Combined tract segmentation and orientation mapping for bundle-specific tractography. Med. Image Anal. **58**, 101559 (2019)

36. Wasserthal, J., Neher, P.F., Isensee, F., Maier-Hein, K.H.: Direct white matter bundle segmentation using stacked u-nets. arXiv:1703.02036, March 2017. https://arxiv.org/abs/1703.02036

37. Yaniv, Z., Lowekamp, B.C., Johnson, H.J., Beare, R.: SimpleITK image-analysis notebooks: a collaborative environment for education and reproducible research. J. Digit. Imag. **31**(3), 290–303 (2018)

38. Zhang, F., Karayumak, S.C., Hoffmann, N., Rathi, Y., Golby, A.J., O'Donnell, L.J.: Deep white matter analysis (DeepWMA): fast and consistent tractography segmentation. Med. Image Anal. **65**, 101761 (2020)

# Complementary Phase Encoding for Pair-Wise Neural Deblurring of Accelerated Brain MRI

Gali Hod[1]([✉]), Michael Green[2], Mark Waserman[3], Eli Konen[3], Shai Shrot[3],
Ilya Nelkenbaum[3], Nahum Kiryati[4], and Arnaldo Mayer[3]

[1] School of Electrical Engineering, Tel Aviv University, Tel Aviv, Israel
galihod@mail.tau.ac.il
[2] School of Computer Sciences, Ben Gurion University, Beersheba, Israel
[3] Diagnostic Imaging Department at Sheba Medical Center Affiliated
with the Beverly Sackler School of Medicine, Tel Aviv University, Tel Aviv, Israel
arnaldo.mayer@sheba.health.gov.il
[4] Klachky Chair of Image Processing, School of Electrical Engineering,
Tel Aviv University, Tel Aviv, Israel
nk@eng.tau.ac.il

**Abstract.** MRI has become an invaluable tool for diagnostic brain imaging, providing unrivalled qualitative and quantitative information to the radiologist. However, due to long scanning times and capital costs, access to MRI lags behind CT. Typical brain protocols lasting over 30 min set a clear limitation to patient experience, scanner throughput, operation profitability, and lead to long waiting times for an appointment.

As image quality, in terms of spatial resolution and noise, is strongly dependent on acquisition duration, significant scanning acceleration must successfully address challenging image degradation. In this work, we consider the scan acceleration scenario of a strongly anisotropic acquisition matrix. We propose a neural approach that jointly deblurs scan pairs acquired with mutually orthogonal phase encoding directions. This leverages the complementarity of the respective phase encoded information as blur directions are also mutually orthogonal between the scans in the pair. The proposed architecture, trained end-to-end, is applied to T1w scan pairs consisting of one scan with contrast media injection (CMI), and one without. Qualitative and quantitative validation is provided against state-of-the-art deblurring methods, for an acceleration factor of 4 beyond compressed sensing acceleration. The proposed method outperforms the compared methods, suggesting its possible clinical applicability for this challenging task.

**Keywords:** Medical imaging · Deep learning · Deblurring · Multi-modal mri · MRI acceleration

## 1 Introduction

The development of fast MRI acquisition techniques is an important field of research, because scanning time is a major pain-point for MRI applicability in

L. Karlinsky et al. (Eds.): ECCV 2022 Workshops, LNCS 13803, pp. 268–280, 2023.
https://doi.org/10.1007/978-3-031-25066-8_13

many clinical indications [11]. Besides poor patient experience, MRI scan duration also affects time to appointment, costs, and overall operational profitability [5].

In the classical process of MRI acquisition, raw data samples are used to fill the k-space. These are discrete data points placed on a finite, usually Cartesian grid in k-space. In this scheme, samples are acquired along the phase encoding direction in a line-by-line ordering, with acquisition duration affected by the phase encoding rather than the frequency encoding direction [3].

In the last ten years, compressed sensing (CS) techniques have been successfully applied to MRI data acquisition, leading to significant scanning time reductions [16,17]. CS is a mathematical framework developed for data reconstruction from highly under-sampled data by leveraging sparse representations. In CS-MRI, the image is recovered from semi-random incomplete sampling of k-space data using an appropriate nonlinear recovery scheme.

Recently, deep convolutional neural networks have achieved superior results in generic computer vision tasks, e.g. segmentation, registration, image deblurring and denoising. Most relevant to this research are [6,13,18,24,26,27,29].

In [18] a fully convolutional network with anisotropic kernels was proposed. The network was optimized for the task of deblurring low resolution (LR) brain MRI using a perceptual cost function. A GAN based architecture was suggested in [27] for the task of MRI reconstruction. The mapping between the low and high resolution images is learnt using a generator with both pixel-wise and perceptual loss components on its output. Another loss, in the frequency domain, is also computed, after applying FFT on the input and the generator's output.

Cross-modal brain MRI enhancement is reported in [24,26,29], where one of the input modalities is enhanced based on the other. Tsiligianni [24] proposes a convolutional network that uses a high resolution (HR) scan of one (source) modality to super-resolve an image of a different (target) modality. Xiang *et al.* [26] present a Unet-based architecture that uses HR T1W scans to reconstruct undersampled T2W scans. Zhou *et al.* [29] also use T1W as prior to reconstruct T2W scans. They use a recurrent network where each block is comprised of both k-space restoration network and image restoration network.

Iwamoto *et al.* [6] use synthesized scans, down sample them to create blurred versions, and use a high quality scan in another modality to learn deblurring. The resulting network is used to enhance an LR scan in one modality using an HR guidance scan in another modality.

Different from natural image processing studies, where many benchmark datasets are available [2,19,28], the medical imaging field is short of such resources [9].

Creating medical image datasets requires resources such as state of the art scanners, informed consent by patients and various regulatory approvals. Another important observation regarding medical image enhancement is that many works are demonstrated only on synthesized data [6,11,21,24]. The synthesis is carried out by simulating sub-sampling of a fully sampled k-space data. Therefore, such datasets may not accurately resemble real LR data. Recently, a dataset of real LR and HR paired scans has appeared [1], but to our knowledge, no published work has yet applied this dataset. Note that some works on MRI acceleration use data that was not acquired using CS protocols. This makes performance evaluation difficult, since the acceleration comes at the expense of potential acceleration by the use of CS [18].

MRI deblurring is closely related to natural image enhancement problems such as super-resolution [14,20] and deblurring [13]. They resemble in their definition as inverse problems and in the applicable deep learning architectures, providing us with an additional comparison basis to SOTA methods [13]. Kupyn et al. [13] present a GAN architecture that uses a conditional GAN with a perceptual loss incorporated in its loss function components. This is a follow up work of [12] reaching SOTA results. A key difference between medical and natural image enhancement is that while in the former a photo-realistic eye-pleasing result is usually sought, in medical image analysis details must be correctly restored to preserve diagnostic value.

In this paper we propose a novel end-to-end framework for deblurring accelerated multi-modal MRI scans. Our contribution is threefold:

- A novel joint deblurring scheme for Low Resolution (LR) MRI scans of different modalities. To the best of our knowledge, this is the first solution to this problem.
- Superior deblurring results in comparison to single-modality methods.
- Creation and application of an actual patient dataset where each entry in the dataset consists of four brain MRI scans of the same patient; two pairs of LR and HR scans, one pair with a contrast media injection (CMI) and one without, where the phase encoding direction is changed between the two LR scans.

The remainder of the paper is organized as follows: Sect. 2 describes the proposed methodology, including pre-processing, network architecture and loss. Section 3 presents the dataset, experiments and results, including comparison to State-of-the-Art (SOTA) methods. Section 4 summarizes the conclusions.

## 2    Proposed Methodology

### 2.1    Registration

Image registration is the alignment of two or more images into the same coordinate system. The image transformed into another coordinate system is referred to as the moving image and the image remaining in its original coordinate system is called the fixed image. Image registration is a common medical imaging task, as it helps to establish correspondence between scans taken at different times, of different subjects, with different modalities. In this research, which is based on actual patients' scans, registration is a crucial pre-processing step consisting of several elements. First, we perform 3D rigid-body registration between each LR-HR pair. In contrast to research based on synthetic data, we must address movements of the patient between the two scans. Otherwise such movements reduce the accuracy of LR-HR pairing and eventually lead to inferior deblurring results. Next, registration between the different modalities takes place. We calculate the 3D rigid-body registration between the two modalities and apply it to all registered scans of the moving modality. At this point, all four (two LR, two HR) scans acquired from a single patient are registered. Finally we apply a normalization process, to help align all four scans together and to eventually have the entire dataset aligned together. The registration is implemented via the SimpleElastix framework [10, 22].

## 2.2 Architecture

The proposed *Multi-Modal MRI Deblurring Network (MMMD-Net)* is a fully convolutional neural network. MMMD-Net takes as input two LR MRI scans of two different modalities, each blurred in a different axial direction ($x$ or $y$) and outputs a deblurred version of both images. The network contains two branches, Fig. 1 demonstrates the branch architecture. The branch input is two registered scans of the two different modalities, stacked one on top of the other. It outputs a deblurred version of the top input scan.

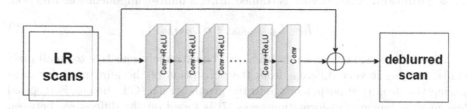

**Fig. 1.** A fully convolutional architecture with 9 convolution layers, 64 filters followed by a ReLU activation function in each layer.

Figure 2 describes the architecture of the proposed MMMD-Net. Anisotropic complementary kernels are used for the two branches, where we specify the kernel size $ks$ by its height and width, i.e., $ks = (k_h \times k_w)$. The upper branch outputs a deblurred version of the T1W scan with $ks = (3 \times 7)$ and the lower branch outputs a deblurred version of the T1W CMI scan with $ks = (7 \times 3)$. We choose anisotropic kernels enlarged in the phase-encoding direction, to take advantage of our prior knowledge regarding the blur direction [18].

At test time, the architecture is fed with two LR brain MRI scans, performs pair-wise deblurring, and outputs deblurred versions of the two input scans.

**Residual Learning.** Residual Learning [4] has proven useful in various image enhancement problems such as deblurring, super-resolution and denoising. With residual learning the network learns the difference between the restored and original images rather than learning the restored image itself. For image enhancement

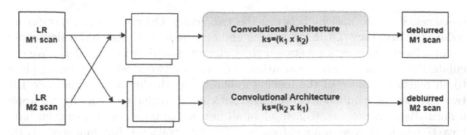

**Fig. 2.** MMMD-Net architecture. M1 and M2 represents the two different modalities and *ks* is the kernel size.

tasks learning the difference is desirable as it promotes similarity between the input and the output images.

$$\hat{y} = F(x; \theta) + x \tag{1}$$

Here, $\hat{y}$ is the reconstructed image, $x$ is the input- in our case the LR image, and $F(x; \theta)$ is the function that represents the learned residual mapping.

**Loss Function.** The network is trained using a multi-component loss function.

$$L = L_{perceptual} + \alpha L_{SSIM} \tag{2}$$

Pixel-wise loss, such as L1 or MSE, can be extremely sensitive to small pixel shifts, making it very different from the perception of the human reader comparing the deblurred output image to the ground truth (GT) image. Perceptual loss used in image transformation tasks [7] is based on the differences between high-level features of the image, rather than between pixel values. Pre-trained convolutional neural networks are often used in order to extract these high-level features.

In our implementation, a pre-trained VGG-16 network [23] was used as a feature extractor. Both the HR image and the output image are fed into the VGG model, where features are extracted from multiple layers. The L1 distance is calculated between each pair of extracted features. Eventually the mean is taken to form the final $L_{perceptual}$ loss component.

$$L_{perceptual} = \frac{1}{|I|} \sum_{i \in I} \|\phi_i(y) - \phi_i(\hat{y}(x; \theta))\|_1 \tag{3}$$

Here $\phi_i$ is feature extracted from the i'th VGG-16 layer and $I = [0, 4, 9, 23, 26, 30]$ Layer 0 refers to the VGG-16 input, i.e. the HR and deblurred output images. Its contribution to the sum is in fact the pixel-wise L1 distance between the HR and deblurred output images. The other layers are numbered according to the sequential model numbering in the Torchvision pre-trained VGG-16 model.

SSIM (Structural Similarity index) [25] is often used for measuring the similarity between two images. It allows for quality comparison between two images where one of them is considered to be of superior quality while the other aspires to match its quality. SSIM, similarly to perceptual loss, is used to analyze the difference between features extracted from the images. The index makes use of three image features: structure, contrast and luminance.

$$SSIM(x,y) = \frac{(2\mu_x\mu_y + c_1)(2\sigma_{xy} + c_2)}{(\mu_x^2 + \mu_y^2 + c_1)(\sigma_x^2 + \sigma_y^2 + c_2)} \tag{4}$$

As it is desired to maximize similarity between the network output and the HR image, corresponding to a high SSIM value, we set $L_{SSIM} = -SSIM$ and multiply by a positive factor $\alpha$.

The loss expression (Eq. 2) is calculated per branch, and the network's total loss is the sum of the two.

## 3 Experiments

### 3.1 Dataset

For this research we assembled a dataset with brain MRI scans collected from 9 distinct subjects, with different medical conditions. Prior to the scanning, each patient provided a written consent in accordance with the IRB clearance delivered by our institution. Each subject provided 4 scans, all acquired using CS protocols [16,17]. In this dataset, the MRI modalities are of two types, T1W scans and T1W CMI-enhanced scans. Both brain scan types are commonly acquired in diagnostic practice.

We thus acquired two HR scans, with acquisition matrix (2D Cartesian k-space sampling) size of 240 × 240, and two complementary LR sagittal scans:

- A T1W scan with a 240 × 60 acquisition matrix, resulting from reduction by a factor of 4 in the phase encoding direction. This reduction translates into an acceleration factor of 4. In the image domain, the acceleration leads to strong blur artifacts along the $x$ axis.
- A T1W CMI-enhanced scan with a 60 × 240 acquisition matrix and the same acceleration factor. This time, the phase encoding direction is swapped, such that in the image domain, the acceleration leads to strong blur artifacts along the $y$ axis.

Eventually the scanner firmware presented the reconstructed HR and LR scans as 512 × 512 × 170 voxel sets.

## 3.2 Training Details

All the experiments were conducted with a NVIDIA GeForce RTX 3070 GPU. We used PyTorch for the implementation. The network was trained for 150 epochs with a batch size of 6. The training duration was 12.5 h. Adam optimizer [8] was used, with betas = 0.9,0.99 and weight decay = 0.001. The initial learning rate was set to 0.001 and a SGDR [15] scheduler was used with minimum lr = 0.0001, 5 iterations for the first restart and a multiplication factor of 2 for the time between restarts.

## 3.3 Results

For evaluating our network in comparison to state-of-the-art methods, it was trained, along with ADMR-Net [18], DeblurGANv2 [13] and DAGAN [27], on our dataset using the leave-one-out validation strategy. The configurations used for training the SOTA methods matched those presented in their original papers. Training times for these algorithms were respectively 117% 102% and 167% compared to ours. Visual results are presented per modality in Figs. 3 and 4. We present patches to allow for the observation of details; nevertheless, zooming in is recommended. In both figures, a strong blur is easily observed in the LR column. While observing Fig. 3 it is evident that all methods but ours struggled to restore details originating from CMI. The images generated by DAGAN demonstrate no such details. ADMR-Net has mostly failed to restore them as well. In DeblurGAN the attempt to restore the CMI-contrast details resulted in strong artifacts. In contrast, our method was able to restore CMI-contrast details in different areas of the brain. In addition to preservation of CMI-contrast details, our method was able to restore fine structural details of the brain and to achieve superior results in this aspect as well. Figure 4 also demonstrates the superior performance of our method. With DAGAN the results are still blurred and have strong patchy looking artifacts. ADMR-Net provides a rather blurry result as well. DeblurGAN produced artifacts in some of the patches along with blurred regions especially noticeable in detailed small regions.

Quantitative performance comparison in terms of the PSNR and SSIM indices is illustrated graphically in Figs. 5 and 6 and summarized in Tables 1 and 2.

**Table 1.** PSNR mean and STD values per method per modality

| Modality | LR | DAGAN | ADMR-Net | DeblurGANv2 | Ours |
|---|---|---|---|---|---|
| $T_1W$ | $16.33 \pm 0.6$ | $16.88 \pm 0.4$ | $24.04 \pm 0.81$ | $23.2 \pm 1.12$ | $\mathbf{24.76 \pm 1}$ |
| $T_1W$ with CMI | $18.15 \pm 1.27$ | $21.01 \pm 1.13$ | $25.69 \pm 0.87$ | $23.91 \pm 0.83$ | $\mathbf{26.41 \pm 1.22}$ |

The graphs are presented per modality and per performance index. The index values are present for the LR scan, for our method and for ADMR-net, DeblurGANv2 and Dagan. In agreement with the comparative visual results, our method scored the highest PSNR and SSIM values for both modalities compared to all other methods. As expected, the LR images obtained the lowest scores in all but one case.

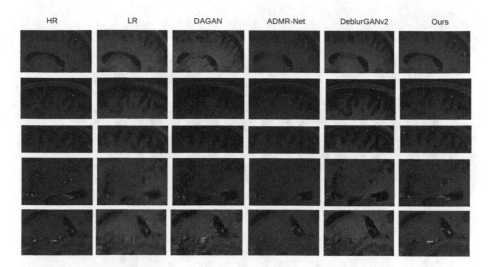

**Fig. 3.** Patches of $T_1W$ MRI scans with CMI. From left to right: HR, LR, DAGAN, ADMR-Net, DeblurGANv2 and Ours

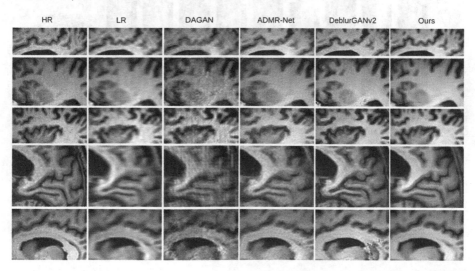

**Fig. 4.** Patches of $T_1W$ MRI scans. From left to right: HR, LR, DAGAN, ADMR-Net, DeblurGANv2 and Ours

**Table 2.** SSIM mean and STD values per method per modality

| Modality | LR | DAGAN | ADMR-Net | DeblurGANv2 | Ours |
|---|---|---|---|---|---|
| $T_1W$ | $0.5 \pm 0.03$ | $0.44 \pm 0.02$ | $0.58 \pm 0.03$ | $0.55 \pm 0.04$ | $\mathbf{0.63 \pm 0.04}$ |
| $T_1W$ with contrast | $0.46 \pm 0.04$ | $0.48 \pm 0.03$ | $0.61 \pm 0.04$ | $0.55 \pm 0.058$ | $\mathbf{0.68 \pm 0.04}$ |

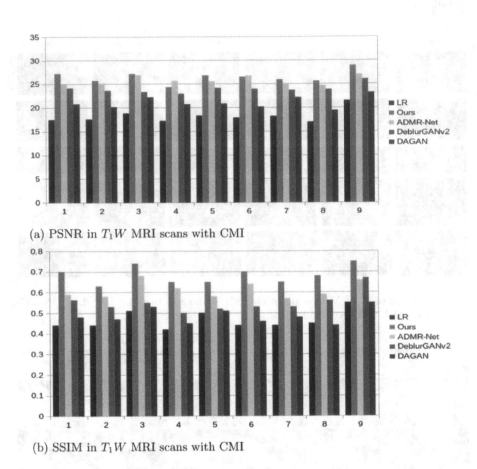

(a) PSNR in $T_1W$ MRI scans with CMI

(b) SSIM in $T_1W$ MRI scans with CMI

**Fig. 5.** SSIM and PSNR indices in $T_1W$ MRI scans with CMI per case number

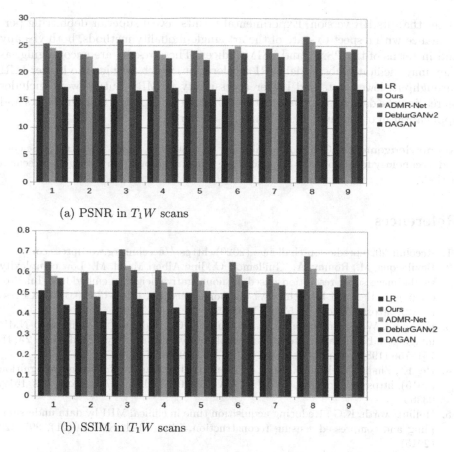

(a) PSNR in $T_1W$ scans

(b) SSIM in $T_1W$ scans

**Fig. 6.** SSIM and PSNR indices in $T_1W$ scans per case number

## 4   Conclusions

We presented a fully convolutional neural network for jointly deblurring multi-modal brain MRI scans, to our knowledge the first of its kind. We applied the network to multi-modal brain scans, where the modalities are T1W with and without contrast media injection (CMI). The scans were acquired with phase encoding directions orthogonal between the two modalities, and with an acceleration factor of 4 along the phase encoding directions. For training and validation, we assembled a unique dataset of actual T1W brain scans with various medical conditions, with four scans per subject: high resolution (HR) and low resolution, with and without CMI where the phase encoding direction is switched. The suggested framework allows to shorten the total scan duration by a factor of 4. Instead of acquiring HR versions of the two scan types, a trained model of the proposed architecture allows to use LR versions, each acquired 4 times faster than its HR version. Experimental results reveal superior deblurring performance with respect to state of the art single-modality methods, both visually and in terms of the PSNR and SSIM indices. These results are encouraging, as they may facilitate substantial MRI acceleration, which may lead to higher MRI throughput, lower costs and better accessibility to MRI. Future work includes increasing the dataset size, systematic evaluation by experienced radiologists and application in other multi-modal imaging scenarios.

**Acknowledgements.** Gali Hod would like to thank the Israeli Ministry of Science and Technology for supporting her under the scholarship program during the research period.

# References

1. Accelmr 2020 prediction challenge (2020). https://accelmrorg.wordpress.com/
2. Bevilacqua, M., Roumy, A., Guillemot, C., line Alberi Morel, M.: Low-complexity single-image super-resolution based on nonnegative neighbor embedding. In: Proceedings of the British Machine Vision Conference, pp. 135.1–135.10. BMVA Press (2012). https://doi.org/10.5244/C.26.135
3. Edelstein, W.A., Hutchison, J.M.S., Johnson, G., Redpath, T.: Spin warp NMR imaging and applications to human whole-body imaging. Phys. Med. Biol. **25**(4), 751–756 (1980). https://doi.org/10.1088/0031-9155/25/4/017
4. He, K., Zhang, X., Ren, S., Sun, J.: Deep residual learning for image recognition (2015). https://doi.org/10.48550/ARXIV.1512.03385, https://arxiv.org/abs/1512.03385
5. Hollingsworth, K.G.: Reducing acquisition time in clinical MRI by data undersampling and compressed sensing reconstruction. Phys. Med. Biol. **60**(21), 297–322 (2015)
6. Iwamoto, Y., Takeda, K., Li, Y., Shiino, A., Chen, Y.W.: Unsupervised MRI super-resolution using deep external learning and guided residual dense network with multimodal image priors (2020)
7. Johnson, J., Alahi, A., Fei-Fei, L.: Perceptual losses for real-time style transfer and super-resolution (2016). https://doi.org/10.48550/ARXIV.1603.08155, https://arxiv.org/abs/1603.08155
8. Kingma, D.P., Ba, J.: Adam: a method for stochastic optimization (2014). https://doi.org/10.48550/ARXIV.1412.6980, https://arxiv.org/abs/1412.6980
9. Kiryati, N., Landau, Y.: Dataset growth in medical image analysis research. J. Imaging **7**, 155 (2021). https://doi.org/10.3390/jimaging7080155
10. Klein*, S., Staring*, M., Murphy, K., Viergever, M.A., Pluim, J.P.: elastix: a toolbox for intensity-based medical image registration. IEEE Trans. Medical Imag. **29**(1), 196–205 (2010)
11. Knoll, F., et al.: Advancing machine learning for MR image reconstruction with an open competition: Overview of the 2019 fastMRI challenge. Magn. Reson. Med. **84**(6), 3054–3070 (2020). https://doi.org/10.1002/mrm.28338
12. Kupyn, O., Budzan, V., Mykhailych, M., Mishkin, D., Matas, J.: Deblurgan: blind motion deblurring using conditional adversarial networks. In: 2018 IEEE/CVF Conference on Computer Vision and Pattern Recognition, pp. 8183–8192 (2018). https://doi.org/10.1109/CVPR.2018.00854

13. Kupyn, O., Martyniuk, T., Wu, J., Wang, Z.: Deblurgan-v2: deblurring (orders-of-magnitude) faster and better. In: The IEEE International Conference on Computer Vision (ICCV) (2019)
14. Ledig, C., et al.: Photo-realistic single image super-resolution using a generative adversarial network (2016). https://doi.org/10.48550/ARXIV.1609.04802, https://arxiv.org/abs/1609.04802
15. Loshchilov, I., Hutter, F.: SGDR: stochastic gradient descent with warm restarts (2016). https://doi.org/10.48550/ARXIV.1608.03983, https://arxiv.org/abs/1608.03983
16. Lustig, M., Donoho, D., Pauly, J.M.: Sparse MRI: the application of compressed sensing for rapid MR imaging. Magn. Reson. Med. **58**(6), 1182–1195 (2007). https://doi.org/10.1002/mrm.21391
17. Lustig, M., Donoho, D.L., Santos, J.M., Pauly, J.M.: Compressed sensing MRI. IEEE Signal Process. Maga. **25**(2), 72–82 (2008). https://doi.org/10.1109/MSP.2007.914728
18. Mayberg, M., et al.: Anisotropic neural deblurring for MRI acceleration. Int. J. Comput. Assist. Radiol. Surg. **17**(2), 315–327 (2021). https://doi.org/10.1007/s11548-021-02535-6
19. Martin, D., Fowlkes, C., Tal, D., Malik, J.: A database of human segmented natural images and its application to evaluating segmentation algorithms and measuring ecological statistics. In: Proceedings Eighth IEEE International Conference on Computer Vision. ICCV 2001, vol. 2, pp. 416–423 (2001). https://doi.org/10.1109/ICCV.2001.937655
20. Mei, Y., et al.: Pyramid attention networks for image restoration (2020). https://doi.org/10.48550/ARXIV.2004.13824, https://arxiv.org/abs/2004.13824
21. Pham, C.H., Tor-Díez, C., Meunier, H., Bednarek, N., Fablet, R., Passat, N., Rousseau, F.: Multiscale brain MRI super-resolution using deep 3D convolutional networks. Comput. Med. Imaging Graph. **77**, 101647 (2019). https://doi.org/10.1016/j.compmedimag.2019.101647
22. Shamonin, D.P., Bron, E.E., Lelieveldt, B.P., Smits, M., Klein, S., Staring, M.: Fast parallel image registration on cpu and gpu for diagnostic classification of alzheimer's disease. Front. Neuroinf. **7**(50), 1–15 (2014)
23. Simonyan, K., Zisserman, A.: Very deep convolutional networks for large-scale image recognition (2014). https://doi.org/10.48550/ARXIV.1409.1556, https://arxiv.org/abs/1409.1556
24. Tsiligianni, E., Zerva, M., Marivani, I., Deligiannis, N., Kondi, L.: Interpretable deep learning for multimodal super-resolution of medical images. In: de Bruijne, M., Cattin, P.C., Cotin, S., Padoy, N., Speidel, S., Zheng, Y., Essert, C. (eds.) MICCAI 2021. LNCS, vol. 12906, pp. 421–429. Springer, Cham (2021). https://doi.org/10.1007/978-3-030-87231-1_41
25. Wang, Z., Bovik, A., Sheikh, H., Simoncelli, E.: Image quality assessment: from error visibility to structural similarity. IEEE Trans. Image Process. **13**(4), 600–612 (2004). https://doi.org/10.1109/TIP.2003.819861
26. Xiang, L., et al.: Deep-learning-based multi-modal fusion for fast mr reconstruction. IEEE Trans. Biomed. Eng. **66**(7), 2105–2114 (2019). https://doi.org/10.1109/TBME.2018.2883958
27. Yang, G., et al.: Dagan: deep de-aliasing generative adversarial networks for fast compressed sensing mri reconstruction. IEEE Trans. Med. Imaging **37**(6), 1310–1321 (2018). https://doi.org/10.1109/TMI.2017.2785879

28. Zeyde, R., Elad, M., Protter, M.: On single image scale-up using sparse-representations. In: Boissonnat, J.-D., Chenin, P., Cohen, A., Gout, C., Lyche, T., Mazure, M.-L., Schumaker, L. (eds.) Curves and Surfaces 2010. LNCS, vol. 6920, pp. 711–730. Springer, Heidelberg (2012). https://doi.org/10.1007/978-3-642-27413-8_47
29. Zhou, B., Zhou, S.K.: Dudornet: learning a dual-domain recurrent network for fast mri reconstruction with deep t1 prior (2020). https://doi.org/10.48550/ARXIV.2001.03799, https://arxiv.org/abs/2001.03799

# Frequency Dropout: Feature-Level Regularization via Randomized Filtering

Mobarakol Islam$^{(\boxtimes)}$ and Ben Glocker

BioMedIA Group, Department of Computing, Imperial College London, London, UK
{m.islam20,b.glocker}@imperial.ac.uk

**Abstract.** Deep convolutional neural networks have shown remarkable performance on various computer vision tasks, and yet, they are susceptible to picking up spurious correlations from the training signal. So called 'shortcuts' can occur during learning, for example, when there are specific frequencies present in the image data that correlate with the output predictions. Both high and low frequencies can be characteristic of the underlying noise distribution caused by the image acquisition rather than in relation to the task-relevant information about the image content. Models that learn features related to this characteristic noise will not generalize well to new data.

In this work, we propose a simple yet effective training strategy, Frequency Dropout, to prevent convolutional neural networks from learning frequency-specific imaging features. We employ randomized filtering of feature maps during training which acts as a feature-level regularization. In this study, we consider common image processing filters such as Gaussian smoothing, Laplacian of Gaussian, and Gabor filtering. Our training strategy is model-agnostic and can be used for any computer vision task. We demonstrate the effectiveness of Frequency Dropout on a range of popular architectures and multiple tasks including image classification, domain adaptation, and semantic segmentation using both computer vision and medical imaging datasets. Our results suggest that the proposed approach does not only improve predictive accuracy but also improves robustness against domain shift.

**Keywords:** Feature-level regularization · Image filtering · Robustness · Domain generalization

## 1 Introduction

The impressive performance of deep convolutional neural networks (CNNs) in computer vision is largely based on their ability to extract complex predictive features from images that correlate well with the prediction targets such as categorical image labels. If the training data, however, contains spurious correlations, for example, between characteristics of image acquisition and image annotations, there is a high risk that the learned features will not generalize to new data acquired under different conditions. A key issue is that features related

© The Author(s), under exclusive license to Springer Nature Switzerland AG 2023
L. Karlinsky et al. (Eds.): ECCV 2022 Workshops, LNCS 13803, pp. 281–295, 2023.
https://doi.org/10.1007/978-3-031-25066-8_14

to such spurious correlations are often much easier to learn and can be trivially picked up via convolution kernels. Specific sensor noise, for example, may manifest itself as high or low frequency patterns in the image signal. When trained on such biased data, CNNs may establish so called 'shortcuts' instead of learning generalizable, task-specific feature representations which may be more difficult to extract.

The issue of shortcut learning [13] and related aspects of texture bias in computer vision have been discussed in great detail in previous work [14,46]. Recently, Wang et al. [43] observe that high-frequency feature components cause a trade-off between accuracy and robustness. Low-level image characteristics are generally easier to pick up and thus lead to a much quicker decrease in the loss function early during training [1].

Several theoretical and empirical studies tried to tackle this issue by synthesizing shape-based representation of the dataset [14], informative dropout [35], pooling geometry [10], smoothing kernels [43], or antialiasing via two step pooling [4]. Curriculum by smoothing (CBS) [37] introduces a curriculum-based feature smoothing approach to reduce the use of high-frequency features during training. The approach controls the high-frequency information propagated through a CNN by applying a Gaussian filter on the feature maps during training. The curriculum consists of decreasing the standard deviation of the Gaussian filter as training progresses. While this avoids the use of high frequency features at the beginning of the training procedure, the CNN may still pick up these features at a later stage.

In this paper, we propose a simple yet effective training strategy, Frequency Dropout (FD), preventing CNNs from learning frequency-specific imaging features by employing randomized feature map filtering. We utilize three different types of filters including Gaussian smoothing, Laplacian of Gaussian, and Gabor filters with randomized parameters. Similar to dropout, these filters are applied randomly during training but act as feature-level regularization instead of dropping activations. FD can be incorporated into any architecture and used across a wide range of applications including image classification and semantic segmentation.

Our main findings are as follows:

- The proposed Frequency Dropout yields consistent improvements in predictive accuracy across a range of popular CNN architectures;
- Besides improved accuracy, we also observe improved robustness for networks trained with FD when tested on corrupted, out-of-distribution data;
- FD improves results in various tasks including image classification, domain adaptation, and semantic segmentation demonstrating its potential value across a wide range of computer vision and medical imaging applications;

Frequency Dropout can be easily implemented in all popular deep learning frameworks. It is complementary to other techniques that aim to improve the robustness of CNNs such as dropout, data augmentation, and robust learning.

## 2    Related Work

### 2.1    Image Filtering in CNNs

Incorporating image filtering such as smoothing, sharpening, or denoising within the training of CNNs has been widely explored and shown potential for improving model robustness and generalization. The main use of image filtering is for input-level data augmentation [11,20,23,28,39] and feature-level normalization [5,27,30,37,47]. SimCLR [8] uses filters such as Gaussian noise, Gaussian blur, and Sobel filters to augment the images for contrastive learning. Taori et al. [39] study model robustness under controlled perturbations using various types of filters to simulate distribution shift. Laplacian networks [25] uses Laplacian smoothing to improve model robustness. Gabor filters have been considered within CNNs to encourage orientation- and scale-invariant feature learning [2,29,33]. Gaussian blurring is also utilized in self-supervised learning [32], and domain adaptation [11]. In terms of CNN feature normalization, smoothing filters are used as anti-aliased max-pooling [47], and anti-alias downsampling [27,30] to improve the internal feature representation. Most recently, anti-aliasing filtering has been used to smooth the CNN feature maps in a curriculum learning manner with a reported increase in performance in various vision tasks [37].

### 2.2    Dropout-Based Regularization

Dropout [19,38] has been widely used as regularization technique to prevent overfitting [38,42], pruning [15,34], spectral transformation [22] and uncertainty estimation [12,31]. Monte Carlo dropout [12] is used to estimate prediction uncertainty at test time. Targeted dropout [15] omits the less useful neurons adaptively for network pruning. Dropout has also been explored for data augmentation by projecting dropout noise into the input space [6]. Spatial dropout [3] proposes 2D dropout to knock out full kernels instead of individual neurons in convolutional layers.

## 3    Background and Preliminaries

In CNNs, a convolution layer is used to extract features from input data $x$ by convolving ($\circledast$) the input with a kernel $w$. A convolution layer in a typical CNN may be followed by a pooling and activation layer (e.g., a ReLU) which can be expressed as

$$out := ReLU(pool(w \circledast x)) \tag{1}$$

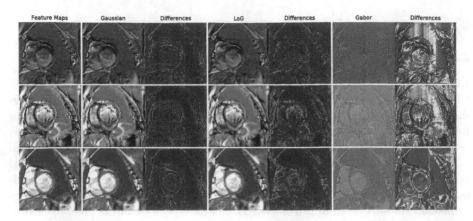

**Fig. 1.** Regularization of feature maps by applying filters such as Gaussian, Laplacian of Gaussian, and Gabor. The differences of feature maps before and after filtering is also illustrated in beside each filter.

### 3.1  Image Filters

Image filters are commonly used in digital image processing, often for smoothing or sharpening of the original image. For example, Gaussian, Laplacian of Gaussian (LoG), Gabor filters are widely used for image smoothing, sharpening, and enhancing. In general, a filter is applied via spatial convolution of the image using a specifically designed filter kernel. We can categorize image filters considered in this work into low-pass and high-pass filters.

The Gaussian filter has a blurring effect by removing high-frequency components of the image signal. The Gaussian filter is characterized by the kernel size and standard deviation. A 2D Gaussian kernel (G) with standard deviation $\sigma$ can be written as

$$G(x, y | \sigma) = -\frac{1}{2\pi\sigma^2} e^{-\frac{x^2+y^2}{2\sigma^2}} \tag{2}$$

Laplacian of Gaussian (LoG) is a hybrid filter with Gaussian smoothing and a Laplacian kernel. It smooths the image and enhances the edges (or regions of strong intensity changes). A zero centered 2D LoG kernel with a standard deviation of $\sigma$ can be expressed as

$$LoG(x, y | \sigma) = -\frac{1}{\pi\sigma^4} \left[ 1 - \frac{x^2 + y^2}{2\sigma^2} \right] e^{-\frac{x^2+y^2}{2\sigma^2}} \tag{3}$$

The Gabor filter is a special case of a band filter where a Gaussian kernel is modulated by a sinusoidal signal of a particular frequency and orientation. It is used in the application of edge detection, texture analysis, and feature extraction in both spatial and frequency domains. There are several parameters such as wavelength ($\lambda$) and phase ($\psi$) of the sinusoidal function, standard deviation ($\sigma$) of Gaussian kernel, and spatial aspect ratio ($\gamma$) of the function, controlled the effect of the filter. The real component of the Gabor kernel (Ga) can be represented as below.

$$Ga(x, y|\sigma) = \exp\left(-\frac{x'^2 + \gamma^2 y'^2}{2\sigma^2}\right) \cos\left(2\pi\frac{x'}{\lambda} + \psi\right) \tag{4}$$

### 3.2  Spatial Dropout

Dropout [19,38] refers to randomly dropping out neurons during the training phase. The probability of dropout can be controlled with a parameter $p$. Based on $p$, dropout generates either 0 or 1 sampled from a Bernoulli distribution to keep or drop the neurons using simple multiplication. If dropout of $p$ probability is applied on a layer $(l)$ of input $y^{(l)}$ then dropout can be formulated as below.

$$r^{(l)} = Bernoulli(p) \tag{5}$$
$$\hat{y}^{(l)} = r^{(l)} * y^{(l)} \tag{6}$$

## 4  Frequency Dropout

We design a feature-level regularization technique by randomizing the choice of filter and its parameters. We utilize three filter types including Gaussian, Laplacian of Gaussian, and Gabor filter within a feature regularization layer that can be incorporated into any CNN architecture. More specifically, a randomized filter from the regularization layer is applied with a certain probability after each convolution operation to post-process the generated feature maps.

A filter is selected randomly in each training iteration and each CNN layer to suppress different frequencies. The randomized filter selection can be formulated as below.

$$RF := \text{rand.choice}[G, \ LoG, \ Ga] \tag{7}$$

### 4.1  FD with Randomized Filtering

Given a randomly selected filter, we also randomly sample filter parameters from specific ranges which affect different frequencies in the input signal. We use dropout to turn off kernels at random locations, adopting the strategy of spatial dropout [3,40] to turn off the entire kernel instead of individual neurons. The dropout probability can vary for each filter type ($p^G$, $p^{LoG}$, $p^{Gabor}$). When a kernel is turned off it means the input feature map in this position will remain unchanged after the frequency dropout layer. If $\sigma^{(n)}$ is the vector of randomly generated frequencies (or standard deviation) to obtain kernels of $n$ channels using a selected filter ($randomized\_filter$ from Eq. 7), then Frequency Dropout by Randomized Filtering (FD-RF) can be formulated as follows

$$\sigma^{(n)} := \text{rand}(n, \, low, \, high) \tag{8}$$

$$\sigma^{(n)}_{fd} := \text{dropout}(\sigma^{(n)}, \, p) \tag{9}$$

$$w^{(n)}_{fd} := RF(\sigma^{(n)}_{fd}) \tag{10}$$

where $low$ and $high$ are the limits of the random frequencies for a filter and $p$ is the dropout probability which can be pre-defined as $p^G$, $p^{LoG}$, $p^{Gabor}$ for Gaussian, LoG and Gabor, respectively. Other notations can be denoted as random frequency $\sigma^{(n)}$, spatial frequency dropout $\sigma^{(n)}_{fd}$ and kernel with frequency dropout $w^{(n)}_{fd}$. Figure 1 illustrates the effect of these filters on feature maps.

An simple example of the layer-wise application of Frequency Dropout with Randomized Filtering (FD-RF) for a few convolutional layers on input data $x$ and convolution kernel $w$ is as follows:

$$w^{(n)}_{fd(i)} := RF(\cdot | \sigma^{(n)}_{fd(i)})$$

$$layer_i := ReLU(pool(w^{(n)}_{fd(i)} \circledast (w \circledast x_i)))$$

$$w^{(n)}_{fd(i+1)} := RF(\cdot | \sigma^{(n)}_{fd(i+1)})$$

$$layer_{i+1} := ReLU(pool(w^{(n)}_{fd} \circledast (w \circledast x_{i+1})))$$

$$w^{(n)}_{fd(i+2)} := RF(\cdot | \sigma^{(n)}_{fd(i+2)})$$

$$layer_{i+2} := ReLU(pool(w^{(n)}_{fd} \circledast (w \circledast x_{i+2})))$$

### 4.2   FD with Gaussian Filtering

To investigate the effectiveness of Frequency Dropout over CBS [37], we additionally define a simplified version of FD-RF with a fixed Gaussian filter. As CBS uses Gaussian filtering in a curriculum manner, Frequency Dropout by Gaussian filtering (FD-GF) correspondingly applies feature map smoothing during training but in a randomized fashion rather than following a curriculum strategy. The example layer-wise application from above would simply change accordingly to:

$$w^{(n)}_{fd(i)} := G(\cdot | \sigma^{(n)}_{fd(i)})$$

$$layer_i := ReLU(pool(w^{(n)}_{fd(i)} \circledast (w \circledast x_i)))$$

$$w^{(n)}_{fd(i+1)} := G(\cdot | \sigma^{(n)}_{fd(i+1)})$$

$$layer_{i+1} := ReLU(pool(w^{(n)}_{fd} \circledast (w \circledast x_{i+1})))$$

$$w^{(n)}_{fd(i+2)} := G(\cdot | \sigma^{(n)}_{fd(i+2)})$$

$$layer_{i+2} := ReLU(pool(w^{(n)}_{fd} \circledast (w \circledast x_{i+2})))$$

# 5  Experiments and Results

We conduct extensive validation of our proposed Frequency Dropout technique including the tasks of image classification and semantic segmentation in the settings of supervised learning and unsupervised domain adaptation. We utilize several state-of-the-art neural network architectures for computer vision and medical imaging applications to demonstrate the effectiveness of our approach. All quantitative metrics are produced from running experiments with two different random seeds. The average is reported as the final metric.

## 5.1  Image Classification

Due to its simplicity and flexibility, we could easily integrate FD into various state-of-the-art classification networks including ResNet-18 [17], Wide-ResNet-52 [45], ResNeXt-50 [44] and VGG-16 [36]. We conduct all experiments on three classification datasets including CIFAR-10, CIFAR-100 [24] and SVHN [16]. To measure the robustness of our method, we test the performance of the trained classification models on CIFAR-10-C and CIFAR-100-C which are corrupted dataset variations [18].

**Table 1.** Classification accuracy of our FD over CBS [37] and Baseline. Boldface indicates the top two models with higher performance and additional underline for the best model. All the experiments are conducted on a common dropout probability of $p^G = 0.4$, $p^{LoG} = 0.5$, $p^{Gabor} = 0.8$

| | | Classification | | | Robustness | |
|---|---|---|---|---|---|---|
| | | CIFAR-100 | CIFAR-10 | SVHN | CIFAR-100-C | CIFAR-10-C |
| ResNet-18 | Baseline | 65.31 ± 0.14 | 89.24 ± 0.23 | 96.26 ± 0.06 | 43.68 ± 0.17 | 70.36 ± 0.75 |
| | CBS | 65.77 ± 0.45 | 89.81 ± 0.09 | 96.27 ± 0.06 | 46.69 ± 0.04 | 74.17 ± 0.29 |
| | FD-GF | **67.45 ± 0.54** | **90.33 ± 0.47** | **96.70 ± 0.30** | **46.78 ± 0.30** | **74.32 ± 3.15** |
| | FD-RF | **68.20 ± 0.52** | **90.53 ± 0.28** | 96.60 ± 0.03 | **48.10 ± 0.07** | **74.43 ± 1.31** |
| VGG-16 | Baseline | 58.54 ± 0.35 | 87.53 ± 0.08 | 95.77 ± 0.05 | 40.10 ± 0.01 | 71.58 ± 0.33 |
| | CBS | **63.67 ± 0.11** | **89.47 ± 0.14** | **96.39 ± 0.02** | 44.53 ± 1.33 | **74.41 ± 0.45** |
| | FD-GF | 63.09 ± 0.14 | 89.32 ± 0.27 | 96.26 ± 0.13 | **44.69 ± 1.22** | 74.01 ± 1.20 |
| | FD-RF | **63.94 ± 0.35** | **89.59 ± 0.07** | 96.26 ± 0.01 | 44.26 ± 0.29 | 73.12 ± 1.97 |
| W-ResNet | Baseline | 68.06 ± 0.07 | **90.98 ± 0.24** | 97.04 ± 0.05 | 39.63 ± 0.52 | 67.39 ± 0.28 |
| | CBS | 65.04 ± 0.45 | 87.00 ± 1.53 | **97.12 ± 0.02** | 37.05 ± 4.22 | 65.45 ± 1.64 |
| | FD-GF | **68.38 ± 0.59** | 90.80 ± 0.45 | 97.05 ± 0.11 | **41.00 ± 0.18** | **69.22 ± 1.17** |
| | FD-RF | **68.11 ± 0.32** | **90.88 ± 0.12** | 97.09 ± 0.16 | **41.39 ± 0.66** | **69.63 ± 0.30** |
| ResNeXt | Baseline | 68.88 ± 0.59 | **90.84 ± 0.30** | 96.31 ± 0.02 | 44.72 ± 0.47 | 68.85 ± 0.86 |
| | CBS | 65.15 ± 1.27 | 89.28 ± 0.40 | 96.15 ± 0.12 | 42.46 ± 1.67 | 69.68 ± 0.37 |
| | FD-GF | 68.35 ± 0.45 | 89.69 ± 0.44 | **96.63 ± 0.01** | 45.12 ± 1.73 | **72.17 ± 1.60** |
| | FD-RF | **68.95 ± 0.66** | **89.86 ± 0.47** | 96.47 ± 0.05 | **46.24 ± 0.56** | 70.36 ± 0.68 |

The original implementation[1] of the closely related technique of Curriculum by Smoothing (CBS) [37] is adopted for comparison with FD and the

---

[1] https://github.com/pairlab/CBS.

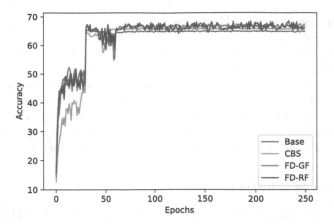

**Fig. 2.** Validation performance over the epochs for ResNet-18 with CIFAR-100.

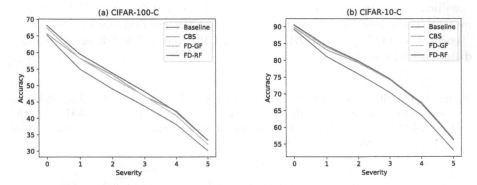

**Fig. 3.** Robustness plots with 0 to 5 severity for (a) CIFAR-100-C and (b) CIFAR-10-C. The result obtains on ResNet-18 for Baseline, CBS, FD-GF, and FD-RF. The proposed method shows constant performance preservation over baseline where CBS preserves robustness with higher severity.

baseline models. In CBS, all classification models are optimized with SGD and the same settings for learning rate schedule, weight decay, and momentum throughout all experiments. We use a common set of dropout parameter ($p^G = 0.4$, $p^{LoG} = 0.5$, $p^{Gabor} = 0.8$) for all the experiments which is tuned with ResNet-18 and CIFAR-100 dataset. Figure 2 shows the validation accuracy versus epochs for ResNet-18 on CIFAR-100. Table 1 presents the classification and robustness performance metrics of our method compared to baselines and CBS. There is a significant improvement in classification accuracy for FD-RF and FD-GF over the baseline and CBS for most of the architectures and datasets. Specifically, FD improves 2–3% accuracy for ResNet-18 with the CIFAR-100 classification dataset. The table also shows the superior performance of our method on corrupted CIFAR-10-C and CIFAR-100-C datasets for the median severity. Robustness performance against severity level is shown in Fig. 3. Both FD-RF and FD-GF show robust prediction accuracy for different severity levels over baseline, while CBS does show competitive performance in higher severity levels of corruption.

## 5.2  Domain Adaptation

**Vision Application.** We investigate the performance of FD with the task of unsupervised domain adaptation (UDA) in visual recognition. We use three datasets of MNIST [26], USPS [21] and SVHN [16] with four architectures of ResNet-18 [17], Wide-ResNet-52 [45], ResNeXt-50 [44] and VGG-16 [36] by following domain setup of [41] where a model train on one dataset and test on different data source for the task of UDA in classification. The datasets consist of 10 classes and arrange into three directions of adaptation. The UDA performance of FD, CBS, and baselines are shown in Table 2. There is 3–4% performance improvement with our method for multiple architectures such as ResNet-18, VGG-16, Wide-ResNet-52 for the adaptation settings of MNIST→USPS, USPS→MNIST, and SVHN→MNIST, respectively. Overall, FD-GF and FD-RF obtain the best performance compared to baseline and CBS for most of the cases.

**Table 2.** Unsupervised domain adaptation (UDA) with classification task. Boldface indicates the top two models with higher accuracy and additional underline for the best model.

| | | MNIST->USPS | USPS->MNIST | SVHN->MNIST |
|---|---|---|---|---|
| ResNet-18 | Baseline | 82.06 ± 0.27 | **77.18 ± 0.08** | 79.88 ± 0.20 |
| | CBS | 84.30 ± 0.46 | 50.82 ± 0.33 | 82.20 ± 0.25 |
| | FD-GF | **85.35 ± 0.27** | 66.48 ± 0.23 | **82.66 ± 0.21** |
| | FD-RF | **86.60 ± 0.20** | **73.76 ± 0.25** | **82.43 ± 0.19** |
| VGG-16 | Baseline | **84.20 ± 0.20** | 52.03 ± 0.33 | 78.68 ± 0.36 |
| | CBS | 78.95 ± 0.12 | 49.61 ± 0.26 | **83.35 ± 0.20** |
| | FD-GF | 82.81 ± 0.35 | **54.07 ± 0.11** | 78.67 ± 0.25 |
| | FD-RF | **84.11 ± 0.14** | **57.67 ± 0.19** | **79.22 ± 0.14** |
| Wide-ResNet-52 | Baseline | 79.32 ± 0.36 | **86.81 ± 0.22** | **83.97 ± 0.50** |
| | CBS | 75.99 ± 0.20 | 85.52 ± 0.26 | 82.26 ± 0.17 |
| | FD-GF | **82.96 ± 0.12** | 86.64 ± 0.07 | 81.71 ± 0.19 |
| | FD-RF | 79.17 ± 0.23 | **88.53 ± 0.11** | **86.02 ± 0.28** |
| ResNeXt-50 | Baseline | **92.62 ± 0.18** | **60.04 ± 0.14** | **83.20 ± 0.38** |
| | CBS | 88.23 ± 0.25 | 51.83 ± 0.13 | 81.97 ± 0.18 |
| | FD-GF | **92.87 ± 0.08** | **59.98 ± 0.11** | **83.12 ± 0.27** |
| | FD-RF | 91.53 ± 0.19 | 52.12 ± 0.19 | 80.83 ± 0.22 |

**Medical Application.** To evaluate the performance of FD with a real-world medical dataset, we utilize the Multi-Centre, Multi-Vendor, and Multi-Disease Cardiac Segmentation (M&MS) MRI dataset [7]. It contains 150 cases each from two vendors of A and B equally. The annotation consists of three cardiac regions including left ventricle (LV), right ventricle (RV), and myocardium (MYO). We split the data vendor-wise for train and validation so that the experiments reflect the domain shift setup. A 3D UNet [9] implementation[2] is adopted as the baseline architecture. We integrate our FD-RF, FD-GF, and closely related work CBS [37]. The filtering effects for all these techniques are only introduced in the encoder part as it plays the role of extracting image features while the decoder operates in the space of segmentation. During training, the Adam optimizer is used with a learning rate of $10^{-4}$ and cross-entropy loss. We use cross-validation by swapping vendor-A and vendor-B as train and validation sets.

**Table 3.** Cross-validation performance of cardiac image segmentation from 3D MRI. All the experiments conduct on a common dropout probability of $p^G = 0.5$, $p^{LoG} = 0.5$, $p^{Ga} = 0.8$. Cross-vendor validation is done to produce the prediction for both vendors A and B. Boldface indicates the top two models with higher DSC and additional underline for the best model.

|  | Vendor-A to Vendor-B | | | | Vendor-B to Vendor-A | | | |
|---|---|---|---|---|---|---|---|---|
|  | LV | RV | MYO | **Mean DSC** | LV | RV | MYO | **Mean DSC** |
| Baseline | 72.19 | **65.43** | 60.28 | 65.97 | 52.98 | 39.30 | **42.08** | 44.79 |
| CBS | 71.60 | 62.28 | 62.24 | 65.37 | **63.11** | **<u>46.40</u>** | 31.33 | 46.95 |
| FD-GF | **<u>72.47</u>** | 63.05 | **64.01** | **66.51** | **<u>63.26</u>** | 44.34 | 40.45 | **49.35** |
| FD-RF | **68.96** | **<u>66.30</u>** | **<u>67.71</u>** | **<u>67.66</u>** | 51.47 | **44.57** | **<u>47.56</u>** | **<u>51.47</u>** |

The results of cross-vendors validation are reported in the Table 3. Dice similarity coefficient (DSC) is used to measure the segmentation prediction for baseline, CBS, and our FD-GF, FD-RF. The left side of the table contains model performances trained on vendor-A and validation on vendor-B and swapping the vendors on the right side. The results suggest a 2–3% increase of the mean DSC for our method over CBS where 4–6% improvement compared to baseline. The prediction visualization is also showing better segmentation for FD-RF compared to other methods in Fig. 4. Overall, the results indicate the better generalization and robustness capacity of the proposed method on dataset shift and domain shift.

## 5.3    Ablation Study

**Dropout Ratio.** To determine the effective dropout ratio for each filter, we perform specific experiments by varying the ratio for the individual filter. Figure 5

---

[2] https://github.com/lescientifik/open_brats2020.

shows the accuracy over dropout ratio in the range 0.1 to 0.9 for ResNet-18 with CIFAR-100. From this, we selected dropout ratios of Gaussian ($p^G = 0.4$), LoG ($p^{LoG} = 0.5$), and Gabor ($p^{Ga} = 0.8$) filters for all further experiments of classification tasks.

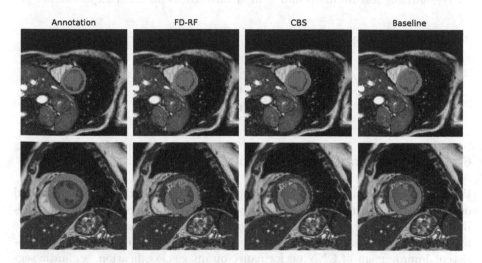

**Fig. 4.** UDA performance in segmentation for Baseline, CBS, and FD-RF. The middle slices of two random MRI scans are visualized for annotation and prediction on different models. The colors of purple, blue and yellow indicate the left ventricle (LV), myocardium (MYO), and right ventricle (RV). (Color figure online)

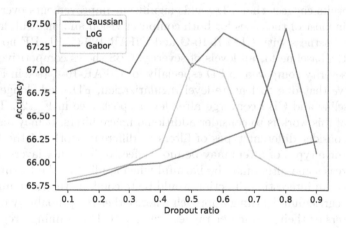

**Fig. 5.** Effect of individual filtering and selecting dropout ratio for our FD technique. The optimal dropout ratios for each filter obtain as $p^G = 0.4$, $p^{LoG} = 0.5$, $p^{Ga} = 0.8$ for ResNet-18 with CIFAR-100.

**Kernel Size.** The considered imaging filters can be used with different kernel sizes. We investigate the FD performance under four different kernel sizes including $[1 \times 1, 3 \times 3, 5 \times 5, 7 \times 7]$. Table 4 presents the performance of FD by varying kernel size. Kernel size of $3 \times 3$ seems most effective in the design of FD for regularizing feature maps and is used throughout all other experiments.

**Table 4.** Effect of different kernel size for FD using ResNet-18 and CIFAR dataset.

|           | $1 \times 1$ | $3 \times 3$ | $5 \times 5$ | $7 \times 7$ |
|-----------|--------------|--------------|--------------|--------------|
| CIFAR-100 | 65.34        | **68.57**    | 68.00        | 67.19        |
| CIFAR-10  | 89.15        | **90.33**    | 89.94        | 89.88        |

## 6   Discussion and Conclusion

We introduced a novel method for feature-level regularization of convolutional neural networks namely Frequency Dropout with Randomized Filtering (FD-RF). We considered Gaussian, Laplacian of Gaussian, and Gabor filters. We provide empirical evidence across a larger number of experiments showing a consistent improvement of CNN performance on image classification, semantic segmentation, and unsupervised domain adaptation tasks on both computer vision and a real-world medical imaging dataset. We also observe improvements for model robustness on the corrupted CIFAR datasets (CIAFR-10-C and CIFAR-100-C). To make a fair comparison with closely related work of CBS [37], we build a simplified version of FD-RF with only Gaussian filters (FD-GF). Our experimental results suggest that FD yields significant improvements over baselines and CBS in most of the cases for both computer vision and medical datasets. In terms of robustness with CIAFR-10-C and CIFAR-100-C, FD-RF notably outperforms the baseline for all levels of severity. CBS shows competitive accuracy in higher severity compared to FD especially for CIFAR-10-C (see in Fig. 3). We also observe that due to feature-level regularization, FD needs longer training where baseline and CBS converge after fewer epoch (see in Fig. 2). The future direction of this work is to consider additional image filters. It may also be beneficial to consider different types of filters at different depths of the CNN. For example, some types of filters may be more effective in earlier layers of the network whereas some filters may be harmful when applied to the feature maps in later layers. An interesting direction would be to combine the randomized filtering with a curriculum learning approach where either the probability of selecting certain filters or their parameter ranges are annealed as training progresses.

**Acknowledgements.** This project has received funding from the European Research Council (ERC under the European Union's Horizon 2020 research and innovation programme (Grant Agreement No. 757173, Project MIRA).

# References

1. Achille, A., Rovere, M., Soatto, S.: Critical learning periods in deep networks. In: International Conference on Learning Representations (2018)
2. Alekseev, A., Bobe, A.: Gabornet: gabor filters with learnable parameters in deep convolutional neural network. In: 2019 International Conference on Engineering and Telecommunication (EnT), pp. 1–4. IEEE (2019)
3. Amini, A., Soleimany, A., Karaman, S., Rus, D.: Spatial uncertainty sampling for end-to-end control. arXiv preprint arXiv:1805.04829 (2018)
4. Azulay, A., Weiss, Y.: Why do deep convolutional networks generalize so poorly to small image transformations? arXiv preprint arXiv:1805.12177 (2018)
5. Azulay, A., Weiss, Y.: Why do deep convolutional networks generalize so poorly to small image transformations? J. Mach. Learn. Res. **20**, 1–25 (2019)
6. Bouthillier, X., Konda, K., Vincent, P., Memisevic, R.: Dropout as data augmentation. arXiv preprint arXiv:1506.08700 (2015)
7. Campello, V.M., et al.: Multi-centre, multi-vendor and multi-disease cardiac segmentation: the m&ms challenge. IEEE Trans. Med. Imaging **40**, 3543–3554 (2021)
8. Chen, T., Kornblith, S., Norouzi, M., Hinton, G.: A simple framework for contrastive learning of visual representations. In: International Conference on Machine Learning, pp. 1597–1607. PMLR (2020)
9. Çiçek, Ö., Abdulkadir, A., Lienkamp, S.S., Brox, T., Ronneberger, O.: 3D U-Net: learning dense volumetric segmentation from sparse annotation. In: Ourselin, S., Joskowicz, L., Sabuncu, M.R., Unal, G., Wells, W. (eds.) MICCAI 2016. LNCS, vol. 9901, pp. 424–432. Springer, Cham (2016). https://doi.org/10.1007/978-3-319-46723-8_49
10. Cohen, N., Shashua, A.: Inductive bias of deep convolutional networks through pooling geometry. arXiv preprint arXiv:1605.06743 (2016)
11. Dai, S., Cheng, Y., Zhang, Y., Gan, Z., Liu, J., Carin, L.: Contrastively smoothed class alignment for unsupervised domain adaptation. In: Proceedings of the Asian Conference on Computer Vision (2020)
12. Gal, Y., Ghahramani, Z.: Dropout as a bayesian approximation: representing model uncertainty in deep learning. In: International Conference on Machine Learning, pp. 1050–1059. PMLR (2016)
13. Geirhos, R., et al.: Shortcut learning in deep neural networks. Nat. Mach. Intell. **2**(11), 665–673 (2020)
14. Geirhos, R., Rubisch, P., Michaelis, C., Bethge, M., Wichmann, F.A., Brendel, W.: Imagenet-trained cnns are biased towards texture; increasing shape bias improves accuracy and robustness. arXiv preprint arXiv:1811.12231 (2018)
15. Gomez, A.N., Zhang, I., Swersky, K., Gal, Y., Hinton, G.E.: Targeted dropout (2018)
16. Goodfellow, I.J., Bulatov, Y., Ibarz, J., Arnoud, S., Shet, V.: Multi-digit number recognition from street view imagery using deep convolutional neural networks. arXiv preprint arXiv:1312.6082 (2013)
17. He, K., Zhang, X., Ren, S., Sun, J.: Deep residual learning for image recognition. In: Proceedings of the IEEE Conference on Computer Vision and Pattern Recognition, pp. 770–778 (2016)
18. Hendrycks, D., Dietterich, T.: Benchmarking neural network robustness to common corruptions and perturbations. arXiv preprint arXiv:1903.12261 (2019)
19. Hinton, G.E., Srivastava, N., Krizhevsky, A., Sutskever, I., Salakhutdinov, R.R.: Improving neural networks by preventing co-adaptation of feature detectors. arXiv preprint arXiv:1207.0580 (2012)

20. Hossain, M.T., Teng, S.W., Sohel, F., Lu, G.: Robust image classification using a low-pass activation function and dct augmentation. IEEE Access **9**, 86460–86474 (2021)
21. Hull, J.J.: A database for handwritten text recognition research. IEEE Trans. Pattern Anal. Mach. Intell. **16**(5), 550–554 (1994)
22. Khan, S.H., Hayat, M., Porikli, F.: Regularization of deep neural networks with spectral dropout. Neural Netw. **110**, 82–90 (2019)
23. Khosla, P., et al.: Supervised contrastive learning. arXiv preprint arXiv:2004.11362 (2020)
24. Krizhevsky, A., Hinton, G., et al.: Learning multiple layers of features from tiny images (2009)
25. Lassance, C., Gripon, V., Ortega, A.: Laplacian networks: bounding indicator function smoothness for neural networks robustness. APSIPA Trans. Signal Inf. Process. **10** (2021)
26. LeCun, Y.: The mnist database of handwritten digits (1998). http://yann.lecun.com/exdb/mnist/
27. Lee, J., Won, T., Lee, T.K., Lee, H., Gu, G., Hong, K.: Compounding the performance improvements of assembled techniques in a convolutional neural network. arXiv preprint arXiv:2001.06268 (2020)
28. Lopes, R.G., Yin, D., Poole, B., Gilmer, J., Cubuk, E.D.: Improving robustness without sacrificing accuracy with patch gaussian augmentation. arXiv preprint arXiv:1906.02611 (2019)
29. Luan, S., Chen, C., Zhang, B., Han, J., Liu, J.: Gabor convolutional networks. IEEE Trans. Image Process. **27**(9), 4357–4366 (2018)
30. Mairal, J.: End-to-end kernel learning with supervised convolutional kernel networks. Adv. Neural Inf. Process. Syst. **29**, 1399–1407 (2016)
31. Nair, T., Precup, D., Arnold, D.L., Arbel, T.: Exploring uncertainty measures in deep networks for multiple sclerosis lesion detection and segmentation. Med. Image Anal. **59**, 101557 (2020)
32. Navarro, F., et al.: Evaluating the robustness of self-supervised learning in medical imaging. arXiv preprint arXiv:2105.06986 (2021)
33. Pérez, J.C., Alfarra, M., Jeanneret, G., Bibi, A., Thabet, A., Ghanem, B., Arbeláez, P.: Gabor layers enhance network robustness. In: Vedaldi, A., Bischof, H., Brox, T., Frahm, J.-M. (eds.) ECCV 2020. LNCS, vol. 12354, pp. 450–466. Springer, Cham (2020). https://doi.org/10.1007/978-3-030-58545-7_26
34. Salehinejad, H., Valaee, S.: Ising-dropout: a regularization method for training and compression of deep neural networks. In: ICASSP 2019–2019 IEEE International Conference on Acoustics, Speech and Signal Processing (ICASSP), pp. 3602–3606. IEEE (2019)
35. Shi, B., Zhang, D., Dai, Q., Zhu, Z., Mu, Y., Wang, J.: Informative dropout for robust representation learning: a shape-bias perspective. In: International Conference on Machine Learning, pp. 8828–8839. PMLR (2020)
36. Simonyan, K., Zisserman, A.: Very deep convolutional networks for large-scale image recognition. arXiv preprint arXiv:1409.1556 (2014)
37. Sinha, S., Garg, A., Larochelle, H.: Curriculum by smoothing. Adv. Neural Inf. Process. Syst. **33** (2020)
38. Srivastava, N., Hinton, G., Krizhevsky, A., Sutskever, I., Salakhutdinov, R.: Dropout: a simple way to prevent neural networks from overfitting. J. Mach. Learn. Res. **15**(1), 1929–1958 (2014)

39. Taori, R., Dave, A., Shankar, V., Carlini, N., Recht, B., Schmidt, L.: Measuring robustness to natural distribution shifts in image classification. arXiv preprint arXiv:2007.00644 (2020)

40. Tompson, J., Goroshin, R., Jain, A., LeCun, Y., Bregler, C.: Efficient object localization using convolutional networks. In: Proceedings of the IEEE Conference on Computer Vision and Pattern Recognition, pp. 648–656 (2015)

41. Tzeng, E., Hoffman, J., Saenko, K., Darrell, T.: Adversarial discriminative domain adaptation. In: Proceedings of the IEEE Conference on Computer Vision and Pattern Recognition, pp. 7167–7176 (2017)

42. Wager, S., Wang, S., Liang, P.S.: Dropout training as adaptive regularization. Adv. Neural Inf. Process. Syst. **26**, 351–359 (2013)

43. Wang, H., Wu, X., Huang, Z., Xing, E.P.: High-frequency component helps explain the generalization of convolutional neural networks. In: Proceedings of the IEEE/CVF Conference on Computer Vision and Pattern Recognition, pp. 8684–8694 (2020)

44. Xie, S., Girshick, R., Dollár, P., Tu, Z., He, K.: Aggregated residual transformations for deep neural networks. In: Proceedings of the IEEE Conference on Computer Vision and Pattern Recognition, pp. 1492–1500 (2017)

45. Zagoruyko, S., Komodakis, N.: Wide residual networks. arXiv preprint arXiv:1605.07146 (2016)

46. Zhang, C., Bengio, S., Hardt, M., Mozer, M.C., Singer, Y.: Identity crisis: memorization and generalization under extreme overparameterization. arXiv preprint arXiv:1902.04698 (2019)

47. Zhang, R.: Making convolutional networks shift-invariant again. In: International Conference on Machine Learning, pp. 7324–7334. PMLR (2019)

# PVBM: A Python Vasculature Biomarker Toolbox Based on Retinal Blood Vessel Segmentation

Jonathan Fhima[1,2]([✉]) [iD], Jan Van Eijgen[3,4]([✉]) [iD], Ingeborg Stalmans[3,4] [iD], Yevgeniy Men[1,5], Moti Freiman[1] [iD], and Joachim A. Behar[1] [iD]

[1] Faculty of Biomedical Engineering, Technion-IIT, Haifa, Israel
jonathanfh@campus.technion.ac.il, jbehar@technion.ac.il
[2] Department of Applied Mathematics Technion-IIT, Haifa, Israel
[3] Research Group Ophthalmology, Department of Neurosciences, KU Leuven, Leuven, Belgium
[4] Department of Ophthalmology, University Hospitals UZ Leuven, Leuven, Belgium
[5] The Andrew and Erna Viterbi Faculty of Electrical and Computer Engineering, Technion-IIT, Haifa, Israel

**Abstract. Introduction:** Blood vessels can be non-invasively visualized from a digital fundus image (DFI). Several studies have shown an association between cardiovascular risk and vascular features obtained from DFI. Recent advances in computer vision and image segmentation enable automatising DFI blood vessel segmentation. There is a need for a resource that can automatically compute digital vasculature biomarkers (VBM) from these segmented DFI. **Methods:** In this paper, we introduce a Python Vasculature BioMarker toolbox, denoted PVBM. A total of 11 VBMs were implemented. In particular, we introduce new algorithmic methods to estimate tortuosity and branching angles. Using PVBM, and as a proof of usability, we analyze geometric vascular differences between glaucomatous patients and healthy controls. **Results:** We built a fully automated vasculature biomarker toolbox based on DFI segmentations and provided a proof of usability to characterize the vascular changes in glaucoma. For arterioles and venules, all biomarkers were significant and lower in glaucoma patients compared to healthy controls except for tortuosity, venular singularity length and venular branching angles. **Conclusion:** We have automated the computation of 11 VBMs from retinal blood vessel segmentation. The PVBM toolbox is made open source under a GNU GPL 3 license and is available on physiozoo.com (following publication).

**Keywords:** Digital fundus images · Digital vascular biomarkers · Glaucoma · Retinal vasculature

## 1 Introduction

According to the National Center for Health Statistics, cardiovascular diseases (CVD), including coronary heart disease and stroke, are the most common cause

© The Author(s), under exclusive license to Springer Nature Switzerland AG 2023
L. Karlinsky et al. (Eds.): ECCV 2022 Workshops, LNCS 13803, pp. 296–312, 2023.
https://doi.org/10.1007/978-3-031-25066-8_15

of death in the USA [35]. Since the beginning of the 20th century, researchers have shown that retinal microvascular abnormalities can be used as biomarkers of CVD [15,16,23]. Retinal vasculature can be non-invasively assessed using DFIs, which can be easily obtained using a fundus camera. Consequently, retinal vascular features obtained from DFI may be used to characterize and analyze vascular health. In order to enable reproducible research, it is necessary to fully automate the computation of these biomarkers from the segmented vasculature. We developed a Python Vasculature Biomarker Toolbox (PVBM), based on DFI segmentations made by expert annotators from the University Hospitals Leuven in Belgium. PVBM enables a quantitative analysis of the vascular geometry thereof with broad application in retinal research. In this paper we illustrate the potential of PVBM by characterizing vascular changes in glaucoma patients.

## 1.1 Prior Works

**Connection Between Retinal Vasculature and Cardiovascular Health:** As early as of beginning of the 20th Century research has been carried out to assess the relationship between the retinal vasculature and cardiovascular health. Marcus Gunn can be seen as one of the first to describe the relation between hypertension and retinal characteristics [15,16], and is followed by the work of H.G Scheie [43] in 1953. In 1974, N.M Keith showed that hypertension and its mortality risk is reflected in the retinal vasculature [23] and in 1999 Sharrett et al. [44] added arterio-venous nicking and arteriole narrowing to the list of pathological findings. Examples of retinal microvascular abnormalities in hypertensive patients can be seen in Fig. 1. Witt et al. [50] concluded that vessel tortuosity significantly distinguished between patients with ischemic heart disease and healthy controls. Over the years retinal vessel calibres were shown to change in hypertension [4], obesity [18], chronic kidney disease [33,42], diabetes mellitus [4], coronary artery disease [17] and glaucoma [22]. Fractal dimensions of the retinal vascular tree are among the newest biomarkers to study cardiovascular risk. Monofractal dimension was shown to change with age, smoking behaviour [24], blood pressure [27], diabetic retinopathy [6], chronic kidney disease [45], stroke [9], and coronary heart disease mortality [25], while Multifractal dimensions were found to be negatively associated with blood pressure and WHO/ISH cardiovascular risk score [49]. The established association between vascular biomarkers and CVD prompts the development of algorithmic solutions for automated computation thereof.

**Automated Biomarker Computation Based on DFIs:** Several attempts have been made to extract meaningful biomarkers to characterize cardiovascular health based on DFI vasculature. In 2000, Martinez-Perez et al. [32] introduced a semi-automated algorithms capable of computing vasculature biomarkers (VBMs) such as vessel diameter, length, tortuosity, area and branching angles. In 2011, Perez-Rovira et al. [38] created Vessel Assessment and Measurement Platform for Images of the REtina (VAMPIRE), a semi-automatic software that can extract the optic disk and compute vessel width, tortuosity, fractal dimension, and branching

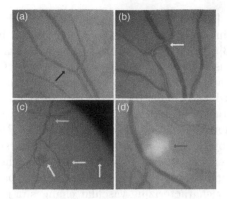

**Fig. 1.** Examples of retinal vascular signs in patients with cardiovascular diseases. Reproduced with permission from Liew et al. [26]. Black arrow: focal arteriolar narrowing, White arrow: arterio-venous nicking, Yellow arrow: haemorrhage, Blue arrow: micro-aneurysm, Red arrow: cotton wool spot. (Color figure online)

coefficient. RetinaCAD was developed in 2014 [8]. This automated system is able to calculate Central Retinal Arteriolar Equivalent (CRAE), the Central Retinal Venular Equivalent (CRVE), and the Arteriolar-to-Venular Ratio (AVR). Lastly, many algorithmic approaches have been developed to estimate the blood vessel tortuosity [14,19,36]. Last year, Provost et al. [40] used the MONA REVA software which semi-automatically segments retinal blood vessels and measures tortuosity and fractal dimension in order to analyze the impact of their changes on children's behaviour.

### 1.2 Research Gap and Objectives

Vasculature biomarkers have been previously proposed and implemented in a semi or fully automated manner. However, these biomarkers were only analyzed individually across multiple groups. Hence the need to use a combined, comprehensive set of VBM within a machine learning (ML) framework for disease diagnosis and risk prediction. The large number of images needed to train ML algorithms require a fully automated computation of these VBM. In this paper we created a computerized toolbox, denoted PVBM, that enables the computation of 11 VBMs engineered from segmented arteriolar or venular networks. A potential application of PVBM is demonstrated by comparison of VBM in glaucoma patients versus healthy controls.

## 2  Methods

### 2.1  Dataset

A database provided by the University Hospitals of Leuven (UZ) in Belgium was used. This database contains 108,516 DFIs, centered around the optic disc. The

**Table 1.** Leuven A/V segmented database (UZFG) summary including the median (Q1–Q3) age, the gender and the diagnosis for the glaucoma (GLA) patient and normal ophthalmic findings (NOR) patient.

|       | № DFIs | Age          | Gender |
|-------|--------|--------------|--------|
| NOR   | 19     | 58 (43–60)   | 42% M  |
| GLA   | 50     | 68 (59–75)   | 42% M  |
| Total | 69     | 61 (57–70)   | 42% M  |

resolution of $1444 \times 1444$ is higher than most public databases, which enables the visualization of smaller blood vessels. Median age was 66 (Q1 and Q3 respectively 54 and 75 years old) and 52% were women. For a subset of the database, the blood vessels were manually segmented by retinal experts using Lirot.ai on Apple iPad Pro 11" and 13" [12]. This subset is denoted UZFG and consists of 69 DFIs. The patients included in UZFG were between 19 and 90 years old with a median age of 61 (Q1 and Q3 were 57 and 70 years old), 58% were female. UZFG has 59% of left and 41% of right eye DFIs. Patients belonging to the UZFG are separated into the classes: Normal ophthalmic findings (NOR) and Glaucoma (GLA) (Table 1).

**Protocol for DFI Vasculature Segmentation by Experts:** Arterioles carry higher concentration of oxyhemoglobin than venules and therefore exhibit a brighter inner part compared to their walls. This feature is known as the central reflex and is more typical for larger arterioles [34]. The exact intensity of this reflection is additionally influenced by the composition of the vessel wall [21], the roughness of the surface, the caliber of the blood column, the indices of refraction of erythrocytes and plasma, the pupil size, the axial length of the eye and the tilt of the blood vessel relative to the direction of incident light. Since the branching of arterioles and venules can be found inside the optic nerve head, two arterioles or venules can be found adjacently at the optic disc rim [5]. The image variability in term of color, contrast, and illumination, challenge an accurate (automatic) arteriole-venule discrimination [3]. Vessel segmentation was manually performed by a pool of ten ophthalmology students experienced in microvascular research, using the software Lirot.ai [12] and afterwards corrected by an UZ retinal expert. Arteriole-venule discrimination was carried out based on the following visual and geometric features [34]:

- Venules are darker than arterioles.
- The central reflex is more recognizable in arterioles.
- Venules are usually thicker than arterioles.
- Venules and arterioles usually alternate near the optic disc.
- It is unlikely that arterioles cross arterioles or that venules cross venules.

### 2.2   Digital Vasculature Biomarkers

A total of 11 VBMs were implemented in PVBM (Table 2). The biomarkers were computed separately for arterioles and venules.

**Table 2.** List of digital vasculature biomarker implemented in PVBM.

| Number | Biomarker | Definition | Unit |
|---|---|---|---|
| 1 | OVLEN [32] | Overall length | Pixel |
| 2 | OVPER [32] | Overall perimeter | Pixel |
| 3 | OVAREA [32] | Overall area | $Pixel^2$ |
| 4 | END [32] | Number of endpoints | – |
| 5 | INTER [32] | Number of intersection points | – |
| 6 | TOR [28,32] | Median tortuosity | % |
| 7 | BA [32] | Branching angles | $°(degree)$ |
| 8 | $D_0$ [7,46] | Capacity dimension | – |
| 9 | $D_1$ [7,46] | Entropy dimension | – |
| 10 | $D_2$ [7,46] | Correlation dimension | – |
| 11 | SL [29] | Singularity Length | – |

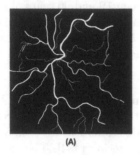 

(A)     (B)

**Fig. 2.** Skeletonization process of a vascular network. (A): Example of a vascular network $y_a$, (B): Corresponding skeleton of $y_a$.

**Overall Length:** The OVLEN biomarker refers to the sum of the length of a vascular network, whether for arterioles or venules. To compute it, the first step is to extract the skeleton of the vascular network, which can be seen in Fig. 2. Then the number of pixels that belong to this skeleton are summed, and divided by the image size ($1444 \times 1444$), then multiplied by 1000 for scaling purposes. It is computed using the following formula:

$$OVLEN = 1e3 * \frac{\Sigma_{p \in S} \sqrt{2} * \mathbb{1}_{|\partial_x p| + |\partial_y p| = 0}(p) + \mathbb{1}_{|\partial_x p| + |\partial_y p| \neq 0}(p)}{1444^2} \quad (1)$$

where $S$ is the set of pixel inside the skeletonized image (Fig. 2), and x/y represent the horizontal/vertical direction.

**Overall Perimeter:** The OVPER refers to the sum of perimeter's length of a vascular network. It is computed as the length of the border of the overall segmentation. It required an edge extraction of the segmentation, which could

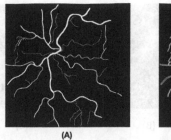

**(A)**                    **(B)**

**Fig. 3.** Border computation process of a vascular network. (A): Vascular network $y_a$, (B): Corresponding computed edge of $y_a$.

be easily found by convolving the original segmentation by a Laplacian filter (Fig. 3). It is computed using the following formula:

$$OVPER = 1e2 * \frac{\Sigma_{p \in E}\sqrt{2} * \mathbb{1}_{|\partial_x p| + |\partial_y p| = 0}(p) + \mathbb{1}_{|\partial_x p| + |\partial_y p| \neq 0}(p)}{1444^2} \qquad (2)$$

where $E$ is set of pixel inside the edge image of the vascular network (Fig. 3) and x/y represent the horizontal/vertical direction.

**Overall Area:** OVAREA is defined as the surface covered by the segmentation. In terms of pixels, it could be represented as the ratio of white pixels in the segmentation to the overall number of pixels. It is computed using the following formula:

$$OVAREA = 1e2 * \frac{\Sigma_{p \in V}\mathbb{1}_1(p)}{1444^2} \qquad (3)$$

where $V$ is set of pixel inside the image of the vascular network (Fig. 3).

**Endpoints and Intersection Points:** The endpoints are the points at the end of the vascular network, which means in the skeleton version of the network, the points which have only one neighbor which belongs to the skeleton. The intersection points are the points where a blood vessel is divided into more than one blood vessel, which means in the skeleton version of the network, the points which have more than two neighbors which belong to the skeleton. Their automatic detection was done using a filter $k$ of size $(3 \times 3)$ where $k_{i,j} = 10$ if $i = j$, or $k_{i,j} = 1$ otherwise. The skeleton is then convolved with this filter to obtain a new image. In this new image, the endpoints will be the pixels with a value of 11, and the intersection points will be the pixels with a value of 13 or larger (Fig. 4). We can represent the endpoints and the intersection points according to the following equation:

$$END = \{p = (i, j) \in Skeleton | (Skeleton \circledast \begin{pmatrix} 1 & 1 & 1 \\ 1 & 10 & 1 \\ 1 & 1 & 1 \end{pmatrix})[i, j] = 11\} \qquad (4)$$

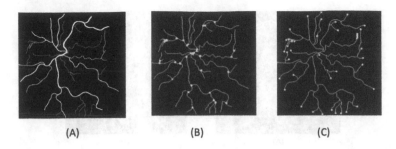

**Fig. 4.** Automatic detection of the particular points of a vascular network. (A): Vascular network $y_a$, (B): Automatic detection of the intersection points (INTER = 32) of $y_a$, (C): Automatic detection of the endpoints (END = 40) of $y_a$.

$$INTER = \{p = (i,j) \in Skeleton | (Skeleton \circledast \begin{pmatrix} 1 & 1 & 1 \\ 1 & 10 & 1 \\ 1 & 1 & 1 \end{pmatrix})[i,j] \geq 13\} \qquad (5)$$

Particular points is the name given to the combined set of points resulting from the union of the endpoints and intersection points. The number of endpoints (END) and the number of intersection points (INTER) were computed as VBMs.

**Median Tortuosity:** The simplest mathematical method to estimate tortuosity is the arc-chord ratio, defined as the ratio between the length of the curve and the distance between its ends [19]. In our work, the median tortuosity was computed using the arc-chord ratio based on the linear interpolation of all the blood vessels. For that purpose, the skeleton is treated as a graph and particular points are extracted. To compute the linear interpolation it is then required to find all the particular points connected to a given particular point. The connection between each particular point was stored in a dictionary according to Algorithm 1; the output of this algorithm is a dictionary where the keys are the particular points, and the values are the list of the connected particular points. Having this dictionary, it is possible to generate the linear interpolation between each connected particular point, as it is shown in Fig. 5. The tortuosity of each blood vessel can be estimated by computing the ratio between the blood vessel's length (yellow curves in Fig. 5) and the length of the interpolation of this blood vessel (red lines in Fig. 5). The median tortuosity will then be the median value of the tortuosity of all the blood vessels.

```
Algorithm 1
program Compute connected pixel dictionary ( S (= skeleton),
P (= particular point list))
    initialize an empty dictionary: connected;
    initialize an empty dictionary: visited;
    for (i,j) in P:
        if S[i,j] == 1 (White):
```

```
                recursive(i,j,S,i,j,visited,P,connected);
end.

program recursive(i_or, j_or, S (= skeleton),
i,j,visited, P(= particular point list) ,connected )
    up = (i-1,j);
    down = (i+1,j);
    left = (i,j-1);
    right = (i,j+1);
    up left = (i-1,j-1);
    up right = (i-1,j+1);
    down left = (i+1,j-1);
    down right = (i+1,j+1);
    if up[0] >= 0 and visited.get(up,default = 0) == 0:
        if up not in P:
            visited[up] = 1;
        if S[up[0]][up[1]] == 1 (White):
        point = up;
        if point is in P:
            connected[i_or,j_or] =
            connected.get((i_or,j_or),default = []) + [up];
        else:
            recursive(i_or, j_or,  S, point[0],
            point[1],visited,P ,connected);

    Do equivalent things for down, left, right, up left, up right,
    down left, down right.
end.
```

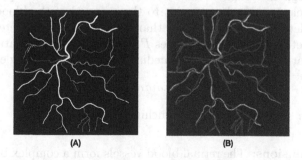

**Fig. 5.** Computing the linear interpolation of a vascular network. (A): Vascular network $y_a$ and (B): Linear interpolation between all the particular points of $y_a$.

**Branching Angles:** A vascular network's branching angles (BA) can be defined as the angle where a blood vessel is divided into smaller blood vessels. The computation of BA is performed using the following steps: starting by extracting all the angles of a vascular network using a simple modification of the linear interpolation algorithm that we developed to compute the tortuosity (Fig. 6). To extract only the branching angles, all other angles need to be discarded. For instance, in Fig. 6 an ellipse has been drawn around a branching angle, giving us the following four points $A$, $B$, $C$ and $D$. These points define the following two segments: $[A, C]$ and $[B, D]$ with their intersection point $C$. Three different angles can be computed from this branching: $\widehat{ACD}$, $\widehat{ACB}$ and $\widehat{BCD}$. The branching angle corresponds to $\widehat{ACB}$. We need to define the centroid of the graph in order to find the only relevant angle.

In a connected graph, we define a centroid as the closest point to any other particular point of the graph. To compute the centroid, we will need to extract a set of points $S$, which will be our particular points in a connected graph;

$$S = \{p, p \in skeleton, N(p) \notin \{2, 0\}\} \tag{6}$$

where $N(p)$ is the number of neighbouring pixels of $p$ which belong to the skeleton.

Then we create a metric $f$ such that for any point $p$ in the skeleton of the segmentation:

$$f(p) = max_{s \in skeleton} dist(p, s) \tag{7}$$

where $dist$ is the distance, measured as the number of pixels required to reach $s$ from $p$ by staying inside the skeleton. The centroid will naturally be the point with the lowest value according to this function. And for a random point $p$, the higher the value of $f(p)$, the farther $p$ is from the centroid. A simple example can be seen in Fig. 7. This also generalizes to segmentation with multiple disconnected parts, assuming that each part has a centroid. The branching angle between $A$, $B$, $C$, and $D$ is the angle between the 3 points that are the farthest from the centroid of the blood vessel in terms of pixel distance when you navigate through the graph of the vascular network which is equivalent to our challenge of deleting the closest point to the centroid of the blood vessel. It is possible to delete the irrelevant point thanks to this centroid detection, and to compute the set of the branching angles $\Gamma = \{BA_i\}_{i=1:n}$ automatically (Fig. 8). The BA biomarker is defined by the median of all the found angles.

$$BA = median\{a \in \Gamma\} \tag{8}$$

where $\Gamma$ is the set of the detected branching angles.

**Fractal Dimensions:** The retinal blood vessels form a complex branching pattern that can be quantified using fractal dimension (monofractal) [11,30,31]. In fractal geometry, the fractal dimension is a measure of the space-filling capacity of a pattern that tells how a fractal scales differently from the space it is embedded in. Fractal dimension can be thought of as an extension of the familiar

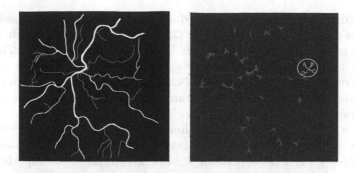

**Fig. 6.** Computation of all the angle of a vascular network.

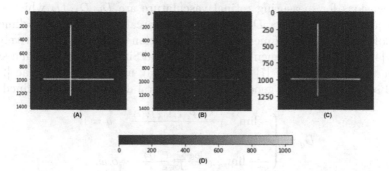

**Fig. 7.** Computation of the centroid of a simple graph. (A): Original graph, (B): Particular point detection, (C) Value of the $f(.)$ function for each pixel of this graph, (D): Heatmap where the centroid is the point with the lowest value according to the function $f(.)$.

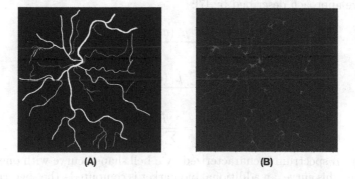

**Fig. 8.** Automated computation of the branching angles of the vascular network. (A): Original segmentation, (B): Branching angles detected automatically.

Euclidean dimensions allowing intermediate states. The fractal dimension of the retinal vascular tree lies between 1 and 2, indicating that its branching pattern fills space more thoroughly than a line, but less than a plane [30]. The fractal dimension measures the global branching complexity, which can be altered by the vessel rarefaction, proliferation and other anatomical changes in a pathological scenario. The simplest and most common method used in the literature for monofractal calculation is the box-counting method [31].

However, a single monofractal is limited in describing human eye retinal vasculature. It has been observed that retinal vasculature has multifractal properties which are a generalized notion of a fractal dimension [46–48]. Multifractal dimensions are characterized by a hierarchy of exponents which can reveal more complex geometrical properties in a structure [47]. The most common multifractal dimensions for measuring retinal vasculature are $D_0, D_1, D_2$ which satisfy the inequality $D_0 > D_1 > D_2$ [39,46] and also called the capacity dimension (a monofractal), entropy dimension and correlation dimension respectively. Following [49], we have implemented $D_0, D_1, D_2$ and Singularity length in a similar manner to Chhabra et al. [7] and the commonly used plugin FracLac [20] for ImageJ software [41]. The generalized multifractal dimensions are defined as:

$$
D_q = \begin{cases} -\lim_{\epsilon \to 1} \frac{\sum_i P_i(\epsilon) \log P_i(\epsilon)}{\log \epsilon} & , q = 1 \\ \frac{1}{q-1} \lim_{\epsilon \to 0} \frac{\log \sum_i P_i^q(\epsilon)}{\log \epsilon} & o.w. \end{cases}
\tag{9}
$$

where $P_i(\epsilon)$ is the pixel probability in the $i^{\text{th}}$ grid box sized $\epsilon$, and $q$ is the order of the moment of the measure. In addition, multifractals can be described by a singularity spectrum $f(\alpha) - \alpha$ which is related to the $D_q - q$ spectrum by the Legendre transformation. In order to calculate $f(\alpha) - \alpha$ spectrum, we use an alternative approach described by [7]:

$$
f(q) = \lim_{\epsilon \to 0} \frac{\sum_i \mu_i(q, \epsilon) \log \mu_i(q, \epsilon)}{\log \epsilon}
\tag{10}
$$

$$
\alpha(q) = \lim_{\epsilon \to 0} \frac{\sum_i \mu_i(q, \epsilon) \log P_i(\epsilon)}{\log \epsilon}
\tag{11}
$$

$$
\mu_i(q, \epsilon) = \frac{P_i(\epsilon)^q}{\sum_i P_i(\epsilon)^q}
\tag{12}
$$

The $f(\alpha) - \alpha$ spectrum is characterized by a bell shaped curve with one maxima point. From this curve, an additional biomarker is computed - the spectrum range $\Delta\alpha$ which is also called Singularity Length (SL).

$$
\alpha_{\min} = \lim_{q \to \infty} \alpha(q), \; \alpha_{\max} = \lim_{q \to -\infty} \alpha(q)
\tag{13}
$$

$$
\Delta\alpha = \alpha_{\max} - \alpha_{\min}
\tag{14}
$$

SL quantifies the multifractality degree [29] of an image.

The calculation of Eqs. 9, 10, 11 was done by linear regression with a linear set of box sizes $\epsilon$ for every $q$. The values of these graphs are sensitive to grid placement on the segmented DFI image, as they depend on pixel distribution across the image. We followed a similar optimization method as used by the FracLac plugin [20], which is to change the sampling grid location by rotating the image randomly, and to choose the measurement which satisfies the inequality $D_0 > D_1 > D_2$ and has the highest value of $D_0$. In addition, to overcome numerical issues saturated grid boxes and grid boxes with low occupancy of pixels were ignored. $\alpha_{min}$ and $\alpha_{max}$ were estimated by $\alpha_{min} \approx \alpha\,(q = 10)$ and $\alpha_{max} \approx \alpha\,(q = -10)$.

**Benchmark Against Existing Software:** In order to validate the implementation of some of the biomarkers we implemented in PVBM we benchmarked their values against two ImageJ plugins, namely AnalyzeSkeleton [2] and FracLac [20]. The benchmark was performed for the arterioles from the entire UZFG dataset. A direct benchmark could be performed for the following biomarkers: OVAREA, TOR, $D_0$, $D_1$, $D_2$ and SL. An extrapolated comparison could be performed for the OVLEN. Indeed, these were not directly outputted by the plugin, but could be derived from the AnalyzeSkeleton [2] plugin. No benchmark could be performed for OVPER, END, INTER and BA because of the lack of open source available benchmark software.

## 3    Results

Table 3 shows that the VBMs benchmarked against reference software had very close values with normalized root mean square errors ranging from 0 to 0.316.

Tables 4 and 5 provide summary statistics and a statistical analysis of the VBMs for the GLA NOR groups. The statistics were presented as median and interquartile (Q1 and Q3), and the p-value from the Wilcoxon signed-rank. The arteriolar OVAREA, OVLEN, and END were the most significant in distinguishing between the two groups. Figure 9 presents qualitative examples of three DFIs with arteriolar OVAREA, OVLEN, BA, END and $D_0$ VBM values.

## 4    Discussion and Future Work

The first contribution of this work is the creation of a toolbox for VBMs, which is made open source under a GNU GPL 3 license and will be made available on physiozoo.com (following publication). In particular, novel algorithms were introduced to estimate the tortuosity and branching angles.

The second contribution of this work is the application of the PVBM toolbox to a new dataset of manually segmented vessels from DFIs. The statistical analysis that we have performed showed that the arterioles-based biomarkers are the most significant in distinguishing between NOR and GLA. For arterioles and venules, all biomarkers were significant and lower in glaucoma patients compared

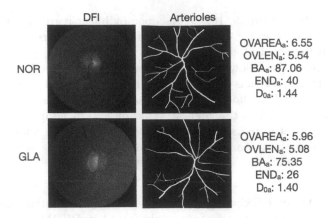

**Fig. 9.** Example of biomarkers computed from the arterioles of two DFIs. The first row is a DFI of a healthy control (NOR) and the second column is from a glaucoma patient (GLA). The OVAREA biomarker shows a larger vascular area in the NOR images. The BA biomarker shows that the branching of the NOR image are bigger than the one of the GLA image. Finally, the END biomarker indicates that the GLA image had less arteriolar branching compared to healthy controls, leading to a lower number of endpoints.

to healthy controls except for tortuosity, venular singularity length and venular branching angles.

A limitation of our experiment is that although the images were taken with the same procedure, which includes the disk being centered, there is some variation in the exact location of the disk due to the non-automated operation. In future work, we need to consider the detection of the disk to delineate a circular frame centered on the disk to engineer the vasculature biomarkers more consistently. Furthermore, other biomarkers may be implemented such as the vessel

**Table 3.** PVBM benchmark against reference ImageJ plugins using the arterioles of the UZFG dataset. $\mu$: mean, $\sigma$: standard deviation, RMSE: root mean square error, NRMSE: normalized root mean square error.

| Biomarker | Benchmark | Benchmark results | | This work | | Difference | |
|---|---|---|---|---|---|---|---|
| | | $\mu$ | $\sigma$ | $\mu$ | $\sigma$ | RMSE | NRMSE |
| OVLEN | AnalyzeSkeleton [2] | 4.687 | 0.928 | 4.868 | 0.974 | 0.194 | 0.060 |
| OVAREA | ImageJ [41] | 4.939 | 0.011 | 4.939 | 0.011 | 0 | 0 |
| TOR | AnalyzeSkeleton [2] | 1.081 | 0.007 | 1.084 | 0.007 | 0.003 | 0.11 |
| $D_0$ | FracLac [20] | 1.373 | 0.035 | 1.425 | 0.026 | 0.054 | 0.301 |
| $D_1$ | FracLac [20] | 1.367 | 0.034 | 1.390 | 0.027 | 0.028 | 0.155 |
| $D_2$ | FracLac [20] | 1.361 | 0.033 | 1.375 | 0.028 | 0.021 | 0.115 |
| $SL$ | FracLac [20] | 0.626 | 0.104 | 0.645 | 0.076 | 0.128 | 0.316 |

**Table 4.** Summary of the biomarkers analyzed for arterioles ($_a$) and extracted with the PVBM toolbox with their median (Q1–Q3). Refer to Table 2 for the definition of the VBM acronyms. P-values are provided using the Wilcoxon signed-rank test.

|  | NOR (n = 19) | GLA (n = 50) | p |
|---|---|---|---|
| OVAREA$_a$ | 5.82 (5.26–6.23) | 4.52 (3.76–5.44) | **4e-4** |
| OVLEN$_a$ | 5.53 (5.28–6.21) | 4.85 (4.33–5.37) | **1e-4** |
| OVPER$_a$ | 3.94 (3.74–4.33) | 3.44 (3.01–3.80) | **2e-4** |
| BA$_a$ | 87.29 (81.88–93.28) | 81.93 (75.41–87.87) | **2e-2** |
| END$_a$ | 37.0 (28.5–43.5) | 27.0 (21.0–32.75) | **7e-4** |
| INTER$_a$ | 36.0 (28.5–43.0) | 24.5 (20.0–30.0) | **8e-4** |
| TOR$_a$ | 1.08 (1.08–1.09) | 1.08(1.08–1.09) | $1e-1$ |
| D$_{0a}$ | 1.44 (1.42–1.45) | 1.42 (1.40–1.44) | **5e-3** |
| D$_{1a}$ | 1.41 (1.38–1.42) | 1.38 (1.37–1.40) | **3e-3** |
| D$_{2a}$ | 1.39 (1.37–1.40) | 1.37 (1.35–1.38) | **4e-3** |
| SL$_a$ | 0.62 (0.59–0.64) | 0.64 (0.62–0.78) | **9e-3** |

**Table 5.** Summary of the biomarkers analyzed for venules ($_v$) and extracted with the PVBM toolbox with their median (Q1–Q3). Refer to Table 2 for the definition of the VBM acronyms. P-values are provided using the Wilcoxon signed-rank test.

|  | NOR (n = 19) | GLA (n = 50) | p |
|---|---|---|---|
| OVAREA$_v$ | 6.3 (5.82–6.71) | 5.28 (4.59–6.01) | **9e-4** |
| OVLEN$_v$ | 5.42 (4.82–5.67) | 4.69 (4.21–5.13) | **1e-3** |
| OVPER$_v$ | 3.77 (3.36–3.99) | 3.26 (2.96–3.59) | **3e-3** |
| BA$_v$ | 84.14 (78.74–86.9) | 85.1 (81.52–91.08) | $1e-1$ |
| END$_v$ | 339.0 (31.5–43.0) | 28.5 (23.0–34.0) | **1e-3** |
| INTER$_v$ | 36.0 (31.5–42.5) | 27.0 (21.0–33.0) | **1e-3** |
| TOR$_v$ | 1.08 (1.08–1.09) | 1.08 (1.08–1.09) | $4e-1$ |
| D$_{0v}$ | 1.43 (1.4–1.44) | 1.40 (1.38–1.42) | **4e-3** |
| D$_{1v}$ | 1.38 (1.36–1.40) | 1.37 (1.35–1.38) | **8e-3** |
| D$_{2v}$ | 1.36 (1.34–1.39) | 1.35 (1.33–1.36) | **1e-2** |
| SL$_v$ | 0.61 (0.58–0.66) | 0.63 (0.59–0.66) | $2e-1$ |

diameter [13], CRAE [37], CRVE [37], branching coefficients [10] which is the ratio of the sum of the cross-sectional areas of the two daughter vessels to the cross-sectional area of the parent vessel at an arteriolar bifurcation [10]. Finally, it is to be studied to what extent VBMs may be evaluated from vasculature obtained using an automated (versus manual) segmentation algorithm as well as the effect of DFI quality [1] on the results.

**Acknowledgment.** The research was supported by Grant No ERANET - 2031470 from the Ministry of Health, by the Israel PBC-VATAT and by the Technion Center for Machine Learning and Intelligent Systems (MLIS).

# References

1. Abramovich, O., Pizem, H., Van Eijgen, J., Stalmans, I., Blumenthal, E., Behar, J.: FundusQ-Net: a regression quality assessment deep learning algorithm for fundus images quality grading. arXiv preprint arXiv:2205.01676 (2022)
2. Arganda-Carreras, I., Fernández-González, R., Muñoz-Barrutia, A., Ortiz-De-Solorzano, C.: 3D reconstruction of histological sections: application to mammary gland tissue. Microsc. Res. Technol. **73**(11), 1019–1029 (2010)
3. Badawi, S.A., Fraz, M.M.: Multiloss function based deep convolutional neural network for segmentation of retinal vasculature into arterioles and venules. In: BioMed Research International 2019 (2019)
4. Betzler, B.K., et al.: Retinal vascular profile in predicting incident cardiometabolic diseases among individuals with diabetes. Microcirculation, p. e12772 (2022)
5. Brinchmann-Hansen, O., Heier, H.: Theoretical relations between light streak characteristics and optical properties of retinal vessels. Acta Ophthalmol. **64**(S179), 33–37 (1986)
6. Cheung, N., et al.: Quantitative assessment of early diabetic retinopathy using fractal analysis. Diabetes Care **32**(1), 106–110 (2009)
7. Chhabra, A., Jensen, R.V.: Direct determination of the f $(\alpha)$ singularity spectrum. Phys. Rev. Lett. **62**(12), 1327 (1989)
8. Dashtbozorg, B., Mendonca, A.M., Penas, S., Campilho, A.: RetinaCAD, a system for the assessment of retinal vascular changes. In: 2014 36th Annual International Conference of the IEEE Engineering in Medicine and Biology Society, pp. 6328–6331. IEEE (2014)
9. Doubal, F.N., MacGillivray, T.J., Patton, N., Dhillon, B., Dennis, M.S., Wardlaw, J.M.: Fractal analysis of retinal vessels suggests that a distinct vasculopathy causes lacunar stroke. Neurology **74**(14), 1102–1107 (2010)
10. Doubal, F.N., et al.: Retinal arteriolar geometry is associated with cerebral white matter hyperintensities on magnetic resonance imaging. Int. J. Stroke **5**(6), 434–439 (2010)
11. Family, F., Masters, B.R., Platt, D.E.: Fractal pattern formation in human retinal vessels. Physica D **38**(1–3), 98–103 (1989)
12. Fhima, J., Van Eijgen, J., Freiman, M., Stalmans, I., Behar, J.A.: Lirot.ai: a novel platform for crowd-sourcing retinal image segmentations. In: Accepted for Proceeding in Computing in Cardiology 2022 (2022)
13. Goldenberg, D., Shahar, J., Loewenstein, A., Goldstein, M.: Diameters of retinal blood vessels in a healthy cohort as measured by spectral domain optical coherence tomography. Retina **33**(9), 1888–1894 (2013)
14. Grisan, E., Foracchia, M., Ruggeri, A.: A novel method for the automatic grading of retinal vessel tortuosity. IEEE Trans. Med. Imaging **27**(3), 310–319 (2008)
15. Gunn, R.M.: Ophthalmoscopic evidence of (1) arterial changes associated with chronic renal disease, and (2) of increased arterial tension. Trans. Ophthalmol. Soc. UK **12**, 124–125 (1892)
16. Gunn, R.M.: Ophthalmoscopic evidence of general arterial disease. Trans. Ophthalmol. Soc. UK **18**, 356–381 (1898)

17. Guo, S., Yin, S., Tse, G., Li, G., Su, L., Liu, T.: Association between caliber of retinal vessels and cardiovascular disease: a systematic review and meta-analysis. Curr. Atheroscler. Rep. **22**(4), 1–13 (2020)

18. Hanssen, H., et al.: Exercise-induced alterations of retinal vessel diameters and cardiovascular risk reduction in obesity. Atherosclerosis **216**(2), 433–439 (2011)

19. Hart, W.E., Goldbaum, M., Côté, B., Kube, P., Nelson, M.R.: Measurement and classification of retinal vascular tortuosity. Int. J. Med. Inform. **53**(2), 239–252 (1999)

20. Karperien, A.: Fraclac for imagej (2013)

21. Kaushik, S., Tan, A.G., Mitchell, P., Wang, J.J.: Prevalence and associations of enhanced retinal arteriolar light reflex: a new look at an old sign. Ophthalmology **114**(1), 113–120 (2007)

22. Kawasaki, R., Wang, J.J., Rochtchina, E., Lee, A.J., Wong, T.Y., Mitchell, P.: Retinal vessel caliber is associated with the 10-year incidence of glaucoma: the Blue Mountains eye study. Ophthalmology **120**(1), 84–90 (2013)

23. Keith, N.M.: Some different types of essential hypertension: their course and prognosis. Am. J. Med. Sci. **197**, 332–343 (1939)

24. Lemmens, S., et al.: Age-related changes in the fractal dimension of the retinal microvasculature, effects of cardiovascular risk factors and smoking behaviour. Acta Ophthalmologica (2021)

25. Liew, G., et al.: Fractal analysis of retinal microvasculature and coronary heart disease mortality. Eur. Heart J. **32**(4), 422–429 (2011)

26. Liew, G., Wang, J.J.: Retinal vascular signs: a window to the heart? Revista Española de Cardiología (English Edition) **64**(6), 515–521 (2011)

27. Liew, G., et al.: The retinal vasculature as a fractal: methodology, reliability, and relationship to blood pressure. Ophthalmology **115**(11), 1951–1956 (2008)

28. Lotmar, W., Freiburghaus, A., Bracher, D.: Measurement of vessel tortuosity on fundus photographs. Albrecht Von Graefes Arch. Klin. Exp. Ophthalmol. **211**(1), 49–57 (1979)

29. Macek, W.M., Wawrzaszek, A.: Evolution of asymmetric multifractal scaling of solar wind turbulence in the outer heliosphere. J. Geophys. Res. Space Phys. **114**(A3) (2009)

30. Mainster, M.A.: The fractal properties of retinal vessels: embryological and clinical implications. Eye **4**(1), 235–241 (1990)

31. Mandelbrot, B.B., Mandelbrot, B.B.: The Fractal Geometry of Nature, vol. 1. WH Freeman, New York (1982)

32. Martínez-Pérez, M.E., et al.: Geometrical and morphological analysis of vascular branches from fundus retinal images. In: Delp, S.L., DiGoia, A.M., Jaramaz, B. (eds.) MICCAI 2000. LNCS, vol. 1935, pp. 756–765. Springer, Heidelberg (2000). https://doi.org/10.1007/978-3-540-40899-4_78

33. Mehta, R., et al.: Phosphate, fibroblast growth factor 23 and retinopathy in chronic kidney disease: the chronic renal insufficiency cohort study. Nephrol. Dial. Transplant. **30**(9), 1534–1541 (2015)

34. Miri, M., Amini, Z., Rabbani, H., Kafieh, R.: A comprehensive study of retinal vessel classification methods in fundus images. J. Med. Sig. Sens. **7**(2), 59 (2017)

35. Murphy, S.L., Kochanek, K.D., Xu, J., Arias, E.: Mortality in the United States, 2020 (2021)

36. Owen, C.G., et al.: Measuring retinal vessel tortuosity in 10-year-old children: validation of the computer-assisted image analysis of the retina (CAIAR) program. Invest. ophthalmol. Vis. Sci. **50**(5), 2004–2010 (2009)

37. Parr, J.C., Spears, G.F.S.: General caliber of the retinal arteries expressed as the equivalent width of the central retinal artery. Am. J. Ophthalmol. **77**(4), 472–477 (1974)
38. Perez-Rovira, A., et al.: VAMPIRE: vessel assessment and measurement platform for images of the REtina. In: 2011 Annual International Conference of the IEEE Engineering in Medicine and Biology Society, pp. 3391–3394. IEEE (2011)
39. Posadas, A.N.D., Giménez, D., Bittelli, M., Vaz, C.M.P., Flury, M.: Multifractal characterization of soil particle-size distributions. Soil Sci. Soc. Am. J. **65**(5), 1361–1367 (2001)
40. Provost, E.B., Nawrot, T.S., Int Panis, L., Standaert, A., Saenen, N.D., De Boever, P.: Denser retinal microvascular network is inversely associated with behavioral outcomes and sustained attention in children. Front. Neurol. **12**, 547033 (2021)
41. Rasband, W.S.: ImageJ, US National Institutes of Health, Bethesda, Maryland, USA (2011)
42. Sabanayagam, C., et al.: Retinal microvascular caliber and chronic kidney disease in an Asian population. Am. J. Epidemiol. **169**(5), 625–632 (2009)
43. Scheie, H.G.: Evaluation of ophthalmoscopic changes of hypertension and arteriolar sclerosis. A.M.A. Arch. Ophthalmol. **49**(2), 117–138 (1953)
44. Sharrett, A.R., et al.: Retinal arteriolar diameters and elevated blood pressure: the atherosclerosis risk in communities study. Am. J. Epidemiol. **150**(3), 263–270 (1999)
45. Sng, C.C.A., et al.: Fractal analysis of the retinal vasculature and chronic kidney disease. Nephrol. Dial. Transplant. **25**(7), 2252–2258 (2010)
46. Stosic, T., Stosic, B.D.: Multifractal analysis of human retinal vessels. IEEE Trans. Med. Imaging **25**(8), 1101–1107 (2006)
47. Ţălu, Ş: Characterization of retinal vessel networks in human retinal imagery using quantitative descriptors. Hum. Veterinary Med. **5**(2), 52–57 (2013)
48. Ţălu, Ş, Stach, S., Călugăru, D.M., Lupaşcu, C.A., Nicoară, S.D.: Analysis of normal human retinal vascular network architecture using multifractal geometry. Int. J. Ophthalmol. **10**(3), 434 (2017)
49. Van Craenendonck, T., et al.: Retinal microvascular complexity comparing mono- and multifractal dimensions in relation to cardiometabolic risk factors in a Middle Eastern population. Acta Ophthalmol. **99**(3), e368–e377 (2021)
50. Witt, N., et al.: Abnormalities of retinal microvascular structure and risk of mortality from ischemic heart disease and stroke. Hypertension **47**(5), 975–981 (2006)

# Simultaneous Detection and Classification of Partially and Weakly Supervised Cells

Alona Golts[1]([⊠])(iD), Ido Livneh[2](iD), Yaniv Zohar[3](iD), Aaron Ciechanover[2](iD), and Michael Elad[1](iD)

[1] Computer Science, Technion – Israel Institute of Technology, Haifa, Israel
{salonaz,elad}@cs.technion.ac.il
[2] Rapaport Faculty of Medicine, Technion – Israel Institute of Technology, Haifa, Israel
aaroncie@technion.ac.il
[3] Pathology Department, Rambam Medical Center, Haifa, Israel
y_zohar@rmc.gov.il

**Abstract.** Detection and classification of cells in immunohistochemistry (IHC) images play a vital role in modern computational pathology pipelines. Biopsy scoring and grading at the slide level is routinely performed by pathologists, but analysis at the cell level, often desired in personalized cancer treatment, is both impractical and non-comprehensive. With its remarkable success in natural images, deep learning is already the gold standard in computational pathology. Currently, some learning-based methods of biopsy analysis are performed at the tile level, thereby disregarding intra-tile cell variability; while others do focus on accurate cell segmentation, but do not address possible downstream tasks. Due to the shared low and high-level features in the tasks of cell detection and classification, these can be treated jointly using a single deep neural network, minimizing cumulative errors and improving the efficiency of both training and inference. We construct a novel dataset of Proteasome-stained Multiple Myeloma (MM) bone marrow slides, containing nine categories with unique morphological traits. With the relative difficulty of acquiring high-quality annotations in the medical-imaging domain, the proposed dataset is intentionally annotated with only 5% of the cells in each tile. To tackle both cell detection and classification within a single network, we model these as a multi-class segmentation task, and train the network with a combination of partial cross-entropy and energy-driven losses. However, as full segmentation masks are unavailable during both training and validation, we perform evaluation on the combined detection and classification performance. Our strategy, uniting both tasks within the same network, achieves a better combined Fscore, at faster training and inference times, as compared to similar disjoint approaches.

**Keywords:** Cell detection · Cell classification · Cell profiling

---

A. Golts, I. Livneh and A. Ciechanover—Equal contribution.

L. Karlinsky et al. (Eds.): ECCV 2022 Workshops, LNCS 13803, pp. 313–329, 2023.
https://doi.org/10.1007/978-3-031-25066-8_16

# 1   Introduction

Immunohistochemistry (IHC) is a widely used method in Histopathology where cellular proteins are stained via the employment of specific antibodies and dyes. The stained proteins serve as markers for different cell types, differentiation levels and tumor classifications of the slide in question. Such staining is crucial for diagnosis, prognosis, treatment guidance, response monitoring, follow up and many other aspects in the clinical life [5, 8]. With the emergence of personalized medicine, IHC staining and screening allow for the identification of the unique features of each patient - and tumor, and subsequent tailoring of the optimal known treatment. A well-known example for such markers that are used to guide treatment, are the estrogen, progesterone and HER2 receptors in breast cancer. Such cellular markers can be stained in a specific biopsy or excised tumor, and their relative abundance – as interpreted by a pathologist – provides valuable information for the management of each patient [43]. Specifically, in an era when cancer treatment has rendered many tumors as curable, we are facing the issue of tumor recurrence, or relapse. In some cases, the recurring cancer cells were in fact present already in the primary tumor, yet in small numbers [7]. It is therefore clear how the identification of unique subpopulations of cells in advance can potentially benefit with patients, as it can predict relapse, dictate the follow-up protocol under remission, and guide treatment may the tumor reoccur. While global slide examination is routine and commonly performed by pathologists, local analysis in the cell level is still out of reach, as such subpopulation of cells may be so rare that the human operator simply cannot spot it.

In the advent of computational pathology, some believe that manual inspection of tissue slides under the microscope will be gradually replaced with high-resolution slide scanning and automatic analysis of Whole Slide Images (WSIs). The field of deep learning has taken the computer vision world by storm, achieving state-of-the-art results in a wide range of applications: object detection [36], image classification [17], segmentation [26], restoration [52], synthesis [18] and more. Inspired by the success of deep learning in natural images, an increasing body of work is directed at solving computational pathology tasks, such as detection, classification and grading of breast [12,23,40], kidney [9] and lung [6,19,46] cancers, mitosis detection [4], detection and classification of cell nuclei [42,44], automatic count of histologic bone marrow samples [47] and many more.

The tasks of detection and classification of cells into multiple categories serve as crucial building blocks in the realization of cell-level analysis of WSIs, and their constituent smaller sized image patches, called 'tiles'. Current cell-level approaches either concentrate on the detection and accurate segmentation of cells [27,33,34], leaving the task of classification to a later stage, or separate the two tasks to two consecutive networks [3,42]. Both approaches do not fully utilize the common low and high-level features of the corresponding neural networks. Envisioning the application of WSI cell-level profiling, we propose simultaneous detection and classification of cells into multiple categories of interest. Apart from minimizing the propagated errors from one network to the other, improving overall robustness of the solution, this reduces training and inference times, both crucial for WSIs. As compared to other works which suggest joint detection and classification [44,47], we do not rely on fully-supervised annotated datasets but use a relatively small and compact set of weak partial annotations.

We propose a novel dataset of Proteasome-stained IHC images of bone marrow biopsies diagnostic for Multiple Myeloma (MM). MM is a malignancy of plasma cells – immune cells which specialize in the production of antibodies, accounting for ≈20% of hematological malignancies, and as much as 2–3% of all human cancers [41]. As part of a vast drug re-purposing project, it has been discovered that proteasome inhibitors, such as Bortezomib [37], are highly efficient for the treatment of MM. The proteasome is the catalytic arm of the ubiquitin-proteasome system, largely responsible for selective cellular protein degradation [25]. Unfortunately, despite the dramatic improvement in survival of many MM patients, the disease is still considered incurable, and virtually all of them experience a relapse of the disease [1]. Notably, it is yet to be unraveled why some patients experience relapse within 2–3 years, while others enjoy a remission as long as 8–9 years. Such a difference may stem from a molecular variability between such patients, which may be determined using IHC staining and analysis at the cell-level. Since MM serves as a suitable prototype and desirable subject for such testing of intra- and inter-patient variability, and since the proteasome is highly abundant in the malignant MM cells, we chose to stain MM biopsies for the proteasome to establish our proposed network.

Our dataset contains point or bounding box annotations with corresponding class information of approximately 5% of the cells in each image. The cells are graded from −4 (most cytosolic) to 4 (most nuclear), amounting to nine distinct classes, with unique morphological signatures. To account for the high level of detail the task of MM biopsy profiling requires, we propose detecting and classifying each and every cell in the WSI for the purpose of aggregating it into a unique histogram representation. We do so by a simultaneous detection and classification neural network, trained end-to-end on the input partial annotations. We model the above tasks as multi-class semantic segmentation, such that the network outputs C (number of classes) segmentation maps. Since both our training and validation sets do not include full segmentation masks, we evaluate the performance of the network using a combined Fscore of both detection and classification.

Inspired by [34], we encode the partial point labels as concentric multi-class circles and surrounding background rings, disregarding all other pixels. We optimize a UNet network [38], with its encoder pre-trained on ImageNet [39]. Our loss function consists of a partial cross-entropy term [45], ensuring fidelity to the extended concentric multi-class labels, and an energy-based smoothness promoting term [10] which enforces the output probability maps to correlate with the edges of the input image. Once training is finished, our network can infer the position and class of each cell in an input tile using a single forward-pass operation. Our experiments show our method can achieve a higher combined Fscore, at faster training and inference speeds, as compared to the approach of training each task disjointly. Our code is available at https://github.com/AlonaGolts/sim-detect-classify-cells

## 2   Related Work

Many works in computational pathology deal with a single task as WSI classification [4,9,46], cell detection [21,28,49] and segmentation [2,20,29,30,33,50]. Classification-centered approaches [4,9,46] are usually global and provide a classification result for each patch or tile. These are then aggregated across the entire WSI using a sliding-window method to create a global classification heatmap. This global approach, however, is less suitable for creating accurate patient-specific cell count histograms. Local approaches, attempting to detect or segment all cells in the image are abundant. Mitosis detection and classification is performed in [28] by morphologically finding cell blobs and classifying them using a Convolutional Neural Network (CNN). In [21,49], a fully-convolutional CNN (FCN) is weakly-supervised with point labels for each cell. The point annotations are encoded either by Gaussians [49] or concentric rings [21] around each cell point. These methods, however, do not account for multiple cell classes and require annotation of all cells in the train set. In the task of cell instance segmentation, full pixelwise masks are incredibly hard to collect, thus weakly-supervised approaches are in favor. In [2,33], Voronoi, cluster-based and repel-code pseudo labels are created to deal with weak cell point annotations.

Another series of works suggests different methods of coping with the relatively low amount of training data [9,14,22,24,27,31,51]. The work of [51] suggests pre-training an encoder-decoder network with unlabelled images to obtain richer representations. The authors of [9,22] propose a semi-supervised approach of training on a smaller dataset and enriching the labels with resulting high-scored predictions. Finally, the works of [24,27] use a cycle-GAN architecture to synthesize corresponding pairs of H&E images and segmentation masks.

An increasing effort is applied at solving both tasks jointly. The authors of [48] introduced a Structured Regression CNN (SR-CNN) which produced a proximity map for each image patch, allowing cell instance detection, classification to cancer/non-cancer cells [42] and tissue phenotyping by grouping individual cells to cohesive neighborhoods [15,16]. These works, however, treat both tasks separately, leading to error accumulation and slower inference times. The authors of [32] propose solving both cell segmentation and classification within the same network, with the help of an additional VGG-assisted perceptual loss, but require the full pixelwise segmentation masks during training. Finally, few papers treat both tasks simultaneously. The work of [47] performs detection and classification of bone marrow smear cells, using a YoloV3 [35] architecture. In [53], a unified network, featuring separate detection and classification branches, allows fine-grained classification of H&E colorectal adenocarcinoma cells. In [44], a synchronized asymmetric hybrid deep autoencoder is used to detect and classify bone marrow stem cells. Note, all of the above methods use cell annotations for the entire image, whereas we utilize only 5% of cells in each image.

# 3   Method

We hereby detail our scheme for simultaneous detection and classification of MM cells. Our approach is inspired by the work of [34], in which partial point annotations of approximately 5% of the total cells in the image, are the driving force for both detection and accurate segmentation for the whole image. As opposed to [34], our final desired outcome is not segmentation, but rather a localization and classification of the cells to nine categories, leading to an accurate and personalized "cell portfolio" for each patient.

## 3.1   Weakly Supervised Segmentation Using Partial Points

For the sake of completeness, we briefly outline the main stages in [34] which aims at cell instance segmentation. The proposed algorithm consists of two stages: (1) cell detection, given partial point annotations, and (2) cell segmentation using pseudo labels. In the detection stage, extended Gaussian masks are created from the partial point annotations in each image, using the following mask

$$M_i = \begin{cases} \exp(-\frac{D_i^2}{2\sigma^2}) & \text{if } D_i < r_1, \\ 0 & \text{if } r_1 \le D_i < r_2, \\ -1 & \text{otherwise,} \end{cases} \tag{1}$$

where $D_i$ is the euclidean distance of pixel $i$ to the closest cell point. The above encoding models each annotated point as a Gaussian, surrounded by a 'background ring' with radii $r_1, r_2$. Pixels located further away from point annotations are ignored during training. A UNet with a ResNet34 pre-trained encoder is trained to regress over the extended Gaussian masks via an MSE loss. To improve the false positive rate, the authors of [34] suggest self-training with background propagation. Once the first network has finished training, a new set of training labels is formed by adding pixels with predicted output probability $p_i < 0.1$ and $p_i > 0.7$ to the background. Another round of training then begins with a newly initialized network and the updated labels. After 2–3 rounds, the predicted cell locations from the trained detection network are used for creating two sets of pseudo labels. In the first, the detected points serve as seeds for Voronoi partitioning, where the resulting partition lines are tagged as background and the detected points as cells. The second pseudo label is formed by clustering the image pixels to 'cell', 'background' and 'ignore' classes using KMeans. A new network is then trained to output 'cell' and 'background' probability maps, via minimization of a cross-entropy loss over the two sets of pseudo-labels.

## 3.2   Multiclass Label Encoding

Inspired by the cell-background encoding in the detection stage of [34], we encode the partial point labels of an input image as a multiclass mask. Denote $X \in \mathbb{R}^{|\Omega| \times 3}$ as the RGB input image, belonging to the discrete image domain $\Omega$, where

$|\Omega|$ is the number of pixels. The corresponding multiclass mask $M \in \mathbb{R}^{|\Omega| \times C}$, where $C$ is the number of classes, is given by

$$M_i^c = \begin{cases} 1 & \text{if } D_i^{c=c_{gt}} < r_1, \\ 0 & \text{if } D_i^{c \neq c_{gt}} \leq r_1, \\ 1 & \text{if } r_1 \leq D_i^{c=c_B} < r_2, \\ 0 & \text{if } r_1 \leq D_i^{c \neq c_B} < r_2, \\ -1 & \text{otherwise,} \end{cases} \tag{2}$$

where $c$ denotes the class, $c_B$ the background class, $c_{gt}$ the ground truth class, and $D_i^c$ the distance of pixel $i$ in class $c$ to the closest point annotation. Each cell is represented as a uniform circle of radius $r_1$ in the corresponding ground truth class, surrounded by a background ring of radii $r_1, r_2$. All other pixels are ignored during training. This representation, shown in Fig. 1, amplifies the weak point annotations and enforces the existence of background around each cell, decreasing cell merging in tight clusters. Note, this representation may not be accurate around the edges of the cell, since cells vary in size and shape, whereas the solution will gravitate towards similar sized circles. This shall be addressed in the energy-based smoothness function, explained in Sect. 3.3. We refer to the above representation as 'ours-ring'.

We propose another cell representation, adding an additional level of supervision to the background, by using a version of the Voronoi labels, introduced in Sect. 3.1. The cell points predicted in the detection stage, are used to compute Voronoi cell boundaries which are now considered as background pixels. As opposed to [34], instead of using all cells as supervision, we use only those that have class annotations (only 5% of cells in the image). We encode these cells as Gaussians as in Sect. 3.1, appearing only in the channel corresponding to the annotated class. We refer to this representation as 'ours-vor'.

### 3.3    Network Architecture

To perform simultaneous detection and classification, we formalize the problem as a multi-class segmentation task. Training the network to solve both tasks simultaneously, as opposed to using an additional cell classifier, is naturally more efficient in both training and inference, since the two share common low- and high-level features. In addition, end-to-end training may reduce accumulated errors in one task that often propagate to the other, improving the accuracy and robustness of the solution. Instead of using pixelwise segmentation masks, we utilize the partial cell encodings, introduced in Sect. 3.2. Given the above masks, a UNet [38] model, shown in Fig. 1, with its encoder – a ResNet34 model [13], pre-trained on ImageNet [39], is trained via a combination of partial cross-entropy [45] and energy-based smoothness [10] losses. The network receives the input RGB image $X \in \mathbb{R}^{|\Omega| \times 3}$, and outputs $f_\theta(X) \triangleq Y \in \mathbb{R}^{|\Omega| \times C}$, the pixelwise $C$-way softmax probabilities.

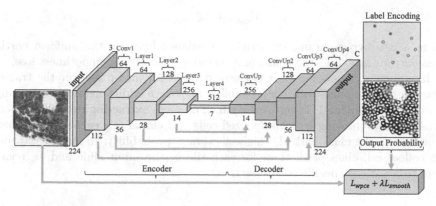

**Fig. 1.** Architecture. A UNet-based network with the number of features and spatial dimensions listed at the top and bottom of each block.

## 3.4   Loss Functions

Our proposed loss function is composed of two terms. The data term is based on the partial cross entropy [45], defined as

$$\mathcal{L}_{wpce} = \sum_{c=1}^{C} \sum_{i \in \Omega_L} -\gamma^c M_i^c \log Y_i^c, \tag{3}$$

where $\Omega_L$ is the set of annotated pixel coordinates, $M_i^c \in \{0,1\}^C$ is the $C$ dimensional one-hot annotation vector at pixel $i$, and $Y_i \in [0,1]^C$ is the $C$-dimensional softmax probability output vector from the network $f_\theta$. We introduce an additional weighting coefficient $\gamma^c$ such that the cell annotations receive stronger weights, as opposed to the background class. The added weight vector is set as $\gamma = [1, \alpha, \alpha, ..., \alpha] \in \mathbb{R}^C, \alpha > 1$ and promotes stronger confidence for the human-provided cell annotations. The smoothness term is based on the random-walker energy-minimization technique suggested in [11] and implemented via DNNs in [10]. The image is represented as a weighted graph in which each pixel is a vertex and adjacent pixels are connected with edges: $w_{ij} = \exp\{-\beta\|X_i - X_j\|^2\}$, where $i, j$ are the pixel coordinates, $X_i, X_j$ are their corresponding RGB values and $\beta$ is a hyper-parameter. The random walker regularization term is given by

$$\mathcal{L}_{smooth} = \sum_{c=1}^{C} \sum_{i,j \in \varepsilon} w_{ij}(Y_i^c - Y_j^c)^2, \tag{4}$$

where $\varepsilon$ is the group of adjoining pixels in a 4-neighborhood connectivity. The above term penalizes variation between adjacent pixel values of the output probability maps. A higher penalty is inflicted on similar colored pixels, as compared to transitional areas with edges and boundaries. This imposes the output probability maps to abide to the given edges and boundaries in the input image. The final loss function, minimized during training is

$$\mathcal{L} = \mathcal{L}_{wpce} + \lambda\mathcal{L}_{smooth}, \tag{5}$$

where $\lambda$ is a hyper-parameter, striking a balance between the uniform partial cross entropy loss, and the shape and boundary preserving smoothness loss.

Inference of a new image is performed via a forward-pass over the trained network, resulting in $C$ softmax probability maps. An additional post-processing, detailed in Algorithm 1, is performed which involves extracting the cell pixels by thresholding, removing irregularly sized cells and classifying each obtained blob by assigning the class with the highest average probability. In such a way, one can collect cell class predictions for each tile in the input slide and aggregate them to a patient-specific cell histogram.

---

**Algorithm 1.** Simultaneous Detection and Classification Inference

---

**Input:** output probability maps $f_\theta(X) = Y \in \mathbb{R}^{|\Omega| \times C}$
**Init:** per-class $scores = 0 \in \mathbb{R}^C$, final prediction $Y_{pred} = 0 \in \mathbb{R}^{|\Omega|}$
$Y_{cell} = \textbf{sum } (Y[:, 1:], dim = 1)$
$Y_{thresh} = \textbf{threshold}(Y_{cell}, t)$
$Y_{comps} = \textbf{connected components}(Y_{thresh})$
$Y_{comps} = \textbf{remove small components}(Y_{comps}, \pi_l)$
$Y_{comps} = \textbf{remove large components}(Y_{comps}, \pi_u)$
$K = \textbf{len}(Y_{comps})$
**for** $k = 0$ **to** $K - 1$ **do**
  $coords = \textbf{component coords}(Y_{comps}[k])$
  **for** $c = 0$ **to** $C - 1$ **do**
    $scores[c] = \textbf{mean}(Y[coords, c])$
  $Y_{pred}[coords] = \textbf{argmax}(scores)$
**Return:** $Y_{pred}$

---

## 4    Experiments

We hereby detail the semi- and weakly-supervised dataset we construct, along with the metrics we use for detection and classification, the experimental setup and the final results of our method.

### 4.1    Dataset

We construct a dataset of IHC images of 12 MM patients' bone-marrow samples, collected in Rambam Medical Center in Israel. Each slide is stained to reveal both MM cells and the sub-cellular Proteasomal pattern, and is then scanned with a 3DHistech scanner. The digitized slides are globally thresholded using Otsu filter and divided up into hundreds of $512 \times 512$ image patches, termed 'tiles', where tiles containing less than 50% tissue are discarded. Each annotated cell in the dataset is ranked for the spread of proteasomal staining, with a score of $-4$ (most cytosolic) to 4 (most nuclear), where the visual characteristics of each consecutive score are slightly different. An additional 'non-relevant' category is assigned to non-malignant cells, appearing in the biopsy. Note that the background class is not annotated, but inferred as part of our solution. See Fig. 2 for different representatives in each class.

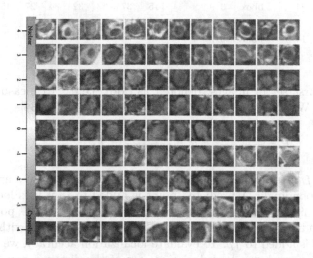

**Fig. 2.** Examples from each class. Each row contains examples of each individual class, which index is shown on the left bar.

In general, we avoid laborious pixelwise annotations of each tile, and instead opt for partial and weakly supervised labels, obtained in several hours of a practiced Pathologist's labor. Our dataset, consisting of 277 512 × 512 tiles, collected from different slides, is annotated with partial annotations of either points, denoting cell centers, or bounding boxes, depicting the location and size of the cell. The images are divided to three groups: 214 partially point/box annotated images for training, 21 fully annotated points for evaluation of detection performance (7 for validation; 14 for test), and partially annotated points/boxes for classification evaluation (18 for validation and 24 for test). All images are partially annotated, with an average of 5% of the cells in each image. To evaluate the detection performance, we annotate all the cells as points in the detection evaluation set, without specifying the exact class, in order to save expensive annotation time. The full description of the dataset is given in Table 1.

### 4.2 Evaluation Metrics

Recall our dataset consists of two types of evaluation sets, one with fully annotated points for each cell location, but without their labels, and the other with partially annotated cells and their corresponding bounding boxes and labels.

**Table 1.** Dataset. Number of examples and annotations in each set and each class, where $-4$ is most cytosolic cell and 4 is most nuclear cell; 5 denotes non-relevant cell.

| Dataset | #images | Type | #annotations | $-4$ | $-3$ | $-2$ | $-1$ | 0 | 1 | 2 | 3 | 4 | 5 |
|---|---|---|---|---|---|---|---|---|---|---|---|---|---|
| 'train' | 214 | Point | 1636 | 63 | 183 | 154 | 167 | 171 | 138 | 234 | 171 | 72 | 283 |
| 'class-val' | 18 | bbox | 220 | 10 | 27 | 25 | 19 | 24 | 16 | 29 | 25 | 9 | 36 |
| 'class-test' | 24 | bbox | 319 | 15 | 37 | 33 | 32 | 34 | 27 | 45 | 34 | 14 | 48 |
| 'det-val' | 7 | Point | 845 | | | | | | | | | | |
| 'det-test' | 14 | Point | 2012 | | | | | | | | | | |

We use the first set of images, denoted as 'det-test' to measure detection performance. We compute the precision, recall and $F_1$ as in [34]

$$P = \frac{tp}{tp + fp}, \quad R = \frac{tp}{tp + fn}, \quad F_1 = \frac{2PR}{P + R}, \tag{6}$$

where $tp, fp, fn$ are the number of true positives, false positives and false negatives, correspondingly. A predicted point is added to $tp$ if it is located within an $r_1$ radius distance from a ground truth (GT) point. A GT point, without any corresponding prediction is added to $fn$, and a prediction without a corresponding GT is added to $fp$. To evaluate localization accuracy, we additionally report the mean and standard deviation $\mu, \sigma$ of the distance between GT and their corresponding $tp$ points, defined as

$$\mu = \frac{1}{N_{tp}} \sum_{i=1}^{N_{tp}} d_i \quad \sigma = \sqrt{\frac{1}{N_{tp}} \sum_{i=1}^{N_{tp}} (d_i - \mu)^2}, \tag{7}$$

where $d_i$ is the euclidean distance between GT and corresponding prediction. To evaluate classification performance, we use the image set 'class-test'. We first compute the relative recall on the partial set of annotations, similarly to the above detection stage. Note that we cannot compute $fp$ since the images are only partially annotated, thus real predictions cannot be distinguished from false positives. The partial recall for 'class-test' is thus $R = tp/(tp + fn)$. To compute the classification performance of the cells tagged as $tp$ in the previous stage, we use the accuracy, precision, recall and F1 classification metrics

$$P = \frac{tp}{tp + fp}, \quad R = \frac{tp}{tp + fn}, \quad F_1 = \frac{2PR}{P + R}, \quad Acc = \frac{tp + tn}{tp + tn + fp + fn}, \tag{8}$$

where $tp, tn$ are positive and negative examples correctly classified, $fn$ are positive examples, incorrectly classified as negative and $fp$ are negative examples, incorrectly classified as positive. We calculate the multi-class scores of each metric by taking the weighted average of the one-vs-all scores of each class. We also compute the mAP (mean Average Precision) score, measuring the area under the precision-recall curve, weighted across all classes.

### 4.3    Implementation Details

The radii and Gaussian width encoding parameters are determined from the validation set and taken as $r_1 = 12, r_2 = 14.5, \sigma = r_1/3$. The post-processing threshold, minimum and maximum areas are $t = 0.85, \pi_l = \pi(r_1/3.5)^2, \pi_u = \pi r_1^2$. The hyper-parameters in our loss function are taken as $\alpha = 10, \beta = 30, \lambda = 1e^{-5}$. During training, we perform data augmentation of the input $512 \times 512$ tiles, including random resize, horizontal and vertical flip, rotation, affine transformation and crop to the input size of the network $224 \times 224$. As suggested in [34], we use ResUNet – a fully-convolutional UNet model, with the ResNet34 encoder pretrained on ImageNet. The network is trained using the Adam optimizer with a learning rate of $10^{-4}$, weight decay of $10^{-3}$ and batch size of 16. We choose the model that gave the best combined $F_1$ metric of both the detection, denoted as $F_1^D$ and classification, denoted as $F_1^C$, represented as a geometric average of both $F_1^{Best} = (2 \cdot F_1^D \cdot F_1^C)/(F_1^D + F_1^C)$. We implement our code in Pytorch and Numpy on a GTX Titan-X Nvidia GPU.

The 'baseline' method is based on the implementation of [34][1]. The detection network is trained for 4 rounds, after which it produces point, Voronoi and cluster labels for the training of the next segmentation network. To produce classifications, we train a CNN with three $3 \times 3$ convolutional layers of width $16, 32, 64$, each followed by BatchNorm, ReLU and $2 \times 2$ MaxPool. The final two linear layers are of size $100, C$, where $C$ is the number of classes. For every segmented component, the surrounding bounding box is extracted and warped to constant size of $64 \times 64$, then fed to the trained classifier. The classifier is trained and validated using bounding boxes in the train and validation sets.

### 4.4    Results

We compare between three approaches of cell detection and classification, the 'baseline' method, which features disjoint detection and classification and 'ours-ring' and 'ours-vor', performing both tasks within the same architecture. The cell versus background detection results on 'det-test' are shown on the left in Table 2. Although one cannot compute precision for 'class-test', we can report the partial recall, denoted as $R_{part}$ in the table. As for classification performance, we treat all detected blobs that received higher scores than the threshold as 'cells', and compute their respective class assignments using either Algorithm 1 for our method, or applying a classifier on the bounding box which surrounds the detected blob, for the baseline. The classification results are listed on the right of Table 2. As can be seen, the 'baseline' method, originally intended for cell detection, receives better detection $P, R, F_1$, as compared to our method. Our method, however, receives better classification results and a higher combined Fscore, suggesting that training simultaneously for both tasks improves the combined performance. Note that the additional Voronoi-based background supervision in 'ours-vor' improves the results, but at the cost of

---

[1] https://github.com/huiqu18/WeaklySegPartialPoints.

**Table 2.** Numeric results. The left and right tables present detection and classification results. $R_{part}$ denotes partial detection recall, and $F1_{comb}$ denotes combined Fscore.

| Method | 'det-test' | | | | | 'class-test' | | | | | | |
|---|---|---|---|---|---|---|---|---|---|---|---|---|
| | $P$ | $R$ | $F_1$ | $\mu$ | $\sigma$ | $R_{part}$ | $Acc$ | $P$ | $R$ | $F_1$ | $mAP$ | $F1_{comb}$ |
| 'baseline' | 0.766 | **0.812** | **0.788** | 3.81 | 2.50 | 0.872 | 0.414 | 0.452 | 0.414 | 0.406 | 0.457 | 0.535 |
| 'ours-ring' | 0.712 | 0.759 | 0.735 | 3.70 | 2.34 | **0.912** | 0.491 | 0.503 | 0.491 | 0.489 | 0.544 | 0.587 |
| 'ours-vor' | **0.923** | 0.629 | 0.748 | **3.51** | **2.33** | 0.897 | **0.539** | **0.550** | **0.539** | **0.530** | **0.577** | **0.621** |

training an additional designated detection network. The qualitative detection and classification results are presented in Figs. 3 and 4. Although the baseline creates tighter segmentations, the predicted classes often do not correspond with the ground truth boxes.

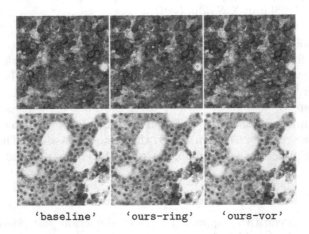

'baseline'          'ours-ring'          'ours-vor'

**Fig. 3.** Qualitative detection results on 'det-test'. Each tile is overlayed with its predicted cells. Blue, red and yellow dots represent $tp, fp, fn$ correspondingly. (Color figure online)

The classification confusion matrix of 'ours-vor' is given in Fig. 5. As can be seen, the confusion matrix is nearly diagonal, up to an offset of ±1 from the main diagonal. This can be expected since neighboring scores, differing by ±1 can be easily confused by the annotating pathologist (see. adjacent rows of Fig. 2) and consequently by our algorithm. We acknowledge this difficulty and show the numeric results given a grace of ±1 between ground truth and predicted scores. We calculate accuracy, precision, recall and $F_1$, based on this extended metric and report the results in Table 3. As can be seen, this considerably increases the classification Fscore to ≈ 85%. As to runtime performance, as shown in Table 3, our joint network is ×4 faster in inference, as it does not require an external classifier over each and every cell, as the 'baseline' method.

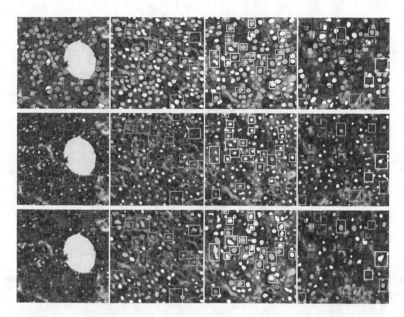

**Fig. 4.** Qualitative classification results on 'class-test'. From top to bottom, the rows feature the 'baseline', 'ours-ring' and 'ours-vor' methods. The annotated ground truths are marked as bounding boxes. Cooler tones represent nuclear cells, whereas warmer tones represent cytosolic cells. The exact colormap is: $-4$: red, $-3$: pink, $-2$: orange, $-1$: yellow, 0: white, 1: green, 2: cyan, 3: blue, 4: purple, 5: grey. (Color figure online)

Returning to the original application, personalized cell histograms, we aggregate the results of our method over three patient' slides (not used for training or validation) and show the results in Fig. 6. We fed the tile images of each patient's WSI into our trained network and obtained the predicted boxes and matching softmax probabilities. In the resulting tile example and histogram, we show cells that achieved a confidence score of 0.35 and higher. The rows (from

**Table 3.** Classification with $\pm 1$ and runtime

|  | 'baseline' | 'ours-ring' | 'ours-vor' |
|---|---|---|---|
| $Acc \pm 1$ | 0.755 | 0.866 | **0.874** |
| $P \pm 1$ | 0.738 | 0.841 | **0.844** |
| $R \pm 1$ | 0.748 | 0.828 | **0.860** |
| $F_1 \pm 1$ | 0.726 | 0.830 | **0.848** |
| $inference[s]$ | 0.0872 | 0.0239 | **0.0213** |

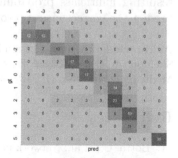

**Fig. 5.** Confusion matrix

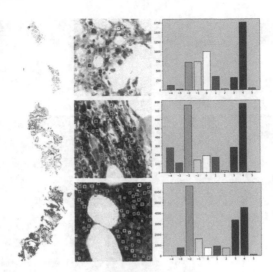

**Fig. 6.** Patient histograms. From left to right: bone marrow sample, example tile from the WSI and accumulated histogram.

top to bottom) in Fig. 6 correspond to patients, tagged globally by a pathologist, as 'nuclear', 'evenly distributed' and 'cytosolic'. This indeed aligns with our obtained histograms, but additional extensive work is needed before finding correlation between a patient's histogram and their suggested treatment.

## 5    Conclusions

We have presented our approach for simultaneous detection and classification of cells in IHC images, derived from biopsies of MM patients. We proposed a new weakly and partially supervised dataset for cell classification, consisting of 9 categories, corresponding to different levels of proteasomal staining. Our DNN, trained to perform both detection and classification, is guided by a loss function, combining fidelity to the given partial labels and smoothness of the resulting multiclass probability maps. We have shown our method is able to predict the correct class up to a margin of $\pm 1$, with an accuracy of $\approx 87\%$.

**Acknowledgements.** The study was supported by grants from the Dr. Miriam and Sheldon Adelson Medical Research Foundation (AMRF) and the Israel Precision Medicine Partnership (IPMP), administered by the Israel Science Foundation (ISF).

## References

1. Botta, C., et al.: Network meta-analysis of randomized trials in multiple myeloma: efficacy and safety in relapsed/refractory patients. Blood Adv. **1**(7), 455–466 (2017)
2. Chamanzar, A., Nie, Y.: Weakly supervised multi-task learning for cell detection and segmentation. In: 2020 IEEE 17th International Symposium on Biomedical Imaging (ISBI), pp. 513–516. IEEE (2020)

3. Chandradevan, R., et al.: Machine-based detection and classification for bone marrow aspirate differential counts: initial development focusing on nonneoplastic cells. Lab. Invest. **100**(1), 98–109 (2020)
4. Cireşan, D.C., Giusti, A., Gambardella, L.M., Schmidhuber, J.: Mitosis detection in breast cancer histology images with deep neural networks. In: Mori, K., Sakuma, I., Sato, Y., Barillot, C., Navab, N. (eds.) MICCAI 2013. LNCS, vol. 8150, pp. 411–418. Springer, Heidelberg (2013). https://doi.org/10.1007/978-3-642-40763-5_51
5. Cordell, J.L., et al.: Immunoenzymatic labeling of monoclonal antibodies using immune complexes of alkaline phosphatase and monoclonal anti-alkaline phosphatase (apaap complexes). J. Histochem. Cytochem. **32**(2), 219–229 (1984)
6. Coudray, N., et al.: Classification and mutation prediction from non-small cell lung cancer histopathology images using deep learning. Nat. Med. **24**(10), 1559–1567 (2018)
7. Dagogo-Jack, I., Shaw, A.T.: Tumour heterogeneity and resistance to cancer therapies. Nat. Rev. Clin. Oncol. **15**(2), 81–94 (2018)
8. Erber, W., Mynheer, L., Mason, D.: APAAP labelling of blood and bone-marrow samples for phenotyping leukaemia. Lancet **327**(8484), 761–765 (1986)
9. Gao, Z., Puttapirat, P., Shi, J., Li, C.: Renal cell carcinoma detection and subtyping with minimal point-based annotation in whole-slide images. In: Martel, A.L., et al. (eds.) MICCAI 2020. LNCS, vol. 12265, pp. 439–448. Springer, Cham (2020). https://doi.org/10.1007/978-3-030-59722-1_42
10. Golts, A., Freedman, D., Elad, M.: Deep energy: task driven training of deep neural networks. IEEE J. Sel. Top. Sig. Process. **15**(2), 324–338 (2021)
11. Grady, L.: Random walks for image segmentation. IEEE Trans. Pattern Anal. Mach. Intell. **28**(11), 1768–1783 (2006)
12. Han, Z., Wei, B., Zheng, Y., Yin, Y., Li, K., Li, S.: Breast cancer multi-classification from histopathological images with structured deep learning model. Sci. Rep. **7**(1), 1–10 (2017)
13. He, K., Zhang, X., Ren, S., Sun, J.: Deep residual learning for image recognition. In: Proceedings of the IEEE Conference on Computer Vision and Pattern Recognition, pp. 770–778 (2016)
14. Hou, L., Agarwal, A., Samaras, D., Kurc, T.M., Gupta, R.R., Saltz, J.H.: Robust histopathology image analysis: to label or to synthesize? In: Proceedings of the IEEE/CVF Conference on Computer Vision and Pattern Recognition, pp. 8533–8542 (2019)
15. Javed, S., et al.: Cellular community detection for tissue phenotyping in colorectal cancer histology images. Med. Image Anal. **63**, 101696 (2020)
16. Javed, S., Mahmood, A., Werghi, N., Benes, K., Rajpoot, N.: Multiplex cellular communities in multi-gigapixel colorectal cancer histology images for tissue phenotyping. IEEE Trans. Image Process. **29**, 9204–9219 (2020)
17. Jia, Y., et al.: Caffe: convolutional architecture for fast feature embedding. In: Proceedings of the 22nd ACM International Conference on Multimedia, pp. 675–678. ACM (2014)
18. Johnson, J., Alahi, A., Fei-Fei, L.: Perceptual losses for real-time style transfer and super-resolution. In: Leibe, B., Matas, J., Sebe, N., Welling, M. (eds.) ECCV 2016. LNCS, vol. 9906, pp. 694–711. Springer, Cham (2016). https://doi.org/10.1007/978-3-319-46475-6_43
19. Kanavati, F., et al.: Weakly-supervised learning for lung carcinoma classification using deep learning. Sci. Rep. **10**(1), 1–11 (2020)

20. Kumar, N., Verma, R., Sharma, S., Bhargava, S., Vahadane, A., Sethi, A.: A dataset and a technique for generalized nuclear segmentation for computational pathology. IEEE Trans. Med. Imaging **36**(7), 1550–1560 (2017)
21. Li, C., Wang, X., Liu, W., Latecki, L.J., Wang, B., Huang, J.: Weakly supervised mitosis detection in breast histopathology images using concentric loss. Med. Image Anal. **53**, 165–178 (2019)
22. Li, J., et al.: Signet ring cell detection with a semi-supervised learning framework. In: Chung, A.C.S., Gee, J.C., Yushkevich, P.A., Bao, S. (eds.) IPMI 2019. LNCS, vol. 11492, pp. 842–854. Springer, Cham (2019). https://doi.org/10.1007/978-3-030-20351-1_66
23. Litjens, G., et al.: Deep learning as a tool for increased accuracy and efficiency of histopathological diagnosis. Sci. Rep. **6**(1), 1–11 (2016)
24. Liu, D., et al.: Unsupervised instance segmentation in microscopy images via panoptic domain adaptation and task re-weighting. In: Proceedings of the IEEE/CVF Conference on Computer Vision and Pattern Recognition, pp. 4243–4252 (2020)
25. Livneh, I., Cohen-Kaplan, V., Cohen-Rosenzweig, C., Avni, N., Ciechanover, A.: The life cycle of the 26s proteasome: from birth, through regulation and function, and onto its death. Cell Res. **26**(8), 869–885 (2016)
26. Long, J., Shelhamer, E., Darrell, T.: Fully convolutional networks for semantic segmentation. In: Proceedings of the IEEE Conference on Computer Vision and Pattern Recognition, pp. 3431–3440 (2015)
27. Mahmood, F., et al.: Deep adversarial training for multi-organ nuclei segmentation in histopathology images. IEEE Trans. Med. Imaging **39**(11), 3257–3267 (2019)
28. Malon, C.D., Cosatto, E.: Classification of mitotic figures with convolutional neural networks and seeded blob features. J. Pathol. Inform. **4**(1), 9 (2013)
29. Naylor, P., Laé, M., Reyal, F., Walter, T.: Segmentation of nuclei in histopathology images by deep regression of the distance map. IEEE Trans. Med. Imaging **38**(2), 448–459 (2018)
30. Nishimura, K., Ker, D.F.E., Bise, R.: Weakly supervised cell instance segmentation by propagating from detection response. In: Shen, D., et al. (eds.) MICCAI 2019. LNCS, vol. 11764, pp. 649–657. Springer, Cham (2019). https://doi.org/10.1007/978-3-030-32239-7_72
31. Peikari, M., Salama, S., Nofech-Mozes, S., Martel, A.L.: A cluster-then-label semi-supervised learning approach for pathology image classification. Sci. Rep. **8**(1), 1–13 (2018)
32. Qu, H., Riedlinger, G., Wu, P., Huang, Q., Yi, J., De, S., Metaxas, D.: Joint segmentation and fine-grained classification of nuclei in histopathology images. In: 2019 IEEE 16th International Symposium on Biomedical Imaging (ISBI 2019), pp. 900–904. IEEE (2019)
33. Qu, H., et al.: Weakly supervised deep nuclei segmentation using points annotation in histopathology images. In: International Conference on Medical Imaging with Deep Learning, pp. 390–400. PMLR (2019)
34. Qu, H., et al.: Weakly supervised deep nuclei segmentation using partial points annotation in histopathology images. IEEE Trans. Med. Imaging **39**(11), 3655–3666 (2020)
35. Redmon, J., Farhadi, A.: Yolov3: An incremental improvement. arXiv preprint arXiv:1804.02767 (2018)
36. Ren, S., He, K., Girshick, R., Sun, J.: Faster R-CNN: towards real-time object detection with region proposal networks. Adv. Neural. Inf. Process. Syst. **28**, 91–99 (2015)

37. Richardson, P.G., et al.: Bortezomib or high-dose dexamethasone for relapsed multiple myeloma. N. Engl. J. Med. **352**(24), 2487–2498 (2005)
38. Ronneberger, O., Fischer, P., Brox, T.: U-net: convolutional networks for biomedical image segmentation. In: Navab, N., Hornegger, J., Wells, W.M., Frangi, A.F. (eds.) MICCAI 2015. LNCS, vol. 9351, pp. 234–241. Springer, Cham (2015). https://doi.org/10.1007/978-3-319-24574-4_28
39. Russakovsky, O., et al.: Imagenet large scale visual recognition challenge. Int. J. Comput. Vision **115**(3), 211–252 (2015)
40. Saha, M., Chakraborty, C.: Her2net: a deep framework for semantic segmentation and classification of cell membranes and nuclei in breast cancer evaluation. IEEE Trans. Image Process. **27**(5), 2189–2200 (2018)
41. Siegel, R.L., Miller, K.D., Fuchs, H.E., Jemal, A.: Cancer statistics, 2022. CA: Cancer J. Clinicians **72**(1), 7–33 (2022). https://doi.org/10.3322/caac.21708
42. Sirinukunwattana, K., Raza, S.E.A., Tsang, Y.W., Snead, D.R., Cree, I.A., Rajpoot, N.M.: Locality sensitive deep learning for detection and classification of nuclei in routine colon cancer histology images. IEEE Trans. Med. Imaging **35**(5), 1196–1206 (2016)
43. Slamon, D.J., et al.: Use of chemotherapy plus a monoclonal antibody against HER2 for metastatic breast cancer that overexpresses HER2. N. Engl. J. Med. **344**(11), 783–792 (2001)
44. Song, T.H., Sanchez, V., Daly, H.E., Rajpoot, N.M.: Simultaneous cell detection and classification in bone marrow histology images. IEEE J. Biomed. Health Inform. **23**(4), 1469–1476 (2018)
45. Tang, M., Perazzi, F., Djelouah, A., Ayed, I.B., Schroers, C., Boykov, Y.: On regularized losses for weakly-supervised CNN Segmentation. In: Ferrari, V., Hebert, M., Sminchisescu, C., Weiss, Y. (eds.) ECCV 2018. LNCS, vol. 11220, pp. 524–540. Springer, Cham (2018). https://doi.org/10.1007/978-3-030-01270-0_31
46. Wei, J.W., Tafe, L.J., Linnik, Y.A., Vaickus, L.J., Tomita, N., Hassanpour, S.: Pathologist-level classification of histologic patterns on resected lung adenocarcinoma slides with deep neural networks. Sci. Rep. **9**(1), 1–8 (2019)
47. Wu, Y.Y., et al.: A hematologist-level deep learning algorithm (BMSNet) for assessing the morphologies of single nuclear balls in bone marrow smears: algorithm development. JMIR Med. Inform. **8**(4), e15963 (2020)
48. Xie, Y., Xing, F., Kong, X., Su, H., Yang, L.: Beyond classification: structured regression for robust cell detection using convolutional neural network. In: Navab, N., Hornegger, J., Wells, W.M., Frangi, A.F. (eds.) MICCAI 2015. LNCS, vol. 9351, pp. 358–365. Springer, Cham (2015). https://doi.org/10.1007/978-3-319-24574-4_43
49. Xie, Y., Xing, F., Shi, X., Kong, X., Su, H., Yang, L.: Efficient and robust cell detection: a structured regression approach. Med. Image Anal. **44**, 245–254 (2018)
50. Xing, F., Xie, Y., Yang, L.: An automatic learning-based framework for robust nucleus segmentation. IEEE Trans. Med. Imaging **35**(2), 550–566 (2015)
51. Xu, J., et al.: Stacked sparse autoencoder (SSAE) for nuclei detection on breast cancer histopathology images. IEEE Trans. Med. Imaging **35**(1), 119–130 (2015)
52. Zhang, K., Zuo, W., Chen, Y., Meng, D., Zhang, L.: Beyond a Gaussian denoiser: residual learning of deep CNN for image denoising. IEEE Trans. Image Process. **26**(7), 3142–3155 (2017)
53. Zhou, Y., Dou, Q., Chen, H., Qin, J., Heng, P.A.: SFCN-OPI: detection and fine-grained classification of nuclei using sibling FCN with objectness prior interaction. In: Proceedings of the AAAI Conference on Artificial Intelligence, vol. 32 (2018)

# Deep-ASPECTS: A Segmentation-Assisted Model for Stroke Severity Measurement

Ujjwal Upadhyay[1]([✉]), Mukul Ranjan[1], Satish Golla[1], Swetha Tanamala[1],
Preetham Sreenivas[1], Sasank Chilamkurthy[1], Jeyaraj Pandian[2],
and Jason Tarpley[3]

[1] Qure.ai, Mumbai, India
{ujjwal.upadhyay,mukul.ranjan,satish.golla,swetha.tanamala,
preetham.sreeniva,sasank.chilamkurthy}@qure.ai
[2] Christian Medical College Ludhiana, Ludhiana, India
[3] Pacific Neuroscience Institute, Santa Monica, USA
jason.tarpley@providence.org

**Abstract.** A stroke occurs when an artery in the brain ruptures and bleeds or when the blood supply to the brain is cut off. Blood and oxygen cannot reach the brain's tissues due to the rupture or obstruction resulting in tissue death. The Middle cerebral artery (MCA) is the largest cerebral artery and the most commonly damaged vessel in stroke. The quick onset of a focused neurological deficit caused by interruption of blood flow in the territory supplied by the MCA is known as an MCA stroke. Alberta stroke programme early CT score (ASPECTS) is used to estimate the extent of early ischemic changes in patients with MCA stroke. This study proposes a deep learning-based method to score the CT scan for ASPECTS. Our work has three highlights. First, we propose a novel method for medical image segmentation for stroke detection. Second, we show the effectiveness of AI solution for fully-automated ASPECT scoring with reduced diagnosis time for a given non-contrast CT (NCCT) Scan. Our algorithms show a dice similarity coefficient of **0.64** for the MCA anatomy segmentation and **0.72** for the infarcts segmentation. Lastly, we show that our model's performance is inline with inter-reader variability between radiologists.

**Keywords:** Stroke · Infarct · Automated ASPECTS scoring framework · Segmentation

## 1 Introduction

A stroke is a medical emergency that requires immediate attention. Brain injury and other consequences can be avoided if intervention is taken early. There are two main types of stroke. **Hemorrhagic Stroke**. When a blood vessel ruptures, it

**Supplementary Information** The online version contains supplementary material available at https://doi.org/10.1007/978-3-031-25066-8_17.

results in hemorrhagic stroke. Aneurysms or arteriovenous malformations (AVM) are the most common causes of hemorrhagic stroke. **Ischemic Stroke**. When the blood flow to a portion of the brain is blocked or diminished, brain tissue is deprived of oxygen and nutrients, resulting in an ischemic stroke. Within minutes, brain cells begin to die. Ischemic and hemorrhagic strokes are managed differently since they have different causes and effects on the body. Rapid diagnosis is critical for minimising brain damage and allowing the doctor to treat the stroke with the most appropriate treatment strategy for the type.

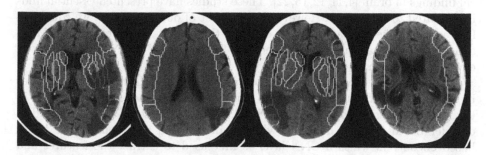

**Fig. 1.** Segmentation output from Deep-ASPECTS. Refer to Fig. 2 for more information on how this is generated.

In this paper, we focus on ischemic strokes. The symptoms of such a stroke vary depending on the brain area affected and the quantity of tissue that has been damaged. The severity of damage can be assessed by ASPECTS. In clinical practice, ASPECTS detects significant Early Ischemic Changes(EIC) in a higher proportion of the early scans [14]. ASPECTS is a topographic scoring system that applies a quantitative approach and does not ask physicians to estimate volumes from two-dimensional images. It is scored out of 10 points. ASPECTS has been one of the recognized scoring scales that serve as key selection criteria on the management of acute stroke in the MCA region, where endovascular therapy in patients with baseline ASPECTS$\geq$ 6 is recommended [6,7,16]. Variations of the ASPECT scoring system is used in the posterior circulation and referred to as pc-ASPECTS [17].

In stroke cases, "time is brain" [19]. The outcomes become progressively worse with time. In the current clinical practice, radiologists must read the NCCT scan to report the ASPECT score, which can take time due to the high volume of cases. We propose an automated ASPECT scoring system, a task currently affected by high inter-reader variability due to manual selection and measurement of the relevant MCA regions scored for ASPECTS. We present Deep-ASPECTS that provides an efficient way to prioritize and detect stroke cases in less than 1 min, making it clinically relevant.

The proposed deep learning solution involves segmenting the acute infarcts and the MCA territory from an NCCT scan. The segmentation step is followed by overlapping these maps to get the affected region across the slices in basal ganglia and corona radiata level. We demonstrate the robustness of our framework by validating both the segmentation maps and the estimated ASPECTS against ground truth clinical data, which contains 150 scans.

## 2   Related Work

Various models have been developed and studied for the segmentation of NCCT scan for qualitative estimation of infarcts, haemorrhage and different other critical findings in brain [4,11,12,15,22]. These studies have presented essential findings related to the NCCT. However, to the best of our knowledge, we could not find any published research which combines the Infarct segmentation and MCA anatomy segmentation to predict the ASPECT score of the NCCT scan in an end-to-end fashion.

Deep learning based semantic segmentation models with encoder and decoder blocks connected through various skip connections have shown to be effective in the medical imaging domain [2,3,8,18,24]. Unet [18] presents a simple encoder-decoder based CNN architecture while TransUnet [3] uses transformer [5,23] based encoder in which a CNN layer is first used to extract the features and then patch embedding is applied to $1 \times 1$ features map. Linknet [2] is another variation of Unet [18] which replaces the CNN block of Unet with Residual connection [9] and instead to stacking the features of encoder blocks to the decoder block in the skip connections it adds them. UNet++ further extends UNet by connecting encoder and decoder blocks through a series of dense Convolutional blocks instead of directly stacking or adding them.

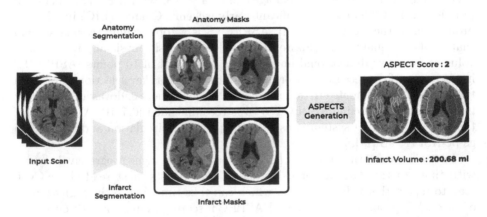

**Fig. 2.** ASPECTS Framework. It has three main components: (1) Infarct segmentation network (2) MCA anatomy segmentation network. (3) ASPECTS generation function.

There has been some prior work on automating ASPECTS. [11] extracted the textures features from ASPECTS regions and then used random forest classifier to classify if the region is affected. [1] used 3 phase architecture, where they preprocessed the NCCT scans for skull striping and then passed it to UNet to find the ASPECTS regions. These ASPECTS regions where then compared contralaterally to find the affected side of the brain. Detailed comparison cannot be drawn between the above methods and ours because of unavailability of the code to reproduce their experiments and unavailability of the dataset used by [1,11].

## 3    Dataset

We have built our dataset containing 50000 studies out of which 8000 have infarcts. Clinicians manually marked pixel-level ground truths for infarcts and MCA anatomy regions on 1500 training examples. A skilled specialist double-checked the annotation findings under stringent quality control. The labelled dataset was separated into three groups at random in training, validation, and testing at a ratio of 8:1:1. The number of slices in each scan in our dataset varies from subject to subject. Our collaborating hospitals have reviewed the data collection process with approval from the local research ethics committee. Details for data collection process is added to supplementary.

## 4    Methods

The ASPECT score is calculated by dividing the MCA territory into ten regions: Caudate, Lentiform Nucleus, Internal Capsule, Insular Cortex, M1, M2, M3 M4 M5, and M6.

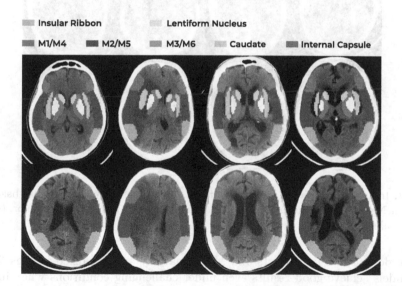

**Fig. 3.** MCA segmentation model results on our test set. These results are based on EfficientNet [21], achieving a DSC of 0.64. Masks are shown in color. (Color figure online)

M1 to M3 is at the level of the basal ganglia. M4 to M6 are at the level of the ventricles immediately above the basal ganglia called corona radiata. 1 point is deducted from the initial score of 10 for every region showing early ischemic signs, such as focal swelling or infarcts. The score was created to aid in identifying patients who were most likely to benefit clinically from intravenous thrombolysis.

Figure 2 illustrates the pipeline of the proposed framework. The UNet [18] segmentation models with EfficientNet backbone are used for both infarct and anatomy segmentation networks. ASPECTS generation function is used to overlap segmentation masks from both segmentation networks and report the ASPECT Score. More qualitative results on segmentation mask can be found in Fig. 1. Infarct volume and masked output are also reported to aid the radiologists in making informed decisions regarding following procedures.

ASPECTS generation function, in Fig. 2, overlays the masks from 2 segmentation over each other and finds the anatomy being overlapped. These overlapped regions are considered affected regions and scored for aspects. This function also finds volume using the 'pixel spacing' attribute, part of dicom metadata. Pixel spacing represents how much volume is enclosed by a voxel.

$$V(\text{ml}) = \sum \text{infarct}_{mask} * spacing_x * spacing_y * spacing_z \qquad (1)$$

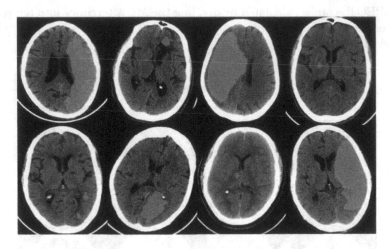

**Fig. 4.** Infarct segmentation model results on our test set. These results are based on EfficientNet [21], achieving a DSC of 0.72. Infarct masks are shown in red color. (Color figure online)

Model outputs for infarcts and MCA anatomy are visualized in Figs. 3 and 4. Models achieve good results even under challenging conditions when infarct is small and hard to identify. Quantitative results are shown in Table 1b.

All the scans were rescaled to 224 × 224 and used as input to our framework. Data augmentation is also applied by introducing random noise, rotation, shift, and flipping. The models were trained on 1 Nvidia 1080Ti GPU with 12 GB RAM. We used SGD optimizer with a cyclic learning rate scheduler for all experiments with a learning rate of 5e−3.

$$L = \alpha * L_1 + \beta * L_2 + \gamma * L_3 \tag{2}$$

In Eq. 2, We used 3 loss functions - focal loss [13] ($L_1$), boundary loss [10] ($L_2$) and dice loss [20] ($L_3$). We found $\alpha = 3$, $\beta = 1$ and $\gamma = 1$, worked best for both infarct segmentation and MCA anatomy segmentation task, after grid search on log linear between 1 and 100. Final evaluation of the acquired data was performed using the dice similarity coefficient (DSC).

# 5   Results

## 5.1   How Does Volume of Infarct Influence the Segmentation Map?

From Table 1a, it can be seen that UNet performed better than other models. We also compared the models based on their ability to segment infarct based on their volume. In Table 1b, it can be seen that UNet was not the best model for infarct volume less than 3 ml. UNet++ was better by 0.04 DSC. However, for other volume categories, UNet was still better. Low volume infarct in terms of ASPECTS translates to a higher ASPECT score i.e. more than 8. In Table 3 and 4, it is evident that the performance is not hampered as even with low dice score. Though in future work, it gives scope of improvement.

**Table 1.** Comparison of different models on infarct segmentation.

(a) Model Evaluation on infarct segmentation task.

| Methods | DSC | Sensitivity | Specificity |
|---|---|---|---|
| UNet | 0.72 | 0.77 | 0.99 |
| TransUNet | 0.57 | 0.61 | 0.98 |
| UNet++ | 0.67 | 0.76 | 0.99 |
| LinkNet | 0.54 | 0.57 | 0.96 |

(b) Infarct volume used to characterise the performance of the models based on DSC.

| Methods | Infarct Volume | | | |
|---|---|---|---|---|
| | <3ml | 3-16ml | 16-66ml | >66ml |
| UNet | 0.45 | 0.72 | 0.78 | 0.01 |
| TransUNet | 0.41 | 0.55 | 0.61 | 0.72 |
| UNet++ | 0.49 | 0.62 | 0.71 | 0.86 |
| LinkNet | 0.31 | 0.54 | 0.52 | 0.79 |

## 5.2   How Do Different Models Segment MCA Territory?

Table 2 shows that UNet performed better than other models when we look at the overall DSC of the MCA territory. Though the smaller regions like Lentiforum nucleus, Internal Capsule, and Insular Ribbon UNet is not the best model, we can see that the variation is not very large between UNet and the best models.

**Table 2.** Comparison of different models on MCA territory segmentation task. The table contains DSC for different regions and across all regions.

| Methods | Overall | Caudate | Lentiform nucleus | Internal capsule | Insular ribbon | M1, M4 | M2, M5 | M3, M6 |
|---|---|---|---|---|---|---|---|---|
| UNet | **0.64** | **0.70** | 0.60 | 0.59 | 0.52 | 0.72 | **0.77** | **0.54** |
| TransUNet | 0.59 | 0.63 | 0.59 | 0.57 | 0.43 | **0.73** | 0.71 | 0.49 |
| UNet++ | 0.62 | 0.69 | **0.61** | 0.58 | 0.50 | 0.69 | 0.75 | 0.54 |
| LinkNet | 0.63 | 0.70 | 0.61 | **0.61** | **0.53** | 0.71 | 0.77 | 0.49 |

## 5.3 Model Performance for Different ASPECT Score

In Table 3, we can see the performance of the best performing model from infarct segmentation and anatomy segmentation tasks. This performance is measured against two readers.

From Table 3, it can be observed that our model has low specificity for some ASPECT scores. However, it is important to notice that specificity is low even between the score of 2 readers. This is attributed to the subjective nature of ASPECTS. Therefore, it is best to judge the performance of model-based on binned ASPECTS, in Table 4. The score was binned based on the treatment outcome given the score. If the score is between 10-8, the treatment outcome is usually favourable and less complex. The ASPECTS cutoff value determined for the prediction of unfavourable outcomes was equal to 7 [7]. Patients with ASPECTS < 4 has even less chance of good functional outcome. The improvement in specificity is evident from Table 4. Here we observe that our agreement with Reader B is more than Reader A. The measure of variability between these two readers is discussed in Sect. 6. For threshold of ASPECTS > 6, we found that sensitivity and specificity was 75.75% and 94.73% respectively.

**Table 3.** Performance of trained model, Deep-ASPECTS against 2 readers and inter-reader difference for each ASPECTS.

| ASPECTS | Model vs Reader A | | Model vs Reader B | | Reader A vs Reader B | |
|---|---|---|---|---|---|---|
| | Sensitivity | Specificity | Sensitivity | Specificity | Sensitivity | Specificity |
| 0 | 0.99 | 0.17 | 0.98 | 0.67 | 0.96 | 0.50 |
| 1 | 1.00 | 0.50 | 1.00 | 0.50 | 0.99 | 0.50 |
| 2 | 1.00 | 0.20 | 0.99 | 0.20 | 0.99 | 1.00 |
| 3 | 0.94 | 0.38 | 0.99 | 0.50 | 0.99 | 0.27 |
| 4 | 0.99 | 0.25 | 0.98 | 0.25 | 0.97 | 0.00 |
| 5 | 0.97 | 0.33 | 0.96 | 0.83 | 0.94 | 0.33 |
| 6 | 0.95 | 0.22 | 0.97 | 0.22 | 0.96 | 0.27 |
| 7 | 0.92 | 0.40 | 0.89 | 0.30 | 0.92 | 0.47 |
| 8 | 0.90 | 0.20 | 0.92 | 0.30 | 0.92 | 0.25 |
| 9 | 0.81 | 0.30 | 0.69 | 0.40 | 0.73 | 0.52 |
| 10 | 0.85 | 0.59 | 0.93 | 0.35 | 0.92 | 0.41 |

**Table 4.** Performance comparison across binned ASPECT score.

| ASPECTS | Model vs Reader A | | Model vs Reader B | | Reader A vs Reader B | |
|---|---|---|---|---|---|---|
| | Sensitivity | Specificity | Sensitivity | Specificity | Sensitivity | Specificity |
| A (0–3) | 0.96 | 0.48 | 0.99 | 0.65 | 0.95 | 0.56 |
| B (4–7) | 0.87 | 0.55 | 0.85 | 0.66 | 0.84 | 0.66 |
| C (8–10) | 0.74 | 0.93 | 0.74 | 0.86 | 0.84 | 0.85 |

## 5.4 Number of Parameters for Models and Their Computation Time

Table 5 shows that other that TransUNet, all of the other three models UNet, UNet++ and LinkNet have a comparable number of parameters since they have the same backbone of EfficientNet [21]. Despite the fact that UNet has 0.9M more parameters compared to LinkNet, its performance on infarct and MCA anatomy segmentation was better than LinkNet. Computational time was also comparable for both models. Therefore we found UNet to be a better option than LinkNet. This evaluation is done on 1 Nvidia 1080Ti GPU with 12 GB RAM.

**Table 5.** Total number of parameters for each model and time taken to process each scan.

| Method | Parameters | Time (seconds) |
|---|---|---|
| UNet | 42.1 M | 28–30 |
| TransUNet | 105 M | 50–63 |
| UNet++ | 42.9 M | 28–32 |
| LinkNet | 41.2 M | 27–29 |

# 6 Inter-reader Variability

Out of 1500 scans, 150 scans were read by two radiologists. This set is the same as the test set to learn what the acceptable ASPECT score might be for a scan. In Tables 3 and 4, we compare how much disagreement there is between 2 readers with regards to ASPECTS from 10 and a binned version of score.

ASPECTS is a very subjective score, and it varies from reader to reader. We found the inter-reader agreement on our data to be 39.45% when we expect an exact score match. However, the agreement increases to 76.87% when the difference of 2 points is allowed. Pearson correlation between the reads of 2 readers came out to be 73.17%.

The ASPECTS reported by our model have an agreement of 42.17% and 36.73% with readers A and B, respectively. If the agreement score is relaxed by 2 points, then the agreement increases to 69.38% and 76.19%. These numbers suggest that our deep learning model performed closely with radiologists maintaining comparable agreement scores.

## 7   Conclusions

We proposed a novel end-to-end system for automated ASPECT scoring. Additionally, a pilot study was conducted, showing the effectiveness of such an AI-based model in getting ASPECTS promptly. Deploying a system that can score an NCCT in less than a minute with reasonable accuracy can aid preliminary diagnosis for a suspected stroke case. Deep-ASPECTS can save precious time from the diagnosis phase and prevent further deterioration in patients' conditions.

Deep-ASPECTS got an agreement of 76.19% with reader B (radiologists), which is the same as reader A (refer Sect. 6). The proposed method consistently performs as good as radiologists. In sum, the method shows a high potential to improve the clinical success rate by alerting the radiologist or neurologists about potential stroke cases with their severity reported as ASPECTS. Furthermore, a more realistic study should be conducted with a bigger sample size. In future, we intend to do a more thorough architecture search for the encoders, investigate more anatomical priors, and improve the model performance even further.

## References

1. Cao, Z., et al.: Deep learning derived automated aspects on non-contrast CT scans of acute ischemic stroke patients. Technical report, Wiley Online Library (2022)
2. Chaurasia, A., Culurciello, E.: LinkNet: exploiting encoder representations for efficient semantic segmentation. In: 2017 IEEE Visual Communications and Image Processing (VCIP), pp. 1–4. IEEE (2017)
3. Chen, J., et al.: TransUNet: transformers make strong encoders for medical image segmentation. arXiv preprint arXiv:2102.04306 (2021)
4. Chilamkurthy, S., et al.: Deep learning algorithms for detection of critical findings in head CT scans: a retrospective study. Lancet **392**(10162), 2388–2396 (2018)
5. Dosovitskiy, A., et al.: An image is worth $16 \times 16$ words: transformers for image recognition at scale. arXiv preprint arXiv:2010.11929 (2020)
6. El Tawil, S., Muir, K.W.: Thrombolysis and thrombectomy for acute Ischaemic stroke. Clin. Med. **17**(2), 161 (2017)
7. Esmael, A., Elsherief, M., Eltoukhy, K.: Predictive value of the Alberta stroke program early CT score (ASPECTS) in the outcome of the acute ischemic stroke and its correlation with stroke subtypes, NIHSS, and cognitive impairment. Stroke Res. Treatment **2021**, 1–10 (2021). https://doi.org/10.1155/2021/5935170
8. Hatamizadeh, A., et al.: UNETR: transformers for 3D medical image segmentation. In: Proceedings of the IEEE/CVF Winter Conference on Applications of Computer Vision, pp. 574–584 (2022)
9. He, K., Zhang, X., Ren, S., Sun, J.: Deep residual learning for image recognition. In: Proceedings of the IEEE Conference on Computer Vision and Pattern Recognition, pp. 770–778 (2016)

10. Kervadec, H., Bouchtiba, J., Desrosiers, C., Granger, E., Dolz, J., Ben Ayed, I.: Boundary loss for highly unbalanced segmentation. Med. Image Anal. **67**, 101851 (2021). https://doi.org/10.1016/j.media.2020.101851

11. Kuang, H., et al.: Automated aspects on noncontrast CT scans in patients with acute ischemic stroke using machine learning. Am. J. Neuroradiol. **40**(1), 33–38 (2019)

12. Liang, K., et al.: Symmetry-enhanced attention network for acute ischemic infarct segmentation with non-contrast CT images. In: de Bruijne, M., et al. (eds.) MIC-CAI 2021. LNCS, vol. 12907, pp. 432–441. Springer, Cham (2021). https://doi.org/10.1007/978-3-030-87234-2_41

13. Lin, T., Goyal, P., Girshick, R.B., He, K., Dollár, P.: Focal loss for dense object detection. CoRR abs/1708.02002 (2017). http://arxiv.org/abs/1708.02002

14. Mokin, M., Primiani, C.T., Siddiqui, A.H., Turk, A.S.: Aspects (Alberta stroke program early CT score) measurement using hounsfield unit values when selecting patients for stroke thrombectomy. Stroke **48**(6), 1574–1579 (2017). https://doi.org/10.1161/STROKEAHA.117.016745

15. Patel, A., et al.: Intracerebral haemorrhage segmentation in non-contrast CT. Sci. Rep. **9**(1), 1–11 (2019)

16. Powers, W.J., et al.: 2015 American heart association/American stroke association focused update of the 2013 guidelines for the early management of patients with acute ischemic stroke regarding endovascular treatment. Stroke **46**(10), 3020–3035 (2015). https://doi.org/10.1161/STR.0000000000000074

17. Puetz, V., et al.: Extent of hypoattenuation on CT angiography source images predicts functional outcome in patients with basilar artery occlusion. Stroke **39**(9), 2485–2490 (2008). https://doi.org/10.1161/STROKEAHA.107.511162

18. Ronneberger, O., Fischer, P., Brox, T.: U-Net: convolutional networks for biomedical image segmentation. In: Navab, N., Hornegger, J., Wells, W.M., Frangi, A.F. (eds.) MICCAI 2015. LNCS, vol. 9351, pp. 234–241. Springer, Cham (2015). https://doi.org/10.1007/978-3-319-24574-4_28

19. Saver, J.L.: Time is brain - quantified. Stroke **37**(1), 263–266 (2006). https://doi.org/10.1161/01.STR.0000196957.55928.ab. https://www.ahajournals.org/doi/abs/10.1161/01.STR.0000196957.55928.ab

20. Sudre, C.H., Li, W., Vercauteren, T., Ourselin, S., Cardoso, M.J.: Generalised dice overlap as a deep learning loss function for highly unbalanced segmentations. CoRR abs/1707.03237 (2017). http://arxiv.org/abs/1707.03237

21. Tan, M., Le, Q.V.: EfficientNet: rethinking model scaling for convolutional neural networks. CoRR abs/1905.11946 (2019). http://arxiv.org/abs/1905.11946

22. Toikkanen, M., Kwon, D., Lee, M.: ReSGAN: intracranial hemorrhage segmentation with residuals of synthetic brain CT scans. In: de Bruijne, M., et al. (eds.) MICCAI 2021. LNCS, vol. 12901, pp. 400–409. Springer, Cham (2021). https://doi.org/10.1007/978-3-030-87193-2_38

23. Vaswani, A., et al.: Attention is all you need. In: Advances in Neural Information Processing Systems, vol. 30 (2017)

24. Zhou, Z., Rahman Siddiquee, M.M., Tajbakhsh, N., Liang, J.: UNet++: a nested U-Net architecture for medical image segmentation. In: Stoyanov, D., et al. (eds.) DLMIA/ML-CDS -2018. LNCS, vol. 11045, pp. 3–11. Springer, Cham (2018). https://doi.org/10.1007/978-3-030-00889-5_1

# ExSwin-Unet: An Unbalanced Weighted Unet with Shifted Window and External Attentions for Fetal Brain MRI Image Segmentation

Yufei Wen[1(✉)], Chongxin Liang[2], Jingyin Lin[2], Huisi Wu[2], and Jing Qin[3]

[1] South China University of Technology, Guangzhou, China
201930034695@mail.scut.edu.cn
[2] Shenzhen University, Shenzhen, China
{2060271074,2110276229}@email.szu.edu.cn, hswu@szu.edu.cn
[3] The Hong Kong Polytechnic University, Hung Hom, Hong Kong
harry.qin@polyu.edu.hk

**Abstract.** Accurate fetal brain MRI image segmentation is essential for fetal disease diagnosis and treatment. While manual segmentation is laborious, time-consuming, and error-prone, automated segmentation is a challenging task owing to (1) the variations in shape and size of brain structures among patients, (2) the subtle changes caused by congenital diseases, and (3) the complicated anatomy of brain. It is critical to effectively capture the long-range dependencies and correlations among training samples to yield satisfactory results. Recently, some transformer-based models have been proposed and achieved good performance in segmentation tasks. However, the self-attention blocks embedded in transformers often neglect the latent relationships among different samples. Model may have biased results due to the unbalanced data distribution in the training dataset. We propose a novel unbalanced weighted Unet equipped with a new ExSwin transformer block to comprehensively address the above concerns by effectively capturing long-range dependencies and correlations among different samples. We design a deeper encoder to facilitate features extracting and preserving more semantic details. In addition, an adaptive weight adjusting method is implemented to dynamically adjust the loss weight of different classes to optimize learning direction and extract more features from under-learning classes. Extensive experiments on a FeTA dataset demonstrate the effectiveness of our model, achieving better results than state-of-the-art approaches.

**Keywords:** Fetal brain MRI images · Transformer · Medical image segmentation

## 1 Introduction

Infancy is the origination stage of everyone's life, but some infants, unfortunately, suffer from congenital diseases and severe congenital diseases may lead to the

L. Karlinsky et al. (Eds.): ECCV 2022 Workshops, LNCS 13803, pp. 340–354, 2023.
https://doi.org/10.1007/978-3-031-25066-8_18

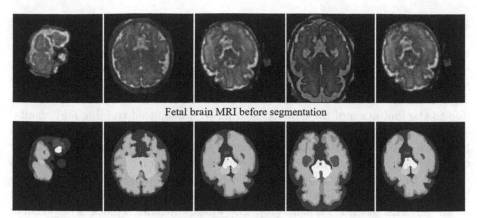

Fetal brain MRI before segmentation

Fetal brain MRI after segmentation

**Fig. 1.** Samples of fetal brain MRI segmentation dataset

death of infants [15]. In this regard, the timely discovery and treatment of infant congenital diseases are significant. For those unfortunate fetuses with congenital diseases, fetal brain MRI results are especially helpful to study the neuro development of the fetus and aid fetal disease diagnosis and treatment [13, 24, 25]. When conducting fetal brain analysis, precise segmentation of crucial structures in MRI images is essential. The fetal brain MRI images are complex and many congenital diseases result in subtle changes in brain tissues [4, 6, 28]. Thus, accurate segmentation of these tissues and structures plays a decisive role in diagnosis and treatment. Manual segmentation is quite laborious, time-consuming, and error-prone. Therefore, automatic segmentation of fetal brain MRI images is highly demanded in practice (Fig. 1).

Recent years, convolutional neural networks (CNNs) based segmentation models have dominated in medical image computing and achieved remarkable success [3, 8, 18, 21, 27]. They apply convolution kernel to perform convolution operations and extract features from local input patches and modularize representations while efficiently utilizing data. However, it is still difficult for these models to precisely segment brain tissues from fetal brain MRI images to capture the subtle changes in different brain tissues. One of the main concerns is the intrinsic locality of convolution operation, which causes CNN-based models difficult to extract long-range semantic information to enhance the segmentation performance in a global view [2]. On the other hand, a newly proposed architecture, namely transformer [23], has achieved great success in the natural language processing domain with its effective self-attention mechanisms. It has been introduced to the vision domain [5] and is widely employed in many computer vision tasks. It performs excellently in the CV area, surpasses the CNN-based model in some areas, and shows that a vigorous model can be constructed with a transformer. Recently, Swin transformer [17] has been proposed and performs well in image classification and detection, while the Swin-Unet [1] has shown its powerful capability in image segmentation. However, the self-attention mechanism embedded in transformers often ignores the correlations among different samples, while correlations between different samples are essential for image segmentation. And Swin transformer structure brings about training instability when changing the window size of the transformer block. In addition, in medical segmentation tasks, the model often confronts limited and biased labeled data due to the limitation of the dataset, leading to unbalanced training and results.

We propose a novel unbalanced weighted Unet equipped with a new ExSwin transformer block for fetal brain MRI image segmentation in order to effectively capture long-range dependencies and correlations among different samples to enhance the segmentation performance. The ExSwin transformer block is composed of the window attention block [16] and the external memory block based on the external attention scheme [10]. The window attention block is responsible for local and global feature representation learning, while the external memory block combines different intra-samples' features with its two external memory units to reduce the information loss due to dimensional reduction and gain inductive bias information of the dataset. Furthermore, we design a special unbalanced Unet structure where we adopt a larger encoder size to facilitate features extracting and preserving deeper semantic information. In addition, an adaptive weight adjusting method is implemented to dynamically adjust the loss weight of different classes, which contributes to optimizing model learning direction and extracting more features from the under-learning classes. Since our dataset is from FeTA 2021 challenge, we implement comparison with several participators' networks, such as Unet, Res-Unet, and Trans-Unet, where our model has a better performance. Quantitative experiments and ablation studies on the dataset demonstrate the effectiveness of the proposed model, achieving better results than state-of-the-art approaches.

## 2   Method

The framework of our unbalanced ExSwin-Unet is as shown in Fig. 2. Our Ex-Swin Unet mainly consists of encoder, bottleneck, decoder and skip connections between encoder and decoder blocks. In our encoder module, input images are divided into non-overlapping patches with patch size $4 \times 4$ and the feature dimension of each patch becomes 16 times. Moreover, the feature dimension is projected to a selected dimension C through a linear embedding layer. After that, we continuously apply ExSwin blocks and patch merging layers alternately where ExSwin blocks grasp feature representation and patch merging layers increase feature dimension for down-sampling. Specifically, ExSwin block size is even since it needs to perform window and shift window attention alternately to capture local and global features of the image. Our ExSwin blocks are able to extract high-level features from input images. Then we apply two ExSwin blocks as the bottleneck block to enhance model convergence ability where the input feature dimension and output feature dimension are the same. On the other hand, in the decoder module, we apply patch expanding layers with multiple ExSwin

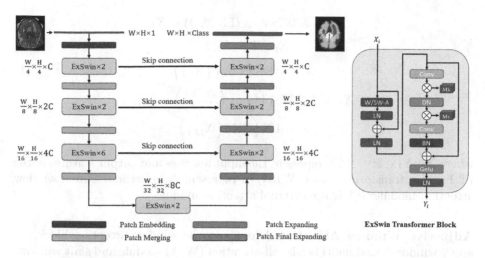

**Fig. 2.** Overview of our proposed ExSwin-Unet. In ExSwin transformer block, W/SW-A is window and shifted window attention module [17]; LN, DN BN represent layer normlization, double normlization [9] and batch normlization respectively; Mk and Mv are external learnable key and value memory respectively; Gelu is the Gaussian error linear unit.

blocks to perform features up-sampling hierarchically. Skip connections between same-level ExSwin blocks are applied to complement detailed information loss during the down-sampling process and retain more high-resolution details contained in high-level feature maps. At the end of the decoder, a particular patch expanding layer is added to conduct 4× up-sampling where feature resolution is mapped to input resolution. In the end, the up-sampled features will be mapped to segmentation predictions through a linear projection layer.

## 2.1 Window-Based Attention Block

Based on the shifted window mechanism and hierarchical structure, the Swin transformer is able to extract both local and global features of the input images. Since our Feta dataset samples are 2D images generated from 3D images, spatial information can be easily lost. In order to make up for the loss of spatial information and improve the feature fusion among different samples, we propose a new transformer block named as ExSwin transformer block. The ExSwin block is constructed with window-based attention and external attention block. The structure of our ExSwin block is shown in Fig. 2. The operation of the ExSwin block can be formulated as follows:

$$X_{i1} = \text{W/SW-A} \left( \text{LN} \left( X_i \right) \right) + X_i,$$
$$\hat{X}_i = \text{LN} \left( X_{i1} \right),$$
$$\hat{X}_{i1} = \text{EA} \left( \text{Conv} \left( \hat{X}_i \right) \right) \tag{1}$$
$$\hat{X}_{i2} = \text{BN} \left( \text{EA} \left( \text{Conv} \left( \hat{X}_{i1} \right) \right) \right) + \hat{X}_{i1},$$
$$Y_i = \text{LN} \left( \text{Gelu} \left( \hat{X}_{i2} \right) \right)$$

where $\mathbf{X_i}, \mathbf{Y_i} \in \mathbb{R}^{C \times H \times W}$ represent the input features and output features of the $i^{\text{th}}$ ExSwin transformer block; W/SW-A represents window and shifted window attention module; EA is the external attention module.

**Adjustive Window Attention Block.** In the window attention block, we apply window-based multi heads self-attention (W-A) module and shift window-based multi-head self-attention (SW-A) [17]. The window-based and shifted window-based multi-head self-attention module are applied in the two successive transformer blocks. Window-based multi-head self attention calculates attention in each window to capture local window features. On the other hand, shifted window-based multi-head self-attention, with its shifting mechanism, calculates attention to mix cross-window features and capture global features. The local self-attention can be formulated as:

$$\text{Attention}(Q, K, V) = \text{Softmax} \left( \frac{QK^T}{\sqrt{d}} + B \right) V \tag{2}$$

where $Q, K, V \in \mathbb{R}^{M^2 \times d}$ represents the query, key and value matrices; $M^2$ denotes the number of patches in a window and $d$ denotes the dimension of the query or key; $B$ is the relative position bias and its values are taken from the bias matrix $\hat{B} \in \mathbb{R}^{(2M-1) \times (2M+1)}$ since the relative position along each axis lies in the range of $[-M+1, M-1]$.

**Post Normalization.** When training the window attention-based model, we may probably encounter training instability since activation values in the network deep layers are quite low [16]. To ease the unstable situation, post normalization, shown in Fig. 3, is applied in attention blocks and adds an additional layer normalization unit before the external attention block.

**Scaled Cosine Attention.** While calculating the self-attention in window attention and shifted window attention module, the attention map in some blocks or heads dominated other features, which leads to biased feature extraction.

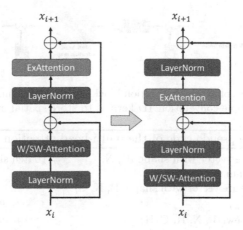

**Fig. 3.** The Pre-norm is transformed to Post-norm

We can replace the inner product similarity with cosine similarity to improve the problem:

$$\text{Similarity}(\mathbf{q}_i, \mathbf{k}_j) = \cos(\mathbf{q}_i, \mathbf{k}_j)/\tau + B_{ij} \tag{3}$$

where $B_{ij}$ is the relative position bias between pixel $i$ and $j$; $\tau$ is a learnable scalar, non-shared across heads and layers. Since the cosine function is equivalently normalized, the substitution can alleviate some inner product domination situation.

## 2.2 External Attention Block

In the external attention block, we design a multi-head external attention module that applies two convolution layers to grasp feature representation and two external learnable memory units to capture spatial information and sample affinity between different samples. The external attention block applies an external attention mechanism, which adopts two external memory units Mk and Mv to restore the spatial information between adjacent slices and store current global information. The external attention module is designed for capturing intra-sample features and it is capable of learning more representative features from input samples. The external attention block structure is as shown in Fig. 4 and the pseudo-code of our multi-head external attention module is as shown in Algorithm 1.

Since, multi-head attention and convolution mechanism are complementary, we apply two convolution layers in the external attention block. The first convolution layer kernel size is $1 \times 1$ in order to aggregate cross-channel features. In order to obtain a useful complement to the attention mechanism, the second convolution layer kernel size is $3 \times 3$ with padding size 1. The $3 \times 3$ convolution layer captures the local information with a larger receptive field and enhances grasping the feature representation [11].

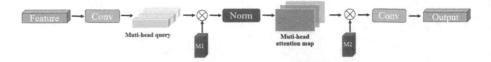

**Fig. 4.** The framework of external attention block which applies multi-head calculation. M1 and M2 are multiple convolution 1D kernels to store spatial information.

---

**Algorithm 1.** The pseudo code of the multi-head attention block.

---

**Input**: $\hat{X}_{in}$, a feature vector with shape [B, N, C]     # (batch size, pixels, channels)
**Parameter**: H, the number of heads
**Output**: $\hat{X}_{out}$, a feature vector with shape [B, N, C]

$Query = Conv(\hat{X}_{in})$                                                        # kernel size $= 1 \times 1$
$Query = Query.view(B , N, H, C/H)$                              # $shape = [B, N, H, M]$
$Query = Query.permute(0 ,2 ,1 ,3 )$                             # $shape = [B, H, N, M]$
$Attn = M_k(Query)$                                                     # $shape = [B, H, N, M]$
# Double normalization
$Attn = Softmax(Attn, dim = 2)$
$Attn = L1Norm(Attn, dim = 3)$
$Out = M_v(Attn)$                                                       # $shape = [B, H, N, M]$
$Out = Out.permute(0 ,2 ,1 ,3 )$                                 # $shape = [B, N, H, M]$
$Out = Out.view(B , N, C)$                                           # $shape = [B, N, C]$
$\hat{X}_{out} = Conv(Out)$                                 # kernel size $= 3 \times 3$, stride $= 1$

---

By utilizing two external memory units to recover and store the spatial information of slices in a 3D sample, our external attention block can be viewed as the dictionary for the whole dataset to calculate attention among 2D slices. The external attention module benefits model learning representative features and alleviates feature loss of dimension reduction process.

## 2.3   Unbalanced Unet Architecture

In the encoder-decoder unet structure, the sizes of Ex-Swin blocks in the encoder and decoder are different. The Ex-Swin blocks size are $[2, 2, 6]$ and $[2, 2, 2]$ for the encoder and decoder module, respectively. The idea of hyper-parameters setting is inspired by Swin-T model and our experimental results also have proven its effectiveness against balanced Unet structure. The encoder block with deeper size Ex-Swin blocks is able to obtain a better feature extraction and enhance model ability of preserving broad contextual information. The decoder block with a thinner size saves calculation resources and also benefits model convergence.

## 2.4   Adaptive Weighting Adjustments

In medical segmentation tasks, we may encounter a biased data segmentation training result due to the limited and unbalanced labeled data situation. To alleviate the biases and increase model performance, we propose an adaptive weighting adjustment strategy on loss function, which conducts model learning on under-learning samples and also prevents the model from overwhelming the well-perform samples during the model training process. In our adaptive weighting adjustment mechanism, the weight value $\mathbf{v_c}$ for the class $c$ is calculated by:

$$v_c = \frac{1}{N_c} \sum_{i=1}^{N_c} \frac{|\text{pre}_c|}{|\text{tru}_c|} \tag{4}$$

where $|\text{pre}_c|$ is prediction pixels that match the ground truth pixels for class $\mathbf{c}$ and the $|\text{tru}_c|$ is the total number of class $\mathbf{c}$ in the corresponding ground truth. And the adaptive weight $w_c$ can be calculated by the weight value:

$$w_c = \text{Softmax}(1 - v_c) \tag{5}$$

where the class-wise weight will be updated for every epoch training process. In that case, a suitable weight is generated to enable model adjusting its learning direction and alleviating the biased segmentation result.

### 2.5  Dual Loss Functions

In order to improve segmentation accuracy and learning speed, we define a dual loss function. Since we adopt an adaptive weight adjusting method, loss is obtained by calculating weighted loss for each class by taking the average value. Assume that $w_c$ is the weight for each class and C is the total number of classes.

**Multi-class Cross Entropy Loss.** Cross entropy loss measures the difference between two probability distributions. It fastens model convergence and reduces model training resource consumption.

$$\mathcal{L}_{ce} = -\frac{1}{C} \times \sum_{i=1}^{C} w_c \times l_c \log (p_c) \tag{6}$$

where $p_c$ is the segmentation probability for class c in the output, $l_c$ is the identification for class c which ranges 0 or 1 and $w_c$ is adaptive weight for class c.

**Square Dice Loss.** Dice loss measures similarity of two distributions and we apply it to calculate the similarity between output prediction and the ground truth. It conduces to improve model performance and increase accuracy.

$$\mathcal{L}_{dice} = \frac{1}{C} \times \sum_{i=1}^{C} w_c \times \left( 1 - \frac{2 \times \sum_{\text{pixels}} y_{\text{pred}} \, y_{\text{true}}}{\sum_{\text{pixels}} \left( y_{\text{pred}}^2 + y_{\text{true}}^2 \right)} \right) \tag{7}$$

where $\sum_{\text{pixels}}$ represents the sum of pixel value, $y_{\text{pixels}}$ and $y_{\text{true}}$ are segmentation prediction and segmentation ground truth respectively and $w_c$ is adaptive weight for class c.

**Table 1.** Quantitative comparison results of segmentation results on feta2021 dataset. The table shows different methods segmentation prediction with % unit. All methods are evaluated by Dice, Jaccard, Sensitivity and Specificity coefficient.

| Method | Year | Dice | Jaccard | SE | SP | Para(M) | Flop(GMac) |
|---|---|---|---|---|---|---|---|
| Attn-Unet [19] | 2018 | 88.6 | 80.5 | 88.9 | 99.7 | **34.9** | 66.6 |
| Segmenter [22] | 2021 | 87.5 | 78.6 | 87.3 | 99.7 | 102.3 | 25.8 |
| Swin-Unet [1] | 2021 | 88.7 | 80.7 | 89.1 | 99.7 | 34.2 | **9.6** |
| Utnet [7] | 2021 | 89.1 | 81.1 | 88.9 | 99.8 | 35.1 | 49.7 |
| Trans-Unet [2] | 2021 | 89.2 | 81.3 | 89.3 | 99.8 | 105.3 | 35.2 |
| **Ours** | 2022 | **90.1** | **82.3** | **90.2** | **99.8** | 50.4 | 11.3 |

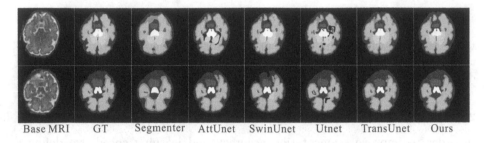

Base MRI    GT    Segmenter    AttUnet    SwinUnet    Utnet    TransUnet    Ours

**Fig. 5.** Segmentation visual results of different methods on FeTA2021 dataset.

**Total Loss.** The total loss is linear combination of average weighted CE loss and average weighted DICE loss with coefficient $\alpha$.

$$\mathcal{L}_{total} = \alpha \times \frac{1}{N} \times \sum_{i=1}^{N} \mathcal{L}_{ce} + (1 - \alpha) \times \frac{1}{N} \times \sum_{i=1}^{N} \mathcal{L}_{dice} \qquad (8)$$

## 3    Experimental Results

### 3.1    Datasets

The dataset is Fetal Brain Tissue Annotation and Segmentation Challenge released in 2021 [20]. The fetal brain MRI was manually segmented into 8 different classes with in-plane resolution of 0.5 mm × 0.5 mm. Dataset includes 80 3D T2-weighted fetal brain and reconstruction methods were used to create a super-resolution reconstruction of the fetal brain. We divided the dataset into 60 training set and 20 testing set. In order to save time and energy consumption, we transform dataset to about 2D images with size 256 × 256.

## 3.2  Implement Details

We train and test our model on a single NVIDIA RTX 2080Ti (11 GB RAM). The ExSwin-Unet model is trained on Python 3.7 and Pytorch 1.7.0. In order to increase data diversity and avoid data overfitting, we applied simple data augmentation flipping and rotation on dataset. We adopt weighted dual loss function and employ lookahead optimizer [26] with Adam optimizer [12] as inner optimizer. Moreover, we experimentally set the coefficient of total loss $\alpha = 0.4$ to obtain a relatively better performance. During the model training period, the initial learning rate is 1e−4 and loss decay for each epoch. We trained the model for 200 epochs with a batch size of 16.

## 3.3  Comparison with SOTA Methods

To evaluate the performances of our method, we compared our network with five state-of-the-art methods including Segmenter [22], Attn-Unet [19], Utnet [7], Swin-Unet [1] and Trans-Unet [2]. The compared models consist of four transformer based model structures and a CNN based models, namely Attention Unet. We implemented the comparison under the same computational environments without using any pre-trained models. Both visual and statistical comparisons are conducted using the same datasets and with same data processing method. The statistical comparison results are shown in Table 1 and visualization results are shown in Fig. 5.

Our model with its unique features generally outperforms other SOTA methods on dice and jaccard score and cost less calculation consumption. We save 50% parameters than Trans-Unet and achieve a better segmentation performance. Visually compared with other segmentation methods in the Fig. 5, our model also outperforms on segmenting fetal tissue with different scales and irregular shapes. Demonstrating that the proposed ExSwin-Unet is capable to improve the segmentation performance.

## 3.4  Ablation Studies

In order to demonstrate the effectiveness of the proposed components, we conduct ablation studies with different components and unbalanced structure. The component ablation experiment results are as shown in Table 2 and the attention hotspots of different methods are as shown in Fig. 6. To illustrate the effectiveness of our unbalanced structure, we conducted ablation studies on the unbalanced structure as shown in Table 4.

As shown in the Table 2, unbalanced Unet structure benefits feature learning process with its larger encoder size. The model with the external attention unit is able to combine different intra-sample features and mitigate spatial information loss. The adaptive loss alleviates model imbalance class under-learning problems.

As shown in Table 3, we further implement comparison with focal loss [14], a typical class imbalance loss, to verify the effectiveness of our adaptive loss

**Table 2.** Ablation studies result on FeTA2021 dataset. The table shows different methods segmentation prediction dice level with % unit. ECF, GM, WM, Ve, Ce, DGM and Br are 7 segmented brain tissues representing External Cerebrospinal Fluid, Grey Matter, White Matter, Ventricles, Cerebellum, Deep Grey Matter, Brainstem, respectively. Specially, except for the first Swin-Unet method is balanced, others structure are unbalanced Unet structure version with encoder size [2, 2, 6] and decoder size [2, 2, 2].

| Method | Mean | ECF | GM | WM | Ve | Ce | DGM | Br |
|---|---|---|---|---|---|---|---|---|
| Swin-Unet(Balanced) | 88.7 | 89.9 | 79.9 | 92.7 | 91.3 | 88.4 | 88.3 | 89.5 |
| Swin-Unet(Unbalanced) | 89.2 | 90.7 | 80.8 | 93.3 | 91.7 | 89.3 | 88.9 | 89.8 |
| Swin-Unet+Adaptive | 89.5 | 90.5 | 82.5 | 93.6 | 91.5 | 89.0 | 89.2 | 89.7 |
| ExSwin-Unet | 89.7 | **91.3** | 81.4 | 93.8 | 92.1 | 89.4 | 89.6 | **90.4** |
| **ExSwin-Unet+Adaptive** | **90.1** | 91.2 | **82.9** | **94.2** | **92.3** | 89.8 | **89.9** | 90.2 |

**Table 3.** Ablation studies result on FeTA2021 dataset. The table shows different methods segmentation prediction dice level with % unit. ECF, GM, WM, Ve, Ce, DGM and Br are 7 segmented brain tissues representing External Cerebrospinal Fluid, Grey Matter, White Matter, Ventricles, Cerebellum, Deep Grey Matter, Brainstem, respectively.

| Method | Mean | ECF | GM | WM | Ve | Ce | DGM | Br |
|---|---|---|---|---|---|---|---|---|
| Swin-Unet+Focal | 88.6 | 90.1 | 81.7 | 93.3 | 90.8 | 87.3 | 87.9 | 89.4 |
| Swin-Unet+Adaptive | 89.5 | 90.5 | 82.5 | 93.6 | 91.5 | 89.0 | 89.2 | 89.7 |
| ExSwin-Unet+Focal | 89.3 | 90.6 | 82.0 | 93.5 | 91.7 | 88.7 | 89.0 | 89.9 |
| **ExSwin-Unet+Adaptive** | **90.1** | **91.2** | **82.9** | **94.2** | **92.3** | **89.8** | **89.9** | **90.2** |

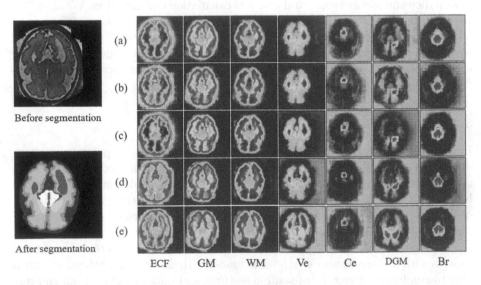

**Fig. 6.** Multiple classes attention hotspot of different methods. (a)–(e) are the ablation experiment methods balanced Swin-Unet, unbalanced Swin-Unet, unbalanced Swin-Unet+Adaptive, unbalanced ExSwin-Unet and unbalanced ExSwin-Unet+Adaptive correspondingly.

**Table 4.** Ablation studies on unbalanced Unet architecture. The mean dice is calculated through five-fold cross-validation to verify our method's effectiveness.

| Method | Encoder size | Decoder size | Mean Dice ± std |
|---|---|---|---|
| ExSwin-Unet | $[2, 2, 6]$ | $[2, 2, 6]$ | $89.5 \pm 0.349$ |
| ExSwin-Unet | $[2, 2, 2]$ | $[2, 2, 2]$ | $89.7 \pm 0.193$ |
| ExSwin-Unet | $[2, 2, 6]$ | $[2, 2, 2]$ | $\mathbf{90.1 \pm 0.252}$ |

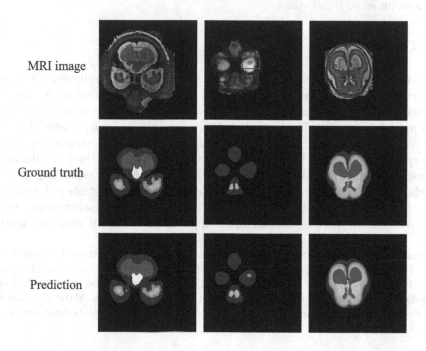

**Fig. 7.** Visual segmentation results of some failure predicting cases.

method. The results demonstrate the our method effectiveness, where our adaptive weighted loss function benefits model learning ability by grasping information of under-learning classes and improving overall performance.

The ablation studies on unbalanced Unet structure is shown in the Table 4, demonstrating the effectiveness of unbalanced Unet structure. With a lager encoder size, our model can achieve a better performance than two other balanced models. The experiment indicates that the unbalanced structure benefit model feature extraction process and improve segmentation result.

## 4    Discussions and Limitations

Through the above ablation studies and comparative experiments, we design an effective 2D-based segmentation network with external attention to implement segmentation tasks on 3D image slices. Our purpose is to discover intra-sample relationships to alleviate spatial information loss and benefit the feature learning process. The external attention module achieves this goal, and experiments demonstrated its effectiveness. Moreover, we discover that balanced Unet structure may not be necessary for Unet framework where unblanced Unet can obtain a better performance than balanced Unet. On the other hand, our method still has some limitations. As shown in Fig. 7, our model fails to achieve correct predictions on some small scales.

## 5    Conclusion

In this paper, we present a novel unbalanced weighted Unet equipped with a new ExSwin transformer block to improve fetal brain MRI segmentation results. The ExSwin transformer is composed of shift-window attention and external attention module. The ExSwin transformer block not only can grasp essential sample features representation, but it also is able to capture intra-sample correlation and spatial information between different 3D slices. And the Unet is unbalanced where the encoder has a larger size to facilitate the feature extracting process. Furthermore, we introduce an adaptive weight adjustment strategy to improve biased data segmentation situations. The quantitative comparison experiments and ablation studies demonstrate the well performance of our proposed model.

**Acknowledgments.** This work was supported partly by National Natural Science Foundation of China (No. 61973221), Natural Science Foundation of Guangdong Province, China (No. 2019A1515011165), the Innovation and Technology Fund-Mainland-Hong Kong Joint Funding Scheme (ITF-MHKJFS) (No. MHP/014/20) and the Project of Strategic Importance grant of The Hong Kong Polytechnic University (No. 1-ZE2Q).

## References

1. Cao, H., et al.: Swin-Unet: Unet-like pure transformer for medical image segmentation. arXiv preprint arXiv:2105.05537 (2021)
2. Chen, J., et al.: TransUNet: transformers make strong encoders for medical image segmentation. arXiv preprint arXiv:2102.04306 (2021)
3. Chen, L.C., Papandreou, G., Kokkinos, I., Murphy, K., Yuille, A.L.: DeepLab: semantic image segmentation with deep convolutional nets, atrous convolution, and fully connected CRFs. IEEE Trans. Pattern Anal. Mach. Intell. **40**(4), 834–848 (2017)
4. Clouchoux, C., et al.: Delayed cortical development in fetuses with complex congenital heart disease. Cereb. Cortex **23**(12), 2932–2943 (2013)
5. Dosovitskiy, A., et al.: An image is worth $16 \times 16$ words: transformers for image recognition at scale. arXiv preprint arXiv:2010.11929 (2020)
6. Egaña-Ugrinovic, G., Sanz-Cortes, M., Figueras, F., Bargalló, N., Gratacós, E.: Differences in cortical development assessed by fetal MRI in late-onset intrauterine growth restriction. Am. J. Obstet. Gynecol. **209**(2), 126-e1 (2013)

7. Gao, Y., Zhou, M., Metaxas, D.N.: UTNet: a hybrid transformer architecture for medical image segmentation. In: de Bruijne, M., et al. (eds.) MICCAI 2021. LNCS, vol. 12903, pp. 61–71. Springer, Cham (2021). https://doi.org/10.1007/978-3-030-87199-4_6

8. Girshick, R.: Fast R-CNN. In: Proceedings of the IEEE International Conference on Computer Vision, pp. 1440–1448 (2015)

9. Guo, M.H., Cai, J.X., Liu, Z.N., Mu, T.J., Martin, R.R., Hu, S.M.: PCT: point cloud transformer. Comput. Vis. Media **7**(2), 187–199 (2021)

10. Guo, M.H., Liu, Z.N., Mu, T.J., Hu, S.M.: Beyond self-attention: external attention using two linear layers for visual tasks. arXiv preprint arXiv:2105.02358 (2021)

11. Guo, R., Niu, D., Qu, L., Li, Z.: SOTR: segmenting objects with transformers. In: Proceedings of the IEEE/CVF International Conference on Computer Vision, pp. 7157–7166 (2021)

12. Kingma, D.P., Ba, J.: Adam: a method for stochastic optimization. arXiv preprint arXiv:1412.6980 (2014)

13. Li, Z., Pan, J., Wu, H., Wen, Z., Qin, J.: Memory-efficient automatic kidney and tumor segmentation based on non-local context guided 3D U-Net. In: Martel, A.L., et al. (eds.) MICCAI 2020. LNCS, vol. 12264, pp. 197–206. Springer, Cham (2020). https://doi.org/10.1007/978-3-030-59719-1_20

14. Lin, T.Y., Goyal, P., Girshick, R., He, K., Dollár, P.: Focal loss for dense object detection. In: Proceedings of the IEEE International Conference on Computer Vision, pp. 2980–2988 (2017)

15. Liu, L., et al.: Global, regional, and national causes of child mortality: an updated systematic analysis for 2010 with time trends since 2000. Lancet **379**(9832), 2151–2161 (2012)

16. Liu, Z., et al.: Swin transformer v2: scaling up capacity and resolution. In: Proceedings of the IEEE/CVF Conference on Computer Vision and Pattern Recognition, pp. 12009–12019 (2022)

17. Liu, Z., et al.: Swin transformer: hierarchical vision transformer using shifted windows. In: Proceedings of the IEEE/CVF International Conference on Computer Vision, pp. 10012–10022 (2021)

18. Long, J., Shelhamer, E., Darrell, T.: Fully convolutional networks for semantic segmentation. In: Proceedings of the IEEE Conference on Computer Vision and Pattern Recognition, pp. 3431–3440 (2015)

19. Oktay, O., et al.: Attention U-Net: learning where to look for the pancreas. arXiv preprint arXiv:1804.03999 (2018)

20. Payette, K., et al.: An automatic multi-tissue human fetal brain segmentation benchmark using the fetal tissue annotation dataset. Sci. Data **8**(1), 1–14 (2021)

21. Ronneberger, O., Fischer, P., Brox, T.: U-Net: convolutional networks for biomedical image segmentation. In: Navab, N., Hornegger, J., Wells, W.M., Frangi, A.F. (eds.) MICCAI 2015. LNCS, vol. 9351, pp. 234–241. Springer, Cham (2015). https://doi.org/10.1007/978-3-319-24574-4_28

22. Strudel, R., Garcia, R., Laptev, I., Schmid, C.: Transformer for semantic segmentation. In: Proceedings of the IEEE/CVF International Conference on Computer Vision, pp. 7262–7272 (2021)

23. Vaswani, A., et al.: Attention is all you need. In: Advances in Neural Information Processing Systems, vol. 30 (2017)

24. Wu, H., Lu, X., Lei, B., Wen, Z.: Automated left ventricular segmentation from cardiac magnetic resonance images via adversarial learning with multi-stage pose estimation network and co-discriminator. Med. Image Anal. **68**, 101891 (2021)

25. Wu, H., Pan, J., Li, Z., Wen, Z., Qin, J.: Automated skin lesion segmentation via an adaptive dual attention module. IEEE Trans. Med. Imaging **40**(1), 357–370 (2020)
26. Zhang, M., Lucas, J., Ba, J., Hinton, G.E.: Lookahead optimizer: k steps forward, 1 step back. In: Advances in Neural Information Processing Systems, vol. 32 (2019)
27. Zhao, H., Shi, J., Qi, X., Wang, X., Jia, J.: Pyramid scene parsing network. In: Proceedings of the IEEE Conference on Computer Vision and Pattern Recognition, pp. 2881–2890 (2017)
28. Zugazaga Cortazar, A., Martín Martinez, C., Duran Feliubadalo, C., Bella Cueto, M.R., Serra, L.: Magnetic resonance imaging in the prenatal diagnosis of neural tube defects. Insights Imaging **4**(2), 225–237 (2013). https://doi.org/10.1007/s13244-013-0223-2

# Contour Dice Loss for Structures with Fuzzy and Complex Boundaries in Fetal MRI

Bella Specktor-Fadida[1]([✉])(iD), Bossmat Yehuda[2,3], Daphna Link-Sourani[2], Liat Ben-Sira[3,4], Dafna Ben-Bashat[2,3], and Leo Joskowicz[1]

[1] School of Computer Science and Engineering, The Hebrew University of Jerusalem, Jerusalem, Israel
bella.specktor@mail.huji.ac.il, josko@cs.huji.ac.il
[2] Sagol Brain Institute, Tel Aviv Sourasky Medical Center, Tel Aviv-Yafo, Israel
[3] Sackler Faculty of Medicine and Sagol School of Neuroscience, Tel Aviv University, Tel Aviv-Yafo, Israel
[4] Division of Pediatric Radiology, Tel Aviv Sourasky Medical Center, Tel Aviv-Yafo, Israel

**Abstract.** Volumetric measurements of fetal structures in MRI are time consuming and error prone and therefore require automatic segmentation. Placenta segmentation and accurate fetal brain segmentation for gyrification assessment are particularly challenging because of the placenta fuzzy boundaries and the fetal brain cortex complex foldings. In this paper, we study the use of the Contour Dice loss for both problems and compare it to other boundary losses and to the combined Dice and Cross-Entropy loss. The loss is computed efficiently for each slice via erosion, dilation and XOR operators. We describe a new formulation of the loss akin to the Contour Dice metric. The combination of the Dice loss and the Contour Dice yielded the best performance for placenta segmentation. For fetal brain segmentation, the best performing loss was the combined Dice with Cross-Entropy loss followed by the Dice with Contour Dice loss, which performed better than other boundary losses.

**Keywords:** Deep learning segmentation · Fetal MRI · Segmentation contour

## 1 Introduction

Fetal MRI has the potential to complement ultrasound (US) imaging and improve fetal development assessment by providing accurate volumetric measurements of the fetal structures [21,22]. Since manual segmentation of the fetal structures is very time consuming and impractical, automatic segmentation is required. However, placental structure and fetal brain have fuzzy and complex contours, hence using standard loss functions may result in inaccuracies on their boundaries even if they demonstrate good performance for the segmentation of the general structure.

© The Author(s), under exclusive license to Springer Nature Switzerland AG 2023
L. Karlinsky et al. (Eds.): ECCV 2022 Workshops, LNCS 13803, pp. 355–368, 2023.
https://doi.org/10.1007/978-3-031-25066-8_19

Recent works propose to address segmentation contour inaccuracies by using additional boundary and contour-based loss functions to train convolutional neural networks (CNN) [3, 11–13, 25, 28]. Contour-based losses aim to minimize directly or indirectly the one-to-one correspondence between points on the predicted and label contour. Therefore, these losses are often complex and their computation cost is high. Recently, Jurdi et al. [11] proposed a computationally efficient loss that optimizes the mean squared error between the predicted and the ground-truth perimeter length. However, a question remains whether the global computation of the contour perimeter is sufficient to regularize very complex boundaries, e.g., the fetal brain contour. Specktor-Fadida et al. [25] proposed a loss function based on the contour Dice metric which performs local regularization of the contour and is efficient. It was reported to perform well on placenta segmentation, but it is not clear how it compares to other boundary losses. Moreover, it was not tested for other structures, e.g. the fetal brain.

Isensee et al. [10] showed that the nnU-Net trained with the Dice loss function in combination with cross-entropy loss surpassed existing approaches on 23 public datasets with various segmentation structures and modalities. Despite the evident success of the compound Dice and Cross-Entropy loss, most of the works on contour losses compare results only to other boundary losses. Ma et al. [16] compared multiple compound loss functions, including the combination of Dice loss with boundary losses, i.e., the Hausdorff distance loss and the Boundary loss, and the combination of Dice loss with regional losses like the Cross Entropy and TopK losses. They reported that the Hausdorff loss yielded the best results for liver tumors segmentation. However, they did not test the losses for structures with fuzzy or complex boundaries. Also, since the paper was published, two additional boundary-based losses, the Perimeter loss and the Contour Dice loss, were introduced [11, 25].

In this paper, we investigate the performance of the Contour Dice loss and other losses for the placenta and fetal brain segmentation in MRI with emphasis on their boundaries and describe a new formulation for the Contour Dice loss function. The main contributions are: 1) a quantitative comparison of the performance of the Contour Dice loss, of other state-of-the-art boundary losses and of the combined Dice and Cross-Entropy loss for the placenta and fetal brain segmentation; 2) a new formulation for the contour Dice loss which is more similar to the Contour Dice metric; 3) quantification of the effect of the contour extraction thresholding on both the Contour Dice and the Perimeter losses.

## 2    Background and Related Work

### 2.1    Placenta Segmentation

The placenta plays an important role in fetal health, as it regulates the transmission of oxygen and nutrients from the mother to the fetus. Placental volume is an important parameter to assess fetal development and identify cases at risk with placental insufficiency [23]. Automatic placenta segmentation in fetal MRI poses numerous challenges. These include MRI related challenges, e.g., varying resolutions and contrasts, intensity inhomogeneity, image artifacts due to the large field of view, and partial volume effect, and fetal scanning related challenges, e.g., motion artifacts due to fetal and maternal movements, high variability in the placenta position, shape, appearance, orientation and fuzzy boundaries between the placenta and uterus.

Most of the existing methods for automatic placenta segmentation are based on deep learning. Alansary et al. [2] describe the first automatic placenta segmentation method. It starts with a coarse segmentation with a 3D multi-scale CNN whose results are refined with a dense 3D Conditional Random Field (CRF). It achieves a Dice score of 0.72 with 4-fold cross validation on 66 fetal MRI scans. Torrents-Barrena et al. [26] present a method based on super-resolution and motion correction followed by Gabor filters based feature extraction and Support Vector Machine (SVM) voxel classification. It achieves a Dice score of 0.82 with 4-fold cross-validation on 44 fetal MRI scans. Han et al. [8] present a method based on a 2D U-Net trained on a small dataset of 11 fetal MRI scans. It achieves a mean IU score of 0.817. Quah et al. [20] compares various methods for placenta segmentation on 68 fetal MRI 3D T2* images. The best method in this study achieves a Dice score of 0.808 using a U-Net with 3D patches. Pietsch et al. [19] describe a network that uses 2D patches and yields a Dice score of 0.76, comparable to expert variability performance. Specktor-Fadida et al. [25] demonstrate good performance for placenta segmentation using a new Contour Dice loss in combination with Soft Dice loss that yields a Dice score of 0.85.

## 2.2 Fetal Brain Segmentation

Fetal brain assessment, relating to its volume and structures, is important to assess fetal development and predict newborn outcome. MRI is often used as a complimentary tool mainly in cases with suspected brain malformations [9]. Automatic segmentation of the fetal brain is important to assess changes of the brain volume with gestational age, and accurate contour segmentation is necessary to assess the cortical folding of the brain, which was found to be an important biomarker for later functional development [4,5].

Multiple works were proposed for automatic fetal brain segmentation [6,7,24, 27]. Recently, a fetal brain tissue segmentation challenge was conducted for seven different fetal brain structures [18]. However, most studies focus on fetal brain segmentation to assess the volume or its structure. In order to assess cortical folding sulcation, a more accurate segmentation of the outer contour is needed.

## 2.3 Boundary Loss Functions

A variety of papers propose to add a constraint to the loss function to improve the accuracy of the segmentation in the structure boundaries. Arif et al. [1] extend the standard cross-entropy term with an average point to curve Euclidean distance factor between predicted and ground-truth contours. This allows the network to take into consideration shape specifications of segmented structures. Caliva et al. [3] use distance maps as weighing factors for a cross-entropy loss term to improve the extraction of shape bio-markers and enable the network to

focus on hard-to-segment boundary regions. As a result, their approach emphasizes the voxels that are in close proximity of the segmented anatomical objects
over those that are far away. Yang et al. [28] use Laplacian filters to develop
a boundary enhanced loss term that invokes the network to generate strong
responses around the boundary areas of organs while producing a zero response
for voxels that are farther from the structures periphery.

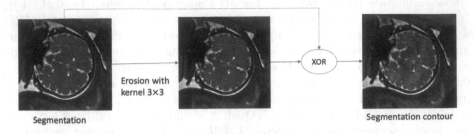

**Fig. 1.** Illustration of the contour extraction method using erosion and XOR operators.
Erosion with a $3 \times 3$ kernel produces a segmentation contour with a width of 2 pixels.

Both the boundary loss [13] and the Hausdorff loss [12] aim to minimize the
distance between the boundaries of the ground truth and the computed segmentation. The boundary loss approximates the distance between the contours
by computing the integrals over the interface between the two boundaries mis-
matched regions. The Hausdorff loss, computed with the distance transform,
approximates the minimization of the Hausdorff distance between the segmentation and ground truth contours. The Hausdorff loss was reported to perform
relatively well compared to multiple other losses on four different segmentation
tasks and yielded the best results for liver tumors segmentation [16]. However,
the loss is computationally costly due to the computation of the distance transform maps, especially for large 3D blocks.

Recently, two loss functions based on the segmentation contour were proposed. One is the Perimeter Loss [11], which optimizes the perimeter length of
the segmented object relative to the ground-truth segmentation using the mean
squared error function. Soft approximation of the contour of the probability map
is performed by specialized, non-trainable layers in the network. The second is
the Contour Dice [25], based on the Contour Dice metric, which was shown to
be highly correlated to the time required to correct segmentation errors [14,17].

Formally, let $\partial T$ and $\partial S$ be the extracted surfaces of the ground truth delineation and the network results, respectively, and let $B_{\partial T}$ and $B_{\partial S}$ be their respective offset bands. The Contour Dice (CD) metric of the offset bands is:

$$CD = \frac{|\partial T \cap B_{\partial S}| + |\partial S \cap B_{\partial T}|}{|\partial T| + |\partial S|} \tag{1}$$

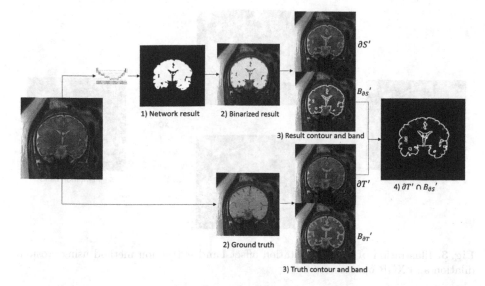

**Fig. 2.** Illustration of the intersection term computation in the Contour Dice loss. 1) Network result; 2) Binary masks of segmentation result and ground truth. To create a binary mask, thresholding is applied on the network result with threshold parameter $t$. The ground truth is already a binary mask; 3) Contours and bands are extracted; 4) Illustration of the computation of $\partial T' \bigcap B_{\partial S}'$. The light green are intersection areas, dark green are ground truth contour areas which do not intersect with the computed band and the white areas are computed band areas which do not intersect with the ground truth contour. The computation of $\partial S' \bigcap B_{\partial T}'$ is performed similarly using the $\partial S'$ and $B_{\partial T}'$ masks. (Color figure online)

To make the Contour Dice loss function differentiable, Specktor-Fadida et al. [25] proposed the following formulation:

$$L_{CD} = -\frac{2|B_{\partial T} \bigcap B_{\partial S}|}{|B_{\partial T}| + |B_{\partial S}|} \tag{2}$$

This formulation makes the function trainable, as we integrate over bands which have a predefined width.

## 3    Methods

For the Contour Dice loss function calculation, we estimate the segmentation contours $\partial T$ and $\partial S$ from Eq. 1 with the erosion and XOR operators (Fig. 1). As a result, the computed contours have a width. We can now integrate over the contour voxels and formulate a loss function that is very similar to the original Contour Dice metric in Eq. 1.

Formally, let $\partial T'$ be the extracted ground truth contour $\partial T$, $\partial S'$ be the extracted segmentation result contour $\partial S$ and $B_{\partial T}'$ and $B_{\partial S}'$ be their respective bands. We formulate the Contour Dice loss $L_{CD}$ function as:

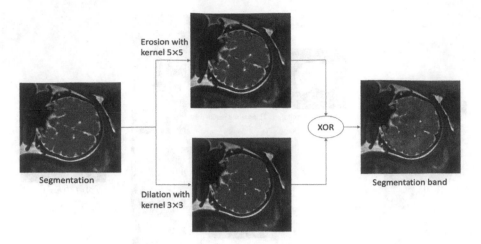

**Fig. 3.** Illustration of the segmentation offset band extraction method using erosion, dilation and XOR operators.

$$L_{CD} = -\frac{|\partial T' \cap B_{\partial S}'| + |\partial S' \cap B_{\partial T}'|}{|\partial T'| + |\partial S'|} \tag{3}$$

The contour Dice loss calculation is performed in two steps: 1) Contour and band extraction for ground truth segmentation and binarized segmentation result; 2) Dice with Contour dice loss computation. Figure 2 illustrates the Contour Dice computation using the extracted contours and bands.

### 3.1 Contours and Bands Extraction

The segmentation contour and the band around it are computed as follows. First, binary thresholding is applied to the network output with a predefined threshold $t \in [0, 1]$. Then, the contours of both the network result and the ground truth segmentation are extracted using erosion and XOR operations (Fig. 1). Finally, the bands are extracted using erosion, dilation and XOR operations (Fig. 3).

### 3.2 Contour Dice Loss Computation

The smoothing term $\epsilon$ is added to the loss function formulation, similar to the Soft Dice:

$$L_{CD} = -\frac{|\partial T' \cap B_{\partial S}'| + |\partial S' \cap B_{\partial T}'| + \epsilon}{|\partial T'| + |\partial S'| + \epsilon} \tag{4}$$

The contour dice loss $L_{CD}$ in Eq. 4 is used in combination with the Soft Dice $L_D$ loss with a constant weighting scheme:

$$L = L_D + \gamma L_{CD} \tag{5}$$

We use the batch implementation for the Soft Dice and Contour Dice loss functions. There, the loss is simultaneously computed for all voxels in a mini-batch instead of a separate computation for each image [15].

## 4  Experimental Results

To evaluate our method, we retrospectively collected fetal MRI scans for placenta segmentation and for fetal brain segmentation.

### 4.1  Datasets and Annotations

We collected fetal MRI scans acquired with the FIESTA, FRFSE and HASTE sequences from the Sourasky Medical Center (Tel Aviv, Israel). Gestational ages (GA) were 22–39 weeks, with most cases 28–39 weeks. For placental segmentation, FIESTA dataset was used, consists of 40 labeled cases acquired on a 1.5T GE Signa Horizon Echo speed LX MRI scanner using a torso coil. Each scan has 50–100 slices, $256 \times 256$ pixels/slice, and resolution of $1.56 \times 1.56 \times 3.0 \, \text{mm}^3$. For fetal brain segmentation, the dataset consists of 48 FRFSE and 21 HASTE cases. The FRFSE cases were acquired on 1.5T MR450 GE scanner. The HASTE cases were acquired on 3T Skyra and Prisma Siemens scanners. Each scan had 11–46 slices, $49 - 348 \times 49 - 348$ pixels/slice and resolution of $0.40 - 1.25 \times 0.40 - 1.25 \times 2.2 - 6 \, \text{mm}^3$.

Ground truth segmentations were created as follows. For placenta segmentation cases, 31 cases were annotated from scratch and 9 additional cases were manually corrected from the network results. For brain segmentation cases, all cases were annotated by correcting network results. Both the annotations and the corrections were performed by a clinical expert.

### 4.2  Experimental Studies

We conducted three studies. The first two studies compare different losses performance: Dice loss, Dice with Cross-Enropy loss, Dice with Boundary loss, Dice with Perimeter loss, Dice with Hausdorff loss and Dice with Contour Dice loss. Study 1 compares the losses for the task of placenta segmentation and Study 2 compares the losses for the task of fetal brain segmentation. Study 3 quantifies the influence of the segmentation result threshold parameter in the contour extraction phase on two contour-based losses, the Perimeter loss and the Contour Dice loss.

In all experiments, the Contour Dice loss weight from Eq. 5 was set to a constant $\gamma = 0.5$; the weight of Cross Entropy loss was set to 1, i.e. the same

**Table 1.** Placenta segmentation results comparison.

| LOSS/METRIC | Dice | Hausdorff | ASSD |
|---|---|---|---|
| Dice | $0.773 \pm 0.117$ | $57.73 \pm 44.24$ | $8.35 \pm 7.43$ |
| Dice + Cross Entropy | $0.807 \pm 0.098$ | $50.48 \pm 40.15$ | $5.83 \pm 3.34$ |
| Dice + Boundary | $0.805 \pm 0.079$ | $52.78 \pm 40.68$ | $6.76 \pm 3.54$ |
| Dice + Perimeter | $0.817 \pm 0.061$ | $50.84 \pm 39.32$ | $5.92 \pm 1.88$ |
| Dice + Hausdorff | $0.829 \pm 0.069$ | $49.40 \pm 43.28$ | $5.63 \pm 4.56$ |
| Dice + Contour Dice | $\mathbf{0.847 \pm 0.058}$ | $\mathbf{44.60 \pm 42.31}$ | $\mathbf{4.46 \pm 2.45}$ |

weight for Dice and Cross-Entropy terms. The weight for the other boundary losses, the Boundary loss, the Perimeter loss and the Hausdorff loss, were set with the dynamic scheme in [11,13]. The schedule was as follows: the initial weight was set to 0.01 and increased by 0.01 every epoch. The network was trained with reducing learning rate on plateau with early stopping of 50 epochs in case there is no reduction of the loss on the validation set.

We used the contour extraction method in Sect. 3.1 for the Perimeter and Contour Dice losses. Since ground truth segmentations for both placenta and fetal brain structures are of a high quality, the band for the Contour Dice loss in all experiments was set to be the estimated contour $B_{\partial T}{}' = \partial T'$ and $B_{\partial S}{}' = \partial S'$ (Eq. 4).

A network architecture based on Dudovitch et al. [6] was used in all experiments. For placenta segmentation we used patch size of $128 \times 128 \times 48$ and for fetal brain segmentation we used a large patch size of $256 \times 256 \times 32$ because of a higher in-plane resolution. The segmentation results were refined by standard post-processing techniques. Hole filling was used only for the placenta segmentation and not for the fetal brain because the brain structure does have holes.

Prior to the use of the placenta segmentation network, the same detection network was used in all experiments to extract the Region of Interest (ROI) around the placenta. The detection network architecture was similar to [6] and was applied on down-scaled scans by $\times 0.5$ in the in-plane axes. Since the fetal brain dataset was constructed after brain ROI extraction, there was no need to apply a fetal brain detection network.

Segmentation quality was evaluated in all studies with the Dice, Hausdorff and 2D ASSD (slice Average Symmetric Surface Difference) metrics. 3D ASSD was not evaluated as it is very dependent on the surface extraction method.

**Study 1: Placenta Segmentation.** The Contour Dice loss method was evaluated on training, validation and test sets of 16, 3 and 21 annotated placenta FIESTA scans respectively and compared to other state-of-the art losses. The segmentation threshold was set to $t = 1$ for both the Perimeter and the Contour Dice losses, as it yielded better results than $t = 0.5$, with a significant performance difference for the Contour Dice loss (see Study 3 for more details).

Table 1 lists the results of various loss functions for placenta segmentation. All compound losses improved upon the Dice loss alone, with the combination

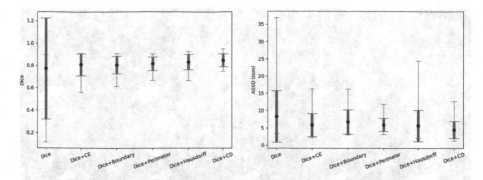

**Fig. 4.** Placenta segmentation results bar plots comparison for the Dice score (left, green) and 2D ASSD (right, blue) metrics. The color bars show the Standard Deviation. The gray bars show minimum and maximum values. X-axis from left to right: Dice loss, Dice with Cross-Entropy loss, Dice with Boundary loss, Dice with Perimeter loss, Dice with Hausdorff loss and Dice with Contour Dice loss. (Color figure online)

of Dice with Contour Dice loss yielding the best Dice score of 0.847. The second best performance was of the combination of Dice with the Hausdorff loss, with a Dice score of 0.829, better than the highly used Dice with Cross-Entropy loss combination. The combination of Dice loss with the Perimeter loss also yielded an improvement upon the Dice with Cross-Entropy loss (Fig. 4).

Figure 5 shows illustrative examples of placenta segmentation results with the boundary-based losses. The Contour Dice loss yields a smooth and relatively accurate segmentation. The perimeter loss also yields a smooth segmentation but sometimes fails to capture the full placenta shape complexity. The Hausdorff loss captures well the placenta shape, but it sometimes misses on the boundary.

**Study 2: Fetal Brain.** The Contour dice loss method was compared to the other state-of-the-art losses on the brain segmentation task using 35/4/30 training/validation/test split. We set the segmentation threshold of $t = 1$ for the Contour Dice loss, and set $t = 0.5$ for the Perimeter loss as it performed significantly better (see Study 3 for more details). All losses were trained with an early stopping of 50 epochs except the Hausdorff loss, which was trained for 42 epochs because of slow training on the large block size of $256 \times 256 \times 32$ and GPU cluster time constraints.

Table 2 lists the results of various loss functions for fetal brain segmentation. Surprisingly, the best performing loss was not a boundary-based loss, but the combination of the Dice loss with the Cross-Entropy loss, with a Dice score of 0.954 and 2D ASSD of 0.68 mm. The combination of the Dice loss with the Contour Dice had slightly worse performance with a Dice score of 0.946 and 2D ASSD of 0.81 mm. This loss performed best compared to all other boundary losses, with the Hausdorff loss having the third best performance with a Dice score of 0.932 and 2D ASSD of 0.93 mm.

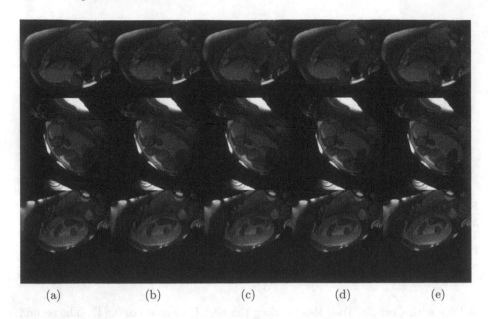

(a)                    (b)                    (c)                    (d)                    (e)

**Fig. 5.** Illustrative placenta segmentation results. (a) Dice and Boundary loss; (b) Dice and Perimeter loss; (c) Dice and Hausdorff loss; (d) Dice with contour Dice loss; (e) ground truth segmentation

**Table 2.** Brain segmentation results comparison.

| LOSS/METRIC | Dice | Hausdorff | ASSD |
|---|---|---|---|
| Dice | $0.924 \pm 0.035$ | $11.05 \pm 4.57$ | $1.12 \pm 0.54$ |
| Dice + cross-entropy | $\mathbf{0.954 \pm 0.019}$ | $\mathbf{7.65 \pm 1.86}$ | $\mathbf{0.68 \pm 0.26}$ |
| Dice + boundary | $0.924 \pm 0.039$ | $10.45 \pm 4.28$ | $1.16 \pm 0.57$ |
| Dice + perimeter | $0.932 \pm 0.032$ | $11.07 \pm 3.27$ | $1.00 \pm 0.43$ |
| Dice + Hausdorff | $0.934 \pm 0.032$ | $10.17 \pm 2.83$ | $0.93 \pm 0.36$ |
| Dice + contour Dice | $0.946 \pm 0.026$ | $8.24 \pm 2.16$ | $0.81 \pm 0.35$ |

Figure 7 shows illustrative results of fetal brain segmentations. While the combined Dice loss and Cross Entropy or Contour Dice losses improved upon the segmentation results of the Dice loss alone, neither captured the full complexity of the brain cortex sulcation.

**Study 3: Effect of the Contour Extraction Threshold Value.** We examined the effect of the threshold parameter $t$ for contour extraction on the Perimeter and the Contour Dice losses. All experimental conditions except for the value of $t$ remained the same. Since the threshold $t$ roughly captures the uncertainty of the network in the output prediction: $t = 0.5$ means that the network has a

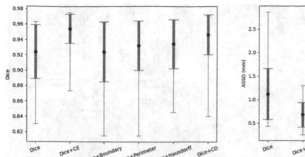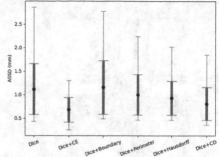

**Fig. 6.** Fetal brain segmentation results bar plots comparison for the Dice score (left in green) and 2D ASSD (right in blue) metrics. The color bars show the Standard Deviation. The gray bars show minimum and maximum values. X-axis from left to right: Dice loss, Dice with Cross-Entropy loss, Dice with Boundary loss, Dice with Perimeter loss, Dice with Haudorff loss and Dice with Contour Dice loss. (Color figure online)

**Table 3.** Ablation results for values of the segmentation threshold $t$ for contour extraction for the Dice with Contour Dice loss (DCD) and Dice with Perimeter loss (DP).

|     | $t$ | Placenta | | | Fetal brain | | |
|-----|-----|----------|-----------|------|-------|-----------|------|
|     |     | Dice | Hausdorff | ASSD | Dice | Hausdorff | ASSD |
| DCD | 0.5 | 0.798 | 52.04 | 6.85 | 0.939 | 9.22 | 0.90 |
|     | 1   | **0.847** | **44.60** | **4.46** | **0.946** | **8.24** | **0.82** |
| DP  | 0.5 | 0.813 | **48.45** | 6.09 | **0.932** | **11.07** | **1.00** |
|     | 1   | **0.817** | 50.85 | **5.92** | 0.895 | 17.19 | 1.59 |

certainty of 50%, while values closer to $t = 1$ mean that the network has a very high certainty about the prediction (Fig. 6).

Table 3 lists the results. For the Perimeter loss, the best performing threshold for fetal brain segmentation task was $t = 0.5$, significantly improving performance compared to a threshold of $t = 1$ from a Dice score of 0.895 to 0.932. For placenta segmentation there was no significant difference between the two. For the Contour Dice loss, the opposite was true. A threshold of $t = 1$ resulted in better performance for both placenta and fetal brain segmentation tasks, improving placenta segmentation from a Dice score of 0.798 to 0.847 and fetal brain segmentation from 0.939 to 0.946.

A looming question is why the difference between the two losses is happening. We hypothesize that for the Perimeter loss it is important to capture the full length of the contour even if the output prediction is not certain because it is a global length property. On the other hand, the Contour Dice loss is applied locally, and it may be beneficial to push the network toward certain contour regions.

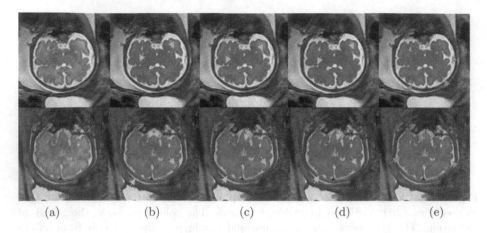

(a)                (b)                (c)                (d)                (e)

**Fig. 7.** Illustrative brain segmentation results: (a) original scan; (b) Dice loss; (c) Dice and Contour Dice loss; (d) Dice and Cross-Entropy loss; (e) ground truth segmentation. Yellow arrows point to segmentation errors of the Dice loss and their correction using either the Dice with Contour Dice or Dice with Cross-Entropy loss. The blue arrows point to cortex sulcation regions that were missed by all losses. (Color figure online)

## 5    Conclusions

This paper demonstrates the effectiveness of the Contour Dice loss compared to other boundary losses and the combined Dice with Cross-Entropy loss for placenta segmentation, a structure with fuzzy boundaries and high manual delineation variability, and for fetal brain segmentation, a structure that has complex boundaries with gyrus and sulcus.

We presented a new formulation for the Contour Dice loss which is more similar to the Contour Dice metric that was found to highly correlate with segmentation correction time. This formulation may be useful for cases with inaccurate boundaries of the ground truth segmentation as it poses relaxation on the contours intersection term.

For contour extraction, we performed thresholding on the segmentation followed by and erosion and XOR operators. We found that the segmentation performance of both the Contour Dice and the Perimeter losses was sensitive to the thresholding parameter $t$. Future work can explore other contour extraction techniques, i.e., min and max pooling operators as proposed in [11].

While the combined Dice with Contour Dice loss resulted in best performance for the task of placenta segmentation, it demonstrated only the second best performance for the task of brain segmentation, with best performance achieved by the compound Dice with Cross-Entropy loss. Maybe a better parameters tuning, a different optimization scheme or a combination with another regional loss function could improve the performance of boundary losses. In any case, the results show the importance of a comparison to the widely used Dice with Cross-Entropy loss even for tasks with fuzzy or complex contours where intuitively boundary losses should improve performance.

**Acknowledgement.** This research was supported in part by Kamin Grants 72061 and 72126 from the Israel Innovation Authority.

# References

1. Al Arif, S.M.M.R., Knapp, K., Slabaugh, G.: Shape-aware deep convolutional neural network for vertebrae segmentation. In: Glocker, B., Yao, J., Vrtovec, T., Frangi, A., Zheng, G. (eds.) MSKI 2017. LNCS, vol. 10734, pp. 12–24. Springer, Cham (2018). https://doi.org/10.1007/978-3-319-74113-0_2
2. Alansary, A., et al.: Fast fully automatic segmentation of the human placenta from motion corrupted MRI. In: Ourselin, S., Joskowicz, L., Sabuncu, M.R., Unal, G., Wells, W. (eds.) MICCAI 2016. LNCS, vol. 9901, pp. 589–597. Springer, Cham (2016). https://doi.org/10.1007/978-3-319-46723-8_68
3. Caliva, F., Iriondo, C., Martinez, A.M., Majumdar, S., Pedoia, V.: Distance map loss penalty term for semantic segmentation. In: International Conference on Medical Imaging with Deep Learning - Extended Abstract Track, pp. 08–10 (2019). https://openreview.net/forum?id=B1eIcvS45V
4. Dubois, J., et al.: Primary cortical folding in the human newborn: an early marker of later functional development. Brain **131**(8), 2028–2041 (2008)
5. Dubois, J., et al.: The dynamics of cortical folding waves and prematurity-related deviations revealed by spatial and spectral analysis of gyrification. Neuroimage **185**, 934–946 (2019)
6. Dudovitch, G., Link-Sourani, D., Ben Sira, L., Miller, E., Ben Bashat, D., Joskowicz, L.: Deep learning automatic fetal structures segmentation in MRI scans with few annotated datasets. In: Martel, A.L., et al. (eds.) MICCAI 2020. LNCS, vol. 12266, pp. 365–374. Springer, Cham (2020). https://doi.org/10.1007/978-3-030-59725-2_35
7. Ebner, M., et al.: An automated framework for localization, segmentation and super-resolution reconstruction of fetal brain MRI. Neuroimage **206**, 116324 (2020)
8. Han, M., et al.: Automatic segmentation of human placenta images with U-Net. IEEE Access **7**, 180083–180092 (2019)
9. Hosny, I.A., Elghawabi, H.S.: Ultrafast MRI of the fetus: an increasingly important tool in prenatal diagnosis of congenital anomalies. Magn. Reson. Imaging **28**(10), 1431–1439 (2010)
10. Isensee, F., Jaeger, P.F., Kohl, S.A., Petersen, J., Maier-Hein, K.H.: nnU-Net: a self-configuring method for deep learning-based biomedical image segmentation. Nat. Methods **18**(2), 203–211 (2021)
11. Jurdi, R.E., Petitjean, C., Honeine, P., Cheplygina, V., Abdallah, F.: A surprisingly effective perimeter-based loss for medical image segmentation. In: Medical Imaging with Deep Learning, pp. 158–167. PMLR (2021)
12. Karimi, D., Salcudean, S.E.: Reducing the Hausdorff distance in medical image segmentation with convolutional neural networks. IEEE Trans. Med. Imaging **39**(2), 499–513 (2019)
13. Kervadec, H., Bouchtiba, J., Desrosiers, C., Granger, E., Dolz, J., Ayed, I.B.: Boundary loss for highly unbalanced segmentation. In: International Conference on Medical Imaging with Deep Learning, pp. 285–296. PMLR (2019)
14. Kiser, K.J., Barman, A., Stieb, S., Fuller, C.D., Giancardo, L.: Novel autosegmentation spatial similarity metrics capture the time required to correct segmentations better than traditional metrics in a thoracic cavity segmentation workflow. J. Digit. Imaging **34**(3), 541–553 (2021)

15. Kodym, O., Španěl, M., Herout, A.: Segmentation of head and neck organs at risk using CNN with batch dice loss. In: Brox, T., Bruhn, A., Fritz, M. (eds.) GCPR 2018. LNCS, vol. 11269, pp. 105–114. Springer, Cham (2019). https://doi.org/10.1007/978-3-030-12939-2_8

16. Ma, J., et al.: Loss odyssey in medical image segmentation. Med. Image Anal. **71**, 102035 (2021)

17. Nikolov, S., et al.: Deep learning to achieve clinically applicable segmentation of head and neck anatomy for radiotherapy. arXiv preprint arXiv:1809.04430 (2018)

18. Payette, K., et al.: Fetal brain tissue annotation and segmentation challenge results. arXiv preprint arXiv:2204.09573 (2022)

19. Pietsch, M., et al.: APPLAUSE: automatic prediction of placental health via U-Net segmentation and statistical evaluation. Med. Image Anal. **72**, 102145 (2021)

20. Quah, B., et al.: Comparison of pure deep learning approaches for placental extraction from dynamic functional MRI sequences between 19 and 37 gestational weeks. In: Proceedings of International Society for Magnetic Resonance in Medicine (2021)

21. Reddy, U.M., Filly, R.A., Copel, J.A.: Prenatal imaging: ultrasonography and magnetic resonance imaging. Obstet. Gynecol. **112**(1), 145 (2008)

22. Rutherford, M., et al.: MR imaging methods for assessing fetal brain development. Dev. Neurobiol. **68**(6), 700–711 (2008)

23. Salavati, N., et al.: The possible role of placental morphometry in the detection of fetal growth restriction. Front. Physiol. **9**, 1884 (2019)

24. Salehi, S.S.M., et al.: Real-time automatic fetal brain extraction in fetal MRI by deep learning. In: 2018 IEEE 15th International Symposium on Biomedical Imaging (ISBI 2018), pp. 720–724. IEEE (2018)

25. Specktor-Fadida, B., et al.: A bootstrap self-training method for sequence transfer: state-of-the-art placenta segmentation in fetal MRI. In: Sudre, C.H., et al. (eds.) UNSURE/PIPPI -2021. LNCS, vol. 12959, pp. 189–199. Springer, Cham (2021). https://doi.org/10.1007/978-3-030-87735-4_18

26. Torrents-Barrena, J., et al.: Fully automatic 3D reconstruction of the placenta and its peripheral vasculature in intrauterine fetal MRI. Med. Image Anal. **54**, 263–279 (2019)

27. Torrents-Barrena, J., et al.: Segmentation and classification in MRI and us fetal imaging: recent trends and future prospects. Med. Image Anal. **51**, 61–88 (2019)

28. Yang, S., Kweon, J., Kim, Y.H.: Major vessel segmentation on X-ray coronary angiography using deep networks with a novel penalty loss function. In: International Conference on Medical Imaging with Deep Learning-Extended Abstract Track (2019)

# Multi-scale Multi-task Distillation for Incremental 3D Medical Image Segmentation

Mu Tian[1], Qinzhu Yang[1], and Yi Gao[1,2,3,4] (✉)

[1] School of Biomedical Engineering, Health Science Center, Shenzhen University, Shenzhen, China
gaoyi@szu.edu.cn
[2] Shenzhen Key Laboratory of Precision Medicine for Hematological Malignancies, Shenzhen, China
[3] Marshall Laboratory of Biomedical Engineering, Shenzhen, China
[4] Pengcheng Laboratory, Shenzhen, China

**Abstract.** Automatic medical image segmentation is the core component for many clinical applications. Substantial number of deep learning based methods have been proposed in past years, but deploying such methods in practice faces certain difficulties, such as the acquisition of massive annotated data for training and the high latency of model iteration. In contrast to the conventional cycle of "data collection, offline training, model update", developing a system that continuously generates robust predictions will be critical. Recently, incremental learning was widely investigated for classification and semantic segmentation on 2D natural images. Existing work showed the effectiveness of data rehearsal and knowledge distillation in counteracting catastrophic forgetting. Inspired by these advances, we propose a multi-scale multi-task distillation framework for incremental learning with 3D medical images. Different from the task-incremental scenario in literature, our proposed strategy focuses on improving robustness against implicit data distribution shift. We introduce knowledge distillation as multi-task regularization to resolve prediction confusions. At each step, the network is instructed to learn towards both the new ground truth and the uncertainty weighted predictions from the previous model. Simultaneously, image features at multiple scales in the segmentation network could participate in a contrastive learning scheme, aiming at more discriminant representations that inherit the past knowledge effectively. Experiments showed that our method improved overall continual learning robustness under the extremely challenging scenario of "seeing each image *once* in a batch of *one*" without any pre-training. In addition, the proposed method could work on top of any network architectures and existing incremental learning strategies. We also showed further improvements by combining our method with data rehearsal using a small buffer.

**Supplementary Information** The online version contains supplementary material available at https://doi.org/10.1007/978-3-031-25066-8_20.

**Keywords:** Multi-scale · Multi-task · Distillation · Incremental learning · 3D medical image segmentation

# 1   Introduction

By exploiting local structures and semantic similarities, elegantly designed deep networks effectively addressed the common issue of noise, low contrast, blurring and shape deformation in medical images [10], leading to state of the arts performances. Generally, these networks should be trained with sufficient amount of high quality annotations provided by clinical experts. Therefore, the common engineering workflow has three steps: data collection, offline training and model update. Labeling images in 3D is a very time consuming task that requires substantial expertise. It is not realistic to expect clinical practitioners to delineate thousands of images within a short time period. Also, it is quite likely that the data distribution could shift over time, due to changes of patient cohort, equipment, imaging protocols and the practitioner's experience. Therefore, the trained model could quickly get outdated and unable to adapt with new data. It is crucial to develop a system that learns with sequentially arriving data while generalizes well to unseen instances.

When learning with streaming data, catastrophic forgetting [7,21] is the most fundamental challenge due to the shifts in learning tasks and data distribution. One naive strategy (referred as "joint training" [6]) would be to re-train the model with all past and present data available, which is obviously not scalable and sustainable. Fine-tuning [6] is a method at the other end of the spectrum, that updates the model parameters solely based on the new task. In this case the model could easily forget the past knowledge. In the past years, multiple online continual/incremental learning (OCL) [1,13] methods have been proposed, aiming at counteracting catastrophic forgetting with constant memory and training time. Two most common successful strategies are data rehearsal and knowledge distillation. The rehearsal strategy, such as Gdumb [20], MIR [13] keeps a memory of constant size to store past data, that participates in training along with new samples. On the other hand, knowledge distillation [2,8,14] is an idea of preserving past experience by learning towards previous model's outputs. The model only need to take the new samples without looking into the past. Both strategies have proved effective in alleviating catastrophic forgetting for classification [9,16,21,24,25] and semantic segmentation tasks [2,14,17,18] .

However, it remains unclear whether these strategies could be directly transferred into medical image segmentation. The majority of existing work attempted to build task-incremental benchmarks. Even for segmentation, the focus was on learning new semantic classes incrementally. However, for medical images, the shift in data distribution, instead of semantic labels, is the dominating source of variations in the stream [19]. Even the images were collected from the same equipment and clinician, these variations could still occur at unknown moments in the stream due to other factors. Therefore, this work investigates a more general setting where no prior information is given about which and when these implicit distribution shifts could occur.

Given a stream of medical images with a fixed segmentation task, the goal is to develop a scalable system that continuously adapts to new data while preserving generalization capability that approaches a deep network trained with the conventional offline fashion. The system should guarantee $O(1)$ training time and memory consumption at each learning step. Assuming that we can only access the new image, the only available source of past knowledge can be queried is from the past model states. Knowledge distillation [2, 8, 14] was originally proposed such that a simpler network could retain the predictive power of a complicated one through learning towards its outputs. Inspired by recent works [6, 26] that extended distillation scheme into task incremental learning, we develop a novel multi-scale multi-task (MSMT) distillation framework particularly for the scenario of medical images.

First, similar to learning without forget (LWF) [6], the prediction head in the network is used to simultaneously learn the new target and previous model's outputs. However, one fundamental difference to LWF [6] is that they use separate classifier heads to distinguish old/new tasks. In the case of segmenting a single target out of the background, we only have one head under the LWF setting. Though this might work as a implicit label smoothing mechanism, this confusion between new ground truth and old model's predictions could break the common configuration of distillation at incremental learning. To resolve this issue, we create a separate head ("dual-head"), whose target is to solely predict previous model's output, and thus formulating a multi-tasking structure. In this setting, the original head is still able to learn the precise ground truth without losing information, while the new head serves as a regularizer that keeps the model's capability of retaining old information, through preventing semantic representations from fluctuating too much. These dual-head structure balances the model's capability of "adapting to the new" and "memorizing the past".

Second, it is known that constructing distillation directly on high level representations with a contrastive loss could further help to counteract catastrophic forgetting. In [26], contrastive pairs were formulated at locations belonging to foreground classes. However, in our case, the extended class map [26] delivers confusion, rather than complement. It is important to carefully design new contrastive schemes that effectively leverages this overlapping map. Here we propose a new contrastive distillation loss, similar but fundamentally different to [26], that promotes internal similarities at true-positive locations while penalizing false positive/negative positions of the old model. At the same time, this design also encourages discriminant representations of the new model by incorporating background positions.

Third, for medical images, U-shaped network architectures [3, 5, 11, 15, 22, 23] worked particularly well due to their capability of capturing multi-scale information efficiently. Our proposed contrastive loss also include multiple scales. The bottleneck layer, the higher resolution visual representations in encoder and decoder at symmetric positions could all contribute to a fusion of spatial similarities.

In summary, this proposed method has four fold contributions.

First, multi-task regularization is used to formulate knowledge distillation for single class segmentation for 3D medical images. This formulation solved the conceptual confusion and balanced the model's capability of learning from new and past tasks.

Second, we designed contrastive learning for single foreground cases in medical images, and introduced multi-scale fusion that enables the network to learn stable representations at both semantic level and visual details.

Third, evaluation on publicly available datasets proved the effectiveness of our framework in 3D medical image segmentation, that produced robust predictions without the depending on pre-training.

Fourth, the proposed framework could be applied on top of any other network architectures and learning strategies. We showed that MSMT could further boost the performance working together with data rehearsal.

## 2   Related Work

Online continual/incremental learning [13] is the line of research that focuses on learning sequentially arriving tasks. The main challenge is to preserve past knowledge while adapting to new information. Data rehearsal, or memory based approaches [13,20] keep a representative set of samples from the past sequence. Despite of its conceptual simplicity, data rehearsal turned out to be very effective for classification and semantic segmentation. However, due to the constant memory buffer size, the amount of knowledge transferred from past will be limited in the long run. Another idea was to customize knowledge distillation to query past information from the model itself [2,6,8,14,26]. Such strategies leverage the output and internal features of the previous model, to encourage identical level of predictive performance on old tasks while intaking new tasks. In particular, [26] proposed a novel approach that combined contrastive learning and knowledge distillation, that enables the model to simultaneously learn discriminant features while counteracting catastrophic forgetting. Their approach achieved new state of the arts in semantic segmentation benchmarks.

The majority of existing work tackle the task-incremental scenario. They assumed a common setting that new tasks (target classes) will be added gradually in a sequence. However, common clinical applications do not fall into this setting. Instead, implicit distribution shift is the dominating variation [19] while the target to be segmented is usually one identical organ/tissue/type of nodules. Domain adaptation [4,27,28] is a related strategy to solve for a transform between source and target domains. However, existing methods assume the existence of both source and target domain samples, which is not realistic in our setting of image stream. Recently, a new memory based method [19] was proposed particularly to deal with domain shifts in medical image segmentation task. They showed that by keeping a diverse set of past samples in the buffer, the model could be robust to domain changes caused by the switch of different equipment. The major limitation of their work might be the reliance on a style network pre-trained on 2D RGB images used for similarity computation, which might limit further extension onto 3D images.

This work was motivated by related works [19, 26]. To the best of our knowledge, this work present the first attempt to discuss and experiment on incremental learning mechanism on *3D medical image segmentation* in a generic setting, without assuming any pre-trained networks and data domain memberships.

## 3   Method

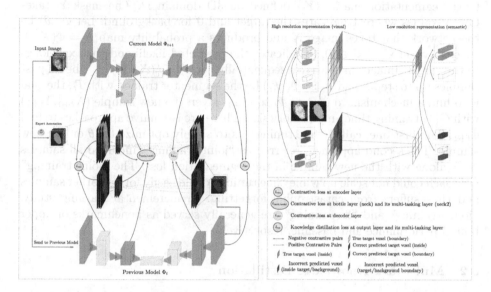

**Fig. 1.** The proposed method overview. In the left panel, a generic encoder-decoder architecture is served as the segmenter network (such as U-Net [5] and V-Net [15]). In practice, the network could have more or less number of layers and does not have to be symmetrical. Three internal layers participate in *Multi-Scale* contrastive learning and the output layer contribute to knowledge distillation. The *Multi-Task* branches are created in parallel with its original feature map and participate in corresponding Multi-Task learning. Upon receiving the input image from the stream, its segmentation mask is acquired from expert annotation. At the same time, the image is fed into the previous step's model $\Phi_t$ and the corresponding visual/semantic feature representations and model output logits could be collected to guide contrastive learning (through losses $L_{enc}$, $L_{neck/neck2}$ and $L_{dec}$, see Sect. 3.3) and knowledge distillation (through loss $L_{kd}$, see Sect. 3.2). The multi-scale contrastive learning scheme is illustrated in the top-right panel, positive and negative contrastive pairs are constructed based on the consistency between previous model's prediction and the ground truth on each voxel (see Sect. 3.3). Note that the voxels along target boundaries were differentiated from those inside the target or background regions. The boundary voxels were used to create edge focused losses through the masking scheme in Sect. 3.4.

The overview of the proposed method is illustrated in Fig. 1. We first describe the general formulation of incremental medical image segmentation, then introduce MSMT in greater detail.

### 3.1  Incremental Medical Image Segmentation

Let $T = \{(X_i, Y_i)\}|_{i=1}^{\infty}$ be a sequence of images $\{X_i\}$ along with their ground truth segmentation masks $\{Y_i\}$ defined on 3D domain $\Omega$. The mask $Y_i$ takes binary values where 1 indicates the target and 0 for background. Let $\Phi(\cdot|\theta)$ be the network that takes image $X$ and produces a probability map $P = \Phi(X|\theta)$. At each voxel $\mathbf{x}$, $P[\mathbf{x}] \in [0, 1]$ indicates the predicted likelihood that $\mathbf{x}$ belongs to the target. Images in $T$ arrives sequentially, let $T_t = \{(X_i, Y_i)\}|_{i=1}^{t}$ be all past samples up to step $t$, and $\Phi_t := \Phi(\cdot|\theta_t)$ be the segmenter trained with $T_t$, the goal is to find a mechanism to update $\Phi_t$ to $\Phi_{t+1}$, given the new sample $(X_{t+1}, Y_{t+1})$ with $O(1)$ training time and memory size. There are two naive approaches to find $\Phi_{t+1}$. The first one, called "fine-tuning", is to simply optimize for $\theta$ on the new sample. The second approach, referred as "joint training", included all samples in $T_t$ along with the new sample in the segmentation loss. The "joint training" approach could not scale up while "fine tuning" could easily forget earlier samples and thus suffers from catastrophic forgetting. In incremental learning study, "joint-training" and "fine-tuning" are generally served as benchmarks of upper bound performance and baseline, respectively.

### 3.2  Multi-task Knowledge Distillation

In task-incremental learning scenarios, distillation is formulated such that the model is learned towards previous model's output at pixel locations other than the new target class. Each pixel is assigned with a unique membership. If we directly apply existing distillation framework into our setting, a confusion could be caused at voxels in the present image where the new ground truth and previous model's prediction do not agree. There are two approaches to solve this confusion. The first is to directly append a distillation loss to the original segmentation loss. This approach is conceptually similar to label smoothing technique and could expect to prevent the model from over-fitting the new sample. In the situation where only minor distribution fluctuation exists, and the model is trained on adequate samples, we could expect $Y_{t+1}$ and $\Phi_t(X_{t+1})$ to converge. However, during early stages in the continual training, or after a sudden unknown switch of data domains, $\Phi_t(X_{t+1})$ could be quite different from the ground truth, causing large regions of false positives/negatives. In this case, a confusion is created and the model is pushed away from the correct information.

Here we provide an alternative choice, to create a parallel "distillation" head $\phi^{kd}$ on top of the network backbone (Fig. 1) while the original head still focuses on segmentation. The existence of $\phi^{kd}$ serves as multi-task regularization that prevents semantic features from fluctuating too far from previous model, and thus helps in mitigating catastrophic forgetting. Formally, let $\Phi^{(l)}, l = 1, \dots, L$

be the network up to layer $l$, where $L$ is the total number of layers; let $\tilde{\Phi} := \phi^{\mathrm{kd}} \circ \Phi^{(L-1)}$, then the distillation loss becomes:

$$L_{\mathrm{kd}}(\theta|X, \Phi_t) = \frac{1}{|\Omega|} [\Phi_t(X) \log \frac{e^{\frac{\tilde{\Phi}(X|\theta)}{\tau}}}{e^{\frac{\tilde{\Phi}(X|\theta)}{\tau}} + 1} + (1 - \Phi_t(X)) \log(\frac{1}{e^{\frac{\tilde{\Phi}(X|\theta)}{\tau}} + 1})] \quad (1)$$

where $\tau > 0$ is the temperature.

## 3.3 Multi-scale Contrastive Learning

The combination of contrastive learning and knowledge distillation was discussed in detail in [26]. The fundamental idea was to contrast on semantic representations based on supposed similarity constraints. However, this idea could not be directly applied for our setting of medical images. First, there's no way to construct negative pairs if we ignore the background class. So we need to include background class in contrastive learning despite its heterogeneity. The motivation is to push positive representations closer together and make better distinctions around target boundaries. Second, it is no longer reasonable to compare current and previous model's representations according to the predicted voxel memberships. Giving a positive reward to a pair of pixels including false positives/negatives from the previous model could just deteriorate the model's discriminant power. We therefore introduce a new configuration below.

Let $G^+ = \{\mathbf{x} : Y[\mathbf{x}] = 1\}$ and $G^- = \{\mathbf{x} : Y[\mathbf{x}] = 0\}$, $S^+ = \{\mathbf{x} : P[\mathbf{x}] > 0.5\}$ and $S^- = \Omega \backslash S^+$. Let $\Phi^{(l)}(X) \in \mathbb{R}^{H^l \times W^l \times D^l \times d^l}$ be internal feature representation extracted from a selected network layer $1 < l < L$ (see Fig. 1), where $H^l, W^l, D^l$ are down-sampled resolution of the input image $X$ of size $H \times W \times D$, and $d^l$ is the feature dimension. Also, the ground truth mask $Y$ and model prediction $P$ are down-sampled along with identical network path into $Y^l, P^l \in \mathbb{R}^{H^l \times W^l \times D^l}$. We therefore have the corresponding sets $G_l^+, G_l^-, S_l^+, S_l^-$ formulated the same way as above. Given $\mathbf{x} \in G_l^+$, we construct two possibly overlap sets covering its "similar" locations: $G_l^+(\mathbf{x}) := G_l^+ \backslash \{\mathbf{x}\}$ and $S_l^p := G_l^+ \cap S_l^+$, where positive pairs will be built on. In addition, its "dissimilar" sets include $G_l^-$ and $S_l^n := G_l^- \cup (G_l^+ \cap S_l^-)$. The idea was to construct a loss that promotes true positives and penalizes false positive/negatives and target-background pairs, as follows:

$$L_{\mathrm{con}}^l(\theta|X, Y, \Phi_t) = \frac{1}{|G_l^+|} \sum_{\mathbf{x} \in G_l^+} [\sum_{\mathbf{y} \in G_l^+(\mathbf{x})} \frac{l_{\mathrm{self}}^l(\mathbf{x}, \mathbf{y})}{|G_l^+(\mathbf{x})|} + \sum_{\mathbf{y} \in S_l^p} \frac{l_{\mathrm{cross}}^l(\mathbf{x}, \mathbf{y})}{|S_l^p|}] \quad (2)$$

with

$$l_{\text{self}}^l(\mathbf{x}, \mathbf{y}) = -\log \frac{\exp(\frac{\langle \Phi^{(l)}(\mathbf{x}), \Phi^{(l)}(\mathbf{y}) \rangle}{\tau})}{\rho}$$

$$l_{\text{cross}}^l(\mathbf{x}, \mathbf{y}) = -\log \frac{\exp(\frac{\langle \Phi^{(l)}(\mathbf{x}), \Phi_t^{(l)}(\mathbf{y}) \rangle}{\tau})}{\rho}$$

$$\rho = \sum_{\mathbf{y}' \in G_l^-} \exp(\frac{\langle \Phi^{(l)}(\mathbf{x}), \Phi^{(l)}(\mathbf{y}') \rangle}{\tau}) + \sum_{\mathbf{y}' \in S_l^n} \exp(\frac{\langle \Phi^{(l)}(\mathbf{x}), \Phi_t^{(l)}(\mathbf{y}') \rangle}{\tau})$$

where $\Phi_t^{(l)}$ is used to denote the feature representation extracted from the previous step's model $\Phi_t$ and $\tau$ is the temperature. It is important to note that the only positive contrast between the current and previous model occurs on locations of true positive predictions $S_l^p$. This design ensures that we distill knowledge only at "correct" positions.

In the case of V-Net, if we pick $l$ to be the bottleneck layer in the encoder (similar to [26]), the feature resolution will be about $1/10$ of the input image. In our setting, input image $X$ has resolution $[30, 55, 55]$, and the size of $\Phi^{(l)}(X)$ becomes $[3, 6, 6]$. Such low resolution might not be sufficient to reveal visual details such as target boundary. Therefore, we include different levels of representation to formulate multi-scale learning. For V-Net, we pick $l_{\text{enc}}, l_{\text{neck}}, l_{\text{dec}}$ to be the second layer of encoder, the second-to-last layer of encoder, and the second-to-last layer of decoder (see Fig. 1). The corresponding spatial resolutions now will be $15 \times 27 \times 27, 3 \times 6 \times 6, 15 \times 27 \times 27$ respectively. Note that $l_{\text{enc}}$ and $l_{\text{dec}}$ locate at symmetric positions and the former was concatenated into the latter with a skip connection. Similar to the multi-task idea in Sect. 3.2, we also introduce a separate branch $\phi^{(l_{\text{neck2}})}$ parallel to $\phi^{(l_{\text{neck}})}$ serving as smoother regularizer within contrastive learning. Eventually, we could combine the contributions from four locations as an integrated contrastive loss, guiding the network to learn representations both at semantic level and more detailed visual structures:

$$L_{\text{con}}(\theta|X, Y, \Phi_t) = \sum_{l \in \{l_{\text{enc}}, l_{\text{neck}}, l_{\text{neck2}}, l_{\text{dec}}\}} \lambda_l L_{\text{con}}^l(\theta|X, Y, \Phi_t) \tag{3}$$

### 3.4   Masking

We introduce two types of masking scheme in our framework. Similar to the idea of [26], we weight the knowledge distillation loss with $\omega_{\text{kd}} = P_t \cdot P_t + (1 - P_t) \cdot (1 - P_t)$ in Eq. 1 to reduce the importance at uncertain locations: $L_{\text{kd}}^\omega := \omega_{\text{kd}} \cdot L_{\text{kd}}$. In addition, with 3D volumetric data, on one hand, we need to construct much more voxel pairs in the contrastive loss; on the other hand, the target boundary, especially for medical images, could be trickier as it could include delicate 3D structures. Therefore, we propose an edge focused mask, $\omega_\delta^l = \mathbf{1}_{\mathbf{x} \in E_\delta^l}$ with $E_\delta^l = \{\mathbf{x} : |G_l^+ \cap \overline{B(\mathbf{x}, \delta)}| > 0, |G_l^- \cap \overline{B(\mathbf{x}, \delta)}| > 0\}$ denoting the "boundary locations" where $B$ represents a 3D ball centered at $\mathbf{x}$ with its

$l$-$\infty$ radius $\delta > 0$. This mask could be applied at visual representation layers $l_{enc}, l_{dec}$ in Eq. 3 to focus on boundary details as well as reducing computational overheads.

### 3.5 Combine with Memory Based Approach

In clinical applications, the past data might not be accessible due to storage constraints, database update and privacy issues, which instructs our work to focus on distillation based approach. Nevertheless, it is beneficial to see how our proposed method could work together with rehearsal, assuming past data can be accessible and we have a quota for constant size buffer.

Let $\mathcal{M} = \{(X_{t_j}, Y_{t_j}, \cdot), t_j \leq t\}|_{j=1}^{m}$ be a memory buffer at step $t$ of size $m$ keeping the input image, label along with necessary auxiliary states inside. Upon arriving of the new sample at $t+1$, the network is trained jointly with $\mathcal{M} \cup \{(X_{t+1}, Y_{t+1})\}$. To enable distillation, a decision must be made which "previous model states" should be aligned with each sample in $\mathcal{M}$. One choice would be to store the model states being used for training at the moment on the fly, i.e. $\mathcal{M} = \{(X_{t_j}, Y_{t_j}, \Phi_{t_j-1})\}$; another option is to re-infer on $\mathcal{M}$ with $\Phi_t$ and use the combined set $\{(X_{t_j}, Y_{t_j}, \Phi_t)\} \cup (X_{t+1}, Y_{t+1}, \Phi_t)$ for training with distillation (Eq. 1) and contrastive (Eq. 3) losses.

The first option could save a constant factor of model inference time, with a cost of increased storage overheads. However, the distillation scheme could only learn towards incomplete past knowledge preceding $t_j$, that will conceptually "degenerate" to a vanilla separate learning task while losing the capability of preserving the past knowledge continuously. Therefore, we should choose the second configuration, that refreshes the auxiliary states in $\mathcal{M}$ each step. Internally, introducing $\mathcal{M}$ enriches representational diversity that promote better performances on both segmentation and distillation tasks.

## 4    Experiments

We evaluate the proposed method on two datasets: NCI-ISBI2013 and BraTS2015 [12]. The NCI-ISBI2013 dataset includes 80 samples for automated segmentation of prostate structures and the target of BraTS2015 is to segment whole brain tumor in 274 FLAIR images. Similar to [12], we use randomly chosen 117 images for incremental training sequence and the 40 images in test set for BraTS2015, and use 64 images for training and the 16 for testing in NCI-ISBI2013. The train-test split is kept identical across our experiments. We assume that images from training set arrives one by one in a sequence, and the network should be trained with proposed strategy on each new sample. Dice score are computed on all evaluation images at each learning step.

### 4.1    Implementation Details

The proposed method can be applied with any segmentation network architectures, we use the V-Net widely adopted in the literature [12,15]. The learning rate is set to $10^{-3}$ throughout training. Since we do not assume any warm-up or pre-training stages, the learning rate, or its changing schedule, should be identical for each training step, as requested by the promised sustainability. For each step, the network is trained with 150 gradient updates. For a fair comparison, we still set batch size to 1 for the case of joint training benchmark and rehearsal approaches. The images are processed in an identical way as in [12]. Each volume was firstly cropped by its non-zero boundaries with a margin of 10 voxels, normalized by the global mean/std values, and eventually resized to [30, 55, 55]. During training, random flipping and rotation transforms was applied on the fly. We set $\lambda_{kd}$ (multiplier on Eq. 1 when combining with segmentation loss) to 0.4 and the values of $\lambda_{l_{enc}}, \lambda_{l_{neck}}, \lambda_{l_{neck2}}, \lambda_{l_{dec}}$ (Eq. 3) to 0.1. We found the performance is not sensitive to temperature $\tau$. We set $\tau$ for both distillation and contrastive losses to 1.0 in NCI-ISBI2013, and 0.5 in BraTS2015. The implementation can be found at https://github.com/kevinmtian/msmt-medseg.

## 4.2   Comparison with Common Strategies

Unlike the flourishing benchmarks for natural image tasks, we did not find a list of publicly available state-of-the-arts for incremental 3D medical image segmentation. Therefore, we implement comparisons with two generic strategies: distillation based and rehearsal-based. For distillation-based methods, we did not found prototypes already implemented for our experiment settings. Therefore, we treat both the single-branch and dual-branch designs as extended versions of learning without forget (LWF). Regarding data rehearsal, we included the fundamental experience replay approach here without looking into more complicated ones. The goal is to see how the proposed MSMT framework works together with rehearsal, not digging into the comparison among different rehearsal-based approaches.

Tables 1 and 2 list evaluation scores on test set. It is expected that model performance could fluctuate across training steps, therefore, we take running averages with a fixed window at several different positions in the sequence. In addition, we only analyze performances during the later stages of training process since they are more representative than earlier stages where few samples have been seen by the model. From both tables, we see that MSMT tends to provide stabilized performance earlier in the training process. Table 1 shows that MSMT significantly outperforms the fine-tuning method at the beginning (the 30th sample) at the second half of training sequence. The last two evaluations appears inferior to the single-head distillation strategy, but MSMT remains to be the most robust approach when summarizing all 5 scores. Similar phenomenon was also observed for "KD2" (dual-head distillation) and Cont neck2 (dual-head bottleneck contrastive learning). This is a demonstration that the proposed multi-task distillation/contrastive learning could promote model robustness. When the rehearsal strategy is involved, we also observed further gain from applying MSMT on top of a rehearsal buffer of size 2 (Mem2) and 4 (Mem4) samples.

For BraTS2015 (Table 2), the improvement brought by MSMT appears subtler, but there's still a clear sign that MSMT dominates at earlier stages when

**Table 1.** Dice scores evaluated on evaluation set of NCI-ISBI2013 dataset vs. learning steps windows. The learning step is equivalent to the number of first samples seen by the network. We compute running average of evaluation set dice scores, across a window of 15 consecutive training steps, with a stride of 5. Here "sn(a)–sn(b)" indicate the window spanning from the $a^{th}$ to the $b^{th}$ sample. Methods were grouped into blocks based on types of strategy and rehearsal memory size, if any (Mem2, Mem4 means we keep a memory buffer capable of storing 2, 4 past samples, respectively). Best score across parts are in **bold**, best inside each part is underlined, *italic* is used to indicate a tie if necessary.

| Method | Average dice on running windows | | | | |
|---|---|---|---|---|---|
|  | sn30–sn44 | sn35–sn49 | sn40–sn54 | sn45–sn59 | sn50–sn64 |
| Fine tuning | 0.5 | 0.47 | 0.54 | 0.63 | 0.65 |
| KD1 | 0.54 | 0.52 | 0.59 | **0.65** | **0.67** |
| KD2 | <u>0.57</u> | <u>0.55</u> | <u>0.6</u> | 0.6 | 0.62 |
| Cont enc | 0.52 | 0.52 | 0.58 | *0.64* | 0.64 |
| Cont neck | 0.54 | 0.52 | 0.57 | *0.64* | <u>0.66</u> |
| Cont neck2 | <u>0.56</u> | <u>0.55</u> | <u>0.59</u> | 0.59 | 0.6 |
| Cont dec | 0.54 | 0.51 | 0.54 | 0.62 | 0.65 |
| MSMT | **0.6** | **0.59** | **0.62** | 0.63 | 0.64 |
| Mem2 | 0.62 | 0.61 | 0.66 | 0.69 | 0.69 |
| Mem2+MSMT | **0.68** | **0.68** | **0.7** | **0.73** | **0.73** |
| Mem4 | 0.7 | 0.7 | 0.73 | 0.74 | 0.75 |
| Mem4+MSMT | 0.7 | **0.72** | **0.74** | **0.76** | **0.76** |
| Joint Training | 0.77 | 0.78 | 0.78 | 0.78 | 0.78 |

no rehearsal is involved. It is also noted that joint training did not provide an absolute performance upper bound. This indicates that even assuming accessibility to all past data, if the training time is required to be constant, it could be less beneficial to loop through all past sequence, that makes the model to over-fit on early samples and under-trained on newer ones. Also, MSMT does not help in Mem2 but provided limited improvements in the case of Mem4. The reason would be that BraTS2015 has a longer training sequence, and we may need more representative data in each round to activate the effect from distillation and contrastive learning.

We also provided trend analysis in Figs. 2 and 3. It shows more straightforwardly, especially for NCI-ISBI2013, that MSMT stays significantly beyond other methods in the long run with superior robustness. In Fig. 3, we see that "joint training" plateaus quickly, indicating the same reasoning above, that it is memorizing old samples without adapting to the new.

## 4.3  Ablation Study

We provide further analysis on how each component contribute and hopefully to shed light on how MSMT could be tailored for different scenarios. Figure 4

**Table 2.** Dice scores evaluated on evaluation set of BraTS2015 dataset vs. learning steps windows. Notations are consistent to Table 1.

| Method | Average dice on running windows | | | | |
|---|---|---|---|---|---|
| | sn83–sn97 | sn88–sn102 | sn93–sn107 | sn98–sn112 | sn103–sn117 |
| Fine Tuning | 0.63 | 0.62 | 0.61 | 0.64 | 0.65 |
| KD1 | 0.64 | 0.63 | *0.62* | <u>0.65</u> | 0.65 |
| KD2 | 0.64 | 0.63 | *0.62* | 0.64 | 0.65 |
| Cont enc | 0.63 | 0.62 | 0.61 | 0.65 | *0.66* |
| Cont neck | 0.64 | 0.63 | 0.61 | 0.64 | 0.65 |
| Cont neck2 | *0.65* | 0.63 | *0.62* | 0.63 | 0.64 |
| Cont dec | 0.64 | 0.63 | *0.62* | *0.66* | *0.66* |
| MSMT | **0.65** | **0.64** | *0.62* | 0.65 | *0.66* |
| Mem2 | 0.71 | 0.7 | 0.68 | 0.69 | 0.7 |
| Mem2+MSMT* | 0.7 | 0.69 | 0.67 | 0.68 | 0.69 |
| Mem4 | 0.73 | 0.73 | 0.71 | 0.7 | 0.68 |
| Mem4+MSMT | 0.73 | 0.73 | 0.71 | **0.71** | **0.71** |
| Joint Training* | 0.7 | 0.69 | 0.69 | 0.69 | 0.7 |

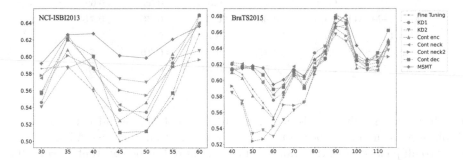

**Fig. 2.** Running averaged Dice scores vs. training samples seen by the model. No rehearsal is included.

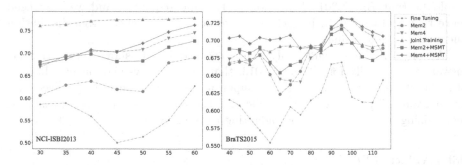

**Fig. 3.** Running averaged Dice scores vs. training samples seen by the model. We focus on analyzing with the case of rehearsal strategies.

(A) Analysis of adding component    (B) Analysis of masking schemes

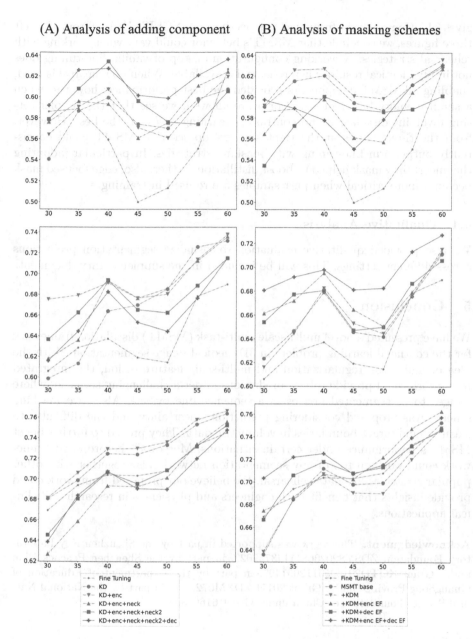

**Fig. 4.** Ablation study on NCI-ISBI2013. Note that the left and right sides give totally different types of analysis, and they do not share common legends. The left three figures analyzes different components in MSMT, and the right three figures analyzes masking schemes. The top two figures are without rehearsal, the middle two figures are under Mem2, rehearsal with a memory of size 2, while the bottom two figures are with Mem4. Here, "KDM" means applying masking scheme to distillation loss, and "EF" means to apply edge focused mask (Sect. 3.4) on contrastive learning.

gives ablation analysis from two perspectives on NCI-ISBI2013. From the left three figures, we conclude that MSMT's behavior could vary when working with rehearsal strategies. A working configuration on top of vanilla fine-tuning does not imply identical results when rehearsal is available. When no rehearsal is used, enabling all MSMT components provided the best performance; however, when a small memory buffer of size 2 or 4 are allowed, we see that the simpler setting involving distillation and contrastive learning on $l_{enc}$ is the best. However, from the three figures at the right side, we do see that MSMT could consistently outperform fine-tuning with masking strategies. In particular, adopting the uncertainty mask helped to boost distillation further, also, edge focused mask becomes more critical when past samples are re-used in training.

### 4.4  Qualitative Analysis

We also provided qualitative evaluation to visualize segmentation predictions across different settings. This will be included in the supplementary document.

## 5   Conclusion

We have presented a novel multi-scale multi-task (MSMT) distillation framework for the continual learning problem in 3D medical image segmentation. With the idea of multi-task regularization and multi-scale feature fusion, the integrated framework could provide robust predictions under a challenging scenario, where images to be segmented arrives in a sequence one by one. Also, two masking scheme was proposed considering prediction uncertainty and the difficulty to learn around target boundaries in volumetric data. They proved to further boost MSMT's performance under certain situations. Moreover, the proposed framework could work on top of any segmentation network, also together with other popular strategies like data rehearsal. We believe our proposed framework could provide insights that benefit both engineers and physicians in research and clinical applications.

**Acknowledgment.** This work was supported in part by the Shenzhen Key Laboratory Foundation ZDSYS20200811143757022, in part by the Shenzhen Peacock Plan under Grant KQTD2016053112051497, in part by the Department of Education of Guangdong Province under Grant 2017KZDXM072, and in part by the National Natural Science Foundation of China under Grant 61601302.

## References

1. Aljundi, R., et al.: Online continual learning with maximally interfered retrieval. arXiv preprint arXiv:1908.04742 (2019)
2. Cermelli, F., Mancini, M., Bulò, S.R., Ricci, E., Caputo, B.: Modeling the background for incremental learning in semantic segmentation. In: 2020 IEEE/CVF Conference on Computer Vision and Pattern Recognition (CVPR), pp. 9230–9239 (2020)

3. Chen, L.C., Papandreou, G., Schroff, F., Adam, H.: Rethinking atrous convolution for semantic image segmentation. arXiv preprint arXiv:1706.05587 (2017)
4. Choudhary, A., Tong, L., Zhu, Y., Wang, M.D.: Advancing medical imaging informatics by deep learning-based domain adaptation. Yearb. Med. Inform. **29**, 129–138 (2020)
5. Çiçek, Ö., Abdulkadir, A., Lienkamp, S.S., Brox, T., Ronneberger, O.: 3D U-net: learning dense volumetric segmentation from sparse annotation. In: Ourselin, S., Joskowicz, L., Sabuncu, M.R., Unal, G., Wells, W. (eds.) MICCAI 2016. LNCS, vol. 9901, pp. 424–432. Springer, Cham (2016). https://doi.org/10.1007/978-3-319-46723-8_49
6. Douillard, A., Chen, Y., Dapogny, A., Cord, M.: Plop: learning without forgetting for continual semantic segmentation. In: Proceedings of the IEEE/CVF Conference on Computer Vision and Pattern Recognition, pp. 4040–4050 (2021)
7. Goodfellow, I.J., Mirza, M., Da, X., Courville, A.C., Bengio, Y.: An empirical investigation of catastrophic forgeting in gradient-based neural networks. CoRR abs/1312.6211 (2014)
8. Hinton, G.E., Vinyals, O., Dean, J.: Distilling the knowledge in a neural network. arXiv abs/1503.02531 (2015)
9. Hou, S., Pan, X., Loy, C.C., Wang, Z., Lin, D.: Learning a unified classifier incrementally via rebalancing. In: 2019 IEEE/CVF Conference on Computer Vision and Pattern Recognition (CVPR), pp. 831–839 (2019)
10. Lei, T., Wang, R., Wan, Y., Du, X., Meng, H., Nandi, A.K.: Medical image segmentation using deep learning: a survey. arXiv abs/2009.13120 (2020)
11. Li, W., Wang, G., Fidon, L., Ourselin, S., Cardoso, M.J., Vercauteren, T.: On the compactness, efficiency, and representation of 3D convolutional networks: brain parcellation as a pretext task. In: Niethammer, M., et al. (eds.) IPMI 2017. LNCS, vol. 10265, pp. 348–360. Springer, Cham (2017). https://doi.org/10.1007/978-3-319-59050-9_28
12. Liao, X., et al.: Iteratively-refined interactive 3D medical image segmentation with multi-agent reinforcement learning. In: 2020 IEEE/CVF Conference on Computer Vision and Pattern Recognition (CVPR), pp. 9391–9399 (2020)
13. Mai, Z., Li, R., Jeong, J., Quispe, D., Kim, H., Sanner, S.: Online continual learning in image classification: an empirical survey. arXiv preprint arXiv:2101.10423 (2021)
14. Michieli, U., Zanuttigh, P.: Incremental learning techniques for semantic segmentation. In: 2019 IEEE/CVF International Conference on Computer Vision Workshop (ICCVW), pp. 3205–3212 (2019)
15. Milletari, F., Navab, N., Ahmadi, S.A.: V-net: fully convolutional neural networks for volumetric medical image segmentation. In: 2016 fourth international conference on 3D vision (3DV), pp. 565–571. IEEE (2016)
16. Ostapenko, O., Puscas, M.M., Klein, T., Jähnichen, P., Nabi, M.: Learning to remember: a synaptic plasticity driven framework for continual learning. 2019 IEEE/CVF Conference on Computer Vision and Pattern Recognition (CVPR), pp. 11313–11321 (2019)
17. Özdemir, F., Fürnstahl, P., Göksel, O.: Learn the new, keep the old: extending pretrained models with new anatomy and images. arXiv abs/1806.00265 (2018)
18. Ozdemir, F., Goksel, O.: Extending pretrained segmentation networks with additional anatomical structures. Int. J. Comput. Assist. Radiol. Surg. **14**(7), 1187–1195 (2019). https://doi.org/10.1007/s11548-019-01984-4
19. Perkonigg, M., et al.: Dynamic memory to alleviate catastrophic forgetting in continual learning with medical imaging. Nat. Commun. **12** (2021)

20. Prabhu, A., Torr, P.H.S., Dokania, P.K.: GDumb: a simple approach that questions our progress in continual learning. In: Vedaldi, A., Bischof, H., Brox, T., Frahm, J.-M. (eds.) ECCV 2020. LNCS, vol. 12347, pp. 524–540. Springer, Cham (2020). https://doi.org/10.1007/978-3-030-58536-5_31
21. Rebuffi, S.A., Kolesnikov, A., Sperl, G., Lampert, C.H.: ICARL: incremental classifier and representation learning. In: Proceedings of the IEEE Conference on Computer Vision and Pattern Recognition, pp. 2001–2010 (2017)
22. Ronneberger, O., Fischer, P., Brox, T.: U-net: convolutional networks for biomedical image segmentation. In: Navab, N., Hornegger, J., Wells, W.M., Frangi, A.F. (eds.) MICCAI 2015. LNCS, vol. 9351, pp. 234–241. Springer, Cham (2015). https://doi.org/10.1007/978-3-319-24574-4_28
23. Shelhamer, E., Long, J., Darrell, T.: Fully convolutional networks for semantic segmentation. IEEE Trans. Pattern Anal. Mach. Intell. **39**, 640–651 (2017)
24. Shin, H., Lee, J.K., Kim, J., Kim, J.: Continual learning with deep generative replay. In: NIPS (2017)
25. Wu, C., Herranz, L., Liu, X., Wang, Y., van de Weijer, J., Raducanu, B.: Memory replay GANs: learning to generate new categories without forgetting. In: NeurIPS (2018)
26. Yang, G., et al.: Uncertainty-aware contrastive distillation for incremental semantic segmentation. IEEE Trans. Pattern Anal. Mach. Intell. (2022)
27. Zhao, S., Li, B., Reed, C., Xu, P., Keutzer, K.: Multi-source domain adaptation in the deep learning era: a systematic survey. arXiv abs/2002.12169 (2020)
28. Zhao, S., et al.: A review of single-source deep unsupervised visual domain adaptation. IEEE Trans. Neural Netw. Learn. Syst. **33**, 473–493 (2022)

# A Data-Efficient Deep Learning Framework for Segmentation and Classification of Histopathology Images

Pranav Singh[1][✉] and Jacopo Cirrone[2]

[1] Department of Computer Science Tandon School of Engineering, New York University New York, New York 11202, USA
ps4364@nyu.edu

[2] Center for Data Science New York University and Colton Center for Autoimmunity NYU Grossman School of Medicine New York, New York 10011, USA
cirrone@courant.nyu.edu

**Abstract.** The current study of cell architecture of inflammation in histopathology images commonly performed for diagnosis and research purposes excludes a lot of information available on the biopsy slide. In autoimmune diseases, major outstanding research questions remain regarding which cell types participate in inflammation at the tissue level, and how they interact with each other. While these questions can be partially answered using traditional methods, artificial intelligence approaches for segmentation and classification provide a much more efficient method to understand the architecture of inflammation in autoimmune disease, holding a great promise for novel insights. In this paper, we empirically develop deep learning approaches that uses dermatomyositis biopsies of human tissue to detect and identify inflammatory cells. Our approach improves classification performance by 26% and segmentation performance by 5%. We also propose a novel post-processing autoencoder architecture that improves segmentation performance by an additional 3%.

**Keywords:** Deep learning · Computer vision · Medical image analysis · Autoimmune diseases · Histopathology images

## 1 Introduction

Our understanding of diseases and their classification has improved multifold with developments in medical science. However, our understanding of autoimmune diseases continues to be incomplete, missing vital information. No mechanism is in place to systematically collect data about the prevalence and incidence of autoimmune diseases (as it exists for infectious diseases and cancers). This deficiency is because we lack a comprehensive and universally acceptable list of

L. Karlinsky et al. (Eds.): ECCV 2022 Workshops, LNCS 13803, pp. 385–405, 2023.
https://doi.org/10.1007/978-3-031-25066-8_21

autoimmune diseases. The most cited study in the epidemiology of autoimmune diseases [12], estimates that autoimmune diseases, combined, affects about 3% of the US population or 9.9 million US citizens. A number comparable to the 13.6 million (US citizens) affected by cancer, which affects almost 4% of the population. Nevertheless, cancer has been widely studied in medical science and at the intersection of medical science and artificial intelligence. The study of autoimmune diseases is not only critical because they affect a sizeable portion of the population, but because we don't have a complete understating of their etiology and treatment. Other compelling reasons to study these diseases is their increasing prevalence [4,13] and their further increase with the recent COVID-19 pandemic [6,8].

Major outstanding research questions exist for autoimmune diseases regarding the presence of different cell types and their role in inflammation at the tissue level. In addition to studying preexisting patterns for different cell interactions, the identification of new cell occurrence and interaction patterns will help us better understand the diseases. While these patterns and interactions can be partially answered using traditional methods, artificial intelligence approaches for segmentation and classification tasks provide a much more efficient and quicker way to understand these architectures of inflammation in autoimmune disease and hold great promise for novel insights. The application of artificial intelligence for medical image analysis has also seen a rapid increase, propelled by the increase in performance and efficiency of such architectures. However, even with these developments mentioned previously, the application of artificial intelligence in autoimmune biopsy analysis has not received the same attention as others. Firstly, autoimmune diseases are highly underrepresented because of significantly fewer data available for aforementioned reasons. Secondly, even within the few existing studies on the application of artificial intelligence for autoimmune disease analysis, dermatomyositis has received significantly less attention. Most research has focused on psoriasis, rheumatoid arthritis, lupus, scleroderma, vitiligo, inflammatory bowel diseases, thyroid eye sisease, multiple sclerosis sisease, and alopecia [22]. We also observe that most of these approaches are based on older techniques and architectures that do not have open-source code to allow more researchers to expand their investigations into this area.

To help bridge this gap, we aim to draw more attention in this paper to autoimmune diseases, specifically to dermatomyositis. With this paper: (i) we improve upon the existing method for classification and segmentation of autoimmune disease images [23] with 26% improvement for classification and 5% for segmentation, (ii) we propose an Autoencoder for Post-Processing (APP) and using image reconstruction loss improve segmentation performance by a further 3% as compared to (i), and (iii) based on these experimentation, we make recommendations for future researchers/practitioners to improve the performance of architectures and understanding of autoimmune diseases. All the aforementioned contributions have been implemented in PyTorch and are publicly available at https://github.com/pranavsinghps1/DEDL.

# 2 Background

## 2.1 Application of Artificial Intelligence for Autoimmune Diseases

Researchers in [22] conducted an in-depth study of the application of computer vision and deep learning in autoimmune disease diagnosis. Based on their study, we found a common trend among the datasets within autoimmune medical imaging - most of the datasets used are extremely small, with a median size of 126 samples. This also correlated with the findings of [20], wherein they also mentioned the median dataset size available for autoimmune diseases ranged between 99–540 samples. Medical imaging datasets tend to be smaller than natural image datasets, and even within medical datasets, autoimmune datasets are comparably smaller than the datasets of diseases with similar prevalence. For example, the prevalence of cancer is around 4% compared to the prevalence of autoimmune diseases at around 3%. However, cancer datasets of sizes ranging from a few thousand samples are readily available as opposed to that of autoimmune, where the median dataset size is between 99–540 samples. Another difference is that most of the autoimmune disease datasets are institutionally restricted.

In addition, there are few studies on the application of artificial intelligence in autoimmune diseases. In 2020, [20] conducted a systematic review of the literature and relevant papers at the intersection of artificial intelligence and autoimmune diseases with the following exclusion criteria: studies not written in English, no actual human patient data included, publication prior to 2001, studies that were not peer-reviewed, non-autoimmune disease comorbidity research and review papers. Only 169 studies met the criteria for inclusion. On further analyzing these 169 studies, only a small proportion of studies 7.7% (13/169) combined different data types. Cross-validation, combined with independent testing set for a more robust model evaluation, occurred only in 8.3% (14/169) of papers.

In 2022, [22] studied the usage of computer vision in autoimmune diseases, its limitations and the opportunities that technology offers for future research. Out of the more than 100 classified autoimmune diseases, research work has mostly focused ten diseases (psoriasis, rheumatoid arthritis, lupus, scleroderma, vitiligo, inflammatory bowel diseases, thyroid eye disease, multiple sclerosis disease and alopecia). [23] is the first, to the best of our knowledge, to apply and study artificial intelligence for medical image analysis of dermatomyositis - an autoimmune disease that has not been studied in much detail. We used the same dataset as used by [23] and propose innovative techniques and architectures to improve performance.

## 2.2 Segmentation

Medical image segmentation, i.e., automated delineation of anatomical structures and other regions of interest (ROIs) paradigms, is an important step in computer-aided diagnosis; for example it is used to extract key quantitative measurements and localize a diseased area from the rest of the slide image. Good segmentation requires the object to see fine picture and intricate details at the same time. The same encoder-decoder architectures have been favored that often use different techniques (e.g., feature pyramid network, dilated networks and atrous networks) to help increase the receptive field of the architecture. When it comes to medical image segmentation, UNet [19] has been the most cited and widely used network architecture. It uses an encoder and decoder architecture with skip connections to learn the segmentation masks. An updated version of UNet was introduced by [27] called UNet++ that is essentially a nested version of UNet. The encoder is a feature extractor that down-samples the input, the decoder then consecutively up-samples to learn the segmentation mask for an input image. We added channel level attention with the help of squeeze and excitation blocks as proposed in [10]. A squeeze and excitation is basically a building block that can be easily incorporated with CNN architecture. Comprised of a bypass that emerges after normal convolution, this is where the squeeze operation is performed. Squeezing basically means compressing each two-dimensional feature map until it becomes a real number. This is followed by an excitation operation that generates a weight for each feature channel to model relevance. Applying these weights to each original feature channel, the importance of different channels can be learned.

### 2.3   Classification

Classification is another important task for medical image analysis. CNNs have been the de facto standard for classification task. The adoption of Transformers for vision tasks from language models have been immensely beneficial. This gain in performance could be the result of Transformer's global receptive field as opposed to limited receptive field on CNNs. Although CNNs have inductive priors that make them more suited for vision tasks, Transformers learn them over the training period. Recently, there has been increased interest in combining the abilities of Transformers and CNNs. Certain CNNs have been trained the way Transformers have been trained as in [16], similarly the introduction of CNN type convolution has been incorporated in Transformers [15]. Within medical image analysis, Transformers with their global receptive field could be extremely beneficial as this can help us learn features that CNNs with their limited receptive field could have missed. We start with the gold-standard of CNNs - ResNet-50 and the ResNet family [9] of architectures. A ResNet-18 model can be scaled up to make ResNet-200. In most cases, this yields a better performance. But this scaling is very random as some models are scaled depthwise and some are scaled width-wise. This problem was addressed with [21], in which the concept of compound scaling was used. They proposed a family of models with balanced scaling that also improved overall model performance. In ResNet-like architectures, batch normalization is often applied at the residual branch. This stabilizes the gradient and enables training of significantly deeper networks. Yet computing batch-level statistics is an expensive operation. To address these issues a set of NFnets (Normalizer-Free Networks) were published [2]. Instead of using batch normalization, NFnets use other techniques to create batch-normalization like effect such as modified residual branches and convolutions with scaled weight standardization and adaptive gradient clipping. Transformers have a global receptive field, as opposed to CNN. [5] introduced Vision Transformers (ViT) adapted from NLP, and since then they have improved upon many benchmarks for vision tasks. Hence, we study the effect of using Transformers as classifiers on the autoimmune dataset. Within Transformers, we start by looking at vision transformers introduced in [5]. Unlike CNNs, Transformer first split each image into patches by a patch module. These patches are then used as "tokens". We then examine the Swin transformer family [15]. Swin transformers use a hierarchical approach with shifted windows. The shifted windows scheme brings greater efficiency by limiting self-attention computation to non-overlapping windows and allows cross-window connection. It also first splits an input RGB image into non-overlapping image patches by a patch splitting module, like ViT [5].

# 3  Methodology

## 3.1  Segmentation

For segmentation, our contribution is twofold. Firstly, we improved upon existing approaches by reducing blank image tiling, adding channel level attention in decoder, by using squeeze and excitation blocks to better map the importance of different channels and by using a pixel normalized cross-entropy loss. This improved performance by 5% over existing approach on the same dataset by [23], secondly, we introduce a novel post-processing Autoencoder which further improved the performance of the segmentation architecture as compared to the first step by 3%.

We used the same dermatomyositis dataset as used in [23]. We started our experimentation by using the same metrics, image tiling and splits as their work, to make comparison easier. Once we surpassed their benchmark, we changed a few things as described in Sect. 5.1.

For segmentation benchmark as mentioned already, we based our study around one of the most widely-cited networks for biomedical image segmentation - UNet and a nested version of UNet - UNet++. These architectures are widely used not only in general segmentation tasks but also in biomedical segmentation. UNet has also been used in previous autoimmune segmentation tasks [3,23].

**Intuition for the Autoencoder Post-processing Architecture.** Traditionally for training segmentation architectures, the output from the decoder is compared against ground truth with a loss function, in our case, we used a cross-entropy loss function. Increasing the model size should help the model with more extracted features from the input but based on our experimentation, this is not the case Table 2. So to provide the segmentation architecture with meaningful insight to improve learning, we provide additional feedback on "How easy is it to reconstruct the ground truth mask with the predicted mask?". To do so, we used simple encoder-decoder architecture with mean-squared error loss. This post-processing autoencoder takes the output of the segmentation architecture as input and then computes this reconstruction loss, which the leading segmentation architecture then uses to improve learning. We expanded more on the experiments and study the results of adding these autoencoders in Sect. 5.1.

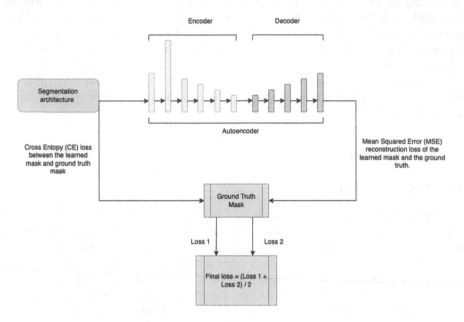

**Fig. 1.** Autoencoder Post-Processing (APP): we introduced the autoencoder labeled with yellow and red colour, after getting the segmentation mask from the segmentation architecture(in our case UNet and UNet++). (Color figure online)

Figure 1 shows the post-processing autoencoder architecture in conjunction with segmentation architectures during training. The feedback from the autoencoder is only used during training and then the trained model is saved for inference. This incurs minimal computational costs during training (as shown in Sect. 5.1); the saved weights are no bigger in size than the saved weights without the autoencoder post-processing architecture and with no change in inference time.

## 3.2   Classification

We used their phenotype as markers to classify the different cell types. Wherein, the presence of a specific phenotype directly correlated to the presence of a cell type - T cell, B Cell, TFH-217, TFH-like cells and other cells. These cells' presence or absence aided us in diagnosing dermatomyositis. As previously mentioned, to develop a better understanding of autoimmune diseases, the study of nonconforming cells to the mapped phenotype-cell classification is extremely important. We classified them as 'others'. This would be potentially helpful in diagnosis and understanding of novel cell patterns present in biopsies, which in-turn could help us understand autoimmune diseases to a better extent by categorizing novel phenotype-cell relationships. We used the auto-fluorescence images of size 352 by 469 and RGB. Since there could be multiple cells present per sample, this would be a multi-label classification. To address this class imbalance, we use the Focal loss [14] function that penalizes the dominant and the underrepresented class in a dataset. We used it to label distribution normalized class weights instead of vanilla cross-entropy, which fails to address this class imbalance. We also used a sixfold approach to train and reported our results on the test set to address any biases that might occur during training.

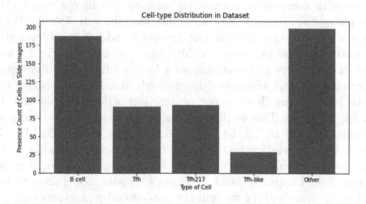

**Fig. 2.** Distribution of different cell types within the dataset. Observing that there is an imbalance in the distribution, we classified cells on the basis of the presence or absence of certain cell phenotype.

## 4 Experimental Details

### 4.1 Dataset

We use the same dataset as used in [23]. This dataset contains 198 TIFF image samples, each containing slides of different protein-stained images - DAPI, CXCR3, CD19, CXCR5, PD1, CD4, CD27 and autofluorescence. Binary thresholds were set for each channel (1- DAPI, 2- CXCR3, 3- CD19, 4- CXCR5, 5- PD1, 6- CD4, 7- CD27, 8- Autofluorescence) to show presence/absence of each representative phenotype. These phenotypes were then classified into B cells and T cells using channels 2–7. The autofluorescence slides are an overlap of all the channels used for classifying cell types.

We use the DAPI stained image for semantic segmentation and the autofluorescence slide images for classification. This approach is a shift from previous work where researchers used DAPI channel images for both tasks as well as using UNet for segmentation and classification.

## 4.2 Segmentation

We use qubvel's implementation of Unet and Unet++ segmentation architectures https://github.com/qubvel/segmentation_models.pytorch. To start, we first convert the TIFF image file into NumPy and then into a PIL Image. We then apply Random Rotation, Random vertical and horizontal flip, and channel normalization before finally converting to tensors. We use the same splits for training, testing and validation as [23] to keep our results comparable. We use cross-entropy loss with component normalized weights, Adam optimizer with 1e-05 decay and cosine learning rate with minimum learning rate 3.4e-04. We also use [10] squeeze and excitation units in the decoder to add channel level attention.

For the Autoencoder processing architecture (as shown in Fig. 1), we use six layers, out of which five are downsampling layers followed by five upsampling layers. The encoder part contains 6 layers with first layer upscaling the input from 480 to 15360. From there on consecutive layers downsample from 15360 to 256, 128, 64, 32 to 16. This is then fed to the decoder which then systematically up scales it from 16, 32, 64, 128, 256 and 480. We use GELU (Gaussian Error Linear Unit) activation in all the layers. This post-processing architecture is only used during training as an added feedback mechanism; this helps the segmentation model improve learning. Times for saved model after training and inference remain same with or without the auto-encoder post-processing. We use Adam optimizer with a constant learning rate of 1e-3. For loss, we use the mean squared error (MSE) loss function.

## 4.3 Classification

To address the class imbalance problem, we use focal loss [14], which is essentially an oscillating cross entropy and this modulation fluctuates with easy and complex examples in the dataset.

We perform an 80/20 split for training to testing and use sixfold cross-validation. We then report the average F1-score across all the folds and different initialization in Sect. 5. We use Adam optimizer without weight decay with a cosine learning schedule and a learning rate of 1e-6 for 16 iterations. For a more generalized result and better optimization, we use Stochastic Weight Averaging [11]. We use timm's implementation [24] of CNN and Transformers for benchmarking.

We conducted all our experiments on a single NVIDIA RTX-8000 GPU with 45 GB of use-able video memory, 64 GB RAM and two cores of CPU. We also used early-stopping with a patience of 5 epochs for both our classification and segmentation training pipeline. We compute average results over 5 runs with different seed values.

# 5    Results and Discussion

## 5.1    Segmentation

**Improvement over Existing Work.** Since the images are 1408 by 1876, we tiled them in $256^2$ and then used blank padding at the edges to make them fit in size $256^2$.

For segmentation, we started by using metrics, architecture and parameters as suggested in [23]. We begin with ImageNet weights as they are readily available for various backbones instead of the brain-MRI segmentation weights used in [23]. By changing the existing learning rate schedule from step to cosine and by using normalized weights of blank pixels and pixels that aren't blank (i.e., hold some information) in the cross-entropy loss, adding channel level attention with squeeze and excitation block to better feature map level channel relationships. We observed that the overall performance improved from 0.933 overall accuracy for [23] to 0.9834 for Unetplusplus with ResNet34 backbone - an improvement of 5%.

We observed that: (i) by using image tiling of size $256^2$, margin padding is primarily empty, and some of the tiles do not have any part of the cell stained on them. This might give false perception of performance as the model achieves higher metric performance from not learning anything. To mitigate this, we tiled the images into $480^2$. This is depicted in Fig. 3; (ii) because accuracy is not the most appropriate metric for segmentation performance, we instead used IoU or Jaccrad Index as our metric. IoU is a better representation if the model has learned meaningful features.

**Fine-Tuning.** For further refining, we use one of the most cited and used architectures for biomedical segmentation, UNet [19], and a nested version of UNet called the UNetplusplus [27]. These two architectures contain an encoder-decoder structure wherein the encoder is a feature extractor, and the decoder learns the segmentation mask on the extracted features. For the choice of the encoder, we used the ResNet family [9] and the newer, more efficient family [21]. These architectures do not have an attention mechanism built-in. So we artificially included channel level attention mechanism using Squeeze and Excitation blocks [10]. In addition to these changes we propose adding a post-processing autoencoder architecture as described in Fig. 1.

**Adding Autoencoder.** As mentioned previously, we also experimented by adding autoencoder after the tuned segmentation architecture. We call this Autoencoder Post Processing or **APP**. We start by using the GELU (Gaussian Error Linear Unit) as our activation function for the autoencoder architecture (AE) as described in Sect. 3.1 and Fig. 1. We report the IoU score on test set for Unet and Unet++ in Table 2. The autoencoder adds additional feedback to the segmentation architecture by computing reconstruction loss between the segmentation architecture's output and ground truth. This adds some extra time during

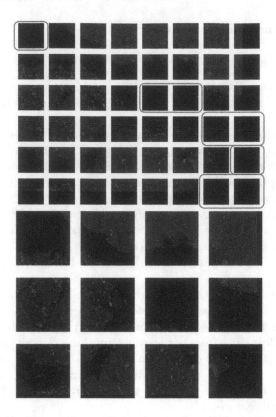

**Fig. 3.** Changing the image tiling reduces the number of blank tiles which are mostly concentrated towards the edges and some of the intermediate tiles. Top images shows $256^2$ tiling while bottom one shows $480^2$ tiling. As we can see for the top image tilled in 256 by 256 image size, the right bottom tiles are almost all empty. Including some of the tiles in the center portion. While for the bottom image tilled in 480 by 480 image size, no tiles are empty. The tiles highlighted in red boxes are empty.

training but the space of saved model and inference time remain constant when compared to networks without the autoencoder post-processing architecture.

**Ablation Study.** In the original implementation, we used GELU activation function with a constant learning rate and Adam optimizer in the autoencoder. In this section, we experimented with the hyperparameters of only the autoencoders to determine its effect on performance.

1. Computational cost: For a training set with 1452 images and validation set with 720 images of size $480^2$ over 50 epochs, we report the following testing time with and without using the GELU autoencoder post processing for UNet with ResNet backbones. We observed that there is an average increase of 1.875% over the ResNet family of encoders Table 1.

**Table 1.** This table shows training time for UNet with and without APP for ResNet family of encoders for 50 epochs. We observed that for smaller encoders like ResNet-18 and 34 the training time increase is greater as opposed to larger encoders.

| Encoder | Without APP | With APP |
|---|---|---|
| Resnet-18 | 1 h 24 m 44 s | 1 h 27 m 40 s |
| Resnet-34 | 1 h 27 m 41 s | 1 h 30 m 39 s |
| Resnet-50 | 1 h 45 m 45 s | 1 h 46 m 24 s |
| Resnet-101 | 1 h 45 m 24 s | 1 h 46 m 25 s |

2. Using ReLU (Rectified Linear Unit) instead of GELU (Gaussian Error Linear Units) for activation function. Table 2 shows the difference in performance for when we changed the activation of the autoencoders layers from ReLU to GELU. And for the same we observed that GELU performed better than ReLU activated autoencoder by around 3% for UNet and UNet++. We observed in most cases for UNet and UNet++, the addition of the autoencoder post processing is highly beneficial. With the addition of autoencoder, GELU activation performed much better than ReLU activation. We further expand upon these results in Appendix 7.1.

**Table 2.** This table shows the IoU score on test set for UNet and UNet++ architectures with and without using cosine learning rate for the autoencoder. Except for ResNet-18 and 101 with UNet++, autoencoder always provide an improvement without cosine learning rate.

| Encoder | UNet | | | UNet++ | | |
|---|---|---|---|---|---|---|
| | Without AE | With ReLU AE | With GELU AE | Without AE | With ReLU AE | With GELU AE |
| Resnet 18 | 0.4347 | 0.4608 | **0.4788** | **0.5274** | 0.4177 | 0.4707 |
| Resnet 34 | 0.4774 | 0.4467 | **0.4983** | 0.3745 | 0.4535 | **0.4678** |
| Resnet 50 | 0.3798 | **0.4187** | 0.3827 | 0.4236 | **0.4685** | 0.4422 |
| Resnet 101 | 0.3718 | 0.4074 | **0.4402** | 0.4311 | 0.4265 | **0.4467** |

3. Using Adam optimizer with cosine learning schedule. For this we kept rest of the setup same. We only added a cosine learning schedule to the autoencoder architecture and ReLU activated layers. From Table 3 we observed that except for two cases in UNet++, autoencoder with constant learning rate perform much better than the one with Adman optimizer and cosine learning rate.

**Table 3.** This table shows the IoU score on test set for UNet and UNet++ architectures. We compared results without, with autoencoder for both ReLU and GELU activations. Except for ResNet-18 with UNet++, autoencoder always provide an improvement. Within autoencoders we see that using GELU activation is much better than ReLU activation.

| Encoder | UNet | | UNet++ | |
|---|---|---|---|---|
| | Without Cosine LR | With Cosine LR | Without Cosine LR | With Cosine LR |
| Resnet 18 | **0.4608** | 0.4106 | 0.4177 | **0.4717** |
| Resnet 34 | **0.4467** | 0.3665 | **0.4535** | 0.4345 |
| Resnet 50 | **0.4187** | 0.3965 | **0.4685** | 0.4268 |
| Resnet 101 | **0.4074** | 0.3846 | 0.4265 | **0.4518** |

## 5.2   Classification

For classification our objective is to classify the different phenotypes in a given image and based on the presence of different cells. Since multiple labels could be assigned to the same image, this is a multi-class classification problem. From Fig. 2, we observed that the classes are highly imbalanced. [23] also reported an F1 score of 0.63 for the classification on this dataset. We improved upon their score with an F1 score of 0.891. This improvement could be attributed to the following, (i) we use focal loss [14] instead of cross entropy as the previous work. Focal loss applies a modulating term to the cross-entropy loss in order to focus learning on hard misclassified examples. It is a dynamically scaled cross-entropy loss, where the scaling factor decays to zero as confidence in the correct class increases. Intuitively, this scaling factor can automatically downweight the contribution of easy examples during training and rapidly focus the model on hard examples. (ii) We used newer and more efficient architectures. We benchmarked pure CNNs, Transformers and newer generation of CNNs trained with Transformer-like techniques Convnext [16]. As mentioned in Sect. 3, in this section we present our results and discuss the effect of different architectures.

**Effect of Architecture.** Recently, the emergence of Transformers for vision tasks has taken the field of computer vision by storm. They have been able to improve upon many benchmarks set by CNNs. They are particularly of interest in the medical imaging because of their global fidelity as opposed to the centered fidelity of CNNs.

More recently, some of the Transformer techniques have also been applied to train CNNs. This has given rise to new architectures like the ConvNeXts family of pure ConvNet models [16]. We studied the effect of using different architectures on performance and hence have found the most suitable architecture for the multiclass classification task at hand. We resized each image to be $384^2$.

Amongst CNNs and Transformers, the peak performance is closely matched with Swin Transformer Base (Patch 4 Window 12) achieving a new state of the art performance for the autoimmune dataset with an F1 Score of 0.891, an improvement of 26.1% over previous work by [23]; compared to nfnet-f3 that provides peak performance for CNNs with an F1 score of 0.8898 Tables 4 and 5.

## 6 Conclusion

Our framework can be adapted to other tissues and diseases datasets. It provides an efficient approach for clinicians to identify and detect cells within histopathology images in order to better comprehend the architecture of inflammation (i.e., which cell types are involved in inflammation at the tissue level, and how cells interact with one other).

Based on our experimentation, we observe that for segmentation of biopsies affected by dermatomyositis, it is better to use Imagenet initialization with a normalized cross-entropy loss. Further performance can be increased by using our proposed autoencoder post-processing architecture (APP). APP gains 3% consistently over architectures without any post-processing segmentation architecture. This addition comes at minimal extra training cost and at no extra space and time to develop inference models.

For classification, using stochastic weight averaging improves generalization and class normalized weights with focal loss and helps to counter the class imbalance problem. In comparing architectures, the performance is relatively similar for CNNs and transformers, but transformers perform slightly better than CNNs. These changes helped us register an improved performance of 8% in segmentation and of 26% in classification performance on our dermatomyositis dataset.

**Acknowledgment.** We would like to thank NYU HPC team for assisting us with our computational needs. We would also like to thank Prof. Elena Sizikova (Moore Sloan Faculty Fellow, Center for Data Science (CDS), New York University (NYU)) for her valuable feedback.

## A  Appendix

### A.1  Expansion of Results

In Tables 6 and 7 we show complete results with mean and standard deviation. These are an expansion of Table 2 in Sect. 5.1 of the main paper. Tables were compressed to save space and only focus on the main results. To provide a complete picture, we added extended results in this section.

**Table 4.** We report the F1 score for with ImageNet initialization for latest set of CNNs and Transformers. We observed the usual trend of increasing ImageNet performance with increasing size of the model is not followed. Overall the best performance is achieved by nfnet-f3 for CNNs. Overall, out of all the tested architectures Swin Transformer Base with Patch 4 and Window 12 performed the best

| Model | Test F1-score |
| --- | --- |
| Resnet-18 | $0.8635 \pm 0.0087$ |
| Resnet-34 | $0.82765 \pm 0.0073$ |
| Resnet-50 | $0.8499 \pm 0.007$ |
| Resnet-101 | $0.871 \pm 0.009$ |
| Efficient-net B0 | $0.8372 \pm 0.0007$ |
| Efficient-net B1 | $0.8346 \pm 0.0026$ |
| Efficient-net B2 | $0.828 \pm 0.00074$ |
| Efficient-net B3 | $0.8369 \pm 0.0094$ |
| Efficient-net B4 | $0.8418 \pm 0.0009$ |
| Efficient-net B5 | $0.8463 \pm 0.00036$ |
| Efficient-net B6 | $0.8263 \pm 0.00147$ |
| Efficient-net B7 | $0.8129 \pm 0.001$ |
| nfnet-f0 | $0.82035 \pm 0.007$ |
| nfnet-f1 | $0.834 \pm 0.007$ |
| nfnet-f2 | $0.8652 \pm 0.0089$ |
| nfnet-f3 | $0.8898 \pm 0.0011$ |
| nfnet-f4 | $0.8848 \pm 0.0109$ |
| nfnet-f5 | $0.8161 \pm 0.0074$ |
| nfnet-f6 | $0.8378 \pm 0.007$ |
| ConvNext-tiny | $0.81355 \pm 0.0032$ |
| ConvNext-small | $0.84795 \pm 0.00246$ |
| ConvNext-base | $0.80675 \pm 0.002$ |
| ConvNext-large | $0.8452 \pm 0.000545$ |
| Swin Transformer large (Patch 4 Window 12) | $0.8839 \pm 0.001$ |
| Swin Transformer Base (Patch 4 Window 12) | $\mathbf{0.891 \pm 0.0007}$ |
| Vit-Base/16 | $0.8426 \pm 0.007$ |
| Vit-Base/32 | $0.8507 \pm 0.0079$ |
| Vit-large/16 | $0.80495 \pm 0.0077$ |
| Vit-large/32 | $0.845 \pm 0.0077$ |

**Table 5.** We compare the best results provided by our algorithm (in bold) as compared to previous benchmark on the same dataset.

| Model | Test F1-score |
| --- | --- |
| Swin transformer large (Patch 4 Window 12) | $\mathbf{0.891 \pm 0.0007}$ |
| nfnet-f3 | $\mathbf{0.8898 \pm 0.0109}$ |
| Vanburen et all [23] | 0.63 |

**Table 6.** This table shows the IoU score on the test set for UNet. We compared results without and with autoencoder for both ReLU and GELU activations for UNet Architecture. These results are averaged over five runs with different seed values. We observed that in all cases addition of APP improved performance. GELU activated APP seems out perform the ReLU activated APP in all cases except for ResNet-50.

| Encoder | UNet | | |
|---------|------|---|---|
| | Without AE | With ReLU AE | With GELU AE |
| ResNet 18 | $0.4347 \pm 0.0006$ | $0.4608 \pm 0.0001$ | $\mathbf{0.4788 \pm 0.0004}$ |
| ResNet 34 | $0.4774 \pm 0.0004$ | $0.4467 \pm 0.0012$ | $\mathbf{0.4983 \pm 0.0008}$ |
| ResNet 50 | $0.3798 \pm 0.00072$ | $\mathbf{0.4187 \pm 0.0006}$ | $0.3827 \pm 0.0003$ |
| ResNet 101 | $0.3718 \pm 0.0001$ | $0.4074 \pm 0.0012$ | $\mathbf{0.4402 \pm 0.00018}$ |

**Table 7.** This table shows the IoU score on the test set for UNet++. These results are averaged over five runs with different seed values. We compare results without and with autoencoder for both ReLU and GELU activations for UNet++ Architecture. We observed that in most cases, APP improves performance except for UNet++ with Resnet-18, where APP segmentation techniques lag by around 5%. However, as a counter for ResNet-34 APP-based segmentation techniques are almost 10% better than UNet++ without APP.

| Encoder | UNet++ | | |
|---------|--------|---|---|
| | Without AE | With ReLU AE | With GELU AE |
| ResNet 18 | $\mathbf{0.5274 \pm 0.0004}$ | $0.4177 \pm 0.0005$ | $0.4707 \pm 0.00067$ |
| ResNet 34 | $0.3745 \pm 0.0006$ | $0.4535 \pm 0.0008$ | $\mathbf{0.4678 \pm 0.0004}$ |
| ResNet 50 | $0.4236 \pm 0.0004$ | $\mathbf{0.4685 \pm 0.0002}$ | $0.4422 \pm 0.0007$ |
| ResNet 101 | $0.4311 \pm 0.0003$ | $0.4265 \pm 0.0002$ | $\mathbf{0.4467 \pm 0.0003}$ |

## A.2   Autoencoder with Efficientnet Encoder for Segmentation

In Tables 8 and 9 we compared the time taken to train and the performance of the respective trained architecture for segmentation using EfficientNet encoders. We observed that with the addition of autoencoder post-processing, training time increased by an average of 3 m 7.3 s over 50 epochs (averaged over the entire efficientnet family). This is an increase of 2.93% in training time over the eight encoders. In other words, an average increase of 0.36% increase in time per encoder over 50 epochs.

Performance wise architecture with autoencoder post-processing consistently outperformed segmentation architectures without them by 2.75%.

Similarly, we compared computational and performance for UNet++ with and without the autoencoder post-processing in Tables 10 and 11 respectively.

**Table 8.** In this table we report the running time averaged over 5 runs with different seeds, for efficient-net encoder family with UNet.The variation is almost negligible(< 6$s$).

| Encoder | Without APP | With APP |
|---------|-------------|----------|
| B0 | 1 h 26 m 27 s | 1h 29 m 04 s |
| B1 | 1 h 31 m 16 s | 1h 33 m 42 s |
| B2 | 1 h 32 m 12 s | 1h 34 m 27 s |
| B3 | 1h 38 m | 1h 40 m 33 s |
| B4 | 1 h 44 m 20 s | 1 h 50 m 02 s |
| B5 | 1 h 55 m 46 s | 1 h 58 m 40 s |
| B6 | 2 h 06 m 55 s | 2 h 10 m 08 s |
| B7 | 2 h 16 m 40 s | 2 h 19 m 59 s |

**Table 9.** In this table we report the IoU averaged over 5 runs with different seeds, for efficient-net encoder family with UNet architecture.

| Encoder | Without APP | With APP |
|---------|-------------|----------|
| B0 | $0.3785 \pm 0.00061$ | **$0.4282 \pm 0.0008$** |
| B1 | $0.3301 \pm 0.0002$ | **$0.4237 \pm 0.0006$** |
| B2 | $0.2235 \pm 0.0007$ | **$0.3735 \pm 0.0009$** |
| B3 | **$0.3982 \pm 0.0007$** | $0.2411 \pm 0.0004$ |
| B4 | $0.3826 \pm 0.0004$ | **$0.3829 \pm 0.0006$** |
| B5 | $0.4056 \pm 0.0008$ | **$0.4336 \pm 0.0008$** |
| B6 | $0.4001 \pm 0.0001$ | **$0.4311 \pm 0.0006$** |
| B7 | $0.3631 \pm 0.0002$ | **$0.3937 \pm 0.0004$** |

In this case, we observed that the gain in performance with autoencoder post-processing is 5% averaged over the efficientnet family of encoders. This also corresponds to a 3 m 7 s increase in training time which is an increase of 2.6%.

**Table 10.** In this table we report the running time averaged over 5 runs with different seeds, for efficient-net encoder family with UNet++.

| Encoder | Without APP | With APP |
|---------|-------------|----------|
| B0 | 1 h 32 m 50 s | 1 h 35 m 31 s |
| B1 | 1 h 37 m 40 s | 1 h 40 m 51 s |
| B2 | 1 h 38 m 30 s | 1 h 40 m 41 s |
| B3 | 1 h 46 m 30 s | 1 h 49 m 34 s |
| B4 | 1 h 54 m 01 s | 1 h 57 m 41 s |
| B5 | 2 h 07 m 54 s | 2 h 11 m 39 s |
| B6 | 2 h 20 m 23 s | 2 h 23 m 41 s |
| B7 | 2 h 29 m 01 s | 2 h 32 m 04 s |

**Table 11.** In this table we report the IoU averaged over 5 runs with different seeds, for efficient-net encoder family with UNet++ architecture.

| Encoder | Without APP | With APP |
|---------|-------------|----------|
| B0 | 0.3584 ± 0.0002 | **0.3751 ± 0.0007** |
| B1 | 0.4260 ± 0.0005 | **0.4269 ± 0.0003** |
| B2 | 0.3778 ± 0.0007 | **0.3942 ± 0.0009** |
| B3 | 0.3928 ± 0.0006 | **0.4174 ± 0.0003** |
| B4 | 0.4138 ± 0.0003 | **0.4273 ± 0.0002** |
| B5 | **0.3884 ± 0.0001** | 0.3875 ± 0.0005 |
| B6 | 0.4090 ± 0.0008 | **0.4214 ± 0.0007** |
| B7 | 0.3784 ± 0.0009 | **0.4002 ± 0.0005** |

## A.3  Metrics Description

For measuring segmentation performance, we use IoU or intersection over union metric. It helps us understand how similar sample sets are.

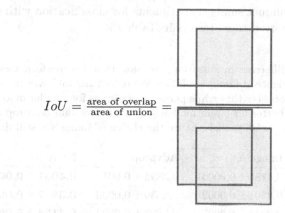

$$IoU = \frac{\text{area of overlap}}{\text{area of union}} =$$

Here the comparison is made between the output mask by segmentation pipeline against the ground truth mask.

For measuring classification performance, we use the F1 score.
Computed as $F1 = \frac{2*\text{Precision}*\text{Recall}}{\text{Precision}+\text{Recall}} = \frac{2*\text{TP}}{2*\text{TP}+\text{FP}+\text{FN}}$

## A.4  Effect of Different Weights

ImageNet initialization has been the defacto norm for most transfer learning tasks. Although in some cases, as in [1] it was observed that noisy student weights performed better than ImageNet initialization. To study the effect in our case, we used advprop and noisy student initialization. ImageNet weights for initialization work for medical data not because of feature reuse but because of better weight scaling and faster convergence [18]. Noisy student training [26] extends the idea of self-training and distillation with the use of equal-or-larger student models, and noise such as dropout, stochastic depth, and data augmentation via RandAugment is added to the student during learning so that the student generalizes better than the teacher. First, an EfficientNet model is trained on labelled images and is used as a teacher to generate pseudo labels for 300M unlabeled images. We then train a larger EfficientNet as a student model on the combination of labelled and pseudo-labelled images. This helps reduce the error rate, increases robustness and improves performance over the existing state-of-the-art on ImageNet.

(ii) AdvProp training, which banks on Adversarial examples, which are commonly viewed as a threat to ConvNets. In [25] they present an opposite perspective: adversarial examples can be used to improve image recognition models. They treat adversarial examples as additional examples to prevent overfitting. It performs better when the models are bigger. This improves upon performance for various ImageNet and its' subset benchmarks.

Since initially all these were developed for the EfficientNet family of the encoders, we used them for benchmarking. We present their results in Table 12.

Similarly, we conduct similar experiments for classification with different initialization. We reported these results in Table 13.

**Table 12.** Using different initialization, we saw that the performance of different encoders of the EfficientNet family on UNet. We report the IoU over the test set in the following table. We observe that while performance gains for smaller models, ImageNet initialisation works better for larger models. Also, the fact that advprop and noisy are not readily available for all models, hence the choice of ImageNet still dominates.

| Encoder | ImageNet | Advprop | Nosiy |
|---|---|---|---|
| B0 | $0.3785 \pm 0.00061$ | $0.3895 \pm 0.001$ | $\mathbf{0.4081 \pm 0.0006}$ |
| B1 | $0.3301 \pm 0.0002$ | $0.2330 \pm 0.0006$ | $\mathbf{0.3817 \pm 0.0008}$ |
| B2 | $0.2235 \pm 0.0007$ | $0.3823 \pm 0.0004$ | $\mathbf{0.4154 \pm 0.0001}$ |
| B3 | $\mathbf{0.3982 \pm 0.0007}$ | $0.3722 \pm 0.0004$ | $0.3509 \pm 0.0007$ |
| B4 | $0.3819 \pm 0.0004$ | $\mathbf{0.4366 \pm 0.0006}$ | $0.4333 \pm 0.0003$ |
| B5 | $\mathbf{0.4056 \pm 0.0008}$ | $0.3910 \pm 0.0001$ | $0.3806 \pm 0.0004$ |
| B6 | $\mathbf{0.4001 \pm 0.0001}$ | $0.3807 \pm 0.0005$ | $0.3941 \pm 0.0006$ |
| B7 | $\mathbf{0.3631 \pm 0.0002}$ | $0.3543 \pm 0.0009$ | $0.3308 \pm 0.0006$ |

**Table 13.** We report the F1 score for different initializations for the EfficientNet family of encoders. We reported the average of 6-fold runs on the test set with five different seed values. We observed that 0.8463 is the peak with ImageNet, 0.843 with advprop and 0.8457 with noisy student initialization.

| Encoder | ImageNet | Advprop | Noisy student |
|---------|----------|---------|---------------|
| B0 | $0.8372 \pm 0.0007$ | **$0.8416 \pm 0.0008$** | $0.839 \pm 0.0008$ |
| B1 | $0.8346 \pm 0.0026$ | $0.8367 \pm 0.0007$ | **$0.8448 \pm 0.0014$** |
| B2 | $0.828 \pm 0.00074$ | **$0.843 \pm 0.00138$** | $0.8430 \pm 0.0012$ |
| B3 | $0.8369 \pm 0.0094$ | $0.84138 \pm 0.00135$ | **$0.8457 \pm 0.007$** |
| B4 | **$0.8418 \pm 0.0009$** | $0.82135 \pm 0.00058$ | $0.8377 \pm 0.00032$ |
| B5 | **$0.8463 \pm 0.00036$** | $0.8133 \pm 0.002$ | $0.8326 \pm 0.00066$ |
| B6 | **$0.8263 \pm 0.00147$** | $0.8237 \pm 0.004$ | $0.8233 \pm 0.0065$ |
| B7 | $0.8129 \pm 0.001$ | $0.8132 \pm 0.004$ | **$0.8257 \pm 0.0008$** |

As we can see for segmentation, ImageNet initialization performed better in most cases. Similarly, in classification, it not only performed better in most cases but also provided the best overall result—these inferences, combined with the fact that advprop and noisy student requires additional computational resources. Hence we decide to stick with ImageNet initialization.

### A.5   Expansion on Experimental Details

**Segmentation.** We used PyTorch lightning's [7] seed everything functionality to seed all the generator values uniformly. For setting the seed value, we randomly generated a set of numbers in the range of 1 and 1000. We did not perform an extensive search of space to optimise performance with seed value as suggested in [17]. We used seed values 26, 77, 334, 517 and 994. For augmentation, we used conversion to PIL Image to apply random rotation (degrees=3), random vertical and horizontal flip, then conversion to tensor and finally channel normalisation. We could have used a resize function to reshape the 1408 by 1876 Whole Slide Images (WSI), but we instead tilled them in 480 square tile images. We then split them into a batch size of 16 before finally passing through the segmentation architecture (UNet/UNet++). We used channel attention only decoder, with ImageNet initialisation and a decoder depth of 3 (256, 128, 64).

We used cross-entropy loss with dark/light pixel normalization, Adam optimizer with LR set to 3.6e-04 and weight decay of 1e-05. We used a cosine scheduling rate with a minimum value set to 3.4e-04.

**APP Segmentation.** When using APP we used GELU activation by default with adam optimizer and lr set to 1e-3.

**Classification.** For Classification, we used the same seed values with PyTorch lightning's [7] seed everything functionality, as described for segmentation above. For augmentation, we resized the images to 384 square images, followed by randomly applying colour jitter (0.2, 0.2, 0.2) or random perspective (distortion scale=0.2) with probability 0.3, colour jittering (0.2, 0.2, 0.2) or random affine (degrees=10) with probability 0.3, random vertical flip and random horizontal flip with probability 0.3 and finally channel normalization.

We used Stochastic weigh averaging with adam optimizer. We used a cosine learning rate starting at 1e-3 and a minimum set to 1e-6. We used focal loss with normalized class weight as our loss function. We used 6-fold validation with each fold of 20 epochs and batch size of 16. We used same parameters for both CNN and Transformers.

# References

1. Agarwal, V., Jhalani, H., Singh, P., Dixit, R.: Classification of melanoma using efficient nets with multiple ensembles and metadata. In: Tiwari, R., Mishra, A., Yadav, N., Pavone, M. (eds.) Proceedings of International Conference on Computational Intelligence. AIS, pp. 101–111. Springer, Singapore (2022). https://doi.org/10.1007/978-981-16-3802-2_8

2. Brock, A., De, S., Smith, S.L., Simonyan, K.: High-performance large-scale image recognition without normalization. In: International Conference on Machine Learning, pp. 1059–1071. PMLR (2021)

3. Dash, M., Londhe, N.D., Ghosh, S., Semwal, A., Sonawane, R.S.: Pslsnet: Automated psoriasis skin lesion segmentation using modified u-net-based fully convolutional network. Biomed. Signal Process. Contr. **52** 226–237 (2019). https://doi.org/10.1016/j.bspc.2019.04.002, https://www.sciencedirect.com/science/article/pii/S1746809419300990

4. Dinse, G.E., et al.: Increasing prevalence of antinuclear antibodies in the united states. Arthritis Rheumatol. **72**(6), 1026–1035 (2020)

5. Dosovitskiy, A., et al.: An image is worth 16x16 words: Transformers for image recognition at scale. arXiv preprint arXiv:2010.11929 (2020)

6. Ehrenfeld, M., et al.: Covid-19 and autoimmunity. Autoimmun. Rev. **19**(8), 102597 (2020)

7. Falcon, W., et al.: Pytorch lightning. GitHub. Note: https://github.com/PyTorchLightning/pytorch-lightning vol. 3(6) (2019)

8. Galeotti, C., Bayry, J.: Autoimmune and inflammatory diseases following covid-19. Nat. Rev. Rheumatol. **16**(8), 413–414 (2020)

9. He, K., Zhang, X., Ren, S., Sun, J.: Deep residual learningfor image recognition. In: ComputerScience (2015)

10. Hu, J., Shen, L., Sun, G.: Squeeze-and-excitation networks. In: Proceedings of the IEEE conference on computer vision and pattern recognition, pp. 7132–7141 (2018)

11. Izmailov, P., Podoprikhin, D., Garipov, T., Vetrov, D., Wilson, A.G.: Averaging weights leads to wider optima and better generalization. arXiv preprint arXiv:1803.05407 (2018)

12. Jacobson, D.L., Gange, S.J., Rose, N.R., Graham, N.M.: Epidemiology and estimated population burden of selected autoimmune diseases in the united states. Clin. Immunol. Immunopathol. **84**(3), 223–243 (1997)

13. Lerner, A., Jeremias, P., Matthias, T.: The world incidence and prevalence of autoimmune diseases is increasing. Int. J. Celiac Disease **3**(4), 151–155 (2015). 10.12691/ijcd-3-4-8, http://pubs.sciepub.com/ijcd/3/4/8

14. Lin, T.Y., Goyal, P., Girshick, R., He, K., Dollár, P.: Focal loss for dense object detection. In: Proceedings of the IEEE International Conference on Computer Vision, pp. 2980–2988 (2017)

15. Liu, Z., et al.: Swin transformer: Hierarchical vision transformer using shifted windows. In: Proceedings of the IEEE/CVF International Conference on Computer Vision, pp. 10012–10022 (2021)

16. Liu, Z., Mao, H., Wu, C.Y., Feichtenhofer, C., Darrell, T., Xie, S.: A convnet for the 2020s. arXiv preprint arXiv:2201.03545 (2022)

17. Picard, D.: Torch.manual_seed(3407) is all you need: On the influence of random seeds in deep learning architectures for computer vision. CoRR abs/2109.08203 (2021). https://arxiv.org/abs/2109.08203

18. Raghu, M., Zhang, C., Kleinberg, J., Bengio, S.: Transfusion: Understanding transfer learning for medical imaging. In: Advances in Neural Information Processing Systems, vol. 32 (2019)

19. Ronneberger, O., Fischer, P., Brox, T.: U-Net: convolutional networks for biomedical image segmentation. In: Navab, N., Hornegger, J., Wells, W.M., Frangi, A.F. (eds.) MICCAI 2015. LNCS, vol. 9351, pp. 234–241. Springer, Cham (2015). https://doi.org/10.1007/978-3-319-24574-4_28

20. Stafford, I., Kellermann, M., Mossotto, E., Beattie, R., MacArthur, B., Ennis, S.: A systematic review of the applications of artificial intelligence and machine learning in autoimmune diseases. NPJ Digital Med. **3**(1), 1–11 (2020)

21. Tan, M., Le, Q.: Efficientnet: Rethinking model scaling for convolutional neural networks. In: International Conference on Machine Learning, pp. 6105–6114. PMLR (2019)

22. Tsakalidou, V.N., Mitsou, P., Papakostas, G.A.: Computer vision in autoimmune diseases diagnosis—current status and perspectives. In: Smys, S., Tavares, J.M.R.S., Balas, V.E. (eds.) Computational Vision and Bio-Inspired Computing. AISC, vol. 1420, pp. 571–586. Springer, Singapore (2022). https://doi.org/10.1007/978-981-16-9573-5_41

23. Buren, V., et al.: Artificial intelligence and deep learning to map immune cell types in inflamed human tissue. Journal of Immunological Methods **505**, 113233 (2022). https://doi.org/10.1016/j.jim.2022.113233, https://www.sciencedirect.com/science/article/pii/S0022175922000205

24. Wightman, R.: Pytorch image models. https://github.com/rwightman/pytorch-image-models (2019). https://doi.org/10.5281/zenodo.4414861

25. Xie, C., Tan, M., Gong, B., Wang, J., Yuille, A.L., Le, Q.V.: Adversarial examples improve image recognition. In: Proceedings of the IEEE/CVF Conference on Computer Vision and Pattern Recognition, pp. 819–828 (2020)

26. Xie, Q., Luong, M.T., Hovy, E., Le, Q.V.: Self-training with noisy student improves imagenet classification. In: Proceedings of the IEEE/CVF Conference on Computer Vision and Pattern Recognition, pp. 10687–10698 (2020)

27. Zhou, Z., Rahman Siddiquee, M.M., Tajbakhsh, N., Liang, J.: UNet++: a nested u-net architecture for medical image segmentation. In: Stoyanov, D., et al. (eds.) DLMIA/ML-CDS -2018. LNCS, vol. 11045, pp. 3–11. Springer, Cham (2018). https://doi.org/10.1007/978-3-030-00889-5_1

# Bounded Future MS-TCN++ for Surgical Gesture Recognition

Adam Goldbraikh[1]([✉]) [ID], Netanell Avisdris[2] [ID], Carla M. Pugh[3] [ID],
and Shlomi Laufer[4] [ID]

[1] Applied Mathematics Department, Technion – Israel Institute of Technology,
3200003 Haifa, Israel
sgoadam@campus.technion.ac.il
[2] School of Computer Science and Engineering, The Hebrew University of Jerusalem,
Jerusalem, Israel
netana03@cs.huji.ac.il
[3] School of Medicine Stanford, Stanford University, Stanford, CA, USA
cpugh@stanford.edu
[4] Faculty of Industrial Engineering and Management,
Technion – Israel Institute of Technology, 3200003 Haifa, Israel
laufer@technion.ac.il

**Abstract.** In recent times there is a growing development of video-based applications for surgical purposes. Part of these applications can work offline after the end of the procedure, other applications must react immediately. However, there are cases where the response should be done during the procedure but some delay is acceptable. In the literature, the online-offline performance gap is known. Our goal in this study was to learn the performance-delay trade-off and design an MS-TCN++-based algorithm that can utilize this trade-off. To this aim, we used our open surgery simulation data-set containing 96 videos of 24 participants that perform a suturing task on a variable tissue simulator. In this study, we used video data captured from the side view. The Networks were trained to identify the performed surgical gestures. The naive approach is to reduce the MS-TCN++ depth, as a result, the receptive field is reduced, and also the number of required future frames is also reduced. We showed that this method is sub-optimal, mainly in the small delay cases. The second method was to limit the accessible future in each temporal convolution. This way, we have flexibility in the network design and as a result, we achieve significantly better performance than in the naive approach.

**Keywords:** Surgical simulation · Surgical gesture recognition · Online algorithms

## 1 Introduction

Surgical data science is an emerging scientific area [21,22]. It explores new ways to capture, organize and analyze data with the goal of improving the quality of

L. Karlinsky et al. (Eds.): ECCV 2022 Workshops, LNCS 13803, pp. 406–421, 2023.
https://doi.org/10.1007/978-3-031-25066-8_22

interventional healthcare. With the increased presence of video in the operating room, there is a growing interest in using computer vision and artificial intelligence (AI) to improve the quality, safety, and efficiency of the modern operating room.

A common approach for workflow analysis is to use a two-stage system. The first stage is a feature extractor such as I3D [4] or ResNet50 [5,29]. The next stage usually includes temporal filtering. The temporal filtering may include recurrent neural networks such as LSTM [6,33,35], and temporal convolutional networks (TCN) such as MS-TCN++ [19] or transformers [25,34].

Automatic workflow analysis of surgical videos has many potential applications. It may assist in an automatic surgical video summarizing [2,20], progress monitoring [26], and the prediction of remaining surgery duration [32]. The development of robotic scrub nurses also depends on the automatic analysis of surgical video data [14,30]. In addition, video data is used for the assessment of surgical skills [3,9,10] and identifying errors [15,23,24].

Traditionally systems are divided into causal and acausal. However, not all applications require this dichotomic strategy. For example, the prediction of the remaining surgery duration may tolerate some delay if this delay ensures a more stable and accurate estimation. On the other hand, a robotic scrub nurse will require real-time information. In general, any acausal system may be transformed into a causal system if a sufficient delay is allowed. Where in the extreme case, the delay would be the entire video. This study aims to find the optimal system, given a constraint on the amount of delay allowed.

Many studies use Multi-Stage Temporal Convolutional Networks (MS-TCN) for workflow analysis [19]. It has both causal and acausal implementations. The network's number of layers and structure defines the size of its receptive field. In the causal case, the receptive field depends on past data. On the other hand, in the acausal case, half of the receptive field depends on future data and half on the past. Assume a fixed amount of delay $T$ is allowed. A naive approach would be to use an acausal network with a receptive field $2 \cdot T$. However, this limits the number of layers in the network and may provide sub-optimal results. In this study, we develop and assess a network with an asymmetric receptive field. Thus we may develop a network with all the required layers while holding the constraint on the delay time $T$. This network will be called Bounded Future MS-TCN++ (BF-MS-TCN++). We will compare this method to the naive approach that reduces the receptive field's size by changing the network's depth. The naive approach will be coined Reduced Receptive Field MS-TCN++ (RR-MS-TCN++). We perform gesture recognition using video from an open surgery simulator to evaluate our method.

The main contribution of our work is the development of an MS-TCN++ with a bounded future window, which makes it possible to improve the causal network performance by delaying the return of output at a predetermined time. In addition, we evaluated a causal and acausal video-based MS-TCN++ for gesture recognition on the open surgery suturing simulation data.

## 2   Related Work

Lea *et al.* [16] was the first to study TCN's ability to identify and segment human actions. Using TCN, they segmented several non-surgical data sets such as 50 Salads, GETA, MERL Shopping, and Georgia Tech Egocentric Activities. They implemented causal and acausal TCNs and compared their performance on the MERL data set. The acausal solution provided superior results. They also out-perform a previous study that uses an LSTM as a causal system and a Bidirectional LSTM as an acausal system. In TeCNO [5] causal Multi-Stage Temporal Convolutional Networks (MS-TCN) were used for surgical phase recognition. Two data-sets of laparoscopic cholecystectomy were used for evaluation. The MS-TCN outperformed various state-of-the-art LSTM approaches. In another study [29], this work was expanded to a multi-task network and was used for step and phase recognition of gastric bypass procedures.

In one study, the performance of an acausal TCN was assessed. The analysis included both non-surgical action segmentation datasets as well as a dataset of a simulator for robotic surgery [18]. Zhang *et al.* [36] used a Convolutional Network to extract local temporal information and an MS-TCN to capture global temporal information. They used acausal implantation to perform Sleeve Gastrectomy surgical workflow recognition. Not all studies use a separate network for capturing temporal information. In Funke *et al.* [8] a 3D convolutional neural networks was used. In this study, they used the sliding window approach to evaluate different look-ahead times.

The use of TCN is not limited to video segmentation. It has been studied in the context of speech analysis as well [27]. In this context, the relationship between delay and accuracy has been assessed [28].

## 3   Methods

### 3.1   Dataset

Eleven medical students, one resident, and thirteen attending surgeons participated in the study. Their task was to place three interrupted instrument-tied sutures on two opposing pieces of the material. Various materials can simulate different types of tissues; for example, in this study, we used tissue paper to simulate a friable tissue and a rubber balloon to simulate an artery. The participants performed the task on each material twice. Thus, the data set contains 100 procedures, each approximately 2–6 min long. One surgeon was left-handed and was excluded from this study. Thus, this study includes a total of 96 procedures. Video data were captured in 30 frames per second, using two cameras, providing top and side views. In addition to video, kinematic data were collected using electromagnetic motion sensors (Ascension, trakSTAR). For this study, we only use the side-view camera. We perform a gesture recognition task, identifying the most subtle surgical activities within the surgical activity recognition task family. Six surgical gestures were defined: G0 - nonspecific motion, G1 - Needle passing, G2 - Pull the suture, G3 - Instrumental tie, G4 - Lay the knot G5 - Cut the suture. The video data were labeled using Behavioral Observation Research Interactive Software (BORIS) [7].

## 3.2 Architecture

MS-TCN++ [19] is a *temporal convolutional network* (TCN) designed for video data activity recognition. The input for this network is a vector of features extracted from the raw video using a CNN, such as I3D [4]. The video length is not predetermined. Let us assume that the video is given in 30 frames per second and contains $T$ frames, namely, T is a parameter of a specific video. In the following sections, we will describe the different components of the MS-TCN++ and the modifications made for the BF-MS-TCN++. It should be noted that the naive approach, RR-MS-TCN++ has the same structure as the acausal MS-TCN++ and was coined with a unique name to emphasize the limitation on the receptive field size.

The architecture of MS-TCN++ is structured from two main modules: the prediction generator and the refinement. For the sake of simplicity, we will first describe the refinement module structure and then the prediction generator.

**Refinement Module:** As shown in Fig. 1, the refinement module includes several refinement stages, where the output of each stage is the input of the next. The refinement stage is a pure TCN that uses *dilated residual layers* (DRL). This allows the module to handle varying input lengths. To match the input dimensions of the stage with the number of feature maps, the input of the refinement stage passes through a $1 \times 1$ convolutional layer. Then these features are fed into several DRLs where the dilation factor is doubled in each layer. The dilation factor determines the distance between kernel elements, such that a dilation of 1 means that the kernel is dense. Formally, the dilation factor is defined as $\delta(\ell) = 2^{\ell-1} : \ell \in \{1, 2, \ldots L\}$, where $\ell$ is the layer number and $L$ is the total number of layers in the stage. In MS-TCN++, the DRL is constructed from an acausal dilated temporal convolutional layer (DTCL), with a kernel size of 3, followed by ReLU activation and then a $1 \times 1$ convolutional layer. The input of the block is then added to the result by a standard residual connection, which is the layer's output. To get the prediction probabilities, the output of the last DRL passes through a prediction head which includes a $1 \times 1$ convolution, to adjust the number of channels to the number of classes, and is activated by a softmax.

**Prediction Generation Module:** The prediction generator consists of only one prediction generation stage. The general structure of this stage is similar to the refinement stage; however, instead of a DRLs, it has a *dual dilated residual layers* (DDRLs). Let's consider layer $\ell \in \{1, 2, \ldots, L\}$. The input of the DDRL is entered into two DTCLs, one with a dilation factor of $\delta_1(\ell) = 2^{\ell-1}$ and the other with a dilation factor of $\delta_2(\ell) = 2^{L-\ell}$. Then, the outputs of the two DTCLs are merged by concatenation of the feature in the channel dimension, followed by a

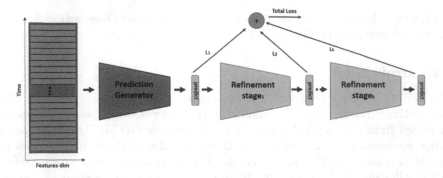

**Fig. 1.** General structure of MS-TCN++. The input is a vector of features (blue). It composes of multiple stages, where each stage predicts the frame segmentation. The first stage is the prediction generator (red) and the other are refinement stages (green), which can be any number ($k \geq 0$) of them. The loss is computed over all stages' predictions. (Color figure online)

1D convolutional layer to reduce the number of channels back to the constant number of feature maps. This output passes through a ReLU activation and an additional 1D convolutional layer before the residual connection. For a formal definition of MS-TCN++ stages and modules, see [19].

**Future Window Size Analysis:** As MS-TCN++ is a temporal convolutional network with different dilatation among the layers, analyzing the temporal dependence is not trivial. In order to determine the number of future frames involved in the output calculation, careful mathematical analysis is required. Calculating the number of future frames required is equivalent to the output delay of our system.

In the naive approach, RR-MS-TCN++, the number of layers of the network governs the receptive field and the future window. The BF-MS-TCN++ is based on the limitation of the accessible future frames in each temporal convolution; hence the name Bounded Future. This section aims to analyze both methods and calculate desired future window.

The input and the output of DRLs and DDRLs (i.e. (D)DRL) have the same dimensions of $N_f \times T$, where $N_f$ represents the number of feature maps in the vector encoding the frame and $T$ is the number of frames in the video. We assume that for every (D)DRL, the vector in the $t$ index represents features that correspond to the frame number $t$. However, in the acausal case, the features of time $t$ can be influenced by a future time point of the previous layer output. In MS-TCN++, (D)DRL has symmetrical DTCLs with a kernel size of 3. The layer's input $\ell$ is padded by $2 \times \delta(\ell)$ zero vectors to ensure the output dimensions are equal to the input dimensions. A symmetric convolution is created by padding the input with $\delta(\ell)$ number of zeros vectors before and after it. The result is that half of each layer's receptive field represents the past and the other half represents the future. To obtain a causal solution, all $2 \times \delta(\ell)$ zero vectors are added before the input. As a result, the entire receptive field is based on the past time. This method has been used in [5,16].

Let $S = \{PG, R, Total\}$ be a set of symbols, where $PG$ represents relation to the prediction generator, $R$ to the refinement stage, and $Total$ to the entire network. Let $L_s : s \in S$ be the number of (D)DRLs in some stage or in the entire network. The number of refinement stages in the network is $N_R$. Note that we assume that all refinement stages are identical, and that $L_{Total} = L_{PG} + N_R \cdot L_R$. Let the ordered set $\mathcal{L}_{total} = (1, 2, \ldots, L_{Total})$, where $\ell \in \mathcal{L}_{total}$ represents the index of $\ell^{th}$ (D)DRL in the order it appears in the network. Given integer $x$, $[x]$ denotes the set of integers satisfying $[x] = \{1, \cdots, x\}$.

**Definition 1.** *Let $DTCL$ $\phi$ with dilation factor of $\delta(\phi)$. The* Direct Future Window *of a $\phi$ is $m \in [\delta(\phi)]$ if and only if the number of the padding vectors after the layers input is $m$ and number of padding vectors before the vector is $2m - \delta(\phi)$. The function $DFW(\phi) = m$ gets a DTCL and returns it's direct future window.*

The definition of a Direct Future Window is shown in Fig. 2.

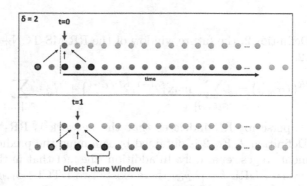

**Fig. 2.** Illustration of Direct Future Window of a dilated temporal convolutional layer (DTCL) with $\delta = 2$, for first (upper part) and second (lower part) timeframes. Blue dots denote input features, yellow dots denote padding needed for DTCL, and orange dots denotes the output of DTCL. (Color figure online)

**Reduced Receptive Field MS-TCN++ Case:** In the DRLs, in the refinement stages, the direct future window is equal to the direct future window of it's DTCL. Formally, let $\ell \in [L_R]$ a DRL, that contains a DTCL $\phi$. The direct future window of this layer is given by $\mathcal{DFW}_R(\ell) = DFW(\phi)$, where the subscript $R$ indicates an association with a refinement stage. However, each DDRL contains two different DTCLs $\phi_1$ and $\phi_2$, as described in Sect. 3.2. Consider a DDRL $\ell \in [L_{PG}]$. The direct future window of this layers is given by Eq. 1.

$$\mathcal{DFW}_{PG}(\ell) = \max\{DFW(\phi_1), DFW(\phi_2)\} \tag{1}$$

**Definition 2.** *The* Future Window *of layer* $\ell \in \mathcal{L}_{total}$ *defined as follows:*

$$FW(\ell) = \sum_{i \in [\ell]} \mathcal{DFW}(i)$$

Figure 3 shows how direct future windows are aggregated to the future window.

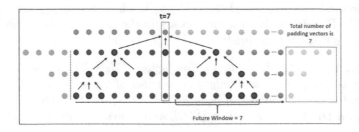

**Fig. 3.** Illustration of Future Window in a refinement stage with three dilated residual layers (DRLs).

Based on Definition 2, the future window of the RR-MS-TCN++ network is defined in Eq. 2.

$$FW^{RR}(L_{Total}) = \sum_{\ell \in [L_{PG}]} \max\{2^{\ell-1}, 2^{L_{PG}-\ell}\} + N_R \cdot \sum_{\ell \in [L_R]} 2^{\ell-1} \qquad (2)$$

Note that superscript RR indicates that the network is RR-MS-TCN++. According to Definition 2, Eq. 2 is obtained by summing the prediction generator and refinement stages separately. In addition, the fact that in the prediction generator, for every DDRL $\ell \in [L_{PG}]$ there exists two DTCLs $\phi_1, \phi_2$ that satisfied $DFW(\phi_1) = 2^{\ell-1}$, and $DFW(\phi_2) = 2^{L_{PG}-\ell}$, yields that the direct future window of the DDRL is the maximum between these terms, as defined in Eq. 3. We take the maximum since the direct future window is determined by the furthest feature in the layer's input that participates in the calculation.

$$\mathcal{DFW}_{PG}(\ell) = \max\{2^{\ell-1}, 2^{L_{PG}-\ell}\} \qquad (3)$$

and for some DRL $\ell \in [L_R]$ in the refinement $\mathcal{DFW}(\ell) = 2^{\ell-1}$.

**Bounded Future MS-TCN++ Case:** Let $w_{max}$ be a bounding parameter that bounds the size of the direct future window. We determine that the direct future window of every DRL $\ell \in [L_R]$, in the refinement stage of the BF-MS-TCN++, is given by Eq. 4.

$$\mathcal{DFW}_R^{BF}(\ell) = \min\{w_{max}, \delta(\ell)\} = \min\{w_{max}, 2^{\ell-1}\} \qquad (4)$$

The superscript $BF$ indicates that the network is BF-MS-TCN++ and the subscript $R$ indicates that this belongs to the refinement stage, where the replacement of $R$ with $PG$ indicates association with the prediction generator.

Figure 4 illustrates how the convolution's symmetry is broken in the case of $\delta(\ell) > w_{max}$.

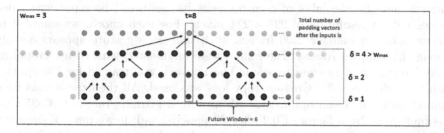

**Fig. 4.** Illustration of a Future Window and asymmetry in padding in a refinement stage of Bounded Future MS-TCN++ with three dilated residual layers (DRLs) and $w_{max} = 3$. In the last layer, since the dilation factor $\delta$ is larger than $w_{max}$, the number of padding vectors (yellow) before and after the sequence is different.

The future window of the refinement stage is given by Eq. 5.

$$FW_R^{BF}(L_R) = \sum_{\ell \in [L_R]} \mathcal{DFW}_R^{BF}(\ell) = \sum_{\ell \in [L_R]} \min\{w_{max}, 2^{\ell-1}\} \qquad (5)$$

For the DDRL $\ell \in [L_{PG}]$ of the prediction generators, the direct future window is given by Eq. 6

$$\mathcal{DFW}_{PG}^{BF}(\ell) = \max\{min\{w_{max}, 2^{\ell-1}\}, min\{w_{max}, 2^{L_{PG}-\ell}\}\} \qquad (6)$$

Hence the future window of the prediction generator for a causal network with delay is given by Eq. 7.

$$FW_{PG}^{BF}(L_{PG}) \sum_{\ell \in [L_{PG}]} \mathcal{DFW}_{PG}^{BF}(\ell) \qquad (7)$$

$$= \sum_{\ell \in [L_{PG}]} \max\{min\{w_{max}, \delta_1(\ell)\}, min\{w_{max}, \delta_2(\ell)\}\}$$

This leads to Eq. 8 which presents the future window for the entire network.

$$FW^{BF}(L_{Total}) = \sum_{\ell \in [L_{PG}]} \max\{min\{w_{max}, 2^{\ell-1}\}, min\{w_{max}, 2^{L_{PG}-\ell}\}\} \qquad (8)$$

$$+ N_R \cdot \sum_{\ell \in [L_R]} \min\{w_{max}, 2^{\ell-1}\}$$

Note that a network with $w_{max} = 0$ is a causal network, which may have a fully online implementation.

### 3.3 Feature Extractor Implementation Details

As a first step, we trained a (2D) EfficientNetV2 medium [31] in a frame-wise manner; namely, the label of each frame is its gesture. The input was video frames with a resolution of $224 \times 224$ pixels. For each epoch, we sampled in a class-balanced manner 2400 frames, such that each gesture appears equally among the sampled frames. The frames were augmented with corner cropping, scale jittering, and random rotation. The network was trained for 100 epochs, with a batch size of 32. Cross-entropy loss was used. All the experiments were trained using an Adam optimizer, with an initial learning rate of 0.00025 that was multiplied by a factor of 0.2 after 50 epochs, and decay rates of $\beta_1 = 0.9$ and $\beta_2 = 0.999$. The code of this part is based on the code provided by Funke *et al.* [8]. After the individual training of each split, the one before the last linear layer was extracted and used as a feature map for the MS-TCN++.

## 4  Experimental Setup and Results

### 4.1  Experimental Setup

In order to evaluate the effect of the delay on the performance of online algorithms we compare two methods, the naive **RR-MS-TCN++** and our **BF-MS-TCN++**. To this aim, we performed a hyperparameter search. For both networks, the number of refinement stages and the number of (D)DRLs inside the stages affect the total receptive field and hence the future window. The uniqueness of the BF-MS-TCN++ is that the future window can be limited by the bounding parameter $W_{max}$ as well, regardless of the values of the other parameters. In our search, we forced the number of DRLs in the prediction generator to be equal to the number of DRLs in the refinement stages, namely $L_{PG} = L_R = L$.

To this end, for the RR-MS-TCN++ network, the number of DRLs $L$ included in the search was in the range of $\{2, 3, 4, 5, 6, 8, 10\}$. The number of refinement stages $N_R$ was in the range of $\{0, 1, 2, 3\}$. For the BF-MS-TCN++, the search included two grids. In the first grid, the values of $w_{max}$ were in the range of $\{0, 1, 2, 3, 6, 7, 8, 10, 12, 13, 14, 15, 16, 17, 20\}$, where 0 represents a online algorithm. The number of DRLs $L$ was in the range of $\{6, 8, 10\}$, and the number of refinement stages $N_R$ was in the of $\{0, 1, 2, 3\}$. In the second grid, the values of $w_{max}$ were in the range of $\{1, 3, 7, 10, 12, 15, 17\}$. The number of DRLs $L$ was in the range of $\{2, 3, 4, 5\}$, and the number of refinement stages $N_R$ was in the of $\{0, 1, 2, 3\}$. In total, we performed 320 experiments. The rest of the hyperparameters remained constant, where the learning rate was 0.001, dropout probability was 0.5, the number of feature maps was 128, batch size of 2 videos, the number of epochs was 40, and the loss function was the standard MS-TCN++ loss with hyperparameters $\tau^2 = 16$ and $\lambda = 1$. All experiments were trained with an Adam optimizer with the default parameters. Training and evaluation were done using a DGX cluster with 8 Nvidia A100 GPUs.

## 4.2 Evaluation Method

We evaluated the models using 5-fold cross-validation. All videos of a specific participant were in the same group (leave-n-users-out approach). The videos assigned to the fold served as the test set of that fold. The remaining participants' videos were divided into train and validation sets.

The validation set for fold $i \in [5]$ consists of 3 participants from fold $(i + 1) \bmod 5$. Namely, for each fold, 12 videos were used as a validation set (3 participants $\times$ 4 repetitions). With this method, we create unique validation sets for each fold. The stopping point during training was determined based on the best F1@50 score on the validation set. The metrics were calculated for each video separately, so the reported results for each metric are the mean and the standard deviation across all 96 videos when they were in the test set.

## 4.3 Evaluation Metrics

We divide the evaluation metrics into two types: segmentation metrics and frame-wise metrics. The segmentation metrics contain F1@k where $k \in \{10, 25, 50\}$ [16], and edit distance [17]. While the frame-wise metrics included Macro-F1 [12] and Accuracy. We calculated each metric for each video and then averaged them across all videos.

**F1@k:** The intersection over union (IoU) between the predicted segments and the ground truth was calculated for each segment. If there is a ground truth segment with IoU greater than $k$, that ground truth segment is marked as true positive and is not available for future use. Otherwise, the predicted segment has been defined as a false positive. Based on these calculations, the F1 score was determined.

**Segmental Edit Distance:** The segmental edit distance is based on the Levenshtein distance, where the role of a single character is taken by segments of the activity. The segmental edit distance was calculated for all gesture segments in the video and normalized by dividing it by the maximum between the ground truth and prediction lengths.

**Frame-Wise Metrics:** We calculated the Accuracy and the multi-class F1-Macro scores as used in [11].

## 4.4 Experimental Studies

We performed four studies: (1) Baseline estimation; (2) Performance comparison; (3) Network hyperparameter importance; and (4) Competitive analysis.

(1) First, we try to estimate the best fully casual and acausal networks, which will serve as a baseline.

(2) In the *Performance comparison* study we compare directly between the performance of *Reduced Receptive Field MS-TCN++* and our *Bounded Future MS-TCN++*, with respect to Future Window size. We defined 12 Future Window intervals: $[0, 0.001, 0.125, 0.25, 0.5, 1, 2, 4, 8, 16, 32, 64, \infty]$ seconds. For each method separately, we divide the networks into these intervals considering the Future Window. A representative value of each interval was selected based on the highest results within each interval. Notice that a few intervals may be left empty.

(3) In the third study, we analyzed the *marginal performance* and *marginal importance* of the investigated hyperparameters on the F1@50 using the fANOVA method [13]. The receptive field is determined by the hyperparameters of our network structure. To understand the advantages and weaknesses of each of our methods, it's essential to assess the importance and trends of the hyperparameters.

(4) In the *Competitive analysis* study, we have two aims. First, it is to try to reveal what is the required delay to approach the best offline performance. Next, we aim to determine which method is more advantageous at which delay values. To this end, we need to define two new metrics: Global and local competitive ratio. In this paper, the *Competitive-Ratio*, inspired by its definition in theoretical analysis of online algorithms [1], will be the ratio between the performance of the causal (with delay) algorithm and the best acausal network. In addition, for each interval, we define the Local Competitive Ratio as the ratio between the best BF-MS-TCN++ and RR-MS-TCN++ networks in that interval.

## 4.5   Results

Table 1 lists the results of the baseline estimation study. The feature extractor performed relatively well in a frame-wise manner with an accuracy of 82.66 and F1-Macro of 79.46. Both the causal and acausal networks improve the accuracy and F1-Macro scores, however, the acausal network has a significant effect in both frame-wise metrics while in the causal case only the accuracy has been improved significantly. The Causal case has the best results in all metrics. Figure 5 shows a performance comparison study, where the trend is similar for all metrics, where BF-MS-TCN++ outperforms RR-MS-TCN++, especially for small delay values.

In the feature importance study (Fig. 6), $w_{max}$ was found to have negligible importance. Other hyperparameters perform better when their values are increased. Lastly, the Competitive analysis results are illustrated in Fig. 7. In the global analysis (left) plot, it is evident that BF-MS-TCN++ has 80% of the performance of the best offline algorithm after only one second and 90% after $2\frac{1}{3}$ s. In the local analysis (right) plot, it is seen clearly that for future windows smaller than one second the BF-MS-TCN++ is significantly preferred over the RR-MS-TCN++.

**Table 1.** The Feature extractor EfficientNet v2, causal and acausal MS-TCN++ results on a gesture recognition task. Bold denotes best results for metric.

| | $F_1$-Macro | Accuracy | Edit distance | F1@10 | F1@25 | F1@50 |
|---|---|---|---|---|---|---|
| EfficientNet v2 | $79.46 \pm 8.10$ | $82.66 \pm 6.03$ | – | – | – | – |
| Causal MS-TCN++ | $80.42 \pm 8.67$ | $85.04 \pm 5.77$ | $64.69 \pm 12.36$ | $74.30 \pm 10.85$ | $72.34 \pm 12.04$ | $64.96 \pm 14.07$ |
| Acausal MS-TCN++ | $\mathbf{83.85 \pm 8.90}$ | $\mathbf{86.94 \pm 6.50}$ | $\mathbf{84.65 \pm 9.25}$ | $\mathbf{88.66 \pm 7.79}$ | $\mathbf{87.13 \pm 9.03}$ | $\mathbf{80.01 \pm 13.21}$ |

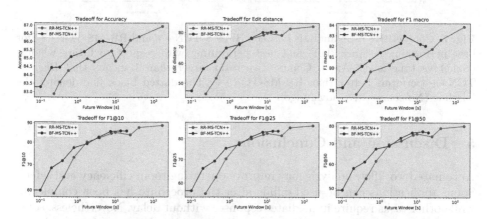

**Fig. 5.** performance-delay trade-off. Comparison of best BF-MS-TCN++ and RR-MS-TCN++ networks, in respect to future window size. The plots show the performance in terms of Accuracy, Edit distance, F1-Macro, and $F1@k$, $k \in \{10, 25, 50\}$.

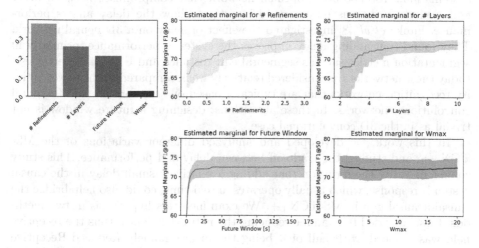

**Fig. 6.** fANOVA analysis of BF-MS-TCN++. Marginal importance (leftmost graph) and estimated marginal F1@50 on architecture hyperparameters: number of refinement stages, number of (D)DRLs in every stage, the Future Window of the network, and the value of $w_{max}$.

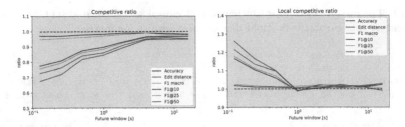

**Fig. 7.** Competitive ratio analysis between BF-MS-TCN++ and RR-MS-TCN++, for best performing network (left) and best performing with respect to future window (right) vs time [s] (X-Axis). Competitive ratio larger than 1 means that BF-MS-TCN++ performs better than RR-MS-TCN++. The dotted black line denotes the baseline (No competitive advantage).

## 5    Discussion and Conclusions

Automated workflow analysis may improve operating room efficiency and safety. Some applications can be used offline after the procedure has been completed, while other tasks require immediate responses without delay. Nevertheless, some applications require real-time yet may allow a delayed response, assuming it improves accuracy.

Studies showed that there is a performance gap between causal and acausal systems [36]. To choose the optimal network that compromises between delay and performance, it is necessary to investigate how the delay affects performance. Funke *et al.* [8] investigated the effect of delay on a 3D neural network. They found that adding delay improves the system's performance, primarily in segmentation metrics such as segmental edit distance and F1@10. Nevertheless, today these networks are considered relatively weak compared to the newer activity recognition networks that are typically based on transformers and temporal convolutional networks. In these algorithms, designing a future window is not trivial as in the 3D convolutional case.

In this work, we developed and analyzed different variations of the MS-TCN++, and studied the trade-off between delay and performance. This study sought to verify the intuition that adding a relatively small delay in the causal system's response, which usually operates in real-time, could also help bridge the causal-acausal gap in MS-TCN++. We examined this hypothesis in two methods. First, we tried to examine reducing the network's depth, thus the receptive field was reduced, with half of it being the future, namely Reduced Receptive field MS-TCN++, RR-MS-TCN++. We expected that this method would not work well for small future window sizes because reducing the window size in this method means a significant decrease in the number of layers or even eliminating of few refinement stages. Li *et al.* [19], showed these parameters have a crucial effect on performance. Therefore we developed the BF-MS-TCN++. This network involved applying a convolution in which the target point in the previous layer is not in the middle of the convolution's receptive field. This way, we bound the future window of the entire network even in large architectures with a small delay. With this method, different architectures can be implemented for approximately the same delay values. Thus in our analysis, several time intervals were defined for both methods, and we selected the network with the best results to represent each interval.

According to Fig. 5, the performance comparison study illustrates the trade-off between performance and delay, considering all metrics in both methods. As seen in Table 1, the gap between causal and acausal networks is much more prominent in the segmentation metrics than in the frame-wise metrics. Similarly, in Fig. 5 and Fig. 7, these metrics exhibit a stronger trade-off effect. Furthermore, BF-MS-TCN++ outperforms RR-MS-TCN++ especially when small delays are allowed. In the left image of Fig. 7, a future window of $2\frac{1}{3}$ s achieves more than 90% of the best offline network. Namely, getting close to an offline network's performance is possible even with a relatively small delay.

The fANOVA test showed, in Fig. 6, that the number of (D)DRLs and the number of refinement stages are the most important hyperparameters for optimizing performance, even more than the total delay. Where increasing these hyperparameter values tends to improve estimated marginal performance. Thus our BF-MS-TCN++, which allows the design of larger networks with that same delay factor, takes advantage of this fact. Another interesting finding is that the value of $w_{max}$ barely affects the outcome (Fig. 6), whereas the total future delay plays a pivotal role. Thus, we conclude that minimizing the value of $w_{max}$ for designing larger networks is an acceptable approach.

This study has a few limitations. First, it has been evaluated using only one data set. In addition, only data from surgical simulators were analyzed. Larger data sets, including data from the operating room, should be analyzed in the future.

The algorithms presented in this study are not limited to the surgical domain. Online with delay activity recognition is relevant to many other applications that evaluate human performance, even outside the surgical field. Therefore, this study lays the foundations for a broader study on the relationship between accuracy and delayed response in video-based human activity recognition.

# References

1. Albers, S.: Online algorithms: a survey. Math. Program. **97**(1), 3–26 (2003)
2. Avellino, I., Nozari, S., Canlorbe, G., Jansen, Y.: Surgical video summarization: multifarious uses, summarization process and ad-hoc coordination. Proc. ACM Hum.-Comput. Interact. **5**(CSCW1), 1–23 (2021)
3. Basiev, K., Goldbraikh, A., Pugh, C.M., Laufer, S.: Open surgery tool classification and hand utilization using a multi-camera system. Int. J. Comput. Assisted Radiol. Surg. **17**, 1497–1505 (2022)
4. Carreira, J., Zisserman, A.: Quo Vadis, action recognition? A new model and the kinetics dataset. In: Proceedings of the IEEE Conference on Computer Vision and Pattern Recognition, pp. 6299–6308 (2017)

5. Czempiel, T., et al.: TeCNO: surgical phase recognition with multi-stage temporal convolutional networks. In: Martel, A.L., et al. (eds.) MICCAI 2020. LNCS, vol. 12263, pp. 343–352. Springer, Cham (2020). https://doi.org/10.1007/978-3-030-59716-0_33

6. Donahue, J., et al.: Long-term recurrent convolutional networks for visual recognition and description. In: Proceedings of the IEEE Conference on Computer Vision and Pattern Recognition, pp. 2625–2634 (2015)

7. Friard, O., Gamba, M.: Boris: a free, versatile open-source event-logging software for video/audio coding and live observations. Methods Ecol. Evol. **7**, 1325–1330 (2016). https://doi.org/10.1111/2041-210X.12584

8. Funke, I., Bodenstedt, S., Oehme, F., von Bechtolsheim, F., Weitz, J., Speidel, S.: Using 3D convolutional neural networks to learn spatiotemporal features for automatic surgical gesture recognition in video. In: Shen, D., et al. (eds.) MICCAI 2019. LNCS, vol. 11768, pp. 467–475. Springer, Cham (2019). https://doi.org/10.1007/978-3-030-32254-0_52

9. Funke, I., Mees, S.T., Weitz, J., Speidel, S.: Video-based surgical skill assessment using 3D convolutional neural networks. Int. J. Comput. Assist. Radiol. Surg. **14**(7), 1217–1225 (2019)

10. Goldbraikh, A., D'Angelo, A.L., Pugh, C.M., Laufer, S.: Video-based fully automatic assessment of open surgery suturing skills. Int. J. Comput. Assist. Radiol. Surg. **17**(3), 437–448 (2022)

11. Goldbraikh, A., Volk, T., Pugh, C.M., Laufer, S.: Using open surgery simulation kinematic data for tool and gesture recognition. Int. J. Comput. Assisted Radiol. Surg. **17**, 965–979 (2022)

12. Huang, C., et al.: Sample imbalance disease classification model based on association rule feature selection. Pattern Recogn. Lett. **133**, 280–286 (2020)

13. Hutter, F., Hoos, H., Leyton-Brown, K.: An efficient approach for assessing hyperparameter importance. In: International Conference on Machine Learning, pp. 754–762. PMLR (2014)

14. Jacob, M.G., Li, Y.T., Wachs, J.P.: A gesture driven robotic scrub nurse. In: 2011 IEEE International Conference on Systems, Man, and Cybernetics, pp. 2039–2044. IEEE (2011)

15. Jung, J.J., Jüni, P., Lebovic, G., Grantcharov, T.: First-year analysis of the operating room black box study. Ann. Surg. **271**(1), 122–127 (2020)

16. Lea, C., Flynn, M.D., Vidal, R., Reiter, A., Hager, G.D.: Temporal convolutional networks for action segmentation and detection. In: Proceedings of the IEEE Conference on Computer Vision and Pattern Recognition, pp. 156–165 (2017)

17. Lea, C., Vidal, R., Hager, G.D.: Learning convolutional action primitives for fine-grained action recognition. In: 2016 IEEE International Conference on Robotics and Automation (ICRA), pp. 1642–1649. IEEE (2016)

18. Lea, C., Vidal, R., Reiter, A., Hager, G.D.: Temporal convolutional networks: a unified approach to action segmentation. In: Hua, G., Jégou, H. (eds.) ECCV 2016. LNCS, vol. 9915, pp. 47–54. Springer, Cham (2016). https://doi.org/10.1007/978-3-319-49409-8_7

19. Li, S.J., AbuFarha, Y., Liu, Y., Cheng, M.M., Gall, J.: MS-TCN++: multi-stage temporal convolutional network for action segmentation. IEEE Trans. Pattern Anal. Mach. Intell. 1 (2020). https://doi.org/10.1109/TPAMI.2020.3021756

20. Lux, M., Marques, O., Schöffmann, K., Böszörmenyi, L., Lajtai, G.: A novel tool for summarization of arthroscopic videos. Multimed. Tools Appl. **46**(2), 521–544 (2010)

21. Maier-Hein, L., et al.: Surgical data science-from concepts toward clinical translation. Med. Image Anal. **76**, 102306 (2022)
22. Maier-Hein, L., et al.: Surgical data science for next-generation interventions. Nat. Biomed. Eng. **1**(9), 691–696 (2017)
23. Mascagni, P., et al.: A computer vision platform to automatically locate critical events in surgical videos: documenting safety in laparoscopic cholecystectomy. Ann. Surg. **274**(1), e93–e95 (2021)
24. Mascagni, P., et al.: Artificial intelligence for surgical safety: automatic assessment of the critical view of safety in laparoscopic cholecystectomy using deep learning. Ann. Surg. **275**(5), 955–961 (2022)
25. Neimark, D., Bar, O., Zohar, M., Asselmann, D.: Video transformer network. In: Proceedings of the IEEE/CVF International Conference on Computer Vision, pp. 3163–3172 (2021)
26. Padoy, N.: Machine and deep learning for workflow recognition during surgery. Minim. Invasive Ther. Allied Technol. **28**(2), 82–90 (2019)
27. Pandey, A., Wang, D.: TCNN: temporal convolutional neural network for real-time speech enhancement in the time domain. In: 2019 IEEE International Conference on Acoustics, Speech and Signal Processing (ICASSP), ICASSP 2019, pp. 6875–6879. IEEE (2019)
28. Peddinti, V., Povey, D., Khudanpur, S.: A time delay neural network architecture for efficient modeling of long temporal contexts. In: Sixteenth Annual Conference of the International Speech Communication Association (2015)
29. Ramesh, S., et al.: Multi-task temporal convolutional networks for joint recognition of surgical phases and steps in gastric bypass procedures. Int. J. Comput. Assist. Radiol. Surg. **16**(7), 1111–1119 (2021). https://doi.org/10.1007/s11548-021-02388-z
30. Sun, X., Okamoto, J., Masamune, K., Muragaki, Y.: Robotic technology in operating rooms: a review. Curr. Robot. Rep. **2**(3), 333–341 (2021)
31. Tan, M., Le, Q.: EfficientNetV2: smaller models and faster training. In: International Conference on Machine Learning, pp. 10096–10106. PMLR (2021)
32. Twinanda, A.P., Yengera, G., Mutter, D., Marescaux, J., Padoy, N.: RSDNet: learning to predict remaining surgery duration from laparoscopic videos without manual annotations. IEEE Trans. Med. Imaging **38**(4), 1069–1078 (2018)
33. Ullah, A., Ahmad, J., Muhammad, K., Sajjad, M., Baik, S.W.: Action recognition in video sequences using deep bi-directional LSTM with CNN features. IEEE Access **6**, 1155–1166 (2017)
34. Yi, F., Wen, H., Jiang, T.: ASFormer: transformer for action segmentation. In: The British Machine Vision Conference (BMVC) (2021)
35. Yue-Hei Ng, J., Hausknecht, M., Vijayanarasimhan, S., Vinyals, O., Monga, R., Toderici, G.: Beyond short snippets: deep networks for video classification. In: Proceedings of the IEEE Conference on Computer Vision and Pattern Recognition, pp. 4694–4702 (2015)
36. Zhang, B., Ghanem, A., Simes, A., Choi, H., Yoo, A., Min, A.: Swnet: surgical workflow recognition with deep convolutional network. In: Medical Imaging with Deep Learning, pp. 855–869. PMLR (2021)

# Anatomy-Aware Contrastive Representation Learning for Fetal Ultrasound

Zeyu Fu[1]([✉])(iD), Jianbo Jiao[1](iD), Robail Yasrab[1](iD), Lior Drukker[2,3](iD),
Aris T. Papageorghiou[2](iD), and J. Alison Noble[1](iD)

[1] Department of Engineering Science, University of Oxford, Oxford, UK
zeyu.fu@eng.ox.ac.uk
[2] Nuffield Department of Women's and Reproductive Health, University of Oxford, Oxford, UK
[3] Department of Obstetrics and Gynecology, Tel-Aviv University, Tel Aviv-Yafo, Israel

**Abstract.** Self-supervised contrastive representation learning offers the advantage of learning meaningful visual representations from unlabeled medical datasets for transfer learning. However, applying current contrastive learning approaches to medical data without considering its domain-specific anatomical characteristics may lead to visual representations that are inconsistent in appearance and semantics. In this paper, we propose to improve visual representations of medical images via anatomy-aware contrastive learning (AWCL), which incorporates anatomy information to augment the positive/negative pair sampling in a contrastive learning manner. The proposed approach is demonstrated for automated fetal ultrasound imaging tasks, enabling the positive pairs from the same or different ultrasound scans that are anatomically similar to be pulled together and thus improving the representation learning. We empirically investigate the effect of inclusion of anatomy information with coarse- and fine-grained granularity, for contrastive learning and find that learning with fine-grained anatomy information which preserves intra-class difference is more effective than its counterpart. We also analyze the impact of anatomy ratio on our AWCL framework and find that using more distinct but anatomically similar samples to compose positive pairs results in better quality representations. Experiments on a large-scale fetal ultrasound dataset demonstrate that our approach is effective for learning representations that transfer well to three clinical downstream tasks, and achieves superior performance compared to ImageNet supervised and the current state-of-the-art contrastive learning methods. In particular, AWCL outperforms ImageNet supervised method by 13.8% and state-of-the-art contrastive-based method by 7.1% on a cross-domain segmentation task.

**Keywords:** Representation learning · Contrastive learning · Ultrasound

Z. Fu, J. Jiao and R. Yasrab—Equal contribution.

L. Karlinsky et al. (Eds.): ECCV 2022 Workshops, LNCS 13803, pp. 422–436, 2023.
https://doi.org/10.1007/978-3-031-25066-8_23

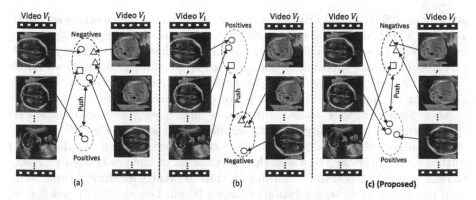

**Fig. 1.** Illustration of different representation learning approaches for fetal ultrasound, (a) self-supervised contrastive learning, (b) contrastive learning with patient metadata, and (c) our proposed anatomy-aware contrastive learning. Icon shapes of circle (o), square (□) and triangle (Δ) denote the anatomical categories of fetal head, profile, and abdomen, respectively. The anchor image is highlighted with a red bounding box, while the red dotted circle means pull together (Best viewed in colored version). (Color figure online)

## 1   Introduction

Semi-supervised and self-supervised representation learning with or without annotations have attracted significant attention across various medical imaging modalities [7,10,26–28]. These learning schemes are able to well exploit large-scale unlabeled medical datasets and learn meaningful representations for downstream task finetuning. In particular, contrastive representation learning based on instance discrimination tasks [6,11] has become the leading paradigm for self-supervised pretraining, where a model is trained to pull together each instance and its augmented views and meanwhile push it away from those of all other instances in the embedding space.

However, directly applying self-supervised contrastive learning (e.g. SimCLR [6] and MoCo [11]) in the context of medical imaging may result in visual representations that are inconsistent in appearance and semantics. We illustrate this issue in Fig. 1(a), which shows that a vanilla contrastive learning approach without considering the domain-specific anatomical characteristics leads to false negatives, i.e. some negative samples having high affinity with the anchor image are "pushed away". To address this, we explore the following question: *Is domain-specific anatomy information helpful in learning better representations for medical data?*

We investigate this question via the proposed anatomy-aware contrastive learning (AWCL), as depicted in Fig. 1(c), where "anatomy-aware" here means that the inclusion of anatomy information is leveraged to augment the positive/negative pair sampling in a contrastive learning manner. In this work, we demonstrate the proposed approach for fetal ultrasound imaging tasks, where a number of different fetal anatomies can be present in a diagnostic scan. Motivated by Khosla et al. [18], we expand the pool of positive samples by grouping images from the same or different ultrasound scans that share common anatomical categories. More importantly, our approach is optimized alternately with both conventional and anatomy-aware contrastive objectives, as shown in Fig. 2(a), given that the anatomy information is not always accessible for each sampling process. Moreover, we consider both coarse- and fine-grained anatomical categories with the availability for data sampling, as shown in Fig. 2(b) and (c). We also empirically investigate their effect on the transferability of the learned feature representations. To assess the effectiveness of our pre-trained representations, we evaluated transfer learning on three downstream clinical tasks: standard plane detection, segmentation of Crown Rump Length (CRL) and Nuchal Translucency (NT), and recognition of first-trimester anatomical structures. In summary, the main contributions and findings are:

- We develop an anatomy-aware contrastive learning approach for medical fetal ultrasound imaging tasks.
- We empirically compare the effect of inclusion of anatomy information with coarse- and fine-grained granularity respectively, within our contrastive learning approach. The comparative analysis suggests that contrastive learning with fine-grained anatomy information which preserves intra-class difference is more effective than its counterpart.
- Experimental evaluations on three downstream clinical tasks demonstrate the better generalizability of our proposed approaches over learning from an ImageNet pre-trained ResNet, vanilla contrastive learning [6], and contrastive learning with patient information [2,7,25].
- We provide an in-depth analysis to show the proposed approach learns high-quality discriminative representations.

## 2   Related Work

**Self-supervised Learning (SSL) in Medical Imaging.** Prior works using SSL for medical imaging typically selecting on designing pre-text tasks, such as solving a Rubik' cube [28], image restoration [14,27], predicting anatomical position [3] and multi-task joint reasoning [17]. Recently, contrastive based SSL [6,11] has been favourably applied to learn more discriminative representations across various medical imaging tasks [7,24,26]. In particular, Sowrirajan et al. [24] successfully adapted a MoCo-contrastive learning method [11] into chest X-rays and demonstrated better transferable representations and initialization for chest X-ray diagnostic tasks. Taher et al. [13] presented a benchmark evaluation study to investigate the effectiveness of several established contrastive learning models pre-trained on ImageNet on a variety of medical imaging tasks. In addition, there have been recent approaches [2,7,25] that leverage patient metadata to improve the medical imaging contrastive learning. These approaches constrain the selection of positive pairs only from the same subject (video), with the assumption that visual representations from the same subject share similar semantic meaning. However, these approaches may not generalize well to a scenario, where different organs or anatomical structures are captured in a single video. For instance, as seen from Fig. 1(b), some positive pairs having low affinity in visual appearance and semantics are pulled together, i.e. false positives, which can degrade the representation learning. The proposed learning scheme, as shown in Fig. 1(c), is advantageous to address the aforementioned limitations by augmenting the sampling process with the inclusion of anatomy information. Moreover, our approach differs from [26] and [7] which combine label information as an additional supervision signal with self supervision for multi-tasking.

**Representation Learning in Fetal Ultrasound.** There are related works exploring representation learning for fetal ultrasound imaging tasks. Baumgartner et al. [4] and Schlemper et al. [22] proposed a VGG-based network and an attention-gated network respectively to detect fetal standard planes. Sharma et al. [23] presented a multi-stream network which combines 2D image and spatio-temporal information to automate clinical workflow description of full-length routine fetal anomaly ultrasound scans. Cai et al. [5] considered incorporating the temporal dimension into visual attention modelling via multi-task learning for standard biometry plane-finding navigation. However, the generalization and transferability of those models to other target tasks remains unclear. Droste et al. [8] proposed to learn transferable representations for fetal ultrasound interpretation by modelling sonographer visual attention (gaze tracking) without manual supervision. More recently, Jiao et al. [16] proposed to derive a meaningful representation from raw data by developing a cross-modal contrastive learning which aligns the correspondence between fetal ultrasound video and narrative speech audio. Our work differs by focusing on learning general image representations without requiring additional data modalities (e.g. gaze tracking and audio) from the domain of interest, and we also perform extensive experimental analysis on three downstream clinical tasks to assess the effectiveness of the learned representations.

## 3   Fetal Ultrasound Imaging Dataset

This study uses a large-scale fetal ultrasound imaging dataset, which was acquired as part of PULSE (Perception Ultrasound by Learning Sonographic Experience) project [9]. The scans were performed by operators including sonographers and fetal medicine specialists using a commercial Voluson E8 version BT18 (General Electric, Zipf, Austria) ultrasound machine. During a routine scanning session, the operator views several fetal or maternal anatomical structures. The frozen views saved by sonographers are referred to as *standard planes* in the paper, following the UK Fetal Anomaly Screening Programme (FASP) nomenclature [1].

Fetal ultrasound videos, recorded from the ultrasound scanner display using a lossless compression and sampled at the rate 30 Hz. We consider a subset of the entire ultrasound dataset for the proposed pre-training approach. This consists of

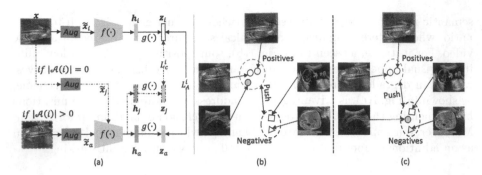

**Fig. 2.** (a) presents the overview of proposed anatomy-aware contrastive learning approach. (b) and (c) illustrate using coarse and fine-grained anatomy categories, respectively for the proposed AWCL framework. Icon shapes of white-circle (○), grey-circle (◉), square (□) and triangle (△) denote the classes of coronal view of spine, sagittal view of spine, profile, and abdomen, respectively.

total number of 2,050,432 frames[1] from 534 second-trimester ultrasound videos. In this sub-dataset, there are 15,384 frames labeled with 13 fine-grained anatomy categories, including four views of heart, three-vessel and trachea (3VT), four-chamber (4CH), right ventricular outflow tract (RVOT), and left ventricular outflow tract (LVOT), two views of brain, transventricular (BrainTv.) and transcerebellum (BrainTc.), two views of spine, coronal (SpineCor.) and sagittal (SpineSag.), abdomen, femur, kidneys, lips, profile and background class. In addition, 69,671 frames are labeled with coarse anatomy categories without dividing the heart, brain and spine into further sub-categories as those of above, but also 3D mode, maternal anatomy including Doppler, abdomen, nose and lips, kidneys, face-side profile, full-body-side profile, bladder including Doppler, femur and "Other" class. All image frames were preprocessed by cropping the ultrasound image region and resizing it to $224 \times 288$ pixels.

## 4   Method

In this section, we first describe the problem formulation of contrastive learning with medical images, and then present our anatomy-aware contrastive learning algorithm design as well as training details.

### 4.1   Problem Formulation

For each input image $\mathbf{x}$ in a mini-batch of $N$ samples, randomly sampled from a pre-training dataset $\mathcal{V}$, a contrastive learning framework (i.e. SimCLR [6]) applies two augmentations to obtain a positive pair $(\tilde{\mathbf{x}}_i, \tilde{\mathbf{x}}_j)$, yielding a set of $2N$

---

[1] Every 8th frame is extracted to reduce temporal redundancy of ultrasound videos.

samples. Let $i$ denote the anchor input index, the contrastive learning objective can be defined as,

$$L_C^i = -\log \frac{\exp\left(\text{sim}\left(\mathbf{z}_i, \mathbf{z}_j\right)/\tau\right)}{\sum_{k=1}^{2N} \mathbf{1}_{[k \neq i]} \exp\left(\text{sim}\left(\mathbf{z}_i, \mathbf{z}_k\right)/\tau\right)}, \tag{1}$$

where $\mathbf{1} \in \{0,1\}$, $\tau$ is a temperature parameter and $sim(\cdot)$ is the pairwise cosine similarity. $\mathbf{z}$ is a representation vector, calculated by $\mathbf{z} = g(f(\mathbf{x}))$, where $f(\cdot)$ denotes a shared encoder modelled by a convolutional neural network (CNN) and $g(\cdot)$ is a multi-layer perception (MLP) projection head.

The above underpins the vanilla contrastive learning. However in some cases (e.g. ultrasound scan as illustrated in this paper), this standard approach, as well as its extended version that leverages patient information [2,7,25], may lead to false negatives and false positives respectively, as seen from Fig. 1(a) and (b). To this end, we introduce a new approach as detailed next.

### 4.2   Anatomy-Aware Contrastive Learning

Figure 1(c) illustrates the main idea of the new anatomy-aware contrastive learning (AWCL) approach, which incorporates additional samples belonging to the same anatomy category from the same or different US scans. In addition to positive sampling from the same image and its augmentation, AWCL is tailored to the case where multiple anatomical structures are present.

As shown in Fig. 2(a), we utilize the available anatomy information as detailed in Sect. 3, forming a positive sample set $\mathcal{A}(i)$ with the same anatomy as sample $i$. The assumption for such a design is that image samples within the same anatomy category should have similar appearances, based on a clinical perspective [9]. Motivated by [18], we design the anatomy-aware contrastive learning objective as follows,

$$L_A^i = -\frac{1}{|\mathcal{A}(i)|} \sum_{a \in \mathcal{A}(i)} \log \frac{\exp\left(\text{sim}\left(z_i, z_a\right)/\tau\right)}{\sum_{k=1}^{2N} \mathbf{1}_{[k \neq i]} \exp\left(\text{sim}\left(z_i, z_k\right)/\tau\right)}, \tag{2}$$

where $|\mathcal{A}(i)|$ denotes the cardinality.

Due to the limited availability of some anatomical categories, $\mathcal{A}(i)$ is not always achievable for each sampling process. In this regard, the AWCL framework is formulated as an alternate optimization combining both learning objectives of Eq. 1 and Eq. 2. This gives a loss function defined as

$$L^i = \begin{cases} L_C^i & \text{if } |\mathcal{A}(i)| = 0 \\ L_A^i & \text{if } |\mathcal{A}(i)| > 0. \end{cases} \tag{3}$$

Furthermore, we consider both coarse- and fine-grained anatomical categories for the proposed AWCL framework, and compare their effect on the transferability of visual representations. Figure 2(b) and (c) shows the motivation of this comparative analysis. For an anatomical structure with different views of visual

---

**Algorithm 1:** Anatomy-aware Contrastive Learning (AWCL)

---

    **Input**   : Sample $x_i$ and its positive set $\mathcal{A}(i)$, pre-training dataset $\mathcal{V}$
    **Output:** The loss value $L$ of the current learning step
1 Sample mini-batch training data $x_i \in \mathcal{V}$
2 **if** $|\mathcal{A}(i)| = 0$ **then**
3     |   Apply data augmentations $\rightarrow$ positive pair $(\tilde{\mathbf{x}}_i, \tilde{\mathbf{x}}_j)$
4     |   Extract representation vectors $z_i = g(f(\tilde{\mathbf{x}}_i)), z_j = g(f(\tilde{\mathbf{x}}_j))$
5     |   $L = -\log \frac{\exp\left(\operatorname{sim}\left(\mathbf{z}_i, \mathbf{z}_j\right)/\tau\right)}{\sum_{k=1}^{2N} \mathbb{1}_{[k \neq i]} \exp(\operatorname{sim}(\mathbf{z}_i, \mathbf{z}_k)/\tau)}$
6 **else**
7     |   Sample data $x_a$ with the same anatomy as $x_i$, where $x_a \in \mathcal{A}(i)$
8     |   Apply data augmentations $\rightarrow$ positive pair $(\tilde{\mathbf{x}}_i, \tilde{\mathbf{x}}_a)$
9     |   Extract representation vectors $z_i = g(f(\tilde{\mathbf{x}}_i)), z_a = g(f(\tilde{\mathbf{x}}_a))$
10    |   $L = -\frac{1}{|\mathcal{A}(i)|} \sum_{a \in \mathcal{A}(i)} \log \frac{\exp(\operatorname{sim}(z_i, z_a)/\tau)}{\sum_{k=1}^{2N} \mathbb{1}_{[k \neq i]} \exp(\operatorname{sim}(z_i, z_k)/\tau)}$
11 **end**
12 Return $L$

---

appearance (e.g. the spine has two views as sub-classes), we observe that AWCL with coarse-grained anatomy information tends to minimize the intra-class difference by pulling together all the instances of the same anatomy. In contrast, AWCL with fine-grained anatomy information tends to preserve the intra-class difference by pushing away images with different visual appearances despite the same anatomy. Both strategies of the proposed learning approach are evaluated and compared in Sect. 6.3. We further study the impact of the ratio of anatomy information used in AWCL pre-training in Sect. 6.4.

### 4.3 Implementation Details

Algorithm 1 provides the pseudo-code of AWCL. Following the prior art [7,24, 25], we use ResNet-18 [12] as our backbone architecture. Further studies on different network architectures are out of scope of this paper. We split the pre-training dataset as detailed in Sect. 3 into training and validation sets (80%/20%), and train the model using the Adam optimizer with a weight decay of $10^{-6}$, and a mini-batch size of 32. We follow [6] for the data augmentations applied to the sampled training data. The output feature dimension of $z$ is set to 128. The temperature parameter $\tau$ is set as 0.5. The models are trained with the loss functions defined earlier (Eq. 2 and Eq. 1) for 10 epochs. The learning rate is set as $10^{-3}$. The whole framework is implemented with the PyTorch [21] framework on a PC with NVIDIA Titan V GPU card. The code is available at https:// github.com/JianboJiao/AWCL.

To demonstrate the effectiveness of AWCL trained models, we compare them with random initialization, ImageNet pre-trained ResNet18 [12], supervised pre-training with coarse labels, supervised pre-training with fine-grained labels, vanilla contrastive learning (SimCLR) [6], and contrastive learning with

**Table 1.** Details of the downstream datasets and imaging tasks.

| Trimester | Task | #Scans | #Images | #Classes |
|---|---|---|---|---|
| 2nd | I- Standard Plane Detection | 58 | 1,075 | 14 |
| 1st | II- Recognition of first-trimester anatomies | 90 | 25,490 | 5 |
| 1st | III- Segmentation of NT and CRL | 128 | 16,093 | 3 |

patient information (CLPI) [2,7,19]. All pre-training methods presented here are pre-trained from scratch on the pre-training dataset with the similar parameter configurations as listed above.

## 5 Experiments on Transfer Learning

In this section, we evaluate the effectiveness of the SSL pre by supervised transfer learning with end-to-end fine-tuning on three downstream clinical tasks, which are second-trimester standard plane detection (Task I), recognition of first-trimester anatomies (Task II) and segmentation of NT and CRL (Task III). The datasets for downstream task evaluation are listed in Table 1, and are independent datasets from [9]. For fair comparison, all compared pre-training models were fine-tuned with the same parameter settings and data augmentation policies within each downstream task evaluation.

### 5.1 Evaluation on Standard Plane Detection

**Evaluation Details.** Here, we investigate how the pre-trained representations generalize to an in-domain second-trimester classification task, which consists of the same fine-grained anatomical categories as detailed in Sect. 3. We fine-tune each trained backbone encoder and attach a classifier head [4] to train the entire network for 70 epochs with a learning rate of 0.01, decayed by 0.1 at epochs 30 and 55. The network training is performed via SGD with momentum of 0.9, weight decay of $5 \times 10^{-4}$, mini-batch size of 16 and a cross-entropy loss, and it is evaluated via a three-fold cross validation. The augmentation policy used is analogous to [8], including random horizontal flipping, rotation (10°), and varying gamma and brightness. We employ precision, recall and F1-scores computed as macro-averages as the evaluation metrics.

**Results and Discussion.** Table 2 shows a quantitative comparison of fine-tuning performance for the three evaluated downstream tasks. From the results of Task I, we observe that AWCL pre-trained models, i.e. *AWCL (coarse)* and *AWCL (fine-grained)*, generally outperform the compared contrastive learning methods SimCLR and CLPI. In particular, *AWCL (coarse)* improves on SimCLR and CLPI by 1.9% and 3.8% in F1-score, respectively. Compared to the supervised pre-training methods, both AWCL approaches achieve better performance in Recall and F1-score than vanilla supervised pre-training with coarse-grained

**Table 2.** Quantitative comparison of fine-tuning performance (mean ± std. [%]) on the tasks of standard plane detection (Task I), first-trimester anatomy recognition (Task II) and CRL / NT segmentation (Task III). Best results are marked in **bold**.

| Pre-training methods | Task I | | | Task II | | | Task III | | |
|---|---|---|---|---|---|---|---|---|---|
| | Precision (↑) | Recall (↑) | F1-score (↑) | Precision (↑) | Recall (↑) | F1-score (↑) | GAA (↑) | MA(↑) | mIoU(↑) |
| Rand.Init. | 70.4±1.7 | 58.3±3.1 | 61.6±3.1 | 81.4±3.4 | 79.2±0.1 | 81.5±0.3 | 67.3±0.2 | 63.0±2.1 | 46.7±0.1 |
| ImageNet | 78.8±4.6 | 73.6±4.1 | 73.6±2.8 | 92.0±0.5 | 93.4±1.5 | 92.1±2.9 | 71.6±1.3 | 64.2±1.0 | 49.0±0.1 |
| Supervised (coarse) | 74.2±2.7 | 67.5±3.4 | 69.0±3.2 | 95.2±0.1 | 93.7±0.2 | 94.1±0.4 | 76.4±0.3 | 67.5±1.1 | 50.1±0.3 |
| Supervised (fine-grained) | **84.6 ± 1.0** | **77.1 ± 2.3** | **78.6 ± 2.1** | 96.1±0.1 | 96.8±1.0 | 96.4±0.9 | 80.0±0.2 | 75.5±0.1 | 62.8±0.4 |
| SimCLR | 71.7±0.3 | 69.6±1.5 | 69.4±0.7 | 96.0±0.5 | 95.2±0.3 | 95.2±0.8 | 77.6±1.4 | 69.2±0.1 | 55.7±0.2 |
| CLPI | 68.6±4.2 | 68.5±3.2 | 67.5±3.7 | 89.2±0.1 | 88.3±0.8 | 89.6±1.1 | 72.7±0.2 | 65.4±1.4 | 48.1±1.2 |
| AWCL (coarse) | 71.4±3.3 | 73.1±1.9 | 71.3±2.2 | 95.6±0.7 | 96.2±1.6 | 95.9±0.2 | 79.8±0.7 | 76.1±0.3 | 61.2±1.3 |
| AWCL (fine-grained) | 71.8±2.7 | 70.0±1.2 | 70.1±1.7 | **96.9 ± 0.1** | **96.8 ± 1.8** | **97.1 ± 0.2** | **80.2 ± 1.1** | **76.0 ± 0.5** | **62.8 ± 0.1** |

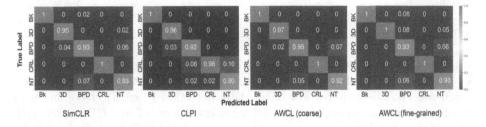

**Fig. 3.** Illustration of the confusion matrix for the first-trimester classification task.

labels. These findings suggest that incorporating anatomy information to select positive pairs from multiple scans can notably improve representation learning.

However, we find that all the contrastive pre-training approaches presented here underperform the supervised pre-training (fine-grained) which has the same form of semantic supervision as Task I. This suggests that without explicitly encoding semantic information, contrastively learned representations may provide limited benefits to the generalization of a fine-grained multi-class classification task, which is line with the findings in [15].

## 5.2   Evaluation on Recognition of First-Trimester Anatomies

**Evaluation Details.** We investigate how the pre-trained representations generalize to a cross-domain classification task using the first-trimester US scans. This first-trimester classification task recognises five anatomical categories: crown rump length(CRL), nuchal translucency (NT), biparietal diameter (BPD), 3D and background (Bk). We split the data into training and testing sets (78%/22%). The trained encoders followed by two fully-connected layers and a softmax layer were fine-tuned for 200 epochs with a learning rate of 0.1 decayed by 0.1 at 150 epochs. The network was trained using SGD with momentum of 0.9. Standard data augmentation was used, including rotation $[-30°, 30°]$, horizontal flip, Gaussian noise, and shear $\leq 0.2$. Batch size was adjusted according to model size and GPU memory restrictions. We use the same metrics as presented in Task I for performance evaluation.

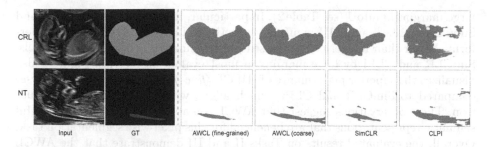

**Fig. 4.** Illustration of the qualitative results for the first-trimester segmentation task.

**Results and Discussion.** For Task II, we see from Table 2, that *AWCL (fine-grained)* achieves the best performance among all the compared solutions. In particular, it achieves a performance gain of 4.9%, 3.4% and 5.0% in Precision, Recall and F1-score compared to ImageNet pre-training, and even improves on supervised pre-training with fine-grained labels (the upper-bound baseline) by 0.7% in F1-score. Moreover, *AWCL (coarse)* also surpasses ImageNet and supervised pre-training with coarse-grained labels by 1.9% and 6.3% in F1-score. For comparison with other contrastive learning methods, we observe a similar improved trend as described in Task I, i.e. *AWCL (coarse)* and *AWCL (fine-grained)* perform better than SimCLR and CLPI. Further evidence is provided in Fig. 3, which shows that both *AWCL (coarse)* and *AWCL (fine-grained)* provide better prediction accuracy than CLPI for all anatomy categories. These experimental results again demonstrate the effectiveness of AWCL approaches and suggest that the inclusion of anatomy information in contrastive learning is a good practice when it is available at hand.

### 5.3   Evaluation on Segmentation of NT and CRL

**Evaluations Details.** In this section, we evaluate how the pre-trained models generalize to a cross-domain segmentation task with the data from the first-trimester US scans. Segmentation of NT and CRL was defined as a three-class segmentation task with the three classes being; mid-sagittal view, nuchal translucency, background. The data is divided into training and testing with 80%/20%. We follow the design of ResNet-18 auto-encoder by attaching additional decoders with the trained encoders, and then fine-tuned the entire model for 50k iterations with a learning rate of 0.001, RMSprop optimization (momentum=0.9) and a weight decay of 0.001. We apply random scaling, random shifting, random rotation, and random horizontal flipping for data augmentation. We use global average accuracy (GAA), mean accuracy (MA), and mean intersection over union (mIoU) metrics for evaluating the segmentation task (Task III).

**Results and Discussion.** For Task III, we find that *AWCL (fine-grained)*, achieves comparable or slightly better performance than supervised pre-training with fine-grained labels and surpasses other compared pre-training methods by large margins in mIoU (see Table 2). In particular, it outperforms ImageNet and SimCLR by 13.8% and 7.1% in mIoU, respectively. Likewise, *AWCL (coarse)* performs better than ImageNet, supervised pre-training with coarse-grained labels, SimCLR and CLPI by large margins in most evaluation metrics. Figure 4 also visualizes the superior performance of *AWCL (fine-grained)* and *AWCL (coarse)* compared to SimCLR and CLPI, which aligns with the quantitative evaluation. These observations suggest that AWCL are able to learn more meaningful semantic representations that are beneficial for this pixel-wise segmentation task. Overall, the evaluated results on Tasks II and III demonstrate that the AWCL models report consistently better performance than the compared pre-trained models, implying the advantage of learning task-agnostic features that can better generalized to the tasks from different domains.

# 6    Analysis

## 6.1    Partial Fine-Tuning

To analyze representation quality, we extract fixed feature representations from the last layer of the ResNet-18 encoder and then evaluate them in two classification target tasks (Task I and Task II). Experimentally, we freeze the entire backbone encoder and attach a classification head [4] for Task I, and a non-linear classifier as mentioned in Sect. 5.2 for Task II. From Table 3, we observe that the AWCL approaches show better representation quality by surpassing the three compared approaches in terms of F1-score for both tasks. This suggests that the learned representations are strong non-linear features which are more generalizable and transferable to the downstream tasks. Comparing Tables 2 and 3, we find that although the reported scores of partial fine-tuning are generally lower than for full fine-tuning, the performance between two implementations of transfer learning is correlated.

**Table 3.** Performance comparison of partial fine-tuning (mean ± std. [%]) on the tasks of standard plane detection (Task I) and first-trimester anatomy recognition (Task II). Best results are marked in **bold**.

| Pre-training methods | Task I | | | Task II | | |
|---|---|---|---|---|---|---|
| | Precision (↑) | Recall (↑) | F1-score (↑) | Precision (↑) | Recall (↑) | F1-score (↑) |
| ImageNet | 65.5 ± 4.9 | 58.1 ± 4.3 | 60.2 ± 4.9 | 84.03 ± 0.13 | 84.25 ± 0.45 | 83.92 ± 0.62 |
| SimCLR | 67.6 ± 3.5 | 67.3 ± 4.1 | 65.9 ± 3.6 | 87.65 ± 0.09 | 86.82 ± 0.11 | 86.12 ± 0.20 |
| CLPI | **71.2 ± 0.6** | 68.6 ± 4.9 | 67.9 ± 5.1 | 82.07 ± 0.62 | 80.28 ± 0.83 | 81.10 ± 1.03 |
| *AWCL (coarse)* | 70.5 ± 2.7 | **71.3 ± 1.7** | **69.5 ± 1.9** | 86.21 ± 1.20 | 87.67 ± 0.32 | 86.14 ± 0.59 |
| *AWCL (fine-grained)* | 69.7 ± 1.0 | 68.8 ± 0.2 | 68.4 ± 0.5 | **88.65 ± 0.49** | **88.14 ± 0.17** | **87.00 ± 0.01** |

## 6.2    Visualization of Feature Representations

In this section we investigate why the feature representations produced with AWCL pre-trained models result in better downstream task performance. We

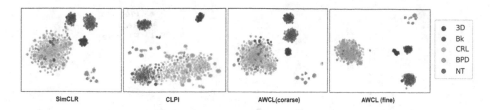

**Fig. 5.** t-SNE feature visualization of the model penultimate layers on Task II.

visualize the image representations of Task II extracted from the penultimate layers using t-SNE [20] in Fig. 5, where different anatomical categories are denoted with different color. We compare the resulting t-SNE embeddings of AWCL models with those as SimCLR and CLPI. We observe that the feature representation by CLPI is not quite separable, especially for classes of NT (orange) and CRL (green). The features embeddings from SimCLR are generally better separated than those in CLPI, while confusion between CRL (green) and Bk (blue) remains. By comparison, *AWCL (fine-grained)* achieves the best separated clusters among five anatomical categories, which means that the learned representations in the embedding space are more distinguishable. These visualization results demonstrate that AWCL approaches are able to learn discriminative feature representations which are better generalized to downstream tasks.

### 6.3   Impact of Data Granularity on AWCL

We analyze how the inclusion of coarse- and fine-grained anatomy information impact the AWCL framework, by comparing the experimental results between *AWCL (coarse)* and *AWCL (fine-grained)* from Sect. 5.1 to Sect. 6.2. Based on the transfer learning results in Table 2, we find that *AWCL (fine-grained)* achieves better performance than *AWCL (coarse)* for Tasks II and III, despite the slight performance drop in Task I. We hypothesize that *AWCL (coarse)* learns more generic representations than *AWCL (fine-grained)*, which leads to better in-domain generalization. Qualitative results in Fig. 3 and Fig. 4 also reveal the advantage of *AWCL (fine-grained)* over its counterpart. Based on the ablation analysis, Table 3 shows a similar finding as seen in Table 2. Figure 6 shows that feature embeddings of *AWCL (fine-grained)* are more discriminative than those of *AWCL (coarse)* thereby resulting in better generalization to downstream tasks. These observations suggest the importance of learning intra-class feature representations for better generalization to downstream tasks especially when there is a domain shift.

### 6.4   Impact of Anatomy Ratio on AWCL

We investigate how varying anatomy ratios impact the AWCL framework. Note that higher anatomy ratio represents that larger number of samples from same or different US scans belonging to the same anatomy category are included

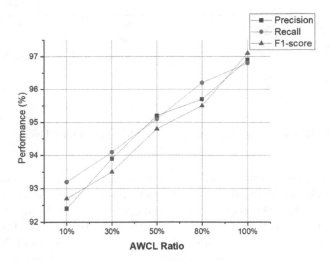

**Fig. 6.** Impact of anatomy ratio on AWCL (fine-grained) evaluated on Task II.

to form positive pairs for contrastive learning. We incorporated the anatomy information with four different ratios: 10%, 30%, 50%, and 80% to train the AWCL (fine-grained) models on the pre-training dataset, respectively. Then, we evaluate these trained models on Task II via full fine-tuning. As shown in Fig. 6, we observe that the performance improves with an increasing anatomy ratio. It suggests that using more distinct but anatomically similar samples to compose positive pairs results in a better quality representation.

## 7   Conclusion

In this paper, we presented a new anatomy-aware contrastive learning (AWCL) approach for fetal ultrasound imaging tasks. The proposed approach is able to leverage more positive samples from the same or different US videos with the same anatomy category and align well with the anatomical characteristics of ultrasound videos. The feature representative analysis shows AWCL approaches learn discriminative representations that can be better generalized to downstream tasks. Through the reported comparative study, AWCL with fine-grained anatomy information which preserves intra-class difference was more effective than its counterpart. Experimental evaluations demonstrate that our AWCL approach provides useful transferable representations for various downstream clinical tasks, especially for cross-domain generalization. The proposed approach can be potentially applied to other medical imaging modalities where such anatomy information is available.

**Acknowledgement.** The authors would like to thank Lok Hin Lee, Richard Droste, Yuan Gao and Harshita Sharma for their help with data preparation. This work is supported by the EPSRC Programme Grants Visual AI (EP/T028572/1) and Seebibyte (EP/M013774/1), the ERC Project PULSE (ERC-ADG-2015 694581), the NIH grant U01AA014809, and the NIHR Oxford Biomedical Research Centre. The NVIDIA Corporation is thanked for a GPU donation.

# References

1. Fetal Anomaly Screen Programme Handbook. NHS Screening Programmes, London (2015)
2. Azizi, S., et al.: Big self-supervised models advance medical image classification. arXiv:2101.05224 (2021)
3. Bai, W., et al.: Self-supervised learning for cardiac MR image segmentation by anatomical position prediction. In: Shen, D., et al. (eds.) MICCAI 2019. LNCS, vol. 11765, pp. 541–549. Springer, Cham (2019). https://doi.org/10.1007/978-3-030-32245-8_60
4. Baumgartner, C.F., et al.: SonoNet: real-time detection and localisation of fetal standard scan planes in freehand ultrasound. IEEE Trans. Med. Imaging **36**(11), 2204–2215 (2017)
5. Cai, Y., et al.: Spatio-temporal visual attention modelling of standard biometry plane-finding navigation. Med. Image Anal. **65**, 101762 (2020)
6. Chen, T., Kornblith, S., Norouzi, M., Hinton, G.: A simple framework for contrastive learning of visual representations. In: International Conference on Machine Learning (ICML), pp. 1597–1607 (2020)
7. Chen, Y., et al.: USCL: pretraining deep ultrasound image diagnosis model through video contrastive representation learning. In: de Bruijne, M., et al. (eds.) MICCAI 2021. LNCS, vol. 12908, pp. 627–637. Springer, Cham (2021). https://doi.org/10.1007/978-3-030-87237-3_60
8. Droste, R., et al.: Ultrasound image representation learning by modeling sonographer visual attention. In: Chung, A.C.S., Gee, J.C., Yushkevich, P.A., Bao, S. (eds.) IPMI 2019. LNCS, vol. 11492, pp. 592–604. Springer, Cham (2019). https://doi.org/10.1007/978-3-030-20351-1_46
9. Drukker, L., et al.: Transforming obstetric ultrasound into data science using eye tracking, voice recording, transducer motion and ultrasound video. Sci. Rep. **11**, 14109 (2021)
10. Haghighi, F., Hosseinzadeh Taher, M.R., Zhou, Z., Gotway, M.B., Liang, J.: Learning semantics-enriched representation via self-discovery, self-classification, and self-restoration. In: Martel, A.L., et al. (eds.) MICCAI 2020. LNCS, vol. 12261, pp. 137–147. Springer, Cham (2020). https://doi.org/10.1007/978-3-030-59710-8_14
11. He, K., Fan, H., Wu, Y., Xie, S., Girshick, R.: Momentum contrast for unsupervised visual representation learning. In: IEEE Conference on Computer Vision and Pattern Recognition (CVPR) (2020)
12. He, K., Zhang, X., Ren, S., Sun, J.: Deep residual learning for image recognition. In: IEEE Conference on Computer Vision and Pattern Recognition (CVPR), pp. 770–778 (2016)
13. Hosseinzadeh Taher, M.R., Haghighi, F., Feng, R., Gotway, M.B., Liang, J.: A systematic benchmarking analysis of transfer learning for medical image analysis. In: Albarqouni, S., et al. (eds.) DART/FAIR 2021. LNCS, vol. 12968, pp. 3–13. Springer, Cham (2021). https://doi.org/10.1007/978-3-030-87722-4_1

14. Hu, S.Y., et al.: Self-supervised pretraining with DICOM metadata in ultrasound imaging. In: Proceedings of the 5th Machine Learning for Healthcare Conference, pp. 732–749 (2020)
15. Islam, A., Chen, C.F.R., Panda, R., Karlinsky, L., Radke, R., Feris, R.: A broad study on the transferability of visual representations with contrastive learning. In: IEEE International Conference on Computer Vision (ICCV), pp. 8845–8855 (2021)
16. Jiao, J., Cai, Y., Alsharid, M., Drukker, L., Papageorghiou, A.T., Noble, J.A.: Self-supervised contrastive video-speech representation learning for ultrasound. In: Martel, A.L., et al. (eds.) MICCAI 2020. LNCS, vol. 12263, pp. 534–543. Springer, Cham (2020). https://doi.org/10.1007/978-3-030-59716-0_51
17. Jiao, J., Droste, R., Drukker, L., Papageorghiou, A.T., Noble, J.A.: Self-supervised representation learning for ultrasound video. In: 2020 IEEE 17th International Symposium on Biomedical Imaging (ISBI), pp. 1847–1850. IEEE (2020)
18. Khosla, P., et al.: Supervised contrastive learning. In: Advances in Neural Information Processing Systems, vol. 33, pp. 18661–18673 (2020)
19. Kiyasseh, D., Zhu, T., Clifton, D.A.: CLOCS: contrastive learning of cardiac signals across space, time, and patients. In: International Conference on Machine Learning (ICML), vol. 139, pp. 5606–5615 (2021)
20. van der Maaten, L., Hinton, G.: Visualizing data using t-SNE. J. Mach. Learn. Res. **9** (2008)
21. Paszke, et al.: PyTorch: an imperative style, high-performance deep learning library. In: Advances in Neural Information Processing Systems, pp. 8024–8035 (2019)
22. Schlemper, J., et al.: Attention-gated networks for improving ultrasound scan plane detection. In: International Conference on Medical Imaging with Deep Learning (MIDL) (2018)
23. Sharma, H., Drukker, L., Chatelain, P., Droste, R., Papageorghiou, A., Noble, J.: Knowledge representation and learning of operator clinical workflow from full-length routine fetal ultrasound scan videos. Med. Image Anal. **69**, 101973 (2021)
24. Sowrirajan, H., Yang, J., Ng, A.Y., Rajpurkar, P.: MoCo-CXR: MoCo pretraining improves representation and transferability of chest X-ray models. In: Medical Imaging with Deep Learning (MIDL) (2021)
25. Vu, Y.N.T., Wang, R., Balachandar, N., Liu, C., Ng, A.Y., Rajpurkar, P.: MedAug: contrastive learning leveraging patient metadata improves representations for chest X-ray interpretation. In: Machine Learning for Healthcare Conference, vol. 149, pp. 755–769 (2021)
26. Zhou, H.-Y., Yu, S., Bian, C., Hu, Y., Ma, K., Zheng, Y.: Comparing to learn: surpassing imagenet pretraining on radiographs by comparing image representations. In: Martel, A.L., et al. (eds.) MICCAI 2020. LNCS, vol. 12261, pp. 398–407. Springer, Cham (2020). https://doi.org/10.1007/978-3-030-59710-8_39
27. Zhou, Z., et al.: Models genesis: generic autodidactic models for 3D medical image analysis. In: Shen, D., et al. (eds.) MICCAI 2019. LNCS, vol. 11767, pp. 384–393. Springer, Cham (2019). https://doi.org/10.1007/978-3-030-32251-9_42
28. Zhuang, X., Li, Y., Hu, Y., Ma, K., Yang, Y., Zheng, Y.: Self-supervised feature learning for 3D medical images by playing a Rubik's cube. In: Shen, D., et al. (eds.) MICCAI 2019. LNCS, vol. 11767, pp. 420–428. Springer, Cham (2019). https://doi.org/10.1007/978-3-030-32251-9_46

# Joint Calibrationless Reconstruction and Segmentation of Parallel MRI

Aniket Pramanik[✉][iD] and Mathews Jacob[iD]

The University of Iowa, Iowa City, IA 52246, USA
{aniket-pramanik,mathews-jacob}@uiowa.edu

**Abstract.** The volume estimation of brain regions from MRI data is a key problem in many clinical applications, where the acquisition of data at high spatial resolution is desirable. While parallel MRI and constrained image reconstruction algorithms can accelerate the scans, image reconstruction artifacts are inevitable, especially at high acceleration factors. We introduce a novel image domain deep-learning framework for calibrationless parallel MRI reconstruction, coupled with a segmentation network to improve image quality and to reduce the vulnerability of current segmentation algorithms to image artifacts resulting from acceleration. Combination of the proposed calibrationless approach with a segmentation algorithm offers improved image quality, while improving segmentation accuracy. The novel architecture with an encoder shared between the reconstruction and segmentation tasks is seen to reduce the need for segmented training datasets. In particular, the proposed few-shot training strategy requires only 10% of segmented datasets to offer good performance.

**Keywords:** Calibrationless · Parallel MRI · Low-rank · Few-shot learning

## 1 Introduction

The degeneration/atrophy of brain structures is a key biomarker that is predictive of progression in several neurodegenerative disorders, including Alzheimer's disease [2–4]. The volume measures of these brain structures are usually estimated in the clinical setting from the segmentation of MR images. High-resolution images can improve the precision of volume measures and thus enable the early detection and improve the prediction of progression [13,17,23]. Unfortunately, MRI is a slow imaging modality; the acquisition of high-resolution images often comes with a proportional increase in scan time, which is challenging for several patient sub-groups, including older adults. In addition, longer scans are vulnerable to motion artifacts [32]. Calibrated [8,22] parallel MRI (PMRI) schemes, which use pre-estimated coil sensitivities, and calibration-free PMRI schemes [9,19,26] have been used to accelerate the acquisition. While calibrationless approaches can eliminate motion artifacts and offer higher acceleration factors, the high computational complexity of these methods is a limitation. While current PMRI schemes use constrained models including compressed

© The Author(s), under exclusive license to Springer Nature Switzerland AG 2023
L. Karlinsky et al. (Eds.): ECCV 2022 Workshops, LNCS 13803, pp. 437–453, 2023.
https://doi.org/10.1007/978-3-031-25066-8_24

sensing [15,18] and deep-learning priors [1,10], the reconstructed images often exhibit residual aliasing and blurring at high acceleration rates. The direct use of current segmentation algorithms on these images may result in segmentation errors, which may offset the benefit of the higher spatial resolution. In particular, we note that the reconstruction and segmentation algorithms are often designed and developed independently, even though there is extensive synergy between segmentation and reconstruction tasks. A challenge that restricts the development of joint recovery-segmentation algorithms is the lack of datasets with segmentation labels. In particular, segmentation often requires extensive time and expert supervision; a semi-supervised segmentation approach that can reduce the number of labelled datasets can significantly improve the widespread adoption of such methods.

The main focus of this paper is to introduce a deep-learning framework for joint calibrationless reconstruction and semi-supervised segmentation of PMRI data. Motivated by k-space deep-learning (DL) strategy [21] for calibrationless PMRI, we introduce a novel image domain DL framework. The proposed framework is motivated by the CLEAR approach [29], which exploits the low-rank structure of the patches obtained from coil sensitivity weighted images. To reduce the computational complexity, we unroll an iterative re-weighted least squares (IRLS) algorithm to minimize the CLEAR cost function and train it end-to-end, similar to the approach in [1,10,25] with shared weights across iterations as in model-based deep-learning (MoDL) [1]. We note that in an IRLS-CLEAR formulation, the iterations alternate between a data consistency (DC) and a projection step, which projects each set of multi-channel patches to a linear signal subspace, thus *denoising* them. The IRLS algorithm estimates the linear subspaces or, equivalently, the linear annihilation relations from the data itself, which requires several iterations that contribute to the high computational complexity. We instead replace the linear projection step with a residual UNET [24] convolutional neural network (CNN). We hypothesize that the pre-learned non-linear CNN acts as a spatially varying annihilation filterbank for the multi-channel patches. The use of the CNN, whose parameters are pre-learned from exemplar data, translates to good performance with much fewer iterations and, hence, significantly reduced computational complexity. We note that significantly more inter- and intra-patch annihilation relations are available in the spatial domain compared to the k-space approach in [21], which translates to improved performance.

To make the segmentation scheme robust to blurring and aliasing artifacts at high acceleration rates, we propose an integrated framework for the joint segmentation and reconstruction. In particular, we attach a segmentation-dedicated decoder to the encoder of the projection CNN at the final iteration of the unrolled reconstruction network. Please see Fig. 1. The shared architecture, which uses the common latent representation for both the segmentation and *denoising* at all the iterations, facilitates the exploitation of the synergies between the tasks. More importantly, this approach enables a semi-supervised training strategy that significantly reduces the number of segmented datasets that are needed to train the joint reconstruction-segmentation framework. We also expect the reconstruction task to benefit from the addition of the segmentation task, which emphasizes the accuracy of the edge locations. In the fully supervised setting, the multi-task architecture is trained end-to-end; the loss is the weighted linear combination of mean squared reconstruction error (MSE) and multi-class cross entropy segmentation loss. We also introduce a semi-supervised setting to reduce the number of segmented datasets. In this setting, we use a weighted linear combination of reconstruction and segmentation losses for the datasets with segmentation labels. By contrast, we only use the reconstruction MSE loss to train the network for datasets without segmentation labels. The shared encoder, which is trained in the reconstruction task on all subjects, enables us to keep the generalization error for segmentation low, even when few labelled segmentation datasets are used. This semi-supervised approach is an attractive option, especially since there are not many large publicly available datasets with both raw k-space data and segmentation labels.

The proposed work has conceptual similarities to the cascade of reconstruction and segmentation networks [12] and the recent JRS [28]. The use of the same latent spaces in our approach facilitates the improved exploitation of the synergies, which offers improved performance and enables the semi-supervised approach rather than the cascade approach [12]. In addition to being restricted to the single-channel MRI setting, the JRS framework has some important distinctions from the proposed approach. The parameters of the auto-encoders at each iteration are not shared in [28]. To constrain the latent variables at each iteration, they are each fed to a segmentation network; the segmentation errors from each iteration are combined linearly to define the segmentation loss. Unlike [28], we use the same reconstruction network at each iteration and only connect the segmentation network to the latent variables in the final iteration. The sharing of the parameters in our setting ensures that the segmentation task will regularize the projection networks at all iterations, thus facilitating the semi-supervised training strategy.

## 2  Background

### 2.1  Forward Model

Parallel MRI acquisition of $\gamma$ can be modelled as $b_i = \mathcal{U}(\mathcal{F}(\underbrace{s_i\,\gamma}_{\gamma_i})) + \eta_i$, $i = 1,\ldots,N$ where $N$ is the number of receiver coils, $s_i$ is the $i^{\text{th}}$ coil sensitivity, $b_i$ represents the undersampled measurements, and $\eta_i$ is zero mean Gaussian noise. Here, $\mathcal{F}$ is a 2D fast Fourier transform (FFT) operator that transforms the spatial domain signal $\gamma_i$ to its k-space, which is sampled at the locations defined by the undersampling operator $\mathcal{U}$.

The model can be compactly represented as $\mathbf{b} = \mathcal{A}(\gamma) + \mathbf{n}$, where $\gamma$ is a 3D volume obtained by stacking the coil images $\gamma_1,..,\gamma_N$ in the spatial domain, $\mathbf{b}$ is a concatenation of corresponding noisy undersampled Fourier measurements

$b_1, .., b_N$ across $N$ channels and $\mathbf{n}$ is a concatenation of corresponding noise $\eta_1, .., \eta_N$. Here, $\widehat{\gamma} = \mathcal{F}(\gamma)$ is multi-channel data in the Fourier domain and, similarly, the Fourier domain image $\widehat{\gamma_i} = \mathcal{F}(\gamma_i)$ for $i^{\text{th}}$ channel.

## 2.2    Calibrationless PMRI Recovery

Parallel MRI recovery methods can be broadly classified into calibrated and calibrationless approaches. Calibrated approaches such as SENSE [22], GRAPPA [8], and ESPiRIT [30] have a separate coil sensitivity estimation step, either through additional calibration scans or from self-calibration data. Recently, MoDL methods have been used to further improve the performance of the above calibrated PMRI schemes [1,10]. These methods pose the recovery as an optimization problem, which minimizes the sum of a DC term and a deep-learned CNN prior. An alternating minimization scheme to solve the above problem is unrolled for a finite number of iterations and trained in an end-to-end fashion [1]. These approaches offer improved performance over classical calibrated PMRI schemes. A challenge with the calibrated schemes is their sensitivity to the estimated sensitivity maps; any imperfection introduces error in reconstructions.

Calibrationless recovery methods were introduced to minimize the above model mismatch. These approaches, often termed as structured low-rank (SLR) methods, exploit the low-rank property of multi-channel k-space patches to recover the multi-channel images from undersampled data [9,19,26]. Another SLR method, CLEAR [29], was introduced to recover PMRI by utilizing relations in the spatial domain instead of k-space. It works on a locally low-rank assumption where a matrix containing patches extracted from images across the coils is low-rank. The spatial relations are obtained based on the smoothness of coil sensitivities. While these methods are powerful, the main challenge is their high computational complexity; it often takes several minutes to recover slices of a subject. Recently, DL methods were introduced for faster calibrationless recovery. For example, K-UNET [11] is a multi-channel UNET used to map undersampled coil images directly to their corresponding fully sampled ones. A similar approach, called Deep-SLR, is used in the model-based setting and offers improved performance over the data-driven method [11].

## 2.3    End-to-End Multi-task Training Approaches

Most of the MRI segmentation algorithms are designed for fully sampled data. When the images are acquired with high acceleration factors, the recovered images often exhibit significant artifacts, including aliasing artifacts and blurring. The direct application of the above segmentation algorithms to these images may result in segmentation errors. Recently, some researchers have looked into minimizing these errors by coupling image denoising or reconstruction tasks with segmentation, thus improving both tasks [16]. [12] considers the cascade of reconstruction and segmentation networks, which are trained end-to-end, while [28] introduces a shared encoder architecture for the single-channel MRI setting. Similar multi-task learning strategies are also used in the MRI setting [20]; these studies show that when the tasks are complementary, the joint learning of them using a single network can facilitate the exploitation of the synergies.

A challenge is scarcity of labelled segmentation datasets that has prompted some researchers to introduce semi-supervised learning strategies. A recent work proposes a few-shot learning approach for segmentation, which couples the segmentation task with few labelled datasets, with an image denoising task [6]. The segmentation and denoising tasks share the encoder; the training of the shared encoder-denoising decoder on all the datasets allows them to keep the generalization error of the segmentation task low, even though few segmentation labels were used for training. Another work proposes few-shot learning for an image classification task, which is coupled with a self-supervised auxiliary task [7].

## 3   Proposed Framework

We will first introduce the image domain deep-SLR (IDSLR) scheme, which offers improved recovery over the k-space deep-SLR in [21]. We will then combine the reconstruction scheme with a semi-supervised segmentation scheme to reduce the sensitivity to alias artifacts and blurring associated with high acceleration factors.

CLEAR [29] exploits the low-rankness of the patches obtained from coil sensitivity weighted images; consider a patch extraction operator $P_c$, which extracts $M \times M$ patches centered at $c$ from the multi-channel data $\gamma_i(\mathbf{r}) = s_i(\mathbf{r})\gamma(\mathbf{r})$, where $\mathbf{r}$ denotes spatial coordinates. The coil sensitivity functions $s_i(\mathbf{r})$ are smooth functions and hence can be safely approximated as $s_i(\mathbf{r}) \approx s_i(\mathbf{c}); \forall \mathbf{r}$ within the $M \times M$ patch neighborhood. The CLEAR relies on this approximation to show that the patch matrices, $\boldsymbol{\Gamma}_c = [P_c(\gamma_1), \ldots, P_c(\gamma_N)] \approx P_c(\gamma) [s_1(\mathbf{c}), s_2(\mathbf{c}), \ldots, s_N(\mathbf{c})]$ can be approximated as low-rank matrices [29]. CLEAR solves for 3-D multi-channel volume $\gamma$ by minimizing the nuclear norm optimization problem [29]:

$$\gamma = \arg\min_{\gamma} \|\mathcal{A}(\gamma) - \mathbf{b}\|_2^2 + \lambda \sum_c \|\boldsymbol{\Gamma}_c\|_* \tag{1}$$

An IRLS approach to (1) results in an alternating minimization algorithm that alternates between the solution of the images:

$$\arg\min_{\gamma} \|\mathcal{A}(\gamma) - \mathbf{b}\|_2^2 + \lambda \sum_c \|\mathcal{Q}_c * \boldsymbol{\Gamma}_c\|_F^2 \tag{2}$$

and deriving the spatially varying multichannel filters $\mathcal{Q}_c$. These filters are estimated from the signal patch matrices $\boldsymbol{\Gamma}_c$. The regularization term in (2) minimizes the projection of the signal energy to the null-space.

### 3.1   Image Domain Deep-SLR Formulation

An alternating minimization scheme, similar to [21], is derived to obtain a recursive network that iterates between a data consistency enforcement (DC) and a denoising step. Motivated by k-space DL framework [21], we replace the linear null-space projection step with a non-linear spatial domain CNN that can learn a set of non-linear spatially varying annihilating filters from the training data and generalize well over unseen test data. The proposed formulation is $\gamma = \arg\min_{\gamma} \|\mathcal{A}(\gamma) - \mathbf{b}\|_2^2 + \lambda\|(\mathcal{I} - \mathcal{D}_{\mathrm{I}})(\gamma)\|_2^2$ where $\mathcal{N}_{\mathrm{I}} = \mathcal{I} - \mathcal{D}_{\mathrm{I}}$ denotes a multi-channel spatial domain CNN and $\mathcal{I}$ is the identity operator. The solution to the formulation alternates between the residual CNN $\mathcal{D}_{\mathrm{I}}$ and the DC block as shown in Fig. 1. The corresponding equations are $\mathbf{z}_n = \mathcal{D}_{\mathrm{I}}(\gamma_n)$ and $\widehat{\gamma}_{n+1} = (\mathcal{A}^H\mathcal{A} + \lambda\mathcal{I})^{-1}(\mathcal{A}^H\mathbf{b} + \lambda\widehat{\mathbf{z}}_n)$. The recursive network structure has weights are shared across iterations. An MSE loss is computed between the predicted image $\gamma$ and the gold standard image $\gamma^{gs}$ obtained from fully sampled measurements. It can be written as $\mathcal{L}_{\mathrm{rec}} = \|\gamma_s - \gamma_s^{gs}\|_2^2$. Here, $\gamma_s$ is the sum-of-squares image obtained from the multi-channel reconstructed image $\gamma$, while $\gamma_s^{gs}$ is the reference (gold standard) sum-of-squares image.

The proposed algorithm does not require any calibration data for estimation of coil sensitivities. Computationally, this approach is orders of magnitude faster than CLEAR since the unrolled algorithm requires fewer iterations to converge. Apart from the multi-channel annihilation relations (2), MR images also satisfy annihilation properties resulting from the redundancy of patches. For instance, smooth regions can be annihilated by derivative filters, while wavelet filters are effective in annihilating texture regions. The proposed CNN framework can exploit these additional annihilation relations, which is a distinction from the k-space approach in [21]; the larger number of annihilation relations can translate to improved performance. Note that the traditional CLEAR approach only exploits the multi-channel relations in (2).

## 3.2    Joint Reconstruction-Segmentation Framework

MRI reconstruction and segmentation are often studied as two different tasks in the MR image analysis pipeline. In the case of a direct cascade of tasks, any aliasing or blurring artifact in reconstructions due to highly accelerated acquisition gets propagated to the segmentation task, leading to degradation of segmentation quality [28]. In this section, we propose a multi-task framework to obtain segmentations that are robust to undersampling artifacts. We also expect the segmentation task to aid the reconstructions. The addition of segmentation network would reduce overfitting errors, when trained with fewer datasets.

The proposed multi-task DL framework is shown in Fig. 1. It consists of an unrolled IDSLR network which is attached to an additional decoder for segmentation at its final iteration. For IDSLR, the CNN $\mathcal{N}_{\mathrm{I}}$ is a UNET with encoder $\mathcal{N}_{\mathrm{IE}}$ and decoder $\mathcal{N}_{\mathrm{ID}}$. The segmentation task dedicated decoder $\mathcal{S}_{\mathrm{ID}}$ is connected to encoder $\mathcal{N}_{\mathrm{IE}}$ from the final iteration. The parameters of encoder $\mathcal{N}_{\mathrm{IE}}$ are shared for both the tasks. We expect that the shared encoder will encourage the learning of common latent variables for both the segmentation and reconstruction tasks, thus exploiting the synergy between them. In particular, we hypothesize that the spatial domain features extracted by the encoder $\mathcal{N}_{\mathrm{IE}}$ from multiple channels would become sharper due to the auxiliary segmentation loss, resulting in improved representation of edges and finer details. The shared encoder architecture also facilitates the learning of the segmentation task from a few labelled datasets as explained below.

A weighted linear combination of reconstruction and segmentation losses is used to train the multi-task network end-to-end. Let $\theta$, $\phi$ and $\psi$ be the weights of $\mathcal{N}_{\text{IE}}$, $\mathcal{N}_{\text{ID}}$, and $\mathcal{S}_{\text{ID}}$, respectively. We used the MSE loss,

$$\mathcal{L}_{\text{rec}}(\gamma_s^{gs}, \gamma_s; \theta, \phi) = \|\gamma_s^{gs} - \gamma_s(\theta, \phi)\|_2^2 \tag{3}$$

as the reconstruction loss. The output of the reconstruction network is only dependent on $\theta$ and $\phi$. The segmentation decoder outputs probability maps of the tissue classes denoted by $p(\theta, \phi, \psi)$; note that the probability maps are dependent on all the network parameters. We compare $p$ against the reference segmentation $z(\mathbf{r})$ using the pixel-level multi-label cross-entropy loss

$$\mathcal{L}_{\text{seg}}(p, z; \theta, \phi, \psi) = -\sum_{\mathbf{r}} z(\mathbf{r}) \log(p(\mathbf{r}; \theta, \phi, \psi)). \tag{4}$$

Here, $p_i(\mathbf{r}; \theta, \phi, \psi)$ is the predicted probabilities at the pixel $\mathbf{r}$ corresponding to the $i^{\text{th}}$ dataset, while $z_i(\mathbf{r})$ is the reference labels obtained from the reference image $\gamma_{s,i}^{gs}(\mathbf{r})$. The total loss is given as

$$\mathcal{L}(\theta, \phi, \psi) = \sum_{i=1}^{N_t} (1 - \alpha) \, \mathcal{L}_{\text{rec}}(\gamma_{s,i}^{gs}, \gamma_{s,i}) + \alpha \, \mathcal{L}_{\text{seg}}(p_i, z_i), \tag{5}$$

where $0 \le \alpha < 1$ is the weight term that controls the strength of each term and $N_t$ is the total number of training datasets. Here, $\mathcal{L}_{\text{rec}}$ and $\mathcal{L}_{\text{seg}}$ are specified by (3) and (4), respectively.

## 3.3    Few-Shot Learning for Semantic Segmentation

Current DL models for medical image segmentation are often trained with large amount of training data. However, obtaining accurate segmentation labels is often a challenging task, requiring high computation time and expert supervision. On the other hand, using a few datasets to train current networks is often associated with the risk of over-fitting. In order to address the limited availability of labelled data, we propose a semi-supervised learning approach capitalizing on our shared network shown in Fig. 1. In particular, we utilize the reconstruction as an auxiliary task for spatial feature learning. We are motivated by [6], where an image denoising block is added as an auxiliary task to keep the generalization error of the segmentation task low.

The proposed multi-task network is trained end-to-end in a supervised fashion with few labelled subjects (10% of training data) for segmentation. We use the combined loss function specified by (5) for datasets with segmentation labels, while we use the reconstruction loss specified by (3) for datasets without segmentation labels. The shared encoder facilitates the learning of latent features

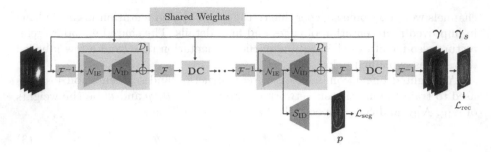

**Fig. 1.** Proposed reconstruction-segmentation architecture.

that are useful for both tasks. The latent variables derived by the shared encoder are learned from all the datasets. Because the segmentation task relies on these latent variables, a few labelled segmentation datasets are sufficient to obtain good generalization performance.

## 4 Implementation Details

Calgary Campinas (CCP) Dataset [27] is used for all the experiments. It consists of raw k-space for T1-weighted brain MRI acquired using a 12-channel coil with a gradient-echo sequence. The acquisition parameters are either TR (repetition time) = 6.3/7.4 ms, TE (echo time) = 2.6/3.1 ms, TI (inversion time) = 650/400 ms. The datasets are split into 40 subjects for training, 7 for validation and 15 for testing. Since, CCP does not provide reference segmentation, we generated reference segmentation using FMRIB's Automated Segmentation Tool (FAST) software [33]. It uses a hidden Markov random field model and solves an associated Expectation-Maximization algorithm to classify the image pixels into CSF, GM and WM. This segmentation task was chosen to demonstrate the technical feasibility of the proposed framework mainly because of the availability of well-established software that is directly applicable to the above dataset with raw multi-channel k-space data. While several segmentation datasets with more challenging tasks are publicly available, most of them rely on post-processed DICOM images that are not appropriate for our setting.

### 4.1 Architecture of the CNNs and Training Details

The IDSLR network in Fig. 1 takes a multi-channel raw k-space $\widehat{\gamma}$ as input that goes through an inverse fast Fourier transform (IFFT) operation before entering the $\mathcal{D}_{\mathrm{I}}$ block. It consists of a residual UNET $\mathcal{N}_{\mathrm{I}}$ [24]. The encoder and decoder of $\mathcal{N}_{\mathrm{I}}$ is denoted as $\mathcal{N}_{\mathrm{IE}}$ and $\mathcal{N}_{\mathrm{ID}}$ respectively. The segmentation decoder $\mathcal{S}_{\mathrm{ID}}$ is attached to $\mathcal{N}_{\mathrm{IE}}$ from the final iteration of the unrolled IDSLR network.

The IDSLR network is unrolled and trained end-to-end for $K = 5$ iterations. We choose $K = 5$ for the other DL models for fair comparisons. For six-fold and eight-fold accelerations, we chose $\alpha = 10^{-4}$ and $\alpha = 10^{-6}$, respectively.

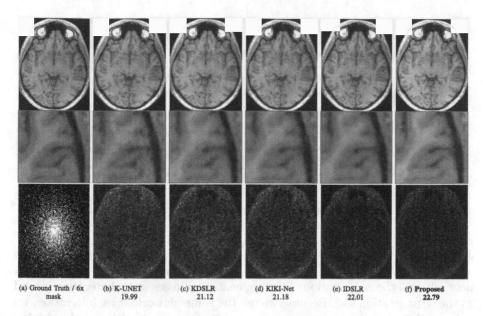

|  (a) Ground Truth / 6x mask | (b) K-UNET 19.99 | (c) KDSLR 21.12 | (d) KIKI-Net 21.18 | (e) IDSLR 22.01 | (f) Proposed 22.79 |

**Fig. 2.** Reconstruction results of 6x accelerated data. SNR (dB) values are reported.

The CNN weights are Xavier initialized and trained for 1000 epochs with the Adam optimizer at a learning rate of $10^{-4}$. We quantitatively evaluate reconstruction quality using the signal-to-noise ratio (SNR) and structural similarity [31] (SSIM) metrics. The segmentation performance is evaluated using Dice coefficients [14].

## 4.2 State-of-the-Art Methods for Comparison

The spatial domain deep-SLR (IDSLR) is compared against KDSLR [21], k-space UNET [11] and a multi-channel KIKI-Net [5] to demonstrate reconstruction performance. The K-UNET approach is a direct inversion scheme that uses a multi-channel UNET in k-space without any DC step. KIKI-Net relies on a cascade of k-space and spatial domain CNNs with a DC step embedded between them. KDSLR is a deep-SLR approach exploiting annihilation relations in k-space. We consider cascading a UNET-based segmentation network SEG, which is pre-trained on fully sampled images, with the above-mentioned recovery methods to study the effect of reconstruction on segmentation quality. KDSLR+SEG, IDSLR+SEG, K-UNET+SEG and KIKI-Net+SEG thus denote the direct cascade of SEG to these recovery methods without any end-to-end training.

We also consider the joint end-to-end training strategies with a cascaded version of our proposed architecture and the JRS scheme [28]. The former is the cascade of a segmentation UNET to the IDSLR network, similar to [12], which is trained end-to-end with the loss function specified by (5). We implemented a multi-channel version of JRS scheme [28] for comparisons. A key distinction of

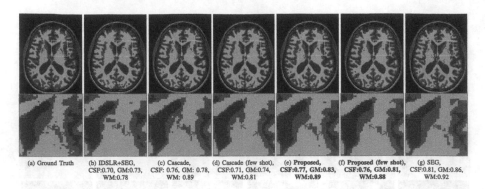

(a) Ground Truth  (b) IDSLR+SEG,  (c) Cascade,  (d) Cascade (few shot),  (e) Proposed,  (f) Proposed (few shot),  (g) SEG,
CSF:0.70, GM:0.73,  CSF: 0.76, GM: 0.78,  CSF:0.71, GM:0.74,  CSF:0.77, GM:0.83,  CSF:0.76, GM:0.81,  CSF:0.81, GM:0.86,
WM:0.78  WM: 0.89  WM:0.81  WM:0.89  WM:0.88  WM:0.92

**Fig. 3.** Segmentation results on 8x accelerated data. Dice coefficients are reported.

JRS with our setting is that the parameters of the autoencoders at each iteration are not shared. The latent variables at each iteration are fed to a segmentation network, and the combination of the segmentation losses at each iteration is used as the segmentation loss. Because we use the same network at each iteration, we only need to connect the segmentation network to the final layer; the sharing of the parameters will ensure that the segmentation task will regularize the projection networks at all iterations.

## 5    Experiments and Results

We now compare the proposed framework against state-of-the-art reconstruction and segmentation algorithms on brain MRI data. The data was retrospectively undersampled in each case using a Cartesian 2D non-uniform variable-density mask. The reconstruction performance figures show magnitude images in gray-scale and their corresponding error images.

### 5.1    Calibration-Free PMRI

Comparisons of proposed IDSLR against other calibrationless methods are reported in Table 1 and Fig. 2, respectively. The experiments were conducted on 12-channel brain images. Comparisons have been made for both six- and eight-fold undersampling. We observe that the KDSLR scheme, which uses interleaved DC steps along with the k-space annihilation relations used in K-UNET, can offer improved performance over K-UNET. KIKI-Net offers a marginally improved performance over KDSLR, while IDSLR offers more than 1 dB improvement over all of the other methods. From Fig. 2, IDSLR is able to minimize alias artifacts while offering sharper edges. The images denoted by the proposed scheme are a combination of the proposed IDSLR approach with a segmentation network as shown in Fig. 1 and were jointly trained end-to-end. This approach offers an additional 0.75 dB improvement in the SNR, which can also be visualized from the zoomed-in images in Fig. 2.

**Table 1.** Quantitative comparison of methods in Fig. 2 with SNR values

| Comparison of reconstruction | | |
|---|---|---|
| Methods | 6-fold | 8-fold |
| IDSLR | 21.99 | 20.53 |
| KDSLR | 21.04 | 20.01 |
| KIKI-Net | 21.12 | 19.93 |
| K-UNET | 19.93 | 19.08 |
| **Proposed** | 22.76 | 21.32 |

## 5.2 Segmentation Quality Comparison

We compare segmentation performances of UNET-based network SEG on the pre-trained reconstruction networks (K-UNET+SEG, KDSLR+SEG, IDSLR+SEG, KIKI-Net+SEG), as shown in Table 2 and Fig. 3, respectively. Among all the methods, the SEG network alone is the best performing since it is trained and tested on fully sampled images. The segmentation performance of other methods is dependent on the reconstruction quality. The IDSLR+SEG provides more accurate labels than other direct cascade methods due to the improved quality of reconstructions (Table 1). The end-to-end training approach, IDSLR-SEG offers better performance than direct cascades, including IDSLR+SEG. IDSLR-SEG is trained with labels for all training datasets. The quantitative results in Table 2 agree with the results in Fig. 3. In particular, the proposed approach offers segmentations that are the closest to SEG. The improved segmentation quality of proposed methods over IDSLR+SEG is evident from Fig. 3.

**Table 2.** Comparison of segmentation performance. Dice coefficient is reported

| Comparison of segmentation | | | | | | |
|---|---|---|---|---|---|---|
| Methods | 6-fold | | | 8-fold | | |
| | CSF | GM | WM | CSF | GM | WM |
| KDSLR+SEG | 0.714 | 0.749 | 0.823 | 0.652 | 0.728 | 0.801 |
| IDSLR+SEG | 0.738 | 0.751 | 0.841 | 0.704 | 0.749 | 0.803 |
| KIKI-Net+SEG | 0.703 | 0.754 | 0.837 | 0.637 | 0.711 | 0.793 |
| KUNET+SEG | 0.642 | 0.741 | 0.788 | 0.611 | 0.692 | 0.747 |
| **Proposed** | 0.793 | 0.841 | 0.898 | 0.767 | 0.826 | 0.889 |
| | CSF | | GM | | WM | |
| SEG | 0.805 | | 0.855 | | 0.913 | |

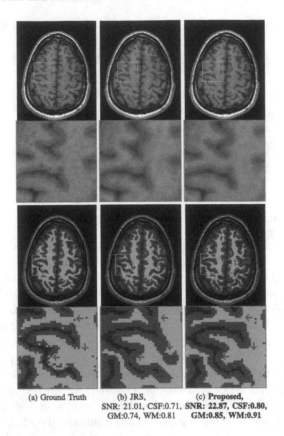

(a) Ground Truth    (b) JRS,    (c) Proposed,
SNR: 21.01, CSF:0.71, SNR: 22.87, CSF:0.80,
GM:0.74, WM:0.81    GM:0.85, WM:0.91

**Fig. 4.** JRS [28] compared with the proposed scheme on 6x accelerated data.

## 5.3   Benefit of Shared Encoder Architecture

We study the performance of different multi-task frameworks in Table 3. The difference in performance between the proposed fully supervised and the few-shot learning approaches is marginal, which is also seen from Fig. 3(e). On the other hand, for cascade architecture, the segmentation performance degrades significantly in the few-shot learning case. Note that in few-shot approaches, segmentation labels are available for only 10% of the data. This trend is mirrored by the reconstruction performances in Table 3. The proposed few-shot approach gives a similar performance compared to the fully supervised one in terms of average SNR. We note from the zoomed-in sections of Fig. 3 that the cascaded network misses some key regions of the gray matter at the center of the slice, denoted by blue arrows. The proposed fully supervised and few-shot learning approaches have preserved those segmentation details and perform at par with the SEG network (segmentation on fully sampled images). In few-shot learning, the multi-task scheme reduces over-fitting errors, thus improving performance.

**Table 3.** Comparison of joint architectures for reconstruction and segmentation quality evaluation. Dice coefficients and SNR values are reported

| Comparison of joint Recon-seg architectures | | | | |
|---|---|---|---|---|
| Methods | SNR | CSF | GM | WM |
| 6-fold undersampling | | | | |
| **Proposed** | 22.76 | 0.793 | 0.841 | 0.898 |
| **Proposed (few shot)** | 22.71 | 0.780 | 0.828 | 0.895 |
| Cascade | 22.75 | 0.795 | 0.838 | 0.891 |
| Cascade (few shot) | 22.39 | 0.726 | 0.773 | 0.840 |
| JRS | 20.89 | 0.705 | 0.721 | 0.796 |
| 8-fold undersampling | | | | |
| **Proposed** | 21.32 | 0.767 | 0.826 | 0.889 |
| **Proposed (few shot)** | 21.28 | 0.761 | 0.817 | 0.886 |
| Cascade | 21.35 | 0.761 | 0.818 | 0.893 |
| Cascade (few shot) | 20.97 | 0.711 | 0.747 | 0.824 |
| JRS | 19.74 | 0.683 | 0.702 | 0.776 |

## 5.4 Comparison with State-of-the-Art

Proposed method is compared against the JRS in Table 3 and Fig. 4. In JRS, the weights of the reconstruction network are not shared across the iterations, while the segmentation decoders are shared. Hence, if $K$ denotes the number of iterations of the reconstruction network, then there are $K$ times more trainable parameters than the proposed method. In the JRS scheme, the segmentation output of decoders from all the iterations are linearly combined to obtain a final segmentation. Both reconstruction and segmentation quality of the proposed scheme are superior to JRS; we note a reduction in blurring and improved preservation of details in the proposed reconstruction. JRS segmentation shows errors in the regions indicated by arrows which are not present in the proposed segmentation. Thus, there is a significant improvement in segmentation quality.

## 6   Discussion

Comparison of the proposed deep-SLR scheme (IDSLR) against other calibrationless methods in Table 1 and Fig. 2 shows the benefit of combining image domain annihilation relations (IDSLR) with the segmentation task. The improved performance of KDSLR over K-UNET can be attributed to the model-based strategy that offers improved data consistency. We observe that the proposed IDSLR scheme offers more than 1 dB of improvement over all of the other methods, which can be evidenced by the lower alias artifacts and sharper edges shown in Fig. 2. The improved performance over KDSLR [21] can be attributed to the use of the spatial domain annihilation relations that result from the multi-channel acquisition scheme, as well as those resulting from the redundancy of patches. The additional improvement in the reconstruction performance offered by the proposed scheme, which uses a linear combination of segmentation and reconstruction losses, may be slightly counter-intuitive from an optimization perspective. However, this improved performance confirms our hypothesis that the exploitation of the synergies between the segmentation and reconstruction tasks will aid both tasks. In particular, the addition of the segmentation task encourages the network to learn improved latent representations, resulting in improved preservation of edges and details in the reconstructed images.

The segmentation results in Table 2 and Fig. 3 clearly show the benefit of the joint training strategy IDSLR-SEG compared to straightforward cascading of a pre-trained segmentation network with reconstruction algorithms, including IDSLR+SEG. In particular, the SEG scheme that was pre-trained on fully sampled images will offer lower performance in the presence of blurring and alias artifacts. By contrast, the proposed joint training strategy IDSLR-SEG offers improved segmentation, evidenced visually and by the lower Dice scores. The slight degradation in the segmentation accuracy of IDSLR-SEG over the SEG approach, where fully sampled images are used, is the price to pay for the acceleration.

The performance of the few-shot training Table 3 shows that the proposed shared encoder network reduces over-fitting and thus, offers minimal degradation in segmentation and reconstruction performance. By contrast, the few-shot cascade learning degrades with reduction in ground truth availability. The shared encoder enables the segmentation network to reduce generalization errors, even though it is trained with few subjects. Since the encoder is trained with a supervised reconstruction loss from several subjects, it learns a very robust feature representation, thus reducing the risk of overfitting in the segmentation network. In contrast to the few-shot learning of the cascade network, the segmentation algorithm fails to learn enough features due to insufficient training data. We also note that this trend is mirrored by the reconstruction performances in Table 3. The proposed few-shot approach gives similar performance to the fully supervised one in terms of average SNR.

# 7    Conclusion

We introduced a novel image-domain deep structured low-rank algorithm (IDSLR) for calibrationless PMRI recovery. The proposed IDSLR scheme is a DL-based non-linear extension of locally low-rank approaches such as CLEAR. The network learns annihilation relations in the spatial domain obtained from the smoothness assumption on coil sensitivities. IDSLR exploits more annihilation relations than the k-space approach KDSLR, thus offering improved performance. To reduce the impact of undersampling on downstream tasks such as MRI segmentation, we proposed a joint reconstruction-segmentation framework trained in an end-to-end fashion. An IDSLR network is combined with a

UNET for segmentation such that the encoder is shared between the tasks. The proposed joint training strategy outperforms the direct cascade of pre-trained reconstruction and segmentation algorithms, in addition to improving the reconstruction performance. Proposed IDSLR-SEG network with a shared encoder for segmentation and reconstruction tasks enables a few-shot learning strategy, which reduces the number of segmented datasets for training by a factor of ten.

# References

1. Aggarwal, H.K., Mani, M.P., Jacob, M.: MoDL: model-based deep learning architecture for inverse problems. IEEE Trans. Med. Imaging **38**(2), 394–405 (2018)
2. Carlesimo, G.A., Piras, F., Orfei, M.D., Iorio, M., Caltagirone, C., Spalletta, G.: Atrophy of presubiculum and subiculum is the earliest hippocampal anatomical marker of Alzheimer's disease. Alzheimer's Dement. Diagn. Assess. Dis. Monit. **1**(1), 24–32 (2015)
3. Chételat, G.: Multimodal neuroimaging in Alzheimer's disease: early diagnosis, physiopathological mechanisms, and impact of lifestyle. J. Alzheimers Dis. **64**(s1), S199–S211 (2018)
4. De Flores, R., La Joie, R., Chételat, G.: Structural imaging of hippocampal subfields in healthy aging and Alzheimer's disease. Neuroscience **309**, 29–50 (2015)
5. Eo, T., Jun, Y., Kim, T., Jang, J., Lee, H.J., Hwang, D.: KIKI-net: cross-domain convolutional neural networks for reconstructing undersampled magnetic resonance images. Magn. Reson. Med. **80**(5), 2188–2201 (2018)
6. Feyjie, A.R., Azad, R., Pedersoli, M., Kauffman, C., Ayed, I.B., Dolz, J.: Semi-supervised few-shot learning for medical image segmentation. arXiv preprint arXiv:2003.08462 (2020)
7. Gidaris, S., Bursuc, A., Komodakis, N., Pérez, P., Cord, M.: Boosting few-shot visual learning with self-supervision. In: Proceedings of the IEEE/CVF International Conference on Computer Vision, pp. 8059–8068 (2019)
8. Griswold, et al.: Generalized autocalibrating partially parallel acquisitions (GRAPPA). Magn. Reson. Med. Official J. Int. Soc. Magn. Reson. Med. **47**(6), 1202–1210 (2002)
9. Haldar, J.P.: Low-rank modeling of local $k$-space neighborhoods (LORAKS) for constrained MRI. IEEE Trans. Med. Imaging **33**(3), 668–681 (2013)
10. Hammernik, et al.: Learning a variational network for reconstruction of accelerated MRI data. Magn. Reson. Med. **79**(6), 3055–3071 (2018)
11. Han, Y., Sunwoo, L., Ye, J.C.: K-space deep learning for accelerated MRI. IEEE Trans. Med. Imaging **39**(2), 377–386 (2019)
12. Huang, Q., Yang, D., Yi, J., Axel, L., Metaxas, D.: FR-Net: joint reconstruction and segmentation in compressed sensing cardiac MRI. In: Coudière, Y., Ozenne, V., Vigmond, E., Zemzemi, N. (eds.) FIMH 2019. LNCS, vol. 11504, pp. 352–360. Springer, Cham (2019). https://doi.org/10.1007/978-3-030-21949-9_38
13. Iglesias, et al.: A computational atlas of the hippocampal formation using ex vivo, ultra-high resolution MRI: application to adaptive segmentation of in vivo MRI. Neuroimage **115**, 117–137 (2015)
14. Jadon, S.: A survey of loss functions for semantic segmentation. In: 2020 IEEE Conference on Computational Intelligence in Bioinformatics and Computational Biology (CIBCB), pp. 1–7. IEEE (2020)

15. Liang, D., Liu, B., Wang, J., Ying, L.: Accelerating SENSE using compressed sensing. Magn. Reson. Med. Official J. Int. Soc. Magn. Reson. Med. **62**(6), 1574–1584 (2009)

16. Liu, D., Wen, B., Liu, X., Wang, Z., Huang, T.S.: When image denoising meets high-level vision tasks: a deep learning approach. In: Proceedings of the 27th International Joint Conference on Artificial Intelligence, pp. 842–848 (2018)

17. Lüsebrink, F., Wollrab, A., Speck, O.: Cortical thickness determination of the human brain using high resolution 3 T and 7 T MRI data. Neuroimage **70**, 122–131 (2013)

18. Lustig, M., Donoho, D., Pauly, J.M.: Sparse MRI: the application of compressed sensing for rapid MR imaging. Magn. Reson. Med. Official J. Int. Soc. Magn. Reson. Med. **58**(6), 1182–1195 (2007)

19. Mani, M., Jacob, M., Kelley, D., Magnotta, V.: Multi-shot sensitivity-encoded diffusion data recovery using structured low-rank matrix completion (MUSSELS). Magn. Reson. Med. **78**(2), 494–507 (2017)

20. Oksuz, et al.: Deep learning-based detection and correction of cardiac MR motion artefacts during reconstruction for high-quality segmentation. IEEE Trans. Med. Imaging **39**(12), 4001–4010 (2020)

21. Pramanik, A., Aggarwal, H.K., Jacob, M.: Deep generalization of structured low-rank algorithms (Deep-SLR). IEEE Trans. Med. Imaging **39**(12), 4186–4197 (2020)

22. Pruessmann, K.P., Weiger, M., Scheidegger, M.B., Boesiger, P.: SENSE: sensitivity encoding for fast MRI. Magn. Reson. Med. Official J. Int. Soc. Magn. Reson. Med. **42**(5), 952–962 (1999)

23. Pruessner, et al.: Volumetry of hippocampus and amygdala with high-resolution MRI and three-dimensional analysis software: minimizing the discrepancies between laboratories. Cereb. Cortex **10**(4), 433–442 (2000)

24. Ronneberger, O., Fischer, P., Brox, T.: U-Net: convolutional networks for biomedical image segmentation. In: Navab, N., Hornegger, J., Wells, W.M., Frangi, A.F. (eds.) MICCAI 2015. LNCS, vol. 9351, pp. 234–241. Springer, Cham (2015). https://doi.org/10.1007/978-3-319-24574-4_28

25. Schlemper, J., Caballero, J., Hajnal, J.V., Price, A.N., Rueckert, D.: A deep cascade of convolutional neural networks for dynamic MR image reconstruction. IEEE Trans. Med. Imaging **37**(2), 491–503 (2017)

26. Shin, et al.: Calibrationless parallel imaging reconstruction based on structured low-rank matrix completion. Magn. Reson. Med. **72**(4), 959–970 (2014)

27. Souza, et al.: An open, multi-vendor, multi-field-strength brain MR dataset and analysis of publicly available skull stripping methods agreement. NeuroImage **170**, 482–494 (2018)

28. Sun, L., Fan, Z., Ding, X., Huang, Y., Paisley, J.: Joint CS-MRI reconstruction and segmentation with a unified deep network. In: Chung, A.C.S., Gee, J.C., Yushkevich, P.A., Bao, S. (eds.) IPMI 2019. LNCS, vol. 11492, pp. 492–504. Springer, Cham (2019). https://doi.org/10.1007/978-3-030-20351-1_38

29. Trzasko, J.D., Manduca, A.: CLEAR: calibration-free parallel imaging using locally low-rank encouraging reconstruction. In: Proceedings of the 20th Annual Meeting of the International Society for Magnetic Resonance in Medicine (ISMRM), vol. 517 (2012)

30. Uecker, et al.: ESPIRiT-an eigenvalue approach to autocalibrating parallel MRI: where SENSE meets GRAPPA. Magn. Reson. Med. **71**(3), 990–1001 (2014)

31. Wang, Z., Bovik, A.C., Sheikh, H.R., Simoncelli, E.P.: Image quality assessment: from error visibility to structural similarity. IEEE Trans. Image Process. **13**(4), 600–612 (2004)

32. Zaitsev, M., Maclaren, J., Herbst, M.: Motion artifacts in MRI: a complex problem with many partial solutions. J. Magn. Reson. Imaging **42**(4), 887–901 (2015)
33. Zhang, Y., Brady, M., Smith, S.: Segmentation of brain MR images through a hidden Markov random field model and the expectation-maximization algorithm. IEEE Trans. Med. Imaging **20**(1), 45–57 (2001)

# Patient-Level Microsatellite Stability Assessment from Whole Slide Images by Combining Momentum Contrast Learning and Group Patch Embeddings

Daniel Shats[1]([✉])[ID], Hadar Hezi[2], Guy Shani[3], Yosef E. Maruvka[3][ID], and Moti Freiman[2][ID]

[1] Faculty of Computer Science, Technion, Haifa, Israel
shats@campus.technion.ac.il
[2] Faculty of Biomedical Engineering, Technion, Haifa, Israel
[3] Faculty of Biotechnology and Food Engineering, Technion, Haifa, Israel

**Abstract.** Assessing microsatellite stability status of a patient's colorectal cancer is crucial in personalizing treatment regime. Recently, convolutional-neural-networks (CNN) combined with transfer-learning approaches were proposed to circumvent traditional laboratory testing for determining microsatellite status from hematoxylin and eosin stained biopsy whole slide images (WSI). However, the high resolution of WSI practically prevent direct classification of the entire WSI. Current approaches bypass the WSI high resolution by first classifying small patches extracted from the WSI, and then aggregating patch-level classification logits to deduce the patient-level status. Such approaches limit the capacity to capture important information which resides at the high resolution WSI data. We introduce an effective approach to leverage WSI high resolution information by momentum contrastive learning of patch embeddings along with training a patient-level classifier on groups of those embeddings. Our approach achieves up to 7.4% better accuracy compared to the straightforward patch-level classification and patient level aggregation approach with a higher stability (AUC, $0.91 \pm 0.01$ vs. $0.85 \pm 0.04$, p-value $< 0.01$). Our code can be found at https://github.com/TechnionComputationalMRILab/colorectal_cancer_ai.

**Keywords:** Digital pathology · Colorectal cancer · Self-supervised learning · Momentum contrast learning

## 1 Introduction

Colorectal cancer is a heterogeneous type of cancer which can be generally classified into one of two groups based on the condition of Short Tandem Repeats (STRs) within the DNA of a patient. These two groups are known as Microsatellite Stable (MSS) and Microsatellite Instable (MSI). When the series

© The Author(s), under exclusive license to Springer Nature Switzerland AG 2023
L. Karlinsky et al. (Eds.): ECCV 2022 Workshops, LNCS 13803, pp. 454–465, 2023.
https://doi.org/10.1007/978-3-031-25066-8_25

of nucleotides in the central cores of the STRs exhibit a highly variable number of core repeating units, the patient sample is referred to as Microsatellite Instable. Otherwise, if the number nucleotides is consistent, that patient sample is referred to as Microsatellite Stable (MSS) [18]. Microsatellite instability (MSI) exists in approximately 15% of colorectal cancer patients. Due to the fact that patients with MSI have different treatment prospects than those without MSI, it is critical to know if a patient exhibits this pathology before determining treatment direction [2].

Currently, it is possible to determine microsatellite status in a patient by performing various laboratory tests. While these methods are effective, they are expensive and take time, resulting in many patients not being tested for it at all. Therefore, there exist many recent efforts aimed toward utilizing deep-learning-based methods to uncover a computational solution for the detection of MSI/MSS status from hematoxylin and eosin (H&E) stained biopsy whole slide images (WSI).

However, the WSIs have an extremely large resolution, often reaching over a billion pixels (Fig. 1). Since neural networks can operate, due to computational constraints, only on relatively low resolution data, the input WSI must be shrunk down in some way to a size that is manageable by today's models and hardware. A relatively naïve approach to do this is by down sampling inputs to approximately the same resolution as Image Net (which also allows one to leverage transfer learning) [10,16]. While this is not detrimental for natural images, where fine detail might not be critical, the same cannot be said for WSI of human tissue where the information of concern may exist in various scales. Our work presents a novel method for effectively down sampling patches into lower dimensional embeddings while preserving features critical to making clinical decisions. Due to our effective dimensional reduction of individual patches, we are then able to concatenate the features of multiple patches together and make clinical decisions using inter-patch information.

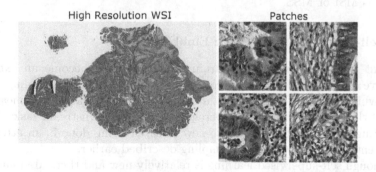

**Fig. 1.** Example of an single patient WSI and a subset of the patches generated from the WSI. The resolution of the WSI is 45815 × 28384 and the corresponding patches are 512 × 512.

## 1.1   CNN-Based MSI/MSS Classification from WSI Data

A simple and straightforward method to alleviate the above issue is by cutting WSI's into many patches and then applying modern deep learning methods to these individual patches as if each of them were a separate image from a data set [10,16]. Once the input has been tessellated, there are a few known ways of training a model from them. One way is to simply assign a label to every patch based on the patient from which they are derived, and then train a model on a patch-by-patch basis. Once the model has been trained, its outputs on all the patches can be averaged for a patient-level classification. Described mathematically, the inference procedure is described below:

Suppose we have some trained classifier $F$ (which returns the probability that a patch belongs to class MSI or MSS e.g. $0 \leq F(x) \leq 1$ for some input patch $x$), a whole slide image $W$, and $n$ patches extracted from $W$ such that $\{p_1, p_2, ...p_n\} \in W$. Then, a patient level probability prediction $P_W$ (on the WSI) can be formulated as such:

$$P_W = \frac{1}{n} \sum F(p_n) \tag{1}$$

And given some classification threshold $0 \leq t \leq 1$, we can arrive at a final classification $C_W$ for the patient:

$$C_W = \begin{cases} MSS & P_W < t \\ MSI & P_W \geq t \end{cases} \tag{2}$$

However, such approaches practically ignore the fact that much of the information critical to making an informed decision on a patient level may reside in the high resolution and inter-patch space. Further, the classification of the patches based on the patient-level data may result in incorrect classification as not necessarily all patches are contributing equally to the classification of a patient as MSI or MSS.

## 1.2   Self-supervision for Patch Embbedings

In recent years self-supervised learning methods have become an extremely attractive replacement for autoencoders as encoding or downsampling mechanisms, while learning features that are much more informative and meaningful [19] and therefore can be easily leveraged for use in downstream tasks. This is an exciting property that has led to new research being done in an attempt to circumvent the issues with downsampling described earlier.

Although self-supervised learning is relatively new and there are many algorithms that attempt to achieve effectively the same goal [3,11,12]. Of the various methods, contrastive learning techniques such as those introduced in SimCLR [7] have become very popular due to their high efficacy and simplicity.

Recently there have been some attempts to use self-supervised contrastive learning, and more specifically the aforementioned SimCLR algorithm, to aid in classification of WSI imagery. Unfortunately SimCLR requires large computational resources to train in a reasonable amount of time. That is why we decided to test the advent of Momentum Contrast Learning with MoCo v2 by Chen et al. [8]. This framework relies on storing a queue of previously encoded samples to overcome the large batch size requirement of SimCLR while seemingly improving downstream classification accuracy as well.

## 2 Related Work

### 2.1 CNN-Based MSS/MSI Classification

Building on top of Echle [10] and the Eqs. 1 and 2 was that of Kather et al. [16]. Here, transfer learning via pretraining with ImageNet [9] was employed to marginally improve results of this straightforward method. Although it has an enormous amount of images, they are not medical, and certainly not pictures of H&E stained biopsy slides. Therefore the degree to which the learned features from ImageNet transfer well to H&E stained WSI imagery is arguably negligible.

It is also important to discuss the particular resolution under which the patches were acquired. Due to the small size of the patches, any individual patch may not be large enough to contain the information required to make a classification (even on the tissue contained within only that patch). One must understand whether or not the task at hand requires intra-cellular information (requiring maximum slide resolution) or tissue-level information (requiring downsampling before patching). Unfortunately in either case, it is also possible that information at multiple levels of resolution is required for optimal results.

Still, more drawbacks can be found tessellating high resolution images into many smaller patches, regardless of patch resolution concerns. For one, the model cannot learn inter-patch information. This is especially important considering that it is very possible that a majority of the patches do not actually contain targeted tissue. Moreover, training the model in such a way is misleading, considering that many patches which have MSS tissue (yet are found on an MSI classified patient) will be marked as MSI for the model. This is likely to result in the model learning less than optimal features.

The work by Hemati et al. [14] gets around this issue by creating a mosaic of multiple patches per batch during training. Unfortunately, this creates other drawbacks. Most notably, they cannot use all the patches per patient, and so they use another algorithm which is mutually exclusive to the learning procedure in order to choose patches, with no guarantee that they contain targeted tissue. Moreover, the mosaic of these selected patches is also still limited by resolution, and so they still must scale down the patches from their original resolution before training.

An improvement on all these previous works was done in the research by Bilal et al. [1]. Most notably, they advanced upon the work from Hemati [14] by learning the patch extraction, or as they call it, patch detection, using a neural network as well. Thereby alleviating an inductive bottleneck. Their process also includes significant work surrounding intermittent detection of known biologically important features to such a problem, such as doing nuclear segmentation, and then providing that information to the next model to make a better-informed decision.

## 2.2   Self Supervision for Patch Embbedings

Due to self-supervised learning being a fairly recent invention, the works similar to ours which cite using it are rather sparse. One of the works which explores the validity of using these methods in the first place is that of Chen et al. [6]. They show that features learned through self-supervised contrastive learning are robust and can be learned from in a downstream fashion. Another paper that uses a similar two stage approach with self-supervised learning being the first stage is DeepSMILE by Schirris et al. [21]. They used a similar approach to the above mentioned contrastive self-supervised learning step with SimCLR, but learned on the features using Deep Multiple Instance Learning (MIL) [15]. While this approach was effective, the computational requirements of SimCLR and the added complexity of MIL may keep the advent of this research out of the hands of many researchers.

Very recently an improvement on the work by Chen [6] was introduced in their research using Hierarchical Vision transformers [5]. Here, the authors apply self-supervised learning through DINO [4] to train 3 stages of vision transformers in achieving entire WSI level classifications. Though seemingly effective, the increased complexity of their approach and the necessity of utilizing transformers makes it relatively inflexible.

## 2.3   Hypothesis and Contributions

*Hypothesis:* Learning effective patch embeddings with self-supervised learning and training a small classifier on groups of those embeddings is more effective than either training on down-sampled WSI's or training on individual patch embeddings and averaging the classification for a patient.

We believe this hypothesis to be true due to the ability of a network to learn inter-patch information at an embedding level. This way, information that is encoded in one patch can impact the decision of the entire WSI. Our contribution is an intuitive and elegant framework that improves patient classification accuracy by up to 7.4%. We argue that this method is very simple to understand and has many avenues of possible improvement. Specifically when considering the initial feature extraction stage, there are many other self-supervised representation learning methods that can be tested and directly compared using our approach.

## 3   Data

The training and validation data used in our method consists of the COAD and READ cohorts of The Cancer Genome Atlas (TCGA) [22]. Out of a total of 360 unique patients, the train set is comprised of 260 patients and the validation set is comprised of 100 patients, where each patient is equivalent to one WSI. Each of these WSIs were tessellated into patches of size $512 \times 512$ pixels at $0.5\,\mu\mathrm{m/px}$ and then downsampled to $224 \times 224$ (this was done only for comparison with Kather et al. [16]). Next the probability of each patch to contain cancerous tissue was computed by a trained CNN and only the patches which were likely to have cancer tissue were kept. Finally, the patches were color normalized. Further detail on the data preprocessing procedure can be found in the paper by Kather et al. [17]. The 260 train patients were tessellated into 46,704 MSS labeled patches, and 46,704 MSIMUT labeled patches. The 100 validation patients were tessellated into 70,569 MSS labeled patches, and 28,335 MSIMUT labeled patches.

Finally, we also ran some final experiments on a more balanced subset of the validation dataset, referred to later in this paper as the "Balanced Validation Set". It comprises of 15 MSS patients and 15 MSIMUT patients which all have a relatively similar distribution of patches extracted from them. 7281 patches were extracted from the MSIMUT patients and 7446 patches were extracted from the MSS patients.

## 4 Method and Model

### 4.1 Overview

Our method comprises of two main training stages (Fig. 2):

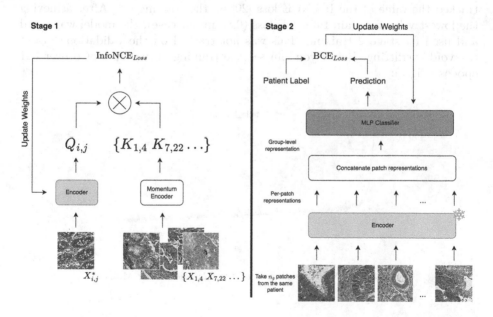

**Fig. 2.** Both stages of our proposed model.

Stage 1 utilizes MoCo and stands to generate robust patch level embeddings that encode patches in a way such that they can be learned from in a downstream fashion. Above, $Q_{i,j}$ represents the encoded query for patient $i$ and patch $j$. Similarly $K_{i,j}$ represents the same for patches encoded by the momentum encoder stored in the queue, and $X_{i,j}$ corresponds to individual samples from the dataset relating to patient $i$ and patch $j$. The stage 1 diagram is very similar to that found in Chen et al. [8].

Stage 2 groups the patches so that their features can be aggregated and the head of our model can learn from a set of patches as opposed to an individual sample. The encoder in stage 2 is a frozen copy of the trained encoder from stage 1. The snowflake indicates that its gradients are not tracked.

### 4.2 Stage 1: Training a Self-supervised Feature Extractor for Patch-Level Embeddings

In stage 1 of training, our feature extractor is trained in exactly the same way as described in the MoCo v2 paper [8]. Data loading and augmentation are unchanged. The main difference is our use of a Resnet18 [13] backbone as opposed to the Resnet50 (C4) backbone which was tested in the original implementation. This was done due to computational constraints and for comparison to the baseline approach from Kather. We also used cosine learning rate annealing, which seems to improve training. The output dimension of our feature extractor is 512 ($n_o$).

To evaluate the ability of MoCo to extract usable features from patches, we tracked the value of the InfoNCE loss [20] on the training set. After achieving the lowest value for train InfoNCE loss (0.88 in our case), the model was saved and used for stage 2 training. This was not tracked on the validation dataset to avoid overfitting. Below you can see the training curve for MoCo over 621 epochs (Fig. 3):

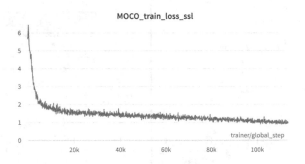

**Fig. 3.** InfoNCE training loss over 621 epochs for MoCo.

The goal of contrastive learning, and thus momentum contrast learning, is to (as stated in the paper by He et al.) learn an encoder for a dictionary look up task. The contrastive loss function InfoNCE exists to return a low loss when encoded queries to the dictionary are similar to keys, and a high loss otherwise.

## 4.3  Stage 2: Training a Supervised Classifier on Patch Embedding Groups

In stage 2, the resnet18 [13] feature extractor trained by MOCO is frozen, and so gradients are not tracked. When making the forward pass, features extracted from patches are grouped by $n_g$ (group size), meaning they are concatenated into one long vector. The length of this vector ($l_g$) will be:

$$l_g = n_g * n_o \tag{3}$$

Meaning that the input dimension of our multi-layer percpetron (MLP) group-level classifier, or model head that we are training in stage 2 must have an input dimension of $l_g$. This brings us to the first issue regarding the $n_g$ parameter. The larger the group size after feature extraction, the larger the first layer of the head must be. This is likely why we found a group size of 4 to be optimal for our dataset. When using a larger group size, the number of parameters for the head of the model increases dramatically, and it tends to overfit much faster.

As an interesting test, we also attempted $n_g = 1$, which performed very similarly to the standard approach. This is what we expected as it indicates the embedding space from Momentum Contrastive Learning is similarly effective to the embedding space of a model trained in a supervised fashion.

## 4.4  Evaluation

Judgement of the algorithm is performed using two main criteria. The first is patch level accuracy and the second is patient level accuracy.

Patch level accuracy ($A_{patch}$) is exactly the same as accuracy in the general context. The only caveat is how the patches are assigned their label. Since our WSI's are labeled on a patient basis, the patches are labeled by inheriting the label of the patient to which they belong.

$$A_{patch} = \frac{\text{Number of Correctly Predicted Patches}}{\text{Total Number of Patches}} \tag{4}$$

Patient level accuracy ($A_{patient}$) is a more crucial and more difficult to improve upon metric. It cannot be trained for directly, as an entire WSI cannot fit on GPU without downsampling. To measure this metric, we must save the models predictions on individual patches (or on groups of patches and extrapolate individual patch predictions) and calculate a final prediction for a patient using the cumulative predictions of its constituent patches. This can be done using a majority vote approach or it is also possible to treat each patches prediction as a probability and average the probabilities before thresholding on a patient level and achieving a final prediction.

$$A_{patient} = \frac{\text{Number of Correctly Predicted Patients}}{\text{Total Number of Patients}} \tag{5}$$

## 5   Experiments and Results

### 5.1   Standard Dataset

The results in this section refer to performance measured on the original dataset processed by Kather et al. [17]. We show the results of our implementation of Kather's method compared to our improvement using momentum contrastive learning and group patch embeddings (Table 1 and Fig. 4).

**Table 1.** Accuracy comparison on standard dataset. Both methods trained to 100 epochs. These are the validation results of an average of 10 runs per method. Our method achieves significantly higher accuracy in both patient (paired t-test, $p < 0.001$), and patch level (paired t-test, $p \ll 1e-6$) evaluation while also having a much more stable result given its smaller standard deviation.

| Method | Patient accuracy | Patch accuracy |
|---|---|---|
| Ours | **0.862** $\pm$ 0.006 | **0.797** $\pm$ 0.005 |
| Kather et al. | 0.837 $\pm$ 0.016 | 0.716 $\pm$ 0.015 |

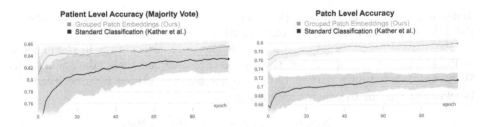

**Fig. 4.** Validation accuracy during training. Average of 10 runs.

Due to the feature extractor already having been learned, our method initially trains much faster than the baseline. We have even noted that for some hyperparameter combinations it may be most effective to stop training after only a few epochs. And below are the ROC curves for the above models (Fig. 5):

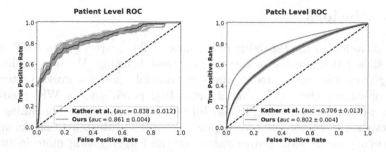

**Fig. 5.** ROC curves on standard validation dataset. Average of 10 runs. Our method achieves a significantly higher AUC in both patient (paired t-test, p < 0.01) and patch level evaluation (paired t-test, p ≪ 1e−7).

## 5.2 Balanced Validation Set

The results in this section refer to performance measured on the balanced subset of the validation dataset processed by Kather et al. [17]. We describe the composition of this dataset in the second paragraph of the data section of this paper. Our method does even better on this balanced validation set compared to the original one from above. The differences were significant for both patient level (paired t-test, p ≪ 1e−4), and patch level (paired t-test, p ≪ 1e−6) classification. This suggest that our method is less prone to bias and overfitting to patients with more or less patches that have been extracted from them (Table 2 and Fig. 6).

**Table 2.** Accuracy comparison on balanced dataset.

| Method | Patient accuracy | Patch accuracy |
|---|---|---|
| Ours | **0.797** ± 0.010 | **0.751** ± 0.006 |
| Kather et al. | 0.723 ± 0.026 | 0.662 ± 0.013 |

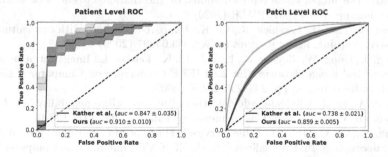

**Fig. 6.** ROC curves on balanced dataset. Average of 10 runs. Our method achieves a significantly higher AUC in both patient (paired t-test, p < 0.01) and patch level evaluation (paired t-test, p ≪ 1e−6).

# 6 Conclusions

Our work validates the feasibility of learning usable features from H&E stained biopsy slide patches using momentum contrast learning. We also qualify that learning from the aggregated features of multiple patches works better than simply averaging the predictions of individual patches for a WSI prediction. Finally, we contribute a simple and intuitive framework for combining these concepts with huge potential for improvement. The future for this domain lies in improving patch level feature extraction and aggregating more features to make global WSI decisions. The advent of a WSI classifier that is as accurate as laboratory testing for microsatellite status can drastically improve the rate at which patients are diagnosed and their treatment prospects.

# References

1. Bilal, M., et al.: Development and validation of a weakly supervised deep learning framework to predict the status of molecular pathways and key mutations in colorectal cancer from routine histology images: a retrospective study. Lancet Digit. Health **3**, e763–e772 (2021)
2. Boland, C.R., Goel, A.: Microsatellite instability in colorectal cancer. Gastroenterology **138**(6), 2073–2087 (2010)
3. Caron, M., Misra, I., Mairal, J., Goyal, P., Bojanowski, P., Joulin, A.: Unsupervised learning of visual features by contrasting cluster assignments. In: Advances in Neural Information Processing Systems, vol. 33, pp. 9912–9924 (2020)
4. Caron, M., et al.: Emerging properties in self-supervised vision transformers. In: Proceedings of the IEEE/CVF International Conference on Computer Vision, pp. 9650–9660 (2021)
5. Chen, R.J., et al.: Scaling vision transformers to gigapixel images via hierarchical self-supervised learning. In: Proceedings of the IEEE/CVF Conference on Computer Vision and Pattern Recognition, pp. 16144–16155 (2022)
6. Chen, R.J., Krishnan, R.G.: Self-supervised vision transformers learn visual concepts in histopathology. arXiv preprint arXiv:2203.00585 (2022)
7. Chen, T., Kornblith, S., Norouzi, M., Hinton, G.: A simple framework for contrastive learning of visual representations. In: International Conference on Machine Learning, pp. 1597–1607. PMLR (2020)
8. Chen, X., Fan, H., Girshick, R., He, K.: Improved baselines with momentum contrastive learning. arXiv preprint arXiv:2003.04297 (2020)
9. Deng, J., Dong, W., Socher, R., Li, L.J., Li, K., Fei-Fei, L.: ImageNet: a large-scale hierarchical image database. In: 2009 IEEE Conference on Computer Vision and Pattern Recognition, pp. 248–255. IEEE (2009)
10. Echle, A., et al.: Clinical-grade detection of microsatellite instability in colorectal tumors by deep learning. Gastroenterology **159**(4), 1406–1416 (2020)
11. Feng, Z., Xu, C., Tao, D.: Self-supervised representation learning by rotation feature decoupling. In: Proceedings of the IEEE/CVF Conference on Computer Vision and Pattern Recognition, pp. 10364–10374 (2019)
12. Grill, J.B., et al.: Bootstrap your own latent-a new approach to self-supervised learning. In: Advances in Neural Information Processing Systems, vol. 33, pp. 21271–21284 (2020)

13. He, K., Zhang, X., Ren, S., Sun, J.: Deep residual learning for image recognition. In: Proceedings of the IEEE Conference on Computer Vision and Pattern Recognition, pp. 770–778 (2016)
14. Hemati, S., Kalra, S., Meaney, C., Babaie, M., Ghodsi, A., Tizhoosh, H.: CNN and deep sets for end-to-end whole slide image representation learning. In: Medical Imaging with Deep Learning (2021)
15. Ilse, M., Tomczak, J., Welling, M.: Attention-based deep multiple instance learning. In: International Conference on Machine Learning, pp. 2127–2136. PMLR (2018)
16. Kather, J.N., et al.: Deep learning can predict microsatellite instability directly from histology in gastrointestinal cancer. Nat. Med. **25**(7), 1054–1056 (2019)
17. Kather, J.: Histological images for MSI vs. MSS classification in gastrointestinal cancer, FFPE samples. ZENODO (2019)
18. Li, K., Luo, H., Huang, L., Luo, H., Zhu, X.: Microsatellite instability: a review of what the oncologist should know. Cancer Cell Int. **20**(1), 1–13 (2020)
19. Liu, X., et al.: Self-supervised learning: Generative or contrastive. IEEE Trans. Knowl. Data Eng. **35**, 857–876 (2021)
20. van den Oord, A., Li, Y., Vinyals, O.: Representation learning with contrastive predictive coding. arXiv preprint arXiv:1807.03748 (2018)
21. Schirris, Y., Gavves, E., Nederlof, I., Horlings, H.M., Teuwen, J.: DeepSmile: self-supervised heterogeneity-aware multiple instance learning for dna damage response defect classification directly from H&E whole-slide images. arXiv preprint arXiv:2107.09405 (2021)
22. Weinstein, J.N., et al.: The cancer genome atlas pan-cancer analysis project. Nat. Genet. **45**(10), 1113–1120 (2013)

# Segmenting Glandular Biopsy Images Using the Separate Merged Objects Algorithm

David Sabban[iD] and Ilan Shimshoni[✉][iD]

University of Haifa, 31905 Haifa, Israel
ishimshoni@is.haifa.ac.il

**Abstract.** The analysis of the structure of histopathology images is crucial in determining whether biopsied tissue is benign or malignant. It is essential in pathology to be precise and, at the same time, to be able to provide a quick diagnosis. These imperatives inspired researchers to automate the process of segmenting and diagnosing biopsies. The main approach is to utilize semantic segmentation networks. Our research presents a post-processing algorithm that addresses one weakness of the semantic segmentation method - namely, the separation of close objects that have been mistakenly merged by the classification algorithm. If two or more objects have been merged, the object can be mis-classified as cancerous. This might require the pathologist to manually validate the biopsy. Our algorithm separates the objects by drawing a line along the points where they touch. We determine whether a line should be passed along the edges to separate the objects according to a loss function that is derived from probabilities based on semantic segmentation (of various classes of pixels) and pixel distances from the contour. This method is general and can be applied to different types of tissue biopsies with glandular objects. We tested the algorithm on colon biopsy images. The newly developed method was able to improve the detection rate on average from 76% to 86%.

**Keywords:** Semantic segmentation · Gland segmentation

## 1 Introduction and Related Work

Cancer is one of the leading causes of death worldwide, no matter the socioeconomic background of the patient. In 2020, there were approximately 19.3 million cancer cases and almost 10 million cancer deaths worldwide. These numbers are expected to grow as the world population increases [18]. One of the most common cancer types, colon cancer, has a particularly high mortality rate. In 2020, there were 1.93 million new cases and 0.94 million deaths [20] from colon cancer. There is general consensus that early detection of the disease can prevent its development and eventually lead to recovery from the disease [20].

**Supplementary Information** The online version contains supplementary material available at https://doi.org/10.1007/978-3-031-25066-8_26.

The diagnosis of disease in biopsy images is often done by the identification of certain histological structures. Extent, size, shape, and other morphological characteristics of these structures are key indicators for the presence or severity of the disease. The first step in quantifying these tissues, and consequently in identifying the associated diseases, is to segment the glandular structures [7].

The colorectal biopsy is composed of glands (crypts) and immune system cells that surround the crypts (lumen, goblet cells, and cytoplasm) with an outer layer of nuclei (see Fig. 1). A colon crypt, is an intestinal gland found in the epithelial layer of the colon. In each human colon biopsy performed, there will be millions of glands [12], which must be classified; however, approx. 80% of the 1 million biopsies performed every year in the US are benign. In other words, 80% of the time, the pathologist is examining benign biopsy tissue [5].

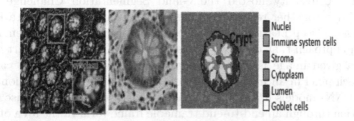

**Fig. 1.** Healthy colon tissue and crypt structure. A healthy crypt is composed of a lumen and goblet cells in cytoplasm, surrounded by a thin layer of nuclei. The crypt and immune system cells are in the stromal intermedium.

This glandular structure is also found in other biopsy tissues such as prostate, breast, and thyroid tissues, as shown in Fig. 2. For all of these types of tissues, gland segmentation is critical for the diagnosis process.

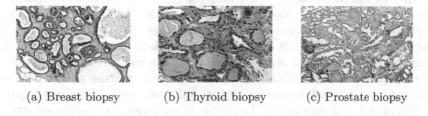

(a) Breast biopsy          (b) Thyroid biopsy          (c) Prostate biopsy

**Fig. 2.** Examples of other types of biopsies that contain glandular structures

It is clear that there is a need for an automated tool to work in tandem with the human pathologist. Such a tool should help focus attention on specific behaviors and make the diagnosis process fast and accurate. Researchers in the fields of image processing and computer vision have developed several approaches in the aim of developing this much needed tool [1,4,5,17].

In [3], a biopsy analysis algorithm was introduced. It is based on the Pixel-Level Classification (PLC) stage, in which each pixel is classified as belonging to one of six classes. Based on that, crypts are segmented using a specially developed active contour approach. The crypts are then classified as false positives, benign or cancerous.

In [15], which continued the research presented in [3], an attempt was made to improve upon the first stage of the previous method, the PLC. First, the previous method is applied. Then the PLC as well as the distance transform between the pixel and the crypt boundary are used to calculate the probability that the pixel is correctly classified. High probability pixels are automatically selected and used to train a new image-specific PLC classifier. This is all done without manual intervention. Using the new PLC classifier, the process described above is repeated, yielding improved segmentation results.

Warwick University hosted the Gland Segmentation Challenge Contest, aimed to bring together researchers interested in the gland segmentation problem. Alongside Deep Learning (DL) methods, teams also used the semantic segmentation approach that deals with this task [16]. An important part of understanding a given image is semantic segmentation, which assigns a categorical label to each pixel in an image. The recent success of deep convolutional neural network (CNN) models has enabled remarkable advances in pixel-wise semantic segmentation through an end-to-end trainable framework. In the area of biomedical image segmentation, U-Net is one of the most recognized Fully Convolutional Networks (FCN) [11]. The team that won the challenge known as "Deep Contour Aware - DCAN" used the Deep Learning semantic segmentation approach [2]. It is based on the FCN algorithm [8]. The DCAN architecture is designed in two steps. The first step uses annotated images of the ground truth to segment the structure. The second path uses the ground truth's shape to segment the contour and ultimately merge the results of the two steps.

From the related work, we can see that an enormous effort [6, 10, 14, 16, 19] has been made in the field of segmentation of glandular objects using DL. We note that for images where there is a nearby object, these models face difficulties and sometimes merge the objects into a single object as can be seen in Fig. 3. Merged objects naturally have different shapes, sizes, and extents. These characteristics, however, are critical in determining whether the biopsy is malignant or benign. Every time several objects are merged, the classifier might mis-classify this biopsy as malignant. Each false positive means that the pathologist must investigate and determine whether this is really cancerous. These investigations are critical and time-consuming. Thus, the primary aim of the proposed study is to develop a framework that segments glandular objects in a biopsy image using a DL semantic segmentation that includes an additional step when objects that were mistakenly merged are searched for and separated.

## 2   Separation of Merged Objects

A challenging aspect of semantic segmentation is the detection of objects that are close to one another. In this research, we outline a generic algorithm framework that can handle such situations. We also demonstrate this algorithm's real-life application on a colon biopsy data set (Sect. 3).

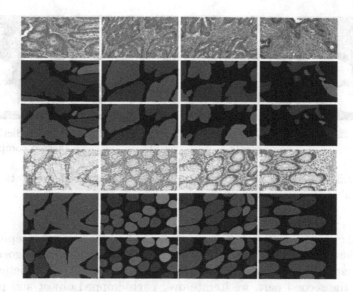

**Fig. 3.** Examples of benign and malignant predictions with merged objects. For each example, the input image, the segmentation result, and the ground truth are given.

The separation of merged objects algorithm is based on three studies that were conducted to deal with the segmentation and classification of colon biopsies. The first study presented the pixel-level classification [3]. The second study [15] used the distance between each pixel and the object boundary as a tool to improve the results of the first study [3]. In the last study, the authors presented an algorithm that won the GlaS challenge. Their algorithm uses a semantic segmentation neural network [2]. Our algorithm utilizes the key parts from each research as a basis on which to solve the merged objects problem.

Our method for separating merged objects is a post-processing algorithm that relies on a number of preconditions. First, the input to the algorithm must be a semantic segmentation prediction image for the object and its components, where each pixel is classified according to the probability of it belonging to one of the components. The second consideration is that the object we wish to analyze is composed of several components, at least one of which is usually close to the object's contour; see Figs. 1 and 7. The algorithm searches for areas where two or more objects were mistakenly merged. The algorithm then separates close objects that have been wrongly merged, thereby making up for one of the weaknesses of semantic segmentation. As soon as the algorithm locates objects that have merged, it draws lines in the object boundaries to separate them. It is assumed that the line mainly goes through pixels of the component mentioned above. In Fig. 4, a demonstration of the merged objects problem is presented.

This figure illustrates the ground truth and the prediction results, which merges three objects into one.

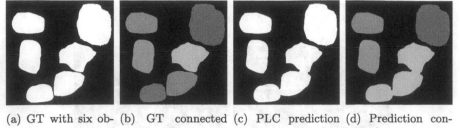

(a) GT with six ob-  (b) GT connected  (c) PLC prediction  (d) Prediction con-
jects                 components        results             nected components

**Fig. 4.** Example of the merged objects problem. According to the prediction results, there are only four objects, not six.

The algorithm consists of several parts. In the first part, we preprocess the classification image objects. We apply the connected components algorithm and crop the image to reflect the number of objects found in the prediction object image. In the second part, we iterate over each cropped object and pass it to the separation algorithm framework. The separation algorithm framework draws lines in the cropped image and evaluates whether the line is a good separation.

The separation of merged objects algorithm is an optimization problem in which we try to find the correct location to place the line in the object image. The line can be placed in many places in the image, and we need a process that will locate the optimal placement. For an optimization process to run, we need a function that we are interested in minimizing (the loss function), as well as the initial starting point, which in our case is a line in the image. The result is the location of the separation line within the image.

### 2.1 Line Representation

In our algorithm, the separation line plays a significant role. There are several ways to represent a line. The standard line equation is:

$$ax + by + c = (a, b, c)^T \cdot (x, y, 1) = 0, \qquad (1)$$

where $a$, $b$, and $c$ are constants. If $ax+by+c > 0$ then the point $(x, y)$ is above the line, otherwise it is below the line. Parameters $a$ and $b$ can also be represented by $\cos\theta$ and $\sin\theta$, where $\theta$ is the angle between the coordinate system to the line.

The main problems with these representations is that they do not deal with lines lying outside the image, and running an optimization function using these representations is non-linear and in many cases and thus will not converge.

As a solution to these problems, we decided to depict the separation line as two locations on the image border. The location of the left upper edge point is

0, and that of the right upper edge point is 0 + width. Likewise, the location of the lower right edge point is 0 + width + height. By applying this approach, we are able to modify the location of the line at every optimization iteration and this line is kept by definition within the image.

## 2.2 Optimization

The optimization procedure consists of two parts. In the first step, we scan the image space to determine the initial line. To do this, we need to avoid local minima and find a point within the basin of attraction of the global minimum. In the second step, we execute the optimization algorithm where we define the function we wish to minimize, the parameters that modify it and select the starting point.

### 2.2.1 Optimization Part 1: Scanning the Image Space

The separation algorithm uses lines to separate objects that have been mistakenly merged. Our initial input is a cropped image of a candidate object to be separated. In the scanning step, we determine where the optimization will start. After the scanning is complete, the optimization will begin with the best candidate, which will lead to convergence to the line at the local minimum [13]. In our algorithm we draw lines horizontally and vertically across the image shape to scan the space. For each line that we draw we calculate the line score using our loss function, which is explained in the next sections. The line with the lowest score determines where optimization will begin.

### 2.2.2 Removing Lines

Separation lines should by definition separate the object into two relatively large parts. When this is not the case they should be eliminated. Examples of such lines can be seen in Fig. 5. Thus, for each potential line, we calculate the number of object pixels above and below the line. If the ratio of the object pixels above the line to those below the line is lower than the given threshold, the line will be discarded.

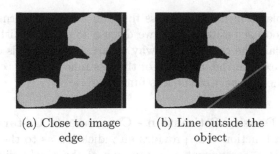

(a) Close to image edge          (b) Line outside the object

**Fig. 5.** Two examples of when our lines will be removed from evaluation

### 2.2.3    Optimization Part 2: F-Min

In the optimization procedure, we have a function that we want to optimize and the line parameters whose values affect the function score. The optimization process finds the line with the lowest loss function value.

In our framework, we use the f-min optimization [9] function. The process minimizes the function using the Nelder-Mead method. This algorithm only uses function values, not derivatives or second derivatives of the function. The Nelder-Mead method is a commonly applied numerical technique for finding the minimum or maximum of an objective function in a multidimensional space.

The f-min optimization search gets as an input a function we want to optimize, and the starting point from where the optimization will start. In our case, we give the loss function that we define and the line equation for which the loss function is computed. In Fig. 6, we can see (a) the merged objects image, (b) the results after the scanning process and, finally, (c) the results after the optimization converged.

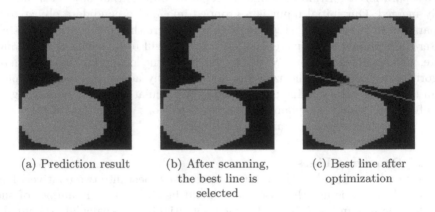

(a) Prediction result        (b) After scanning,        (c) Best line after
                                 the best line is           optimization
                                   selected

**Fig. 6.** The process of selecting the best line

### 2.3    Loss Function

This section explains in detail the loss function used to determine whether or not a line is a good separator. First, we discuss component distances from the contour and explain the rationale for why we are analyzing this component. We then proceed to explain the first part of the loss function and conclude with an explanation of the last part of the loss function.

### 2.3.1    Kernel Density Estimation - Components Distance

We now define a function that provides an indication as to the likelihood of a pixel belonging to a component as a function of the pixel's distance from the object's contour. Additionally, we meet one of our preconditions by getting the distance function as shown in Fig. 7: determining whether or not the component is close to the contour. The intuition is that we want to perform a separation at pixels that belong to a component whose pixels are usually close to the contour.

For the loss calculation, we need to prepare a probability function indicating how likely it is that there will be an object or a component at a given distance. This will be based on the ground truth of the gland's shape and the classification of the pixels.

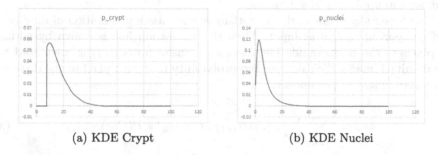

(a) KDE Crypt                    (b) KDE Nuclei

**Fig. 7.** The probability in the y axis and the distance in the x axis

In Fig. 7 we can see the KDE value for a colon biopsy analysis. The component that is close to the contour is the nucleus and the object that we are analyzing is the crypt. We can see when the distance to the contour is small, the probability of a nucleus is high. In the crypt probability calculation, when we are close to the contour, the probability is almost zero. The conclusion from Fig. 7 is that we want to separate the objects near the contours that are in the locations of the nuclei.

### 2.3.2 Loss 1: Calculating the Points in the Line

There are two parts to our loss function. Here we discuss calculating the points in the line. Then we present the second part of the loss function, the probability function based on a given distance.

The first part of the loss function focuses on the suggested separation line. For each component of the object that we are analyzing, we assign a weight of importance, where components whose pixels are usually close to the object's contour receive a higher value. The idea is to lay the line in a position such that most of its pixels belong to such a component. Based on Fig. 7, we can see that for the nucleus component when the distance is small, the probability is higher. By giving a higher weight value to the component that is close to the contour, the score will be higher for lines whose pixels belong to that component. For each component, we sum up the prediction values, i.e., the probabilities, $p_i(component_j)$, of each point on the line. The loss 1 equation is thus:

$$Loss\,1 = \sum_{i=1}^{n}\sum_{j=1}^{m} component\ weight_j * p_i(component_j). \qquad (2)$$

### 2.3.3    Loss 2: Probability Condition Based on a Given Distance

According to [15], the distance of a pixel to an object can provide information about the component to which it belongs. By using the ground truth, for each component we can calculate the distance transform, as shown in Sect. 2.3.1.

Once we have a conditional probability for a given distance $P(dist(i,j) \mid Comp_k)$ as computed by the KDE process, and the probability of belonging to a component based on the semantic segmentation, $P(Comp_k, i, j)$ and a suggested line, we can apply our loss 2 function.

The loss 2 algorithm works by getting a semantic segmentation of an object and component images as input. Then the separation line is drawn by adding its pixels to the background. The distance transform $dist(i,j)$ is calculated for the modified image. We calculate the probability that each pixel in the distance transform image is a component as:

$$Loss\, 2 = \sum_{i,j=0,0}^{n,m} \sum_{k=1}^{K} P(dist(i,j) \mid Comp_k) * P(Comp_k, i, j). \tag{3}$$

We sum up the conditional probabilities for each pixel and for each component. We combine loss 1 and loss 2, to give us the total loss function. Both components give us a good estimate of whether the line we evaluated is on the boundary of two objects.

### 2.4    Recursive Cutting Decision

We use our separation touching object algorithm as a post-processing step to improve object classification accuracy. This is done by running our algorithm recursively. At each step, the optimization is run on the object which returns the selected line. The object is separated using this line only if the loss 2 value of the original image is higher than the loss 2 of the image with the line. When this happens the algorithm is applied to the two sub-objects. The process continues until no object needs to be separated.

## 3    Use Case: Colon Biopsy Crypt Segmentation

In this section, we present the implementation of our algorithm framework based on a real-life use case. The platform we develop consists of a neural network that takes as input an image and performs semantic segmentation using DL, followed by the application of our post-processing algorithm for separating the merged objects. Our framework comprises three consecutive processes. In the first step, we perform data preparation for the semantic segmentation models. Our second process executes two semantic segmentation models, one for the crypt and the other for the nuclei. In the third step, we apply our post-processing algorithm as discussed in Sect. 2. Lastly, we evaluate the segmentation's accuracy before and after applying the separation algorithm.

When examining a microscopical image of a colon biopsy, the goal is to locate the crypts in the image. Depending on the structure of the crypts, a pathologist can determine if the tissue is malignant. In our framework, we concentrate on the nuclei and the crypt (the object). As shown in Fig. 1, the nuclei are in proximity to the crypt contour. Therefore, it meets our algorithm requirements. The next sections provide a detailed description of each process.

## 3.1   Data Set

We, tested our algorithm on a colon biopsy data set, which met our separation algorithm requirements. The crypt is composed of several components, one of which (the nucleus) usually lies near the crypt contour.

The biopsies for the database were randomly chosen by Dr. E. Sabo, a medical pathologist from the Gyneco-oncology Unit in Rambam Health Care Campus. This data set has been approved and used in a number of research studies [3,15]. The database was created by scanning the biopsies under a microscope with a 200x magnification scale. From each scanned biopsy image, sub-images were taken at 4x zoom. The average size of a sub-image is 800 × 500 pixels. There were 109 sub-images of healthy colons taken from 33 biopsies and 91 sub-images of cancerous colons taken from 21 biopsies. The classification of these sub-images was confirmed by the pathologist.

## 3.2   Data Preparation

Data were prepared in three steps: image preprocessing, ground truth preparation, and applying semantic segmentation.

To be able to feed them into the segmentation models properly, the images have to be preprocessed. First, all images must have the same shape. We, therefore, resized all the images to 480 × 480 pixels. In addition, the images were converted to gray scale.

As a first step in creating a supervised ML model, we need annotated data on which the model can be trained. Our data set for this application is a colon biopsy image that has been labeled with the six crypt components as shown in Fig. 1. Since we had a small number of images, we applied data augmentation to them to produce a larger training set. For this use case, we decided to predict two components, the nuclei that are located near the crypt contour and the crypt itself.

One requirement of the separation algorithm is the ability to classify on a pixel level. We therefore used the well-known and widely used U-Net algorithm [11]. Other methods for semantic segmentation could also be used. We implemented the same neural network for the crypt and the nucleus classes. Once the training was complete, we saved the model and used it to predict new images (the test set) so we could evaluate its performance.

## 3.3 Separation Algorithm

In this section we describe the implementation of the separation of merged objects algorithm for a colon biopsy data set, prepared as described in Sect. 2. The optimization algorithm is run as described in Sect. 2. A typical example of the distance transform on the crypt image is shown in Fig. 8. Figure 8(a-b) show the result of the semantic segmentation for the crypt and nucleus classes. Figure 8(c) shows the distance image of the merged crypts. At the center of the object, the intensity is strong as the distance is relatively large, but as we get closer to the contour, the intensity gets weaker. Concentrating on the red circle in Fig. 8(c), it can be seen that the intensity is strong, meaning that it is far from the contour. The probability of finding a crypt in this area is greater if we translate the distance to a probability based on Fig. 7. Meanwhile, in Fig. 8(b), we see that it should be a nucleus. Essentially, we are experiencing a mismatch between the distance calculation and the semantic segmentation probability of the object being a nucleus.

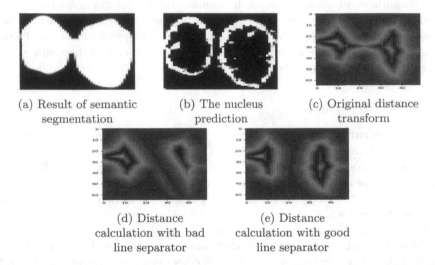

(a) Result of semantic segmentation     (b) The nucleus prediction     (c) Original distance transform

(d) Distance calculation with bad line separator     (e) Distance calculation with good line separator

**Fig. 8.** Distance transform on two merged objects. (c) Within the circle, we can see the intensities within the merging region. In (d) and (e) we can see examples of the effect of good and bad separation lines.

Figure 8(d-e) illustrates how the line can change the loss 2 score. Figure 8(d) shows the distance transform of the merged objects after placing a bad separation line. In Fig. 8(d), we can see that the intensity is weak across the line, which indicates that we are near the contour, which indicates a greater probability of being nucleus. Below the line it is a nucleus which also matches with the distance. However, above this line, the intensity of all of the regions is weaker, meaning that the probability of being a nucleus is greater, but it is a crypt. By selecting this line we have negatively affected the value of the loss 2. In Fig. 8(e), on the other hand, when the line passes in the merged zone the intensity becomes weaker, meaning that the probability of a nucleus is higher, as it is similar to Fig. 8(b).

## 4    Experimental Results

Throughout this Section, we present a number of results derived from our algorithm and compare them with the results obtained from semantic segmentation classification using the deep learning U-Net network. Our method was applied to 402 crypt objects from seven different biopsy images. Three images are of malignant tissue, containing 81 crypts objects, and four images are of benign tissue, containing 324 crypts objects. A typical result of the algorithm is shown in Fig. 9. All the results are presented in the supplementary material.

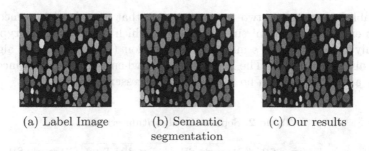

(a) Label Image          (b) Semantic          (c) Our results
                         segmentation

**Fig. 9.** The red lines in figure (c) indicate the separation lines place in the prediction image (Color figure online)

A segmented glandular object with an Intersection over Union (IOU) of at least 50% of its ground truth object is considered a true positive. Otherwise, the object is considered a false positive. A ground truth object that has no corresponding segmented object or has less than 50% of its area overlapped by its corresponding segmented object is considered a false negative. Our goal in separating merged objects is to solve the under-segmentation problem. In this scenario, if there is only one segmented object in the semantic segmentation and we expected two or more segments, the IoU scores will be low.

In Table 1, we present the average precision of IoU scores for each image that we evaluated, which are above 50% ($AP@[0.5]$). The table is divided into malignant and benign tissue images. The $AP@[0.5]$ that was generated by the U-Net results for benign images is 75%. After applying the separation post-processing algorithm on the same images, the average score is 86%, indicating an improvement of 11%. The improvement is almost the same for malignant images, whose shapes are much more complicated. For U-Net, the $AP@[0.5]$ is 78% for our separation algorithm, the average score is 87%, indicating an improvement of 9%.

**Table 1.** $AP@[0.5]$ for benign and malignant tissue images. Image ID refers to the figure ID in the supplementary material.

| Image Fig # | # Crypts in GT | U-Net $AP@[0.5]$ | Sep Alg $AP@[0.5]$ | Change Rate |
|---|---|---|---|---|
| $AP@[0.5]$ Benign Score | | | | |
| supp:1 | 55 | 74% | 82% | 10% |
| supp:4 | 81 | 69% | 78% | 10% |
| supp:5 | 120 | 85% | 95% | 11% |
| supp:7 | 68 | 75% | 88% | 17% |
| $AP@[0.5]$ Malignant Score | | | | |
| supp:6 | 21 | 85% | 95% | 11% |
| supp:2 | 37 | 75% | 82% | 9% |
| supp:3 | 23 | 75% | 83% | 9% |

In Table 2, we present two types of errors that our algorithm encounters. The first error is cases of missing separation, which indicates that crypts were mistakenly merged by the semantic segmentation (U-Net) and our algorithm was not able to detect it. The second error is when our algorithm incorrectly performs a separation when no separation is necessary.

**Table 2.** Separation algorithm errors.

| Image ID (ref) | # Missed SP | # Wrongly SP | %Missed of Total | %Wrongly SP of total |
|---|---|---|---|---|
| Separation algorithm errors - benign | | | | |
| supp:1 | 2 | 1 | 3.6% | 1.8% |
| supp:4 | 7 | 0 | 8.6% | 0% |
| supp:5 | 0 | 0 | 0% | 0% |
| supp:7 | 3 | 0 | 4.4% | 0% |
| Separation algorithm errors - Malignant | | | | |
| supp:6 | 0 | 1 | 0% | 4.8% |
| supp:2 | 2 | 0 | 5.4% | 0% |
| supp:3 | 1 | 0 | 4.3% | 0% |

As can be seen from the Table 2, the times when the object should have been separated and was not is more dominant than the incorrectly separated objects. The former does not affect the final score when compared to the U-Net since we are not making any changes. As a result, there were only two occasions where our algorithm separated objects unnecessarily. Overall, we observe that our algorithm improves the $AP@[0.5]$ for malignant and benign tissue images by at least 10% from 76% to 86%.

**Limitations:** When studying the errors mentioned above, there are two causes for them. The first cause is low quality of the nuclei prediction stage. When nuclei pixels were misclassified the object which should have been separated was not as can be seen in Fig. 10(a). On the other hand, when pixels were misclassified as nuclei, objects were separated by mistake as can be seen in Fig. 10(b-c). This problem can be alleviated by improving the quality of the semantic segmentation. The main limitation of our method is that we assume that the separator between crypts is a straight line. Although this is usually true there are cases such as the one shown in Fig. 11, where this is not the case. Thus, one of the suggested future works is to deal with this special case.

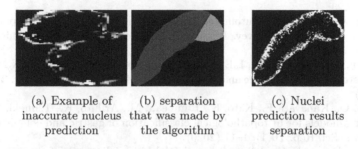

(a) Example of inaccurate nucleus prediction　　(b) separation that was made by the algorithm　　(c) Nuclei prediction results separation

**Fig. 10.** Examples of errors due to pixel misclassification.

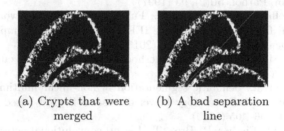

(a) Crypts that were merged　　(b) A bad separation line

**Fig. 11.** An object where a straight line is not a good separator.

## 5    Conclusions and Future Work

In this paper we developed a novel algorithm to improve semantic segmentation results. This method is general and applicable to a wide range of objects with structures that contain components, where at least one component is located near the object's contour. We demonstrated our novel algorithm in the framework of examining colon biopsy images containing glandular objects (crypts). Our method could also be used on other types of glandular tissues. Analyzing 402 different objects, the proposed algorithm method improved the $AP@[0.5]$ by 10%. This improvement was noted both in cases of malignant and benign biopsies. In addition, we were able to improve the IoU scores in images where they were originally relatively low.

Future work will be devoted to testing our algorithm on other types of glandular tissues. In addition, we plan to develop a more general algorithm in which the separator will not be a straight line but a more general curve.

# References

1. Belsare, A., Mushrif, M.: Histopathological image analysis using image processing techniques: an overview. Signal Image Process. **3**(4), 23 (2012)
2. Chen, H., Qi, X., Yu, L., Heng, P.A.: DCAN: deep contour-aware networks for accurate gland segmentation. In: Proceedings of the IEEE Conference on Computer Vision and Pattern Recognition, pp. 2487–2496 (2016)
3. Cohen, A., Rivlin, E., Shimshoni, I., Sabo, E.: Colon biopsy classification using crypt architecture. In: Wu, G., Zhang, D., Zhou, L. (eds.) MLMI 2014. LNCS, vol. 8679, pp. 182–189. Springer, Cham (2014). https://doi.org/10.1007/978-3-319-10581-9_23
4. Demir, C., Yener, B.: Automated cancer diagnosis based on histopathological images: a systematic survey. Rensselaer Polytechnic Institute, Technical report (2005)
5. Gurcan, M.N., Boucheron, L.E., Can, A., Madabhushi, A., Rajpoot, N.M., Yener, B.: Histopathological image analysis: a review. IEEE Rev. Biomed. Eng. **2**, 147–171 (2009)
6. Iizuka, O., Kanavati, F., Kato, K., Rambeau, M., Arihiro, K., Tsuneki, M.: Deep learning models for histopathological classification of gastric and colonic epithelial tumours. Sci. Rep. **10**(1), 1–11 (2020)
7. Jenkins, D., et al.: Guidelines for the initial biopsy diagnosis of suspected chronic idiopathic inflammatory bowel disease. The British society of gastroenterology initiative. J. Clin. Pathol. **50**(2), 93 (1997)
8. Long, J., Shelhamer, E., Darrell, T.: Fully convolutional networks for semantic segmentation. In: Proceedings of the IEEE Conference on Computer Vision and Pattern Recognition, pp. 3431–3440 (2015)
9. Nelder, J.A., Mead, R.: A simplex method for function minimization. Comput. J. **7**(4), 308–313 (1965)
10. Pradhan, P., et al.: Semantic segmentation of non-linear multimodal images for disease grading of inflammatory bowel disease: a SegNet-based application. In: ICPRAM, pp. 396–405 (2019)
11. Ronneberger, O., Fischer, P., Brox, T.: U-Net: convolutional networks for biomedical image segmentation. In: Navab, N., Hornegger, J., Wells, W.M., Frangi, A.F. (eds.) MICCAI 2015. LNCS, vol. 9351, pp. 234–241. Springer, Cham (2015). https://doi.org/10.1007/978-3-319-24574-4_28
12. Rubin, R., Strayer, D.S., Rubin, E., et al.: Rubin's Pathology: Clinicopathologic Foundations of Medicine. Lippincott Williams & Wilkins (2008)
13. Ruder, S.: An overview of gradient descent optimization algorithms. arXiv preprint arXiv:1609.04747 (2016)
14. Rundo, L., et al.: CNN-based prostate zonal segmentation on T2-weighted MR images: a cross-dataset study. In: Esposito, A., Faundez-Zanuy, M., Morabito, F.C., Pasero, E. (eds.) Neural Approaches to Dynamics of Signal Exchanges. SIST, vol. 151, pp. 269–280. Springer, Singapore (2020). https://doi.org/10.1007/978-981-13-8950-4_25
15. Shahin, S.A.: Colon Biopsy Classification by Improving the Pixel Level Classification Stage. University of Haifa (Israel) (2018)
16. Sirinukunwattana, K., et al.: Gland segmentation in colon histology images: the GlaS challenge contest. Med. Image Anal. **35**, 489–502 (2017)
17. Smochină, C., Herghelegiu, P., Manta, V.: Image processing techniques used in microscopic image segmentation. In: Technical Report. Gheorghe Asachi Technical University of Iași (2011)

18. Sung, H., et al.: Global cancer statistics 2020: GLOBOCAN estimates of incidence and mortality worldwide for 36 cancers in 185 countries. CA: Cancer J. Clin. **71**(3), 209–249 (2021)

19. Talo, M.: Convolutional neural networks for multi-class histopathology image classification. arXiv preprint arXiv:1903.10035 (2019)

20. Xi, Y., Xu, P.: Global colorectal cancer burden in 2020 and projections to 2040. Transl. Oncol. **14**(10), 101174 (2021)

# qDWI-Morph: Motion-Compensated Quantitative Diffusion-Weighted MRI Analysis for Fetal Lung Maturity Assessment

Yael Zaffrani-Reznikov[1]([✉])[iD], Onur Afacan[2][iD], Sila Kurugol[2][iD], Simon Warfield[2][iD], and Moti Freiman[1][iD]

[1] Faculty of Biomedical Engineering, Technion, Haifa, Israel
yael.rez@campus.technion.ac.il
[2] Boston Children's Hospital, Boston, MA, USA

**Abstract.** Quantitative analysis of fetal lung Diffusion-Weighted MRI (DWI) data shows potential in providing quantitative imaging biomarkers that indirectly reflect fetal lung maturation. However, fetal motion during the acquisition hampered quantitative analysis of the acquired DWI data and, consequently, reliable clinical utilization. We introduce qDWI-morph, an unsupervised deep-neural-network architecture for motion compensated quantitative DWI (qDWI) analysis. Our approach couples a registration sub-network with a quantitative DWI model fitting sub-network. We simultaneously estimate the qDWI parameters and the motion model by minimizing a bio-physically-informed loss function integrating a registration loss and a model fitting quality loss. We demonstrated the added-value of qDWI-morph over: 1) a baseline qDWI analysis without motion compensation and 2) a baseline deep-learning model incorporating registration loss solely. The qDWI-morph achieved a substantially improved correlation with the gestational age through in-vivo qDWI analysis of fetal lung DWI data ($R^2 = 0.32$ vs. $0.13, 0.28$). Our qDWI-morph has the potential to enable motion-compensated quantitative analysis of DWI data and to provide clinically feasible bio-markers for non-invasive fetal lung maturity assessment. Our code is available at: https://github.com/TechnionComputationalMRILab/qDWI-Morph.

**Keywords:** Motion compensation · Quantitative DWI · Fetal imaging

## 1 Introduction

Fetal lung parenchyma maldevelopment may lead to life-threatening physiologic dysfunction due to pulmonary hypoplasia and pulmonary hypertension [10]. Accurate assessment of lung maturation before delivery is critical in determining pre-natal and post-natal care and potential interventions [2]. In current

This research was supported in part by a grant from the United States-Israel Binational Science Foundation (BSF), Jerusalem, Israel.

clinical practice, the non-invasive assessment of fetal lung parenchyma development relies upon fetal lung volume estimation from either ultrasonography [12], or anatomical magnetic resonance imaging (MRI) [5] data. However, these modalities are limited in providing an insight into lung function and are therefore suboptimal in assessing fetal lung maturity and in providing relevant biomarkers for fetal lung parenchyma maldevelopment.

Diffusion-weighted MRI (DWI) is a non-invasive imaging technique sensitive to the random movement of individual water molecules. The displacement of individual water molecules results in signal attenuation in the presence of a magnetic field encoding gradient pulses. This attenuation increases with the degree of sensitization-to-diffusion of the MRI pulse sequence (the "b-value") [1]. Quantitative analysis of the DWI data (qDWI) has been suggested previously for functional assessment of fetal lung parenchyma development [1,13]. The qDWI analysis is performed by a least-squares fitting of a signal decay model describing the DWI signal attenuation as a function of the b-value to the acquired DWI data. Commonly, a mono-exponential signal decay model of the form of:

$$S_i = S_0 \cdot e^{-b_i \cdot ADC} \tag{1}$$

where $S_i$ is the signal at b-value $b_i$, $S_0$ is the signal without sensitizing the diffusion gradients, and $ADC$ is the apparent diffusion coefficient, which represents the overall diffusivity, is used in clinical practice [8].

However, qDWI analysis is intrinsically highly susceptible to gross motion between the acquisition of the different b-value images [1]. Hence, the irregular and unpredictable motion of the fetus, in addition to the maternal respiratory and abdominal motion, causes misalignment between image volumes, acquired at multiple b-values, and impairs the accuracy and robustness of the signal decay parameter estimation [9].

Specifically, Afacan et al. [1] examined the influence of the presence of motion during the DWI data acquisition on the correlation between the overall diffusivity characterized by the ADC parameter and the gestational age (GA). They found that the dependence of the ADC parameter on the GA is of an exponential saturation type, with an R-squared ($R^2$) value of 0.71 for fetal DWI data that was not affected by fetus motion. However, when fetal DWI data with gentle motion were included, the $R^2$ dropped to 0.232, and when fetal DWI data with severe motion were included, the $R^2$ dropped to 0.102.

We introduce qDWI-Morph, an unsupervised deep-neural-network model for motion-compensated qDWI analysis. Our specific contributions are: 1) unsupervised deep-neural-network (DNN) model for simultaneous motion compensation and qDWI analysis, 2) Introduction of a bio-physically-informed loss function incorporating registration loss with qDWI model fitting loss, and 3) improved correlation between qDWI analysis by means of the ADC parameter and GA demonstrated using in-vivo DWI data of fetal lungs.

## 1.1   Background

**Fetal Lung Development.** The normal fetal lung parenchyma development starts in the second trimester and progresses through multiple phases before becoming fully functional at full term. The first phase of development, the embryonic and pseudo glandular stage, is followed by the canalicular phase, which starts at 16 weeks, then the saccular stage, which starts at 24 weeks, and ends with the alveolar stage, which starts at 36 weeks gestation. [14]. The formation of a dense capillary network and a progressive increase in pulmonary blood flow, leading to reduced extra-cellular space on the one hand and increased perfusion on the other hand, characterizes the progression through these phases [6]. Hence, changes in the overall tissue diffusivity are thought to serve as a functional indicator of lung development.

**Diffusion-Weighted MRI.** DWI is a non-invasive imaging technique sensitive to the random movement of individual water molecules. The movement of water molecules depends on tissue micro-environments and results from two phenomena: diffusion and perfusion [6]. The water molecule mobility attenuates the DWI signal according to the degree of sensitivity-to-diffusion, the "b-value", used in the acquisition. Typically, DWI images are acquired at multiple b-values, and the signal decay rate parameters are quantified per voxel by least-squares model fitting [9]. DWI can potentially be a functional biomarker for lung maturity since it results from diffusion and perfusion. According to the mono-exponential model, the signal of a particular voxel decreases exponentially with an increasing b-value (Eq. 1), with a decay rate that depends on the overall diffusivity of the voxel, the ADC, encapsulating both pure diffusion and perfusion influence on the DWI signal. Quantifying the ADC out of the measurements is done by applying $log(\cdot)$ on both sides of Eq. 1:

$$log(S_i) = log(S_0) - b_i \cdot ADC \tag{2}$$

A set of B equations is obtained by scanning with B different b-values. This linear set of equations can be solved by linear least-squares regression (LLS) [15]. The solution is given by:

$$\hat{x}_{LS} = \operatorname*{argmin}_{S_0, ADC} \sum_{i \in B} |\log(S_i) - \log(S_0) + b_i ADC|^2 \tag{3}$$

$$= (A^T A)^{-1} A^T \beta \tag{4}$$

where:

$$A = \begin{pmatrix} 1 & -b_0 \\ \vdots & \vdots \\ 1 & -b_B \end{pmatrix}, \quad \beta = [log(S_0), \cdots, log(S_B)]^T, \quad x = [log(S_0), ADC]^T \tag{5}$$

In order to be stable against outliers in the acquired DWI signals, a more robust solution can be obtained by using an iterative-least squares algorithm [4]. This is an iterative method in which each step involves solving a weighted least squares problem:

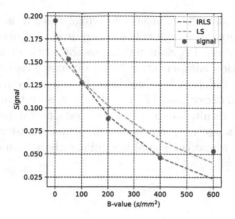

**Fig. 1.** LLS Vs. IRLS fitting of DWI Signal.

$$\hat{x}_{IRLS}^{(t)} = \operatorname*{argmin}_{S_0, ADC} \sum_{i \in B} w_i^{(t)} |\log(S_i) - \log(S_0) + b_i ADC| \tag{6}$$

$$= (A^T W^{(t)} A)^{-1} A^T W^{(t)} b \tag{7}$$

where $W^{(t)}$ is a diagonal matrix of weights, with all elements set initially to:

$$\left\{ w_i^{(0)} \right\}_{i \in B} = 1 \tag{8}$$

and updated after each iteration to:

$$\left\{ w_i^{(t)} \right\}_{i \in B} = \frac{1}{\max \left\{ |Ax^{(t-1)} - b|_i, 0.0001 \right\}} \tag{9}$$

As shown In Fig. 1, the IRLS is more accurate and ignores outliers by setting them a low weight. On the other hand, the IRLS algorithm is computationally expensive and time-consuming.

## 1.2  Fetal Lung Maturity Assessment with qDWI

Assessing fetal lung maturation using qDWI was first suggested by Moore et al. [13] in 2001. More recently, Afacan et al. [1] demonstrated a strong association between ADC and GA in fetuses with normal lung development. The ADC is quantified per voxel by LLS regression (Eq. 4). Then, the averaged ADC in the lung can indicate the GA of the fetus according to the saturation-exponential model suggested by [1]:

$$ADC = ADC_{sat}(1 - e^{-\alpha \cdot GA}) \tag{10}$$

Since the ADC quantification is done voxel-wise, it is sensitive to potential misalignment between image volumes acquired at different b-values, which accrue from fetal motion. Previous techniques for motion compensation include post-acquisition motion compensation based on image registration to bring the volumes acquired at different b-values into the same physical coordinate space before fitting a signal decay model [7,11]. However, each b-value image has a different contrast. As a result, independent registration of different b-value images to a b $= 0\,\mathrm{s/mm^2}$ image may result in suboptimal registration, especially for high b-value images where the signal is significantly attenuated and the signal-to-noise ratio is low [9].

## 2  Method

We hypothesize that simultaneous qDWI analysis and fetal motion compensation will result in improved quantification of the ADC, which will correlate better with fetal lung maturity compared to qDWI analysis without compensating for motion and to motion compensated approach which does not account for the signal decay model fit quality.

We define the simultaneous qDWI analysis and motion compensation problem as follows:

$$\widehat{\Phi}, \widehat{ADC} = \underset{\Phi, ADC}{\mathrm{argmin}} \sum_{i \in B} \|\phi_i \circ S_i - S_0 \exp(-b_i \cdot ADC)\|^2 \tag{11}$$

where $\Phi = \{\phi\}_{i \in B}$ is the set of transformations that align the different b-value images to the images predicted by the model, and $B$ is the number of b-values used during the DWI acquisition.

Taking an unsupervised DNN-based perspective, the optimization problem seeks to find the DNN weights that minimize the following:

$$\widehat{\Theta} = \underset{\Theta}{\mathrm{argmin}} \sum_{i \in B} \|f_\Theta(\{b_i, S_i\}_{i \in B})[0] \circ S_i - S_0 \exp(-b_i \cdot f_\Theta(\{b_i, S_i\}_{i \in B})[1])\|^2 \tag{12}$$

where $\Theta$ are the DNN parameters to be optimized, $f_\Theta(\{b_i, S_i\}_{i=0}^{B})[0]$ is the first output of the DNN model, represents the set of transformations $\Phi$ between the different b-value images $\{S_i\}_{i \in B}$, and $f_\Theta(\{b_i, S_i\}_{i \in B})[1])$ is the second output of the DNN model represents the predicted ADC.

### 2.1  DNN Architecture

Figure 2 presents our qDWI-Morph architecture used for the optimization of Eq. 12. The qDWI-Morph model is composed of two sub-networks: a qDWI model fitting sub-network and an image registration sub-network. Our method is iterative, where the motion-corrected images from the previous iteration enter as input to the next iteration until convergence. We normalize the input in each

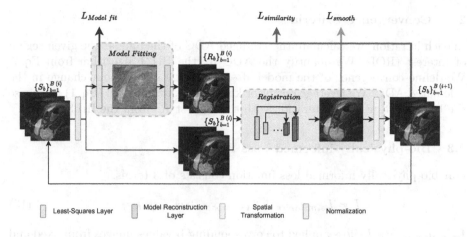

**Fig. 2.** Overall architecture of our proposed framework.

iteration $k$ by the maximal value of the signal at the previous iteration $S_0^{(k-1)}$. The input to the model at each iteration $k$ is a set of 3D images from the same patient, each scanned with a different b-value along with their corresponding b-values $\{b_i, S_i\}_{i \in B}^{(k)}$. The output of our model is composed of the set of motion-corrected images forced pixel-wise to the mono-exponential model (Eq. 1).

**Quantitative DWI Model Fitting Sub-network.** first layer is a LLS layer for quantifying the ADC at each pixel using Eq. 4. The second layer is a recon-struction layer that generates a new set of 3D images, $\{R_b\}_{b \in B}^{(k)}$, by calculating the model pixel-wise using Eq. 1. The reconstructed set, $\{R_b\}_{b \in B}^{(k)}$, is retreated as fixed images and is an input for the second sub-network, the registration network.

**Registration Sub-network.** is based on the publicly available deep learn-ing framework for deformable medical image registration, VoxelMorph [3]. The moving images are $\{S_b\}_{b \in B}^{(k)}$, and the fixed one are $\{R_b\}_{b \in B}^{(k)}$. The output of the VoxelMorph network is a registration field, $\phi$, that can be applied to the moving images to get the next motion-compensated images, $\{S_b\}_{b \in B}^{(k+1)}$. The registra-tion is done between corresponding b-values images from the moving and fixed images. For example, the moving image $S_0$ will be registered to the fixed image $R_0$. This way, we overcome the differences in contrast between b-values and use prior knowledge that the resulting compensated image needs to follow the mono-exponential model.

## 2.2    Convergence Criteria

In each iteration, we calculate the averaged ADC of the lung at the given region of interest (ROI). We quantify the ADC by the IRLS algorithm from Eq. 7. We define convergence of the model after five iterations without change in the calculated ADC. We leverage our prior knowledge of the model (Eq. 1) to choose the best iteration as the iteration with the highest $R^2$ from the IRLS fitting.

## 2.3    Bio-physically-informed Loss Function

Our bio-physically-informed loss function consists of 3 terms.

$$L = L_{similarity} + \alpha_1 L_{smooth} + \alpha_2 L_{Model\ fit} \tag{13}$$

**L**$_{similarity}$ is a $L_1$ loss applied to corresponding b-values images from fixed and wrapped moving images, averaged over the $B$ b-values images:

$$L(R^{(k)}, S^{(k)} \circ \phi_i^{(k)}) = \frac{1}{B} \frac{1}{|\Omega|} \sum_{i \in B} \sum_{p \in \Omega} |R_i^{(k)}(p) - [S_i^{(k)} \circ \phi_i^{(k)}](p)| \tag{14}$$

where $\Omega$ is the image dimension. Minimizing $L_{similarity}$ will encourage $\{S_i\}_{b \in B}^{(k)} \circ \phi_i^{(k)}$ to approximate $\{R_i\}_{b \in B}^{(k)}$, but may generate a non-smooth $\phi$ that is not physically realistic [3].

**L**$_{smooth}$ is a regularization term introduced by Balakrishnan et al. [3]. This loss term encourages a smooth displacement field $\phi$ using a diffusion regularize on the spatial gradients of displacement $u$:

$$L_{smooth}(\Phi) = \sum_{p \in \Omega} \|\nabla u(p)\|^2 \tag{15}$$

**L**$_{model\ fit}$ is the loss that responsible for forcing each voxel to act according to the mono-exponential model from Eq. 2. This loss is an MSE loss on the least square residual:

$$L_{model\ fit}(ADC) = \frac{1}{B} \frac{1}{|\Omega^*|} \sum_{i \in B} \sum_{p \in \Omega^*} (log(S_0) - b_i \cdot ADC - log(S_i))^2 \tag{16}$$

where $\Omega^*$ is a subset of the image, including only voxels from the lung, as we assume the mono-exponential model can describe them, contrary to the background, which doesn't present a mono-exponential decay of the DWI signal.

# 3    Experiments

## 3.1    Clinical Data-Set

Legacy fetal DWI data was used in the study. The data acquisition was performed on a Siemens 3T Skyra scanner equipped with an 18-channel body matrix coil. Each patient was scanned with a multi-slice, single shot, echo-planar imaging (EPI) sequence that was used to acquire diffusion-weighted scans of the lungs. The in-plane resolution was 2.5 mm × 2.5 mm for each study, and the slice thickness was set at 3 mm. Echo time (TE) was 60 ms, whereas repetition time (TR) ranged from 2 s to 4.4 s depending on the number of slices required to cover the lungs. Each patient was scanned with 6 different b-values (0, 50, 100, 200, 400, 600 s/mm$^2$) in axial and coronal planes. A region of interest (ROI) was manually drawn for each case in the right lung [1]. We cropped each image to a shape: $96 \times 96 \times 16$ and normalized it by the maximal value at $S_0$. The data includes 38 cases with a range of minor misalignment between the different b-values image volumes.

## 3.2    Experiment Goals

The objectives of the experiment are: 1) To analyze the effect of quantitative DWI motion-compensation on the correlation between the ADC parameter and GA. 2) To analyze the contribution of our proposed model fitting quality loss in addition to the registration loss in improving the observed correlation between the ADC parameter and GA.

## 3.3    Experimental Setup

Our baseline method is ADC quantification without motion compensation. We compared the baseline method to our suggested method with: 1) Registration loss solely, and 2) Our hybrid loss combining both registration loss and a model fitting quality loss. For each of the three methods, we quantified the averaged ADC in the right lung of each subject by IRLS algorithm. We used an exponential saturation model suggested by Afacan et al. [1] described in Eq. 10 above.

## 3.4    Implementation Details

We implemented our models on PyCharm 2021.2.3, Python 3.9.12 with PyTorch 1.11.0.

We applied our suggested methods with a batch size of one, meaning that each batch is data from one patient with size: $n_b \times n_x \times n_y \times n_z$, where $n_b$ is the number of b-values used for scanning the patient, and $n_x \times n_y \times n_z$ is the image shape. We trained the networks in two epochs so that the earlier cases will be contributed from the later ones. We experimented the affect of training in a wide range of epochs and found that the network converged after the second epoch, and there was no further improvement after that. We used an Adam optimizer

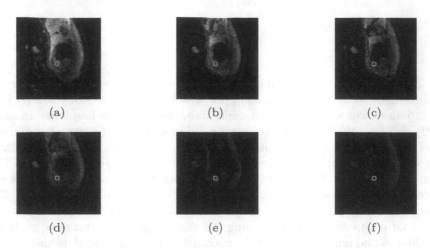

**Fig. 3.** Sample case with ROI. Top row left to right: b = 0, 50, 100 s/mm². Bottom row left to right: b-value = 200, 400, 600 s/mm².

with an initial learning rate of 10-4. If the total loss increases compared to the previous iteration, we reduce the learning rate by 10. We limited the number of iterations for each case to 50. The hyper-parameter we used are: $\alpha_1 = 0.01$, $\alpha_2 = 1000$. We choose $\alpha_2$ by a search-grid of the best $\alpha_2$ in terms of correlation between ADC and GA. For testing the qDWI-Morph method without fit model loss term, we set $\alpha_2 = 0$.

## 4    Results

Figure 3 presents an example of DWI data before accounting for motion by the registration. Figure 4 presents our iterative motion-compensation and model estimation. The top row is the motion-compensation b = 0 s/mm² images from different iterations, where the top-left image is the input image. The second row is the ADC map result from the LLS layer, and the bottom row is the corresponded reconstructed b = 0 s/mm² image. The right column is the convergence iteration. The ADC map (middle row) becomes sharper and more clinically-meaningful as the motion is correct during the iterations. However, the final motion-compensated b-value images become more blurry as a result of the registration wrapping process by the linear interpolation.

Figure 5 demonstrates the averaged signal in the ROI as a function of the b-value at different iterations. The lines are the corresponding fit to Eq. 1. At the last iteration, after the motion compensation process (5c), the dots behave according to the expected model and are less scattered.

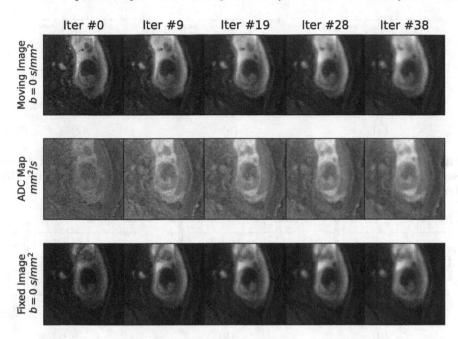

**Fig. 4.** The process done by our method, qDWI-Morph, demonstrated over time. Top row: the motion-compensation $b = 0 \, s/mm^2$ images. Second row: ADC map result from the LLS layer. Bottom row: the model-reconstructed $b = 0 \, s/mm^2$ images.

Figure 6 shows the ADC parameter plotted against GA for all three methods. The data is color-coded by the degree of conformity to the mono-exponential model from Eq. 1 in terms of $R^2$. The Darker the dot, the higher $R^2$. Our approach achieved the best correlation ($R^2 = 0.32$) between the ADC and the GA compared to both motion compensation by our qDWI-Morph without adding the model fit loss ($R^2 = 0.28$) and to baseline approach without motion compensation ($R^2 = 0.13$).

Table 1 summarizes the fitted ADC saturation model (Eq. 10) using the different approaches along with the correlation between the ADC and GA in terms of $R^2$. The fitted model parameters are bout the same for all methods. However, our qDWI-Morph achieved the best correlation between the ADC and the GA, since the average ADC vs. GA is less sparse, as seen in Fig. 6.

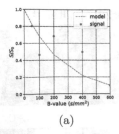 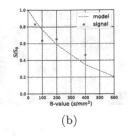 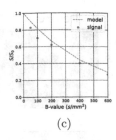

(a)                    (b)                    (c)

**Fig. 5.** The averaged signal in the ROI for one case, as a function of b-value at different iterations, with a curve fit to the mono-exponential model: (a) Original data, (b) iteration 19, and (c) iteration 38.

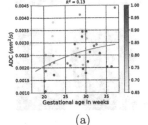 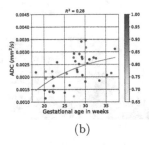 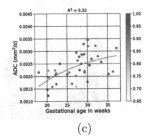

(a)                    (b)                    (c)

**Fig. 6.** Averaged ADC in the right lung vs. GA with the three methods: (a) Baseline; W.O motion compensation. (b) qDWI-Morph W.O model fit loss. (c) qDWI-Morph. The data is color-coded by the degree of conformity to the mono-exponential model from Eq. 1 in terms of $R^2$.

**Table 1.** The correlation of the averaged ADC in the right lung vs. GA by the three methods: W.O M.C; the baseline method, ADC quantification without motion compensation. W.O F.L; qDWI-Morph without model fit loss, and our method: q-DWI-Morph.

| Method | $R^2$ | A $\times 10^{-3}$ $(\mathrm{mm}^2/\mathrm{s})$ | B $\times 10^{-3}$ $(\mathrm{mm}^2/\mathrm{s})$ | C |
|---|---|---|---|---|
| W.O. M.C | 0.13 | 3.2 | 0.005 | 0.07 |
| W.O. F.L | 0.28 | 3.17 | 0.006 | 0.07 |
| qDWI-Morph | **0.32** | 3.19 | 0.006 | 0.07 |

## 5    Conclusions

We introduced qDWI-Morph, an unsupervised deep-neural-networks approach for Motion-compensated quantitative Diffusion-Weighted MRI analysis with application in fetal lung maturity assessment. Our model coupled a registration sub-network with a model fitting sub-network to simultaneously estimate the motion between the different b-value images and the signal decay model parameters. The optimization of the model weights is driven by minimizing a bio-physically-informed loss representing both the registration loss and the fit quality loss. The integration of the fit quality loss encourages the model to produce deformation fields that will lead to bio-physically expected behavior of the signal along the b-value axis, in addition to the standard registration loss, which encourages similarity and smoothness within the deformation field of each b-value image. Our experiments demonstrated an added-value of adding the model-fitting loss in addition to the registration loss for fetal lung maturity assessment. The proposed approach can potentially improve our ability to quantify DWI signal decay model parameter in cases with motion.

# References

1. Afacan, O., et al.: Fetal lung apparent diffusion coefficient measurement using diffusion-weighted MRI at 3 Tesla: correlation with gestational age. J. Magn. Reson. Imaging **44**(6), 1650–1655 (2016). https://doi.org/10.1002/jmri.25294
2. Ahlfeld, S.K., Conway, S.J.: Assessment of inhibited alveolar-capillary membrane structural development and function in bronchopulmonary dysplasia. Birth Def. Res. Part A: Clin. Molec. Teratol. **100**(3), 168–179 (2014)
3. Balakrishnan, G., Zhao, A., Sabuncu, M.R., Guttag, J., Dalca, A.V.: Voxelmorph: a learning framework for deformable medical image registration. IEEE Trans. Med. Imaging **38**(8), 1788–1800 (2019)
4. Burrus, S.: Iterative reweighted least squares * c (2018)
5. Deshmukh, S., Rubesova, E., Barth, R.: Mr assessment of normal fetal lung volumes: a literature review. Am. J. Roentgenol. **194**(2), W212–W217 (2010)
6. Ercolani, G., et al.: IntraVoxel incoherent motion (IVIM) MRI of fetal lung and kidney: can the perfusion fraction be a marker of normal pulmonary and renal maturation? Eur. J. Radiol. **139**, 109726 (2021). https://doi.org/10.1016/J.EJRAD.2021.109726
7. Guyader, J.M., Bernardin, L., Douglas, N.H., Poot, D.H., Niessen, W.J., Klein, S.: Influence of image registration on apparent diffusion coefficient images computed from free-breathing diffusion mr images of the abdomen. J. Magn. Reson. Imaging **42**(2), 315–330 (2015)
8. Koh, D.M., Collins, D.J.: Diffusion-weighted mri in the body: applications and challenges in oncology. Am. J. Roentgenol. **188**(6), 1622–1635 (2007)
9. Kurugol, S., et al.: Motion-robust parameter estimation in abdominal diffusion-weighted MRI by simultaneous image registration and model estimation. Med. Image Anal. **39**, 124–132 (2017). https://doi.org/10.1016/J.MEDIA.2017.04.006
10. Lakshminrusimha, S., Keszler, M.: Persistent pulmonary hypertension of the newborn. Neoreviews **16**(12), e680–e692 (2015)
11. Mazaheri, Y., Do, R.K., Shukla-Dave, A., Deasy, J.O., Lu, Y., Akin, O.: Motion correction of multi-b-value diffusion-weighted imaging in the liver. Acad. Radiol. **19**(12), 1573–1580 (2012)
12. Moeglin, D., Talmant, C., Duyme, M., Lopez, A.C.: Fetal lung volumetry using two- and three-dimensional ultrasound. Ultrasound Obstet. Gynecolo. **25**(2), 119–127 (2005). https://doi.org/10.1002/UOG.1799, https://onlinelibrary.wiley.com/doi/full/10.1002/uog.1799https://onlinelibrary.wiley.com/doi/abs/10.1002/uog.1799, https://obgyn.onlinelibrary.wiley.com/doi/10.1002/uog.1799
13. Moore, R., Strachan, B., Tyler, D., Baker, P., Gowland, P.: In vivo diffusion measurements as an indication of fetal lung maturation using echo planar imaging at 0.5 t. Magn. Reson. Med. Off. J. Int. Soc. Magn. Reson. Med. **45**(2), 247–253 (2001)

14. Schittny, J.C.: Development of the lung. Cell Tissue Res. **367**(3), 427–444 (2017). https://doi.org/10.1007/s00441-016-2545-0
15. Watson, G.S.: Linear least squares regression. Ann. Math. Stat. **38**(6), 1679–1699 (1967). http://www.jstor.org/stable/2238648

# Estimating Withdrawal Time
# in Colonoscopies

Liran Katzir[1], Danny Veikherman[1], Valentin Dashinsky[2], Roman Goldenberg[3], Ilan Shimshoni[4], Nadav Rabani[1], Regev Cohen[3(✉)], Ori Kelner[3], Ehud Rivlin[3], and Daniel Freedman[3]

[1] Google, Mountain View, USA
[2] DRW, Mountain View, USA
[3] Verily Life Sciences, San Francisco, USA
regevcohen@google.com
[4] University of Haifa, Haifa, Israel

**Abstract.** The Colonoscopic Withdrawal Time (CWT) is the time required to withdraw the endoscope during a colonoscopy procedure. Estimating the CWT has several applications, including as a performance metric for gastroenterologists, and as an augmentation to polyp detection systems. We present a method for estimating the CWT directly from colonoscopy video based on three separate modules: egomotion computation; depth estimation; and anatomical landmark classification. Features are computed based on the modules' outputs, which are then used to classify each frame as representing forward, stagnant, or backward motion. This allows for the optimal detection of the change points between these phases based on efficient maximization of the likelihood; from which the CWT follows directly. We collect a dataset consisting of 788 videos of colonoscopy procedures, with the CWT for each annotated by gastroenterologists. Our algorithm achieves a mean error of 1.20 min, which nearly matches the inter-rater disagreement of 1.17 min.

**Keywords:** Colonoscopy · Detection · Visual odometry · Withdrawal time

## 1 Introduction

Colorectal Cancer (CRC) claims many lives per year [1,2]; however, as is well known, CRC may be prevented via early screening. In particular, the colonoscopy procedure is able to both detect polyps in the colon while they are still precancerous, and to resect them. There is, however, some variation in the quality of colonoscopy procedures, as performed by different gastroenterologists. This paper is concerned with one aspect which underlies this variation, namely the time spent by the endoscopist during the withdrawal phase of the colonoscopy.

V. Dashinsky and I. Shimshoni—Work performed while at Alphabet.

**Background and Motivation.** By way of background, we briefly describe the structure of a colonoscopy. When the procedure commences, the goal of the physician is to insert the colonoscope all the way to the end the colon, known as the cecum; see Fig. 1. This is known as the colonoscopic insertion. The time it takes to reach the cecum is known as the Cecal Intubation Time (CIT). During this process the physician moves the endoscope both forwards and backwards. From time to time only the wall of the colon is seen by the camera.

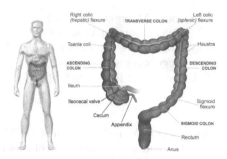

**Fig. 1.** The map of the colon, adapted with minor modifications from wikipedia. Our goal is to estimate the Colonoscopic Withdrawal Time (CWT), the duration elapsed from the time the colonoscope reaches the cecum until it has been entirely withdrawn.

Once the cecum has been reached the physician starts to slowly extract the colonoscope; the goal in this stage is to detect the polyps.[1] This phase is easier from the navigation point of view and the motion is usually backwards. Forward motion generally only occurs when a polyp is detected, examined and extracted. Thus, there is a clear distinction between colonoscopic insertion and colonoscopic withdrawal phases. The duration of the colonoscopic withdrawal is referred to as the Colonoscopic Withdrawal Time (CWT).

In this paper, we are particularly interested in measuring the CWT, as the CWT can impinge directly upon the successful detection and removal of polyps. Specifically, success in polyp detection is often measured by the Adenoma Detection Rate (ADR), defined as the fraction of procedures in which a physician discovers at least one adenomatous polyp. Several studies have found a positive correlation between the CWT and rates of neoplasia detection [7,41]. As a result, current guidelines recommend that the CWT be at least 6–7 min in order to achieve the desired higher ADR [7,25,32]. Higher ADR is directly linked to lower rates of interval CRC (a CRC which develops within 60 months of a negative colonoscopy screening) [23]; thus, ensuring a sufficiently high CWT is of paramount importance.

**Overview of the Proposed Method.** Our goal is to estimate the CWT. In order to do so, we seek to find the first time point in the procedure where

---

[1] Cheng *et al.* [10] studied the effect of detecting polyps during colonoscopic insertion and found it did not improve ADR.

the operator stops inserting the endoscope deeper into the colon and starts the colonoscopic withdrawal phase. Typically, this happens at the cecum, see Fig. 1.

Our technique relies on the extraction of three key quantities: the egomotion of the camera, depth estimates of the colon, and detection of anatomical landmarks. The use of the egomotion is clear: it allows us to assess in which direction we are moving, which is an obvious differentiator between the colonoscopic insertion phase and the colonoscopic withdrawal phase. The use of the depth maps is more subtle: they help to distinguish between frames in which the camera is adjacent to the colon (which occur more often in colonoscopic insertion) and frames which see an unobstructed view of the colon (which occur more often in colonoscopic withdrawal). The use of landmark detection is straightforward: the presence of the relevant landmarks near the cecum – namely the appendiceal orifice, ileocecal valve, and triradiate fold – constitutes strong evidence the withdrawal phase has begun.

Our method then learns the optimal way of combining features based on these three key quantities in order to best estimate the probability that any given frame is in one of the following three phases: {forward, stagnant, backward}. Precise definitions of these three phases are given in Sect. 3; in brief, forward corresponds to the time moving forward from the rectum to the cecum; stagnant to the time spent inspecting the cecum; and backward to the time moving backward from the cecum to the rectum. Based on the per frame probabilities, we propose an optimization problem for determining the optimal temporal segmentation of the entire video into phases, which in turn gives the CWT. The main reason that this problem is challenging is that the accuracy of the estimated egomotion and depth images is not always high. We perform ablation studies to carefully show the role each of these features plays in the estimate.

**Applications.** There are several uses of this segmentation procedure. The first use is as a performance metric for GIs. As we have already mentioned, the standard guidelines require that the physician spend at least six minutes after the ileocecal valve has been detected. In such scenarios the withdrawal time should not be measured manually nor by chronometric instruments. However, the amount of time the physician has actually spent can be easily estimated from the segmentation results, and used as a performance metric. Second, the system could be used in combination with existing automatic polyp detection systems, such as those described in [35,37]. In particular, some physicians prefer to detect and remove polyps only in the colonoscopic withdrawal, thus the polyp detection could be optionally turned off during the colonoscopic insertion phase. Indeed, given that our method is based on visual odometry computations, it could in principle store the approximate locations of any polyps detected in the colonoscopic insertion; then in the colonoscopic withdrawal when the physician returns to these locations she can be alerted to the existence of the polyp and take more care in finding it. Third, systems for detecting deficient colon coverage such as [16] could similarly be turned off during the colonoscopic insertion.

A final use case for such a system is as a tool for training novice endoscopists. In [24], it was shown empirically that as training proceeds, the CIT becomes shorter. This shows that the trainee becomes more proficient in navigation; however, this is not necessarily correlated to an improvement in ADR. Thus, navigation and polyp detection are two different capabilities which have to be mastered and the segmentation of the procedure can help in analyzing their proficiency separately.

**Contributions and Paper Outline.** The main contributions of the paper are:

1. We propose a novel approach to withdrawal time estimation based on the combination of egomotion, depth, and landmark information.
2. We collect a gastroenterologist-annotated dataset of 1,447 colonoscopy videos for the purposes of training and validation.
3. We validate our algorithm on this dataset, showing that the algorithm leads to high quality segmentations. Specifically, our error is smaller in magnitude than inter-physician disagreement.

The remainder of the paper is organized as follows. Section 2 reviews related work. Section 3 describes our proposed method, focusing on egomotion and depth computation; detection of anatomical landmarks; derivation of features from the foregoing; a technique for combining these features into a per-frame phase classifier; and a method for video phase segmentation based on this classifier. Section 4 describes our dataset and presents experimental results, including ablation studies. Section 5 concludes the paper and discusses future work.

## 2    Related Work

**Withdrawal Time Estimation.** We begin by discussing [11], which has addressed a very similar problem. The goal of this work is to detect the Cecal Intubation Time (CIT) based on the motion of the endoscope. The assumption made is that when the cecum is reached the magnitude of the motion is small. The relative motion is estimated between consecutive frames using a network which estimates the optical flow between them based on the Horn-Schunk algorithm [22]. Each motion is then classified as +1 for insertion, −1 for withdrawal, and 0 for stop. A section of the video where the sum of these values is lowest is assumed to be the turning point. One of the problems that must be dealt with is that in such videos, many of the frames are of low quality. Thus, at first frames are classified as informative or non-informative [5,12] and the optical flow algorithm is run only on the informative frames. This is due to the fact that on non-informative frames the estimated motion is usually classified as stop. In [27], the authors present a two-stage method for detecting the withdrawal point, i.e., the moment when the endoscope begins to withdraw. First, a deep network is trained to classify each frame whether it is an image of the ileocecal valve, the opening of the appendix or it contains background. Second, the trained classifier is used to generate a time series consisting of the per-frame ileocecal valve class probabilities. This temporal signal is then processed with sliding windows to identify the first window with a sufficient number of frames recognized as the ileocecal valve. Given this window, the withdrawal point is estimated as the last frame of the window.

**Egomotion and Depth Estimation.** The first step of our algorithm runs a visual odometry and depth estimation procedure on a single colonoscopy video. In practice it is also possible to run these procedures separately. Visual odometry algorithms only recover the relative motion between consecutive frames. Classical methods extract and match feature points using descriptors such as SIFT or ORB and estimate from them the relative motion. A review of these methods can be found in [30,31]. Deep learning visual odometry also exists, for example [38]; and specialized visual odometry for endoscopy videos have also been developed [3, 29,36].

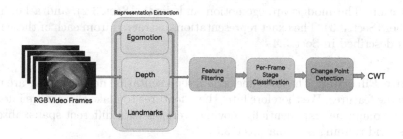

**Fig. 2.** The main blocks of the estimation pipeline.

Since relative motion and depth images are closely related as will be described below, deep learning methods which recover them together have been developed yielding superior results. In initial methods the training was performed in a supervised manner, where the ground truth depth and relative motion were available [4,39,43]. Following this unsupervised methods were developed [20,26, 40,42,44]. The possibility of learning in an unsupervised manner is extremely useful in our case since for colonoscopy videos, ground truth depth and motion are not available. We obtain our depth images and relative motion using a state of the art method of that type [20], which will be reviewed below.

**Phase Detection.** Our work may be thought as a type of phase detection on colonoscopy procedures, where there are three phases. There has been some related work on phase detection in medical procedures. In [33], the video frames are analyzed using a network based on AlexNet. The output of the network are detected tools and the phase to which the frame belongs. An HMM is then used to classify the phase of the frame taking into account temporal constraints. In a more recent work [15], the backbone is replaced with more modern networks and the HMM is replaced with a multi-stage convolutional network. This algorithm was used for analyzing surgery stages in [13]. A very recent validation of surgical phase detection on a much larger dataset is presented in [6].

# 3    Methods

## 3.1    The Estimation Pipeline

Our pipeline is illustrated in Fig. 2. It consists of three deep neural network modules which take as inputs RGB frames, the outputs of which are combined to generate a low dimensional representation. Following this stage, the pipeline has three sequential blocks which take the processed low dimensional representation to generate an estimate of the CWT. More specifically, the stages are as follows:

**Representation Extraction:** The RGB frames are first passed (in parallel) through deep neural network modules to transform them into a succinct representation. The models are egomotion and depth (Sect. 3.2), and a landmark classifier (Sect. 3.3). The exact representations extracted from each of these modules is described in Sect. 3.4.

**Feature Filtering:** The low dimensional representation derived above are noisy per-frame features. We therefore filter these features to make them more informative, by computing exponentially moving averages with different spans; absolute values; and running maxima (Sect. 3.5).

**Per-frame Phase Classifier:** Given the robust features thus computed, this block combines the features into per-frame algorithmic phase probability estimates (Sect. 3.6).

Change-Point Detector: Given the per-frame probability estimates, we compute the change point which induces maximum likelihood (Sect. 3.7).

We now expand upon each of these stages of the pipeline.

## 3.2    Unsupervised Visual Odometry

From an RGB video of a colonoscopy, our goal is to estimate the relative motion between consecutive frames; and additionally, for each frame its corresponding depth image. Given that the vast majority of current endoscopes are monocular we consider the monocular setting. We adopt the *struct2depth* method [8,9,20], which is unsupervised. This is beneficial in the colonoscopy setting, as neither the ground truth position of the colonoscope nor the depth image are available for training. Furthermore, the method does not assume that the camera is calibrated, which can be useful in cases where the camera parameters are unknown.

Our method, like many algorithms for unsupervised depth and motion estimation [19,20,28,34,40,44], is based on the following property: corresponding points in two frames usually have very similar RGB values. This is especially true when the relative motion between the frames is small, as is the case in consecutive video frames. This property can be used to define a loss function known as the *view synthesis loss*, which combines both the egomotion and the depth estimated by the algorithm.

Concretely, the network consists of several sub-networks, as shown in Fig. 3. The depth estimation network is given as input $I_t$, the RGB frame of time $t$ and produces the depth image $D_t$. In addition, given the image pair $I_{t-1}$ and $I_t$, the pose network produces the relative pose / egomotion. Specifically, the pose is defined as the rigid transformation (rotation matrix $R$ and translation vector $t$) from the current frame $t$ to the previous frame $t - 1$. An additional intrinsics network can be used to produce an estimate of the internal calibration $K$; alternatively, a pre-learned $K$ can be given as input.

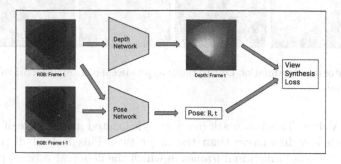

**Fig. 3.** The view synthesis loss and corresponding network architecture. See accompanying description in the text.

The relationship between the geometric location of corresponding pixels in the two frames may be expressed by combining the depth, pose, and intrinsics information as:

$$z'p' = KRK^{-1}zp + Kt \tag{1}$$

where $p$ and $p'$ are the corresponding pixels in homogeneous coordinates and $z$ and $z'$ are their corresponding depth values. The view synthesis loss then compares the RGB values of the pixels at $p$ in image $I_{t-1}$ and $p'$ in image $I_t$. In particular, the loss function is the $L_1$ loss of the RGB difference combined with an analogous loss based on structural similarity (SSIM).

In practice, there may sometimes be pairs of pixels for which Equation (1) does not hold, for example due to self occlusion in one frame or due to non-rigid motion of the colon. In such cases, when the relative transformation is applied to one depth image, there will be a difference in the corresponding depth in the other depth image. These depth differences are incorporated into the loss function, effectively reducing the weight for these pixel pairs.

## 3.3   The Landmark Prediction Module

As shown in Fig. 1, the cecum has several distinctive features which when detected may indicate the end of the insertion phase and beginning of withdrawal.

**Appendiceal Orifice and Triradiate Fold.** Both of these landmarks (see Fig. 4) reside inside the cecum and may therefore be used as indicators of arrival at the cecum. We train a dual-head binary classification model to predict the presence of these two landmarks within the frame.

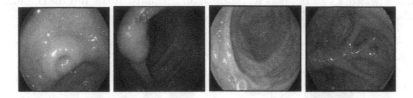

**Fig. 4.** Anatomical landmarks. Left to right: appendiceal orifice; ileocecal valve; cecum; triradiate fold.

**Ileocecal Valve.** This landmark (see Fig. 4) is located just outside of the cecum and is much less distinctive than the other two. This lack of distinctiveness makes annotation of individual frames in which the ileocecal valve appears quite

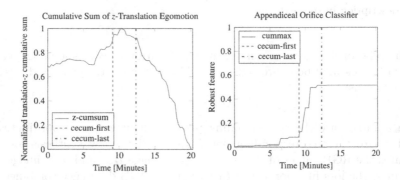

**Fig. 5.** Illustration of the role of various features in computing CWT. In both graphs, the triradiate fold is first observed at the green vertical line and last observed at the red vertical line; this denotes the boundaries of the time spent in the cecum. Left: egomotion. The cumulative sum of the $z$-translation egomotion (blue) can be seen to increase while the scope is inserted (before the green line), to remain constant in the cecum (between green and red), and to decrease during withdrawal (after the red line). Right: landmarks. The cumulative maximum of the appendiceal orifice classifier (blue) increases significantly in the cecum. See accompanying description in the text. (Color figure online)

challenging. We deal with this by labelling the overall temporal region in which the ileocecal value is located; that is, we annotate an initial frame before which the ileocecal valve does not appear, and a final frame after which it does not appear. We then use this temporal region as weak supervision in a Multiple Instance Learning (MIL) [14] scheme where the model learns to predict if a frame has any distinctive features that are common to the cecum region.

Both of the networks - the dual head classification network for the appendiceal orifice and triradiate fold, and the MIL classifier for the ileocecal valve – share a common feature extractor backbone, a Resnet-50 CNN [21]. Each then has a separate fully-connected layer mounted on top of the resulting embedding to yield the probabilities for each of the three landmark features. All networks are trained together in an end-to-end fashion.

### 3.4   Representation Extraction

Thus far, we have described three conceptually different sources of feature representation: egomotion and depth are generated from a module described in Sect. 3.2, while the landmark detection module is described in Sect. 3.3. We now summarize the actual representations:

**Egomotion:** For each pair of consecutive frames, the egomotion estimate has 3 translation coordinates $(x, y, z)$ and 3 rotation coordinates $(\phi_x, \phi_y, \phi_z)$, which are Euler angles. The egomotion estimate reflects the camera motion with respect to the camera coordinates.

**Depth:** For each frame, the depth map estimate consists of an entire image, where each pixel contains the depth estimate corresponding to that pixel.

**Landmarks:** For each frame and landmark (ileocecal valve; appendiceal orifice; and triradiate fold), the landmark feature is the classifier's probability estimate of the landmark's presence in the frame.

We now give some intuition as to why each of these features can play a role in the estimation of the Colonoscopic Withdrawal Time, beginnning with egomotion. Logically, a positive $z$-axis egomotion should indicate forward motion, while negative $z$-axis motion should indicate backward motion. This idea is illustrated on the left side of Fig. 5, which shows a graph of the cumulative sum of the $z$-translation egomotion, overlaid with a ground truth annotation of the triradiate fold (cecum area). The cumulative sum of the $z$-translation egomotion, which we call the $z$-cumsum, can be seen to have the following rough characteristics: (1) there is a small positive $z$-cumsum while the scope moves forward; (2) there is roughly zero $z$-cumsum while the scope is in the cecum area; and (3) there is a large negative $z$-cumsum while the scope moves backward.

The depth maps can be useful in estimating the CWT, in that they help to distinguish between (1) frames in which the camera is adjacent to the colon, which occurs more often in colonoscopic insertion; and (2) frames which see an unobstructed view of the colon implying more pixels with high depth values, which occurs more often in colonoscopic withdrawal.

Finally, by definition the landmark features are extremely indicative of having reached the cecum. A clear view of the appendiceal orifice or the triradiate fold is strong evidence the withdrawal phase has started. The right side of Fig. 5 shows that the landmark classifier for the appendiceal orifice is high around the cecum area. The cumulative maximum of the classification probabilities creates a very robust feature, which reaches its maximum once the navigation away from cecum area has begun.

### 3.5   Feature Filtering

Since raw per-frame features are extremely noisy on their own, we apply smoothing filters to aggregate the values of multiple frames. Specifically, the exponential-weighted-moving-average of a discrete signal $s$, denoted by $\text{ewma}(s)$, is calculated as follows:

$$\text{ewma}(s)[0] = 0$$
$$\text{ewma}(s)[i] = (1 - \alpha) \cdot \text{ewma}(s)[i - 1] + \alpha \cdot s[i]$$

Here $\alpha$ determines the effective memory span: a span of $m$ steps uses $\alpha = 2/m$. Given the above definition, we define smoothed versions of each of the features described in Sect. 3.4 as follows:

**Smoothed $z$-Translation:** Exponentially weighted moving average of egomotion's $z$-translation, where the moving average is taken with a variety of length spans: 1, 2, 3, and 4 min.

**Smoothed Depth Quantiles:** For each depth map, first the $(0.1, 0.25, 0.5, 0.75, 0.9)$-quantiles are extracted. Then, the quantiles are smoothed using exponentially weighted moving average with two different length spans: 2 and 4 min.

**Peaked Smoothed Landmarks:** The running maximum (cumulative maximum) of exponentially weighted moving average of a landmark's probability estimates. For each of the ileocecal valve, appendiceal orifice, and triradiate fold landmarks, the moving average is taken with a variety of length spans: 8, 15, and 30 s.

Note that all filtered features at time $t$ are computed using only representations observed up to time $t$. Moreover, all filtered features for frame $t$ require only frames $t - 1$ and $t$ as well as the filtered features for frame $t - 1$. Therefore, the computation is well suited for an online setup.

## 3.6   Per-frame Algorithmic Phase Classification

As the penultimate stage of our algorithm, we learn a per-frame classifier for the phases of the colonoscopy procedure. One of the useful side benefits of learning a per-frame classifier is that doing so greatly simplifies the pipeline, in an engineering sense. We define the following three phases:

1. forward marks the navigation from the rectum to cecum area. This phase is characterized by forward motion (positive $z$-translation); many frames see the colon's walls; and no landmarks are detected.
2. stagnant marks the start of the screening process (reaching the cecum area). This phase is characterized by little movement ($z$-translation is approximately 0) and possibly one or more landmarks have been detected.
3. backward marks the withdrawal from the cecum area. This phase is characterized by strong backward movement (negative $z$-translation); many frames view deep regions (the lumen of the colon); and one or more landmarks have been detected.

Thus, the colonoscopic insertion phase consists of the forward phase, while the colonoscopic withdrawal phase consists of the union of the stagnant phase and the backward phase.

For the classification model we use gradient boosted decision trees [17,18] on the filtered features from Sect. 3.5. The classifier is trained by minimizing the standard cross-entropy loss. The resulting model generates a probability estimate for each of the three algorithmic phases (which sum up to 1). The reason for choosing a gradient boosted decision trees over a deep-learning model is that at this point we are left with a small number of features and a small number of (relatively independent) samples. Moreover, the separation into modules greatly simplifies the training.

To generate the training data each procedure video is sampled at a fixed rate. Each frame's weight is set to account for duration variability. Specifically, the total weight of samples for a specified video is fixed: longer videos do not influence the training more than shorter videos. At inference time, a probability estimate is generated for every frame.

## 3.7   Change-Point Detection

The phases in an actual colonoscopy procedure form a sequence of contiguous segments with known order: forward → stagnant → backward. Using this order constraint, we seek a segmentation of the per-frame probability estimate which yields the maximum likelihood solution.

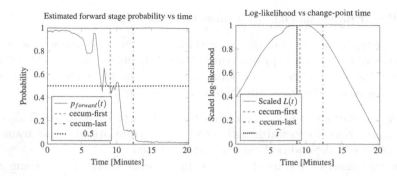

**Fig. 6.** Illustration of the per-frame classifier and the log likelihood of the change point. In both graphs, the triradiate fold is first observed at the green vertical line and last observed at the red vertical line; this denotes the boundaries of the time spent in the cecum. Left: the per-frame classifier probability of being in the `forward` phase. Observe that around the entrance to the cecum area the probability is very close to 0.5 Right: the log likelihood $L(t)$ of the change point vs. time. The estimated change point, which occurs at the maximum of $L(t)$, is shown with a black vertical line, and is very close to the manually annotated change point shown in green. (Color figure online)

Let $\widehat{p}_{c,t}$ denote the probability estimate for phase $c$ at frame $t$. For two phases, denote by $L(t)$ the log-likelihood of the change-point occurring at time $t$. Then we may write

$$L(t) = \sum_{t' \leq t} \log \widehat{p}_{1,t'} + \sum_{t' > t} \log \widehat{p}_{2,t'}$$

The log-likelihood (and the rest of the analysis) can be similarly extended into two splits and three phases.

The optimal change point, $\hat{t}$, is chosen such that $\hat{t} = \arg\max_t L(t)$. (If there are multiple such points, we take the earliest one.) We now present an online computational method for $\hat{t}$. To this end, we define $V[c, t]$ as the value of the optimal log-likelihood for the $1, 2, \ldots, t$ frames which end with phase $c$. We have $V[1, 0] = V[2, 0] = 0$ and

$$V[1, t] = V[1, t-1] + \log \widehat{p}_{1,t}$$
$$V[2, t] = \max \left( V[1, t-1], V[2, t-1] \right) + \log \widehat{p}_{2,t} \tag{2}$$

Finally, $L(\hat{t}) = V[2, T]$ where $T$ is the index of last frame. The value $\hat{t}$ can be retrieved by setting $\hat{t} = t$ where $t$ is the last index for which $V[1, t-1] > V[2, t-1]$. Note that Equation (2) can be easily extended to accommodate 3 (or any) number of phases.

$L(t)$ is visualized for a given sequence in Fig. 6. In this figure, we used the colonoscopic insertion phase as the first segment and the colonoscopic withdrawal phase (union of the **stagnant** phase and **backward** phase) as the second segment.

# 4    Results

## 4.1    The Dataset

The dataset consists of real de-identified colonoscopy videos (acquired from Orpheus Medical) of procedures performed at an academic hospital. In total, there are 788 videos. All videos were recorded at 30 frames per second, with a compression rate of 16 mbps. The distribution of frames in each phase is: forward - 45.2%; stagnant - 12.1%; backward - 42.7%. To maximize the data usage, we employ 5-fold cross validation on the entire set of videos.

The videos were annotated offline by gastroenterologist annotators, drawn from a pool of four with 4, 7, 7, and 9 years of experience. For every landmark, the annotators were asked to carefully mark the first and last frame in which the landmark is visible (for every contiguous period separately), as well as the time point which marks the start of the withdrawal phase.

## 4.2    Hyperparameters

The hyperparameters for Sect. 3.2 and Sect. 3.3 were chosen separately (each for its own sub-task). We focus here on the hyperparameters for the per-frame classifier Sect. 3.6. The video was sub-sampled at a fixed rate of a frame every 15 s. Technically, computational resources allow for the training of the per-frame classifier without sub-sampling. However, consecutive frames are highly correlated and their inclusion does not improve overall metrics. Moreover, sub-sampling is useful in stochastic gradient boosting models to create diversity. The other hyperparameters relate to gradient boosting, namely: the number of trees; the maximum tree depth; the learning rate (0.03); and the subsample probability (0.5). The effect on performance of the choice of both the number of trees and maximum tree depth is presented in a separate discussion in Sect. 4.4.

## 4.3    Results

We report various statistics for the absolute error of the CWT. Specifically, if the ground truth time is $t$ and the estimated time is $\hat{t}$, then the absolute error is denoted $\Delta t = |\hat{t} - t|$. To capture the error distribution we use: (1) the mean absolute error (MAE) and (2) the $i^{th}$ percentile/quantile, which we denote as $q_i(\Delta t)$. To reduce variability, each estimate was taken as the median of 9 bootstrap runs. The results reported are attained using the hyperparameter settings described in Sect. 4.2, along with 1000 trees with maximum depth 1. (The role of the latter hyperparameters is analyzed in Sect. 4.4.)

The results are reported in Table 1, where times are reported in minutes. The MAE is 1.20 min, while the median absolute error is 0.58 min, and the $75^{th}$ percentile is 1.32 min. In order to calibrate the size of these errors, we compared them with the disagreement between gastroenterologist experts. We provided 45 videos to be analyzed by 3 gastroenterologists and measured the difference between the earliest and latest of the 3 annotations, which we refer

**Table 1.** Statistics for the CWT error in minutes: mean and various percentiles. Notice that the algorithm error and the annotator spread are roughly the same.

|  | Mean | $50^{th}$ | $75^{th}$ |
|---|---|---|---|
| $|\Delta t|$ = Algorithm Error | 1.20 | 0.58 | 1.32 |
| Annotator Spread | 1.17 | 0.62 | 1.38 |

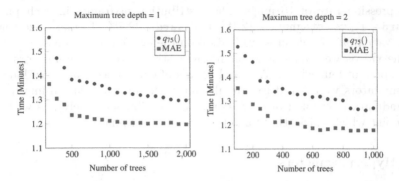

**Fig. 7.** Sensitivity of the algorithm to hyperparameters: number of trees and maximum tree depth. In each case, both MAE and the $75^{th}$ percentile error are plotted vs. number of trees. Left: maximum tree depth of 1. Right: maximum tree depth of 2.

to as the *annotator spread.* The mean annotator spread is 1.17 min, while the median annotator spread is 0.62 min; and the $75^{th}$ percentile is 1.38 min. Note that the algorithm error and the annotator spread are quite similar in value. More specifically, in the case of the median and $75^{th}$ percentile, the annotator spread is higher than the algorithm error; while in the case of the mean, the algorithm's error is slightly higher.

## 4.4 Sensitivity Analysis

We now study the sensitivity of the estimator to the choice of hyperparameters, specifically the number of trees used in the gradient boosting algorithm, as well as the maximum depth of these trees. Figure 7 shows graphs of the performance - as measured by both MAE as well as the $75^{th}$ percentile error $q_{75}(\Delta t)$ - vs. the number of trees. There are two separate graphs, corresponding to maximum tree depths of 1 and 2. We note that the graphs are not entirely smooth due to the optimization performed in the change-point detection algorithm (see Sect. 3.7). Nonetheless, overall there is little variability in the metrics, especially once the number of trees has increased past 500 (for depth 1) or 300 (for depth 2). Thus, it is strongly evident that the estimate is quite robust to the choice of these hyperparameters.

## 4.5    Analysis of Feature Importance

We now turn to analyzing the role of each of the features in the algorithm, in particular their role in the per-frame classifier. To assess the role of each feature, we use the relative importance heuristic described in [17], which we report in Table 2. We see that the egomotion is quite important, accounting for 40.7% of the total contribution. Furthermore, the landmarks are also very important, accounting for a total of 56.3% between the three of them; the triradiate fold is the most important, followed by the ileocecal valve and the appendiceal orifice. Finally, the depth appears to be less important, accounting for only 3.0%.

## 4.6    Ablation Studies

We continue our analysis of the algorithm by performing an ablation study. In particular, we test the effect on performance of removing each of the features one by one, and then retraining the per-frame classifier. The features we study are: (1) egomotion (2) depth (3) all landmarks considered together. In the case of each retraining, we used maximum tree depth equal to 1 and chose the number of trees which minimized $q_{75}(\Delta t)$ for each ablation test.

The results are shown in Table 3. Note that by removing the landmarks, the performance suffers the most: the MAE increases from 1.20 to 1.85, with a concomitant increase in $q_{75}(\Delta t)$. Next is the egomotion, followed by the depth. Interestingly, the ablation study shows that the depth is still quite important: removing it increases the MAE from 1.20 to 1.33, a 10.8% relative increase.

**Table 2.** Feature importance Analysis.

| Feature | Importance |
| --- | --- |
| Egomotion | 0.407 |
| Depth | 0.030 |
| Appendiceal orifice | 0.090 |
| Ileocecal value | 0.331 |
| Triradiate fold | 0.143 |

**Table 3.** Ablation studies. Performance is shown when each feature is removed.

| Ablation | MAE | $q_{75}(\Delta t)$ |
| --- | --- | --- |
| None | 1.20 | 1.32 |
| z-translation | 1.42 | 1.68 |
| Depth | 1.33 | 1.42 |
| Landmarks | 1.85 | 2.08 |

## 5    Conclusions

We presented a system for estimating the colonoscopic withdrawal time in colonoscopy procedures. The method is based on combining features from three disparate sources: egomotion, depth, and landmark classification. The resulting algorithm has been validated on a GI-annotated dataset, and has demonstrated an error which is smaller than the inter-rater disagreement. As a result, the algorithm shows promise in a variety of applications, including as a performance metric for GIs, as an add-on to existing polyp detection systems, and as part of a training system for novice endoscopists.

**Acknowledgment.** The authors would like to thank Gaddi Menahem and Yaron Frid of Orpheus Medical Ltd. for helping in the provision of data.

# References

1. Cancer Facts & Figures (2019). https://www.cancer.org/research/cancer-facts-statistics/all-cancer-facts-figures/cancer-facts-figures-2019.html. Accessed 26 Nov 2019
2. Colorectal Cancer Fact Sheet (2018). https://gco.iarc.fr/today/data/factsheets/cancers/10_8_9-Colorectum-fact-sheet.pdf. Accessed 08 Jan 2020
3. Aghanouri, M., Ghaffari, A., Serej, N.D., Rabbani, H., Adibi, P.: New image-guided method for localisation of an active capsule endoscope in the stomach. IET Image Process. **13**(12), 2321–2327 (2019)
4. Almalioglu, Y., Saputra, M.R.U., de Gusmao, P.P., Markham, A., Trigoni, N.: GANVO: unsupervised deep monocular visual odometry and depth estimation with generative adversarial networks. In: 2019 International Conference on Robotics and Automation (ICRA), pp. 5474–5480. IEEE (2019)
5. Ballesteros, C., Trujillo, M., Mazo, C., Chaves, D., Hoyos, J.: Automatic classification of non-informative frames in colonoscopy videos using texture analysis. In: Beltrán-Castañón, C., Nyström, I., Famili, F. (eds.) CIARP 2016. LNCS, vol. 10125, pp. 401–408. Springer, Cham (2017). https://doi.org/10.1007/978-3-319-52277-7_49
6. Bar, O., et al.: Impact of data on generalization of AI for surgical intelligence applications. Sci. Rep. **10**(1), 1–12 (2020)
7. Barclay, R.L., Vicari, J.J., Doughty, A.S., Johanson, J.F., Greenlaw, R.L.: Colonoscopic withdrawal times and adenoma detection during screening colonoscopy. New Engl. J. Med. **355**(24), 2533–2541 (2006)
8. Casser, V., Pirk, S., Mahjourian, R., Angelova, A.: Depth prediction without the sensors: leveraging structure for unsupervised learning from monocular videos. In: Thirty-Third AAAI Conference on Artificial Intelligence (AAAI-2019) (2019)
9. Casser, V., Pirk, S., Mahjourian, R., Angelova, A.: Unsupervised monocular depth and ego-motion learning with structure and semantics. In: CVPR Workshop on Visual Odometry and Computer Vision Applications Based on Location Cues (VOCVALC) (2019)
10. Cheng, C.L., et al.: Comparison of polyp detection during both insertion and withdrawal versus only withdrawal of colonoscopy: a prospective randomized trial. J. Gastroent. Hepatol. **34**(8), 1377–1383 (2019)
11. Cho, M., Kim, J.H., Hong, K.S., Kim, J.S., Kong, H.J., Kim, S.: Identification of cecum time-location in a colonoscopy video by deep learning analysis of colonoscope movement. PeerJ **7** (2019). https://doi.org/10.7717/peerj.7256
12. Cho, M., Kim, J.H., Kong, H.J., Hong, K.S., Kim, S.: A novel summary report of colonoscopy: timeline visualization providing meaningful colonoscopy video information. Int. J. Colorectal Dis. **33**(5), 549–559 (2018). https://doi.org/10.1007/s00384-018-2980-3
13. Czempiel, T., et al.: TeCNO: surgical phase recognition with multi-stage temporal convolutional networks. arXiv preprint arXiv:2003.10751 (2020)
14. Dieterich, T.G., Lathrop, R.H., Lozano-Pérez, T.: Solving the multiple instance problem with axis-parallel rectangles. Artif. Intell. **89**(1–2), 31–71 (1997)

15. Farha, Y.A., Gall, J.: MS-TCN: multi-stage temporal convolutional network for action segmentation. In: Proceedings of the IEEE Conference on Computer Vision and Pattern Recognition, pp. 3575–3584 (2019)
16. Freedman, D., et al.: Detecting deficient coverage in colonoscopies. IEEE Trans. Med. Imaging **39**(11), 3451–3462 (2020). https://doi.org/10.1109/TMI.2020.2994221
17. Friedman, J.H.: Greedy function approximation: a gradient boosting machine. Ann. Stat., 1189–1232 (2001)
18. Friedman, J.H.: Stochastic gradient boosting. Comput. Stat. Data Anal. **38**(4), 367–378 (2002)
19. Garg, R., B.G., V.K., Carneiro, G., Reid, I.: Unsupervised CNN for single view depth estimation: geometry to the rescue. In: Leibe, B., Matas, J., Sebe, N., Welling, M. (eds.) ECCV 2016. LNCS, vol. 9912, pp. 740–756. Springer, Cham (2016). https://doi.org/10.1007/978-3-319-46484-8_45
20. Gordon, A., Li, H., Jonschkowski, R., Angelova, A.: Depth from videos in the wild: unsupervised monocular depth learning from unknown cameras. In: Proceedings of the IEEE International Conference on Computer Vision (2019)
21. He, K., Zhang, X., Ren, S., Sun, J.: Deep residual learning for image recognition. In: Proceedings of the IEEE Conference on Computer Vision and Pattern Recognition, pp. 770–778 (2016)
22. Horn, B.K., Schunck, B.G.: Determining optical flow. In: Techniques and Applications of Image Understanding, vol. 281, pp. 319–331. International Society for Optics and Photonics (1981)
23. Kaminski, M.F.: Increased rate of adenoma detection associates with reduced risk of colorectal cancer and death. Gastroenterology **153**(1), 98–105 (2017)
24. Lee, S.H., et al.: An adequate level of training for technical competence in screening and diagnostic colonoscopy: a prospective multicenter evaluation of the learning curve. Gastrointest. Endosc. **67**(4), 683–689 (2008)
25. Lee, T., et al.: Longer mean colonoscopy withdrawal time is associated with increased adenoma detection: evidence from the Bowel Cancer Screening Programme in England. Endoscopy **45**(01), 20–26 (2013)
26. Li, R., Wang, S., Long, Z., Gu, D.: UndeepVO: monocular visual odometry through unsupervised deep learning. In: 2018 IEEE International Conference on Robotics and Automation (ICRA), pp. 7286–7291. IEEE (2018)
27. Li, Y., Ding, A., Cao, Y., Liu, B., Chen, S., Liu, X.: Detection of endoscope withdrawal time in colonoscopy videos. In: IEEE International Conference on Machine Learning and Applications (ICMLA), pp. 67–74 (2021)
28. Mahjourian, R., Wicke, M., Angelova, A.: Unsupervised learning of depth and ego-motion from monocular video using 3D geometric constraints. In: Proceedings of the IEEE Conference on Computer Vision and Pattern Recognition, pp. 5667–5675 (2018)
29. Pinheiro, G., Coelho, P., Salgado, M., Oliveira, H.P., Cunha, A.: Deep homography based localization on videos of endoscopic capsules. In: 2018 IEEE International Conference on Bioinformatics and Biomedicine (BIBM), pp. 724–727. IEEE (2018)
30. Scaramuzza, D., Fraundorfer, F.: Visual odometry [tutorial]. IEEE Rob. Autom. Maga. **18**(4), 80–92 (2011)
31. Scaramuzza, D., Fraundorfer, F.: Visual odometry part ii: matching, robustness, optimization and applications. IEEE Rob. Autom. Maga. **19**(2), 78–90 (2012)
32. Simmons, D.T., et al.: Impact of endoscopist withdrawal speed on polyp yield: implications for optimal colonoscopy withdrawal time. Aliment. Pharmacol. Therapeut. **24**(6), 965–971 (2006)

33. Twinanda, A.P., Shehata, S., Mutter, D., Marescaux, J., De Mathelin, M., Padoy, N.: EndoNet: a deep architecture for recognition tasks on laparoscopic videos. IEEE Trans. Med. Imaging **36**(1), 86–97 (2016)
34. Ummenhofer, B., et al.: DeMoN: depth and motion network for learning monocular stereo. In: CVPR (2017)
35. Urban, G., et al.: Deep learning localizes and identifies polyps in real time with 96% accuracy in screening colonoscopy. Gastroenterology **155**(4), 1069–1078 (2018)
36. Wang, M., Shi, Q., Song, S., Hu, C., Meng, M.Q.H.: A novel relative position estimation method for capsule robot moving in gastrointestinal tract. Sensors **19**(12), 2746 (2019)
37. Wang, P., et al.: Development and validation of a deep-learning algorithm for the detection of polyps during colonoscopy. Nat. Biomed. Eng. **2**(10), 741–748 (2018)
38. Wang, S., Clark, R., Wen, H., Trigoni, N.: DeepVO: towards end-to-end visual odometry with deep recurrent convolutional neural networks. In: 2017 IEEE International Conference on Robotics and Automation (ICRA), pp. 2043–2050. IEEE (2017)
39. Yang, N., Wang, R., Stuckler, J., Cremers, D.: Deep virtual stereo odometry: leveraging deep depth prediction for monocular direct sparse odometry. In: Proceedings of the European Conference on Computer Vision (ECCV), pp. 817–833 (2018)
40. Yin, Z., Shi, J.: GeoNet: unsupervised learning of dense depth, optical flow and camera pose. In: Proceedings of the IEEE Conference on Computer Vision and Pattern Recognition, pp. 1983–1992 (2018)
41. Young, Y.G., et al.: Colonoscopic withdrawal time and adenoma detection in the right colon. Medicine **97** (2018)
42. Zhan, H., Garg, R., Saroj Weerasekera, C., Li, K., Agarwal, H., Reid, I.: Unsupervised learning of monocular depth estimation and visual odometry with deep feature reconstruction. In: Proceedings of the IEEE Conference on Computer Vision and Pattern Recognition, pp. 340–349 (2018)
43. Zhou, H., Ummenhofer, B., Brox, T.: DeepTAM: deep tracking and mapping. In: Proceedings of the European Conference on Computer Vision (ECCV), pp. 822–838 (2018)
44. Zhou, T., Brown, M., Snavely, N., Lowe, D.G.: Unsupervised learning of depth and ego-motion from video. In: Proceedings of the IEEE Conference on Computer Vision and Pattern Recognition, pp. 1851–1858 (2017)

# Beyond Local Processing: Adapting CNNs for CT Reconstruction

Bassel Hamoud[1][(✉)] (iD), Yuval Bahat[2] (iD), and Tomer Michaeli[1] (iD)

[1] Technion–Israel Institute of Technology, Haifa, Israel
bassel164@campus.technion.ac.il
[2] Princeton University, Princeton, NJ, USA

**Abstract.** Convolutional neural networks (CNNs) are well suited for image restoration tasks, like super resolution, deblurring, and denoising, in which the information required for restoring each pixel is mostly concentrated in a small neighborhood around it in the degraded image. However, they are less natural for highly non-local reconstruction problems, such as computed tomography (CT). To date, this incompatibility has been partially circumvented by using CNNs with very large receptive fields. Here, we propose an alternative approach, which relies on the rearrangement of the CT projection measurements along the CNN's $3^{rd}$ (channels') dimension. This leads to a more local inverse problem, which is suitable for CNNs. We demonstrate our approach on sparse-view and limited-view CT, and show that it significantly improves reconstruction accuracy for any given network model. This allows achieving the same level of accuracy with significantly smaller models, and thus induces shorter training and inference times.

**Keywords:** CT reconstruction · Machine learning · ConvNets

## 1 Introduction

Deep learning has led to major leaps in our ability to solve complex inverse problems. In the context of image restoration, the architectures of choice are typically convolutional neural networks (CNNs). CNNs have pushed the state-of-the-art in tasks like super-resolution [11,13,33], denoising [12,16,31], deblurring [21,25], dehazing [30] and deraining [19]. These are tasks that are well suited for CNNs because of their local nature. Namely, good restoration in those problems can be achieved by predicting the value of each pixel from a local neighborhood around it in the input image, which is precisely how CNNs operate. However, there exist many important inverse problems that do not possess this locality

---

Y. Bahat—Part of the work was done while the author was affiliated with the Technion.

---

**Supplementary Information** The online version contains supplementary material available at https://doi.org/10.1007/978-3-031-25066-8_29.

L. Karlinsky et al. (Eds.): ECCV 2022 Workshops, LNCS 13803, pp. 513–526, 2023.
https://doi.org/10.1007/978-3-031-25066-8_29

property, and are thus a-priori less natural for CNNs. One particularly important example is computed tomography (CT), where cross-sections of the human body are reconstructed from partial measurements of their Radon transform (or sinogram).

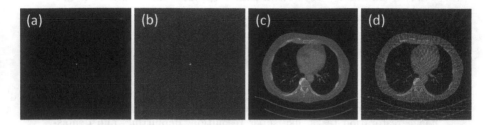

**Fig. 1. Non-local nature of CT reconstruction.** Ground-truth images (a & c) and their corresponding reconstructions (b & d) based on 30 projections, using FBP. Reconstruction artifacts exhibit a global structure, resembling streaks or rays, making them hard to tackle by CNNs, which operate locally.

CT reconstruction has seen significant progress over the years. Theoretically, infinitely many projection angles and noise-free measurements allow perfect reconstruction, *e.g.* via the *filtered back-projection* (FBP) method [6]. However, under practical capturing conditions, the reconstruction problem is ill-posed and the naive FBP method leads to poor reconstructions (Fig. 1). The effect becomes more severe when using low radiation doses and/or a small number of projections, which are desired for reducing scan times and exposure to ionizing X-ray radiation, but lead to noisy and under-sampled sinograms. Several works suggested to use CNNs for learning to reconstruct CT images from degraded sinograms. These methods first convert the low-quality sinogram into an image, and then feed it into the CNN. The conversion stage is done either with the FBP method [3,9] or using learned operations [8]. However, in both cases, the CNN is typically left with solving a very nonlocal reconstruction problem. This can be appreciated from Figs. 1(b), (d), which depict FBP reconstructions of a delta function and a chest scan, from 30 projection angles. As can be seen, the FBP-reconstructed images contain global streaking artifacts. Therefore, CNNs applied on such inputs, must have very large receptive fields [9] to ensure that they can access the information relevant for recovering each pixel. This comes at the cost of many learned parameters, and networks that are harder to train.

In this paper, we present an approach for adapting CNNs to the highly nonlocal CT reconstruction task, by modifying their input such that the information relevant for reconstructing each pixel is available in a small spatial neighborhood around it, across different channels. To this end, we begin by converting the captured sinogram into a series of per-projection images, which we then stack along the channels dimension, and feed into the CNN. The process is depicted in Fig. 2 and described in detail in Sect. 3.

We bring our approach to bear on several sparse-view and limited-view CT reconstruction tasks, including under low dose and patient motion settings. These tasks are all highly relevant for the ongoing effort to reduce radiation exposure in patients undergoing CT scans, without compromising image quality and impairing medical diagnosis.

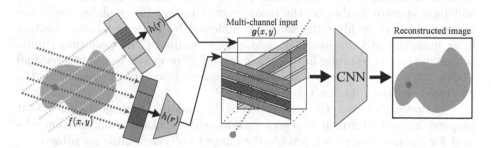

**Fig. 2. Sketch of the proposed method.** After data is acquired from $N$ projection angles (left), we propose to arrange the filtered projections along the channel axis of a 3D tensor $g(x, y)$ (middle), before feeding it to a reconstruction CNN to yield the reconstructed image (right).

## 2   Related Work

As mentioned above, the simple FBP method is not well suited for sparse-view and low-dose CT settings. Several more advanced techniques have been developed over the years. These can be broadly put in several categories, as follows.

*Iterative Algorithms.* These reconstruct the CT image by iteratively minimizing an objective function [2,26,32]. The algebraic reconstruction technique (ART) [7] and simultaneous ART (SART) [1] minimize a fidelity term that pushes the reconstructed image to be consistent with the sinogram. Other methods incorporate image priors over the reconstructed image by adding a regularization term, *e.g.* penalizing the $\ell_1$ [20] or $\ell_0$ [29] norm of the image gradients.

*Radon Space.* These methods operate on the sinogram before transforming it to an image. Some low-dose reconstruction methods proposed to denoise the sinogram using a MAP estimator based on low-dose CT noise statistics [24], or by using a total variation based denoising model tailored for Poisson noise [34]. For sparse-view CT, Lee et al. [14] trained a U-Net-like residual CNN on downsampled and interpolated sinogram patches, while Li et al. [15] learned a dictionary to fill in missing sinogram projections. However, common to all Radon space methods is the need to eventually transform the reconstructed sinogram into image space (e.g. using FBP), which typically results in introducing new artifacts.

*Image Space.* These techniques attempt to enhance a naïvely reconstructed image (obtained e.g. using FBP). Y. Chen et al. [4] employed dictionary learning to reduce the noise and streaking artifacts in low-dose CT images. Kang et al.

[10] applied a CNN to the wavelet coefficients of the reconstructed image, while an auto-encoder design was used in [3]. Yang et al. [28] trained a WGAN to improve perceptual quality. In terms of reconstruction performance, Jin et al. [9] achieved state of the art results by training a U-Net network to suppress the global streaking artifacts that arise in sparse-view reconstruction using FBP.

**End-to-End.** These methods process the sinogram, transform it into an image, and then operate further on the reconstructed image. He et al. [8] used fully connected layers to filter the sinogram, followed by a learned backprojection layer mapping into the image domain, and a residual CNN performing post-processing on the resulting image. Wang et al. [22] proposed to use two residual CNNs, one on the sinogram and one on the image obtained by transforming the processed sinogram into an image using FBP.

Our approach falls in the intersection of the Radon and image domains, as it proposes a novel intermediate representation, which makes the information relevant for reconstructing each pixel in the image locally accessible for subsequent processing. As such, it can be paired with almost any method operating in the image domain, including end-to-end methods.

## 3    Adapting Radon Space Representation for CNNs

CT scans are closely related to the Radon transform. Specifically, when using the simple setting of *parallel-beam* CT, the measurements correspond precisely to the Radon transform of the imaged slice, at the measured angles. When using the more popular *fan-beam* geometry, a narrow X-ray source illuminates a wide section of the body, and as a result the individual projections do not form horizontal lines in Radon space. However, those measurements can be converted into a regular sinogram by using a simple closed-form computation [5,23]. Therefore, in the interest of simplicity, we will describe our method for the parallel-beam CT setting, keeping in mind that it is applicable also to fan-beam CT in a straight-forward manner. We begin by describing the FBP method, and then explain how sinograms can be rearranged to better fit the local nature of CNNs.

### 3.1    The Filtered Backprojection Algorithm

Let $f : \mathbb{R}^2 \to \mathbb{R}$ be a real-valued image. Its Radon transform [17] is the function $s : [0, \pi) \times \mathbb{R} \to \mathbb{R}$ obtained from the linear transformation

$$s(\theta, r) = \int_{-\infty}^{\infty} f(z \sin \theta + r \cos \theta, -z \cos \theta + r \sin \theta) \, dz. \tag{1}$$

That is, each point $(\theta, r)$ in the sinogram corresponds to an integral of $f$ over the straight line $\{(x, y) : x \cos \theta + y \sin \theta = r\}$, which lies at a distance $r$ from the origin and forms an angle $\theta$ with the vertical axis. In the context of CT, $f$ corresponds to the attenuation coefficients across a slice within the imaged object, and $s$ is obtained by measuring the amount of radiation that passes through lines within $f$ (see Fig. 2). Under mild assumptions, $f$ can be recovered from $s$ using the inverse Radon transform, which can be expressed as [17]

$$f(x,y) = \frac{1}{2\pi} \int_0^\pi \tilde{s}(\theta, x\cos\theta + y\sin\theta) \, d\theta, \tag{2}$$

where $\tilde{s}(\theta, \cdot) = s(\theta, \cdot) * h$. Here, $*$ denotes 1D convolution and $h(r)$ is the Ram-Lak filter [18] (ramp filter). In words, each projection is filtered with $h(r)$ and then backprojected onto the image space at its corresponding angle.

In practical settings, the sinogram $s(\theta, r)$ is measured only on a discrete set of angles $\theta$. In such cases, $f$ generally cannot be perfectly reconstructed. One approach to obtain an approximation of $f$ is to discretize (2). Specifically, assuming $N$ uniformly spaced angles $\{\theta_n\}_{n=1}^N$ over the interval $[0, \pi)$, one can approximate $f$ as

$$f(x,y) \approx \frac{1}{2N} \sum_{n=1}^N \tilde{s}(\theta_n, x\cos\theta_n + y\sin\theta_n), \tag{3}$$

This is the filtered back-projection (FBP) algorithm of Feldkamp et al. [6].

The FBP method produces satisfactory reconstructions when the angles are sufficiently densely spaced. However, it leads to highly nonlocal artifacts when the projections are sparse, as illustrated in Fig. 1. To gain intuition into the nonlocal nature of the problem, consider the setting of Figs. 1(a), (b). Here, we measure $N$ projections along angles $\{\theta_n\}_{n=1}^N$ of a Dirac delta function that is located at $(x_0, y_0)$ in the image domain,

$$f(x,y) = \delta(x - x_0, y - y_0). \tag{4}$$

The resulting sinogram is given by (see Supp. A.1)

$$s(\theta_n, r) = \delta(r - (x_0 \cdot \cos\theta_n + y_0 \cdot \sin\theta_n)), \quad n = 1, 2, ..., N. \tag{5}$$

Namely, it is a sinusoid in Radon space. As we show in Supp. A.2, reconstructing the image using FBP yields an approximation of $f$ in the form of a star-like pattern with $N$ rays that are inclined at angles $\{\theta_n\}_{n=1}^N$ and intersect at $(x_0, y_0)$,

$$f(x,y) \approx \frac{1}{2N} \sum_{n=1}^N g\big((x - x_0)\cos\theta_n + (y - y_0)\sin\theta_n\big). \tag{6}$$

Here, $g(r) = k_{\max}^2 \left(\text{sinc}(k_{\max}r) - \frac{1}{4}\text{sinc}^2\left(\frac{k_{\max}r}{2}\right)\right)$. These artifacts can be suppressed using a CNN, however their non-local nature necessitates architectures with very large receptive fields.

### 3.2   Localizing the Inverse Transformation

To be able to effectively exploit the power of CNNs for improving CT reconstruction, here we propose to rearrange the sinogram data into a new multi-channel image-domain representation $g(x, y)$, before feeding it into the CNN. The key idea is to make each (vector-valued) pixel in $g(x, y)$ contain all the sinogram

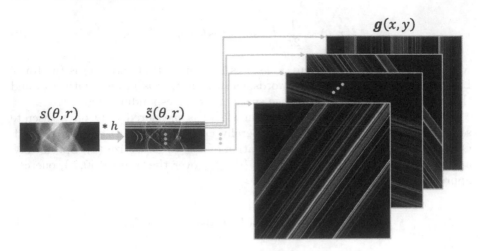

**Fig. 3. Our proposed method (CBP).** The captured sinogram's rows are first filtered with the Ram-Lak filter $h$ to yield $\tilde{s}$. The multi-channel tensor $g(x, y)$ is then constructed by smearing each row of the filtered sinogram $\tilde{s}$ along its corresponding angle to form a 2D image. The smeared images are finally arranged along the channel dimension to form $g$.

entries affected by the corresponding pixel in $f(x, y)$. Specifically, note from (1) that a pixel $f(x_0, y_0)$ affects the sinogram entries at all locations $(\theta, r)$ satisfying $x_0 \cos\theta + y_0 \sin\theta = r$. Since we have $N$ discrete angles, this affected set contains precisely $N$ sinogram entries (one $r$ for each angle $\theta_n$). We collect those $N$ entries from the filtered sinogram $\tilde{s}$ and arrange them along the channel dimension of $g(x, y)$. Namely, for each pixel location $(x, y)$ and channel $n \in \{1, ..., N\}$, we set

$$[g(x, y)]_n = \tilde{s}(\theta_n, r_n), \tag{7}$$

where $r_n = x \cos\theta_n + y \sin\theta_n$. Accordingly, we term our method *Channel Back Projection* (CBP).

It is instructive to note that the $n$th channel of $g(x, y)$ contains a pattern that changes only along the direction $\theta_n$. Particularly, for a fixed $\theta_n$, if we walk along any line of the form $(x, y) = (r \cos\theta_n, r \sin\theta_n)$ by varying $r \in \mathbb{R}$, then the pattern we encounter is precisely $\tilde{s}(r, \theta_n)$. Thus, each channel of $g(x, y)$ can be formed by taking all entries of $\tilde{s}$ corresponding to a single angle, and "smearing" them along that angle to form a 2D image. This is illustrated in Fig. 3.

An additional important observation is that averaging the entries of $g(x, y)$ along the channel direction yields the naive FBP reconstruction (3) (up to a factor of $1/2$). This suggests that our representation $g(x, y)$ generally contains more information than the FBP representation. As we now illustrate, this turns out to allow a significant reduction in the CNN's receptive field (and thus number of parameters) while retaining good reconstruction accuracy.

# 4   Experiments

To demonstrate the advantages of the proposed approach, we evaluate it on the task of sparse-view CT reconstruction in noiseless, noisy, and motion blur settings (we study both sensor blur and patient movement blur). These settings are often encountered when working with low radiation doses. We also evaluate our method on the task of limited-view CT reconstruction, where the projection angles are spread over $[0, \theta_{max})$ with a $\theta_{max}$ that is strictly smaller than $180°$. Such settings arise, $e.g.$ when imaging an object with periodic motion such as coronary vasculature.

We use the DeepLesion [27] dataset, which contains clean full-dose ground-truth (GT) CT images spanning the whole body. We used 5152 CT images for training and 960 for testing. The train and test sets contain CT scans of different patients (74 and 13 patients, respectively). Our approach can be paired with any method operating in the image domain. Here, we demonstrate it with the deep U-Net architecture proposed in [9], which achieves state of the art results in sparse-view CT reconstruction. The architecture is shown in Fig. 4. We next elaborate on each of the reconstruction tasks we consider.

## 4.1   CT Reconstruction Tasks

We simulate four sparse-view CT reconstruction tasks, in which we use 30 projection directions uniformly spread in $[0°, 180°)$, and two limited-view CT reconstruction tasks where the projections are limited to $[0°, 120°)$ and $[0°, 90°)$.

1. **Sparse-view CT:**
   (a) **Noise-free:** The 30-view sinogram $s(\theta, r)$ serves as the measurements.
   (b) **Noisy:** A noisy version of the 30-view $s(\theta, r)$ serves as the measurements. This is done by converting $s(\theta, r)$ into intensity measurements as $I(\theta, r) = I_0 \exp\{-s(\theta, r)\}$, contaminating it by Poisson noise as $I_n(\theta, r) \sim \text{Poisson}(I(\theta, r))$, and converting it back into a noisy sinogram $s_n(\theta, r) = -\log(I_n(\theta, r)/I_0)$. Here $I_0$ is the X-ray source's intensity, which we set to $10^5$.
   (c) **Sensor-motion blur:** Here we simulate motion blur resulting from the rotation of the X-ray tube and the sensor array during the scan. Each of the 30 projections in $s(\theta, r)$ is calculated as the average of 12 projections spanning $6°$, without overlap.
   (d) **Patient-motion blur:** We consider a simplified case of rigid patient motion. Assuming short exposure times, we focus on motion in straight lines. To simulate this, we draw a random direction and a random length $l$ for each GT image. We then sequentially compute the 30 projections of the corresponding degraded sinogram, while uniformly translating the image between $-\ell/2$ and $\ell/2$, along the chosen direction. Thus, each projection within the sinogram corresponds to a different shift of the image (projection $j \in \{0, 1, ..., 29\}$ corresponds to a shift of $-\ell/2 + j\ell/30$).
2. **Limited-view CT:** The measurements here consist of 120 or 90 projections spread uniformly over $[0°, 120°)$ or $[0°, 90°)$, respectively.

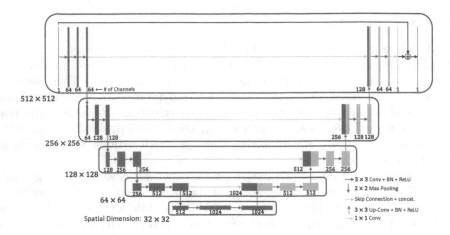

**Fig. 4. U-Net architecture.** The original architecture of [9], comprising five scales consisting of $34.5 \times 10^6$ parameters, and inducing a receptive field of $203 \times 203$.

## 4.2   Reconstruction Methods

We compare our approach to other *CNN based methods*. These achieve far better reconstruction accuracy than classical iterative techniques (see comparison in Supp. C. Particularly, in all tasks, we compare the performance of our method (CBP) to that of feeding the CNN with the image reconstructed using FBP. Moreover, to evaluate the significance of allowing *deep* network processing over the localized input, we additionally compare to a degenerate variant of our method, that learns a weighted linear combination of channels $[\boldsymbol{g}(x, y)]_n$, $n = 1, \ldots, N$, before feeding the resulting *single-channel* image to the CNN. This variant is equivalent to learning a *sinusoidal back projection* (SBP) layer, as proposed in [8], which allows a degree of freedom over using the FBP as input, but does not allow subsequent network layers to access the localized information.

We experiment with different model sizes corresponding to different receptive field sizes, to show how using our method for localizing the input's arrangement alleviates the need for large receptive fields. To this end, we use the original U-Net architecture of [9] (Fig. 4), as well as variants having reduced number of scales, and thus smaller receptive fields and fewer learned parameters. The full architectures and parameters of the variants are provided in Supp. B.

In all experiments, we train the network until convergence by minimizing the mean square error (MSE) between the network's outputs and the corresponding GT CT images, as proposed in [9]. We employ the Adam optimizer with the default settings and batch size of 32. We initialize the learning rate to $10^{-4}$ and automatically decrease it by half whenever the loss reaches a plateau.

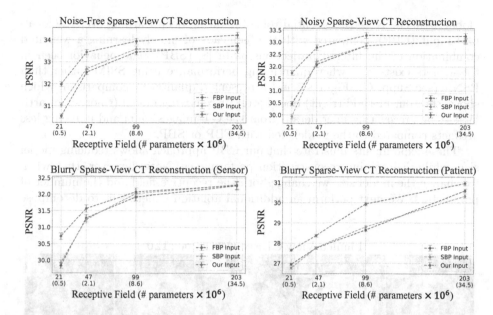

**Fig. 5. Reconstruction accuracy vs. receptive field size in four sparse-view CT reconstruction tasks.** Our CBP input (green) consistently improves performance over the FBP and SBP inputs, and the effect is stronger for smaller models (with smaller receptive fields). These models struggle to handle global artifacts. Error bars correspond to standard error of the mean (SEM) and horizontal axes correspond to receptive field sizes and number of parameters (in parentheses). Please refer to Supp. C for SSIM comparisons, which reveal the same behavior.

### 4.3 Results

We present quantitative comparisons in Figs. 5 and 6, demonstrating the advantage of our method across all tasks and model sizes. This advantage is typically bigger with smaller models, whose smaller receptive field does worse at handling

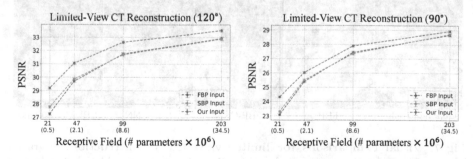

**Fig. 6. Reconstruction accuracy vs. receptive field in limited-view CT.** Our method is consistently advantageous over operating on FBP or SBP inputs, and the effect is stronger for smaller models, whose small receptive fields cannot handle global artifacts. Error bars correspond to standard error of the mean (SEM). Please refer to Supp. C for SSIM comparisons, which reveal the same behavior.

global artifacts. Note the essential role of allowing deep learning over the localized input (CBP), compared with the inefficacy of merely learning a weighted combination of the filtered projections as in [8] (SBP). The behavior seen in these plots exists also when quantifying performance using SSIM rather than PSNR (see Supp. C). Figures 7 and 8 present a qualitative comparison, which shows that our method (right) does better at removing global (smear-like) artifacts, while recovering finer details from the GT images (left) and inducing less artifacts compared to the models fed with FBP or SBP.

These experiments illustrate that our CBP approach allows reducing model size, while maintaining reconstruction performance. Note, however, that when we change the model size we change both the receptive field and the number of parameters. We next perform an additional ablation experiment to decompose

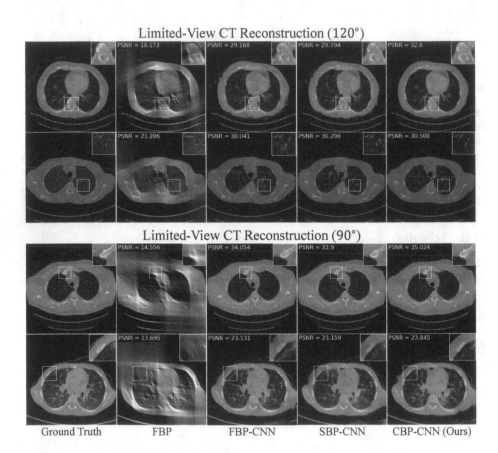

**Fig. 7. Visual comparisons for limited-view CT reconstruction.** Degraded images ($2^{nd}$ col.) are reconstructed using a $21 \times 21$ receptive field U-Net, which is fed with FBP-reconstructed images ($3^{rd}$ col.), with the outputs of the SBP layer of [8] ($4^{th}$ col.), or with our proposed tensor $g$ (last col.). Our method leads to more reliable reconstructions, with less artifacts.

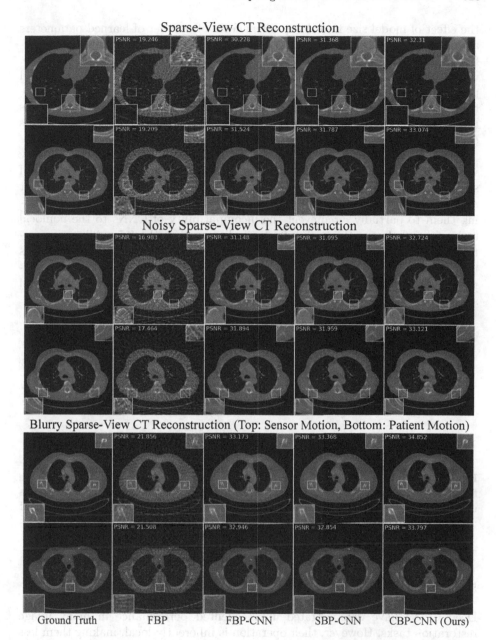

**Fig. 8. Visual comparisons for sparse-view CT reconstruction.** Degraded images (2$^{nd}$ col.) are reconstructed using a $47 \times 47$ receptive field U-Net, which is fed with the FBP-reconstructed images (3$^{rd}$ col.), with the outputs of the SBP layer of [8] (4$^{th}$ col.), or with our proposed tensor $g$ (last col.). Our method leads to more reliable reconstructions, with less artifacts.

the effect of model size into the specific effects of number of learned parameters and receptive field size. To this end, we train six network variants on noise-free sparse-view CT reconstruction, keeping a fixed (21×21) receptive field size, while varying the number of channels to induce different numbers of learned parameters. In particular, we use our 0.5 million parameter (21 × 21 receptive field) variant of U-Net (denoted U-Net-S2, see Supp. B) and five sub-variants thereof, modified to have 0.5, 1.5, 2, 2.5, or 3 times the number of channels (resulting in 0.1, 1.1, 1.9, 3, or 4.3 million parameters, respectively).

Figure 9 shows that the advantage of our approach is unchanged when merely varying the number of learned parameters, without changing the receptive field size. This supports our hypothesis that the improved performance of larger U-net models in Figs. 5 and 6 should be attributed to their larger receptive fields, allowing them to partially circumvent the incompatibility of CNNs to the nonlocal task at hand. Interestingly, Fig. 9 also indicates that the 0.5 million parameter model that uses our CBP performs better than 4.3 million parameter models that use the FBP and SBP representations. Namely, CBP achieves better performance with 8.6× less parameters, illustrating its effectiveness in reducing model sizes.

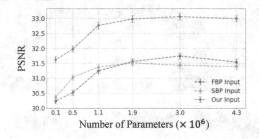

**Fig. 9. Reconstruction performance vs. model size, using a fixed receptive field.** With a fixed receptive field size, the advantage of our approach (green) remains unchanged as the number of learned model parameters increases. This suggests that the improved performance of larger U-net models (right points in Fig. 5) is primarily due to their larger receptive fields, which compensate for the non-local nature of the CT reconstruction problem. Error bars correspond to standard error of the mean (SEM).

## 5   Conclusion

Deep CNNs have demonstrated unprecedented performance in many image restoration tasks. However, their operation is inherently local, making them less suitable for handling CT reconstruction tasks, which suffer from global artifacts. While existing works circumvent this incompatibility using deeper networks with very large receptive fields, we inherently address the problem by proposing a new locality-preserving representation for the CNN's input data. We evaluate our approach across different CT reconstruction tasks, showing how it can improve a network's performance, or alternatively allow reducing the model size, while maintaining reconstruction quality.

**Acknowledgement.** This research was partially supported by the Ollendorff Miverva Center at the Viterbi Faculty of Electrical and Computer Engineering, Technion. Yuval Bahat is funded by the European Union's Horizon 2020 research and innovation programme under the Marie Skłodowska-Curie grant agreement No 945422.

# References

1. Andersen, A., Kak, A.: Simultaneous algebraic reconstruction technique (sart): a superior implementation of the art algorithm. Ultrasonic Imaging **6**(1), 81–94 (1984)
2. Cai, J.F., Jia, X., Gao, H., Jiang, S.B., Shen, Z., Zhao, H.: Cine cone beam ct reconstruction using low-rank matrix factorization: algorithm and a proof-of-princple study (2012)
3. Chen, H., et al.: Low-dose CT with a residual encoder-decoder convolutional neural network. IEEE Trans. Med. Imaging **36**(12), 2524–2535 (2017)
4. Chen, Y.: Artifact suppressed dictionary learning for low-dose CT image processing. IEEE Trans. Med. Imaging **33**(12), 2271–2292 (2014)
5. Dreike, P., Boyd, D.P.: Convolution reconstruction of fan beam projections. Comput. Graph. Image Process. **5**(4), 459–469 (1976)
6. Feldkamp, L.A., Davis, L.C., Kress, J.W.: Practical cone-beam algorithm. J. Opt. Soc. Am. A-optics Image Sci. Vision **1**, 612–619 (1984)
7. Gordon, R., Bender, R., Herman, G.T.: Algebraic reconstruction techniques (art) for three-dimensional electron microscopy and x-ray photography. J. Theor. Biol. **29**(3), 471–481 (1970)
8. He, J., Wang, Y., Ma, J.: Radon inversion via deep learning. IEEE Trans. Med. Imaging **39**(6), 2076–2087 (2020)
9. Jin, K.H., McCann, M.T., Froustey, E., Unser, M.: Deep convolutional neural network for inverse problems in imaging. IEEE Trans. Image Process. **26**(9), 4509–4522 (2017)
10. Kang, E., Min, J., Ye, J.C.: A deep convolutional neural network using directional wavelets for low-dose x-ray CT reconstruction. Med. Phys. **44**(10), e360–e375 (2017)
11. Lai, W.S., Huang, J.B., Ahuja, N., Yang, M.H.: Fast and accurate image superresolution with deep laplacian pyramid networks. IEEE Trans. Pattern Anal. Mach. Intell. **41**(11), 2599–2613 (2018)
12. Laine, S., Lehtinen, J., Aila, T.: Self-supervised deep image denoising. CoRR abs/1901.10277 (2019)
13. Ledig, C., et al.: Photo-realistic single image super-resolution using a generative adversarial network. CoRR abs/1609.04802 (2016)
14. Lee, H., Lee, J., Kim, H., Cho, B., Cho, S.: Deep-neural-network-based sinogram synthesis for sparse-view CT image reconstruction. IEEE Trans. Radiat. Plasma Med. Sci. **3**(2), 109–119 (2019)
15. Li, S., Cao, Q., Chen, Y., Hu, Y., Luo, L., Toumoulin, C.: Dictionary learning based sinogram inpainting for ct sparse reconstruction. Optik - Int. J. Light Electron Optics **125**, 2862–2867 (2014)
16. Mao, X., Shen, C., Yang, Y.: Image denoising using very deep fully convolutional encoder-decoder networks with symmetric skip connections. CoRR abs/1603.09056 (2016)
17. Radon, J.: On the determination of functions from their integral values along certain manifolds. IEEE Trans. Med. Imaging **5**(4), 170–176 (1986)

18. Ramachandran, G.N., Lakshminarayanan, A.V.: Three-dimensional reconstruction from radiographs and electron micrographs: application of convolutions instead of fourier transforms. Proc. Natl. Acad. Sci. USA **68**(9), 2236–40 (1971)

19. Ren, D., Zuo, W., Hu, Q., Zhu, P., Meng, D.: Progressive image deraining networks: a better and simpler baseline. In: Proceedings of the IEEE/CVF Conference on Computer Vision and Pattern Recognition (CVPR) (2019)

20. Sidky, E.Y., Pan, X.: Image reconstruction in circular cone-beam computed tomography by constrained, total-variation minimization. Phys. Med. Biol. **53**, 4777 (2008)

21. Tao, X., Gao, H., Wang, Y., Shen, X., Wang, J., Jia, J.: Scale-recurrent network for deep image deblurring. CoRR abs/1802.01770 (2018)

22. Wang, W., et al.: An end-to-end deep network for reconstructing CT images directly from sparse sinograms. IEEE Trans. Comput. Imaging **6**, 1548–1560 (2020)

23. Wecksung, G.W., Kruger, R.P., Morris, R.A.: Fan-to parallel-beam conversion in cat by rubber sheet transformation. Appl. Digital Image Process. **III**(0207), 76–83 (1979)

24. Xie, Q., et al.: Robust low-dose CT sinogram preprocessing via exploiting noise-generating mechanism. IEEE Trans. Med. Imaging **36**(12), 2487–2498 (2017)

25. Xu, L., Ren, J.S., Liu, C., Jia, J.: Deep convolutional neural network for image deconvolution. Adv. Neural Inf. Process. Syst. **27**, 1790–1798 (2014)

26. Xu, Q., Yu, H., Mou, X., Zhang, L., Hsieh, J., Wang, G.: Low-dose x-ray ct reconstruction via dictionary learning. IEEE Trans. Med. Imaging **31**, 1682–1697 (2012)

27. Yan, K., Wang, X., Lu, L., Summers, R.M.: Deeplesion: automated deep mining, categorization and detection of significant radiology image findings using large-scale clinical lesion annotations. CoRR abs/1710.01766 (2017)

28. Yang, Q., et al.: Low-dose CT image denoising using a generative adversarial network with wasserstein distance and perceptual loss. IEEE Trans. Med. Imaging **37**(6), 1348–1357 (2018)

29. Yu, W., Wang, C., Nie, X., Huang, M., Wu, L.: Image reconstruction for few-view computed tomography based on l0 sparse regularization. Procedia Comput. Sci. **107**, 808–813 (2017)

30. Zhang, H., Sindagi, V., Patel, V.M.: Joint transmission map estimation and dehazing using deep networks. IEEE Trans. Circ. Syst. Video Technol. **30**(7), 1975–1986 (2019)

31. Zhang, K., Zuo, W., Chen, Y., Meng, D., Zhang, L.: Beyond a gaussian denoiser: residual learning of deep CNN for image denoising. IEEE Trans. Image Process. **26**(7), 3142–3155 (2017)

32. Zhang, Y., Zhang, W.H., Chen, H., Yang, M., Li, T.Y., Zhou, J.L.: Few-view image reconstruction combining total variation and a high-order norm. Int. J. Imaging Syst. Technol. **23**, 249–255 (2013)

33. Zhang, Y., Li, K., Li, K., Wang, L., Zhong, B., Fu, Y.: Image super-resolution using very deep residual channel attention networks. CoRR abs/1807.02758 (2018)

34. Zhu, Y., Zhao, M., Zhao, Y., Li, H., Zhang, P.: Noise reduction with low dose CT data based on a modified ROF model. Opt. Express **20**(16), 17987–18004 (2012)

# CL-GAN: Contrastive Learning-Based Generative Adversarial Network for Modality Transfer with Limited Paired Data

Hajar Emami[1], Ming Dong[1(✉)], and Carri Glide-Hurst[2]

[1] Department of Computer Science, Wayne State University, Detroit, MI, USA
{hajar.emami.gohari,mdong}@wayne.edu
[2] Department of Human Oncology, University of Wisconsin Madison,
Detroit, MI, USA
glidehurst@humonc.wisc.edu

**Abstract.** Separate acquisition of multiple modalities in medical imaging is time-consuming, costly and increases unnecessary irradiation to patients. This paper proposes a novel deep learning method, contrastive learning-based Generative Adversarial Network (CL-GAN) for modality transfer with limited paired data. We employ CL-GAN to generate synthetic PET (synPET) images from MRI data, and it has a three-phase training pipeline: 1) intra-modality training for separate source (MRI) and target (PET) domain encoders, 2) cross-modality encoder training with MRI-PET pairs and 3) GAN training. As obtaining paired MRI-PET training data in sufficient quantities is often very costly and cumbersome in clinical practice, we integrate contrastive learning (CL) in all three training phases to fully leverage paired and unpaired data, leading to more accurate and realistic synPET images. Experimental results on benchmark datasets demonstrate the superior performance of CL-GAN, both qualitatively and quantitatively, when compared with current state-of-the-art methods.

**Keywords:** Medical image synthesis · Contrastive learning · Generative adversarial network · Modality transfer · Self-supervised learning · Magnetic resonance imaging

## 1 Introduction

Multimodal medical imaging is essential in various clinical practices including early diagnosis and treatment planning. However, separate acquisition of multiple modalities is time-consuming, costly and increases unnecessary irradiation to patients. Thus, a need exists for an automated imaging modality transfer approach to synthesize missing modalities from the available ones. For example, Positron emission tomography (PET), is a key imaging modality for the diagnosis and treatment of a number of diseases (e.g., Alzheimer, Epilepsy, and Head

L. Karlinsky et al. (Eds.): ECCV 2022 Workshops, LNCS 13803, pp. 527–542, 2023.
https://doi.org/10.1007/978-3-031-25066-8_30

and Neck Cancer) [16,22,32,37,45]. However, high-quality PET acquisition procedure is costly (associated with specialized hardware and tools, logistics, and expertise) and involves side effects of radioactive injection. Clinically, PET imaging is often only considered much further down the pipeline, after information from other non-invasive approaches has been collected. Compared with PET, MRI is a safer imaging modality that does not involve radiation exposure side effects and provides excellent soft tissue contrast. Therefore, estimating synPETs from MRIs has emerged as a viable approach to circumvent the issues in PET acquisition, reducing unnecessary radiation exposure and cost, and to streamline clinical efficiency [32,36,37].

In this direction, many deep learning models have been successfully applied to learn the mapping between two paired image domains [17,19]. However, clinically, acquiring paired training data from different modalities in sufficient quantities is often very costly and cumbersome as patients have to go through different scans. Additionally, there might be inconsistency between images of different modalities even with the same patient, typically caused by internal status/tissue changes as different modalities are usually obtained at a different time. On the other hand, large amount of unpaired data from (different) patients are typically available from various medical imaging studies.

Therefore, there are motivations towards unsupervised approaches of modality transfer, in which, source and target image domains are completely independent with no paired examples between them. To this end, the need for paired training images is removed by introducing the cycle consistency loss in unsupervised image-to-image translation [44,46]. However, the cycle-consistency assumption in these methods could cause a bijection relationship between the two domains [21], which can be too restrictive. Recently, contrastive learning (CL) between multiple views of data has been used as an effective tool in visual representation learning [3,11, 31,43]. Contrastive unpaired translation (CUT) [33] introduced CL for unpaired image-to-image translation by encouraging the corresponding patches from the input and output images to share mutual information.

In this paper, we propose a novel deep learning method, a contrastive learning-based generative adversarial network (CL-GAN), for modality transfer with limited paired data. We employ CL-GAN to generate synthetic PET (synPET) images from MRI data, and it has a three-phase training pipeline: 1)intra-modality training for separate source (MRI) and target (PET) domain encoders, 2)cross-modality encoder training with MRI-PET pairs and 3)GAN training. We integrate contrastive learning (CL) in all three training phases to fully leverage paired and unpaired data. From a computing perspective, the novelty and contribution of this work are as follows:

- To our knowledge, CL-GAN is the first framework that has integrated CL in both the generator and discriminator of a GAN model. This helps CL-GAN use the information available in augmented paired and unpaired data to improve the quality of the synthesized images.
- Different from [33] where only one embedding is employed for both the source and target domains, CL-GAN learns two separate embeddings for the source

and target domains so that both domain-invariant (content) and domain-specific (texture) features can be learned by encoders.

- CL is seamlessly integrated in all three training phases of CL-GAN: In intra-modality training, CL-GAN captures the information from unpaired data to train both source and target domain encoders using CL; CL in cross-modality training encourages corresponding patches from source and target domains to share high mutual information. The number of paired data is augmented by additional positive pairs identified using corresponding image patches; CL is also employed in GAN training to help the discriminator better discriminate between the generated and real images, which, in turn, will help the training of the generator as well. Overall, CL-enhanced training leads to more accurate and realistic synthetic images, especially when the paired training data is very limited.

- We use CL-GAN to generate synPET images from MRI data and evaluate it using MRI/PET images from the ADNI database [18]. Through extensive experiments, we show that, both qualitatively and quantitatively, CL-GAN significantly outperforms other state-of-the-art modality transfer models. CL-GAN offers strong potential for modality transfer applications in a clinical setting when only limited paired data are available.

## 2   Related Work

To date, many conventional machine learning approaches have been employed for synthesizing missing medical imaging modalities from the available ones [15,35]. Recently, many deep learning models have been proposed to learn the mapping between images from a source domain and images from a target domain [4, 14,17,23,28,46,47] and been successfully employed for medical image synthesis, e.g., generating synthetic CT images from MRI images [7,10,29] or estimating PET images from MRI [1,22,32,36,37]. Convolutional neural networks (CNNs) were used in medical image synthesis by capturing nonlinear mapping between the source and target image domains. Once trained, the CNN-based models were found to outperform conventional machine learning approaches for medical image synthesis. A CNN-based model was used in [29] to generate synthetic CT images from the corresponding MRI inputs.

[22] proposed a deep learning based approach using 3D-CNN for synthesizing the missing PET images from the input MRI images. U-net, a well-established fully convolutional network (FCN) [24], is a deep learning model that was first introduced in [34] for biomedical image segmentation. Han developed a deep learning-based approach [10] and employed U-net architecture to generate synthetic CT images from T1-weighted MRI on brain dataset.

Recently, Generative Adversarial Networks (GAN) [8] have achieved state of the art image generation results, and have been shown to generate realistic images in both supervised [17] and unsupervised settings [46]. The adversarial training in GANs can retain fine details in medical image synthesis and has further improved the performance over CNN models [1,6,26,30,32,32]. Typically, in image-to-image translation tasks, a model learns the mapping between two domains using a training set of paired data. However, acquiring a large amount of paired training data is very expensive and sometimes impossible for many real-world tasks including medical imaging applications. To this end, unsupervised approaches [14, 44, 46] were introduced to model the mapping between unpaired image domains. For example, CycleGAN [46] were used in [20, 42] to generate the missing modality images using unpaired data.

More recently, CL between multiple views of data was shown as an effective tool in visual representation learning [3, 9, 11, 13, 31, 41, 43] and achieved state-of-the-art performance in the field of self-supervised representation learning. CL generally learns an embedding for associating "positive" samples and dissociating "negative" samples in a training dataset through maximizing mutual information. In particular, positive samples can be an image with itself [5, 11, 43], a patch image and its neighbour [13, 31], multiple views of an input image [38], or an image with the augmented views of the same image [3, 27]. CUT [33] first introduced CL to unpaired image-to-image translation.

Different from CUT that uses only one encoder, our proposed CL-GAN model employs two separate encoders for the source and target domains to learn both domain-invariant (content) and domain-specific (texture) features. Furthermore, CL-GAN integrates CL in both generator and discriminator for successful modality transfer with limited paired data while CUT integrates CL only in the encoder of the generator.

## 3   Method

State-of-the-art medical image generation models typically train on either paired or unpaired data. To leverage both paired and unpaired data, CL-GAN consists of three training phases: 1) intra-modality training on unpaired data using separate source and target encoders, 2) cross-modality embedding learning using augmented MRI-PET pairs, and 3) GAN training with CL. More specifically, the goal of CL-GAN is to estimate a mapping $G_{MRI \rightarrow PET}$ from the source domain (MRI) to the target domain (PET). CL-GAN is trained using unpaired training data $\{mri_i\}_{i=1}^{N}$ and $\{pet_j\}_{j=1}^{M}$ in the intra-modality training phase and then paired training data $\{(mri_k, pet_k) | mri_k \in MRI, pet_k \in PET, k = 1, 2, ..., K\}$ (K is the number of MRI-PET pairs) in the cross-modality training phase. Finally, we have GAN training (both the generator and discriminator) using CL on MRI-PET pairs.

Figure 1 illustrates the overall architecture of CL-GAN, which includes the MRI encoder, the PET encoder, the decoder and the discriminator classifier. The generator function $G$ includes two components, the pre-trained MRI encoder $Enc_m$ followed by a decoder $Dec$, which are applied sequentially to produce the synPET image $synpet = G(mri) = Dec(Enc_m(mri))$. The discriminator $D$ consists of the pre-trained PET encoder $Enc_p$ and the classifier $C$.

We use CL in all three training phases to better learn the $G_{MRI \rightarrow PET}$ mapping and generate more accurate synPETs. CL-GAN jointly optimizes three

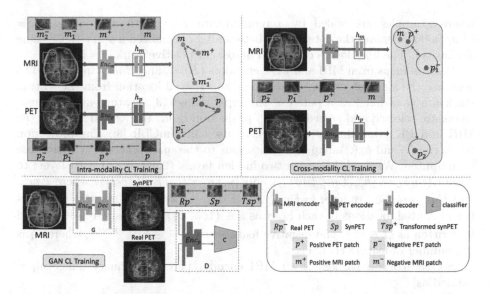

**Fig. 1.** Our CL-GAN framework to generate synPET images from MRI. CL-GAN has a CL-enhanced three-phase training pipeline: intra-modality training on separate MRI and CT domains (encoders) using unpaired data, cross-modality training using augmented MRI-PET pairs, MRI → PET GAN training using the (pre-trained) MRI encoder and decoder as the generator $G$, and the (pre-trained) PET encoder and a small classifier network as the discriminator $D$.

loss functions including the adversarial loss, the patch contrastive loss, and the reconstruction loss as detailed in the following sections.

### 3.1  Intra-modality Training

CL-GAN first learns a generic representation (i.e., the encoders) of images with unpaired MRI and PET data using CL, and then fine-tunes it with a small amount of paired image data. The idea of CL is to associate a single query sample and its related samples referred to as "positive" samples, in contrast to other samples within the dataset, referred to as "negatives" samples. In the intra-modality training phase, each patch with its augmented views (replacement or intensity changing) of the same patch are considered as positive pairs while other patches as negatives.

The query, positive, and N negative samples are mapped to K-dimensional vectors $z, z^+ \in \mathbb{R}^k$ and $z^- \in \mathbb{R}^{N \times k}$, respectively. The contrastive loss for MRI and PET encoders is formulated as follows:

$$
\ell(z, z^+, z^-)
$$
$$
= -log[\frac{exp(sim(z, z^+)/\tau)}{exp(sim(z, z^+)/\tau) + \sum_{n=1}^{N} exp(sim(z, z_n^-)/\tau)}] \tag{1}
$$

where examples are scaled by a temperature $\tau = 0.07$, and $sim(u,v) = u^T v / \|u\| \|v\|$ denotes the dot product between L2 normalized $u$ and $v$ (i.e., cosine similarity). The final loss is computed across all positive pairs in a batch.

Feature maps from MRI and PET encoders ($Enc_m$ and $Enc_p$) are selected to generate the stack of features. Each layer and spatial location from the feature stack are considered as a patch of the input image [3]. Feature maps from $L$ layers are selected and passed through projection heads ($h_m(.)$ and $h_p(.)$ for the MRI and PET domain, respectively), as used in SimCLR [3]. The projection heads ($h_m(.)$ and $h_p(.)$) map representations to the space where contrastive loss is computed. We use MLPs with two hidden layers (256 units at each layer) to obtain $\{m_l\}_L = \{h_m^l(Enc_m^l(mri))\}$ and $\{p_l\}_L = \{h_p^l(Enc_p^l(pet))\}$ where $Enc_m^l$ and $Enc_p^l$ are the $l$-th layer in $Enc_m$ and $Enc_p$, respectively. Denoting the number of spatial locations in each layer as $s \in \{1, ..., S_l\}$, the positive features can be written as $z_l^s \in \mathbb{R}^{C_l}$, and negative features as $z_l^{S \backslash s} \in \mathbb{R}^{(S_l - 1) \times C_l}$, where $C_l$ is the number of feature channels.

So, the contrastive loss of an MRI encoder patch in a query image $\hat{m}_l$ is defined as:

$$\mathcal{L}_{PCL_m} = \mathbb{E}mri \sim MRI \sum_{l=1}^{L} \sum_{s=1}^{S_l} \ell(\hat{m}_l^s, m_l^s, m_l^{S \backslash s}) \tag{2}$$

Similarly, the contrastive loss of a PET encoder patch is defined as:

$$\mathcal{L}_{PCL_p} = \mathbb{E}pet \sim PET \sum_{l=1}^{L} \sum_{s=1}^{S_l} \ell(\hat{p}_l^s, p_l^s, p_l^{S \backslash s}) \tag{3}$$

We randomly sample a minibatch of $N$ examples and calculate contrastive loss on query, positive, and negative samples derived from the minibatch. Following [3], we do not sample negative examples explicitly. Instead, we treat the other patches within a minibatch as negative examples. The final objective of the intra-modality training phase is defined as:

$$\begin{aligned} \mathcal{L}_{intra}(Enc_m, Enc_p, h_m, h_p) \\ = \mathcal{L}_{PCL_m}(Enc_m, h_m, MRI) + \mathcal{L}_{PCL_p}(Enc_p, h_p, PET) \end{aligned} \tag{4}$$

Both $Enc_m$ and $Enc_p$ encoders are trained in this phase. In the testing phase, we only use the MRI encoder $Enc_m$, and the projection heads $h_m(.)$ and $h_p(.)$ are not needed anymore.

### 3.2 Cross-modality Training

After intra-modality training on unpaired data, the $Enc_m$ encoder is fine-tuned in cross-modality training on MRI-PET pairs. As shown in Fig. 1, positive and negative pairs are corresponding and non-corresponding MRI-PET patches in the cross-modality training phase. The augmented MRI patches and the corresponding PET patches of the original MRI patches are also considered as positive pairs to increase the number of training pairs. Using CL in cross-modality training results in representations of corresponding patches to be more closely associated with each other than other patches. For example, a MRI patch containing bone regions would be more closely associated with a corresponding PET patch containing bone regions than other patches containing air, tissue regions or the background. Note that the inputs to the PET encoder $Enc_p$ in the next stage (GAN training) are generated synPETs and real PETs. So, it is not necessary for $Enc_p$ to go through the cross-modality training. Additionally, once CL-GAN is fully trained, $Enc_p$ is not used for synthesising PET images. Only $Enc_m$ is required in the generator $G$ for synPET generation.

Specifically, the objective of the cross-modality training is to improve the MRI encoder $Enc_m$ using augmented MRI-PET pairs:

$$\mathcal{L}_{cross} = \mathcal{L}_{PCL_{mp}}(Enc_m, h_m, MRI, PET) \tag{5}$$

where $\mathcal{L}_{PCL_{mp}}$ is defined similar to $\mathcal{L}_{PCL_m}$ with patches from MRI and PET domains. Different from [33] where only one embedding is used for both source and target domains which may not efficiently capture the domain gap, in CL-GAN, we train a separate encoder for each domain so that the MRI encoder $Enc_m$ is able to capture both domain-invariant features such as organ shapes and domain-specific features such as texture and style of the organs.

## 3.3  CL-GAN Training

After intra-modality and cross-modality pre-training of encoders, CL-GAN's generator $G$ and discriminator $D$ are trained on paired training data. $G$ tries to generate synPETs as close as possible to real PET images while $D$ tries to distinguish between the generated synPETs and real PETs. Following [25], we replace the negative log likelihood objective by a least square loss for more stable training:

$$\mathcal{L}_{GAN}(G, D) = \mathbb{E}_{pet \sim P(pet)}[(D(pet) - 1)^2] \\ + \mathbb{E}_{mr \sim P(mr)}[D(G(mr))^2] \tag{6}$$

Moreover, $L_1$ norm is also used as the regularization in the reconstruction error:

$$\mathcal{L}_{L_1}(G) = \mathbb{E}_{mr \sim P(mr), pet \sim P(pet)}[||G(mr) - pet||_1] \tag{7}$$

In GAN, the discriminator tries to classify the input to either fake or real images. *One novel contribution in CL-GAN is that we deploy CL in the discriminator network* to train a more powerful discriminator. Our idea is to associate the synthetic PET patch and its augmented view as positives in contrast to corresponding synthetic PET-real PET patches as negatives. This helps the discriminator to better discriminate between the real PETs and synPETs. Improving the discriminator, in turn, will also improve the generator performance, ultimately leading to more accurate and realistic synPETs. Specifically, we use the pre-trained PET encoder $Enc_p$ and the classifier $C$ as the discriminator $D$ in

CL-GAN. Again, the projection head $h_p(.)$ is used to map the representations obtained from $Enc_p$ to the space where contrastive loss is computed.

The objective of this training phase is defined as:

$$G^*_{MR \to PET}, D^* = arg \min_G \max_D \mathcal{L}_{GAN}(G_{MR \to PET}, D)$$
$$+ \lambda_{L_1}\mathcal{L}_{L_1} + \lambda_{L_{PCL_p}}\mathcal{L}_{PCL_p} \tag{8}$$

where the loss hyper-parameters $\lambda_{L_1}$ and $\lambda_{L_{PCL_p}}$ are set to 1 throughout our experiments.

## 4 Experiments

### 4.1 Dataset and Implementation Details

Following the MRI to PET translation literature [22,32,36,37], we used Alzheimer's Disease Neuroimaging Initiative (ADNI)[1] database in our experiment. The ADNI was launched in 2003 as a public-private partnership, led by Principal Investigator Michael W. Weiner, MD. The primary goal of ADNI has been to test whether serial magnetic resonance imaging (MRI), positron emission tomography (PET), other biologicalmarkers, and clinical and neuropsychological assessment can be combined to measure the progression of mild cognitive impairment (MCI) and early Alzheimers disease (AD). For up-to-date information, see https://www.adni-info.org/.

The FDG-PET and T1- weighted MRI brain scans of 50 subjects from ADNI database was acquired. First, PET images were rigidly co-registered to their corresponding MRI images of the same subject using SPM12 [2] and then all MRI and PET image slices were cropped to the size of 256 × 256 for training. To evaluate the performance, 20 cases were randomly selected as unpaired data for the intra-modality training, and the remaining cases were used for cross-modality training, GAN training and testing (within cross-validation scheme).

We use a ResNet-based [12] generator with 9 residual blocks. The pre-trained PET encoder and a 3-layer neural network classifier are used as the discriminator. The weights in CL-GAN were all initialized from a Gaussian distribution with parameter values of 0 and 0.02 for mean and standard deviation. The model is trained with ADAM optimizer with an initial learning rate of 0.0002. CL-GAN is trained for 100 epochs in the intra-modality and cross-modality training, and then 300 epochs in CL-GAN training. The model was implemented using PyTorch. The source code of this work will be made publicly available.

---

[1] Data used in the experiments for this work were obtained directly from the Alzheimers Disease Neuroimaging Initiative (ADNI) database (adni.loni.usc.edu). As such, the investigators within the ADNI contributed to the design and implementation of ADNI and/or provided data but did not participate in analysis or writing of this report. A complete listing of ADNI investigators can be found at: http://adni.loni.usc.edu/wp-content/uploads/how_to_apply/ADNI_Acknowledgement_List.pdf.

**Table 1.** Ablation study of key components in CL-GAN. MAE, PSNR, and SSIM metrics computed on MRI → PET translation for ablations of our proposed approach by removing CL in the discriminator (CL-GAN-wo-$CL_D$), removing CL in the generator (CL-GAN-wo-$CL_G$), using one embedding for both domains (CL-GAN-wo-2Emb), removing the cross-modality training phase (CL-GAN-wo-cross), CL-GAN, and including more unpaired cases (CL-GAN+) are listed in Table 1.

| Method | MAE | PSNR | SSIM |
|---|---|---|---|
| cGAN | $43.8 \pm 6.6$ | $28.3 \pm 1.5$ | $0.80 \pm 0.04$ |
| CL-GAN-wo-$CL_D$ | $39.6 \pm 7.1$ | $28.9 \pm 1.4$ | $0.81 \pm 0.03$ |
| CL-GAN-wo-$CL_G$ | $40.6 \pm 5.9$ | $28.8 \pm 1.9$ | $0.82 \pm 0.04$ |
| CL-GAN-wo-2Emb | $39.1 \pm 6.3$ | $29.1 \pm 2.0$ | $0.83 \pm 0.05$ |
| CL-GAN-wo-cross | $38.0 \pm 5.6$ | $29.3 \pm 2.1$ | $0.84 \pm 0.05$ |
| CL-GAN | $36.2 \pm 4.7$ | $29.8 \pm 1.3$ | $0.86 \pm 0.05$ |
| CL-GAN+ | $\mathbf{34.7 \pm 5.3}$ | $\mathbf{30.2 \pm 1.8}$ | $\mathbf{0.87 \pm 0.05}$ |

**Evaluation Metrics.** Three commonly-used quantitative measures are adopted for evaluation: Mean Absolute Error (MAE), Peak Signal-to-Noise Ratio (PSNR), and Structure Similarity Index (SSIM). For PSNR and SSIM, higher values indicate better results. For MAE, the lower the value, the better is the generation.

### 4.2 Experimental Results

**Ablation Study.** We first performed model ablation on MR to PET translation to evaluate each component of CL-GAN. In Table 1, we report MAE, PSNR and SSIM for different configurations of our model. First, we removed CL in CL-GAN including the generator CL and the discriminator CL. In this case, our model is reduced to the cGAN architecture (cGAN) with the regular loss function.

Next, we removed CL in the discriminator of the CL-GAN but kept the CL in the generator (CL-GAN-wo-$CL_D$). Our results show that this leads to a higher MAE and lower PSNR and SSIM when compared with the full version of CL-GAN. This ablation study showed the effectiveness of introducing CL in the discriminator to better discriminate between the real PETs and generated synPETs. Further, we removed CL in the generator of the CL-GAN. As a consequence, the intra-modality and cross-modality training phases were not performed. This leads to worse performance of CL-GAN-wo-$CL_G$ reported in Table 1. Clearly, using the information from unpaired data in the intra-modality training and maximizing the mutual information between corresponding MRI and PET patches in the cross-modality training help us achieve more accurate results.

In our model ablation, we also compared a different configuration of CL-GAN using one embedding for both domains (Cl-GAN-wo-2Emb) similar to CUT [33]. This leads to worse results (reported in Table 1) than our model with two separate embeddings, one for each domain. The result is mainly attributed to the fact that one embedding fails to capture both domain-invariant (content) and

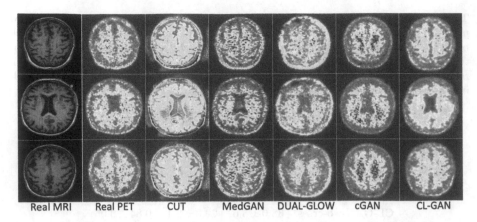

Real MRI    Real PET    CUT    MedGAN    DUAL-GLOW    cGAN    CL-GAN

**Fig. 2.** Qualitative comparison of synPETs generated with CL-GAN and state-of-the-art models. The input MRI, real PET, synPET generated by CUT, MedGAN, DUAL-GLOW, cGAN and CL-GAN, are shown in the columns, respectively.

**Table 2.** Quantitative performance comparison of CL-GAN with other models for synPET generation.

| Method | MAE | PSNR | SSIM |
|---|---|---|---|
| cGAN [17] | $43.82 \pm 6.61$ | $28.32 \pm 1.57$ | $0.80 \pm 0.04$ |
| DUAL-GLOW [37] | $42.68 \pm 5.85$ | $28.71 \pm 2.14$ | $0.83 \pm 0.05$ |
| CUT [33] | $49.52 \pm 10.24$ | $26.53 \pm 1.85$ | $0.75 \pm 0.05$ |
| MedGAN [1] | $43.34 \pm 7.26$ | $28.38 \pm 1.95$ | $0.80 \pm 0.05$ |
| CL-GAN | $\mathbf{36.20 \pm 4.72}$ | $\mathbf{29.89 \pm 1.30}$ | $\mathbf{0.86 \pm 0.05}$ |

domain-specific (texture) features, ultimately leading to less accurate synPETs. Additionally, if we remove the cross-modality training phase, the model (CL-GAN-wo-*cross*) also performs worse.

Clearly, including the generator and discriminator CL in the full version of CL-GAN (CL-GAN), helps achieve the most accurate generated synPETs. Finally, if we include more unpaired cases (e.g., doubling unpaired data by adding 20 more cases) in the intra-modality training phase (CL-GAN+), we get a lower MAE and a slightly higher PSNR and SSIM. This result indicates that the three-phase training of CL-GAN can effectively leverage unpaired data for better modality transfer. In the remaining experiments, we use the CL-GAN with all its components in our evaluation.

**Comparison with the State-of-the-Arts.** Extensive experiments were performed to compare CL-GAN with the current state-of-the-art synPET generation model: DUAL-GLOW [37], the most recent model with CL for image-to-image translation: CUT [33], the state-of-the-art medical image synthesis model: MedGAN [1], and the image-to-image translation baseline cGAN reported in recent medical image translation literature [39]. Based on the same training and testing splits, cGAN, DUAL-GLOW, CUT and MedGAN are included in our empirical evaluation and comparison.

Figure 2 shows qualitative synPET results provided by different models for three testing cases, where color is used to better illustrate the differences. The images in the columns are the input MRI, real PET, the generated synPETs using CUT, MedGAN, DUAL-GLOW, cGAN, and CL-GAN, respectively. Clearly, the synPET images generated by CL-GAN are quite similar to the ground truth PET images. The synPETs generated by DUAL-GLOW, MedGAN and cGAN show less consistency with the ground truth image when compared with CL-GAN results. Shown in the third column, CUT was not able to capture the style of PET domain in the generated synPETs. Overall, our model generates more accurate synPET images and competes favorably against all other methods.

The average MAE, PSNR, and SSIM metrics computed based on the real and synthetic PETs for all test cases are listed in Table 2 for each of the aforementioned methods. CL-GAN achieved the best performance in generating synPET images when compared with other state-of-the-art models, with an average MAE of $36.20 \pm 4.72$ over all test cases. We believe the improvement in synPET results is due to CL in the generator of CL-GAN that helps generate accurate results even with limited training pairs. Additionally, the CL in the discriminator allows our model to better differentiate between the synPETs and real PETs, ultimately leading to a more powerful generator through adversarial learning.

Finally, it is worth mentioning that the small number of MRI-PET pairs presents a significant challenge for synPET methods such as cGAN, MedGAN and DUAL-GLOW. Although CUT was introduced for unpaired image-to-image translation in the absence of paired data, as shown in Fig. 2, it was unable to maintain the style of PET domain in the generated synPETs. This is probably due to using only one encoder for both source and target domains.

**Comparison on Limited Number of Paired Data.** To better evaluate the performance of our CL-GAN model on a smaller number of MRI-PET training pairs, we performed additional experiments with different numbers of MRI-PET pairs using both cGAN and CL-GAN. Specifically, we repeated the same experiment with 10, 20 and 30 paired cases. For CL-GAN, we use 20 cases as unpaired training data in all experiments regardless the number of paired training cases. The visual examples in Fig. 3 shows the clear advantage of CL-GAN over cGAN when handling the challenges associated with very limited paired data.

The average MAE, PSNR, and SSIM metrics computed based on the real and synthetic PETs for all test cases are listed in Table 3 for each of the aforementioned experiments. The performance of CL-GAN degrades slightly as the number of paired data decreases while cGAN degrades much quicker. The effectiveness of our proposed CL-GAN model becomes more evident when a smaller number of paired training data is available.

**Other Potential Applications.** While not a key focus of our work, we also evaluated CL-GAN on two standard computer vision image-to-image translation tasks in order to show our model's generality on various applications. Using the Facades dataset of 400 training images from [40], we constructed photo images from labels. Facades dataset only includes paired data, and thus we did not use any unpaired data for intra-modlity training.

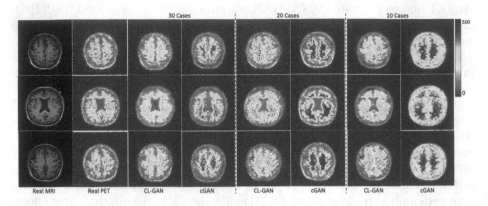

**Fig. 3.** Qualitative comparison of generated synPETs using CL-GAN and cGAN with different numbers of paired data.

**Table 3.** Quantitative performance comparison of CL-GAN and cGAN with different number of paired data.

|      | 10 cases | | 20 cases | | 30 cases | |
|------|----------|----------|----------|----------|----------|----------|
|      | cGAN | CL-GAN | cGAN | CL-GAN | cGAN | CL-GAN |
| MAE  | $53.78 \pm 10.14$ | $41.54 \pm 5.37$ | $48.04 \pm 8.90$ | $38.15 \pm 5.43$ | $43.82 \pm 6.61$ | $\mathbf{36.20 \pm 4.72}$ |
| PSNR | $27.31 \pm 1.83$ | $29.03 \pm 1.67$ | $27.93 \pm 2.03$ | $29.51 \pm 1.26$ | $28.32 \pm 1.57$ | $\mathbf{29.89 \pm 1.30}$ |
| SSIM | $0.73 \pm 0.06$ | $0.82 \pm 0.04$ | $0.76 \pm 0.04$ | $0.84 \pm 0.05$ | $0.80 \pm 0.04$ | $\mathbf{0.86 \pm 0.05}$ |

Figure 4 shows the results generated by GAN models with CL in both generator and discriminator (CL-GAN), with CL only in the generator (wo-$CL_D$), without CL (wo-$CL$) and ground truth images for three example images. Again, these results show the benefits of introducing CL in both the generator and discriminator of a GAN model, and also suggest that CL-GAN is general and applicable to other image-to-image translation tasks even though a more specialized network design could yield better results for this task. We also trained CL-GAN on edges to shoes dataset and generated realistic shoe images from their edges (see Fig. 5).

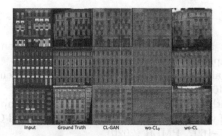

**Fig. 4.** Examples of CL-GAN on facades label → photo.

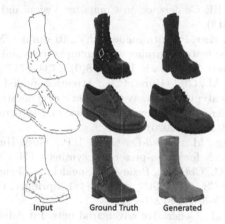

**Fig. 5.** Examples of CL-GAN on edge → shoe.

## 5    Conclusion

In this paper, we proposed CL-GAN for clinical modality transfer. CL-GAN has a three-phase training pipeline: intra-modality training, cross-modality training and GAN training. CL-GAN employs CL in all three training phases to fully leverage paired and unpaired data, and thus improves the quality of generated images. Our experimental results show that CL-GAN achieved superior performance of synthesizing PET images from MRI, both qualitative and quantitative, over current state-of-the-arts. The effectiveness of CL-GAN is more evident when only a small number of paired training data is available. In the future, we plan to extend CL-GAN from 2D to 3D and use it in 3D image-to-image translation applications.

**Acknowledgment.** Work reported in this publication was supported in part by the National Cancer Institute of the National Institutes of Health under award number R01CA204189 and the National Heart, Lung, and Blood Institute of the National Institutes of Health under award number R01HL153720. The content is solely the responsibility of the authors and does not necessarily represent the official views of the National Institutes of Health.

# References

1. Armanious, K., et al.: Medgan: medical image translation using gans. Comput. Med. Imaging Graph. **79**, 101684 (2020)
2. Ashburner, J., et al.: Spm12 manual. Wellcome Trust Centre for Neuroimaging, London, UK, p. 2464 (2014)
3. Chen, T., Kornblith, S., Norouzi, M., Hinton, G.: A simple framework for contrastive learning of visual representations. In: International Conference on Machine Learning, pp. 1597–1607. PMLR (2020)
4. Choi, Y., Choi, M., Kim, M., Ha, J.W., Kim, S., Choo, J.: Stargan: unified generative adversarial networks for multi-domain image-to-image translation. In: Proceedings of the IEEE Conference on Computer Vision and Pattern Recognition, pp. 8789–8797 (2018)
5. Dosovitskiy, A., Fischer, P., Springenberg, J.T., Riedmiller, M., Brox, T.: Discriminative unsupervised feature learning with exemplar convolutional neural networks. IEEE Trans. Pattern Anal. Mach. Intell. **38**(9), 1734–1747 (2015)
6. Emami, H., Dong, M., Glide-Hurst, C.K.: Attention-guided generative adversarial network to address atypical anatomy in synthetic ct generation. In: 2020 IEEE 21st International Conference on Information Reuse and Integration for Data Science (IRI), pp. 188–193. IEEE (2020)
7. Emami, H., Dong, M., Nejad-Davarani, S.P., Glide-Hurst, C.K.: SA-GAN: structure-aware GAN for organ-preserving synthetic CT generation. In: de Bruijne, M., Cattin, P.C., Cotin, S., Padoy, N., Speidel, S., Zheng, Y., Essert, C. (eds.) MICCAI 2021. LNCS, vol. 12906, pp. 471–481. Springer, Cham (2021). https://doi.org/10.1007/978-3-030-87231-1_46
8. Goodfellow, I., et al.: Generative adversarial nets. In: Advances in Neural Information Processing Systems, pp. 2672–2680 (2014)
9. Han, J., Shoeiby, M., Petersson, L., Armin, M.A.: Dual contrastive learning for unsupervised image-to-image translation. In: Proceedings of the IEEE/CVF Conference on Computer Vision and Pattern Recognition, pp. 746–755 (2021)
10. Han, X.: Mr-based synthetic ct generation using a deep convolutional neural network method. Med. Phys. **44**(4), 1408–1419 (2017)
11. He, K., Fan, H., Wu, Y., Xie, S., Girshick, R.: Momentum contrast for unsupervised visual representation learning. In: Proceedings of the IEEE/CVF Conference on Computer Vision and Pattern Recognition, pp. 9729–9738 (2020)
12. He, K., Zhang, X., Ren, S., Sun, J.: Deep residual learning for image recognition. In: Proceedings of the IEEE Conference on Computer Vision and Pattern Recognition, pp. 770–778 (2016)
13. Henaff, O.: Data-efficient image recognition with contrastive predictive coding. In: International Conference on Machine Learning, pp. 4182–4192. PMLR (2020)
14. Huang, X., Liu, M.Y., Belongie, S., Kautz, J.: Multimodal unsupervised image-to-image translation. In: Proceedings of the European Conference on Computer Vision (ECCV), pp. 172–189 (2018)
15. Huynh, T., et al.: Estimating CT image from MRI data using structured random forest and auto-context model. IEEE Trans. Med. Imaging **35**(1), 174–183 (2015)
16. Islam, J., Zhang, Y.: Gan-based synthetic brain pet image generation. Brain Inf. **7**(1), 1–12 (2020)
17. Isola, P., Zhu, J.Y., Zhou, T., Efros, A.A.: Image-to-image translation with conditional adversarial networks. In: Proceedings of the IEEE Conference on Computer Vision and Pattern Recognition, pp. 1125–1134 (2017)

18. Jack Jr, C.R., et al.: The alzheimer's disease neuroimaging initiative (adni): Mri methods. J. Magn. Reson. Imaging: Off. J. Int. Soc. Magn. Reson. Med. **27**(4), 685–691 (2008)

19. Ledig, C., et al.: Photo-realistic single image super-resolution using a generative adversarial network. In: Proceedings of the IEEE Conference on Computer Vision and Pattern Recognition, pp. 4681–4690 (2017)

20. Lei, Y.: Mri-only based synthetic ct generation using dense cycle consistent generative adversarial networks. Med. Phys. **46**(8), 3565–3581 (2019)

21. Li, C.: Alice: towards understanding adversarial learning for joint distribution matching. Adv. Neural Inf. Process. Syst. **30**, 5495–5503 (2017)

22. Li, R., et al.: Deep learning based imaging data completion for improved brain disease diagnosis. In: Golland, P., Hata, N., Barillot, C., Hornegger, J., Howe, R. (eds.) MICCAI 2014. LNCS, vol. 8675, pp. 305–312. Springer, Cham (2014). https://doi.org/10.1007/978-3-319-10443-0_39

23. Liu, M.Y., Breuel, T., Kautz, J.: Unsupervised image-to-image translation networks. In: Advances in Neural Information Processing Systems, pp. 700–708 (2017)

24. Long, J., Shelhamer, E., Darrell, T.: Fully convolutional networks for semantic segmentation. In: Proceedings of the IEEE Conference on Computer Vision and Pattern Recognition, pp. 3431–3440 (2015)

25. Mao, X., Li, Q., Xie, H., Lau, R.Y., Wang, Z., Paul Smolley, S.: Least squares generative adversarial networks. In: Proceedings of the IEEE International Conference on Computer Vision, pp. 2794–2802 (2017)

26. Maspero, M., et al.: Dose evaluation of fast synthetic-ct generation using a generative adversarial network for general pelvis mr-only radiotherapy. Phys. Med. Biol. **63**(18), 185001 (2018)

27. Misra, I., Maaten, L.v.d.: Self-supervised learning of pretext-invariant representations. In: Proceedings of the IEEE/CVF Conference on Computer Vision and Pattern Recognition, pp. 6707–6717 (2020)

28. Murez, Z., Kolouri, S., Kriegman, D., Ramamoorthi, R., Kim, K.: Image to image translation for domain adaptation. In: Proceedings of the IEEE Conference on Computer Vision and Pattern Recognition, pp. 4500–4509 (2018)

29. Nie, D., Cao, X., Gao, Y., Wang, L., Shen, D.: Estimating CT image from MRI data using 3D fully convolutional networks. In: Carneiro, G., et al. (eds.) LABELS/DLMIA -2016. LNCS, vol. 10008, pp. 170–178. Springer, Cham (2016). https://doi.org/10.1007/978-3-319-46976-8_18

30. Nie, D., et al.: Medical image synthesis with context-aware generative adversarial networks. In: Descoteaux, M., Maier-Hein, L., Franz, A., Jannin, P., Collins, D.L., Duchesne, S. (eds.) MICCAI 2017. LNCS, vol. 10435, pp. 417–425. Springer, Cham (2017). https://doi.org/10.1007/978-3-319-66179-7_48

31. Oord, A.v.d., Li, Y., Vinyals, O.: Representation learning with contrastive predictive coding. arXiv preprint arXiv:1807.03748 (2018)

32. Pan, Y., Liu, M., Lian, C., Zhou, T., Xia, Y., Shen, D.: Synthesizing missing PET from MRI with cycle-consistent generative adversarial networks for alzheimer's disease diagnosis. In: Frangi, A.F., Schnabel, J.A., Davatzikos, C., Alberola-López, C., Fichtinger, G. (eds.) MICCAI 2018. LNCS, vol. 11072, pp. 455–463. Springer, Cham (2018). https://doi.org/10.1007/978-3-030-00931-1_52

33. Park, T., Efros, A.A., Zhang, R., Zhu, J.-Y.: Contrastive learning for unpaired image-to-image translation. In: Vedaldi, A., Bischof, H., Brox, T., Frahm, J.-M. (eds.) ECCV 2020. LNCS, vol. 12354, pp. 319–345. Springer, Cham (2020). https://doi.org/10.1007/978-3-030-58545-7_19

34. Ronneberger, O., Fischer, P., Brox, T.: U-Net: convolutional networks for biomedical image segmentation. In: Navab, N., Hornegger, J., Wells, W.M., Frangi, A.F. (eds.) MICCAI 2015. LNCS, vol. 9351, pp. 234–241. Springer, Cham (2015). https://doi.org/10.1007/978-3-319-24574-4_28

35. Roy, S., Wang, W.T., Carass, A., Prince, J.L., Butman, J.A., Pham, D.L.: Pet attenuation correction using synthetic CT from ultrashort echo-time MR imaging. J. Nucl. Med. **55**(12), 2071–2077 (2014)

36. Sikka, A., Peri, S.V., Bathula, D.R.: MRI to FDG-PET: cross-modal synthesis using 3D U-Net for multi-modal alzheimer's classification. In: Gooya, A., Goksel, O., Oguz, I., Burgos, N. (eds.) SASHIMI 2018. LNCS, vol. 11037, pp. 80–89. Springer, Cham (2018). https://doi.org/10.1007/978-3-030-00536-8_9

37. Sun, H., et al.: Dual-glow: conditional flow-based generative model for modality transfer. In: Proceedings of the IEEE/CVF International Conference on Computer Vision, pp. 10611–10620 (2019)

38. Tian, Y., Krishnan, D., Isola, P.: Contrastive multiview coding. In: Vedaldi, A., Bischof, H., Brox, T., Frahm, J.-M. (eds.) ECCV 2020. LNCS, vol. 12356, pp. 776–794. Springer, Cham (2020). https://doi.org/10.1007/978-3-030-58621-8_45

39. Tie, X., Lam, S.K., Zhang, Y., Lee, K.H., Au, K.H., Cai, J.: Pseudo-ct generation from multi-parametric mri using a novel multi-channel multi-path conditional generative adversarial network for nasopharyngeal carcinoma patients. Med. Phys. **47**(4), 1750–1762 (2020)

40. Tyleček, R., Šára, R.: Spatial pattern templates for recognition of objects with regular structure. In: Weickert, J., Hein, M., Schiele, B. (eds.) GCPR 2013. LNCS, vol. 8142, pp. 364–374. Springer, Heidelberg (2013). https://doi.org/10.1007/978-3-642-40602-7_39

41. Wang, W., Zhou, W., Bao, J., Chen, D., Li, H.: Instance-wise hard negative example generation for contrastive learning in unpaired image-to-image translation. arXiv preprint arXiv:2108.04547 (2021)

42. Wolterink, J.M., Dinkla, A.M., Savenije, M.H.F., Seevinck, P.R., van den Berg, C.A.T., Išgum, I.: Deep MR to CT synthesis using unpaired data. In: Tsaftaris, S.A., Gooya, A., Frangi, A.F., Prince, J.L. (eds.) SASHIMI 2017. LNCS, vol. 10557, pp. 14–23. Springer, Cham (2017). https://doi.org/10.1007/978-3-319-68127-6_2

43. Wu, Z., Xiong, Y., Yu, S.X., Lin, D.: Unsupervised feature learning via nonparametric instance discrimination. In: Proceedings of the IEEE Conference on Computer Vision and Pattern Recognition, pp. 3733–3742 (2018)

44. Yi, Z., Zhang, H., Tan, P., Gong, M.: Dualgan: unsupervised dual learning for image-to-image translation. In: Proceedings of the IEEE International Conference on Computer Vision, pp. 2849–2857 (2017)

45. Zhou, B., et al.: Synthesizing multi-tracer PET images for alzheimer's disease patients using a 3D unified anatomy-aware cyclic adversarial network. In: de Bruijne, M., et al. (eds.) MICCAI 2021. LNCS, vol. 12906, pp. 34–43. Springer, Cham (2021). https://doi.org/10.1007/978-3-030-87231-1_4

46. Zhu, J.Y., Park, T., Isola, P., Efros, A.A.: Unpaired image-to-image translation using cycle-consistent adversarial networks. In: Proceedings of the IEEE International Conference on Computer Vision, pp. 2223–2232 (2017)

47. Zhu, J.Y., et al.: Toward multimodal image-to-image translation. In: Advances in Neural Information Processing Systems, pp. 465–476 (2017)

# IMPaSh: A Novel Domain-Shift Resistant Representation for Colorectal Cancer Tissue Classification

Trinh Thi Le Vuong[1], Quoc Dang Vu[2], Mostafa Jahanifar[2], Simon Graham[2], Jin Tae Kwak[1(✉)], and Nasir Rajpoot[2]

[1] School of Electrical Engineering, Korea University, Seoul, Korea
jkwak@korea.ac.kr
[2] Tissue Image Analytics Centre, University of Warwick, Coventry, UK
n.m.rajpoot@warwick.ac.uk

**Abstract.** The appearance of histopathology images depends on tissue type, staining and digitization procedure. These vary from source to source and are the potential causes for domain-shift problems. Owing to this problem, despite the great success of deep learning models in computational pathology, a model trained on a specific domain may still perform sub-optimally when we apply them to another domain. To overcome this, we propose a new augmentation called PatchShuffling and a novel self-supervised contrastive learning framework named IMPaSh for pre-training deep learning models. Using these, we obtained a ResNet50 encoder that can extract image representation resistant to domain-shift. We compared our derived representation against those acquired based on other domain-generalization techniques by using them for the cross-domain classification of colorectal tissue images. We show that the proposed method outperforms other traditional histology domain-adaptation and state-of-the-art self-supervised learning methods. Code is available at: https://github.com/trinhvg/IMPash.

**Keywords:** Domain generalization · Self-supervised learning · Contrastive learning · Colon cancer

## 1 Introduction

Although Deep learning (DL) models have been shown to be very powerful in solving various computational pathology (CPath) problems [20], they can be very fragile against the variations in histology images [5]. One of the main challenges in CPath is *domain-shift* where the distribution of the data used for training and testing of the models varies significantly. There are many sources that can cause domain-shift in CPath, such as variation in sample preparation and staining protocol, the colour palette of different scanners and the tissue type itself.

T. T. L. Vuong and Q. D. Vu—First authors contributed equally.
J. T. Kwak and N. Rajpoot—Last authors contributed equally.

In CPath, there have been several efforts attempting to solve the domain-shift problem by using stain normalization, stain-augmentation, domain adaptation or domain generalization techniques [10,13]. The objective of stain normalization (SN) is to match the colour distribution between the training and testing domains [2,23]. On the other hand, stain augmentation (SA) tries to artificially expand the training set with stain samples that could potentially be within the unseen testing set. This is often achieved by randomly changing the stain/colour information of source images [19]. In certain problems, radically altering the image colour from its usual distribution may also be beneficial, such as by using medically-irrelevant style transfer augmentation [25].

On the other hand, both domain adaptation and domain generalization families try to directly reinforce the model's ability to represent images, such as via the loss functions, to achieve robustness across training and testing domains. The former relies on data from the unseen domain, whereas the latter does not. And due to such reliance on the data of target domains, domain adaptation techniques are thus also highly dependent on the availability and quality of curated data in such domains. Abbet et al. [1] recently proposed a training scheme that we can consider a prime example of a domain adaptation technique. In particular, they employed in-domain and cross-domain losses for colorectal tissue classification.

As for domain generalization, recent proposals mostly focus on pre-training models. In fact, self-supervised algorithms such as MoCoV2 [4] or Self-Path [13] are particularly attractive because they can leverage a huge number of unlabelled histology images. However, in order to effectively train such a model based on self-supervised contrastive learning (SSCL) methods, a careful selection of augmentations and their corresponding parameters is of crucial importance.

In this research, we propose training a DL-based model in self-supervised contrastive learning (SSCL) manner so that the resulting model can extract robust features for colorectal cancer classification across *unseen* datasets. We propose a new domain generalization method inspired by [4,22] that does not rely on data in the domains to be evaluated. Our contributions include:

- We propose PatchShuffling augmentation to pre-train the model such that they can extract invariant representation.
- We propose a new SSCL framework that combines InfoMin [22] augmentations and PatchShuffling for model pre-training called **IMPaSh** based on contrastive learning with momentum contrast.
- We provide a comparative evaluation to demonstrate the effectiveness of IMPaSh.

## 2    The Proposed Method

### 2.1    Self-supervised Contrastive Learning

An overview of our proposed IMPaSh method is presented in Fig. 1. As can be seen in the figure, encoders and projectors that are of the same color have

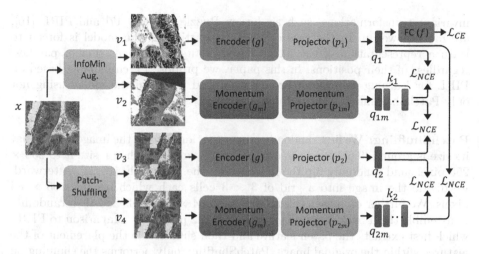

**Fig. 1.** Overview of the proposed self-supervised contrastive learning method. *FC* denotes a fully connected layer that acts as a classifier (i.e., the classification head). In the figure, encoders that are of the same color have their weights shared. (Color figure online)

their weights shared. The proposed method includes two types of augmentation: InfoMin [22] and our proposed PatchShuffling (Sect. 2.2). For each approach, an input image $x$ is augmented twice, resulting into 4 different views $\{v_1, v_2, v_3, v_4\}$ of $x$. An encoder $g$ and two multilayer perceptron (MLP) projectors, called $p_1$ and $p_2$, are applied on $v_1$ and $v_3$ to respectively extract feature vector $q_1$ and $q_2$ both in $\mathbb{R}^{128}$. As for $v_2$ and $v_4$, in a slightly different manner, their corresponding feature vector $q_{1m}$ and $q_{2m}$ are extracted by the *momentum* version of $g$ and $p$, which we subsequently denote as $g_m$, $p_{1m}$ and $p_{2m}$. These $q_{1m}$ and $q_{2m}$ features are further used to update 2 queues of feature vectors $k_1$ and $k_2$.

For the SSCL task, query $q_1$ and query $q_2$ are then respectively compared to $k_1$ and $k_2$ to allow the model to learn the similarity between the different views. Training is thereby achieved by optimizing the InfoNCE loss ($\mathcal{L}_{NCE}$) [18]. Specifically, new $q_1$ and $q_2$ treat existing features that are from different images in the queue as negative samples for optimization.

For the transfer learning task, we first freeze the encoder $g$ and projector $p_1$ and then add another classifier on top of projector $p_1$. We further describe the implementation of the PatchShuffling augmentation and momentum contrast in the following sections.

## 2.2 Learning Pretext-Invariant Representation

Self-supervised learning endows the model with the ability to extract robust representation by maximizing the similarity across variations of the same image. Traditionally, such variations are obtained by covariant transformations like translation or rotation [6,7]. However, recent investigations have shown that invariant transformations, such as Jigsaw Puzzle Solving [17] and PIRL [16], are more powerful. In the case of PIRL in particular, the model is forced to learn a representation of the image based on its constituent smaller patches, regardless of their positions. In this paper, we propose PatchShuffling based on PIRL for learning invariant representation and pre-train the model using not only PatchShuffling but also InfoMin.

**PatchShuffling:** We first randomly crop a portion of from the image $x$ such that its size is around $[0.6, 1.0]$ of the original image area. We then resize it to $255 \times 255$ pixels and randomly flip the image using the settings in $[16, 22]$. Afterward, we divide the image into a grid of $3 \times 3$ cells each which occupies $85 \times 85$ pixels. We further crop each cell randomly to $64 \times 64$ pixels and then randomly re-assemble them back to an image of $192 \times 192$ pixels. In comparison to PIRL, which first extracts the patch feature and then shuffles the the placement of the features within the original image, PatchShuffling only performs the shuffling on the original image itself.

**InfoMin:** We construct views $v_1$ and $v_2$ of a given image $x$ by using the augmentation setup in [22]. In particular, InfoMin augmentation is specifically designed so that the mutual information of the original image and the augmented images is as low as possible while keeping any task-relevant information intact.

### 2.3 Momentum Contrast

In contrastive learning, the most common approach for end-to-end learning is by using only the sample in the current iteration [3]. However, in order to obtain a good image representation, contrastive learning requires a large set of negative samples. Thus a large batch size is required for the training processes (i.e., high GPU memory demand). To handle this memory problem, Wu et al. [24] proposed a memory bank mechanism that stores all the features obtained from previous iterations. Then, from this memory bank, a set of negative samples are randomly selected for the current training iteration. As a result, a large number of samples can be obtained without relying on back-propagation, which in turn dramatically decreases the required training time. However, because the selected samples may come from different training iterations (i.e., from vastly different encoders), there may exist a large discrepancy between them which can severely hinder the training process. To alleviate this problem, MoCo [8] introduced momentum contrast which allows the construction of a consistent dictionary of negative samples in near linear scaling. Inspired by this, to best exploit contrastive learning, we utilize momentum contrast for both InfoMin and PatchShuffling by constructing two dedicated momentum branches as introduced in Fig. 1.

The first momentum contrast branch encodes and stores a dictionary of image representations from an image augmented based on InfoMin. Meanwhile, the second one handles the representation of PatchShuffling. Parameters of these momentum encoders and projectors are updated following the momentum principle. Formally, we denote the parameters of the momentum branch $\{g_m, p_{1m}, p_{2m}\}$ as $\theta_m$ and $\{g, p_1, p_2\}$ as $\theta_q$, we update $\theta_m$ by:

$$\theta_m \leftarrow \alpha\theta_m + (1-\alpha)\theta_q. \tag{1}$$

Here, $\alpha$ is a momentum coefficient to condition the training process to update $\theta_m$ more than $\theta_q$. We empirically set $\alpha = 0.9999$ to the value as used in MoCo [8].

**Loss Function:** Our proposed loss function is an extended version of InfoNCE loss [18]. In essence, the loss maximizes the mutual information between positive pair obtained from an encoder and its momentum version. At the same time, the loss also tries to minimize the similarity in the representation of the current view of the image compared to other $K = 65536$ negative samples from the momentum encoder.

We denote the encoded query as $q$, where $q_1 = p(g(v_1))$, and $q_2 = p(g(v_3))$, and denote the set of keys from momentum branch as $k$, where $k_1 = p_m(g_m(v_2))$, and $k_2 = p_m(g_m(v_4))$. Objective function $\mathcal{L}_{NCE}$ for each pair of query $q$ and queue $k$ is defined as

$$\mathcal{L}_{NCE}(q, k) = -\mathbb{E}\left[\log \frac{\exp(q_i \cdot k_i/\tau)}{\sum_{j=1}^{K} \exp(q_i \cdot k_j/\tau)}\right] \tag{2}$$

where the temperature hyper-parameter $\tau = 0.07$. In summary, the objective function for our contrastive learning framework is as follows,

$$\mathcal{L}_{NCE} = \mathcal{L}_{NCE}(q_1, k_1) + \mathcal{L}_{NCE}(q_1, k_2) + \mathcal{L}_{NCE}(q_2, k_1) + \mathcal{L}_{NCE}(q_2, k_2) \tag{3}$$

### 2.4   Transfer Learning Task

After pre-training the self-supervised task, we obtain encoder $g$ and projector $p$. Instead of discarding the momentum branch and all projectors, we keep and freeze both the projector $p$ and the encoder $g$ to embed the input image to 128-dimensional features. Then, in a supervised manner, we can train another classifier $f$ on top of these features using cross-entropy loss ($\mathcal{L}_{CE}$) and the labels from the corresponding training dataset.

## 3   Experiment

### 3.1   The Datasets

We employed two publicly available datasets of colorectal histology images to evaluate our method: 1) K19 [11] dataset, which includes 100,000 images of size $244 \times 224$ pixels from 9 tissue classes as the source domain, and 2) K16 [12] containing 5,000 images of size $150 \times 150$ pixels from 8 classes as the target domain. Example images from these two datasets are shown in Fig. 2. Since

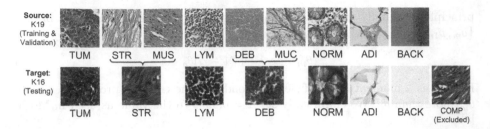

**Fig. 2.** Example images from the two used datasets of colorectal histology images: K19 and K16. Since these two datasets have different class definitions, stroma/muscle and debris/mucus in K19 are grouped into the stroma (STR) and debris (DEB) in K16. On the other hand, complex stroma (COMP) in K16 is excluded from the analyses.

these two datasets have different class labels, we followed [1] to group 9 classes from the training dataset (K19) into 7 classes that are best matched to the 7 classes in the test set (K16). In particular, stroma/muscle and debris/mucus are grouped as stroma (STR) and debris (DEB), respectively. Additionally, for K16, we excluded 625 "complex stroma" images due to the lack of that group in the training domain, leaving us with a test set of 4,375 images. In total, the 7 classes that we evaluate are: adipole (ADI), background (BACK), debris (DEB), lymphocyte (LYM), normal (NORM), stroma (STR) and tumour (TUM).

### 3.2 Experimental Settings

In this study, we adopted ResNet-50 [9] feature extractor as our backbone network (i.e. the encoder). All the projectors in the self-supervised training stage consist of 2 fully-connected layers. Meanwhile, the classifier $f$ in the transfer-learning stage consists of only one fully-connected layer. Both pre-trained encoder $g$ and classifier $f$ were trained using 4 GPUs with a batch size of 256 and optimized with SGD default parameters. We trained the encoder $g$ following MoCo settings which utilized 65,536 negatives samples. The encoder was trained for 200 epochs with an initial learning rate of 0.03 and decayed based on a cosine annealing schedule. On the other hand, the linear classifier $f$ was trained for 40 epochs. Its learning rate started at 30 and then reduced by 5 at epoch 30. For evaluating the performance, we measured the accuracy (Acc), recall (Re), precision (Pre) and F1 of each class, then we took the averaged and reported the results.

### 3.3 Comparative Experiments

We compare our methods with several existing domain generalization methods.

**Domain-Specific Methods.** 1) (SN Macenko) [14] is a stain normalization method proposed by Macenko, which is used to stain normalized the whole training dataset. 2) SN Vahadane [23] is another stain normalization method aiming to preserve histology images' structure.

**Self-supervised Methods.** 1) InsDis [24] formulate the supervised problem as instance-level discrimination, which stores feature vectors in a discrete memory bank and directly compares distances (similarity) between instances. 2)PIRL [16] is a self-supervised method that employs the pre-text task representation, and memory bank [24] to store the negative sample in the self-supervised contrastive task. There are two views in PIRL; one is the original image while the other is cropped into nine patches, encoding these nine patches into nine 128 dimension vectors and then concatenating them in random order. 3) MocoV2 [4] is similar to InsDis in constructing multiple views, but in MoCoV2, negative samples are stored in the momentum updated manner. 4) InfoMin [22] is a combination of InfoMin augmentation and PIRL [16]. InfoMin constructs two views while PIRL constructs one more view; there is only one momentum encoder that constructs a dictionary of negative samples, i.e., negative samples only come from the InfoMin augmentation.

### 3.4   Ablation Study

We conducted experiments to investigate the contribution of the projection heads and the momentum encoders on the model's overall performance when testing on a different domain. In addition to that, we also compared the benefits of shuffling only on the original images (PatchShuffling) against doing the shuffling in feature space (PIRL).

## 4   Results and Discussion

Table 1 reports the performance of ResNet50 when training and validating on K19 and testing on the unseen dataset K16. It is clear that using pre-trained ImageNet weights for the ResNet50 feature extractor and training the classification head $f$ on a dataset is enough to achieve good results on the test data from the same domain with a 0.942 accuracy (*ImageNet - Upper Bound*). However, when we trained the same model on the K19 dataset and tested it on the K16 dataset, it performed poorly, with a drop of more than 25% in accuracy (*ImageNet - Lower Bound*). Despite the simplicity of the task at hand and the well-known capacity of deep neural networks, these results demonstrate that when a testing dataset is of different distribution compared to the training set, the deep learning model can still fail to generalize. As such, our main target is to identify techniques that can improve the performance of any model trained on K19 (*ImageNet - Lower Bound*) to the extent that they can be comparable to those trained directly on K16 (*ImageNet - Upper Bound*).

**Table 1.** Results of the domain generalization experiments between a different source domain (K19) and target domains (K16) using various domain-adaptation techniques

| Method | Training set | Acc. | Re | Pre | F1 |
|---|---|---|---|---|---|
| ImageNet - Upper Bound | K16 (Target) | 0.942 | 0.942 | 0.941 | 0.941 |
| ImageNet - Lower Bound | K19 | 0.654 | 0.654 | 0.741 | 0.626 |
| SN Macenko [14] | K19 | 0.660 | 0.660 | 0.683 | 0.645 |
| SN Vahadane [23] | K19 | 0.683 | 0.683 | 0.696 | 0.656 |
| InsDis [24] | K19 | 0.694 | 0.694 | 0.766 | 0.659 |
| PIRL [16] | K19 | 0.818 | 0.818 | 0.853 | 0.812 |
| MocoV2 [4] | K19 | 0.675 | 0.675 | 0.816 | 0.642 |
| InfoMin [22] | K19 | 0.750 | 0.750 | 0.824 | 0.752 |
| IMPaSh (Ours) | K19 | **0.868** | **0.868** | **0.887** | **0.865** |

**Table 2.** Ablation results when we trained each component of IMPaSh on K19 (the source domain) and then independently tested them on the K16 (the target domain). *ME* denotes using extra Momentum Encoder while *Head* denotes using an additional projector

| Method | Head | Add ME | Acc. | Re | Pre | F1 |
|---|---|---|---|---|---|---|
| InfoMin + PIRL | ✓ | | 0.862 | 0.862 | 0.875 | 0.859 |
| InfoMin + PIRL | ✓ | ✓ | 0.838 | 0.838 | 0.867 | 0.836 |
| InfoMin + PatchShuffling (IMPaSh—ours) | ✓ | | 0.855 | 0.855 | 0.870 | 0.852 |
| InfoMin + PatchShuffling (IMPaSh—ours) | ✓ | ✓ | **0.868** | **0.868** | **0.887** | **0.865** |

Given the fact that K16 and K19 were obtained via different protocols, i.e., their color distribution is different, the above premise and results suggest that utilizing simple domain-specific adaptation techniques such as stain normalization could improve the model performance. In this work, we evaluated this hypothesis by employing the Macenko method [14] and Vandadane method [23]. From Table 1, we demonstrate that such conjecture is plausible as the model performance on F1 scores on the unseen K16 dataset was improved by about 2–3%.

We further evaluated the effectiveness of domain generalization techniques. Table 1 shows that not all self-supervised learning approaches are noticeably superior to the more simple stain normalization methods. In fact, out of all comparative self-supervised learning methods, only PIRL and InfoMin achieved higher performance compared to the baseline ImageNet encoder and stain normalization techniques with more than 10% difference in F1 and at least 7% difference in accuracy.

**Table 3.** Quantitative measurements on the degree of separations across clusters in term of class label and domain alignment for each method

(a) Silhouette scores when measuring clusters of each class (higher is better)

| Domains | ImageNet | SN Vahadane | InsDis | PIRL | MoCov2 | InfoMin | IMPaSh |
|---------|----------|-------------|--------|------|--------|---------|--------|
| Target | 0.402 | 0.461 | 0.465 | 0.485 | 0.410 | 0.471 | **0.553** |
| All | 0.144 | 0.199 | 0.246 | **0.360** | 0.171 | 0.264 | 0.311 |

(b) Silhouette scores when measuring clusters of each domain *within* each class cluster (lower is better)

| Class | ImageNet | SN Vahadane | InsDis | PIRL | MoCov2 | InfoMin | IMPaSh |
|-------|----------|-------------|--------|------|--------|---------|--------|
| ADI | 0.557 | 0.559 | 0.497 | 0.580 | 0.689 | 0.544 | 0.667 |
| BACK | 0.808 | 0.671 | 0.617 | 0.614 | 0.687 | 0.707 | 0.580 |
| DEB | 0.214 | 0.139 | 0.137 | 0.132 | 0.067 | 0.056 | 0.057 |
| LYM | 0.815 | 0.719 | 0.652 | 0.517 | 0.500 | 0.436 | 0.332 |
| NORM | 0.304 | 0.181 | 0.282 | 0.291 | 0.335 | 0.384 | 0.320 |
| STR | 0.448 | 0.499 | 0.405 | 0.418 | 0.413 | 0.442 | 0.418 |
| TUM | 0.647 | 0.512 | 0.448 | 0.426 | 0.582 | 0.580 | 0.545 |
| All | 0.542 | 0.469 | 0.434 | 0.425 | 0.468 | 0.450 | **0.417** |

From Table 1, it can also be observed that our proposed IMPaSh out-performed all other domain-specific techniques and state-of-the-art SSCL approaches such as stain normalization and MoCoV2 [4,16,21,22,24]. In particular, the F1 score on the unseen K16 dataset increases by 24% using our proposed SSCL approach compared to using ImageNet weights without using any stain normalization during test time. In comparison with MoCoV2 [4] which has a single momentum branch and PIRL [16] which only uses Jigsaw Puzzle Solving augmentation, our method respectively achieved 22% and 5% higher F1-score.

We conducted an ablation study and reported the results in Table 2. The results suggest that PatchShuffling provided better performance compared to PIRL when used in conjunction with additional projection heads and momentum branches. On the other hand, using momentum encoders can be detrimental to model overall performance when jointly utilizing InfoMin and PIRL.

To qualitatively assess the impact of each technique on the actual image representation, we utilized UMAP (Uniform Manifold Approximation and Projection) [15] for visualizing the distribution of samples (i.e., their ResNet50 features) from both source (K19) and target domain (K16) in Fig. 3. For NORM (purple) and STR (brown) classes, we observed that the top three methods (PIRL, InfoMin and our proposed IMPaSh) were able to noticeably increase the distance between these two clusters while keeping the samples from the source (lime green) and target domains (cyan) that belong to each of these clusters close to each other. On the other hand, image features obtained using methods with less competitive results like the baseline or stain normalization are highly

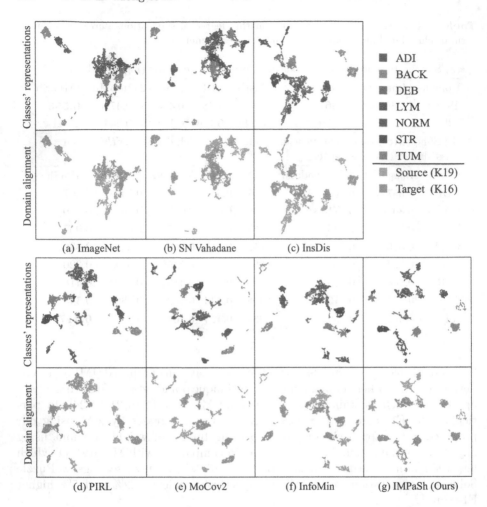

**Fig. 3.** The UMAP [15] visualization of the source (K19) and target (K16) domain feature representation vectors for domain alignment (top row) and the different classes' representations (bottom row). We compare our (g) IMPaSh methods and the a) baseline ImageNet encoder, b) ImageNet encoder+SN and 2 other self-supervised methods e) MoCoV2 and f) InfoMin.

clumped together or visibly closer. These observations suggest that the features from IMPaSh are highly resistant when switching the domain from K19 to K16 for NORM and STR. In the same vein, IMPaSh seems to provide better representation compared to PIRL and InfoMin on LYM (red), given the small area occupied by their samples. Nonetheless, all methods have little success in making NORM (purple) and DEB (green) categories more distinguishable. In our case, while IMPaSh features for NORM and DEB samples are nicely separated into clusters that are far from each other, the samples from the source (lime green) and target (cyan) domains for the same label are not close, thus indicating a strong shift in distribution.

In addition to the qualitative results, we further utilized silhouette scores to measure the degree of separation of the clusters obtained from each method in terms of the classes and the domain alignment. Intuitively, from a better domain generalization technique, we would be able to obtain clusters that satisfy: a) the clusters of each class across all domains will get further apart from each other while b) the clusters of each domain *within* each class cluster get pulled closer. In other words, a better method has higher silhouette scores when measuring the clusters of class labels across domains (class-level scores). At the same time, for methods that have close class-level scores, one that has a smaller silhouette score when measuring for domain alignment within each class label (domain-level score) is the better. We present the class-level scores in Table 3a and domain-level scores in Table 3b. Consistent with our intuition, we observe that methods that have better performance in Table 1 achieved higher scores than others in Table 3a. Interestingly, PIRL has a higher score than IMPaSh while having noticeably lower classification performance. When measuring domain-level separability in 3b, IMPaSh achieved the lowest score when averaging across all classes, but PIRL is the closest in terms of overall performance. All in all, results in 3a and 3b further demonstrate the benefits of IMPaSh framework. However, contradictory observations for PIRL and IMPaSh suggest further investigations in the future are required.

## 5    Conclusion

We proposed a new augmentation named PatchShuffling and a new SSCL framework called IMPaSh that utilize momentum, PatchShuffling and InfoMin to pretrain a neural network encoder. We demonstrated that the resulting encoder was able to extract discriminative image representation while being highly robust against distribution shift. However, our research stops short of evaluating only colorectal tissue classification. Further investigations are required to identify if our IMPaSh can scale for other tasks and/or with more data. In addition, as PatchShuffling is quite modular, investigations on using it in combination with other domain-specific augmentations may also be beneficial for improving IMPaSh.

**Acknowledgement.** This work was funded by the Medical Research Council (MRC) UK and South Korea biomedical and health researcher exchange scheme grant No. MC/PC/210-14, and the National Research Foundation of Korea (NRF) grant funded by the Korean government (MSIP) (No. 2021K1A3A1A88100920 and No. 2021R1A2C2-014557).

## References

1. Abbet, C., et al.: Self-rule to adapt: Learning generalized features from sparsely-labeled data using unsupervised domain adaptation for colorectal cancer tissue phenotyping. In: Medical Imaging with Deep Learning (2021)

2. Alsubaie, N., Trahearn, N., Ahmed Raza, S., Rajpoot, N.M.: A discriminative framework for stain deconvolution of histopathology images in the maxwellian space. In: MIUA, pp. 132–137 (2015)
3. Chen, T., Kornblith, S., Norouzi, M., Hinton, G.: A simple framework for contrastive learning of visual representations. In: International Conference on Machine Learning, pp. 1597–1607. PMLR (2020)
4. Chen, X., Fan, H., Girshick, R.B., He, K.: Improved baselines with momentum contrastive learning. CoRR abs/2003.04297 (2020). https://arxiv.org/abs/2003.04297
5. Foote, A., Asif, A., Rajpoot, N., Minhas, F.: REET: robustness evaluation and enhancement toolbox for computational pathology. Bioinformatics (Oxford, England) 38, 3312–3314 (2022)
6. Gidaris, S., Singh, P., Komodakis, N.: Unsupervised representation learning by predicting image rotations. arXiv preprint arXiv:1803.07728 (2018)
7. Goyal, P., Mahajan, D., Gupta, A., Misra, I.: Scaling and benchmarking self-supervised visual representation learning. In: Proceedings of the IEEE/CVF International Conference on Computer Vision, pp. 6391–6400 (2019)
8. He, K., Fan, H., Wu, Y., Xie, S., Girshick, R.: Momentum contrast for unsupervised visual representation learning. In: Proceedings of the IEEE/CVF Conference on Computer Vision and Pattern Recognition, pp. 9729–9738 (2020)
9. He, K., Zhang, X., Ren, S., Sun, J.: Deep residual learning for image recognition. In: Proceedings of the IEEE Conference on Computer Vision and Pattern Recognition, pp. 770–778 (2016)
10. Jahanifar, M., et al.: Stain-robust mitotic figure detection for the mitosis domain generalization challenge. In: Aubreville, M., Zimmerer, D., Heinrich, M. (eds.) MICCAI 2021. LNCS, vol. 13166, pp. 48–52. Springer, Cham (2022). https://doi.org/10.1007/978-3-030-97281-3_6
11. Kather, J.N., et al.: Predicting survival from colorectal cancer histology slides using deep learning: A retrospective multicenter study. PLoS Med. 16(1), e1002730 (2019)
12. Kather, J.N., et al.: Multi-class texture analysis in colorectal cancer histology. Sci. Rep. 6(1), 1–11 (2016)
13. Koohbanani, N.A., Unnikrishnan, B., Khurram, S.A., Krishnaswamy, P., Rajpoot, N.: Self-path: self-supervision for classification of pathology images with limited annotations. IEEE Trans. Med. Imaging 40(10), 2845–2856 (2021)
14. Macenko, M., et al.: A method for normalizing histology slides for quantitative analysis. In: 2009 IEEE International Symposium on Biomedical Imaging: From Nano to Macro, pp. 1107–1110. IEEE (2009)
15. McInnes, L., Healy, J., Melville, J.: Umap: uniform manifold approximation and projection for dimension reduction. arXiv preprint arXiv:1802.03426 (2018)
16. Misra, I., Maaten, L.V.D.: Self-supervised learning of pretext-invariant representations. In: Proceedings of the IEEE/CVF Conference on Computer Vision and Pattern Recognition, pp. 6707–6717 (2020)
17. Noroozi, M., Favaro, P.: Unsupervised learning of visual representations by solving jigsaw puzzles. In: Leibe, B., Matas, J., Sebe, N., Welling, M. (eds.) ECCV 2016. LNCS, vol. 9910, pp. 69–84. Springer, Cham (2016). https://doi.org/10.1007/978-3-319-46466-4_5
18. Oord, A.V.D., Li, Y., Vinyals, O.: Representation learning with contrastive predictive coding. arXiv preprint arXiv:1807.03748 (2018)
19. Pocock, J., et al.: Tiatoolbox: an end-to-end toolbox for advanced tissue image analytics. bioRxiv (2021)

20. Srinidhi, C.L., Ciga, O., Martel, A.L.: Deep neural network models for computational histopathology: a survey. Med. Image Anal. **67**, 101813 (2021)
21. Tian, Y., Krishnan, D., Isola, P.: Contrastive multiview coding. In: Vedaldi, A., Bischof, H., Brox, T., Frahm, J.-M. (eds.) ECCV 2020. LNCS, vol. 12356, pp. 776–794. Springer, Cham (2020). https://doi.org/10.1007/978-3-030-58621-8_45
22. Tian, Y., Sun, C., Poole, B., Krishnan, D., Schmid, C., Isola, P.: What makes for good views for contrastive learning? Adv. Neural Inf. Process. Syst. **33**, 6827–6839 (2020)
23. Vahadane, A., et al.: Structure-preserving color normalization and sparse stain separation for histological images. IEEE Trans. Med. Imaging **35**(8), 1962–1971 (2016)
24. Wu, Z., Xiong, Y., Yu, S.X., Lin, D.: Unsupervised feature learning via non-parametric instance discrimination. In: Proceedings of the IEEE Conference on Computer Vision and Pattern Recognition, pp. 3733–3742 (2018)
25. Yamashita, R., Long, J., Banda, S., Shen, J., Rubin, D.L.: Learning domain-agnostic visual representation for computational pathology using medically-irrelevant style transfer augmentation. IEEE Trans. Med. Imaging **40**(12), 3945–3954 (2021)

# Surgical Workflow Recognition: From Analysis of Challenges to Architectural Study

Tobias Czempiel[1]([⊠]), Aidean Sharghi[2], Magdalini Paschali[3], Nassir Navab[1,4], and Omid Mohareri[2]

[1] Computer Aided Medical Procedures, Technical University of Munich, Munich, Germany
tobias.czempiel@tum.de
[2] Intuitive Surgical Inc., Sunnyvale, USA
[3] Department of Psychiatry and Behavioral Sciences, Stanford University School of Medicine, Stanford, USA
[4] Computer Aided Medical Procedures, Johns Hopkins University, Baltimore, USA

**Abstract.** Algorithmic surgical workflow recognition is an ongoing research field and can be divided into laparoscopic (Internal) and operating room (External) analysis. So far, many different works for the internal analysis have been proposed with the combination of a frame-level and an additional temporal model to address the temporal ambiguities between different workflow phases. For the External recognition task, Clip-level methods are in the focus of researchers targeting the local ambiguities present in the operating room (OR) scene. In this work, we evaluate the performance of different combinations of common architectures for the task of surgical workflow recognition to provide a fair and comprehensive comparison of the methods for both settings, Internal and External. We show that the methods particularly designed for one setting can be transferred to the other mode and discuss the architecture effectiveness considering the main challenges for both Internal and External surgical workflow recognition.

**Keywords:** Surgical workflow analysis · Surgical phase recognition · Concept ablation · Cholecystectomy · Benchmarking · Analysis

## 1 Introduction

Automatic recognition of surgical workflow is essential for the Operating Room (OR) of the future, increasing the patient's safety through early detection of surgical workflow variations and improving the surgical results [16]. By supplying surgeons with the necessary information for each surgical step, a cognitive OR can reduce the stress induced by an overload of information [20] and build the foundation for more efficient surgical scheduling and reporting systems [3].

Research in Surgical workflow analysis is separated into two fundamental direction (modes). The first mode is analyzing the internal surgical scene captured by

L. Karlinsky et al. (Eds.): ECCV 2022 Workshops, LNCS 13803, pp. 556–568, 2023.
https://doi.org/10.1007/978-3-031-25066-8_32

laparoscopic or robotic cameras, which has been the main research focus in the past (Internal). Procedures such as laparoscopic cholecystectomy, colorectal and laparoscopic sleeve gastrectomy have been widely analyzed [14].

The other mode is closely related to activity and action recognition, where cameras are placed inside the OR to capture human activities and external processes (External) [29]. These external cameras are rigidly installed to the ceiling or are attached to a portable cart. The goal of external OR workflow analysis is to capture the events happening inside the OR such as patient roll-in or docking of a surgical robot [27].

Automatic surgical workflow recognition remains challenging due to the limited amount of publicly available annotated training data, visual similarities of different phases, and visual variability of frames among the same phase. The External OR scene is a complex environment with many surgical instruments and people working in a dense and cluttered environment. Compared to popular action recognition datasets, such as the Breakfast Action [22] or the GTEA [11] dataset, the duration of a surgery is considerably longer. Therefore, the amount of information to be analyzed is substantially higher. Additionally, factors like variation of patient anatomy or surgeon style and personal preferences impose further challenges for the automatic analysis.

Our primary goal in this study is to identify the advantages and disadvantages of the fundamental building blocks used for surgical phase recognition in a structured and fair manner. We do not aim to establish a new state-of-the-art model or directly compare out-of-the-box methods, similar to endoscopic challenges. Our main objective and contribution is, considering the challenges of surgical phase recognition, to compare different network architecture blocks and main conceptual components that have been widely used for surgical phase recognition, analysing their advantages and disadvantages.

## 1.1 Related Work

The analysis of the internal surgical workflow is an ongoing research topic [24] that gained additional attention with the introduction of convolutional neural networks (CNNs) for computer vision tasks [30]. The basic building blocks of such approaches consisted of a frame-level method in combination with a temporal method to analyze the temporal context. Even though the frame-level methods got upgraded to more recent architectures [4,18] with more learning capabilities, the majority of research tried to improve the results with more capable temporal models such as Recurrent Neural Networks (RNNs) [31]. Jin et al. [19] used Long short-term memory (LSTM) Networks [4] to temporally refine the surgical phase results with the prediction of the surgical tool and showed that both tasks profited from this combination. Czempiel et al. [7] proposed to replace the frequently used LSTMs with a multi-stage temporal convolution network (TCN) [10] analyzing the long temporal relationships more efficiently. Additionally, attention-based transformer architectures [32] have been proposed [8,13] to refine the temporal context even further and increase model interpretability.

**Fair Evaluation.** One of the biggest challenges in this domain is the limited benchmarking between existing methods. New architectures are proposed, incrementally improving the results on the datasets. However, tracing improvements back to each fundamental architectural changes remains a challenging task.

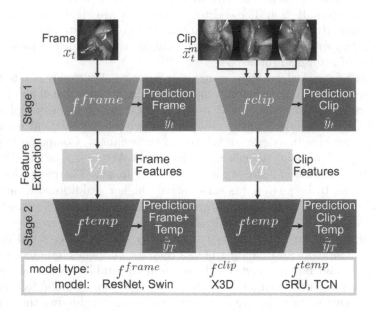

**Fig. 1.** Stage 1 consists of the backbone models on both frame- and clip-level. Stage 2 describes the additional temporal training on extracted features

Researchers have to choose from a plethora of additional settings and methods that can influence the results such as data augmentation, choice of optimizers, learning rate schedules, or even dataset splits. The comparison gets even more challenging for 2-Stage methods that split the training into an image to feature and feature to prediction part which allows for long temporal modelling which otherwise would not be possible due to computational hardware limitations. In Fig. 1 an overview of the different model combinations and configurations as Stage 1 and Stage 2 models is visualized. The novelty of some works is only limited to the Stage 2, however the quality of the extracted features from Stage 1 is essential for a method's overall prediction. Hence, the final results generated in the Stage 2 heavily rely on the Stage 1 features making it difficult to pinpoint the exact point of improvement. Finally, publicly available datasets usually provide limited training data, compromising model performance on unseen data. Thus, the conclusions we get can be misleading or only applicable to a particular dataset.

A very effective way to make a comparison fair for all participants is through the creation of challenges with an unpublished test set that is used to evaluate all submissions. The computer-assisted intervention community releases different challenges every year such as the Endovis[1] grand challenge. We strongly believe that challenges are a fundamental tool to identify the best solution for a surgical workflow task. However, even in public challenges the choice of learning strategies, learning rates, frameworks, optimizers vary among participants making the impact of each methodological advancement hard to identify. Nevertheless, architectural studies can provide additional insights to identify the methodological advancements in a structured way.

To this end, we conduct a fair and objective evaluation on multiple architectures for both Stages 1 and 2. We evaluate the various architectural components on two datasets for internal and external surgical workflow recognition.

## 2   Analysis of Challenges: Local vs. Global

**Descriptive Frames.** In Fig. 2 we visualized the four main architectural differences for video classification that are discussed in this paper: end-to-end Image and Clip level methods and 2-Stage Image and Clip methods with an additional temporal model operating on extracted features. For descriptive frames such as the ones visualized in Fig. 2 all of the aforementioned main architectural directions are likely to produce correct results. A single descriptive image is enough in this case to correctly identify the Phase as clipping&cutting (IP3) mainly due to the presence of the clipping tool that only appears in that particular phase.

**Local Ambiguities.** In the second row of Fig. 2 two examples of frames with Local Ambiguities are shown. In the external scene several occlusions of the camera appear, caused by bulky OR equipment or medical staff operating in a crowded environment. In the internal scene, the view can be impeded by smoke generated during tissue coagulation or pollution of the endoscopic lens through body fluids. Surgical frames do not receive a phase label purely based on their visual properties but also based on their semantic meaning within the OR. Surgical workflow analysis methods should be able to categorize ambiguous frames, such as ones captured with a polluted lens, based on this global semantic context. However, purely image-level methods lack the temporal context and the semantic information of previous time points to resolve ambiguities reliably. Clip-level methods are able to understand the context of a frame neighbourhood, which is mostly sufficient as these Local Ambiguities in the internal and external OR scene often only persist for a few seconds before the person in the scene moves to a different location or the surgeon cleans the lens to continue the surgery safely.

**Global Ambiguities.** There are many activities in the external and internal analysis with high visual similarities belonging to different phases (high inter-class similarities). For instance, the phases of patient roll-in and roll-out in the external dataset contain frames that look almost identical (Fig. 2). To correctly differentiate these two phases only based on single images could be challenging even for an experience surgeon. Considering a limited temporal context could also be insufficient for a correct classification, since the activity can be considerably

---

[1]  http://endovissub-workflow.grand-challenge.org.

longer. However, the rich temporal context modeled in a 2-Stage approach, can alleviate this confusion more effectively as it is clear that the roll-out phase appears always after the roll-in phase.

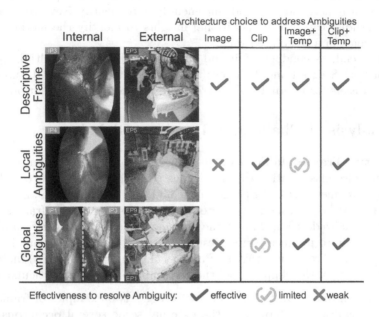

**Fig. 2.** Different types of Surgical Workflow Datasets Internal (IE) and External (EP) along with three frames from each datasets that highlight Local and Global Ambiguities. On the right we highlight different architectural choices regarding the input type and addition of a temporal component (+Temp) that can be used to address the ambiguities

## 3   Methodology

We conduct a fair and objective evaluation on multiple architectures under the same settings. We evaluate our results on two different datasets for internal and external surgical workflow recognition. Each surgical workflow video consists of different amount of frames $T$. For each frame $x_t \in \{x_1, x_2, \ldots x_T\}$ the target is to create an accurate prediction corresponding to the label $y_t$ of the frame-level model $f^{frame}(x_t) = \hat{y}_t$. For the clip-level model the input is not a single frame but a sequence of frames $\overrightarrow{x}_t^n = (x_{t-n}, x_{t-n+1}, \ldots, x_t)$ were $n$ is the size of the input clip to generate one prediction $f^{clip}(\overrightarrow{x}_t^n) = \hat{y}_t$. In the feature extraction process we extract for each time point $t$ of a video a feature vector $\overrightarrow{v}_t$. The feature vectors from an entire video $\overrightarrow{V}_T = \{\overrightarrow{v}_1, \overrightarrow{v}_2, \ldots, \overrightarrow{v}_T\}$ are then used as input for the temporal models of the second Stage that predict a corresponding output probability for all feature vectors $f^{temp}(\overrightarrow{V}_T) = \hat{\overrightarrow{y}}_T$, where $\hat{\overrightarrow{y}}_T = (\hat{y}_1, \hat{y}_2, \ldots, \hat{y}_T)$.

### 3.1   Visual Backbone - Stage 1

**Frame-level.** Frame-level methods operate on a single image using the visual contend to create a classification (Fig. 1 - Prediction Frame). For the task of surgical workflow recognition a purely frame-level model has problems in resolving Local Ambiguities as discussed in Sect. 2. The one exception to this is when a particular local ambiguity only appears in a single phase of the surgery which would make the local ambiguity a descriptive frame for both algorithm and clinician. A surgeon would still be able to resolve local ambiguities by considering the frames proceeding the ambiguous one.

**Clip-level.** We can imitate the strategy to also consider proceeding frames using Clip-level models that take as input multiple frames to refine the frame-level prediction. Clip-level models usually have a limited temporal knowledge but have the ability to detect motion patterns and directions which is critical when predicting activities. In contrast to frame-level models, clip-level models are more resilient to local ambiguities by design. However, both frame and clip level models have trouble handling Global Ambiguities. Thus, using a Temporal Model in an additional training Stage can be used to address this challenge.

**Models.** For the frame-level methods we build up on the many previous works in this domain by using the CNN based ResNet-50 [15] model. Additionally, we utilize a novel visual transformer network Swin [23] into our study, that does not perform any convolutions but purely relies on self-attention, with state-of-the-art performance on the large scale Imagenet image classifcation task [9]. Both models have a similar amount of parameters and memory requirements. For the Clip-level methods we choose the expanding architectures for efficient video recognition (X3D) [12] based on 3D CNNs and residual layers, which outperform many methods with a substantially higher number of learnable parameter on the challenging kinetics human action dataset [28].

**Feature Extraction.** Due to hardware memory limitations, training the Visual Backbone end-to-end with a temporal model is often impossible. Hence, training the temporal models is done in a second Stage using the feature embedding $v$ extracted from the Stage 1 models.

### 3.2   Temporal Models - Stage 2

To address the Global Ambiguities discussed in Sect. 2 an additional temporal model can be used with the capabilities to analyse the entire surgical procedure at once.

**Models.** For the temporal models of the Stage 2 we first utilize commonly used architectures, such as RNNs and specifically Gated Recurrent Units (GRU) [1] which have been shown to achieve comparable results to LSTMs with a simplified architecture [5]. RNNs follow an autoregressive pattern for the inference where each prediction builds up on the previous one. Temporal convolutions [10] (TCN), on the other hand, follow a non-autoregressive structure taking all the past time points as input for the current prediction. Temporal transformer architectures follow a similar non-autoregressive pattern but tend to overfit more easily due to their increase amount of learable parameters. TCNs are lightweight and fast to train which is why we selected them as the architectural choice for the non-autoregressive group. In Fig. 1 the different combinations of Stage 1 frame- and clip-based models with a Stage 2 temporal model is visualized.

**Table 1.** Phase definitions for the datasets used in this study Cholec80 as the Internal dataset with Internal Phases (IP) and External DS as External dataset with External Phases (EP).

| Cholec80 phase names | | External DS phase names | |
|---|---|---|---|
| IP1 | Preparation | EP1 | Sterile preparation |
| IP2 | Calot triangle dissection | EP2 | Patient roll-in |
| IP3 | Clipping&cutting | EP3 | Patient preparation |
| IP4 | Gallbladder dissection | EP4 | Robot roll-up |
| IP5 | Gallbladder packaging | EP5 | Robot docking |
| IP6 | Cleaning & coagulation | EP6 | Surgery |
| IP7 | Gallbladder retraction | EP7 | Robot undocking |
| | | EP8 | Robot rollback |
| | | EP9 | Patient close |
| | | EP10 | Patient roll-out |

# 4   Experimental Setup

In the following we conducted an ablation of the aforementioned methods while keeping the hyperparameters of the architecture comparable to conduct a fair and comprehensive study on two dataset for surgical workflow recognition. The phase names of both datasets are summarized in Table 1. The Cholec80 dataset contains surgery specific phases while the External dataset phases are less specific to an intervention.

## 4.1   Datasets

**Cholec80.** The internal workflow recognition capabilities are evaluated with the publicly available Cholec80 [30] dataset that includes 80 cholecystectomy surgeries were every RGB-frame belongs to one out of seven surgical phases (Table 1). We split the dataset into half training and half testing as described in [30] For a fair comparison between datasets we did not use the additional surgical tool labels in Cholec80.

**External DS.** The external dataset (External DS) includes 400 videos from 103 robotic surgical cases. The dataset includes 10 different phases describing the workflow in the OR (Table 1). Every surgery is recorded from four different angles. We treat each angle as a separate input for our model and strictly separate the recordings on a procedural level. Recordings of the same surgery from different views were always in the same train or test split. In this work, we do not consider the combination of multiple views of one surgery as this is out of scope for our work. We used time of flight cameras (ToF) not only for improved privacy compared to RGB but ToF also provides geometrical rich information about the scene. External DS contains 10 general, surgery type agnostic, classes for robotic surgery in a multi-label setting. The dataset contains 28 different surgery types all recorded using an daVinci Xi surgical system. We used 80% of all videos for training and the remaining 20% of all videos for testing. To ensure generalizability of our approach to various minimally invasive surgical procedures, all training videos originated from 16 surgery types and all testing videos from the remaining 12 surgery types with no overlap.

**Model Training.** For our training setup we keep the training configuration options constant between the different methods to establish a fair evaluation. This way, we try to develop a comparison focusing on the main architectural design choices without unwanted metric advantages by using, e.g., a more sophisticated optimizer or training scheme. For the learning on the multi-class Cholec80 dataset we are using the Cross-Entropy and the Binary Cross-Entropy loss for the multi-label External DS. For all our experiments we used the Adam optimizer with a learning rate of 1e–4 for the Stage 1 models and 1e–3 for the Stage 2 models with a step learing rate scheduler (beta: 0.1, interval: 10 epochs). For the training of the Stage 1 models we used the RandAugment [6] which combines a multitude of different augmentation techniques in an optimized manner. For both datasets, we resize all the frames from each video to $224 \times 224$ pixels which allowed us to use pretrained weights on ImageNet [9] (ResNet, Swin) and Kinetics [21] (X3D) to accelerate the convergence. We further follow the related work and subsample both datasets to 1 frame per second. All of our experiments are performed using python and the deep learning training library pytorch.

### 4.2 Architecture Settings

For all our models we adapted the size of the output fully-connected layer to match the number of surgical phases for each dataset (Output layer - Internal dataset: 7, Output layer - External dataset: 10).

**ResNet.** We choose the ResNet-50 model with pretrained ImageNet weights.

**Swin.** For the visual transformer architecture we choose the Swin-T version with initialized weights from Imagenet.

**X3D.** For the X3D architecture we choose X3D-M a efficient compromise between model size and performance. We set the size of the input clip $n$ to 16 frames resulting in a temporal receptive window of 16 s. The weights are initialized from Kinetics.

**GRU** For GRU we used a hidden dimensionality matching the dimensionality of the features extrated from the Stage 1 models. Additionally we selected 2 GRU layers.

**TCN.** For the TCN model we used 15 layers over 2 Stages and ensured to set the models in causal inference mode for online phase recognition results without temporal leakage.

### 4.3 Evaluation Metrics and Baselines

To comprehensively measure the results we report the harmonic mean (F1) of Precision and Recall [24]. Precision and Recall are calculated by averaging the results for each class over all samples followed by an average over all class averages. In that way the metrics are reported on a video level to make sure that a long and short intervention contribute equally to the results. For the multiclass cholec80 dataset we also used the mean accuracy following Endonet [30], averaged over all videos of the test set, since it is a commonly used metric for surgical workflow recognition. For the multi-label external surgical workflow dataset we chose the mean average precision (mAP [17]) as an objective metric similar to other works in the field of activity and action recognition [27]. All of the models are run with 3 different random seeds and the mean and standard variation across the runs are reported.

**Table 2.** Comparative study of architectural components for surgical phase recognition using different model backbones for Stages 1 and 2 on our selected surgical workflow datasets.

| Architecture | | Cholec80 | | External DS | |
|---|---|---|---|---|---|
| Stage 1 | Stage 2 | Acc | F1 | mAP | F1 |
| ResNet | | $81.23 \pm 2.5$ | $71.27 \pm 3.0$ | $67.50 \pm 1.9$ | $58.66 \pm 0.7$ |
| Swin | | $82.16 \pm 1.7$ | $71.86 \pm 2.3$ | $68.89 \pm 1.8$ | $59.89 \pm 0.4$ |
| X3D | | $82.62 \pm 2.6$ | $77.55 \pm 1.8$ | $69.33 \pm 1.4$ | $68.42 \pm 0.9$ |
| ResNet | GRU | $85.62 \pm 2.4$ | $76.68 \pm 1.7$ | $70.8 \pm 1.9$ | $65.12 \pm 1.5$ |
| Swin | GRU | $\mathbf{87.73 \pm 2.2}$ | $80.65 \pm 1.2$ | $67.81 \pm 2.0$ | $63.65 \pm 1.3$ |
| X3D | GRU | $85.53 \pm 1.7$ | $78.99 \pm 1.4$ | $\mathbf{80.35 \pm 1.5}$ | $\mathbf{76.70 \pm 1.1}$ |
| ResNet | TCN | $87.37 \pm 1.4$ | $82.48 \pm 1.8$ | $67.34 \pm 1.9$ | $65.35 \pm 0.9$ |
| Swin | TCN | $87.55 \pm 0.7$ | $\mathbf{81.70 \pm 1.6}$ | $73.02 \pm 3.1$ | $71.08 \pm 2.0$ |
| X3D | TCN | $85.81 \pm 1.4$ | $80.41 \pm 1.3$ | $77.15 \pm 1.5$ | $74.96 \pm 1.6$ |

## 5    Results and Discussion

In Table 2 the models with temporal components in Stage 2 (GRU, TCN) have been trained with extracted features from the respective visual backbones of

Stage 1 (ResNet, Swin, X3D). On both Cholec80 and External DS the image-level Swin backbone improves the results of the ResNet architecture on all metrics. The clip-level modeling with local temporal context in X3D further improves the results which is especially noticeable in the F1 score for both Cholec80 (+5.69%) and the External DS (+8.62%).

Furthermore, with the addition of a temporal model we can refine the results, leading in an increase in both Accuracy and F1 on Cholec80 for both the TCN and GRU. On Cholec80, the TCN improves the results more than GRU on most settings (ResNet, X3D) but the combination of Swin and GRU achives the highest accuracy. For the External DS although the F1 results are generally more stable with the addition of TCN, the best result is achieved with a combination of X3D with GRU (80.35 mAP, 76.70 F1).

In the discussion we want to bring together the implications of the Analysis of Challenges (Sect. 2) and the practical results of the architectural study. In Fig. 1 we see that the addition of the temporal models should indeed improve the results as it is more capable to establish a longer temporal context and resolving the global ambiguities. In fact, the results (Sect. 5) showcase that the addition of a temporal model such as GRU or TCN improves the metrics across the board.

Additionally, as we reasoned in Sect. 2, Clip-level models would be able to resolve Local Ambiguities more effectively in comparison to Image-level models. Indeed, our results highlight that the clip-level model achieves higher Acc (82.62) compared to both image-level models and outperforms them regarding their F1-score on both datasets (+6% and +8%).

Interestingly, all metrics on Cholec80 highlight that the addition of a temporal model is beneficial, as expected. However, for External DS SwinGRU and ResTCN achieve marginally lower mAPs (67.81 and 67.34 respectively) compared to Swin (68.89) and ResNet (67.50). However, the F1-score for the External DS is still consistently improved by the addition of a Stage 2 model, which shows the need for evaluating multiple metrics to determine which model would be better suited for computer-aided surgical workflow systems.

Moreover, our results show that the clip-level backbone could not reach the performance of the frame-level backbones when combined with a temporal model on Cholec80. This could be attributed to the limited amount of training data in comparison to External DS and to the increased amount of learnable parameters, that could cause overfitting on the clip-level backbone. We believe, that since clip-level models are capable of overcoming Local Ambiguities and have shown to outperform image-level architectures on the External DS, they will be more widely used and preferred once the internal surgical workflow recognition datasets, as soon as they expand further.

# 6    Conclusion

In this work we presented a fair analysis of models for the task of surgical workflow recognition on two datasets. We conducted an analysis of the challenges related to surgical workflow recognition and provided an intuition on how architectures can be chosen specifically to address them.

Our results show that utilizing a Temporal Model is a critical component of a model for surgical workflow and can help overcome Global Ambiguities. TCNs are particularly suitable as temporal models, combining a low number of trainable parameters and a large receptive field. Furthermore, we showed that clip-level models can alleviate Local Ambiguities on the External DS and have the potential of benefiting Internal Datasets as they expand in training data size. Future work includes utilizing recently introduced transformer architectures [8,13] as temporal backbones, since they have the potential to increase the model performance. However, using larger datasets is critical when training such complex architectures or analyzing procedures including more surgical phases with longer duration and higher surgical complexity.

Furthermore, as has been recently shown [26] the choice of the evaluation metric is crucial and should carefully consider for each task. This is in line with the results of our study, where models' mAP, Acc and F1-score were critical for the fair evaluation and comparison of each architecture. Finally, when comparing model architectures for surgical workflow analysis, the model robustness [25] and tolerance to outliers [2] could additionally be taken into account.

**Acknowledgements.** The majority of this work was carried out during an internship by Tobias Czempiel at Intuitive Surgical Inc.

# References

1. Bahdanau, D., Cho, K.H., Bengio, Y.: Neural machine translation by jointly learning to align and translate. In: 3rd International Conference on Learning Representations, ICLR 2015 - Conference Track Proceedings, pp. 1–15 (2015)
2. Berger, C., Paschali, M., Glocker, B., Kamnitsas, K.: Confidence-based out-of-distribution detection: a comparative study and analysis. In: Sudre, C.H., et al. (eds.) UNSURE/PIPPI -2021. LNCS, vol. 12959, pp. 122–132. Springer, Cham (2021). https://doi.org/10.1007/978-3-030-87735-4_12
3. Berlet, M., et al.: Surgical reporting for laparoscopic cholecystectomy based on phase annotation by a convolutional neural network (CNN) and the phenomenon of phase flickering: a proof of concept. Int. J. Comput. Assit. Radiol. Surg. **17**, 1991–1999 (2022)
4. Bodenstedt, S., et al.: Unsupervised temporal context learning using convolutional neural networks for laparoscopic workflow analysis (February 2017). http://arxiv.org/1702.03684arxiv.org/abs/1702.03684
5. Bodenstedt, S., et al.: Prediction of laparoscopic procedure duration using unlabeled, multimodal sensor data. Int. J. Computer Assit. Radiol. Surg. **14**(6), 1089–1095 (2019). https://doi.org/10.1007/s11548-019-01966-6
6. Cubuk, E.D., Zoph, B., Shlens, J., Le, Q.V.: Randaugment: practical automated data augmentation with a reduced search space. In: IEEE Computer Society Conference on Computer Vision and Pattern Recognition Workshops 2020-June, 3008–3017 (2020). https://doi.org/10.1109/CVPRW50498.2020.00359

7. Czempiel, T., et al.: TeCNO: surgical phase recognition with multi-stage temporal convolutional networks. In: Martel, A.L., Abolmaesumi, P., Stoyanov, D., Mateus, D., Zuluaga, M.A., Zhou, S.K., Racoceanu, D., Joskowicz, L. (eds.) MICCAI 2020. LNCS, vol. 12263, pp. 343–352. Springer, Cham (2020). https://doi.org/10.1007/978-3-030-59716-0_33

8. Czempiel, T., Paschali, M., Ostler, D., Kim, S.T., Busam, B., Navab, N.: OperA: attention-regularized transformers for surgical phase recognition. In: de Bruijne, M., et al. (eds.) MICCAI 2021. LNCS, vol. 12904, pp. 604–614. Springer, Cham (2021). https://doi.org/10.1007/978-3-030-87202-1_58

9. Deng, J., Dong, W., Socher, R., Li, L.J., Li, K., Fei-Fei, L.: Imagenet: a large-scale hierarchical image database. In: 2009 IEEE Conference on Ccomputer Vision and Pattern Recognition, pp. 248–255. IEEE (2009)

10. Farha, Y.A., Gall, J.: MS-TCN: multi-stage Temporal Convolutional Network for Action Segmentation. In: 2019 IEEE/CVF Conference on Computer Vision and Pattern Recognition (CVPR). vol. 2019-June, pp. 3570–3579. IEEE (June 2019). https://doi.org/10.1109/CVPR.2019.00369, https://ieeexplore.ieee.org/document/8953830

11. Fathi, A., Ren, X., Rehg, J.M.: Learning to recognize objects in egocentric activities. In: Proceedings of the IEEE Computer Society Conference on Computer Vision and Pattern Recognition, pp. 3281–3288 (2011). https://doi.org/10.1109/CVPR.2011.5995444

12. Feichtenhofer, C.: X3D: expanding architectures for efficient video recognition. In: 2020 IEEE/CVF Conference on Computer Vision and Pattern Recognition (CVPR), pp. 200–210. IEEE (June 2020). https://doi.org/10.1109/CVPR42600.2020.00028

13. Gao, X., Jin, Y., Long, Y., Dou, Q., Heng, P.-A.: Trans-SVNet: accurate phase recognition from surgical videos via hybrid embedding aggregation transformer. In: de Bruijne, M., et al. (eds.) MICCAI 2021. LNCS, vol. 12904, pp. 593–603. Springer, Cham (2021). https://doi.org/10.1007/978-3-030-87202-1_57

14. Garrow, C.R., et al.: Machine learning for surgical phase recognition: a systematic review. Ann. Surg. 273(4), 684–693 (2021). https://journals.lww.com/annalsofsurgery/pages/default.aspxhttps://pubmed.ncbi.nlm.nih.gov/33201088/

15. He, K., Zhang, X., Ren, S., Sun, J.: Deep residual learning for image recognition. In: 2016 IEEE Conference on Computer Vision and Pattern Recognition (CVPR). pp. 770–778. IEEE (June 2016). https://doi.org/10.1109/CVPR.2016.90, https://image-net.org/challenges/LSVRC/2015/,https://ieeexplore.ieee.org/document/7780459

16. Huaulmé, A., Jannin, P., Reche, F., Faucheron, J.L., Moreau-Gaudry, A., Voros, S.: Offline identification of surgical deviations in laparoscopic rectopexy. Artiff. Intell. Med. 104(2019) (2020). https://doi.org/10.1016/j.artmed.2020.101837

17. Idrees, H., et al.: The THUMOS challenge on action recognition for videos "in the wild". Comput. Vis. Image Underst. 155, 1–23 (2017)

18. Jin, Y., Dou, Q., Chen, H., Yu, L., Qin, J., Fu, C.W., Heng, P.A.: SV-RCNet: workflow recognition from surgical videos using recurrent convolutional network. IEEE Trans. Med. Imaging 37(5), 1114–1126 (2018). https://doi.org/10.1109/TMI.2017.2787657

19. Jin, Y., et al.: Multi-task recurrent convolutional network with correlation loss for surgical video analysis. Med. Image Anal. 59 (2020). https://doi.org/10.1016/j.media.2019.101572

20. Katić, D., et al.: Bridging the gap between formal and experience-based knowledge for context-aware laparoscopy. Int. J. Comput. Assist. Radiol. Surg. **11**(6), 881–888 (2016). https://doi.org/10.1007/s11548-016-1379-2
21. Kay, W., et al.: The Kinetics Human Action Video Dataset (may 2017). http://arxiv.org/1705.06950
22. Kuehne, H., Arslan, A., Serre, T.: The language of actions: Recovering the syntax and semantics of goal-directed human activities. In: Proceedings of the IEEE Computer Society Conference on Computer Vision and Pattern Recognition, pp. 780–787 (2014). https://doi.org/10.1109/CVPR.2014.105
23. Liu, Z., et al.: Swin transformer: hierarchical vision transformer using shifted windows. In: IEEE/CVF International Conference on Computer Vision (ICCV), pp. 9992–10002 (2021)
24. Padoy, N., Blum, T., Ahmadi, S.A., Feussner, H., Berger, M.O., Navab, N.: Statistical modeling and recognition of surgical workflow. Med. Image Anal. **16**(3), 632–641 (2012)
25. Paschali, M., Conjeti, S., Navarro, F., Navab, N.: Generalizability *vs.* robustness: investigating medical imaging networks using adversarial examples. In: Frangi, A.F., Schnabel, J.A., Davatzikos, C., Alberola-López, C., Fichtinger, G. (eds.) MICCAI 2018. LNCS, vol. 11070, pp. 493–501. Springer, Cham (2018). https://doi.org/10.1007/978-3-030-00928-1_56
26. Reinke, A., et al.: Metrics reloaded-a new recommendation framework for biomedical image analysis validation. In: Medical Imaging with Deep Learning (2022)
27. Sharghi, A., Haugerud, H., Oh, D., Mohareri, O.: Automatic operating room surgical activity recognition for robot-assisted surgery. In: Martel, A.L., et al. (eds.) MICCAI 2020. LNCS, vol. 12263, pp. 385–395. Springer, Cham (2020). https://doi.org/10.1007/978-3-030-59716-0_37
28. Smaira, L., Carreira, J., Noland, E., Clancy, E., Wu, A., Zisserman, A.: A Short Note on the Kinetics-700-2020 Human Action Dataset. arXiv (i) (2020). http://arxiv.org/2010.10864
29. Srivastav, V., Issenhuth, T., Kadkhodamohammadi, A., de Mathelin, M., Gangi, A., Padoy, N.: MVOR: a multi-view RGB-D operating room dataset for 2D and 3D human pose estimation. In: MICCAI-LABELS, pp. 1–10 (2018). http://arxiv.org/1808.08180
30. Twinanda, A.P., Shehata, S., Mutter, D., Marescaux, J., De Mathelin, M., Padoy, N.: EndoNet: a Deep Architecture for Recognition Tasks on Laparoscopic Videos. IEEE Trans. Med. Imaging **36**, 86–97 (2017).https://doi.org/10.1109/TMI.2016.2593957
31. Twinanda, A.P., Padoy, N., Troccaz, M.J., Hager, G.: Vision-based Approaches for surgical activity recognition using laparoscopic and RBGD Videos. Thesis (7357) (2017), https://theses.hal.science/tel-01557522/document
32. Vaswani, A., et al.: Attention is all you need. In: Proceedings of Advances in Neural Information Processing Systems 2017-Decem(Nips), pp. 5999–6009 (2017)

# RVENet: A Large Echocardiographic Dataset for the Deep Learning-Based Assessment of Right Ventricular Function

Bálint Magyar[1]([✉]), Márton Tokodi[2], András Soós[1,2], Máté Tolvaj[2],
Bálint Károly Lakatos[2], Alexandra Fábián[2], Elena Surkova[3], Béla Merkely[2],
Attila Kovács[2], and András Horváth[1]

[1] Faculty of Information Technology and Bionics, Pázmány Péter Catholic
University, Budapest, Hungary
magyar.balint@itk.ppke.hu
[2] Heart and Vascular Center, Semmelweis University, Budapest, Hungary
tokmarton@gmail.com, attila.kovacs@med.semmelweis-univ.hu
[3] Harefield Hospital, Royal Brompton and Harefield Hospitals,
Part of Guy's and St Thomas' NHS Foundation Trust, London, UK

**Abstract.** Right ventricular ejection fraction (RVEF) is an important indicator of cardiac function and has a well-established prognostic value. In scenarios where imaging modalities capable of directly assessing RVEF are unavailable, deep learning (DL) might be used to infer RVEF from alternative modalities, such as two-dimensional echocardiography. For the implementation of such solutions, publicly available, dedicated datasets are pivotal.

Accordingly, we introduce the RVENet dataset comprising 3,583 two-dimensional apical four-chamber view echocardiographic videos of 831 patients. The ground truth RVEF values were calculated by medical experts using three-dimensional echocardiography. We also implemented benchmark DL models for two tasks: (i) the classification of RVEF as normal or reduced and (ii) the prediction of the exact RVEF values. In the classification task, the DL models were able to surpass the medical experts' performance. We hope that the publication of this dataset may foster innovations targeting the accurate diagnosis of RV dysfunction.

**Keywords:** Echocardiography · Right ventricle · Right ventricular ejection fraction · Deep learning

B. Magyar and M. Tokodi—These authors contributed equally to this work and are joint first authors.
B. Merkely, A. Kovács and A. Horváth—These authors contributed equally to this work and are joint last authors.

**Supplementary Information** The online version contains supplementary material available at https://doi.org/10.1007/978-3-031-25066-8_33.

# 1   Introduction

Echocardiography is an ultrasound-based imaging modality that aims to study the physiology and pathophysiology of the heart. Important indicators that describe the cardiac pump function can be calculated based on the annotations of echocardiographic recordings. One of these indicators is the EF. This ratio indicates the amount of blood pumped out by the examined ventricle during its contraction. In other words, EF defines the normalized difference between the end-diastolic volume (EDV) which is the largest volume during the cardiac cycle, and the end-systolic volume (ESV) which is the smallest volume during the cardiac cycle.

$$EF(\%) = \frac{EDV - ESV}{EDV} * 100$$

Two-dimensional (2D) echocardiographic images acquired from standardized echocardiographic views can be used to approximate left ventricular volumes, and thus, the left ventricular ejection fraction (LVEF) with sufficient accuracy. Due to the more complex three-dimensional (3D) shape of the RV (see Fig. 1), there is no accurate and clinically used method for estimating RV ejection fraction (RVEF) from 2D recordings [15]. Nevertheless, the availability of this indicator in daily clinical practice would be highly desirable. RVEF can be calculated using 3D echocardiography, which is validated against the gold-standard cardiac magnetic resonance imaging; however, it requires additional hardware and software resources along with significant human expertise to maximize its accuracy and reproducibility [11].

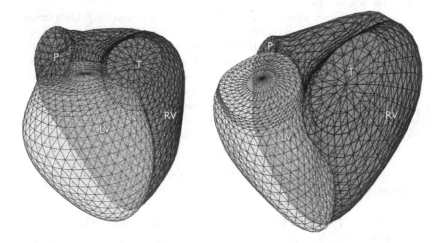

**Fig. 1.** This figure depicts the left and right ventricles of a normal (left) and a diseased heart (right), as well as the geometric differences between the two ventricles. It can be seen that the estimation of the left ventricular volume from a single 2D plane is feasible due to its regular shape. In contrast, the right ventricle's more complex shape requires 3D evaluation. RV - right ventricle, LV - left ventricle, P - pulmonary valve, T - tricuspid valve

In this work, we publish a dataset that contains 2D echocardiographic videos from 831 patients suitable for RV function assessment. 3D recordings were also collected from the same patients for ground truth generation (i.e., RVEF). We assume that deep learning methods can find relevant patterns on the 2D recordings to detect RV dysfunction and to predict the exact value of RVEF.

To test our assumption, we used an off-the-shelf video classification model and a custom spatiotemporal convolutional neural network for the classification of reduced/normal RVEF and for the prediction of RVEF. RV function can be classified as normal if the RVEF is equal or above 45%, and reduced if it is below 45%. Even if the prediction of RVEF seems to be a more comprehensive method, the aforementioned binary classification is the most widely used evaluation in clinical practice having a clear association with the risk of future adverse clinical events [6,10].

Our contributions can be summarized as follows:

- We publish a large-scale echocardiographic dataset for the assessment of RV function. According to our knowledge, this is the first dedicated dataset aiming RV evaluation. Its uniqueness lies in the calculation of the ground truth RVEF which was done using 3D recordings.
- Baseline deep learning models were developed and applied to classify RV reduced/normal EF and to predict the RVEF value. Based on our literature review, there is no solution that solves the same clinical problem.
- We compared the models' performance with two experienced medical doctors' performance (one from the center from which the dataset originates and one external expert). This was a unique comparison, as the ground truth values were created using another modality (3D echocardiography).

## 2   Related Work

### 2.1   Datasets

To the best of our knowledge, there is no dataset for RV function assessment. Based on an exhaustive literature review the following open-source datasets for the assessment of LV function were identified.

The first one called CETUS [3] which contains 45 3D echocardiogram recordings (each from a different patient). The dataset was collected from three different hospitals using different machines, and it was annotated (3D LV segmentation) by three expert cardiologists based on a pre-defined protocol. The main purpose of this dataset is to compare 3D segmentation algorithms, therefore different segmentation metrics (e.g. Dice similarity index and 3D Hausdorff distance) are used for evaluation. To evaluate the clinical performance of the methods they calculate mean absolute error (MAE) and means squared error (MSE) for the ESV, EDV and LVEF values.

CAMUS represents another important dataset [7]. 500 2D echocardiogram recordings (each from a different patient) were collected from a single hospital for this data set. LV segmentation mask that was created by an experienced cardiologist (except in the test set where 2 other cardiologist were involved) and LVEF values are provided for the recordings. The estimation of the LVEF values based on the Simpson's biplane method of discs [5]. The same measures were applied for evaluation as in the CETUS dataset (except the 2D version of segmentation metrics instead of 3D).

The third dataset is the EchoNet-Dynamic [13] which contains 10036 2D echocardiogram recordings (each from a different patient). Each recording was captured from apical-4-chamber view. LVEF, EDV and ESV measurements were obtained by experienced cardiologists on the standard clinical workflow. The original recordings were filtered based on image quality and resized to $112 \times 112$ gray-scale image sequences.

They also used MAE and MSE as well as $R^2$ to evaluate the LVEF, EDV and ESV predictions.

## 2.2   Methods

The assessment of cardiac functions using machine learning and particularly deep learning algorithms is a commonly applied approach nowadays due to the superior performance of these systems compared to traditional computer vision algorithms. These methods can even surpass human performance in certain cases [1, 2, 22].

Some of the earlier methods focused on individual frames to predict echocardiographic view, ventricular hypertrophy, EF, and other metrics. Madani et al. implemented a method for view classification, which is usually the first step before further analysis [9].

Another group of researchers developed a more complex system that identifies the view, applies segmentation on the 2D recordings and predicts the LVEF as well as one of 3 diagnostic classes [23].

Leclerc et al. used a more advanced encoder-decoder style segmentation network on the CAMUS dataset to calculate the EDV and ESV and to predict the LVEF using these values [7].

Application of anatomical atlas information as a segmentation constraint was successfully applied by Oktay et al. to create a more accurate segmentation method for LV segmentation and LVEF prediction [12].

In order to predict measures like EF that can be estimated only using multiple frames, the key frames has to be selected manually for these methods.

More advanced methods use video input and spatiotemporal convolutional networks to provide an end-to-end solution.

Shat et al. applied 3D convolutional layers along with optical flow to detect temporal changes along the video and to predict post-operative RV failure [18].

The effect of temporally consistent segmentation has been investigated previously [4]. In this study, the authors used a custom convolutional layer to obtain bi-directional motion fields. The motion detection was combined with the segmentation results to obtain precise LV segmentation.

Ouyang et al. [14] presented a two stage convnet applying atrous convolution to first segment the LV, and then another stage of spatiotemporal convolutions to predict the LVEF.

Compared to existing approaches, we proposed a single stage method that aims to predict the RVEF directly from the input videos using satiotemporal convolutional networks. We assume that in contrast to segmentation based methods, different regions of the input image can also contribute to the RVEF prediction, and therefore this task is feasible.

## 3    Overview of the RVENet Dataset

### 3.1    Data Collection

To create the RVENet dataset, we retrospectively reviewed the transthoracic echocardiographic examinations performed between November 2013 and March 2021 at the Heart and Vascular Center of Semmelweis University (Budapest, Hungary). We aimed to identify those examinations that included one or more 2D apical four-chamber view echocardiographic videos and an electrocardiogram (ECG)-gated full-volume 3D echocardiographic recording (with a minimum volume rate of 15 volumes/s, acquired from an RV-optimized apical view, and reconstructed from four cardiac cycles) suitable for 3D RV analysis and RVEF assessment. The 2D apical four-chamber view videos were exported as Digital Imaging and Communications in Medicine (DICOM) files, whereas the 3D recordings were used for generating labels (see the detailed description of data labeling in Sect. 3.2). Protected health information was removed from all exported DICOM files. 2D videos with (i) invalid heart rate or frame per second (FPS) values in the DICOM tags, (ii) acquisition issues comprising but not limited to severe translational motion, gain changes, depth changes, view changes, sector position changes, (iii) duration shorter than one cardiac cycle, or (iv) less than 20 frames per cardiac cycle were discarded. All transthoracic echocardiographic examinations were performed by experienced echocardiographers using commercially available ultrasound scanners (Vivid E95 system, GE Vingmed Ultrasound, Horten, Norway; iE33, EPIQ CVx, 7C, or 7G systems, Philips, Best, The Netherlands).

### 3.2    Data Labeling

The exported 2D echocardiographic videos were reviewed by a single experienced echocardiographer who (i) assessed the image quality using a 5-point Likert scale (1 - non-diagnostic, 2 - poor, 3 - moderate, 4 - good, 5 - excellent), (ii) labeled them as either standard or RV-focused, (iii) determined LV/RV orientation (Mayo - RV on the right side and LV on the left side; Stanford - LV on the right side and RV on the left side), and (iv) ascertained that none of them meet the exclusion criteria.

The 3D echocardiographic recordings were analyzed by expert readers on desktop computers using a commercially available software solution (4D RV-Function 2, TomTec Imaging, Unterschleissheim, Germany) to compute RV end-diastolic and end-systolic volumes, as well as RVEF. These parameters were calculated only once for each echocardiographic examination. However, an examination may contain multiple 2D apical four-chamber view videos; thus, the same label was linked to all 2D videos within that given examination.

A comprehensive list and description of the generated labels are provided in Table 1.

**Table 1.** Description of the labels. RV - right ventricular

| Variable | Description |
| --- | --- |
| FileName | Hashed file name used to link videos and labels |
| PatientHash | Hashed patient name |
| PatientGroup | Patient subgroup referring to the primary diagnosis |
| Age | Age in years, rounded to nearest year |
| Sex | Sex reported in medical record (M - male, F - female) |
| UltrasoundSystem | Ultrasound system used for video acquisition |
| FPS | Frames per second (1/s) |
| NumFrames | Number of frames in the whole video |
| VideoViewType | Standard or RV-focused apical four-chamber view |
| VideoOrientation | LV/RV orientation (Mayo or Stanford) |
| VideoQuality | 2D video quality on a 5-point scale (1 - non-diagnostic, 2 - poor, 3 - moderate, 4 - good, 5 - excellent) |
| RVEDV | 3D echocardiography-derived RV end-diastolic volume (mL) |
| RVESV | 3D echocardiography-derived RV end-systolic volume (mL) |
| RVEF | 3D echocardiography-derived RV ejection fraction (%) |
| Split | Train-test splitting used for benchmarking |

### 3.3   Composition of the Dataset

The RVENet dataset contains 3,583 2D apical four-chamber view echocardiographic videos from 944 transthoracic echocardiographic examinations of 831 individuals. It comprises ten distinct subgroups of subjects: (i) healthy adult volunteers (without history and risk factors of cardiovascular diseases, n = 192), (ii) healthy pediatric volunteers (n = 54), (iii) elite, competitive athletes (n = 139), (iv) patients with heart failure and reduced EF (n = 98), (v) patients with LV non-compaction cardiomyopathy (n = 27), (vi) patients with aortic valve disease (n = 85), (vii) patients with mitral valve disease (n = 70), (viii) patients who underwent orthotopic heart transplantation (n = 87), (ix) pediatric patients who underwent kidney transplantation (n = 23), and (x) others (n = 56). Beyond the primary diagnosis and the labels mentioned in Sect. 3.2, we also provided the age (rounded to the nearest year) and the biological sex (as reported in medical records) for each patient, and train-test splitting (80:20 ratio) that we used for the training and the evaluation of the benchmark models (see Sect. 4). In addition, the ultrasound system utilized for video acquisition, the frame rate (i.e., FPS), and the total number of frames are also reported for each video among the labels.

### 3.4 Data De-identification and Access

Before publication of the RVENet dataset, all DICOM files were processed to remove any protected health information. We also ensured that no protected health information is included among the published labels. Thus, the RVENet dataset complies with the General Data Protection Regulation of the European Union. The dataset with the corresponding labels is available at https://rvenet. github.io/dataset/.

The RVENet dataset is available only for personal, non-commercial research purposes. Any commercial use, sale, or other monetization is prohibited. Re-identification of individuals is strictly prohibited. The RVENet dataset can be used only for legal purposes.

## 4  Benchmark Models

### 4.1  Methodology

**Ethical Approval.** The study conforms to the principles outlined in the Declaration of Helsinki, and it was approved by the Semmelweis University Regional and Institutional Committee of Science and Research Ethics (approval No. 190/2020). Methods and results are reported in compliance with the Proposed Requirements for Cardiovascular Imaging-Related Machine Learning Evaluation (PRIME) checklist (Supplementary Table 3) [17].

**Data Preprocessing.** The RVENet dataset can be used for various purposes within the realm of cardiovascular research. In this section, we describe data preprocessing that proceeded both deep learning tasks, namely (i) the prediction of the exact RVEF values (i.e. regression task) and (ii) and the classification of reduced/normal RVEF (i.e. binary classification task).

All echocardiographic recordings were exported as DICOM files. Each DICOM file contains a series of frames depicting one or more cardiac cycles. This arrangement of the data had to be modified to achieve a representation that is more suitable for neural networks. The three main steps of preprocessing can be described as follows: (1) frame selection, (2) image data preparation, and (3) handling imbalance in the train set.

Frame selection refers to the preprocessing step in which 20 frames are selected to represent a cardiac cycle (20 frames per cardiac cycle proved to be the appropriate number in [14] for left ventricle EF prediction). Recordings may contain multiple cardiac cycles and may differ in length and frame rate (FPS - frames per second). We applied the following formula to estimate the length of a cardiac cycle (L) based on heart rate (HR) and FPS extracted from the DICOM file tags:

$$L = \frac{60}{HR} * FPS$$

Since all the recordings are ECG-gated and start with the end-diastolic frame, the split of the videos into consecutive, non-overlapping fragments (depicting exactly one cardiac cycle) was feasible. Fragments containing less than L frames were excluded. Then, a predefined number of frames (N = 20) were sampled from the fragments based on the sampling frequency (SF) which was calculated using the following formula:

$$SF = \frac{L}{N}$$

A subset of randomly selected videos underwent a manual verification process by an experienced physician to evaluate the cardiac cycle selection.

The next step is the image data generation which is shown in Fig. 2. The selected frames contain multiple components that are unnecessary for the neural network training, such as ECG signal, color-scale and other signals and texts. These unwanted items were removed using motion-based filtering. Our algorithm tracks the changes frame by frame and set the pixels to black if they change fewer times than a predefined threshold (in our case 10). We also cropped the relevant region of the recordings and generate a binary mask for training.

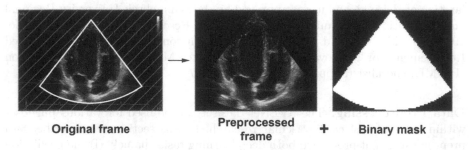

**Original frame**          **Preprocessed frame**    **+**    **Binary mask**

**Fig. 2.** Schematic illustration of the preprocessing. First, static objects (e.g., technical markers and the ECG signal) were removed from the area marked by red diagonal lines using motion-based filtering, while the region of interest (enclosed by the white contour) was left intact. Second, the region of interest was cropped from the filtered image and a binary mask was also generated.

The removal of unwanted components is performed along with the binary mask generation. This mask is created for every video fragment with the consideration of all the frames. The aim of this additional binary image is to prevent the network from extracting features from the outside of the region of interest.

The closest enclosing rectangle is applied to both the preprocessed frames and the binary image. After that they are resized to $224 \times 224$ pixels. Previously, Madani et al. examined the effect of the input image size on the system's accuracy in a echocardiography view classification task, and they found out that the accuracy saturated if they used higher resolution than this [9].

**Train-Test Splitting.** The dataset was split in an approximately 80:20 ratio into train and test sets. Splitting was performed at the patient level to avoid data leakage (i.e., we assigned all videos of a given patient either to the train or the test set).

**Dataset Balancing.** An optional step in pre-processing is the dataset balancing which aims to compensate the high imbalance between negative (normal EF) and positive (reduced EF) cases. In binary classification this means the oversampling of the positive cases and the undersampling of negative cases. In case of a regression problem, the EF values are assigned to discrete bins, and the algorithm aims to balance the number of samples in these bins. The method takes the number of videos and heart cycles from a certain patient into account. It aims to keep at least one video from every patient in the undersampling phase, and oversamples the videos from patients with reduced EF uniformly.

**Spatiotemporal Convolutional Neural Networks.** As it was mentioned in the Related Work section, spatiotemporal processing of the echocardiographic videos provide a more accurate approach for EF prediction [4,14].

Based on these results, we used two neural network models. The first one is composed of R(2+1)D spatiotemporal convolutional blocks [20], and a PyTorch implementation (called R2Plus1D_18) of such a model is available off-the-shelf. We refer to this model as "R2+1D" in the text. We also designed a more efficient, single stage neural network called EFNet (Ejection Fraction Network) that consist of a feature extractor backbone (ShuffleNet V2 [8] or ResNext50 [21]) a temporal merging layer and two fully connected layers. The architecture of our custom model is visualized in (Fig. 3).

Both networks predict the RVEF directly from an input image sequence (and the corresponding binary mask). The same architectures can be used for classification or regression by changing the number of outputs.

**Model Training and Evaluation.** Several experiments were performed to find the best training parameters. In the followings, the final parameter sets are introduced.

The backbone model of the EFNet was ShuffleNet V2 [8] for binary classification and ResNext50 [23] for regression, which is a more challenging task and therefore needs a more complex architecture. To distinguish these two versions of EFNet, we refer to the classification model as EFNet_class and to the regression model as EFNet_reg.

ImageNet pre-trained weights can improve the performance of deep learning models applied in medical datasets [19]. In our case, only the EFNet_reg model

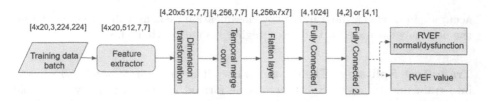

**Fig. 3.** The training batch contains batch size × video frames images with resolution of 224 × 224 pixels. The feature extractor is a ShuffleNet V2 [8] or a ResNext50 [23] model. The dimension transformation layer groups the frame features corresponding to the videos in the batch, then these group of frames are processed using the spatiotemporal convolutional layer to extract dynamic features. The final features are downscaled using fully connected layers and forwarded to either a classification or a regression head.

was initialized using ImageNet pre-trained weights, the weights of the R2+1D and EFNet_reg models were initialized randomly.

As it was described in Sect. 4.1, the videos were split into distinct cardiac cycles, and 20 frames were sampled from each of them.

Dataset balancing (described above) also improved the accuracy of the system as well as the F1 and $R^2$ scores. This is mainly due the substantial imbalance in the dataset.

Augmentation techniques were also applied, namely vertical flipping and rotation (+/−10°). Normalization was not applied.

The models were trained for maximum 30 epochs with a batch size of 4. We used Adam optimizer (initial learning rate = 0.003, momentum = 0.9), and the PyTorch cyclic learning rate scheduler (lambda = 0.965).

Cross-entropy loss was used in the classification experiments, and MAE in the regression experiments.

The parameter search and model selection was done applying a four fold cross validation using the whole training set. For the final experiments, the training set was split in 75%–25% ratio into training and validation set. A balanced version of the training set was used for training, and the best model was selected using the validation set results based on f1 score in case of classification and $R^2$ in case of regression training.

For both the classification and regression tasks, the deep learning models were evaluated on the test set. As a classification model predicts a class for a single cardiac cycle, these predictions were averaged for each video (taking the majority vote). This way the results can be compared with the human experts' performance as they also saw the whole video during evaluation. In the regression task, the models' prediction were averaged for each test video similarly to the classification task. In this case no human expert comparison was performed as the exact prediction of the RVEF value is not part of the clinical evaluation of 2D echocardiographic recordings (it is done only using 3D recordings).

**Human Expert Evaluation.** Videos of the test set were evaluated by two expert cardiologists: one from the same center where the echocardiographic videos were acquired (referred to as Expert$_{Internal}$) and one from an external center (referred to as Expert$_{External}$). Although patient identification information, medical history, and diagnosis were hidden from both of them during evaluation, the first cardiologist might have seen some of the videos previously, as he performs echocardiographic examinations at a daily basis. On the other hand, the second cardiologist has not seen any of the videos previously, enabling a completely unbiased comparison.

Both evaluator used the same custom desktop application for evaluation, which displayed the original videos one by one in a random order and the evaluating expert had to decide based on visual estimation whether the video belongs to patient with normal (RVEF is equal or greater than 45) or reduced (RVEF is less than 45) RV function.

## 4.2 Results

Table 2 shows the results of the deep learning models and the human experts in the detection of RV dysfunction (i.e., binary classification task). Both deep learning models achieved a numerically higher accuracy, specificity, sensitivity, and F1 score than the medical experts. We also confirmed these differences using McNemar's tests. EFNet_class model exhibited an accuracy, specificity, and sensitivity comparable to those of the internal expert, whereas it had higher accuracy and sensitivity than the external expert (Table 3). The R2+1D model achieved a higher sensitivity than the internal expert, and it also outperformed the external expert in terms of accuracy and sensitivity (Table 3).

**Table 2.** Performance of the deep learning models and the cardiologists in the binary classification task.

|  | Accuracy | Specificity | Sensitivity | F1 score |
|---|---|---|---|---|
| EFNet_class | 0.911 | 0.942 | 0.688 | 0.655 |
| R2+1D | 0.920 | 0.940 | 0.775 | 0.705 |
| Expert$_{Internal}$ | 0.897 | 0.940 | 0.588 | 0.584 |
| Expert$_{External}$ | 0.859 | 0.923 | 0.400 | 0.410 |

In the regression task (i.e. prediction of the exact RVEF value), the two deep learning models performed similarly (Supplementary Fig. 1). The EFNet_reg model predicted RVEF with an $R^2$ of 0.411, a mean absolute error of 5.442% points, and a mean squared error of 47.845% points$^2$, whereas the R2+1D model achieved an $R^2$ of 0.417, a mean absolute error of 5.374% points, and a mean squared error of 47.377% points$^2$.

**Table 3.** P-values of the McNemar's tests comparing the accuracy, specificity, and sensitivity of the deep models with those of the cardiologists in the binary classification task.

|  | Accuracy | Specificity | Sensitivity |
|---|---|---|---|
| EFNet_class vs. Expert$_{\text{Internal}}$ | 0.362 | 1.000 | 0.170 |
| EFNet_class vs. Expert$_{\text{External}}$ | 0.001 | 0.229 | <0.001 |
| R2+1D vs. Expert$_{\text{Internal}}$ | 0.106 | 1.000 | 0.004 |
| R2+1D vs. Expert$_{\text{External}}$ | <0.001 | 0.275 | <0.001 |

The Bland-Altman analysis showed a significant bias between the deep learning-predicted and the 3D echocardiography-based ground truth RVEF values (EFNet_reg: $-2.496\%$ points, $p < 0.001$; R2+1D: $0.803\%$ points, $p < 0.001$; Supplementary Fig. 1).

Table 4 shows the comparison of the R2+1D and the two EFNet models in terms of size, and inference speed. Even if the R2+1D model performed better in the classification task and slightly better in the regression task, EFNet is a more efficient model, and its speed can be a huge advantage in model training and inference both in experimentation and in clinical applications.

**Table 4.** Size and inference speed results of the baseline models. Inference speed was measured by averaging 100 iterations with batch size of 1 on an Nvidia V100 GPU.

|  | Feature extractor | Inference time [ms] | Model size [MB] |
|---|---|---|---|
| EFNet_class | ShuffleNet V2 | 16 | 61 |
| EFNet_reg | ResNext50 | 31 | 177 |
| R2+1D | R(2+1)D | 53 | 119 |

## 5  Discussion

### 5.1  Potential Clinical Application

RV dysfunction is significantly and independently associated with symptomatology and clinical outcomes (e.g. all-cause mortality and/or adverse cardiopulmonary outcomes) in different cardiopulmonary diseases irrespective of which side of the heart is primarily affected. Among echocardiographic parameters, 3D echocardiography-derived quantification of RVEF provides the highest predictive value for future adverse events [16]. However, there are several issues that prevent RVEF to be a standard measure in the daily clinical routine. 3D echocardiography-based quantification of RVEF requires advanced hardware and software environment along with experienced cardiology specialists. First, a high-end ultrasound system equipped with a 3D-capable matrix transducer is required. Compared to a conventional acquisition of an apical four-chamber view video (included in the routine protocols and takes no more than one minute irrespective of the level of expertise), an RV-focused, modified four-chamber view is needed with the 3D option enabled. The investigator needs to ensure the capture of the entire RV endocardial surface, which can be troublesome with distinct anatomical features of the patient. Moreover, to enable higher temporal resolution, multi-beat reconstruction should be used that can be limited in the cases of irregular heart rhythm, transducer motion, in patients who are not able to breathe-hold, and again, user experience is of significant importance to acquire a high-quality 3D dataset free of artifacts. To acquire such a measurement feasible for RVEF measurement takes about 2 to 4 min for an expert user, which can go up to 5–8 min for users not having extensive experience in 3D image acquisition. Then, the 3D DICOM file should be post-processed using standalone software (running on a separate PC or embedded in the high-end ultrasound machine). One vendor enables fully automatic 3D reconstruction of the RV endocardial surface and calculation of RVEF values (which takes about 30 s). However, in the vast majority of the cases (over 90%), correction of endocardial contours is needed in multiple long- and short-axis views both at the end-diastolic and end-systolic frames. Changes in the automatically traced contours made by the human reader can result in notable interobserver and even intraobserver variability. Here again, the experience of the user is a major factor in terms of accurate measurements and also, analysis time. For an experienced reader, the manual correction of the initial contours takes about 4 to 10 min and up to 15 min for an inexperienced user. Overall, from image acquisition through image transfer, preprocessing and finally RVEF calculation, the entire process is generally taking 10 to 25 min in clinical practice for this single parameter.

Due to significant human and also hardware/software resources needed, RVEF calculation by echocardiography is rarely performed in the clinical routine despite its clear value. This can be circumvented by an automated system, which utilizes routinely acquired echocardiographic videos and does not require a high-end ultrasound system or significant human experience either. In the clinical routine, echocardiography is often performed by other medical disciplines (i.e. emergency physicians, cardiac surgeons) with mobile, even handheld machines to answer focused yet important clinical questions (so-called point-of-care ultrasound examinations). These medical professionals generally do not have any experience with 3D echocardiography and either high-end ultrasound equipment. However, in these disciplines, the detection of RV dysfunction is a critical clinical issue. As it may be applied even to handheld devices and allow the fast detection of RV dysfunction using simple, routine 2D echocardiographic videos, our system could be of high clinical interest. It can run in a cloud environment, and provide results within a few seconds. Also, its use does not require deep technical knowledge.

## 5.2  Summary

In this paper, we presented a large dataset for the deep learning-based assessment of RV function. We made publicly available 3,583 two-dimensional echocardiographic apical four-chamber view videos from 831 patients to researchers and medical experts. These videos are labelled with RVEF values (the single best echocardiographic parameter for RV function quantification) derived from 3D echocardiography. We also introduced benchmark models, which were able to outperform an external expert human reader in terms of accuracy and sensitivity to detect RV dysfunction. We foresee further performance improvement through collaborations, definition of RV-related specific clinical tasks, addition of further echocardiographic views or imaging modalities. Our current database and model development may serve as a reference point to foster such innovations.

**Acknowledgement.** This project was also supported by a grant from the National Research, Development and Innovation Office (NKFIH) of Hungary (FK 142573 to Attila Kovács).

# References

1. Akkus, Z., et al.: Artificial intelligence (AI)-empowered echocardiography interpretation: A state-of-the-art review. J. Clin. Med. **10**(7), 1391 (2021)
2. Alsharqi, M., Woodward, W., Mumith, J., Markham, D., Upton, R., Leeson, P.: Artificial intelligence and echocardiography. Echo Res. Pract. **5**(4), R115–R125 (2018)
3. Bernard, O., et al.: Standardized evaluation system for left ventricular segmentation algorithms in 3D echocardiography. IEEE Trans. Med. Imaging **35**(4), 967–977 (2015)
4. Chen, Y., Zhang, X., Haggerty, C.M., Stough, J.V.: Assessing the generalizability of temporally coherent echocardiography video segmentation. In: Medical Imaging 2021: Image Processing, vol. 11596, pp. 463–469, SPIE (2021)
5. Folland, E., Parisi, A., Moynihan, P., Jones, D.R., Feldman, C.L., Tow, D.: Assessment of left ventricular ejection fraction and volumes by real-time, two-dimensional echocardiography. A comparison of cineangiographic and radionuclide techniques. Circulation **60**(4), 760–766 (1979)
6. Lang, R.M., et al.: Recommendations for cardiac chamber quantification by echocardiography in adults: an update from the American society of echocardiography and the European association of cardiovascular imaging. Eur. Heart Jo. Cardiovasc. Imaging **16**(3), 233–271 (2015)
7. Leclerc, S., et al.: Deep learning for segmentation using an open large-scale dataset in 2D echocardiography. IEEE Trans. Med. Imaging **38**(9), 2198–2210 (2019)
8. Ma, N., Zhang, X., Zheng, H.-T., Sun, J.: ShuffleNet V2: practical guidelines for efficient cnn architecture design. In: Ferrari, V., Hebert, M., Sminchisescu, C., Weiss, Y. (eds.) Computer Vision – ECCV 2018. LNCS, vol. 11218, pp. 122–138. Springer, Cham (2018). https://doi.org/10.1007/978-3-030-01264-9_8
9. Madani, A., Ong, J.R., Tibrewal, A., Mofrad, M.R.: Deep echocardiography: data-efficient supervised and semi-supervised deep learning towards automated diagnosis of cardiac disease. NPJ Digit. Med. **1**(1), 1–11 (2018)

10. Muraru, D., et al.: Development and prognostic validation of partition values to grade right ventricular dysfunction severity using 3D echocardiography. Eur. Heart J. Cardiovasc. Imaging **21**(1), 10–21 (2020)

11. Muraru, D., et al.: New speckle-tracking algorithm for right ventricular volume analysis from three-dimensional echocardiographic data sets: validation with cardiac magnetic resonance and comparison with the previous analysis tool. Eur. J. Echocardiogr. **17**(11), 1279–1289 (2015)

12. Oktay, O., et al.: Anatomically constrained neural networks (ACNNs): application to cardiac image enhancement and segmentation. IEEE Trans. Med. Imaging **37**(2), 384–395 (2017)

13. Ouyang, D., et al.: Echonet-dynamic: a large new cardiac motion video data resource for medical machine learning. In: NeurIPS ML4H Workshop: Vancouver, BC, Canada (2019)

14. Ouyang, D., et al.: Video-based AI for beat-to-beat assessment of cardiac function. Nature **580**(7802), 252–256 (2020)

15. Porter, T.R., et al.: Guidelines for the use of echocardiography as a monitor for therapeutic intervention in adults: a report from the American society of echocardiography. J. Am. Soc. Echocardiogr. **28**(1), 40–56 (2015)

16. Sayour, A.A., Tokodi, M., Celeng, C., Takx, R.A.P., Fábián, A., Lakatos, B.K. et al.: Association of right ventricular functional parameters with adverse cardiopulmonary outcomes - a meta-analysis. J. Am. Soc. Echocardiogr. https://doi.org/10.1016/j.echo.2023.01.018. in press

17. Sengupta, P.P., et al.: Proposed requirements for cardiovascular imaging-related machine learning evaluation (prime): a checklist. JACC: Cardiovasc. Imaging **13**(9), 2017–2035 (2020)

18. Shad, R., et al.: Predicting post-operative right ventricular failure using video-based deep learning. Nat. Commun. **12**(1), 1–8 (2021)

19. Tajbakhsh, N., et al.: Convolutional neural networks for medical image analysis: full training or fine tuning? IEEE Trans. Med. Imaging **35**(5), 1299–1312 (2016)

20. Tran, D., Wang, H., Torresani, L., Ray, J., LeCun, Y., Paluri, M.: A closer look at spatiotemporal convolutions for action recognition. In: Proceedings of the IEEE Conference on Computer Vision and Pattern Recognition, pp. 6450–6459 (2018)

21. Xie, S., Girshick, R., Dollár, P., Tu, Z., He, K.: Aggregated residual transformations for deep neural networks. In: Proceedings of the IEEE Conference on Computer Vision and Pattern Recognition, pp. 1492–1500 (2017)

22. Zamzmi, G., Hsu, L.Y., Li, W., Sachdev, V., Antani, S.: Harnessing machine intelligence in automatic echocardiogram analysis: current status, limitations, and future directions. IEEE Rev. Biomed. Eng. (2020)

23. Zhang, J., et al.: Fully automated echocardiogram interpretation in clinical practice: feasibility and diagnostic accuracy. Circulation **138**(16), 1623–1635 (2018)

# W08 - Computer Vision for Metaverse

# W08 - Computer Vision for Metaverse

Computer Vision (CV) research plays an essential role in enabling the future applications of Augmented Reality (AR), Virtual Reality (VR), and Mixed Reality (MR), which are nowadays referred to as the Metaverse. Building the Metaverse requires CV technologies to better understand people, objects, scenes, and the world around us, and to better render contents in more immersive and realistic ways. This brings new problems to CV research and inspires us to look at existing CV problems from new perspectives. As the general public interest is growing and industry is putting more efforts into the Metaverse, we thought it would be a good opportunity to organize a workshop for the computer vision community to showcase our latest research, discuss new directions and problems, and influence the future trajectory of Metaverse research and applications.

October 2022

Bichen Wu
Peizhao Zhang
Xiaoliang Dai
Tao Xu
Hang Zhang
Peter Vajda
Fernando de la Torre
Angela Dai
Bryan Catanzaro

# Initialization and Alignment for Adversarial Texture Optimization

Xiaoming Zhao$^{(\boxtimes)}$, Zhizhen Zhao, and Alexander G. Schwing

University of Illinois, Urbana-Champaign, USA
{xz23,zhizhenz,aschwing}@illinois.edu
https://xiaoming-zhao.github.io/projects/advtex_init_align

**Abstract.** While recovery of geometry from image and video data has received a lot of attention in computer vision, methods to capture the texture for a given geometry are less mature. Specifically, classical methods for texture generation often assume clean geometry and reasonably well-aligned image data. While very recent methods, *e.g.*, adversarial texture optimization, better handle lower-quality data obtained from hand-held devices, we find them to still struggle frequently. To improve robustness, particularly of recent adversarial texture optimization, we develop an explicit initialization and an alignment procedure. It deals with complex geometry due to a robust mapping of the geometry to the texture map and a hard-assignment-based initialization. It deals with misalignment of geometry and images by integrating fast image-alignment into the texture refinement optimization. We demonstrate efficacy of our texture generation on a dataset of 11 scenes with a total of 2807 frames, observing 7.8% and 11.1% relative improvements regarding perceptual and sharpness measurements.

**Keywords:** Scene analysis · Texture reconstruction

## 1 Introduction

Accurate scene reconstruction is one of the major goals in computer vision. Decades of research have been devoted to developing robust methods like 'Structure from Motion,' 'Bundle Adjustment,' and more recently also single view reconstruction techniques. While reconstruction of geometry from image and video data has become increasingly popular and accurate in recent years, recovered 3D models remain often pale because textures aren't considered.

Given a reconstructed 3D model of a scene consisting of triangular faces, and given a sequence of images depicting the scene, texture mapping aims to find for each triangle a suitable texture. The problem of automatic texture mapping has been studied in different areas since late 1990 and early 2000. For instance, in the graphics community [12, 29, 30], in computer vision [27, 43, 45], architecture [19]

**Supplementary Information** The online version contains supplementary material available at https://doi.org/10.1007/978-3-031-25066-8_34.

and photogrammetry [14]. Many of the proposed algorithms work very well in a controlled lab-setting where geometry is known perfectly, or in a setting where accurate 3D models are available from a 3D laser scanner.

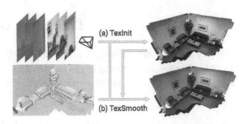

**Fig. 1.** We study texture generation given RGBD images with associated camera parameters as well as a reconstructed mesh. **(a) TexInit**: we initialize the texture using an assignment-based texture generation framework. **(b) TexSmooth**: a data-driven adversarial loss is utilized to optimize out artifacts incurred in the assignment step.

However, applying texture mapping techniques to noisy mesh geometry obtained *on the fly* from a recent LiDAR equipped iPad reveals missing robustness because of multiple reasons: 1) images and 3D models are often not perfectly aligned; 2) 3D models are not accurate and the obtained meshes aren't necessarily manifold-connected. Even recent techniques for mesh flattening [41] and texturing [16,24,54] result in surprising artifacts due to streamed geometry and pose inaccuracies as shown later.

To address this robustness issue we find equipping of the recently-proposed adversarial texture optimization technique [24] with classical initialization and alignment techniques to be remarkably effective. Without the added initialization and alignment, we find current methods don't produce high-quality textures. Concretely and as illustrated in Fig. 1, we aim for texture generation which operates on a sequence of images and corresponding depth maps as well as their camera parameters. Moreover, we assume the 3D model to be given and fixed. Importantly, we consider a streaming setup, with all data obtained *on the fly*, and not further processed, *e.g.*, via batch structure-from-motion. The setup is ubiquitous and the form of data can be acquired easily from consumer-grade devices these days, *e.g.*, from a recent iPad or iPhone [1]. We aim to translate this data into texture maps. For this, we first flatten the triangle mesh using recent advances [41]. We then use a Markov Random Field (MRF) to resolve overlaps in flattened meshes for non-manifold-connected data. In a next step we determine the image frame from which to extract the texture of each mesh triangle using a simple optimization. We refer to this as TexInit, which permits to obtain a high-quality initialization for subsequent refinement. Next we address inaccuracies in camera poses and in geometry by automatically shifting images using the result of a fast Fourier transformation (FFT) [6]. The final optimized texture is obtained by integrating this FFT-alignment component into adversarial optimization [24]. We dub this stage TexSmooth. The obtained texture can be used in any 3D engine for downstream applications.

To study efficacy of the proposed framework we acquire 11 complex scenes using a recent iPad. We establish accuracy of the proposed technique to generate and use the texture by showing that the quality of rendered views is superior to prior approaches on these scenes. Quantitatively, our framework improves prior work by 7.8% and 11.1% relatively with respect to perceptual (LPIPS [53]) and sharpness ($S_3$ [47]) measurements respectively. Besides, our framework improves over baselines on ScanNet [11], demonstrating the ability to generalize.

## 2   Related Work

We aim for accurate recovery of texture for a reconstructed 3D scene from a sequence of RGBD images. For this, a variety of techniques have been proposed, which can be roughly categorized into four groups: 1) averaging-based; 2) warping-based; 3) learning-based; and 4) assignment-based. Averaging-based methods find all views within which a point is visible and combine the color observations. Warping-based approaches either distort or synthesize source images to handle mesh misalignment or camera drift. Learning-based ones learn the texture representation. Assignment-based methods attempt to find the best view and 'copy' the observation into a texture. We review these groups next:

**Averaging-Based:** Very early work by Marshner [29] estimates the parameters of a bidirectional reflectance distribution function (BRDF) for every point on the texture map. To compute this estimation, all observations from the recorded images where the point is visible are used. Similar techniques have been investigated in subsequent work [9].

Similarly, to compute a texture map, [30] and [12] perform a weighted blending of all recorded images. The weights take visibility and other factors into account. The developed approaches are semi-automatic as they require an initial estimate of the camera orientations which is obtained from interactively selected point correspondences or marked lines. Multi-resolution textures [33], face textures [36] and blending [8, 31] have also been studied.

**Warping-Based:** Aganj *et al.* [4] morph each source image to align to the mesh. Furthermore, [23, 54] propose to optimize camera poses and image warping parameters jointly. However, this line of vertex-based optimization has stringent requirements on the mesh density and cannot be applied to a sparse mesh. More recently, Bi *et al.* [10] follow patch-synthesis [7, 50] to re-synthesize aligned target images for each source view. However, such methods require costly multiscale optimization to avoid a large number of local optima. In contrast, the proposed approach does not require those techniques.

**Learning-Based:** Recently, learning-based methods have been introduced for texture optimization. Some works focus on specific object and scene categories [18, 40] while we do not make such assumptions. Moreover, learned representations, *e.g.*, neural textures, have also been developed [5,44,46]. Meanwhile, generative models are developed to synthesize a holistic texture from a single image or pattern [21,32] while we focus on texture reconstruction. AtlasNet [20] and NeuTex [52] focus on learning a 3D-to-2D mapping, which can be utilized in texture editing, while we focus on reconstructing realistic textures from source images. The recently-proposed adversarial texture optimization [24] utilizes adversarial training to reconstruct the texture. However, despite advances, adversarial optimization still struggles with misalignments. We improve this shortcoming via an explicit high-quality initialization and an efficient alignment module.

**Assignment-Based:** Classical assignment-based methods operated within controlled environments [14,15,38,39] or utilized special camera rigs [15,19]. These works suggest computing for each vertex a set of 'valid' images, which are subsequently refined by iterating over each vertex and adjusting the assignment to obtain more consistency. Finally, texture data is 'copied' from the images. In contrast, we aim to create a texture in an uncontrolled setting. Consequently, 3D geometry is not accurate and very noisy. Other early work [13,22,28,39,51] focuses on closed surfaces and small-scale meshes, making them not applicable to our setting. More recently, upon finding the best texture independently for each face using cues like visibility, orientation, resolution, and distortion, refinement techniques like texture coordinate optimization, color adjustments, or scores-based optimization have been discussed [3,34,48].

Related to our approach are methods that formulate texture selection using a Markov Random Field (MRF) [16,27,42]. Shu *et al.* [42] suggest visibility as the data term and employ texture smoothness to reduce transitions. Lempitsky *et al.* [27] study color-continuity which integrates over face seams. Fu *et al.* [16] additionally use the projected 2D face area to select a texture assignment for each face. However, noisy geometry like the one shown in Fig. 2, makes it difficult for assignment-based methods to yield high quality results, which we will show later. Therefore,

**Fig. 2.** Noisy geometry. The wall has two layers.

different from these methods, we address texture drift in a data-driven refinement procedure rather than in an assignment stage.

## 3    Approach

We want to automatically create the texture $\mathcal{T}$ from a set of RGBD images $I = \{I_1, \ldots, I_T\}$, for each of which we also know camera parameters $\{p_t\}_{t=1}^{T}$, *i.e.*, extrinsics and intrinsics. We are also given a triangular scene mesh $M = \{\text{Tri}_i\}_{i=1}^{|M|}$, where $\text{Tri}_i$ denotes the $i$-th triangle. This form of data is easily accessible from commercially available consumer devices, *e.g.*, a recent iPhone or iPad.

We construct the texture $\mathcal{T}$ in two steps that combine advantages of assignment-based and learning-based techniques: 1) TexInit: we generate a texture initialization $\mathcal{T}_{init} \in \mathbb{R}^{H \times W \times 3}$ of height $H$, width $W$ and 3 color channels in an assignment-based manner (Sect. 3.1); 2) TexSmooth: we then refine $\mathcal{T}_{init}$ with an improved data-driven adversarial optimization that integrates an efficient alignment procedure (Sect. 3.2). Formally, the final texture $\mathcal{T}$ is computed via

$$\mathcal{T} = \text{TexSmooth}\left(\mathcal{T}_{init}, \{I_t\}_{t=1}^T, \{p_t\}_{t=1}^T, M\right),$$

$$\text{where } \mathcal{T}_{init} = \text{TexInit}\left(\{I_t\}_{t=1}^T, \{p_t\}_{t=1}^T, M\right). \tag{1}$$

We detail each component next.

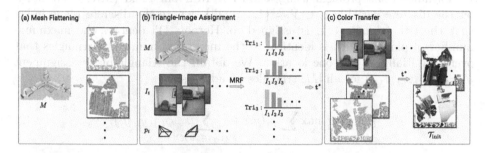

**Fig. 3. Texture initialization TexInit** (Sect. 3.1). **(a)** Mesh flattening: we flatten a 3D mesh into the 2D plane using overlap detection. **(b)** Triangle-image assignment: we develop a simple formulation to compute the triangle-image assignment $\mathbf{t}^*$ from mesh $M$, frames $I_t$ and camera parameters $p_t$. We assign frames to each triangle $\text{Tri}_i$ based on $\mathbf{t}^*$. **(c)** Color transfer: based on the flattened mesh in **(a)** and the best assignment $\mathbf{t}^*$ from **(b)**, we generate the texture $\mathcal{T}_{init}$.

## 3.1 Texture Initialization (TexInit)

The proposed approach to obtain the texture initialization $\mathcal{T}_{init}$ is outlined in Fig. 3 and consists of following three steps: 1) We flatten the provided mesh $M$. For this we detect overlaps within the flattened mesh, which may happen due to the fact that we operate with general meshes that are not guaranteed to have a manifold connectivity. Overlap detection ensures that every triangle is assigned a unique position in the texture. 2) We identify for each triangle the 'best' texture index $\mathbf{t}^*$. Hereby, 'best' is defined using cues like visibility and color consistency. 3) After identifying the index $\mathbf{t}^* = (t_1^*, \ldots, t_{|M|}^*)$ for each triangle, we create the texture $\mathcal{T}_{init}$ by transferring for all $(u,v) \in [1,\ldots,W] \times [1,\ldots,H]$ locations in the texture, the RGB data from the corresponding location $(a,b)$ in image $I_t$.

**1) Mesh Flattening:** In a first step, as illustrated in Fig. 3 (left), we flatten the given mesh $M$. For this we use the recently proposed boundary first flattening (BFF) technique [41]. The flattening is fully automatic, with distortion mathematically guaranteed to be as low or lower than any other conformal mapping.

However, despite those guarantees, BFF still requires meshes to have a manifold connectivity. While we augment work by [41] using vertex duplication to circumvent this restriction, flattening may still result in overlapping regions as illustrated in Fig. 4. To fix this and uniquely assign a triangle to a position in the texture, we perform overlap detection as discussed next.

**Overlap Detection:** Overlap detection operates on flattened and possibly overlapping triangle meshes like the one illustrated in Fig. 4a. Our goal is to assign triangles to different planes. Upon re-packing the triangles assigned to different planes, we obtain the non-overlapping triangle mesh illustrated in Fig. 4b.

In order to not break the triangle mesh at a random position and end up with many individual triangles, $i.e.$, in order to maintain large triangle connectivity, we formulate this problem using a Markov Random Field (MRF). Formally, let the discrete variable $y_i \in \mathcal{Y} = \{1, \ldots, |\mathcal{Y}|\}$ denote the discrete plane index that the $i$-th triangle $\mathtt{Tri}_i$ is assigned to. Hereby, $|\mathcal{Y}|$ denotes the maximum number of planes which is identical to the maximum number of triangles that overlap initially at any one location. We obtain the triangle-plane assignment $y^* = (y_1^*, \ldots, y_{|M|}^*)$ for all $|M|$ triangles by addressing

$$y^* = \arg\max_y \sum_{i=1}^{|M|} \phi_i(y_i) + \sum_{(i,j) \in \mathcal{A} \cup \mathcal{O}} \phi_{i,j}(y_i, y_j), \tag{2}$$

where $\mathcal{A}$ and $\mathcal{O}$ are sets of triangle index pairs which are adjacent and overlapping respectively. Here, $\phi_i(\cdot)$ denotes triangle $\mathtt{Tri}_i$'s priority over $\mathcal{Y}$ when considering only its $local$ information, while $\phi_{i,j}(\cdot)$ refers to $\mathtt{Tri}_i$ and $\mathtt{Tri}_j$'s joint preference on their assignments. Equation (2) is solved with belief propagation [17].

Intuitively, by addressing the program given in Eq. (2) we want a different plane index for overlapping triangles, while encouraging mesh $M$'s adjacent triangles to be placed on the same plane. To achieve this we use

$$\phi_i(y_i) = \begin{cases} 1.0, & \text{if } y_i = \min \mathcal{Y}_{i,\text{non-overlap}} \\ 0.0, & \text{otherwise} \end{cases}, \text{ and} \tag{3}$$

$$\phi_{i,j}(y_i, y_j) = \begin{cases} \mathbb{1}\{y_i = y_j\}, & \text{if } (i,j) \in \mathcal{A} \\ \mathbb{1}\{y_i \neq y_j\}, & \text{if } (i,j) \in \mathcal{O} \end{cases}. \tag{4}$$

Here, $\mathbb{1}\{\cdot\}$ denotes the indicator function and $\mathcal{Y}_{i,\text{non-overlap}}$ contains all plane indices where $\mathtt{Tri}_i$ has no overlap with others. Intuitively, Eq. (3) encourages to assign the minimum of such indices to $\mathtt{Tri}_i$.

As fast MRF optimizers remove most overlaps but don't provide guarantees, we add a light post-processing to manually assign the remaining few overlapping triangles to separate planes. This guarantees overlap-free results. As mentioned before, after having identified the plane assignment $y^*$ for each triangle we use a bin packing to uniquely assign each triangle to a position in the texture. Conversely, for every texture coordinate $u, v$ we obtain a unique triangle index

$$i = G(u, v). \tag{5}$$

A qualitative result is illustrated in Fig. 4b. Next, we identify the image which should be used to texture each triangle.

**2) Textures from Triangle-Image Assignments:** Our goal is to identify a suitable frame $I_{t_i}$, $t_i \in \{1, \ldots, T + 1\}$, for each triangle $\mathsf{Tri}_i$, $i \in \{1, \ldots, |M|\}$. Note that the $(T + 1)$-th option $I_{T+1}$ refers to an empty texture. We compute the texture assignments $\mathbf{t}^* = (t_1^*, \ldots, t_{|M|}^*)$ using a purely local optimization:

$$\mathbf{t}^* = \underset{\mathbf{t}}{\mathrm{argmax}} \sum_{i=1}^{|M|} \psi_i(t_i). \tag{6}$$

Here $\psi_i$ captures *unary* cues. Note, we also studied *pairwise* cues but did not observe significant improvements. Please see the Appendix for more details. Due to better efficiency, we therefore only consider unary cues. Intuitively, we want

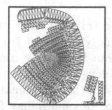

(a) Flattened mesh overlaps.

(b) Overlap-free.

**Fig. 4. Flattening. (a)** Red triangles indicate where overlap happens. **(b)** The proposed method (Sect. 3.1) resolves this issue while keeping connectivity of areas. (Color figure online)

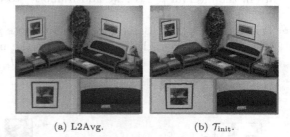

(a) L2Avg.                    (b) $\mathcal{T}_{\mathrm{init}}$.

**Fig. 5. Initialization comparison. (a)** We use PyTorch3D's rendering pipeline [37] to project each pixel of every RGB image back to the texture. The color of each pixel in the texture is the average of all colors that project to it. This texture minimizes the $\mathcal{L}_2$ loss of the difference between the rendered and the ground truth images. We dub it L2Avg. **(b)** $\mathcal{T}_{\mathrm{init}}$ from Sect. 3.1 permits to maintain details. The seam artifacts will be optimized out using TexSmooth (Sect. 3.2). Besides over-smoothness, without taking into account misalignments of geometry and camera poses, L2Avg produces texture that overfits to available views, *e.g.*, the sofa's blue colors are painted onto the wall. (Color figure online)

the program given in Eq. (6) to encourage triangle-image assignment to be 'best' for each triangle $\text{Tri}_i$. We describe the unary cues to do so next.

**Unary Potentials** $\psi_i(t_i)$ for each pair of triangle $\text{Tri}_i$ and frame $I_{t_i}$ are

$$\psi_i(t_i) = \begin{cases} \psi_i^{\text{C}}(t_i), & \text{if } \psi_i^{\text{V}}(t_i) = 1 \\ -\infty, & \text{otherwise} \end{cases}, \tag{7}$$

where $\psi_i^{\text{V}}(t_i)$ and $\psi_i^{\text{C}}(t_i)$ represent validity check and potentials from cues respectively. Concretely, we use

$$\psi_i^{\text{V}}(t_i) = \mathbb{1}\{I_{t_i} \in \mathcal{S}_i^{\text{V}}\}, \tag{8}$$

$$\psi_i^{\text{C}}(t_i) = \omega_1 \cdot \psi_i^{\text{C}_1}(t_i) + \omega_2 \cdot \psi_i^{\text{C}_2}(t_i) + \omega_3 \cdot \psi_i^{\text{C}_3}(t_i), \tag{9}$$

where $\mathcal{S}_i^{\text{V}}$ denotes the set of valid frames for $\text{Tri}_i$ and $\omega_1, \omega_2, \omega_3$ represent weights for potentials $\psi_i^{\text{C}_1}, \psi_i^{\text{C}_2}, \psi_i^{\text{C}_3}$. We discuss each one next:

- **Validity** (V). To assess whether frame $I_{t_i}$ is valid for $\text{Tri}_i$, we check the visibility of $\text{Tri}_i$ in $I_{t_i}$. We approximate this by checking visibility of $\text{Tri}_i$'s three vertices as well as its centroid. Concretely, we transform the vertices and centroid from world coordinates to the normalized device coordinates of the $t_i$-th camera. If all vertices and centroid are visible, *i.e.*, their coordinates are in the interval $[-1, 1]$, we add frame $I_{t_i}$ to the set $\mathcal{S}_i^{\text{V}}$ of valid frames for triangle $\text{Tri}_i$.
- **Triangle area** (C$_1$). Based on a camera's pose $p_{t_i}$, a triangle's area changes. The larger the area, the more detailed is the information for $\text{Tri}_i$ in frame $I_{t_i}$. We encourage to assign $\text{Tri}_i$ to frames $I_{t_i}$ with large area by defining $\psi_i^{\text{C}_1}(t_i) = \text{Area}_{t_i}(\text{Tri}_i)$ and set $\omega_1 > 0$.

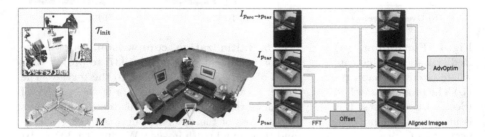

**Fig. 6. Texture smoothing TexSmooth** (Sect. 3.2). We utilize adversarial optimization (AdvOptim) [24] to refine the texture $\mathcal{T}_{\text{init}}$ from Sect. 3.1. Differently: 1) We initialize with $\mathcal{T}_{\text{init}}$. 2) To resolve the issue of misalignment between rendering and ground truth (GT), we integrate an alignment module based on the fast Fourier transform (FFT).

- **Discrepancy between $z$-buffer and actual depth** ($C_2$). For a valid frame $I_{t_i} \in \mathcal{S}_i^V$, a triangle's vertices and its centroid project to valid image coordinates. We compute the discrepancy between: 1) the depth from frame $I_{t_i}$ at the image coordinates of the vertices and centroid; 2) the depth of vertices and centroid in the camera's coordinate system. We set $\psi_i^{C_2}(t_i)$ to be the sum of absolute value differences between both depth estimates while using $\omega_2 < 0$.

- **Perceptual consistency** ($C_3$). Due to diverse illumination, triangle $\text{Tri}_i$'s appearance changes across frames. Intuitively, we don't want to assign a texture to $\text{Tri}_i$ using a frame that contains colors that deviate drastically from other frames. Concretely, we first average all triangle's three vertices color values across all valid frames, *i.e.*, across all $I_{t_i} \in \mathcal{S}_i^V$. We then compare this global average to the local average obtained independently for the three vertices of every valid frame $I_{t_i} \in \mathcal{S}_i^V$ using an absolute value difference. We require $\omega_3 < 0$.

**3) Color Transfer:** Given the inferred triangle-frame assignments $t^*$ we complete the texture $\mathcal{T}_{\text{init}}$ by transferring RGB data from image $I_{t_i^*}$ for $\text{Tri}_i, i \in \{1, \ldots, |M|\}$. For this we leverage the camera pose $p_{t_i^*}$ which permits to transform texture coordinates $(u, v)$ of locations within $\text{Tri}_i$ to corresponding image coordinates $(a, b)$ in texture $I_{t_i^*}$ via the mapping $F : \mathbb{R}^2 \to \mathbb{R}^2$, *i.e.*,

$$(a, b) = F(u, v, t_i^*, p_{t_i^*}). \tag{10}$$

Intuitively, given the $(u, v)$ coordinates on the texture in a coordinate system which is local to the triangle $\text{Tri}_i$, and given the camera pose $p_{t_i^*}$ used to record image $I_{t_i^*}$, the mapping $F$ retrieves the image coordinates $(a, b)$ corresponding to texture coordinate $(u, v)$. Using this mapping, we obtain the texture $\mathcal{T}_{\text{init}}$ at location $(u, v)$, *i.e.*, $\mathcal{T}_{\text{init}}(u, v)$, from the image data $I_{t_i^*}(a, b) \in \mathbb{R}^3$ via

$$\mathcal{T}_{\text{init}}(u, v) = I_{t_i^*}(F(u, v, t_i^*, p_{t_i^*})). \tag{11}$$

Note, because of the overlap detection, we obtain a unique triangle index $i = G(u, v)$ for every $(u, v)$ coordinate from Eq. (5). Having transferred RGB data for all coordinates within all triangles results in the texture $\mathcal{T}_{\text{init}} \in \mathbb{R}^{H \times W \times 3}$, which we compare to standard L2 averaging initialization in Fig. 5. We next refine this texture via adversarial optimization. We observe that this initialization $\mathcal{T}_{\text{init}}$ is crucial to obtain high-quality textures, which we will show in Sect. 4.

### 3.2 Texture Smoothing (TexSmooth)

As can be seen in Fig. 5b, the texture $\mathcal{T}_{\text{init}}$ contains seams that affect visual quality. To reconstruct a seamless texture $\mathcal{T}$, we extend recent adversarial optimization (AdvOptim). Different from prior work [24] which initializes with blank (paper) or averaged (code release[1]) textures, we initialize with $\mathcal{T}_{\text{init}}$. Also, we find AdvOptim doesn't handle common camera pose and geometry misalignment well. To resolve this, we develop an efficient alignment module. This is depicted in Fig. 6 and will be detailed next.

**Smoothing with Adversarial Optimization:** To optimize the texture, AdvOptim iterates over camera poses. When optimizing for a specific target camera pose $p_{\text{tar}}$, AdvOptim uses three images: 1) the ground truth image $I_{p_{\text{tar}}}$ of the target camera pose $p_{\text{tar}}$; 2) a rendering $\hat{I}_{p_{\text{tar}}}$ for the target camera pose $p_{\text{tar}}$ from the texture map $\mathcal{T}$; and 3) a re-projection from another camera pose $p_{\text{src}}$'s ground truth image, which we refer to as $I_{p_{\text{src}} \to p_{\text{tar}}}$. It then optimizes by minimizing an $\mathcal{L}_1$ plus a conditional adversarial loss. However, we find AdvOptim to struggle with alignment errors due to inaccurate geometry. Therefore, we integrate an efficient alignment operation into AdvOptim. Instead of directly using the input images, we first compute a 2D offset $(\Delta h_{p_{\text{tar}}}, \Delta w_{p_{\text{tar}}})$ between $I_{p_{\text{tar}}}$ and $\hat{I}_{p_{\text{tar}}}$, which we apply to align $I_{p_{\text{tar}}}$ and $\hat{I}_{p_{\text{tar}}}$ as well as $I_{p_{\text{src}} \to p_{\text{tar}}}$ via

$$I^{\mathsf{A}} \doteq \text{Align}(I, (\Delta h, \Delta w)), \tag{12}$$

where $I^{\mathsf{A}}$ marks aligned images. We then use the three aligned images as input:

$$\mathcal{L} = \lambda \| I^{\mathsf{A}}_{p_{\text{tar}}} - \hat{I}^{\mathsf{A}}_{p_{\text{tar}}} \|_1 + \mathbb{E}_{I^{\mathsf{A}}_{p_{\text{tar}}}, I^{\mathsf{A}}_{p_{\text{src}} \to p_{\text{tar}}}} \left[ \log D(I^{\mathsf{A}}_{p_{\text{src}} \to p_{\text{tar}}} | I^{\mathsf{A}}_{p_{\text{tar}}}) \right]$$
$$+ \mathbb{E}_{I^{\mathsf{A}}_{p_{\text{tar}}}, \hat{I}^{\mathsf{A}}_{p_{\text{tar}}}} \left[ \log(1 - D(\hat{I}^{\mathsf{A}}_{p_{\text{tar}}} | I^{\mathsf{A}}_{p_{\text{tar}}})) \right]. \tag{13}$$

Here, $D$ is a convolutional deep-net based discriminator. When using the unaligned image $I$ instead of $I^{\mathsf{A}}$, Eq. (13) reduces to the vanilla version in [24]. We now discuss a fast way to align images.

**Alignment with Fourier Transformation:** To align ground truth $I_{p_{\text{tar}}}$ and rendering $\hat{I}_{p_{\text{tar}}}$, one could use naïve grid-search to find the offset which results in the minimum difference of the shifted images. However, such a grid-search is prohibitively costly during an iterative optimization, especially with high-resolution images (*e.g.*, we use a resolution up to $1920 \times 1440$). Instead, we use the fast Fourier transformation (FFT) to complete the job [6]. Specifically, given a misaligned image pair of $I \in \mathbb{R}^{h \times w \times 3}$ and $\hat{I} \in \mathbb{R}^{h \times w \times 3}$, we compute for every channel the maximum correlation via

---

[1] https://github.com/hjwdzh/AdversarialTexture.

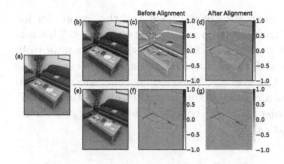

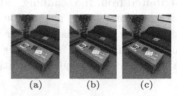

Fig. 7. Alignment with fast Fourier transformation (FFT) (Sect. 3.2). We show results for $\mathcal{T}_{\text{init}}$ (Sect. 3.1) and L2Avg (Fig. 5) in top and bottom rows. (a): ground-truth (GT); (b) and (e): texture rendering with (a)'s corresponding camera; (c) and (d): difference between (a) and (b); (f) and (g): difference between (a) and (e). The top row: FFT successfully aligns GT image and rendering from $\mathcal{T}_{\text{init}}$. Within expectation, there is almost no misalignment for the texture L2Avg as it overfits to available views (Fig. 5).

Fig. 8. Alignment is important for evaluation (Sect. 4.1). Clearly, (c) is more desirable than (b). However, before alignment, LPIPS yields 0.3347 and 0.4971 for (a)–(b) and (a)–(c) pairs respectively. This is misleading as lower LPIPS indicates higher quality. After alignment, LPIPS produces 0.3343 and 0.2428 for the same pairs, which provides correct signals for evaluation.

$$\operatorname*{argmax}_{(i,j)} \ \mathsf{FFT}^{-1}\left(\mathsf{FFT}(I) \cdot \overline{\mathsf{FFT}(\hat{I})}\right)[i,j]. \tag{14}$$

Here, $\mathsf{FFT}(\cdot)$ represents the fast Fourier transformation while $\mathsf{FFT}^{-1}(\cdot)$ denotes its inverse and $\overline{\mathsf{FFT}(\hat{I})}$ refers to the complex conjugate. After decoding the maximum correlation response and averaging across channels, we obtain the final offset $(\Delta h, \Delta w)$. As can be seen in Fig. 7(d), the offset $(\Delta h, \Delta w)$ is very accurate. Moreover, the computation finishes in around 0.4 s even for 1920×1440-resolution images. Note, we don't need to maintain gradients for $(\Delta h, \Delta w)$, since the offset is only used to shift images and not to backpropagate through it.

## 4  Experiments

### 4.1  Experimental Setup

**Data Acquisition.** We use a 2020 iPad Pro and develop an iOS app to acquire the RGBD images $I_t$, camera pose $p_t$, and scene mesh $M$ via Apple's ARKit [1].

**UofI Texture Scenes.** We collect a dataset of 11 scenes: four indoor and seven outdoor scenes. This dataset consists of a total of 2807 frames, of which 91, 2052, and 664 are of resolution 480 × 360, 960 × 720, and 1920 × 1440 respectively.

For each scene, we use 90% of its views for optimization and the remainder for evaluation. In total, we have 2521 training frames and 286 test frames. This setting is more challenging than prior work where [24] "select(s) 10 views uniformly distributed from the scanning video" for evaluation while using up to thousands of frames for texture generation. On average, the angular differences between test set view directions and their nearest neighbour in the training sets are 2.05° (min 0.85°/max 13.8°). Angular distances are computed following [25]. Please see Appendix for scene-level statistics.

**Implementation.** We compare to five baselines for texture generation: L2Avg, ColorMap [54], TexMap [16], MVSTex [48], and AdvTex [24]. For ColorMap, TexMap, and MVSTex, we use their official implementations.[2] For AdvOptim (Sect. 3.2) used in both AdvTex and ours, we re-implement a PyTorch [35] version based on their official release in TensorFlow [2] (See footnote 1). We evaluate AdvTex with two different initializations: 1) blank textures as stated in the paper (AdvTex-B); 2) the initialization used in the official code release (AdvTex-C). We run AdvOptim using the Adam optimizer [26]. See Appendix for more details. For our TexInit (Sect. 3.1), we use a generic set of weights across all scenes: $\omega_1 = 1e^{-3}$ (triangle area), $\omega_2 = -10$ (depth discrepancy), and $\omega_3 = -1$ (perception consistency), which makes cue magnitudes roughly similar.

On a 3.10 GHz Intel Xeon Gold 6254 CPU, ColorMap takes less than two minutes to complete while TexMap's running time ranges from 40 min to 4 h. MVSTex can be completed in no more than 10 min. Our $\mathcal{T}_{init}$ (Sect. 3.1) completes in two minutes. Additionally, the AdvOptim takes around 20 min for 4000 iterations to complete with an Nvidia RTX A6000 GPU.

**Evaluation Metrics.** To assess the efficacy of the method, we study the quality of the texture from two perspectives: perceptual quality and sharpness. 1) For perceptual quality, we assess the similarity between rendered and collected ground-truth views using the Structural Similarity Index Measure (SSIM) [49] and the Learned Perceptual Image Patch Similarity (LPIPS) [53]. 2) For sharpness, we consider measurement $S_3$ [47] and the norm of image gradient (Grad) following [24]. Specifically, for each pixel, we compute its $S_3$ value, whose difference between the rendered and ground truth (GT) is used for averaging across the whole image. A similar procedure is applied to Grad. For all four metrics, we report the mean and standard deviation across 11 scenes.

**Alignment in Evaluation.** As can be seen in Fig. 8, evaluation will be misleading if we do not align images during evaluation. Therefore, we propose the following procedure: 1) for each method, we align the rendered image and the GT using an FFT (Sect. 3.2); 2) to avoid various resolutions caused by different methods, we crop out the maximum common area across methods. 3) we then

---

[2] ColorMap: https://github.com/intel-isl/Open3D/pull/339; TexMap: https://github.com/fdp0525/G2LTex; MVSTex: https://github.com/nmoehrle/mvs-texturing.

compute metrics on those cropped regions. The resulting comparison is fair as all methods are evaluated on the same number of pixels and aligned content.

## 4.2   Experimental Evaluation

**Quantitative Evaluation.** Table 1 reports aggregated results on all 11 scenes. The quality of our texture $\mathcal{T}$ (Row 6) outperforms baselines on LPIPS, $S_3$ and Grad, confirming the effectiveness of the proposed pipeline. Specifically, we improve LPIPS by 7.8% from 0.335 (2nd-best) to 0.309, indicating high perceptual similarity. Moreover, $\mathcal{T}$ maintains sharpness as we improve $S_3$ by 11.1% from 0.135 (2nd-best) to 0.120 and Grad from 7.171 (2nd-best) to 6.871. Regarding SSIM, we find it to favor L2Avg in almost all scenes (see Appendix) which aligns with the findings in [53].

**Table 1. Aggregated quantitative evaluation on UofI Texture Scenes.** We report results in the form of mean±std. Please see Fig. 9 for qualitative texture comparisons and Appendix for scene-level quantitative results.

|     |          | SSIM↑ | LPIPS↓ | $S_3$ ↓ | Grad↓ |
| --- | -------- | ----- | ------ | ------- | ----- |
| 1   | L2Avg    | **0.610** ± 0.191 | 0.386 ± 0.116 | 0.173 ± 0.105 | 7.066 ± 4.575 |
| 2   | ColorMap | 0.553 ± 0.193 | 0.581 ± 0.132 | 0.234 ± 0.140 | 7.969 ± 5.114 |
| 3   | TexMap   | 0.376 ± 0.113 | 0.488 ± 0.097 | 0.179 ± 0.062 | 8.918 ± 4.174 |
| 4   | MVSTex   | 0.476 ± 0.164 | 0.335 ± 0.086 | 0.139 ± 0.047 | 8.198 ± 3.936 |
| 5-1 | AdvTex-B | 0.495 ± 0.174 | 0.369 ± 0.092 | 0.148 ± 0.047 | 8.229 ± 4.586 |
| 5-2 | AdvTex-C | 0.563 ± 0.191 | 0.365 ± 0.096 | 0.135 ± 0.067 | 7.171 ± 4.272 |
| 6   | Ours     | 0.602 ± 0.189 | **0.309** ± 0.086 | **0.120** ± 0.058 | **6.871** ± 4.342 |

**Table 2. Ablation study.** We report results in the form of mean±std.

|   | Adv Optim | FFT Align | $\mathcal{T}_{init}$ | SSIM↑ | LPIPS↓ | $S_3$ ↓ | Grad↓ |
| --- | --- | --- | --- | --- | --- | --- | --- |
| 1 |   |   | ✓ | 0.510 ± 0.175 | 0.342 ± 0.060 | 0.141 ± 0.052 | 8.092 ± 4.488 |
| 2 | ✓ | ✓ |   | 0.592 ± 0.192 | 0.332 ± 0.102 | 0.130 ± 0.066 | **6.864** ± 4.211 |
| 3 | ✓ |   | ✓ | 0.559 ± 0.196 | 0.346 ± 0.082 | 0.125 ± 0.057 | 7.244 ± 4.359 |
| 4 | ✓ | ✓ | ✓ | **0.602** ± 0.189 | **0.309** ± 0.086 | **0.120** ± 0.058 | 6.871 ± 4.342 |

**Ablation Study.** We verify the design choices of TexInit and TexSmooth in Table 2. **1) TexSmooth is required:** we directly evaluate $\mathcal{T}_{init}$ and 1st *vs.* 4th row confirms the performance drop: −0.092 (SSIM), +0.033 (LPIPS), +0.021 ($S_3$), and +1.221 (Grad). **2) $\mathcal{T}_{init}$ is needed:** we replace $\mathcal{T}_{init}$ with L2Avg as it performs better than ColorMap and TexMap in Table 1 and still incorporate FFT into AdvOptim. We observe inferior performance: −0.010 (SSIM), +0.023 (LPIPS), +0.010 ($S_3$) in 2nd *vs.* 4th row. **3) Alignment is important:** we use the vanilla AdvOptim but initialize with $\mathcal{T}_{init}$. As shown in Table 2's 3rd *vs.* 4th row, the texture quality drops by −0.043 (SSIM), +0.037 (LPIPS), +0.005 ($S_3$), and +0.373 (Grad).

**Qualitative Evaluation.** We present qualitative examples in Fig. 9. Figure 9a and Fig. 9b demonstrate that L2Avg and ColorMap produce overly smooth texture. Meanwhile, due to noise in the geometry, *e.g.*, Fig. 2, TexMap fails to

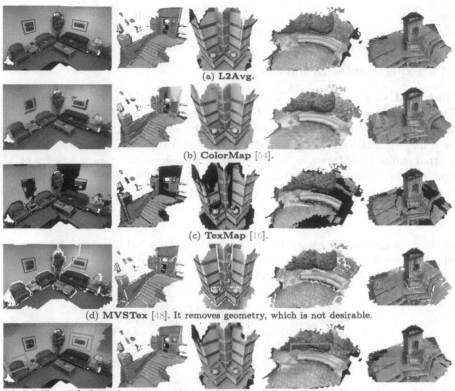

(a) **L2Avg.**

(b) **ColorMap** [54].

(c) **TexMap** [16].

(d) **MVSTex** [48]. It removes geometry, which is not desirable.

(e) **AdvTex-C** [24]. We highlight artifacts with boxes. 1) Scene 1: sofa's texture is mapped to the wall and the figure on the wall is broken; 2) Scene 2: the door's color is mapped to the floor and the brick wall's pattern is mapped to light; 3) Scene 3: ball's color is projected to brick walls; 4) Scene 4: bench's color is added to the bush and ground; 5) Scene 5: the crack breaks and stair's color is on the ground.

(f) **Ours.** Compared to Fig. 9e, our method reduces artifacts.

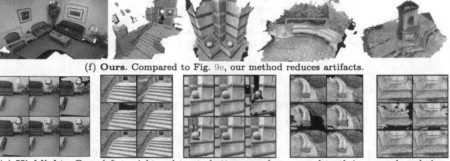

(g) **Highlights.** From left to right and top to bottom, we show ground-truth image and renderings at the same camera pose with texture from ColorMap, TexMap, MVSTex, AdvTex-C, and ours respectively. Compared to AdvTex-C: 1) Scene 1: ours produces much more complete and sharper pattern for the sofa; 2) Scene 2: ours generates sharper cracks on brick walls; 3) Scene 3: our balls are more complete while AdvTex-C maps the top of the ball to the left brick wall; 4) Scene 4: AdvTex-C maps the bench to the bush while ours is much cleaner; 5) Scene 5: the pattern on the ground from ours is much sharper.

**Fig. 9. Qualitative results on UofI Texture Scenes.** For each method, we show results for Scene 1 to 5 from left to right. Best viewed in color and zoomed-in. (Color figure online)

**Table 3. ScanNet results.** We report results in the form of mean±std. Note, this can't be directly compared to [23]'s Table 2: while we reserve 10% views for evaluation, [23] reserves only 10/2011 (≈0.5%), where 2011 is the number of average views per scene.

|  |  | SSIM↑ | LPIPS↓ | $S_3$ ↓ | Grad↓ |
|---|---|---|---|---|---|
| 1-1 | AdvTex-B | $0.534 \pm 0.074$ | $0.557 \pm 0.071$ | $0.143 \pm 0.028$ | $3.753 \pm 0.730$ |
| 1-2 | AdvTex-C | $0.531 \pm 0.074$ | $0.558 \pm 0.075$ | $0.161 \pm 0.044$ | $4.565 \pm 1.399$ |
| 2 | Ours | $\mathbf{0.571} \pm 0.069$ | $\mathbf{0.503} \pm 0.090$ | $\mathbf{0.127} \pm 0.031$ | $\mathbf{3.324} \pm 0.826$ |

**Fig. 10. Remaining six scenes with our textures.** See Appendix for all methods' results on these scenes.

**Fig. 11. Results on ScanNet.** Left to right: AdvTex-B/C and ours. Ours alleviates artifacts: colors from box on the cabinet are mapped to the backpack and wall.

resolve texture seams and cannot produce a complete texture (Fig. 9c). MVS-Tex results in Fig. 9d are undesirable as geometries are removed. This is because MVSTex requires ray collision checking to remove occluded faces. Due to the misalignment between geometries and cameras, artifacts are introduced. We show results of AdvTex-C in Fig. 9e as it outperforms AdvTex-B from Table 1. Artifacts are highlighted. Our method can largely mitigate such seams, which can be inferred from Fig. 9f. In Fig. 9g, we show renderings, which demonstrate the effectiveness of the proposed method. Please see Appendix for complete qualitative results of scenes in Fig. 10.

**On ScanNet** [11]. Following [24], we study scenes with ID ≤ 20 (Fig. 11, Table 3). We improve upon baselines (AdvTex-B/C) by a margin on SSIM ($0.534 \rightarrow 0.571$), LPIPS ($0.557 \rightarrow 0.503$), $S_3$ ($0.143 \rightarrow 0.127$), and Grad ($3.753 \rightarrow 3.324$).

## 5   Conclusion

We develop an initialization and an alignment method for fully-automatic texture generation from a given scene mesh, and a given sequence of RGBD images and their camera parameters. We observe the proposed method to yield appealing results, addressing robustness issues due to noisy geometry and misalignment of prior work. Quantitatively we observe improvements on both perceptual similarity (LPIPS from 0.335 to 0.309) and sharpness ($S_3$ from 0.135 to 0.120).

**Acknowledgements.** Supported in part by NSF grants 1718221, 2008387, 2045586, 2106825, MRI #1725729, and NIFA award 2020-67021-32799.

# References

1. Augmented Reality - Apple Developer. https://developer.apple.com/augmented-reality/ (2021). Accessed 14 Nov 2021
2. Abadi, M., et al: TensorFlow: a system for large-scale machine learning. In: OSDI (2016)
3. Abdelhafiz, A., Mostafa, Y.G.: Automatic texture mapping mega-projects. J. Spatial Sci. (2020)
4. Aganj, E., Monasse, P., Keriven, R.: Multi-view texturing of imprecise mesh. In: Zha, H., Taniguchi, R., Maybank, S. (eds.) ACCV 2009. LNCS, vol. 5995, pp. 468–476. Springer, Heidelberg (2010). https://doi.org/10.1007/978-3-642-12304-7_44
5. Aliev, K.-A., Sevastopolsky, A., Kolos, M., Ulyanov, D., Lempitsky, V.: Neural point-based graphics. In: Vedaldi, A., Bischof, H., Brox, T., Frahm, J.-M. (eds.) ECCV 2020. LNCS, vol. 12367, pp. 696–712. Springer, Cham (2020). https://doi.org/10.1007/978-3-030-58542-6_42
6. Anuta, P.E.: Spatial registration of multispectral and multitemporal digital imagery using fast Fourier Transform techniques. IEEE Trans. Geosci. Electron. **8**(4), 353–368 (1970)
7. Barnes, C., Shechtman, E., Finkelstein, A., Goldman, D.B.: PatchMatch: a randomized correspondence algorithm for structural image editing. In: SIGGRAPH (2009)
8. Baumberg, A.: Blending images for texturing 3D models. In: BMVC (2002)
9. Bernardini, F., Martin, I., Rushmeier, H.: High quality texture reconstruction from multiple scans. TVCG **7**(4), 318–332 (2001)
10. Bi, S., Kalantari, N.K., Ramamoorthi, R.: Patch-based optimization for image-based texture mapping. TOG **36**(4), 106-1 (2017)
11. Dai, A., Chang, A.X., Savva, M., Halber, M., Funkhouser, T.A., Nießner, M.: ScanNet: richly-annotated 3D reconstructions of indoor scenes. In: CVPR (2017)
12. Debevec, P., Taylor, C., Malik, J.: Modeling and rendering architecture from photographs: a hybrid geometry and image-based approach. In: SIGGRAPH (1996)
13. Duan, Y.: Topology adaptive deformable models for visual computing. Ph.D. thesis, State University of New York (2003)
14. El-Hakim, S., Gonzo, L., Picard, M., Girardi, S., Simoni, A.: Visualization of Frescoed surfaces: Buonconsiglio Castle - Aquila Tower, "Cycle of the Months". IAPRS (2003)
15. Früh, C., Sammon, R., Zakhor, A.: Automated texture mapping of 3D city models with oblique aerial imagery. In: 3DPVT (2004)
16. Fu, Y., Yan, Q., Yang, L., Liao, J., Xiao, C.: Texture mapping for 3D reconstruction with RGB-D sensor. In: CVPR (2018)
17. Globerson, A., Jaakkola, T.: Fixing max-product: convergent message passing algorithms for MAP LP-relaxations. In: NIPS (2007)
18. Goel, S., Kanazawa, A., Malik, J.: Shape and viewpoint without keypoints. In: Vedaldi, A., Bischof, H., Brox, T., Frahm, J.-M. (eds.) ECCV 2020. LNCS, vol. 12360, pp. 88–104. Springer, Cham (2020). https://doi.org/10.1007/978-3-030-58555-6_6

19. Grammatikopoulos, L., Kalisperakis, I., Karras, G., Petsa, E.: Automatic multi-view texture mapping of 3D surface projections. In: International Workshop 3D-ARCH (2007)
20. Groueix, T., Fisher, M., Kim, V.G., Russell, B.C., Aubry, M.: AtlasNet: a Papier-Mâché approach to learning 3D surface generation. In: CVPR (2018)
21. Henzler, P., Mitra, N.J., Ritschel, T.: Learning a neural 3D texture space from 2D exemplars. In: CVPR (2020)
22. Hernández-Esteban, C.: Stereo and Silhouette fusion for 3D object modeling from uncalibrated images under circular motion. Ph.D. thesis, École Nationale Supérieure des Télécommunications (2004)
23. Huang, J., Dai, A., Guibas, L., Nießner, M.: 3DLite: towards commodity 3D scanning for content creation. ACM TOG 36, 1–14 (2017)
24. Huang, J., et al.: Adversarial texture optimization from RGB-D scans. In: CVPR (2020)
25. Huynh, D.: Metrics for 3D rotations: comparison and analysis. J. Math. Imaging Vision 35, 155–164 (2009)
26. Kingma, D.P., Ba, J.: Adam: a method for stochastic optimization. ArXiv (2015)
27. Lempitsky, V., Ivanov, D.: Seamless Mosaicing of image-based texture maps. In: CVPR (2007)
28. Lensch, H., Heidrich, W., Seidel, H.P.: Automated texture registration and stitching for real world models. In: Graphical Models (2001)
29. Marshner, S.R.: Inverse rendering for computer graphics. Ph.D. thesis, Cornell University (1998)
30. Neugebauer, P.J., Klein, K.: Texturing 3D models of real world objects from multiple unregistered photographic views. In: Eurographics (1999)
31. Niem, W., Wingbermühle, J.: Automatic reconstruction of 3D objects using a mobile camera. In: IVC (1999)
32. Oechsle, M., Mescheder, L.M., Niemeyer, M., Strauss, T., Geiger, A.: Texture fields: learning texture representations in function space. In: ICCV (2019)
33. Ofek, E., Shilat, E., Rappoport, A., Werman, M.: Multiresolution textures from image sequences. Comput. Graph. Appl. 17(2), 18–29 (1997)
34. Pan, R., Taubin, G.: Color adjustment in image-based texture maps. Graph. Models 79, 39–48 (2015)
35. Paszke, A., et al.: PyTorch: an imperative style, high-performance deep learning library. ArXiv abs/1912.01703 (2019)
36. Pighin, F., Hecker, J., Lischinski, D., Szeliski, R., Salesin, D.H.: Synthesizing realistic facial expressions from photographs. In: CGIT (1998)
37. Ravi, N., et al.: Accelerating 3D deep learning with PyTorch3D. arXiv:2007.08501 (2020)
38. Rocchini, C., Cignoni, P., Montani, C., Scopigno, R.: Multiple textures stitching and blending on 3D objects. In: Eurographics Workshop on Rendering (1999)
39. Rocchini, C., Cignoni, P., Montani, C., Scopigno, R.: Acquiring, stitching and blending diffuse appearance attributes on 3D models. Visual Comput. 18, 186–204 (2002)
40. Saito, S., Wei, L., Hu, L., Nagano, K., Li, H.: Photorealistic facial texture inference using deep neural networks. In: CVPR (2017)
41. Sawhney, R., Crane, K.: Boundary first flattening. ACM TOG 37, 1–14 (2018)
42. Shu, J., Liu, Y., Li, J., Xu, Z., Du, S.: Rich and seamless texture mapping to 3D mesh models. In: Tan, T., et al. (eds.) IGTA 2016. CCIS, vol. 634, pp. 69–76. Springer, Singapore (2016). https://doi.org/10.1007/978-981-10-2260-9_9

43. Sinha, S.N., Steedly, D., Szeliski, R., Agrawala, M., Pollefeys, M.: Interactive 3D architectural modeling from unordered photo collections. In: SIGGRAPH 2008 (2008)

44. Sitzmann, V., Thies, J., Heide, F., Nießner, M., Wetzstein, G., Zollhöfer, M.: Deep-Voxels: learning persistent 3D feature embeddings. In: CVPR (2019)

45. Thierry, M., David, F., Gorria, P., Salvi, J.: Automatic texture mapping on real 3D model. In: CVPR (2007)

46. Thies, J., Zollhöfer, M., Nießner, M.: Deferred neural rendering. ACM TOG **38**, 1–12 (2019)

47. Vu, C.T., Phan, T.D., Chandler, D.M.: $S_3$: a spectral and spatial measure of local perceived sharpness in natural images. IEEE Trans. Image Process. **21**(3), 934–945 (2012)

48. Waechter, M., Moehrle, N., Goesele, M.: Let there be color! Large-scale texturing of 3D reconstructions. In: Fleet, D., Pajdla, T., Schiele, B., Tuytelaars, T. (eds.) ECCV 2014. LNCS, vol. 8693, pp. 836–850. Springer, Cham (2014). https://doi.org/10.1007/978-3-319-10602-1_54

49. Wang, Z., Bovik, A., Sheikh, H.R., Simoncelli, E.P.: Image quality assessment: from error visibility to structural similarity. IEEE Trans. Image Process. **13**(4), 600–612 (2004)

50. Wexler, Y., Shechtman, E., Irani, M.: Space-time video completion. In: CVPR (2004)

51. Wuhrer, S., Atanassov, R., Shu, C.: Fully automatic texture mapping for image-based modeling. Technical report, Institute for Information Technology (2006)

52. Xiang, F., Xu, Z., Havsan, M., Hold-Geoffroy, Y., Sunkavalli, K., Su, H.: NeuTex: neural texture mapping for volumetric neural rendering. In: CVPR (2021)

53. Zhang, R., Isola, P., Efros, A.A., Shechtman, E., Wang, O.: The unreasonable effectiveness of deep features as a perceptual metric. In: CVPR (2018)

54. Zhou, Q.Y., Koltun, V.: Color map optimization for 3D reconstruction with consumer depth cameras. ACM TOG **33**(4), 1–10 (2014)

# SIGNet: Intrinsic Image Decomposition by a Semantic and Invariant Gradient Driven Network for Indoor Scenes

Partha Das[1,3]($\boxtimes$) ⓘ, Sezer Karaoğlu[1,3], Arjan Gijsenij[2], and Theo Gevers[1,3]

[1] CV Lab, University of Amsterdam, Amsterdam, The Netherlands
{p.das,th.gevers}@uva.nl
[2] AkzoNobel, Amsterdam, The Netherlands
arjan.gijsenij@akzonobel.com
[3] 3DUniversum, Amsterdam, The Netherlands
s.karaoglu@3duniversum.com

**Abstract.** Intrinsic image decomposition (IID) is an under-constrained problem. Therefore, traditional approaches use hand crafted priors to constrain the problem. However, these constraints are limited when coping with complex scenes. Deep learning-based approaches learn these constraints implicitly through the data, but they often suffer from dataset biases (due to not being able to include all possible imaging conditions).

In this paper, a combination of the two is proposed. Component specific priors like semantics and invariant features are exploited to obtain semantically and physically plausible reflectance transitions. These transitions are used to steer a progressive CNN with implicit homogeneity constraints to decompose reflectance and shading maps.

An ablation study is conducted showing that the use of the proposed priors and progressive CNN increase the IID performance. State of the art performance on both our proposed dataset and the standard real-world IIW dataset shows the effectiveness of the proposed method. Code is made available here.

**Keywords:** Priors · Semantic segmentation · Intrinsic image decomposition · CNN · Indoor dataset

## 1 Introduction

An image can be defined as the combination of an object's colour and the incident light on it projected on a plane. Inverting the process of image formation is useful for many downstream computer vision tasks such as geometry estimation [18], relighting [34], colour edits [5] and Augmented Reality (AR) insertion and interactions for applications like the Metaverse. The process of recovering the object colour (reflectance or albedo) and the incident light (shading) is known

---

**Supplementary Information** The online version contains supplementary material available at https://doi.org/10.1007/978-3-031-25066-8_35.

as Intrinsic Image Decomposition (IID). As the problem is ill-defined (with only one known), constraint-based approaches are explored to limit the solution space. For example, as an explicit gradient assumption, softer (or smoother) gradient transitions are attributed to shading transitions, while stronger (or abrupt) ones are related to reflectance transitions [21]. Colour palette constraints in the form of sparsity priors and piece-wise consistency are also employed for reflectance estimation [1,15]. However, these approaches are based on strong assumptions of the imaging process and hence are limited in their applicability.

Implicit constraints, by means of deep learning-based methods, are proposed to expand previous approaches [26]. For these methods, the losses implicitly formulate the constraints and are dependent on the training data. These methods learn a flexible representation based on training data which may lead to dataset biases. [23] integrates multiple datasets to manage the dataset bias problem. However, introducing more datasets only acts as an expansion of the imaging distribution. Additionally, multiple purpose-built losses are needed to train the network. An alternative approach of combining constraints and deep learning is explored in [12] where edges are used as an additional constraint to guide the network. However, edges at image locations with strong illumination effects, like pronounced cast shadows, may lead to edge misclassification resulting in undesirable effects like shading-reflectance leakages.

On the other hand, [2] forgoes priors and specialised losses to leverage joint learning of related modalities. They explore semantic segmentation as a closely related task to IID, arguing that jointly learning the semantic maps provides the network information to jointly correct for reflectance-shading transitions. However, no explicit guidance or constraint between the semantics and reflectance

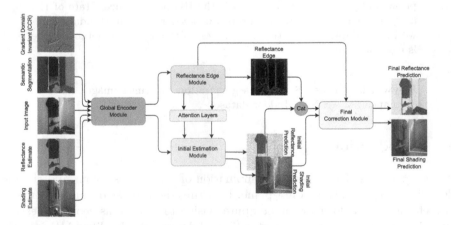

**Fig. 1.** The proposed network overviews. The network consists of i) the global encoder module, ii) the reflectance edge module, iii) the initial estimation module, and iv) the final correction module. The final reflectance and shading outputs are used for all the evaluations. Please refer to the supplementary for more details. Images shown here are ground truth images, for illustrative purposes.

are imposed. The network thus relies on learning the constraints from the ground truth semantic, reflectance and shading jointly. Moreover, only outdoor gardens are considered, where most natural classes (e.g., bushes, trees, and roses) contain similar colours (i.e., constrained colour distributions).

This paper exploits physical and statistical image properties for IID of indoor scenes. Illumination and geometry invariant descriptors [16] yield physics-based cues to detect reflectance transitions, while statistical grouping of pixels in an image provides initial starting estimates for IID components. To this end, a combination of semantic and invariant transition constraints is proposed. Semantic transitions provide valuable information about reflectance transitions i.e., a change in semantics most likely matches a reflectance transition but not always the other way around (objects may consist of different colours). Illumination invariant gradients provide useful information about reflectance transitions but can be unstable (noisy) due to low intensity. Exploiting reflectance transition information on these two levels compensates each other and ensures a stronger guidance for IID. In addition, indoor structures, like walls and ceilings, are often homogeneously coloured. To this end, the semantic map can be used as an explicit homogeneous prior. This allows for integrating an explicit sparsity/piece-wise consistency (homogeneity) prior in the form of constant reflectance colour.

In this paper, a progressive CNN is employed, consisting of two stages. The first stage of the network exploits the prior information to arrive at an initial estimation. This estimation is based on the semantics, the invariant guided boundaries, and sparsity constraints. The second stage of the network takes the initial estimation and fine-tunes it using the original image cues to disentangle the reflectance and shading maps while being semantically correct. This allows the network to separate the problem into two distinct solution spaces that build progressively on each other. In addition, it allows the network to learn a continuous representation that can extrapolate even when the priors contain errors. An overview of the proposed network is shown in the Fig. 1.

While deep learning networks have shown very good performance, they require high quality datasets. Traditional physical-based rendering methods are often time and resource intensive. Recently, these methods are more efficient i.e., real time on consumer hardware. Hence, a dataset of physical-based and photo-realistic rendered indoor images is provided. The synthetic dataset is used to train the proposed method.

In summary, our contributions are as follows:

- **Algorithm**: An end-to-end semantic and physically invariant edge transition driven hybrid network is proposed for intrinsic image decomposition of indoor scenes.
- **Insight**: The use of component specific priors outperforms learning from a single image.
- **Performance**: The proposed algorithm is able to achieve state-of-the-art performance on both synthetic and real-world datasets.
- **Dataset**: A new ray-traced and photo-realistic indoor dataset is provided.

## 2  Related Works

A considerable amount of effort has been put in exploring hand-crafted prior constraints for the problem of IID. [21] pioneered the field by assuming reflectance changes to be related to sharper gradient changes, while smoother gradients correspond to shading changes. Other priors have been explored like piece-wise constancy for the reflectance, and smoothness priors for shading [1], textures [15]. Constraints in the form of additional inputs have also been explored. [22] explores the use of depth as an additional input, while [19] explores surface normals. Near infrared priors are used by [10] to decompose non-local intrinsics. Humans in the loop is also studied by [8] and [27]. However, these works mostly focus on single objects and do not generalise well to complete scenes.

In contrast to the use of explicit (hand-crafted) constraints, deep learning methods that implicitly learn (data-driven) specific constraints are also explored [26]. [3] explores disentangling the shading component into direct and indirect shading. [39] differentiates shading into illumination and surface normals in addition to reflectance. [6] uses a piece-wise constancy property of reflectances and employs Conditional Random Fields to perform IID. [12] shows that image edges contain information about reflectance edges and uses them as a guidance for the IID problem. [23] reduces the solution space by using multiple task specific losses. [31] directly learns the inverse of the rendering function. Finally, [2] forgoes losses and jointly learns semantic segmentation to implicitly learn a posterior on the IID, while [30] uses estimated semantic features as a support for an iterative competing formulation for IID. However, the above approaches do not explicitly integrate the physics-based image formation information and rely on the datasets containing a large set of imaging conditions. Hence, they may fall short for images containing extreme imaging conditions such as strong shadows or reflectance transitions. Large datasets [23,25,29] are proposed to train networks. Unfortunately, they are limited in their photo-realistic appearance.

Unlike IID, physics-based image formation priors have been explored in other tasks. [13] introduces Colour Ratios which are illumination invariant descriptors for objects. [16] then introduces Cross Colour Ratios which are both geometric and illumination invariant reflectance descriptors. [4] shows the applicability of the descriptors to the problem of IID. In contrast to previous methods, in this paper, a combination of explicit image formation-based priors and implicit intrinsic component property losses are explored.

## 3  Methodology

### 3.1  Priors

***Semantic Segmentation:*** [2] shows that semantic segmentation provides useful information for the IID problem. However, components are jointly learned and hence their method lacks any explicit influence of the component's property. Since object boundaries correspond to reflectance changes such boundary information can serve as a useful global reflectance transition guidance for the network. Furthermore, homogeneous colour (i.e., reflectance) constraints (e.g., a wall has a uniform colour) can be imposed on the segmentation explicitly. To this end, in this paper, an off-the-self segmentation algorithm Mask2Former [9] is used to obtain segmentation maps.

*Invariant Gradient Domain:* Solely using semantic regions as priors may cause the network to be biased to the regions generated by the segmentation method. To prevent such a bias, an invariant (edge) map is included as an additional prior to the network. In this work, Cross Colour Ratios (CCR) [16] are employed. These are illumination invariants i.e., reflectance descriptors. Given an image $I$ with channels Red $(R)$, Green $(G)$ and Blue $(B)$ and neighbouring pixels $p_1$ and $p_2$, CCR is defined by $M_{RG} = \frac{R_{p_1} G_{p_2}}{R_{p_2} G_{p_1}}$ , $M_{RB} = \frac{R_{p_1} B_{p_2}}{R_{p_2} B_{p_1}}$ and $M_{GB} = \frac{G_{p_1} B_{p_2}}{G_{p_2} B_{p_1}}$ where, $R_{p_1}$, $G_{p_1}$ and $B_{p_1}$ are the red, green, and blue channel for pixel $p_1$. Descriptors $M_{RG}$, $M_{RB}$ and $M_{GB}$ are illumination free and therefore solely depending on reflectance transitions. Using the reflectance gradient as an additional prior allows the network to be steered by reflectance transitions.

*Reflectance and Shading Estimates:* Consider the simplified Lambertian [32] image formation model: $I = R \times S$, where shading $(S)$ is the scaling term on the reflectance component $(R)$. Hence, for a given constant reflectance region, all the pixels are different shades of the same colour. In this way, the reflectance colour becomes a scale optimisation for which the pixel mean of a segment can be used: $\mathcal{M}_c = \sum_{n=1}^{N} I_n^c$ where, $\mathcal{M}_c$ is the channel-specific mean of the pixels. $\mathcal{M}_R$, $\mathcal{M}_G$ and $\mathcal{M}_B$ values are then spread within the region to obtain an initial starting point for reflectance colour based on the homogeneity constraint. Conversely, these values can be inverted using the image formation to obtain the corresponding scaled shading estimates. A CNN is then employed to implicitly learn the scaling for both priors. Additionally, since the mean of the segment does not consider textures, a deep learning method is proposed to compensate it by means of a dedicated correction module, see Sect. 3.2. The supplementary material provides more visuals for these priors.

## 3.2   Network Architecture

The network consists of 4 components: i) Global encoder blocks, ii) Reflectance edge Decoder, iii) Initial estimation decoder and iv) Final correction module. The network is trained end-to-end. The input to the network is an image and its corresponding segmentation obtained by Mask2Former [9]. The CCR, Reflectance and Shading estimates are computed from the input image for the respective encoder blocks. Additional details and visuals for the modules can be found in the supplementary materials.

*Global Encoder Module:* The input image, the segmentation image, the average reflectance estimate, inverse shading estimate and the CCR images are encoded through their respective encoders. The encoders share the same configuration, but the intermediate features are independent of each other. The semantic features $(\mathcal{F}_S)$ provide guidance for the general outlines of object boundaries, while the CCR features $(\mathcal{F}_G)$ focus on local reflectance transitions, possibly

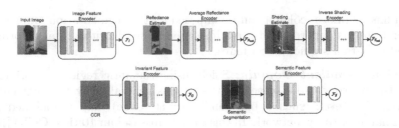

**Fig. 2.** Overview of the global encoder module. Each of the inputs are provided with their independent encoders to enable modality specific feature learning. The respective features are used in the downstream decoders to provide component specific information for the network.

including textures. Correspondingly, the average reflectance estimate features ($\mathcal{F}_{R_{est}}$) and the inverse shading estimate features ($\mathcal{F}_{S_{est}}$) provide a starting point for the reflectance and the shading estimation, respectively. Finally, the image features ($\mathcal{F}_I$) provide the network a common conditioning to learn the scaling and boundary transitions for the intrinsic components. Figure 2 shows the overview of the module.

*Reflectance Edge Module:* This sub-network decodes the reflectance edges of the given input. The decoded reflectance and edges are used as an attention mechanism to the initial estimation module to provide (global) region consistency. The features, $\mathcal{F}_S$ and $\mathcal{F}_G$ are concatenated with the image features $\mathcal{F}_I$ and passed on to the edge decoder. The semantic and CCR features provide object and reflectance transitions, respectively. The image features allow the network to disentangle reflectance from illumination edges. Corresponding skip connections from $\mathcal{F}_I$, $\mathcal{F}_{R_{est}}$ and $\mathcal{F}_G$ encoders are used to generate high frequency details. Scale space supervision, following [36], is provided by a common deconvolution layer for the last 2 layers, for scales of $64 \times 64$ and $128 \times 128$, yielding a scale consistent reflectance edge prediction. The ground truth edges are calculated by using a Canny Edge operation on the ground truth reflectance. Figure 3 shows an overview of the module.

*Initial Estimation Module:* The initial estimation decoder block focuses on learning the IID from the respective initial estimates of the intrinsic (Fig. 3). It consists of two parallel decoders. The Reflectance decoder learns to predict the first estimation from $\mathcal{F}_I$ and $\mathcal{F}_{R_{est}}$. The features are further augmented with the learned boundaries from the reflectance edge decoder passed through an attention layer [35]. $\mathcal{F}_S$ is also passed to the decoder to guide global object transitions and acts as an additional attention. Similarly, the Shading decoder only receives $\mathcal{F}_I$ and $\mathcal{F}_{S_{est}}$, focusing on properties like smoother (shading) gradient changes. The reflectance and shading decoders are interconnected to provide an additional cue to learn an inverse of each other. Skip connections from the respective encoders to the decoders are also given. This allows the network to learn an implicit scaling on top of the average reflectance and the inverse shading estimation. The output at this stage is guided by transition and reflectance boundaries and may suffer from local inconsistencies like shading-reflectance leakages.

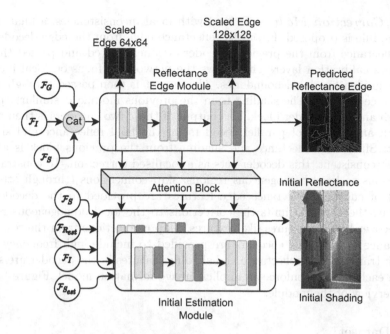

**Fig. 3.** Overview of the reflectance edge and the attention guided initial estimation module. The edge module takes the image encoder, semantic encoder, and the invariant encoder feature to learn a semantically and physically guided reflectance transition. The edge features are then transferred through an attention block to the initial estimation decoder module. The reflectance decoder in this module takes the semantic encoder, image encoder and the average reflectance estimation features and input. The shading decoder correspondingly takes the image encoder along with the average shading estimation feature. Interconnections in the decoder allows the network to use reflectance cues for shading and vice versa.

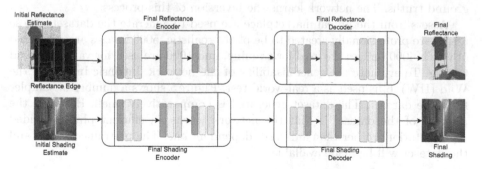

**Fig. 4.** The final decoder module. The initial reflectance and shading estimates from the previous step are further corrected to obtain the final reflectance and shading. The encoder consists of an independent parallel reflectance and shading encoder. The reflectance encoder takes receives the initial reflectance and the reflectance edge as an input, while the shading encoder receives the initial shading. Two parallel decoders are used for reflectance and shading to obtain the final IID outputs.

*Final Correction Module:* To deal with local inconsistencies, a final correction module is proposed. First, the reflectance edge from the edge decoder and the reflectance from the previous decoder is concatenated and passed through a feature calibration layer. This allows the network to focus on local inconsistencies guided by global boundaries. The output is then passed through a final reflectance encoder. The shading from the previous module is similarly passed through another encoder block. The output of these two encoders is then passed through another set of parallel decoders for the final reflectance and shading output. Since the reflectance and shading from the previous block is already globally consistent, this decoder acts as a localised correction. To constrain the corrections to local homogeneous regions, skip connections (through attention layers) of encoded reflectance edge features are provided to the decoders. In this way, the network limits the corrections to the local homogeneous regions and recover local structures like textures. Skip connections from the respective reflectance and shading encoders are provided to include high frequency information transfer. The reflectance and shading features in the decoder are shared within each other to enforce an implicit image formation model. Figure 4 shows the overview of the module.

### 3.3   Dataset

Unreal Engine [11] is used to generate a dataset suited for the task. The rendering engine supports physically based rendering, with real-time raytracing (RTX) support. The engine first calculates the intrinsic components from the various material and geometry property of the objects making up the scene. Then, the illumination is physically simulated through ray tracing and lighting is calculated. Finally, all these results are combined to render the final image. Since the engine calculates the intrinsic components, ground truth intrinsic is recovered using the respective buffer. The dataset consists of dense reflectance and shading ground-truths. The network learns the inversion of this process.

Assets from the unreal marketplace are used to generate the dataset. These assets are professionally created to be photo realistic. 5000 images are generated of which 4000 images are used for training, and 1000 are used for validation and testing. To evaluate the generalisability of the network, Intrinsic Images in the Wild (IIW) [6] is used as a real-world test. Figure 5 shows a number of samples from the dataset. The dataset generated is comparatively small. However, the purpose of the dataset is that the network learns an efficient physics guided representation, rather than a dataset dependent one. The pretrained model and the dataset will be made available.

### 3.4   Loss Functions and Training Details

MSE loss is applied for each output of the network: (i) Initial estimation loss ($\mathcal{L}_e$ & $\mathcal{L}_i$) and (ii) Final correction loss ($\mathcal{L}_f$). $\mathcal{L}_e$ is the loss applied on the scale space reflectance edge. $\mathcal{L}_i$ is the loss on the reflectance and shading output from the initial estimation module. Additional losses are also applied on the reflectance and

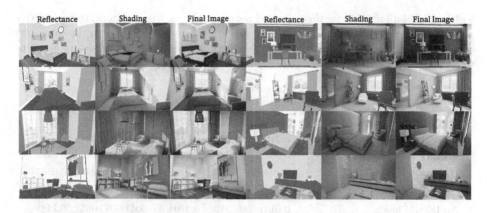

**Fig. 5.** Samples from the proposed dataset. The dataset comes with the corresponding dense reflectance and shading maps. The dataset consists of various everyday objects and lighting, containing both near local light sources, like lamps, and more global light sources like sunlight and windows.

shading output from the final correction module. This reflectance and shading are also combined and compared with the input image for a reconstruction loss. These 3 losses are collected in the term $\mathcal{L}_f$. An invariance loss $\mathcal{L}_{Norm}$ is added between the normalised $RGB$ and the prediction of the network for each segment. A Total Variation (TV) loss ($\mathcal{L}_{TV}$) is included to deal with the assumption that large indoor classes like walls and ceilings are homogeneously coloured. This loss is only applied to ceilings and wall pixels and minimises the TV between the prediction and the ground truth reflectance. Finally, to encourage perceptually consistent and sharper textures, a perceptual and dssim loss are included and grouped as $\mathcal{L}_\delta$. The final loss term to minimise for the network thus becomes:

$$
\begin{aligned}
\mathcal{L} = \lambda_e\, \mathcal{L}_e + \lambda_i\, \mathcal{L}_i + \mathcal{L}_f \\
+ \mathcal{L}_{Norm} + \mathcal{L}_{TV} + \mathcal{L}_\delta
\end{aligned}
\tag{1}
$$

where $\lambda_e$ and $\lambda_i$ are weighting terms for the edge and initial estimation losses. They are empirically set to 0.4 and 0.5, respectively. The network is trained for 60 epochs, with a learning rate of $2e-4$ and the Adam [20] optimiser. Please refer to the supplementary materials for more details.

## 4  Experiments

### 4.1  Ablation Study

To study the influence of different architecture components and losses, an ablation study is conducted. For a fair evaluation, the ablation study is performed on the test-set of the rendered dataset. For all the ablations, all hyper-parameters are kept constant. The results of the ablation study are presented in Table 1.

**Table 1.** Ablation study for the proposed network. For each experiment, the respective parts of the network are modified. All the experiments are conducted on the same test and train split of the proposed dataset. All the applicable hyper-parameters are kept constant.

| | Reflectance | | | Shading | | |
|---|---|---|---|---|---|---|
| | MSE | LMSE | DSSIM | MSE | LMSE | DSSIM |
| w/o final correction | 0.0029 | 0.0020 | 0.0225 | 0.0044 | 0.0035 | 0.0276 |
| w/o priors | 0.0105 | 0.0047 | 0.0444 | 0.0054 | 0.0034 | 0.0399 |
| w canny edges | 0.0032 | 0.0037 | 0.0229 | 0.0031 | 0.0049 | 0.0293 |
| w/o average estimates | 0.0030 | 0.0023 | 0.0232 | 0.0041 | 0.0043 | 0.0267 |
| w/o reflectance edge module | 0.0097 | 0.0156 | 0.3254 | 0.0033 | 0.0061 | 0.0270 |
| No DSSIM loss | 0.0131 | 0.0240 | 0.3704 | 0.0041 | 0.0055 | 0.1488 |
| No perceptual loss | 0.0032 | 0.0022 | 0.0289 | 0.0032 | 0.0038 | 0.0285 |
| No invariant & homogeneity loss | 0.0032 | 0.0027 | 0.0288 | **0.0024** | **0.0024** | 0.0318 |
| Proposed | **0.0026** | **0.0018** | **0.0219** | 0.0030 | 0.0033 | **0.0252** |

***Influence of Final Correction Module:*** In this experiment, the influence of the final correction module is studied. The output from the initial estimation decoder is taken as the final output.

From the results, it is shown that the final correction module helps in improving the outputs. The improvement in the DSSIM metric for both components shows that the final correction module is able to deal with structural artefacts.

***Influence of Priors:*** The influence of all the priors is studied in this ablation. The additional priors are removed, and the network only receives the image as an input. All network structures are kept the same. This setup studies if the network can disentangle the additional information from the input image without any specific priors.

Removing all priors makes the network to perform worse for all metrics. In this setting, the network only uses the image to derive both the reflectance and shading changes. This is challenging for strong illumination effects. This shows that the priors are an important source of information enabling a better disentanglement between intrinsic components.

***Influence of Specialised Edges:*** This experiment studies the need of specialised edges obtained from the semantic transition boundaries and invariant features. The edges obtained from the input image are provided to the network. The study focuses on whether the network can distinguish between reflectance, geometry, and shadow edges directly from the image.

From the results, it is shown that using image edges is not sufficient. Image edges can be ambiguous due to the presence of shadow edges. However, the performance is still better than using the image as the only input, showing that edges yield, to a certain extent, useful transition information.

*Influence of Reflectance and Shading Estimate Priors:* In this experiment, the efficacy of the statistic-based homogeneous reflectance and the inverted shading estimate is studied.

Removing the average reflectance and shading estimates degrades the performance. With the priors of the estimates, the network can use its learning capacity to deal with the scaling of the initial estimation to obtain the correct IID. The network needs to learn the colour as well as the scaling within the same learning capacity.

*Influence of Reflectance Edge Guidance Module:* For this experiment, the edge guidance module is removed. As such, the network is then forced to learn the attention and the reflectance transition boundaries implicitly as part of the solution space.

Removing the reflectance edge module results in the second worse result. This shows that, apart from the priors, the ability to use those features to learn a reflectance transition, is useful. It is shown that without such a transition guidance, the network is susceptible of misclassifying shadow edges as reflectance transitions. Furthermore, it is shown that without this module, the reflectance performance suffers more than the shading performance. Hence, using a learned edge guidance allows the network to be more flexible and better able to distinguish between true reflectance transitions.

*Influence of Different Losses:* The influence of the different losses is studied in this experiment. For each sub-experiment, the same proposed structure is used, and the respective losses are selectively turned off.

From the results, it is shown that the DSSIM loss contributes to a large extend, to the performance, because this loss penalises perceptual variations like contrast, luminance, and structure. As such, by removing the supervision, the network learns an absolute difference which is not expressive to smaller spatial changes. Similar trend of performance decrease is shown when removing the perceptual and homogeneity losses. This is expected since both losses contribute to region consistency. With the addition of the losses on the reflectance, the shading values suffer slightly. However, structurally they perform better when including the losses, as shown by the DSSIM metric. This indicates that applying such a loss helps not only to achieve a better reflectance value, but it also jointly improves shading, resulting in sharper outputs.

## 4.2  Comparison to State of the Art

*On the Proposed Dataset*: To study the influence of the dataset, the proposed network is compared to baseline algorithm's performance. For these experiments, the standard, MSE, LMSE and the DSSIM metric are used. The baselines are chosen based on their performance of the Weighted Human Disagreement Rate (WHDR), widely used in the literature. Hence, [23] is chosen as a baseline. [39] does not provide any publicly available code, hence is not included. Although [12] is the state of the art, their provided code generates errors when trying to run on custom datasets and hence is not used for comparison. For completeness, [37] and [33] is also compared. [37] uses an optimization-based method based on the pioneering Retinex model. Since it is a purely physical constraint-based model, it is included for comparison. For a fair comparison, methods focusing on indoors are used. [2] assumes outdoor settings and requires semantic ground truths to train and hence is not included. For all the networks, they are retrained on the dataset that is proposed in this paper, using the optimum hyperparamters as mentioned in the respective publication. The results are shown in Table 2 and Fig. 6

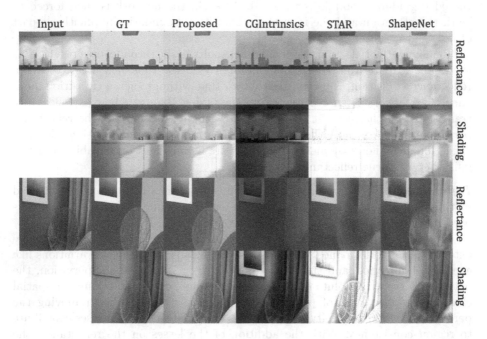

**Fig. 6.** Comparison of the proposed to baseline methods. It is shown that the proposed method is able to better disentangle the illumination effect. In comparison, CGIntrinsics, which has comparable results on the WHDR SoTA, suffers from discolouration. STAR misses the illumination while ShapeNet suffers from artefacts.

From the table it is shown that our proposed model is able to provide the highest scores. From the figure, the baselines suffer from strong illumination effects. CGIntrinsics discolours the regions while STAR mostly fails. ShapeNet, suffers from artefacts and colour variations around the illumination regions. In comparison, the proposed network is able to recover from such effects.

*On IIW* [6]: The proposed network is finetuned on the IIW dataset and compared to the baselines. The training and testing splits are used as specified in the original publication. For the baselines, the numbers are obtained from the respective original publications. The results are shown in Table 3 and visuals in Fig. 7.

**Table 2.** Comparison to the baseline methods on the proposed dataset. It is shown that the proposed method outperforms all other methods.

|                   | Reflectance | | | Shading | | |
|-------------------|--------|--------|--------|--------|--------|--------|
|                   | MSE    | LMSE   | DSSIM  | MSE    | LMSE   | DSSIM  |
| ShapeNet [33]     | 0.0084 | 0.0133 | 0.1052 | 0.0065 | 0.0129 | 0.1862 |
| STAR [37]         | 0.0304 | 0.0166 | 0.1180 | 0.0290 | 0.0128 | 0.1572 |
| CGIntrinsics [23] | 0.0211 | 0.0156 | 0.0976 | 0.0848 | 0.0577 | 0.2180 |
| Proposed          | **0.0026** | **0.0018** | **0.0219** | **0.0030** | **0.0033** | **0.0252** |

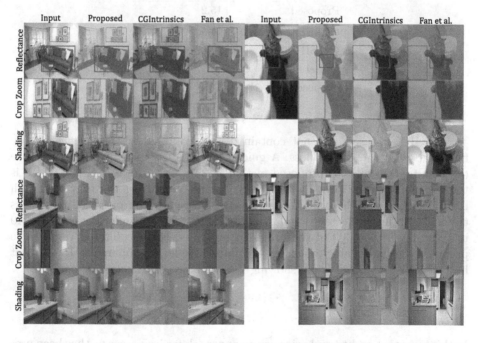

**Fig. 7.** Visual results on the IIW test set. Compared to CGIntrinsics [23] and Fan *et al.* [12], the proposed method disentangles better the shading and highlights (highlighted in red boxes), showing a smoother reflectance. Both CGIntrinsics and [12] are unable to remove the highlights from the reflectance, resulting in discolouration. They are also susceptible to reflectance colour change as be seen on the cat and furniture (highlighted green boxes). The proposed method is able to better retain the original colour in the reflectance. (Color figure online)

**Table 3.** Baseline comparison for the IIW dataset. Results marked with * are post-processed with a guided filter [28]

| Methods | WHDR (mean) |
|---|---|
| Direct intrinsics [26] | 37.3 |
| Color retinex [17] | 26.9 |
| Garces et al. [14] | 25.5 |
| Zhao et al. [38] | 23.2 |
| IIW [6] | 20.6 |
| Nestmeyer et al. [28] | 19.5 |
| Bi et al. [7] | 17.7 |
| Sengupta et al. [31] | 16.7 |
| Li et al. [24] | 15.9 |
| CGIntrinsics [23] | 15.5 |
| GLoSH [39] | 15.2 |
| Fan et al. [12] | 15.4 |
| Proposed | 15.2 |
| CGIntrinsics [23]* | 14.8 |
| GLoSH [39]* | 14.6 |
| Fan et al. [12]* | 14.5 |
| Proposed* | **13.9** |

The IIW dataset does not contain dense ground truth and hence is only finetuned with the ordinal loss. A guided filter [28] is used to further improve the results. Overall, our proposed method is on par with GLoSH [39] which is the best performing method without any post filtering. However, they need both lighting and normal information as supervision, while the proposed method is trained with just reflectance and shading, along with a smaller dataset (58, 949 images of [39] vs. 5000 of the proposed method). For the filtered results, the proposed method is able to achieve a comfortable lead compared to the current best of 14.5 obtained by [12], showing the efficiency of the current model.

## 5    Conclusions

In this paper, an end-to-end prior driven approach for indoor scenes has been proposed for the task of intrinsic image decomposition. Reflectance transitions and invariant illuminant descriptors has been used to guide the reflectance decomposition. Image statistics-based priors have been used to provide the network a starting point for learning. To integrate explicit homogeneous constraints, a progressive CNN was used. To train the network, a custom physically rendered dataset was proposed.

An extensive ablation was performed to validate the proposed network showing that: i) using explicit reflectance transition priors helps the network to
achieve an improved intrinsic image decomposition, ii) image statistics-based
priors are helpful for simplifying the problem and, iii) the proposed method
attains sota performance for the standardised real-world dataset IIW.

# References

1. Barron, J.T., Malik, J.: Shape, illumination, and reflectance from shading. IEEE
   TPAMI **37**, 1670–1687 (2015)
2. Baslamisli, A.S., Groenestege, T.T., Das, P., Le, H.A., Karaoglu, S., Gevers, T.:
   Joint learning of intrinsic images and semantic segmentation. In: ECCV (2018)
3. Baslamisli, A.S., Das, P., Le, H., Karaoglu, S., Gevers, T.: Shadingnet: image
   intrinsics by fine-grained shading decomposition. IJCV **129**, 2445–2473 (2021)
4. Baslamisli, A.S., Liu, Y., Karaoglu, S., Gevers, T.: Physics-based shading reconstruction for intrinsic image decomposition. Comput. Vis. Image Understanding,
   1–14 (2020)
5. Beigpour, S., van de Weijer, J.: Object recoloring based on intrinsic image estimation. In: ICCV, pp. 327–334 (2011)
6. Bell, S., Bala, K., Snavely, N.: Intrinsic images in the wild. ACM TOG **33**, 1–12
   (2014)
7. Bi, S., Han, X., Yu, Y.: An l1 image transform for edge-preserving smoothing and
   scene-level intrinsic decomposition. ACM TOG **34**(4) (2015). https://doi.org/10.
   1145/2766946
8. Bonneel, N., Sunkavalli, K., Tompkin, J., Sun, D., Paris, S., Pfister, H.: Interactive
   intrinsic video editing. ACM TOG **33**, 197:1–197:10 (2014)
9. Cheng, B., Misra, I., Schwing, A.G., Kirillov, A., Girdhar, R.: Masked-attention
   mask transformer for universal image segmentation. arXiv (2021)
10. Cheng, Z., Zheng, Y., You, S., Sato, I.: Non-local intrinsic decomposition with
    near-infrared priors. In: ICCV (2019)
11. Epic Games: Unreal engine. https://www.unrealengine.com
12. Fan, Q., Yang, J., Hua, G., Chen, B., Wipf, D.: Revisiting deep intrinsic image
    decompositions. In: CVPR (2018)
13. Finlayson, G.D.: Colour Object Recognition. Master's thesis, Simon Fraser Uni-
    versity (1992)
14. Garces, E., Munoz, A., Lopez-Moreno, J., Gutierrez, D.: Intrinsic images by cluster-
    ing. Comput. Graph. Forum **31**(4) (2012). https://www-sop.inria.fr/reves/Basilic/
    2012/GMLG12
15. Gehler, P.V., Rother, C., Kiefel, M., Zhang, L., Schölkopf, B.: Recovering intrinsic
    images with a global sparsity prior on reflectance. In: NeurIPS (2011)
16. Gevers, T., Smeulders, A.: Color-based object recognition. PR **32**, 453–464 (1999)
17. Grosse, R., Johnson, M.K., Adelson, E.H., Freeman, W.T.: Ground truth dataset
    and baseline evaluations for intrinsic image algorithms. In: ICCV (2009)
18. Henderson, P., Ferrari, V.: Learning single-image 3D reconstruction by generative
    modelling of shape, pose and shading. Int. J. Comput. Vis. **128**, 835–854 (2019)
19. Jeon, J., Cho, S., Tong, X., Lee, S.: Intrinsic image decomposition using structure-
    texture separation and surface normals. In: Fleet, D., Pajdla, T., Schiele, B., Tuyte-
    laars, T. (eds.) ECCV 2014. LNCS, vol. 8695, pp. 218–233. Springer, Cham (2014).
    https://doi.org/10.1007/978-3-319-10584-0_15

20. Kingma, D.P., Ba, J.: Adam: A method for stochastic optimization (2014). https://arxiv.org/abs/1412.6980, arxiv:1412.6980Comment: Published as a conference paper at the 3rd International Conference for Learning Representations, San Diego, 2015

21. Land, E.H., McCann, J.J.: Lightness and retinex theory. J. Opt. Soc. Am. **61**, 1–11 (1971)

22. Lee, K.J., et al.: Estimation of intrinsic image sequences from image+depth video. In: Fitzgibbon, A., Lazebnik, S., Perona, P., Sato, Y., Schmid, C. (eds.) ECCV 2012. LNCS, vol. 7577, pp. 327–340. Springer, Heidelberg (2012). https://doi.org/10.1007/978-3-642-33783-3_24

23. Li, Z., Snavely, N.: Cgintrinsics: better intrinsic image decomposition through physically-based rendering. In: ECCV (2018)

24. Li, Z., Shafiei, M., Ramamoorthi, R., Sunkavalli, K., Chandraker, M.: Inverse rendering for complex indoor scenes: shape, spatially-varying lighting and svbrdf from a single image. In: CVPR, pp. 2472–2481 (2020)

25. Li, Z., et al.: Openrooms: an end-to-end open framework for photorealistic indoor scene datasets. CoRR abs/2007.12868 (2020). https://arxiv.org/abs/2007.12868

26. Narihira, T., Maire, M., Yu, S.X.: Direct intrinsics: learning albedo-shading decomposition by convolutional regression. In: ICCV (2015)

27. Narihira, T., Maire, M., Yu, S.X.: Learning lightness from human judgement on relative reflectance. In: CVPR, pp. 2965–2973 (2015). https://doi.org/10.1109/CVPR.2015.7298915

28. Nestmeyer, T., Gehler, P.V.: Reflectance adaptive filtering improves intrinsic image estimation. CoRR abs/1612.05062 (2016). https://arxiv.org/abs/1612.05062

29. Roberts, M., et al.: Hypersim: a photorealistic synthetic dataset for holistic indoor scene understanding. In: International Conference on Computer Vision (ICCV) 2021 (2021)

30. Saini, S., Narayanan, P.J.: Semantic hierarchical priors for intrinsic image decomposition. CoRR abs/1902.03830 (2019). https://arxiv.org/abs/1902.03830

31. Sengupta, S., Gu, J., Kim, K., Liu, G., Jacobs, D.W., Kautz, J.: Neural inverse rendering of an indoor scene from a single image. CoRR abs/1901.02453 (2019). https://arxiv.org/abs/1901.02453

32. Shafer, S.: Using color to separate reflection components. Color Res. App. **10**, 210–218 (1985)

33. Shi, J., Dong, Y., Su, H., Yu, S.X.: Learning non-lambertian object intrinsics across shapenet categories. In: CVPR (2017)

34. Shu, Z., Yumer, E., Hadap, S., Sunkavalli, K., Shechtman, E., Samaras, D.: Neural face editing with intrinsic image disentangling. CoRR abs/1704.04131 (2017). https://arxiv.org/abs/1704.04131

35. Tang, H., Qi, X., Xu, D., Torr, P.H.S., Sebe, N.: Edge guided gans with semantic preserving for semantic image synthesis. CoRR (2020)

36. Xie, S., Tu, Z.: Holistically-nested edge detection. In: ICCV (2015)

37. Xu, J., et al.: Star: a structure and texture aware retinex model. IEEE TIP **29**, 5022–5037 (2020)

38. Zhao, Q., Tan, P., Dai, Q., Shen, L., Wu, E., Lin, S.: A closed-form solution to retinex with nonlocal texture constraints. IEEE TPAMI **34**(7), 1437–1444 (2012). https://doi.org/10.1109/TPAMI.2012.77

39. Zhou, H., Yu, X., Jacobs, D.W.: Glosh: global-local spherical harmonics for intrinsic image decomposition. In: ICCV (2019)

# Implicit Map Augmentation
# for Relocalization

Yuxin Hou[1,2], Tianwei Shen[3(✉)], Tsun-Yi Yang[3], Daniel DeTone[3],
Hyo Jin Kim[3], Chris Sweeney[3], and Richard Newcombe[3]

[1] Aalto University, Espoo, Finland
[2] Niantic, San Francisco, USA
[3] Reality Labs, Meta, Redmond, USA
tianweishen@fb.com

**Abstract.** Learning neural radiance fields (NeRF) has recently revo-
lutionized novel view synthesis and related topics. The fact that the
implicit scene models learned via NeRF greatly extend the representational capability compared with sparse maps, however, is largely overlooked. In this paper, we propose implicit map augmentation (IMA) that
utilizes implicit scene representations to augment the sparse maps and
help with visual relocalization. Given a sparse map reconstructed by
structure-from-motion (SfM) or SLAM, the method first trains a NeRF
model conditioned on the sparse reconstruction. Then an augmented
sparse map representation can be sampled from the NeRF model to render better relocalization performance. Unlike the existing implicit mapping and pose estimation methods based on NeRF, IMA takes a hybrid
approach by bridging the sparse map representation with MLP-based
implicit representation in a non-intrusive way. The experiments demonstrate that our approach achieves better relocalization results with the
augmented maps on challenging views. We also show that the resulting
augmented maps not only remove the noisy 3D points but also bring
back missing details that get discarded during the sparse reconstruction,
which helps visual relocalization in wide-baseline scenarios.

**Keywords:** Scene representation · Visual relocalization · View
synthesis

## 1 Introduction

Visual relocalization addresses the challenge of estimating the position and orientation of the given query images by analyzing correspondence between query
images and the map. It is an essential task for many applications like autonomous

---

Y. Hou—Work done during internship at Reality Labs, Meta.

---

**Supplementary Information** The online version contains supplementary material
available at https://doi.org/10.1007/978-3-031-25066-8_36.

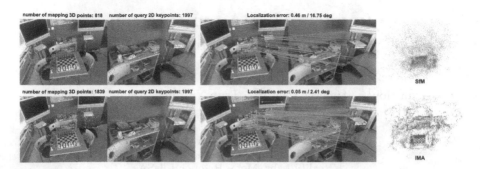

**Fig. 1.** Localization with Implicit Map Augmentation (IMA). Compared to the sparse point cloud from SfM, IMA re-samples more accurate 3D points in weak regions via neural implicit representations. For wide-baseline query images, the augmented map leads to more 2D–3D matches and smaller localization errors.

driving [11,62] and augmented reality [18,32]. Most existing methods represent scenes as sparse 3D points clouds obtained from Structure-from-Motion (SfM) reconstruction [15,42]. 6-Degree-of-Freedom (6-DoF) camera poses can be estimated by matching 2D keypoints in query images and 3D points in the point cloud. The 2D-3D matches can be established via 2D-2D keypoints matching [13,39] or direct 2D-3D matching [42–44]. Researchers have advanced the pipeline by applying better matching strategy [40,43] or better descriptors [4,5,37]. However, it is important to note the impact of the quality of the scene representations on localization performance. When the mapping images only cover a part of the scene, the 3D point cloud may be too sparse for wide-baseline query images to be registered. Figure 1 shows examples of sparse point clouds that build upon images of a partial scene. For rarely observed weak regions, the 3D points are sparse and noisy. As a consequence, for query images that capture the weak regions, the registration will fail or have poor localization performance.

On the other hand, implicit scene representations such as neural radiance fields (NeRF) [29] have attracted the great interest of the scene reconstruction community recently. NeRF and its follow-ups shows impressive novel view synthesis results by using implicit continuous functions to encode volumetric density and colors. However, it is relatively overlooked that NeRF introduces a scene representation that has greater expressiveness compared with sparse maps, in the sense that an arbitrary viewpoint can be rendered from these implicit scene models. The implicit scene models have been under-utilized in traditional geometric tasks such as visual localization.

There exists some attempts that use NeRF to estimate camera poses. iNeRF [60] computes the pixel-level appearance differences between the synthesis result and the query image, and uses gradient back-propagation repeatedly until the synthesis result and the query image are aligned. iMap [49] builds a real-time SLAM system using multi-layer perceptron (MLP) as the underlying scene representation. Most of these methods address the pose estimation problem via iterative optimization, and ignore the traditional explicit 2D-3D matching process, making it hard to integrate these methods into the traditional localization pipelines. In addition, some limitations hinder the application of NeRF to real-world scenes. Estimating both geometry and appearance can be ill-posed when there are not enough input views. Inaccurate geometries can still produce accurate rendering results, which is known as shape-radiance ambiguity [57,61].

In contrast to previous NeRF-based methods, this paper incorporates implicit representations for localization in a non-intrusive way. Rather than using the implicit representation to estimate pose directly, we focus on the scenes representation itself such that it improves the localization performance by improving the map quality, yet can be readily plugged into a traditional visual localization pipeline. To improve the quality of geometry learned by NeRF, we propose to use the given original sparse point clouds to produce per-view monocular depth priors as additional pseudo depth supervision to regularize the density prediction, inspired by [3,38,57].

In summary, we propose a method, termed as *implicit map augmentation (IMA)*, that learns the implicit scene representations conditioned on the given sparse point clouds, and utilizes the implicit representations to produce a better 3D map for localization. Our contributions are listed as follows:

- We show that pseudo-depth supervised NeRF that can learn accurate and consistent depth from multi-view images.
- We propose a novel pipeline that utilizes learned implicit representation to augment sparse point clouds from SfM. We reconcile the NeRF-based models with sparse maps, enabling interoperability between traditional and learning-based map representations for the first time.
- It is demonstrated that after the implicit map augmentation, the localization results get improved significantly without changing the localization pipeline.

## 2   Related Work

*Neural Implicit Representations.* Recent works demonstrate the potential effectiveness of representing a scene with neural networks [27–29,33,35,58]. NeRF [29] represents the radiance field and density distribution of a scene as a continuous implicit function of 3D location and orientations and uses volume rendering to synthesize novel view images. Though NeRF shows impressive high quality synthesis results, since it purely relies on photometric consistency, it suffer from shape-radiance ambiguity and fail in challenging regions like textureless regions. UNISURF [34] proposes a unified framework for implicit surfaces and radiance fields that enable more efficient sampling procedures and solve the ambiguities during early iterations. Since the sparse 3D points can be a cheap source of supervision, to overcome the inherent ambiguity of the correspondence, some methods incorporate sparse depth observations into NeRF reconstruction [3,38,57]. DS-NeRF [3] use the sparse point cloud from COLMAP [46] as an extra supervision to regularize the learned geometry directly. Since the sparse supervision is not enough, NerfingMVS [57] use depths from MVS to finetune a monocular depth network, and the resulting dense depth predictions are used to compute adaptive sample ranges along rays. [38] convert the sparse depth maps into dense depth maps via a depth completion network. Some methods also solve the pose estimation during the NeRF optimization [22,49]. The iNeRF [60] apply analysis-by-synthesis that estimate pose by iterative alignment. In contrast, our method uses NeRF to augment the sparse point clouds from SfM to improve relocalization results without modifying the traditional localization pipeline.

*Multi-view Reconstruction.* Recovering 3D geometry from multiple images is a fundamental problem in computer vision [10]. The key of 3D reconstruction is to solve the correspondence matching. Classical multi-view stereo (MVS) methods use hand-crafted features and metric to match features across input views and produce depth maps [1,8,9]. These methods need post-processing steps like depth fusion [26]. Some volumetric methods divide the 3D space into discretized grids and perform global optimization based on photo-consistency [20,47,55], but these methods suffer from high computational consumption and discretization artifacts. Though classical methods show good results under most ideal scenes, they are not robust to challenging regions like texture-less regions and non-lambertian suarfaces. Instead of using hand-crafted features and metrics, recently many learning-based methods apply learned feature maps [12,14,25,31,50,56,59]. Many methods build a cost volume to encode geometry information [14,56,59] and use deep neural networks for cost volume regularization to predict per-view depth maps. Some methods predict TSDF representation directly via feature unprojection [31,50]. Our work also predict per-view depth maps, but different from learning-based MVS methods, we do not build the cost volume explicitly but use NeRF [29] model to optimize the implicit volumes.

*Structure-Based Relocalization.* Structure-based localization is the task of estimating the 6-DoF poses of query images w.r.t a pre-computed 3D model. Some methods perform direct matching of local descriptors (e.g. SIFT [24], D2-Net [5], R2D2 [37]) between 2D keypoints of query images and 3D points from SfM [23,41,42,44,51]. Some methods merge image retrieval into the pipeline to deal with larger-scale models, performing coarse image retrieval followed by the fine geometric matching [13,15,39]. Kapture establishes 2D-2D correspondences between query images and retrieved similar database images. Since many 2D keypoints of database images correspond to 3D points of the map, there are 2D-3D matches established. HF-Net [39] clusters similar database images based on the co-visibility and successively matches the 2D keypoints in query images to the 3D points in the cluster. When the 2D-3D matches are found, the camera pose can be estimated using a Perspective-n-Point (PnP) solver [19] within a RANSAC loop [7]. Our work demonstrates the augmented map with the Kapture pipeline [13]. Compared to the original 3D point cloud from SfM, our augmented 3D map will assign more 3D points to the 2D keypoints of database images, which lead to more potential 2D-3D matches during localization.

## 3   Method

Figure 2 shows the overview of our proposed implicit map augmentation pipeline. Our method aims to obtain an augmented 3D map to facilitate relocalization

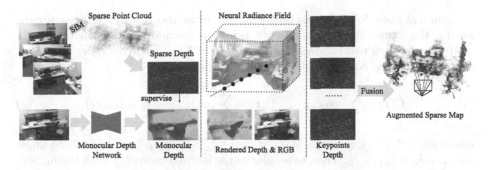

Sparse Point Cloud    Neural Radiance Field

Sparse Depth

supervise

Fusion

Augmented Sparse Map

Monocular Depth    Monocular    Rendered Depth & RGB    Keypoints
Network    Depth    Depth

**Fig. 2.** Overview of the implicit map augmentation pipeline. Given a set of posed RGB images and a sparse point cloud from the SfM method, we project the 3D points to get sparse depth maps, which are used to finetune the monocular depth network and get dense depth predictions. Then we use the dense monocular depth estimation as pseudo ground-truth to regularize the neural radiance field. After NeRF optimization, we obtain the augmented sparse map by fusing the rendered depth maps guided by dense 2D features, which can be plugged into a structure-based localization pipeline.

based on a given collection of posed RGB images and original sparse point clouds from SfM methods (e.g. COLMAP [46]). We obtain depth maps from the NeRF [29] model, then re-sample and fuse the per-view depth maps into a novel sparse point cloud. To get more accurate depth estimations from NeRF models, inspired by NerfingMVS [57], we use a monocular depth network to generate dense depth prior. After obtaining the augmented 3D map, we apply the 3D map into the traditional coarse-to-fine relocalization pipeline directly.

### 3.1 Monocular Depth Pseudo Supervision

To make use of the given sparse point cloud, similar to NerfingMVS [57], our method utilizes a monocular depth network to predict per-view dense depth maps prior. However, different from [57] that use sparse depth maps from MVS results, we use the sparse 3D points from SfM directly to supervise the monocular depth network, so we do not need the expensive MVS preprocessing. By projecting the 3D points in the point cloud, we can get sparse depth maps. To fine-tune the monocular depth network, we employ the scale-invariant loss [6] as the monocular estimations have scale ambiguity.

$$\mathcal{L}_{prior} = \frac{1}{n} \sum_{j=1}^{n} |\log D^i_{prior}(j) - \log D^i_{sparse}(j) + \alpha(D^i_{prior}, D^i_{sparse})|, \quad (1)$$

where $\alpha(D^i_{prior}, D^i_{sparse}) = \frac{1}{n} \sum_j (\log D^i_{prior}(j) - \log D^i_{sparse}(j))$ is the value that minimizes the error for the given predicted depth map $D^i_{prior}$ and the sparse depth map from SfM $D^i_{sparse}$. After obtaining the per-view monocular depth estimations, we align the predicted monocular depth maps with the projected sparse depth maps via the median value.

Different from [57] that uses the monocular estimations to guide the sampling, we use the scale-aligned dense depth priors as pseudo supervision of density prediction for each ray. Following the original NeRF model, given a ray $\mathbf{r} = \mathbf{o} + t\mathbf{d}$, we can render the color and depth using the formula:

$$\hat{\mathbf{C}}(\mathbf{r}) = \sum_{k=1}^{K} w_k \mathbf{c}_k, \qquad \hat{D}(\mathbf{r}) = \sum_{k=1}^{K} w_k t_k, \qquad t_k \in [t_n, t_f] \qquad (2)$$

where $w_k = T_k(1 - \exp(-\sigma_k \delta_k)), T_k = \exp(-\sum_{k'=1}^{k} \sigma_{k'} \delta_{k'})$ and the inter-sample distance $\delta_k = t_{k+1} - t_k$. Then to regularize the learned geometry with the monocular depth prior, we have extra loss function:

$$\mathcal{L}_{depth} = ||D_{prior}(\mathbf{r}) - \hat{D}(\mathbf{r})||^2 \qquad (3)$$

Though the monocular prior predictions are not accurate enough and may suffer from multi-view inconsistency, after NeRF optimization, our method can produce more accurate consistent depth maps. To train the NeRF model, we use the combined loss function:

$$\mathcal{L} = \mathcal{L}_{color} + \lambda \mathcal{L}_{depth}, \qquad (4)$$

where $\lambda$ is the hyperparameter that controls the weight for depth loss.

## 3.2    Active Sampling

Rendering and optimizing all pixels of mapping images can be expensive, especially when the scale of database images is large. Considering the aim of our method is to augment the given sparse map for better localization rather than obtain the accurate dense 3D reconstruction, all pixels are not equally important (e.g. texture-less regions like the white wall in indoor environments are less important for localization). In that case, we can pay more attention to those more important regions with active sampling.

For structure-based localization, to compute the 6-DoF camera poses for query images, we detect 2D keypoints from query images and match them with 3D points in the map to establish 2D-3D matches, so we need to obtain accurate 3D points for detectable keypoints in the augmented map. During training, we assign a probability map for each mapping image based on D2-Net[5] detected keypoints. For each detected 2D keypoint with coordinate $(u, v)$, we assign higher weight for pixels around it:

$$w(i, j) = \begin{cases} 2, \text{if } i \in [u - s, u + s], j \in [v - s, v + s] \\ 1, \text{otherwise} \end{cases} \qquad (5)$$

where $s$ is the hyperparameter that controls the size of window around the keypoints.

For each batch of images, we get the probability distribution inside the batch by normalizing the assigned weights:

$$p(i) = \frac{w(i)}{\sum_j^n w(j)}, \qquad (6)$$

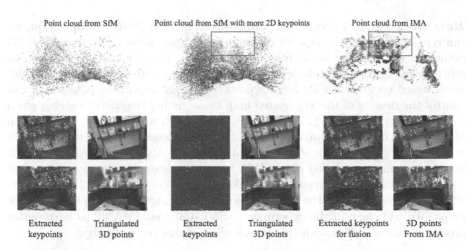

| Point cloud from SfM | | Point cloud from SfM with more 2D keypoints | | Point cloud from IMA | |
|---|---|---|---|---|---|
| Extracted keypoints | Triangulated 3D points | Extracted keypoints | Triangulated 3D points | Extracted keypoints for fusion | 3D points From IMA |

**Fig. 3.** For rarely observed regions (e.g. see red boxes in figures), the 2D features are often unstable across different views. Therefore, these keypoints would not have corresponding 3d points in the SfM maps, even if we increase the number of 2D keypoints. The left two examples show that the dense regions get denser after increasing the number of keypoints for SfM, while the sparse regions remain sparse. In contrast, the point cloud from IMA is more balanced. (Color figure online)

where $n$ is the total number of pixels in the batch. We use this distribution to sample a sparser set of random pixels for NeRF optimization, so those detectable keypoints will have more chance to be sampled and more likely to have better render quality.

### 3.3 Mapping and Relocalization

*Why Map Augmentation?* As most of the visual (re-)localization algorithms depend on a sparse map representation, the quality of the base map plays a decisive role in the success of relocalization queries. Mapping algorithms [32, 46, 52] first detect 2D keypoints in database images, and then compute the 3D points based on pairwise matches via incremental triangulation and bundle adjustment.

There are several factors that would harm the representational ability of sparse maps because of the map building process. For rarely observed regions, the 2D features are often unstable across different views. Therefore, these keypoints would not have corresponding 3d points in the SfM maps, leading to the 3D points in these weak regions being sparse and noisy. In this case, even if we increase the number of 2D keypoints, the 3D points in those weak regions remain sparse. However, these regions are important for wide-baseline query images that have limited overlap with the mapping images. Figure 3 shows an example that when we increase the number of 2D keypoints, point clouds in those dense regions get denser, while those weak regions are still sparse.

*How Does IMA Address the Limitations?* After NeRF optimization, we can render dense depth maps for all mapping images, so we can obtain more 3D points for the detected keypoints in weak regions. To get a unified sparse scene representation from multiple depth maps, we first generate the per-view keypoint mask based on detected D2-Net keypoints to select important points. We can control the density of the augmented map by adjusting the size of patches when generating keypoint-based masked depth maps. Then we fuse these sparse depth maps into a new 3D point cloud with the image-based depth fusion methods in COLMAP [45], and the visibility of 3D points is also recorded during depth fusion. In summary, the output of IMA is still a sparse map representation but with better sampling of the scene to make it capable of localizing difficult views, while being agnostic to the camera relocalization algorithms based on sparse maps. We will show in the experiments (Sect. 4.3) that IMA maps have better representational ability with the same level of sparse point density.

*Localization.* We adopt kapture [13] to demonstrate the improvement brought by the better IMA map representation. In essence, kapture is a coarse-to-fine algorithm based on COLMAP [45]. It first queries the database images to find similar ones that could possibly be matched with the query images. The pairwise matches are computed against the top-$k$ database images and elevated to 2D-3D matches for PnP camera estimation. To adapt to the kapture pipeline, we utilize the recorded visibility to project 3D points in the fused 3D point cloud back to the database images to get new 2D keypoints and extract the descriptor features from dense D2-Net feature maps via bilinear interpolation.

## 4   Experiments

### 4.1   Datasets

We experiment on two indoor datasets, 7Scenes [48] and ScanNet [2]. Though these datasets provide RGB-D data, we do not use the depth data for training and inference, only RGB images and given camera poses for mapping and localization evaluation.

*7Scenes.* The 7Scenes dataset is a collection of tracked Kinect RGB-D camera frames, which is commonly used for evaluation of relocalization methods. We follow the original train/test splits as mapping/query images. For scene *Kitchen*, since there are 7 sequences with 7,000 images for mapping, to train the NeRF more efficiently, we only use 500 images per sequence (frame IDs are even number) for training and mapping.

*7Scenes-sub.* The standard 7Scenes have very dense set-up for most scenes, so there are not many rarely observed regions and coverage of the point cloud from SfM can be already good enough. To further demonstrate the performance for more difficult situations (e.g. wide-baseline relocalization), apart from the original train/test splits for each scene, we create more challenging sequences for evaluation. Instead of using all training images to build the 3D map, we pick a single sequence from each scene and only use the first 500 images as mapping

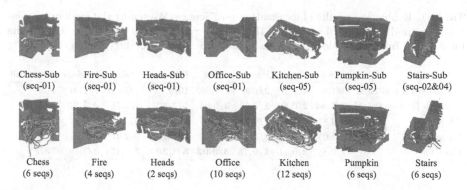

| Chess-Sub | Fire-Sub | Heads-Sub | Office-Sub | Kitchen-Sub | Pumpkin-Sub | Stairs-Sub |
| --- | --- | --- | --- | --- | --- | --- |
| (seq-01) | (seq-01) | (seq-01) | (seq-01) | (seq-05) | (seq-05) | (seq-02&04) |

| Chess | Fire | Heads | Office | Kitchen | Pumpkin | Stairs |
| --- | --- | --- | --- | --- | --- | --- |
| (6 seqs) | (4 seqs) | (2 seqs) | (10 seqs) | (12 seqs) | (6 seqs) | (6 seqs) |

**Fig. 4.** The 7Scenes and 7Scenes-sub sequences. The red and green trajectories show the positions of the cameras in the mapping and query sequences. Compared to the full 7Scenes dataset, these subsets of sequences using much fewer images to build the 3D map and the query images have less overlap with mapping images. (Color figure online)

images, and the remain 500 images as query images. For scene *stairs*, since each sequence only have 500 images, we take *seq-02* as mapping images and *seq-04* as query images. We denote these challenging sets as 7scenes-sub. Compared to the original 7scenes splits that use thousands densely-sampled images for mapping, as 7scenes-sub only build 3D map with observed images from partial scene, the query images generally have less overlap with database images and make the localization task harder. Figure 4 shows examples of mapping/query sequences in 7scenes-sub, and the comparison with the full 7Scene sequences.

***ScanNet.*** For ScanNet dataset, we randomly pick five scene sequences, and for each scene we pick 200 to 300 images for mapping and 200 images for query.

## 4.2 Implementation Details

***Monocular Depth Estimation.*** For each scene, we first get the original sparse point cloud from COLMAP [46]. Then to train the implicit scene representation conditioned on the sparse point cloud, similar to NerfingMVS [57] and CVD [25], we use the pretrained weights and the Mannequin Challenge depth network architecture in [21] as our monocular depth network. We finetune the pretrained weights with the sparse point cloud for 500 iterations.

***NeRF Optimization.*** The NeRF architecture is the same as [29]. Instead of using hierarchical sampling strategy, we only use the coarse sampling so there is only one network. For all datasets we sampled 128 points per ray. The weight for depth loss is set to $\lambda = 0.1$ for Eq. 4. For 7Scenes, all images keep their original resolution as $480 \times 640$, while for ScanNet all images are resized as $484 \times 648$ during NeRF training. The window size in Eq. 5 is set to $s = 1$ to assign weights for active sampling. We use Adam optimizer [17] to train each scene for 200,000 iterations, and each iteration takes 1024 rays. The learning rate starts with $5 \times 10^{-4}$ and decays exponentially to $5 \times 10^{-5}$.

**Table 1.** Evaluation results of localization on 7Scenes. We report the median translation (cm) and rotation (°) errors and the average recall (%) at (5 cm, 5°). The results for AS [44], InLoc [53], hloc [39] are excerpted from corresponding papers.

| | Chess | Fire | Heads | Office | Pumpkin | Stairs | Kitchen | Avg recall |
|---|---|---|---|---|---|---|---|---|
| D2-Net+SfM+APGeM | 3/1.09 | 4/1.51 | 2/1.98 | 5/1.58 | 10/2.78 | 13/3.21 | 7/1.96 | 47.9 |
| D2-Net+IMA+APGeM | **2/0.84** | **2/0.89** | 1/0.90 | 3/0.94 | **5/1.30** | 8/1.71 | 4/1.29 | 70.7 |
| AS [44] | | 3/0.87 | **2/1.01** | 1/0.82 | 4/1.15 | 7/1.69 | **4/1.01** | 5/1.72 | 68.7 |
| InLoc [53] | | 3/1.05 | 3/1.07 | 2/1.16 | 3/1.05 | 5/1.55 | 9/2.47 | 4/1.31 | 66.3 |
| hloc [39] | | 2/0.85 | 2/0.94 | **1/0.75** | **3/0.92** | **5/1.30** | 5/1.47 | 4/1.40 | **73.1** |

*Mapping.* We use the COLMAP library to fuse the masked depth maps to get the augmented 3D point cloud. To get the masked depth maps, we manually adjust the window size within the range $[1, 1.5]$ to get a similar level of number of 3D points as the baseline SfM point clouds. After we get the augmented 3D map, we then obtain 2D keypoints for each mapping image by projecting the visible 3D points back to the mapping images, and extracting the corresponding descriptors from the dense D2-Net feature maps via bilinear interpolation.

*Localization.* To compute the 6-DoF camera poses for query images, we use kapture [13] benchmark pipeline. The image retrieval uses APGeM [36] as global feature to obtain top 5 database images. The 6-DoF camera poses are estimated by solving the PnP [19] problem within the RANSAC [7] loop, which is implemented by COLMAP.

### 4.3   Evaluation

*Baselines.* Since we extract descriptors for new 2D keypoints in database images from dense D2-Net feature maps, we mainly compare with the baseline that also use D2-Net as local descriptors with kapture (D2-Net+SfM+APGeM). The maximum threshold for number of detected keypoints is 10,000. As we use the same global features as the baseline, the retrieved similar database images are the same, and the performance depends on the 2D-3D matching results only. For standard 7Scenes benchmark, we also compare the results with other 3 feature-matching based methods. Active Search (AS) [44] performs direct 2D-to-3D matching with SIFT [24] and perform actively 3D-to-2D search. Both InLoc [53] and hloc [39] a perform image retrieval first. The InLoc uses dense learned descriptors and dense 3D models, while hloc use SuperPoint features and matches with SuperGlue [40].

*Results.* Experiment results show that using the augmented 3D map from our IMA model leads to consistent better localization results than using the sparse 3D point cloud from SfM for all datasets.

The evaluation results for standard 7Scenes sequences are presented in Table 1. We report the median translation (cm) and rotation (°) errors, as well as the average localization recall at (5 cm, 5°). Compared to these state-of-the-art methods that use better descriptors or better matching strategy, the kapture

**Table 2.** Evaluation results of localization on 7Scenes-sub. We report the median translation (cm) and rotation (°) errors and the average recall at (20 cm, 5°).

| | Chess | Fire | Heads | Office | Pumpkin | Stairs | Kitchen | Recall |
|---|---|---|---|---|---|---|---|---|
| D2-Net+SfM+APGeM | 46/19.16 | **3**/1.26 | 4/2.91 | **10**/2.38 | 37/9.63 | 12/3.07 | 25/5.68 | 54.0 |
| D2-Net+mono+APGeM | 11/**3.60** | 4/1.19 | **2**/2.02 | **10**/2.47 | 36/10.22 | 11/2.71 | 16/3.48 | 62.2 |
| D2-Net+IMA+APGeM | **8**/**3.60** | **3**/**0.83** | **2**/**1.51** | **10**/**2.01** | **20**/**4.99** | **8**/**1.99** | **15**/**2.50** | **69.1** |

**Table 3.** Evaluation results of localization on Scannet. We report the median translation (cm) and rotation (°) errors and the average recall at (10 cm, 5°).

| | Scene0079 | Scene0207 | Scene0653 | Scene0707 | Scene0737 | Recall |
|---|---|---|---|---|---|---|
| D2-Net+SfM+APGeM | 4/1.10 | 19/5.31 | 10/2.33 | 8/2.17 | 6/3.21 | 55.2 |
| D2-Net+mono+APGeM | 4/1.24 | **10**/2.91 | 9/1.84 | 9/1.38 | **3**/1.31 | 61.8 |
| D2-Net+IMA+APGeM | **2**/**1.05** | **10**/**2.56** | **4**/**1.36** | **5**/**1.18** | **3**/1.37 | **76.5** |

baseline with D2-Net features perform slightly worse. However, after replacing SfM map with our augmented 3D map, we get on par performance with the SOTA methods. For scenes like *Fire*, *Kitchen* and *Chess*, our methods achieved the best localization results **without increasing the density of 3D maps dramatically**. For scene *Chess* the number of 3D points in our IMA map is 381,116, which is even smaller than the number for SfM map (439,159 points), while using the IMA map still have lower localization errors and higher recall.

For the 7scenes-sub set, since only part of the scene is observed, the quality of the built sparse 3D map will be worse than using the full database images. Also, for NeRF training, less input images can lead to ambiguities easier. Moreover, the query images will have limited overlap with the built 3D map, so accurate localization is more difficult. We report the median translation and rotation errors in Table 2. Though the overall performance of the baseline is worse than the full setting, localizing with the augmented 3D maps still improves the performance consistently for all scenes, which confirms that our method can be used to alleviate wide-baseline localization. We also compute the average recall at (20 cm, 5°), and the augmented map significantly increases the recall from 54.0% to 69.1%. The Fig. 6 shows the qualitative comparison results of the 3D maps on both full 7Scenes and 7Scenes-sub. Compared to the 3D point cloud from SfM with D2-Net, our augmented map is more accurate and much less noisy.

The similar results are observed with the ScanNet sequences. Table 3 shows that the average recall at (10 cm, 5°) increased from 55.2% to 76.5% by replacing the SfM 3D maps with our augmented maps. Qualitative examples are shown in Fig. 5.

### 4.4  Ablation Study

To further understand the components in our pipeline, we performed ablation studies. To demonstrate the effectiveness of NeRF optimization, we experimented

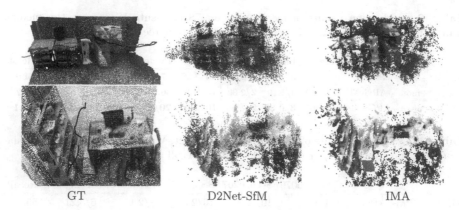

GT                    D2Net-SfM                    IMA

**Fig. 5.** Qualitative examples of 3D maps on ScanNet. The left columns shows the GT point cloud and the camera sequences. The red and green trajectories show the positions of cameras in mapping sequences and query sequences respectively. (Color figure online)

with fusing the monocular estimations directly. To compare the benefit of the pseudo depth supervision, we experiments with two variants of NeRF methods, vanilla NeRF and DS-NeRF [3], which corresponds to no depth supervision and sparse depth supervision respectively.

***NeRF Optimization.*** After aligning the dense depth maps from the monocular depth network, we can also produce a unified 3D map via depth fusion. We report the localization performance with the 3D map generated by monocular depth maps (D2-Net+mono+APGeM) in Table 2, Table 3 and Table 4. Since the inconsistency across views, when we use the same window size to extracting keypoints from monocular depth estimation, we get much less 3D points than IMA, so we manually adjust the window size per scene within range [2, 2.5] to make the fused point cloud have similar number of points as IMA. According to the recall results, using the fused point cloud from monocular depth estimation can also improve the localization performance, though the median errors get worse for several scenes. Moreover, results shows both NeRF-base 3D maps out perform the monocular depths based 3D maps, and localizing with our implicit augmented map has the best results. Different from the monocular depth estimation that did not consider the multi-view geometry consistency, NeRF optimization integrates multi-view constraints well, which lead to higher quality of the 3D point cloud.

***Pseudo Depth Supervision.*** To further study the effectiveness of the depth pseudo supervision during the NeRF optimization, we compared our methods with two other baselines: vanilla NeRF have no depth supervision, while the other being DS-NeRF that only utilizes the sparse point cloud as supervision. For all methods, we use the same window size when extracting depth maps sampled via dense 2D features. Table 4 shows the comparison results on scene *Heads-sub*, which reports the median translation and rotation errors and the

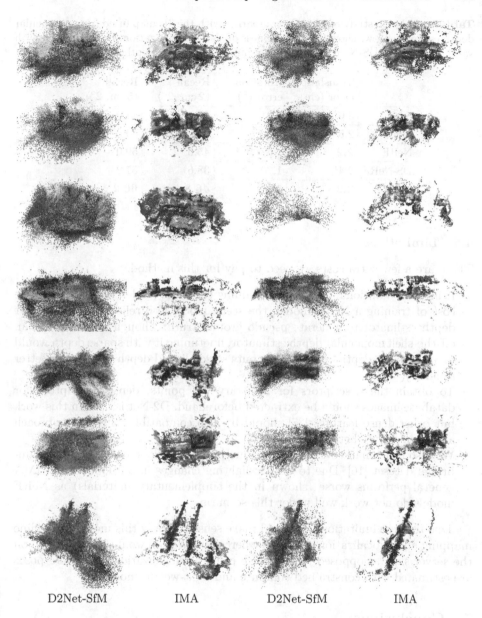

| D2Net-SfM | IMA | D2Net-SfM | IMA |

**Fig. 6.** Qualitative results of 3D maps on 7Scenes and 7SceneS-sub. The left two columns show the comparison on 7Scenes while the right two columns show the comparison on 7Scenes-sub. Our augmented maps are more accurate and neat.

localization recall at (2 cm, 2°) and (5 cm, 2°). Both NeRF-based maps show significant improvement compared to the SfM map, and our methods achieve the best results for all metrics, which indicates the benefit of using dense monocular estimation as pseudo supervision.

**Table 4.** Ablation study results. By comparing with the 3D map fused from monocular depth estimations, we show the effectiveness of NeRF optimization. By comparing with vanila NeRF and DS-NeRF, we show the impact of pseudo depth supervision.

|         | Translation error (cm) | Rotation error (°) | Recall (2 cm, 2°) | Recall (5 cm, 2°) |
|---------|------------------------|--------------------|-------------------|-------------------|
| SfM     | 3.6                    | 2.91               | 26.2              | 38.0              |
| mono    | 2.5                    | 2.02               | 37.6              | 49.0              |
| NeRF    | **2.2**                | 1.67               | 45.6              | 59.0              |
| DS-NeRF | 2.4                    | 1.70               | 38.6              | 57.2              |
| IMA     | **2.2**                | **1.51**           | **46.8**          | **66.2**          |

### 4.5   Limitations

There are a few extra costs we need to pay for this method:

- It involves additional overhead in training the scene model. In addition to the cost of training a NeRF model, the scene model also relies on a monocular depth estimator to generate pseudo ground-truth. Though we could use an off-the-shelf monocular depth estimator, finetuning it with sparse depth would make sure the depth estimator outputs scale-aligned depth maps with better accuracy.
- To obtain the descriptors for arbitrary 3D points, dense descriptors for database images should be extracted beforehand. D2-Net is used in this work but other dense features [37] should be also compatible with this approach and even deliver better results.
- We also benchmarked the proposed method on outdoor scenes such as Cambridge dataset [16]. Due to drastic lighting changes in outdoor scenes, IMA general performs worse (shown in the supplementary materials) as NeRF models do not work well under this scenario.

Due to these limitations, it would make sense to apply this method to offline mapping and relocalization scenarios where computation would be conducted on the server side, as opposed to on-device camera relocalization [54] where poses are estimated with constrained compute and light-weight models.

## 5   Conclusions

In this paper, we propose a non-intrusive method that utilizes implicit scene models to enrich the expressiveness of sparse maps, which is demonstrated via the visual localization task. Empirical results show that IMA enriches the scene models not only because of higher sparse point density, but it can also exploit more critical information embedded in the input. The computational overhead brought by IMA, however, is non-trivial due to the NeRF training. Yet, recent work [30] has shed light on the efficiency end of the problem. We believe that bridging the gap between traditional sparse points and neural-network-based representations is of great significance which could enable many promising directions to explore in the future, such as map updates through online learning.

# References

1. Bleyer, M., Rhemann, C., Rother, C.: Patchmatch stereo-stereo matching with slanted support windows. In: Proceedings of the British Machine Vision Conference (BMVC), vol. 11, pp. 1–11 (2011)
2. Dai, A., Chang, A.X., Savva, M., Halber, M., Funkhouser, T., Nießner, M.: Scannet: richly-annotated 3D reconstructions of indoor scenes. In: Proceedings of the IEEE Conference on Computer Vision and Pattern Recognition (CVPR) (2017)
3. Deng, K., Liu, A., Zhu, J.Y., Ramanan, D.: Depth-supervised nerf: Fewer views and faster training for free. arXiv preprint arXiv:2107.02791 (2021)
4. DeTone, D., Malisiewicz, T., Rabinovich, A.: Superpoint: self-supervised interest point detection and description. In: Proceedings of the IEEE Conference on Computer Vision and Pattern Recognition Workshops (CVPRW), pp. 224–236 (2018)
5. Dusmanu, M., et al.: D2-Net: a trainable CNN for joint detection and description of local features. In: Proceedings of the IEEE/CVF Conference on Computer Vision and Pattern Recognition (CVPR) (2019)
6. Eigen, D., Puhrsch, C., Fergus, R.: Depth map prediction from a single image using a multi-scale deep network. In: Advances in Neural Information Processing Systems, vol. 27 (2014)
7. Fischler, M.A., Bolles, R.C.: Random sample consensus: a paradigm for model fitting with applications to image analysis and automated cartography. Commun. ACM **24**(6), 381–395 (1981)
8. Gallup, D., Frahm, J.M., Mordohai, P., Yang, Q., Pollefeys, M.: Real-time plane-sweeping stereo with multiple sweeping directions. In: Proceedings of the IEEE Conference on Computer Vision and Pattern Recognition (CVPR), pp. 1–8. IEEE (2007)
9. Goesele, M., Curless, B., Seitz, S.M.: Multi-view stereo revisited. In: Proceedings of the 2006 IEEE Computer Society Conference on Computer Vision and Pattern Recognition (CVPR), vol. 2, pp. 2402–2409. IEEE (2006)
10. Hartley, R.I., Zisserman, A.: Multiple View Geometry in Computer Vision, 2nd edn. Cambridge University Press, Cambridge (2004). ISBN 0521540518
11. Heng, L., et al.: Project autovision: localization and 3D scene perception for an autonomous vehicle with a multi-camera system. In: International Conference on Robotics and Automation (ICRA), pp. 4695–4702. IEEE (2019)
12. Hou, Y., Kannala, J., Solin, A.: Multi-view stereo by temporal nonparametric fusion. In: Proceedings of the IEEE/CVF International Conference on Computer Vision (ICCV), pp. 2651–2660 (2019)
13. Humenberger, M., et al.: Robust image retrieval-based visual localization using kapture. arXiv preprint arXiv:2007.13867 (2020)
14. Im, S., Jeon, H., Lin, S., Kweon, I.S.: DPSnet: end-to-end deep plane sweep stereo. In: 7th International Conference on Learning Representations (ICLR) (2019)
15. Irschara, A., Zach, C., Frahm, J.M., Bischof, H.: From structure-from-motion point clouds to fast location recognition. In: 2009 IEEE Conference on Computer Vision and Pattern Recognition, pp. 2599–2606. IEEE (2009)

16. Kendall, A., Grimes, M., Cipolla, R.: Posenet: a convolutional network for real-time 6-DOF camera relocalization. In: Proceedings of the IEEE International Conference on Computer Vision (ICCV), pp. 2938–2946 (2015)
17. Kingma, D.P., Ba, J.: Adam: a method for stochastic optimization. In: International Conference for Learning Representations (ICLR) (2015)
18. Klein, G., Murray, D.: Parallel tracking and mapping for small AR workspaces. In: 2007 6th IEEE and ACM International Symposium on Mixed and Augmented Reality, pp. 225–234. IEEE (2007)
19. Kneip, L., Scaramuzza, D., Siegwart, R.: A novel parametrization of the perspective-three-point problem for a direct computation of absolute camera position and orientation. In: Proceedings of the 2001 IEEE Computer Society Conference on Computer Vision and Pattern Recognition (CVPR), pp. 2969–2976. IEEE (2011)
20. Kutulakos, K.N., Seitz, S.M.: A theory of shape by space carving. Int. J. Comput. Vision **38**(3), 199–218 (2000)
21. Li, Z., et al.: Learning the depths of moving people by watching frozen people. In: Proceedings of the IEEE/CVF Conference on Computer Vision and Pattern Recognition (CVPR), pp. 4521–4530 (2019)
22. Lin, C.H., Ma, W.C., Torralba, A., Lucey, S.: Barf: bundle-adjusting neural radiance fields. In: Proceedings of the IEEE/CVF International Conference on Computer Vision (ICCV), pp. 5741–5751 (2021)
23. Liu, L., Li, H., Dai, Y.: Efficient global 2D–3D matching for camera localization in a large-scale 3D map. In: Proceedings of the IEEE International Conference on Computer Vision (ICCV), pp. 2372–2381 (2017)
24. Lowe, D.G.: Distinctive image features from scale-invariant keypoints. Int. J. Comput. Vision **60**(2), 91–110 (2004)
25. Luo, X., Huang, J.B., Szeliski, R., Matzen, K., Kopf, J.: Consistent video depth estimation. ACM Trans. Graph. (ToG) **39**(4), 71–1 (2020)
26. Merrell, P., et al.: Real-time visibility-based fusion of depth maps. In: Proceedings of the IEEE International Conference on Computer Vision (ICCV), pp. 1–8. IEEE (2007)
27. Mescheder, L., Oechsle, M., Niemeyer, M., Nowozin, S., Geiger, A.: Occupancy networks: learning 3D reconstruction in function space. In: Proceedings of the IEEE/CVF Conference on Computer Vision and Pattern Recognition (CVPR), pp. 4460–4470 (2019)
28. Michalkiewicz, M., Pontes, J.K., Jack, D., Baktashmotlagh, M., Eriksson, A.: Implicit surface representations as layers in neural networks. In: Proceedings of the IEEE/CVF International Conference on Computer Vision (ICCV), pp. 4743–4752 (2019)
29. Mildenhall, B., Srinivasan, P.P., Tancik, M., Barron, J.T., Ramamoorthi, R., Ng, R.: NeRF: representing scenes as neural radiance fields for view synthesis. In: Vedaldi, A., Bischof, H., Brox, T., Frahm, J.-M. (eds.) ECCV 2020. LNCS, vol. 12346, pp. 405–421. Springer, Cham (2020). https://doi.org/10.1007/978-3-030-58452-8_24
30. Müller, T., Evans, A., Schied, C., Keller, A.: Instant neural graphics primitives with a multiresolution hash encoding. arXiv preprint arXiv:2201.05989 (2022)
31. Murez, Z., van As, T., Bartolozzi, J., Sinha, A., Badrinarayanan, V., Rabinovich, A.: Atlas: end-to-end 3D scene reconstruction from posed images. In: Vedaldi, A., Bischof, H., Brox, T., Frahm, J.-M. (eds.) ECCV 2020. LNCS, vol. 12352, pp. 414–431. Springer, Cham (2020). https://doi.org/10.1007/978-3-030-58571-6_25

32. Newcombe, R.A., Lovegrove, S.J., Davison, A.J.: Dtam: dense tracking and mapping in real-time. In: Proceedings of the IEEE International Conference on Computer Vision (ICCV), pp. 2320–2327. IEEE (2011)
33. Niemeyer, M., Mescheder, L., Oechsle, M., Geiger, A.: Differentiable volumetric rendering: learning implicit 3D representations without 3D supervision. In: Proceedings of the IEEE/CVF Conference on Computer Vision and Pattern Recognition (CVPR) (2020)
34. Oechsle, M., Peng, S., Geiger, A.: Unisurf: unifying neural implicit surfaces and radiance fields for multi-view reconstruction. In: Proceedings of the IEEE/CVF International Conference on Computer Vision (ICCV), pp. 5589–5599 (2021)
35. Park, J.J., Florence, P., Straub, J., Newcombe, R., Lovegrove, S.: DeepSDF: learning continuous signed distance functions for shape representation. In: Proceedings of the IEEE/CVF Conference on Computer Vision and Pattern Recognition (CVPR), pp. 165–174 (2019)
36. Revaud, J., Almazán, J., Rezende, R.S., Souza, C.R.d.: Learning with average precision: Training image retrieval with a listwise loss. In: Proceedings of the IEEE/CVF International Conference on Computer Vision (ICCV), pp. 5107–5116 (2019)
37. Revaud, J., et al.: R2d2: repeatable and reliable detector and descriptor. arXiv preprint arXiv:1906.06195 (2019)
38. Roessle, B., Barron, J.T., Mildenhall, B., Srinivasan, P.P., Nießner, M.: Dense depth priors for neural radiance fields from sparse input views. arXiv preprint arXiv:2112.03288 (2021)
39. Sarlin, P.E., Cadena, C., Siegwart, R., Dymczyk, M.: From coarse to fine: robust hierarchical localization at large scale. In: Proceedings of the IEEE/CVF Conference on Computer Vision and Pattern Recognition (CVPR), pp. 12716–12725 (2019)
40. Sarlin, P.E., DeTone, D., Malisiewicz, T., Rabinovich, A.: Superglue: learning feature matching with graph neural networks. In: Proceedings of the IEEE/CVF Conference on Computer Vision and Pattern Recognition, pp. 4938–4947 (2020)
41. Sattler, T., Havlena, M., Radenovic, F., Schindler, K., Pollefeys, M.: Hyperpoints and fine vocabularies for large-scale location recognition. In: Proceedings of the IEEE International Conference on Computer Vision (ICCV), pp. 2102–2110 (2015)
42. Sattler, T., Leibe, B., Kobbelt, L.: Fast image-based localization using direct 2D-to-3D matching. In: Proceedings of the IEEE International Conference on Computer Vision (ICCV), pp. 667–674. IEEE (2011)
43. Sattler, T., Leibe, B., Kobbelt, L.: Improving image-based localization by active correspondence search. In: Fitzgibbon, A., Lazebnik, S., Perona, P., Sato, Y., Schmid, C. (eds.) ECCV 2012. LNCS, vol. 7572, pp. 752–765. Springer, Heidelberg (2012). https://doi.org/10.1007/978-3-642-33718-5_54
44. Sattler, T., Leibe, B., Kobbelt, L.: Efficient & effective prioritized matching for large-scale image-based localization. IEEE Trans. Pattern Anal. Mach. Intell. 39(9), 1744–1756 (2016)
45. Schönberger, J.L., Zheng, E., Frahm, J.-M., Pollefeys, M.: Pixelwise view selection for unstructured multi-view stereo. In: Leibe, B., Matas, J., Sebe, N., Welling, M. (eds.) ECCV 2016. LNCS, vol. 9907, pp. 501–518. Springer, Cham (2016). https://doi.org/10.1007/978-3-319-46487-9_31
46. Schönberger, J.L., Frahm, J.M.: Structure-from-motion revisited. In: Conference on Computer Vision and Pattern Recognition (CVPR) (2016)
47. Seitz, S.M., Dyer, C.R.: Photorealistic scene reconstruction by voxel coloring. Int. J. Comput. Vision 35(2), 151–173 (1999)

48. Shotton, J., Glocker, B., Zach, C., Izadi, S., Criminisi, A., Fitzgibbon, A.: Scene coordinate regression forests for camera relocalization in RGB-D images. In: Proceedings of the IEEE Conference on Computer Vision and Pattern Recognition (CVPR), pp. 2930–2937 (2013)
49. Sucar, E., Liu, S., Ortiz, J., Davison, A.J.: imap: Implicit mapping and positioning in real-time. In: Proceedings of the IEEE/CVF International Conference on Computer Vision (ICCV), pp. 6229–6238 (2021)
50. Sun, J., Xie, Y., Chen, L., Zhou, X., Bao, H.: Neuralrecon: real-time coherent 3D reconstruction from monocular video. In: Proceedings of the IEEE/CVF Conference on Computer Vision and Pattern Recognition, pp. 15598–15607 (2021)
51. Svarm, L., Enqvist, O., Oskarsson, M., Kahl, F.: Accurate localization and pose estimation for large 3D models. In: Proceedings of the IEEE Conference on Computer Vision and Pattern Recognition (CVPR), pp. 532–539 (2014)
52. Sweeney, C., Sattler, T., Hollerer, T., Turk, M., Pollefeys, M.: Optimizing the viewing graph for structure-from-motion. In: Proceedings of the IEEE International Conference on Computer Vision, pp. 801–809 (2015)
53. Taira, H., et al.: Inloc: indoor visual localization with dense matching and view synthesis. In: Proceedings of the IEEE Conference on Computer Vision and Pattern Recognition (CVPR), pp. 7199–7209 (2018)
54. Tran, N.T., et al.: On-device scalable image-based localization via prioritized cascade search and fast one-many RANSAC. IEEE Trans. Image Process. **28**(4), 1675–1690 (2018)
55. Vogiatzis, G., Torr, P.H., Cipolla, R.: Multi-view stereo via volumetric graph-cuts. In: Proceedings of the 2005 IEEE Computer Society Conference on Computer Vision and Pattern Recognition (CVPR), vol. 2, pp. 391–398. IEEE (2005)
56. Wang, K., Shen, S.: Mvdepthnet: real-time multiview depth estimation neural network. In: 2018 International conference on 3D vision (3DV), pp. 248–257. IEEE (2018)
57. Wei, Y., Liu, S., Rao, Y., Zhao, W., Lu, J., Zhou, J.: NerfingMVS: guided optimization of neural radiance fields for indoor multi-view stereo. In: ICCV (2021)
58. Xu, Q., Wang, W., Ceylan, D., Mech, R., Neumann, U.: DISN: deep implicit surface network for high-quality single-view 3D reconstruction. In: Advances in Neural Information Processing Systems (NeurIPS), vol. 32 (2019)
59. Yao, Y., Luo, Z., Li, S., Fang, T., Quan, L.: MVSNet: depth inference for unstructured multi-view stereo. In: Ferrari, V., Hebert, M., Sminchisescu, C., Weiss, Y. (eds.) ECCV 2018. LNCS, vol. 11212, pp. 785–801. Springer, Cham (2018). https://doi.org/10.1007/978-3-030-01237-3_47
60. Yen-Chen, L., Florence, P., Barron, J.T., Rodriguez, A., Isola, P., Lin, T.Y.: iNeRF: inverting neural radiance fields for pose estimation. In: IEEE/RSJ International Conference on Intelligent Robots and Systems (IROS) (2021)
61. Zhang, K., Riegler, G., Snavely, N., Koltun, V.: Nerf++: analyzing and improving neural radiance fields. arXiv preprint arXiv:2010.07492 (2020)
62. Zhou, Y., et al.: DA4AD: end-to-end deep attention-based visual localization for autonomous driving. In: Vedaldi, A., Bischof, H., Brox, T., Frahm, J.-M. (eds.) ECCV 2020. LNCS, vol. 12373, pp. 271–289. Springer, Cham (2020). https://doi.org/10.1007/978-3-030-58604-1_17

# Social Processes: Self-supervised Meta-learning Over Conversational Groups for Forecasting Nonverbal Social Cues

Chirag Raman[1(✉)] , Hayley Hung[1] , and Marco Loog[1,2]

[1] Delft University of Technology, Delft, The Netherlands
{c.a.raman,h.hung,m.loog}@tudelft.nl
[2] University of Copenhagen, Copenhagen, Denmark

**Abstract.** Free-standing social conversations constitute a yet underexplored setting for human behavior forecasting. While the task of predicting pedestrian trajectories has received much recent attention, an intrinsic difference between these settings is how groups form and disband. Evidence from social psychology suggests that group members in a conversation explicitly self-organize to sustain the interaction by adapting to one another's behaviors. Crucially, the same individual is unlikely to adapt similarly across different groups; contextual factors such as perceived relationships, attraction, rapport, etc., influence the entire spectrum of participants' behaviors. A question arises: how can we jointly forecast the mutually dependent futures of conversation partners by modeling the dynamics unique to every group? In this paper, we propose the *Social Process* (SP) models, taking a novel meta-learning and stochastic perspective of group dynamics. Training group-specific forecasting models hinders generalization to unseen groups and is challenging given limited conversation data. In contrast, our SP models treat interaction sequences from a single group as a meta-dataset: we condition forecasts for a sequence from a given group on other observed-future sequence pairs from the same group. In this way, an SP model learns to adapt its forecasts to the unique dynamics of the interacting partners, generalizing to unseen groups in a data-efficient manner. Additionally, we first rethink the task formulation itself, motivating task requirements from social science literature that prior formulations have overlooked. For our formulation of *Social Cue Forecasting*, we evaluate the empirical performance of our SP models against both non-meta-learning and meta-learning approaches with similar assumptions. The SP models yield improved performance on synthetic and real-world behavior datasets.

**Keywords:** Social interactions · Nonverbal cues · Behavior forecasting

**Supplementary Information** The online version contains supplementary material available at https://doi.org/10.1007/978-3-031-25066-8_37.

# 1 Introduction

Picture a conversing group of people in a free-standing social setting. To conduct such exchanges, we transfer high-order social signals across space and time through explicit low-level behavior cues—examples include our pose, gestures, gaze, and floor control actions [1–3]. Evidence suggests that we employ anticipation of these and other cues to navigate daily social interactions [1,4–8]. Consequently, for machines to truly develop adaptive social skills, they need to have the ability to forecast the future. For instance, foreseeing the upcoming behaviors of partners in advance can enable interactive agents to choose more fluid interaction policies [9], or contend with uncertainties in imperfect real-time inferences surrounding cues [3].

In literature, behavior forecasting works mainly consider data at two representations with an increasing level of abstraction: low-level cues or features that are extracted manually or automatically from raw audiovisual data, and manually labeled high-order events or actions. The forecasting task has primarily been formulated to predict future event or action labels from observed cues or other high-order event or action labels [5,6,9–13]. Moreover, identifying patterns predictive of certain semantic events has been a long-standing topic of focus in the social sciences, where researchers primarily employ a top-down workflow. First, the events of interest are selected for consideration. Then their relationship to preceding cues or other high-order actions are studied in isolation through

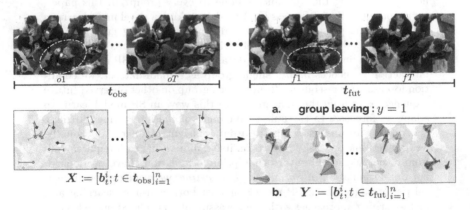

**Fig. 1.** Conceptual illustration of forecasting approaches on an in-the-wild conversation from the MatchNMingle dataset [16]. **Top.** A *group leaving* event [10]: the circled individual has moved from one group in the observed window $t_{\mathrm{obs}} := [o1 \ldots oT]$ to another in a future window $t_{\mathrm{fut}} := [f1 \ldots fT]$. **Bottom.** Input behavioral cues $b_t^i$: head pose (solid normal), body pose (hollow normal), and speaking status (speaker in orange). **a.** The top-down approach entails predicting the event label from such cues over $t_{\mathrm{obs}}$, from only 200 instances of group leaving in over 90 min of interaction [10]. **b.** Our proposed bottom-up, self-supervised formulation of *Social Cue Forecasting* involves regressing a future distribution for the same low-level input cues over $t_{\mathrm{fut}}$ (shaded spread). This enables utilizing the full 90 min of event-unlabeled data.

exploratory or confirmatory analysis [14,15]. Examples of such semantic events include speaker turn transitions [5,6], mimicry episodes [13], the termination of an interaction [9,10], or high-order social actions [11,12].

One hurdle in such a top-down paradigm is data efficiency. The labeled events often occur infrequently over the interaction, reducing the effective amount of labeled data. This, combined with the fact that collecting behavior data is cost and labor-intensive, precludes the effective application of neural supervised learning techniques that tend to be data demanding. More recently, some approaches have adopted a more bottom-up formulation for dyadic conversations. The task entails predicting event-independent future cues for a single target participant or virtual avatar from the preceding observed cues of both participants [17,18]. Since training sequences are not limited to windows around semantic events, such a formulation is more data-efficient. Figure 1 illustrates the top-down and bottom-up approaches conceptually.

In practice, however, the concrete formulations within the bottom-up paradigm [17,18] suffer from several conceptual problems: (i) predictions are made for a single individual using cues from both individuals as input; since people behave differently, this entails training one forecasting model per person; (ii) even so, predicting a future for one individual at a time is undesirable as these futures are not independent; and (iii) the prediction is only a single future, despite evidence that the future is not deterministic, and the same observed sequence can result in multiple socially-valid continuations [19–21].

To address all these issues, we introduce a self-supervised forecasting task called Social Cue Forecasting: predicting a *distribution* over future multimodal cues *jointly for all group members* from their same preceding multimodal cues. Note that we use *self-supervised* here to simply distinguish from the formulations where the predicted quantity (e.g. event-labels) is of a different representation than the observed input (e.g. cues). Given the cue data, the inputs and outputs of our formulation are both cues, so we *obtain the supervisory signal from the data itself.*

Furthermore, a crucial characteristic of free-standing conversations is that people sustain the interaction by explicitly adapting to one another's behaviors [1]. Moreover, the way a person adapts to their partners is a function of several complex factors surrounding their interpersonal relationships and the social setting [22, Chap. 1]; [1, p. 237]. The social dynamics guiding such behavior are embedded in the constellation of participant cues and are distinct for every unique grouping of individuals. As such, a model should adapt its forecasts to the group under consideration. (Even in the pedestrian setting where coordination is only implicit, Rudenko et al. [23, Sec. 8.4.1] observe that failing to adapt predictions to different individuals is still a limitation). For our methodological contribution, we propose the probabilistic Social Processes models, viewing each conversation group as a meta-learning *task*. This allows for capturing social dynamics unique to each group without learning group-specific models and generalizing to unseen groups at evaluation in a data-efficient manner. We believe that this framing of SCF as a *few-shot* function estimation problem is especially suitable for conversation forecasting—a limited data regime where good uncertainty estimates are desirable. Concretely, we make the following contributions:

- We introduce and formalize the novel task of Social Cue Forecasting (SCF), addressing the conceptual drawbacks of past formulations.
- For SCF, we propose and evaluate the family of socially aware probabilistic Seq2Seq models we call Social Processes (SP).

## 2  Related Work

To aid readers from different disciplines situate our work within the broader research landscape, we categorize behavior-forecasting literature by interaction focus [24]. In a focused interaction, such as conversations, participants explicitly coordinate their behaviors to sustain the interaction. In unfocused interactions, coordination is implicit, such as when pedestrians avoid collisions.

*Focused Interactions.* The predominant interest in conversation forecasting stems from the social sciences, with a focus on identifying patterns that are predictive of upcoming speaking turns [5–8], disengagement from an interaction [9,10], or the splitting or merging of groups [25]. Other works forecast the time-evolving size of a group [26] or semantic social action labels [11,12]. More recently, there has also been a growing interest in the computer vision community for tasks related to inferring low-level cues of participants either from their partners' cues [27] or raw multimodal sensor data [28]. Here there has also been some interest in forecasting nonverbal behavior, mainly for dyadic interactions [17,18,29]. The task involves forecasting the future cues of a target individual from the preceding cues of both participants.

*Unfocused Interactions.* Early approaches for forecasting pedestrian or vehicle trajectories were heuristic-based, involving hand-crafted energy potentials to describe the influence pedestrians and vehicles have on each other [30–37]. Recent approaches build upon the idea of encoding relative positional information directly into a neural architecture [38–45]. Some works go beyond locations, predicting keypoints in group activities [46,47]. Rudenko et al. [23] provide a survey of approaches within this space.

*Non-interaction Settings.* Here, the focus has been on forecasting individual poses from images [48] and video [49,50], or synthesizing poses using high-level control parameters [51,52]. The self-supervised aspects of our task formulation are related to visual forecasting, where the goal has been to predict non-semantic low-level pixel features or intermediate representations [34,50,53–57]. Such learned representations have been utilized for other tasks like semi-supervised classification [58], or training agents in immersive environments [59].

For the interested reader, we further discuss practical considerations distinguishing forecasting in conversation and pedestrian settings in Appendix E.

# 3    Social Cue Forecasting: Task Formalization

While self-supervision has shown promise for learning representations of language and video data, is this bottom-up approach conceptually reasonable for behavior cues? The crucial observation we make is that the semantic meaning transferred in interactions (the so-called *social signal* [60]) is already embedded in the low-level cues [61]. So representations of this high-level semantic meaning that we associate with actions and events (e.g. *group leaving*) can be learned from the low-level dynamics in the cues.

## 3.1    Formalization and Distinction from Prior Task Formulations

The objective of SCF is to predict future behavioral cues of *all* people involved in a social encounter given an observed sequence of their behavioral features. Formally, let us denote a window of monotonically increasing observed timesteps as $t_{\text{obs}} := [o1, o2, ..., oT]$, and an unobserved future time window as $t_{\text{fut}} := [f1, f2, ..., fT]$, $f1 > oT$. Note that $t_{\text{fut}}$ and $t_{\text{obs}}$ can be of different lengths, and $t_{\text{fut}}$ need not immediately follow $t_{\text{obs}}$. Given $n$ interacting participants, let us denote their social cues over $t_{\text{obs}}$ and $t_{\text{fut}}$ as

$$X := [b_t^i; t \in t_{\text{obs}}]_{i=1}^n, \quad Y := [b_t^i; t \in t_{\text{fut}}]_{i=1}^n. \tag{1a, b}$$

The vector $b_t^i$ encapsulates the multimodal cues of interest from participant $i$ at time $t$. These can include head and body pose, speaking status, facial expressions, gestures, verbal content—any information streams that combine to transfer social meaning.

*Distribution Over Futures.* In its simplest form, given an $X$, the objective of SCF is to learn a single function $f$ such that $Y = f(X)$. However, an inherent challenge in forecasting behavior is that an observed sequence of interaction does not have a deterministic future and can result in multiple socially valid ones—a window of overlapping speech between people may and may not result in a change of speaker [19, 20], a change in head orientation may continue into a sweeping glance across the room or a darting glance stopping at a recipient of interest [21]. In some cases, certain observed behaviors—intonation and gaze cues [5, 62] or synchronization in speaker-listener speech [63] for turn-taking—may make some outcomes more likely than others. Given that there are both supporting and challenging arguments for how these observations influence subsequent behaviors [63, p. 5]; [62, p. 22], it would be beneficial if a data-driven model expresses a measure of uncertainty in its forecasts. We do this by modeling the distribution over possible futures $p(Y|X)$, rather than a single future $Y$ for a given $X$, the latter being the case for previous formulations for cues [18, 27, 46] and actions [11, 12].

*Joint Modeling of Future Uncertainty.* A defining characteristic of focused interactions is that the participants sustain the shared interaction through explicit, cooperative coordination of behavior [1, p. 220]—the futures of interacting individuals are not independent given an observed window of group behavior. It is therefore essential to capture uncertainty in forecasts at the *global* level—jointly forecasting one future for all participants at a time, rather than at a *local* output level—one future for each individual independent of the remaining participants' futures. In contrast, applying the prior formulations [17,18,27] requires the training of separate models treating each individual as a target (for the same group input) and then forecasting an independent future one at a time. Meanwhile, other prior pose forecasting works [48–52] have been in non-social settings and do not need to model such behavioral interdependence.

*Non-contiguous Observed and Future Windows.* Domain experts are often interested in settings where $t_{obs}$ and $t_{fut}$ are offset by an arbitrary delay, such as forecasting a time lagged synchrony [64] or mimicry [13] episode, or upcoming disengagement [9,10]. We therefore allow for non-contiguous $t_{obs}$ and $t_{fut}$. Operationalizing prior formulations that predict one step into the future [11,12,27,46] would entail a sliding window of autoregressive predictions over the offset between $t_{obs}$ and $t_{fut}$ (from $oT$ to $f1$), with errors cascading even before decoding is performed over the window of interest $t_{fut}$.

Our task formalization of SCF can be viewed as a social science-grounded generalization of prior computational formulations, and therefore suitable for a wider range of cross-disciplinary tasks, both computational and analytical.

# 4   Method Preliminaries

*Meta-Learning.* A supervised learning algorithm can be viewed as a function mapping a dataset $C := (\boldsymbol{X}_C, \boldsymbol{Y}_C) := \{(\boldsymbol{x}^i, \boldsymbol{y}^i)\}_{i \in [N_C]}$ to a predictor $f(\boldsymbol{x})$. Here $N_C$ is the number of datapoints in $C$, and $[N_C] := \{1, \ldots, N_C\}$. The key idea of meta-learning is to learn how to learn from a dataset in order to adapt to unseen supervised tasks; hence the name *meta*-learning. This is done by learning a map $C \mapsto f(\cdot, C)$. In meta-learning literature, a *task* refers to each dataset in a collection $\{\mathcal{T}_m\}_{m=1}^{N_{tasks}}$ of related datasets [65]. Training is episodic, where each task $\mathcal{T}$ is split into subsets $(C, D)$. A meta-learner then fits the subset of target points $D$ given the subset of context observations $C$. At meta-test time, the resulting predictor $f(\boldsymbol{x}, C)$ is adapted to make predictions for target points on an unseen task by conditioning on a new context set $C$ unseen during meta-training.

*Neural Processes (NPs).* Sharing the same core motivations, NPs [66] can be viewed as a family of latent variable models that extend the idea of meta-learning to situations where uncertainty in the predictions $f(\boldsymbol{x}, C)$ are desirable. They do this by meta-learning a map from datasets to stochastic processes, estimating a distribution over the predictions $p(\boldsymbol{Y}|\boldsymbol{X}, C)$. To capture this distribution, NPs model the conditional latent distribution $p(\boldsymbol{z}|C)$ from which a task representation $\boldsymbol{z} \in \mathbb{R}^d$ is sampled. This introduces stochasticity, constituting what is called the model's *latent path*. The context can also be directly incorporated through a *deterministic path*, via a representation $r_C \in \mathbb{R}^d$ aggregated over $C$. An observation model $p(\boldsymbol{y}^i | \boldsymbol{x}^i, r_C, \boldsymbol{z})$ then fits the target observations in $D$. The generative process for the NP is written as

$$p(\boldsymbol{Y} | \boldsymbol{X}, C) := \int p(\boldsymbol{Y} | \boldsymbol{X}, C, \boldsymbol{z}) p(\boldsymbol{z} | C) d\boldsymbol{z} = \int p(\boldsymbol{Y} | \boldsymbol{X}, r_C, \boldsymbol{z}) q(\boldsymbol{z} | \boldsymbol{s}_C) d\boldsymbol{z}, \quad (2)$$

where $p(\boldsymbol{Y} | \boldsymbol{X}, r_C, \boldsymbol{z}) := \prod_{i \in [N_D]} p(\boldsymbol{y}^i | \boldsymbol{x}^i, r_C, \boldsymbol{z})$. The latent $\boldsymbol{z}$ is modeled by a factorized Gaussian parameterized by $\boldsymbol{s}_C := f_s(C)$, with $f_s$ being a deterministic function invariant to order permutation over $C$. When the conditioning on context is removed ($C = \varnothing$), we have $q(\boldsymbol{z} | \boldsymbol{s}_\varnothing) := p(\boldsymbol{z})$, the zero-information prior on $\boldsymbol{z}$. The deterministic path uses a function $f_r$ similar to $f_s$, so that $r_C := f_r(C)$. In practice this is implemented as $r_C = \sum_{i \in [N_C]} \text{MLP}(\boldsymbol{x}_i, \boldsymbol{y}_i)/N_C$. The observation model is referred to as the *decoder*, and $q, f_r, f_s$ comprise the *encoders*. The parameters of the NP are learned for random subsets $C$ and $D$ for a task by maximizing the evidence lower bound (ELBO)

$$\log p(\boldsymbol{Y} | \boldsymbol{X}, C) \geq \mathbb{E}_{q(\boldsymbol{z} | \boldsymbol{s}_D)} [\log p(\boldsymbol{Y} | \boldsymbol{X}, C, \boldsymbol{z})] - \mathbb{KL}(q(\boldsymbol{z} | \boldsymbol{s}_D) \| q(\boldsymbol{z} | \boldsymbol{s}_C)). \quad (3)$$

## 5 Social Processes: Methodology

Our core idea for adapting predictions to a group's unique behavioral dynamics is to condition forecasts on a context set $C$ of the same group's observed-future sequence pairs. By *learning to learn*, i.e., *meta-learn* from a context set, our model can generalize to unseen groups at evaluation by conditioning on an unseen context set of the test group's behavior sequences. In practice, a social robot might, for instance, observe such an evaluation context set before approaching a new group.

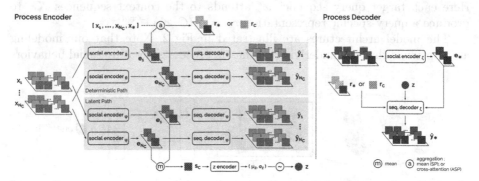

**Fig. 2.** Architecture of the SP and ASP family.

We set up by splitting the interaction into pairs of observed and future sequences, writing the context as $C := (\boldsymbol{X}_C, \boldsymbol{Y}_C) := (\boldsymbol{X}_j, \boldsymbol{Y}_k)_{(j,k) \in [N_C] \times [N_C]}$, where every $X_j$ occurs before the corresponding $Y_k$. Since we allow for non-contiguous $t_{obs}$ and $t_{fut}$, the $j$th $t_{obs}$ can have multiple associated $t_{fut}$ windows for prediction, up to a maximum offset. Denoting the set of target window pairs as $D := (X, Y) := (X_j, Y_k)_{(j,k) \in [N_D] \times [N_D]}$, our goal is to model the distribution $p(Y|X, C)$. Note that when conditioning on context is removed ($C = \varnothing$), we simply revert to the non-meta-learning formulation $p(Y|X)$.

The generative process for our Social Process (SP) model follows Eq. 2, which we extend to social forecasting in two ways. We embed an observed sequence $x^i$ for participant $p_i$ into a condensed encoding $e^i \in \mathbb{R}^d$ that is then decoded into the future sequence using a Seq2Seq architecture [67,68]. Crucially, the sequence decoder only accesses $x^i$ through $e^i$. So after training, $e^i$ must encode the *temporal* information that $x^i$ contains about the future. Further, social behavior is interdependent. We model $e^i$ as a function of both, $p_i$'s own behavior as well as that of partners $p_{j,j\neq i}$ from $p_i$'s perspective. This captures the *spatial* influence partners have on the participant over $t_{obs}$. Using notation we established in Sect. 3, we define the observation model for $p_i$ as

$$p(y^i|x^i, C, z) := p(b^i_{f1}, \ldots, b^i_{fT}|b^i_{o1}, \ldots, b^i_{oT}, C, z) = p(b^i_{f1}, \ldots, b^i_{fT}|e^i, r_C, z). \tag{4}$$

If decoding is carried out in an auto-regressive manner, the right hand side of Eq. 4 simplifies to $\prod_{t=f1}^{fT} p(b^i_t|b^i_{t-1}, \ldots, b^i_{f1}, e^i, r_C, z)$. Following the standard NP setting, we implement the observation model as a set of Gaussian distributions factorized over time and feature dimensions. We also incorporate the cross-attention mechanism from the Attentive Neural Process (ANP) [69] to define the variant Attentive Social Process (ASP). Following Eq. 4 and the definition of the ANP, the corresponding observation model of the ASP for a single participant is defined as

$$p(y^i|x^i, C, z) = p(b^i_{f1}, \ldots, b^i_{fT}|e^i, r^*(C, x^i), z). \tag{5}$$

Here each target query sequence $x^i_*$ attends to the context sequences $X_C$ to produce a query-specific representation $r_* := r^*(C, x^i_*) \in \mathbb{R}^d$.

The model architectures are illustrated in Fig. 2. Note that our modeling assumption is that the underlying stochastic process generating social behaviors

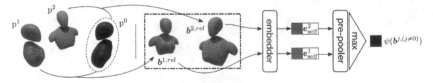

**Fig. 3.** Encoding partner behavior for participant $p^0$ for a single timestep. To model the influence partners $p^1$ and $p^2$ have on the behavior of $p^0$, we transform the partner features to capture the interaction from $p^0$'s perspective, and learn a representation of these features invariant to group size and partner-order permutation using the symmetric max function.

does not evolve over time. That is, the individual factors determining how participants coordinate behaviors—age, cultural background, personality variables [22, Chap. 1]; [1, p. 237]—are likely to remain the same over a single interaction. This is in contrast to the line of work that deals with *meta-transfer learning*, where the stochastic process itself changes over time [70–73]; this entails modeling a different $z$ distribution for every timestep.

*Encoding Partner Behavior.* To encode partners' influence on an individual's future, we use a pair of sequence encoders: one to encode the temporal dynamics of participant $p^i$'s features, $e^i_{self} = f_{self}(x^i)$, and another to encode the dynamics of a transformed representation of the features of $p^i$'s partners, $e^i_{partner} = f_{partner}(\psi(x^{j,(j \neq i)}))$. Using a separate network to encode partner behavior enables sampling an individual's and partners' features at different sampling rates.

How do we model $\psi(x^{j,(j \neq i)})$? We want the partners' representation to possess two properties: *permutation invariance*—changing the order of the partners should not affect the representation, and *group-size independence*—we want to compactly represent all partners independent of the group size. Intuitively, to model partner influence on $p^i$, we wish to *capture a view of the partners' behavior as $p^i$ perceives it.* Figure 3 illustrates the underlying intuition. We do this by computing pooled embeddings of relative behavioral features, extending Gupta et al. [40]'s approach for pedestrian positions to conversation behavior. Note that our partner-encoding approach is in contrast to that of Tan et al. [28], which is order and group-size dependent, and Yao et al. [46], who do not transform the partner features to an individual's perspective.

Since the most commonly considered cues in literature are pose (orientation and location) and binary speaking status [28,74,75], we specify how we transform them. For a single timestep, we denote these cues for $p^i$ as $b^i = [q^i; l^i; s^i]$, and for $p^j$ as $b^j = [q^j; l^j; s^j]$. We compute the relative partner features $b^{j,rel} = [q^{rel}; l^{rel}; s^{rel}]$ by transforming $b^j$ to a frame of reference defined by $b^i$:

$$q^{rel} = q^i * (q^j)^{-1}, \quad l^{rel} = l^j - l^i, \quad s^{rel} = s^j - s^i. \qquad \text{(6a-c)}$$

Note that we use unit quaternions (denoted $q$) for representing orientation due to their various benefits over other representations of rotation [76, Sec. 3.2]. The operator $*$ denotes the Hamilton product of the quaternions. These transformed features $b^{j,rel}$ for each $p^j$ are then encoded using an *embedder* MLP. The outputs are concatenated with their corresponding $e^j_{self}$ and processed by a *pre-pooler* MLP. Assuming $d_{in}$ and $d_{out}$ pre-pooler input and output dims and $J$ partners, we stack the $J$ inputs to obtain $(J, d_{in})$ tensors. The $(J, d_{out})$-dim output is element-wise max-pooled over the $J$ dim, resulting in the $d_{out}$-dim vector $\psi(b^{j,(j \neq i)})$ for any value of $J$, per timestep. We capture the temporal dynamics in this pooled representation over $t_{obs}$ using $f_{partner}$. Finally, we combine $e^i_{self}$ and $e^i_{partner}$ for $p^i$ through a linear projection (defined by a weight matrix $W$) to obtain the individual's embedding $e^i_{ind} = W \cdot [e^i_{self}; e^i_{partner}]$. Our intuition is that with information about both $p^i$ themselves, and of $p^i$'s partners from $p^i$'s point-of-view, $e_{\text{ind}}^i$ now contains the information required to predict $\mathbf{p}^i$'s future behavior.

*Encoding Future Window Offset.* Since we allow for non-contiguous windows, a single $t_{\text{obs}}$ might be associated to multiple $t_{\text{fut}}$ windows at different offsets. Decoding the same $e_{\text{ind}}^i$ into multiple sequences (for different $t_{\text{fut}}$) in the absence of any timing information might cause an averaging effect in either the decoder or the information encoded in $e_{\text{ind}}^i$. One option would be to immediately start decoding after $t_{\text{obs}}$ and discard the predictions in the offset between $t_{\text{obs}}$ and $t_{\text{fut}}$. However, auto-regressive decoding might lead to cascading errors over the offset. Instead, we address this one-to-many issue by injecting the offset information into $e_{\text{ind}}^i$. The decoder then receives a unique encoded representation for every $t_{\text{fut}}$ corresponding to the same $t_{\text{obs}}$. We do this by repurposing the idea of sinusoidal positional encodings [77] to encode window offsets rather than relative token positions in sequences. For a given $t_{\text{obs}}$ and $t_{\text{fut}}$, and $d_e$-dim $e_{\text{ind}}^i$ we define the offset as $\Delta t = f1 - oT$, and the corresponding offset encoding $OE_{\Delta t}$ as

$$OE_{(\Delta t, 2m)} = \sin(\Delta t/10000^{2m/d_e}),\, OE_{(\Delta t, 2m+1)} = \cos(\Delta t/10000^{2m/d_e}). \quad \text{(7a, b)}$$

Here $m$ refers to the dimension index in the encoding. We finally compute the representation $e^i$ for Eq. 4 and Eq. 5 as

$$e^i = e_{\text{ind}}^i + OE_{\Delta t}. \quad (8)$$

*Auxiliary Loss Functions.* We incorporate a geometric loss function for each of our sequence decoders to improve performance in pose regression tasks. For $\mathbf{p}_i$ at time $t$, given the ground truth $b_t^i = [q; l; s]$, and the predicted mean $\hat{b}_t^i = [\hat{q}; \hat{l}; \hat{s}]$, we denote the tuple $(b_t^i, \hat{b}_t^i)$ as $B_t^i$. We then have the location loss in Euclidean space $\mathcal{L}_l(B_t^i) = \|l - \hat{l}\|$, and we can regress the quaternion values using

$$\mathcal{L}_q(B_t^i) = \left\| q - \frac{\hat{q}}{\|\hat{q}\|} \right\|. \quad (9)$$

Kendall and Cipolla [76] show how these losses can be combined using the homoscedastic uncertainties in position and orientation, $\hat{\sigma}_l^2$ and $\hat{\sigma}_q^2$:

$$\mathcal{L}_\sigma(B_t^i) = \mathcal{L}_l(B_t^i)\exp(-\hat{s}_l) + \hat{s}_l + \mathcal{L}_q(B_t^i)\exp(-\hat{s}_q) + \hat{s}_q, \quad (10)$$

where $\hat{s} := \log \hat{\sigma}^2$. Using the binary cross-entropy loss for speaking status $\mathcal{L}_s(B_t^i)$, we have the overall auxiliary loss over $t \in t_{\text{fut}}$:

$$\mathcal{L}_{\text{aux}}(Y, \hat{Y}) = \sum_i \sum_t \mathcal{L}_\sigma(B_t^i) + \mathcal{L}_s(B_t^i). \quad (11)$$

The parameters of the SP and ASP are trained by maximizing the ELBO (Eq. 3) and minimizing this auxiliary loss.

# 6    Experiments and Results

## 6.1    Experimental Setup

*Evaluation Metrics.* Prior forecasting formulations output a single future. However, since the future is not deterministic, we predict a future distribution. Consequently, needing a metric that accounts for probabilistic predictions, we report the log-likelihood (LL) $\log p(Y|X, C)$, commonly used by all variants within the NP family [66,69,70]. The metric is equal to the log of the predicted density evaluated at the ground-truth value. (Note: the fact that the vast majority of forecasting works even in pedestrian settings omit a probabilistic metric, using only geometric metrics, is a limitation also observed by Rudenko et al. [23, Sec. 8.3].) Nevertheless, for additional insight beyond the LL, we also report the errors in the predicted means—geometric errors for pose and accuracy for speaking status—and provide qualitative visualizations of forecasts.

*Models and Baselines.* In keeping with the task requirements and for fair evaluation, we require that all models we compare against forecast a distribution over future cues.

- To evaluate our core idea of viewing conversing groups as meta-learning tasks, we compare against non-meta-learning methods: we adapt variational encoder-decoder (VED) architectures [78,79] to output a distribution.
- To evaluate our specific modeling choices within the meta-learning family, we compare against the NP and ANP models (see Sect. 5). The original methods were not proposed for sequences, so we adapt them by collapsing the timestep and feature dimensions in the data.

Note that in contrast to the SP models, these baselines have direct access to the future sequences in the context, and therefore constitute a strong baseline. We consider two variants for both NP and SP models: *-latent* denoting only the stochastic path; and *-uniform* containing both the deterministic and stochastic paths with uniform attention over context sequences. We further consider two attention mechanisms for the cross-attention module: *-dot* with dot attention, and *-mh* with wide multi-head attention [69]. Finally, we experiment with two choices of backbone architectures: multi-layer perceptrons (MLP), and Gated Recurrent Units (GRU). Implementation and training details can be found in Appendix D. Code, processed data, trained models, and test batches for reproduction are available at https://github.com/chiragraman/social-processes.

## 6.2    Evaluation on Synthesized Behavior Data

To first validate our method on a toy task, we synthesize a dataset simulating two glancing behaviors in social settings [21], approximated by horizontal head rotation. The sweeping *Type I* glance is represented by a 1D sinusoid over 20 timesteps. The gaze-fixating *Type III* glance is denoted by clipping the amplitude for the last six timesteps. The task is to forecast the signal over the last 10

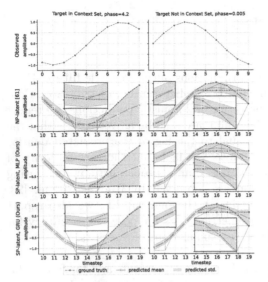

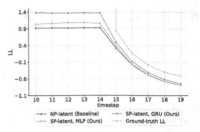

**Fig. 5.** Mean per timestep LL over the sequences in the synthetic glancing dataset. Higher is better.

**Table 1. Mean (Std.) Metrics on the Synthetic Glancing Behavior Dataset.** The metrics are averaged over timesteps; mean and std. are then computed over sequences. Higher is better for LL, lower for MAE.

**Fig. 4.** Ground truths and predictions for the toy task of forecasting simulated glancing behavior. Our SP models learn a better fit than the NP model, SP-GRU being the best (see zoomed insets).

|                  | LL             | Head Ori. MAE (°) |
|------------------|----------------|-------------------|
| NP-latent        | 0.28 (0.24)    | 19.63 (7.26)      |
| SP-latent (MLP)  | 0.36 (0.20)    | 19.46 (7.05)      |
| SP-latent (GRU)  | **0.55 (0.23)**| **18.55 (7.11)**  |

timesteps ($t_{\text{fut}}$) by observing the first 10 ($t_{\text{obs}}$). Consequently, the first half of $t_{\text{fut}}$ is certain, while the last half is uncertain: every observed sinusoid has two ground truth futures in the data (clipped and unclipped). It is impossible to infer from an observed sequence alone if the head rotation will stop partway through the future. Figure 4 illustrates the predictions for two sample sequences. Table 1 provides quantitative metrics and Fig. 5 plots the LL per timestep. The LL is expected to decrease over timesteps where ground-truth futures diverge, being ∞ when the future is certain. We observe that all models estimate the mean reasonably well, although our proposed SP models perform best. More crucially, the SP models, especially the SP-GRU, learn much better uncertainty estimates compared to the NP baseline (see zoomed regions in Fig. 4). We provide additional analysis, alternative qualitative visualizations, and data synthesis details in Appendices A to C respectively.

## 6.3   Evaluation on Real-World Behavior Data

*Datasets and Preprocessing.* With limited behavioral data availability, a common practice in the domain is to solely train and evaluate methods on synthesized behavior dynamics [12,80]. In contrast, we also evaluate on two real-world behavior datasets: the MatchNMingle (MnM) dataset of in-the-wild mingling behavior [16], and the Haggling dataset of a triadic game where two sellers compete to sell a fictional product to a buyer [27]. For MnM, we treat the 42 groups from Day 1

**Table 2. Mean (Std.) Log-Likelihood (LL) on the MatchNMingle and Haggling Test Sets.** For a single sequence, we sum over the feature and participant dimensions, and average over timesteps. The reported mean and std. are over individual sequences in the test sets. Higher is better. Underline indicates best LL within family.

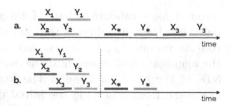

**Fig. 6.** Context Regimes. For a target sequence pair $(X_*, Y_*)$, context pairs (here 3) are sampled either **a.** randomly across the lifetime of the group interaction (*random*), or **b.** from a fixed initial duration (*fixed-initial*).

| | MatchNMingle | | Haggling | |
|---|---|---|---|---|
| | Random | Fixed-initial | Random | Fixed-initial |
| VED family [78,79] | | | | |
| VED-MLP | 8.1 (7.2) | 7.9 (7.0) | 4.0 (8.3) | 4.1 (8.2) |
| VED-GRU | 25.4 (18.0) | 25.1 (19.1) | 60.3 (2.2) | 60.3 (2.1) |
| NP Family [66,69] | | | | |
| NP-latent | 22.1 (17.8) | 21.6 (18.5) | 27.2 (17.3) | 27.9 (16.3) |
| NP-uniform | 21.4 (18.8) | 20.5 (17.8) | 24.8 (22.9) | 25.0 (22.2) |
| ANP-dot | 22.8 (18.6) | 21.0 (18.3) | 26.7 (21.4) | 24.7 (20.8) |
| ANP-mh | 23.6 (15.6) | 20.0 (23.9) | 25.1 (23.1) | 24.8 (22.4) |
| Ours (SP-MLP) | | | | |
| SP-latent | 102.1 (29.9) | 101.5 (29.2) | 136.6 (7.0) | 136.7 (7.0) |
| SP-uniform | 112.8 (34.1) | 111.4 (33.8) | 138.3 (8.0) | 137.6 (8.4) |
| ASP-dot | 109.9 (32.9) | 107.6 (32.1) | 137.8 (7.5) | 136.4 (7.6) |
| ASP-mh | 112.9 (34.7) | 111.3 (33.6) | 146.0 (10.9) | 145.7 (10.2) |
| Ours (SP-GRU) | | | | |
| SP-latent | 86.4 (37.2) | 85.4 (37.2) | 66.7 (27.4) | 66.2 (30.7) |
| SP-uniform | 87.0 (38.4) | 85.5 (38.3) | 79.9 (50.5) | 78.6 (52.2) |
| ASP-dot | 87.6 (39.1) | 83.9 (38.1) | 38.4 (60.4) | 27.2 (93.4) |
| ASP-mh | 85.8 (37.1) | 82.3 (36.0) | 66.3 (30.3) | 59.3 (32.4) |

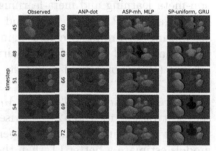

**Fig. 7.** Forecasts over selected timesteps from the Haggling group *170224-a1-group1*. Speaking status is interpolated between orange (speaking) and blue (listening). Translucent models denote the predicted mean ± std. (Color figure online)

as test sets and a total of 101 groups from the other two days as train sets. For Haggling, we use the same split of 79 training and 28 test groups used by Joo et al. [27]. We consider the following cues: *head pose* and *body pose*, described by the location of a keypoint and an orientation quaternion; and binary *speaking status*. These are the most commonly considered cues in computational analyses of conversations [28,74,75] given how crucial they are in sustaining interactions [1,20,61]. For orientation, we first convert the normal vectors (provided in the horizontal direction in both datasets) into unit quaternions. Since the quaternions $\mathbf{q}$ and $-\mathbf{q}$ denote an identical rotation, we constrain the first quaternion in every sequence to the same hemisphere and interpolate subsequent quaternions to have the shortest distance along the unit hypersphere. We then split the interaction data into pairs of $t_{\text{obs}}$ and $t_{\text{fut}}$ windows to construct the samples for forecasting. We specify dataset-specific preprocessing details in Appendix C.

*Context Regimes.* We evaluate on two context regimes: *random*, and *fixed-initial* (see Fig. 6). In the *random* regime, context samples (observed-future pairs) are selected as a random subset of target samples, so the model is exposed to behaviors from any phase of the interaction lifecycle. Here we ensure that batches contain unique $t_{obs}$ to prevent any single observed sequence from dominating the aggregation of representations over the context split. At evaluation, we take 50% of the batch as context. The *fixed-initial* regime investigates how models can learn from observing the initial dynamics of an interaction where certain gestures and patterns are more distinctive [1, Chap. 6]. Here we treat the first 20% of the entire interaction as context, treating the rest as target.

*Conversation Groups as Meta-learning Tasks?* While our core idea of viewing groups as meta-learning tasks is grounded in social science literature (see Sect. 5), does it help to improve empirical performance? Comparing the LL of non-meta-learning and meta-learning models in Table 2 by architecture—VED-MLP against NP and SP-MLP, and VED-GRU against SP-GRU—we find that accounting for group-specific dynamics through meta-learning yields improved performance. All best-in-family pairwise model differences are statistically significant (Wilcoxon signed rank test, $p < 10^{-4}$).

*Comparing Within Meta-learning Methods.* While our SP-MLP models perform the best on LL in Table 2 (pairwise differences are significant), they fare the worst at estimating the mean (Appendix A.2). On the other hand, the SP-GRU models estimate a better LL than the NP models with comparable errors in the mean forecast. The NP models attain the lowest errors in predicted means, but also achieve the worst LL. Why do the models achieving better LL also tend to predict worse means? Upon inspecting the metrics for individual features, we found that the models, especially the MLP variants, tend to improve LL by making the variance over constant features exceedingly small, often at the cost of errors in the means. Note that since the rotation in the data is in the horizontal plane, the $qx$ and $qy$ quaternion dimensions are zero throughout. We do not observe such model behavior in the synthetic data experiments, which do not involve constant features. Figure 7 visualizes forecasts for an example sequence from the Haggling dataset where a turn change has occurred just at the end of the observed window. Here, the SP-GRU model forecasts an interesting continuation to the turn. It anticipates that the buyer (middle) will interrupt the last observed speaker (right seller), before falling silent and looking from one seller to another, both of whom the model expects to then speak simultaneously (see Appendix B for the full sequence). We believe that the forecast indicates that the model is capable of learning believable haggling turn dynamics from different turn continuations in the data. From the visualizations also we observe that the models seem to maximize LL at the cost of orientation errors; in the case of SP-MLP seemingly by predicting the majority orientation in the triadic setting. Also, the NP models forecast largely static futures. In contrast, while being more dynamic, the SP-GRU forecasts contain some smoothing. Overall, the SP-GRU models achieve the best trade-off between maximizing LL and forecasting plausible human behavior.

## 6.4  Ablations

*Encoding Partner Behavior.* Modeling the interaction from the perspective of each individual is a central idea in our approach. We investigate the influence of encoding partner behavior into individual representations $e_{ind}^i$. We train the SP-uniform GRU variant in two configurations: *no-pool*, where we do not encode any partner behavior; and *pool-oT* where we pool over partner representations only at the last timestep (similar to [40]). Both configurations lead to worse LL and location errors (Table 3 and Appendix A).

**Table 3. Mean (Std.) LL for the Ablation Experiments with the SP-uniform GRU Model.** The reported mean and std. are over individual sequences in the test sets. Higher is better.

|  |  | MatchNMingle | | Haggling | |
|---|---|---|---|---|---|
|  |  | Random | Fixed-initial | Random | Fixed-initial |
| Full model | | 87.0 (38.4) | 85.5 (38.3) | 79.9 (50.5) | 78.6 (52.2) |
| Encoding partner behavior | no-pool | 77.8 (31.2) | 76.9 (31.0) | 54.5 (75.5) | 50.1 (97.5) |
| | pool-oT | 82.3 (33.3) | 81.0 (33.6) | 66.9 (26.0) | 66.8 (25.7) |
| No deterministic decoding | Shared social encoders | 88.5 (40.7) | 87.6 (39.6) | 93.1 (39.3) | 91.9 (40.4) |
| | Unshared social encoders | 81.4 (38.1) | 80.2 (37.8) | 66.6 (24.0) | 64.8 (23.4) |

*Deterministic Decoding and Social Encoder Sharing.* We investigate the effect of the deterministic decoders by training the SP-uniform GRU model without them. We also investigate sharing a single social encoder between the Process Encoder and Process Decoder in Fig. 2. Removing the decoders only improves log-likelihood if the encoders are shared, and at the cost of head orientation errors (Table 3 and Appendix A).

## 7  Discussion

The setting of social conversations remains a uniquely challenging frontier for state-of-the-art low-level behavior forecasting. In the recent forecasting challenge involving dyadic interactions, none of the submitted methods could outperform the naive *zero-velocity* baseline [17, Sec. 5.5]. (The baseline propagates the last observed features into the future as if the person remained static.) Why is this? The predominant focus of researchers working on social human-motion prediction has been pedestrian trajectories [23] or actions such as *punching, kicking, gathering, chasing, etc.* [46,47]. In contrast to such activities which involve pronounced movements, the postural adaptation for regulating conversations is far more subtle (also see the discussion in Appendix E). At the same time, the social intelligence required to understand the underlying dynamics that drive a conversation is comparatively more sophisticated than for an action such as a kick. We hope that the social-science considerations informing the design of SCF (joint probabilistic forecasting for all members) and the SP models (groups as meta-learning tasks) constitute a meaningful foundation for future research in this space to build upon. Note that for our task formulation, even the performance of our baseline models constitutes new results.

*Cross-Discipline Impact and Ethical Considerations.* While our work here is an *upstream* methodological contribution, the focus on human behavior entails ethical considerations for downstream applications. One such application involves assisting social scientists in developing predictive hypotheses for specific behaviors by examining model predictions. In these cases, such hypotheses must be verified in subsequent controlled experiments. With the continued targeted development of techniques for recording social behavior in the wild [81], evaluating forecasting models in varied interaction settings would also provide further insight. Another application involves helping conversational agents achieve smoother interactions. Here researchers should be careful that the ability to forecast does not result in nefarious manipulation of user behavior.

**Acknowledgements.** This research was partially funded by the Netherlands Organization for Scientific Research (NWO) under the MINGLE project number 639.022.606. Chirag would like to thank Amelia Villegas-Morcillo for her input and the innumerable discussions, and Tiffany Matej Hrkalovic for feedback on parts of the manuscript.

# References

1. Kendon, A.: Conducting Interaction: Patterns of Behavior in Focused Encounters. Number 7 in Studies in Interactional Sociolinguistics. Cambridge University Press, Cambridge (1990). ISBN 978-0-521-38036-2, 978-0-521-38938-9
2. Vinciarelli, A., Pantic, M., Bourlard, H.: Social signal processing: survey of an emerging domain. Image Vis. Comput. **27**(12), 1743–1759 (2009)
3. Bohus, D., Horvitz, E.: Models for multiparty engagement in open-world dialog. In: Proceedings of the SIGDIAL 2009 Conference on The 10th Annual Meeting of the Special Interest Group on Discourse and Dialogue - SIGDIAL 2009, pp. 225–234. Association for Computational Linguistics, London (2009). ISBN 978-1-932432-64-0. https://doi.org/10.3115/1708376.1708409
4. Ishii, R., Kumano, S., Otsuka, K.: Prediction of next-utterance timing using head movement in multi-party meetings. In: Proceedings of the 5th International Conference on Human Agent Interaction, HAI 2017, pp. 181–187. Association for Computing Machinery, New York, October 2017. ISBN 978-1-4503-5113-3, https://doi.org/10.1145/3125739.3125765
5. Keitel, A., Daum, M.M.: The use of intonation for turn anticipation in observed conversations without visual signals as source of information. Front. Psychol. **6**, 108 (2015)
6. Garrod, S., Pickering, M.J.: The use of content and timing to predict turn transitions. Front. Psychol. **6**, 751 (2015)
7. Rochet-Capellan, A., Fuchs, S.: Take a breath and take the turn: how breathing meets turns in spontaneous dialogue. Philos. Trans. Roy. Soc. B Biol. Sci. **369**(1658), 20130399 (2014)

8. Wlodarczak, M., Heldner, M.: Respiratory turn-taking cues. In: INTERSPEECH (2016)
9. Bohus, D., Horvitz, E.: Managing human-robot engagement with forecasts and... um... hesitations. In: Proceedings of the 16th International Conference on Multimodal Interaction, p. 8 (2014)
10. van Doorn, F.: Rituals of leaving: predictive modelling of leaving behaviour in conversation. Master of Science thesis, Delft University of Technology (2018)
11. Airale, L., Vaufreydaz, D., Alameda-Pineda, X.: SocialInteractionGAN: multi-person interaction sequence generation. arXiv:2103.05916 [cs, stat], March 2021
12. Sanghvi, N., Yonetani, R., Kitani, K.: MGPI: a computational model of multiagent group perception and interaction. arXiv preprint arXiv:1903.01537 (2019)
13. Bilakhia, S., Petridis, S., Pantic, M.: Audiovisual detection of behavioural mimicry. In: 2013 Humaine Association Conference on Affective Computing and Intelligent Interaction, pp. 123–128. IEEE, Geneva, September 2013. ISBN 978-0-7695-5048-0. https://doi.org/10.1109/ACII.2013.27
14. Liem, C.C.S., et al.: Psychology meets machine learning: interdisciplinary perspectives on algorithmic job candidate screening. In: Escalante, H.J., et al. (eds.) Explainable and Interpretable Models in Computer Vision and Machine Learning. TSSCML, pp. 197–253. Springer, Cham (2018). https://doi.org/10.1007/978-3-319-98131-4_9
15. Nilsen, E., Bowler, D., Linnell, J.: Exploratory and confirmatory research in the open science era. J. Appl. Ecol. **57** (2020). https://doi.org/10.1111/1365-2664. 13571
16. Cabrera-Quiros, L., Demetriou, A., Gedik, E., van der Meij, L., Hung, H.: The matchnmingle dataset: a novel multi-sensor resource for the analysis of social interactions and group dynamics in-the-wild during free-standing conversations and speed dates. IEEE Trans. Affect. Comput. (2018)
17. Palmero, C., et al.: Chalearn lap challenges on self-reported personality recognition and non-verbal behavior forecasting during social dyadic interactions: dataset, design, and results. In: Understanding Social Behavior in Dyadic and Small Group Interactions, pp. 4–52. PMLR (2022)
18. Ahuja, C., Ma, S., Morency, L.-P., Sheikh, Y.: To react or not to react: end-to-end visual pose forecasting for personalized avatar during dyadic conversations. arXiv:1910.02181 [cs], October 2019
19. Heldner, M., Edlund, J.: Pauses, gaps and overlaps in conversations. J. Phonet. **38**(4), 555–568 (2010). ISSN 0095-4470. https://doi.org/10.1016/j.wocn.2010.08. 002
20. Duncan, S.: Some signals and rules for taking speaking turns in conversations. J. Person. Soc. Psychol. **23**(2), 283–292(1972). ISSN 1939-1315 (Electronic), 0022-3514 (Print). https://doi.org/10.1037/h0033031
21. Moore, M.M.: Nonverbal courtship patterns in women: context and consequences. Ethol. Sociobiol. **6**(4), 237–247 (1985). ISSN 0162-3095. https://doi.org/10.1016/ 0162-3095(85)90016-0
22. Moore, N.-J., Mark III, H., Don, W.: Stacks. Nonverbal Commun. Stud. Appl. (2013)
23. Rudenko, Palmieri, L., Herman, M., Kitani, K.M., Gavrila, D.M., Arras, K.O.: Human motion trajectory prediction: a survey. Int. J. Robot. Res. **39**(8), 895–935 (2020)
24. Goffman, E.: Behavior in Public Places: Notes on the Social Organization of Gatherings. The Free Press, 1. paperback edn, 24. printing edition, 1966. ISBN 978-0-02-911940-2

25. Wang, A., Steinfeld, A.: Group split and merge prediction with 3D convolutional networks. IEEE Robot. Autom. Lett. **5**(2), 1923–1930, April 2020. ISSN 2377-3766. https://doi.org/10.1109/LRA.2020.2969947
26. Mastrangeli, M., Schmidt, M., Lacasa, L.: The roundtable: an abstract model of conversation dynamics. arXiv:1010.2943 [physics], October 2010
27. Joo, H., Simon, T., Cikara, M., Sheikh, Y.: Towards social artificial intelligence: nonverbal social signal prediction in a triadic interaction. In: 2019 IEEE/CVF Conference on Computer Vision and Pattern Recognition (CVPR), pp. 10865–10875. IEEE, Long Beach, June 2019. ISBN 978-1-72813-293-8. https://doi.org/10.1109/CVPR.2019.01113
28. Tan, S., Tax, D.M.J., Hung, H.: Multimodal joint head orientation estimation in interacting groups via proxemics and interaction dynamics. In: Proceedings of the ACM on Interactive, Mobile, Wearable and Ubiquitous Technologies, vol. 5, no. 1, pp. 1–22, March 2021. ISSN 2474-9567. https://doi.org/10.1145/3448122
29. Tuyen, N.T.V., Celiktutan, O.: Context-aware human behaviour forecasting in dyadic interactions. In: Understanding Social Behavior in Dyadic and Small Group Interactions, pp. 88–106. PMLR (2022)
30. Helbing, D., Molnar, P.: Social force model for pedestrian dynamics. Phys. Rev. E, **51**(5), 4282–4286 (1995). ISSN 1063-651X, 1095-3787. https://doi.org/10.1103/PhysRevE.51.4282
31. Jarosław Wąs, Bartłomiej Gudowski, and Paweł J. Matuszyk. Social Distances Model of Pedestrian Dynamics. In Cellular Automata, volume 4173, pages 492–501. Springer, Berlin Heidelberg, Berlin, Heidelberg, 2006. ISBN 978-3-540-40929-8 978-3-540-40932-8. https://doi.org/10.1007/11861201_57
32. Antonini, G., Bierlaire, M., Weber, M.: Discrete choice models for pedestrian walking behavior. Transport. Res. Part B Methodol. **40**, 667–687 (2006). https://doi.org/10.1016/j.trb.2005.09.006
33. Treuille, A., Cooper, S., Popović, Z.: Continuum crowds. ACM Trans. Graph./SIGGRAPH 2006 **25**(3), 1160–1168 (2006)
34. Robicquet, A., Sadeghian, A., Alahi, A., Savarese, S.: Learning social etiquette: human trajectory understanding in crowded scenes. In: Leibe, B., Matas, J., Sebe, N., Welling, M. (eds.) ECCV 2016. LNCS, vol. 9912, pp. 549–565. Springer, Cham (2016). https://doi.org/10.1007/978-3-319-46484-8_33
35. Wang, J.M., Fleet, D.J., Hertzmann, A.: Gaussian process dynamical models for human motion. IEEE Trans. Pattern Anal. Mach. Intell. **30**(2), 283–298, February 2008. ISSN 1939-3539. https://doi.org/10.1109/TPAMI.2007.1167
36. Tay, C., Laugier, C.: Modelling smooth paths using gaussian processes. In: Proceedings of the International Conference on Field and Service Robotics (2007)
37. Patterson, A., Lakshmanan, A., Hovakimyan, N.: Intent-aware probabilistic trajectory estimation for collision prediction with uncertainty quantification. arXiv:1904.02765 [cs, math], April 2019
38. Alahi, A., Goel, K., Ramanathan, V., Robicquet, A., Fei-Fei, L., Savarese, S.: Social LSTM: human trajectory prediction in crowded spaces. In: 2016 IEEE Conference on Computer Vision and Pattern Recognition (CVPR), pp. 961–971. IEEE, Las Vegas, June 2016. ISBN 978-1-4673-8851-1. https://doi.org/10.1109/CVPR.2016.110
39. Zhang, P., Ouyang, W., Zhang, P., Xue, J., Zheng, N.: SR-LSTM: state refinement for LSTM towards pedestrian trajectory prediction. arXiv:1903.02793 [cs], March 2019

40. Gupta, A., Johnson, J., Fei-Fei, L., Savarese, S., Alahi, A.: Social GAN: socially acceptable trajectories with generative adversarial networks. arXiv:1803.10892 [cs], March 2018

41. Hasan, I., et al.: Forecasting people trajectories and head poses by jointly reasoning on tracklets and vislets. arXiv:1901.02000 [cs], January 2019

42. Huang, Y., Bi, H., Li, Z., Mao, T., Wang, Z.: STGAT: modeling spatial-temporal interactions for human trajectory prediction. In: 2019 IEEE/CVF International Conference on Computer Vision (ICCV), pp. 6271–6280. IEEE, Seoul, October 2019. ISBN 978-1-72814-803-8. https://doi.org/10.1109/ICCV.2019.00637

43. Mohamed, A., Qian, K., Elhoseiny, M., Claudel, C.: Social-STGCNN: a social spatio-temporal graph convolutional neural network for human trajectory prediction. arXiv:2002.11927 [cs], February 2020

44. Zhao, H., et al.: TNT: Target-driveN trajectory prediction. arXiv:2008.08294 [cs], August 2020

45. Gilles, T., Sabatini, S., Tsishkou, D., Stanciulescu, B., Moutarde, F.: THOMAS: trajectory heatmap output with learned multi-agent sampling. arXiv:2110.06607 [cs], January 2022

46. Yao, T., Wang, M., Ni, B., Wei, H., Yang, X.: Multiple granularity group interaction prediction. In: 2018 IEEE/CVF Conference on Computer Vision and Pattern Recognition, pp. 2246–2254. IEEE, Salt Lake City, June 2018. ISBN 978-1-5386-6420-9. https://doi.org/10.1109/CVPR.2018.00239

47. Vida Adeli, Ehsan Adeli, Ian Reid, Juan Carlos Niebles, and Hamid Rezatofighi. Socially and contextually aware human motion and pose forecasting. IEEE Robotics and Automation Letters, 5 (4): 6033–6040, 2020

48. Chao, Y.-W., Yang, J., Price, B., Cohen, S., Deng, J.: Forecasting human dynamics from static images. arXiv:1704.03432 [cs], April 2017

49. Fragkiadaki, K., Levine, S., Felsen, P., Malik, J.: Recurrent network models for human dynamics. arXiv:1508.00271 [cs], September 2015

50. Walker, J., Marino, K., Gupta, A., Hebert, M.: The pose knows: video forecasting by generating pose futures. arXiv:1705.00053 [cs], April 2017

51. Habibie, I., Holden, D., Schwarz, J., Yearsley, J., Komura, T.: A recurrent variational autoencoder for human motion synthesis. In Procedings of the British Machine Vision Conference 2017, p. 119. British Machine Vision Association, London (2017). ISBN 978-1-901725-60-5. https://doi.org/10.5244/C.31.119

52. Pavllo, D., Grangier, D., Auli, M.: QuaterNet: a quaternion-based recurrent model for human motion. arXiv:1805.06485 [cs], July 2018

53. Ranzato, M.A., Szlam, A., Bruna, J., Mathieu, M., Collobert, R., Chopra, S.: Video (language) modeling: a baseline for generative models of natural videos. arXiv:1412.6604 [cs], December 2014

54. Walker, J., Gupta, A., Hebert, M.: Dense optical flow prediction from a static image. In: 2015 IEEE International Conference on Computer Vision (ICCV), pp. 2443–2451. IEEE, Santiago, December 2015. ISBN 978-1-4673-8391-2. https://doi.org/10.1109/ICCV.2015.281

55. Dosovitskiy, A., et al.: FlowNet: learning optical flow with convolutional networks. In: 2015 IEEE International Conference on Computer Vision (ICCV), pp. 2758–2766. IEEE, Santiago, December 2015. ISBN 978-1-4673-8391-2. https://doi.org/10.1109/ICCV.2015.316

56. Walker, J., Gupta, A., Hebert, M.: Patch to the future: unsupervised visual prediction. In: 2014 IEEE Conference on Computer Vision and Pattern Recognition, pp. 3302–3309. IEEE, Columbus, June 2014. ISBN 978-1-4799-5118-5. https://doi.org/10.1109/CVPR.2014.416

57. Vondrick, C., Pirsiavash, H., Torralba, A.: Anticipating visual representations from unlabeled video. In: 2016 IEEE Conference on Computer Vision and Pattern Recognition (CVPR), pp. 98–106. IEEE, Las Vegas, June 2016. ISBN 978-1-4673-8851-1. https://doi.org/10.1109/CVPR.2016.18

58. Srivastava, N., Mansimov, E., Salakhutdinov, R.: Unsupervised learning of video representations using LSTMs. arXiv:1502.04681 [cs], February 2015

59. Dosovitskiy, A., Koltun, V.: Learning to act by predicting the future. arXiv:1611.01779 [cs], November 2016

60. Ambady, N., Bernieri, F.J., Richeson, J.A.: Toward a histology of social behavior: judgmental accuracy from thin slices of the behavioral stream. In: Advances in Experimental Social Psychology, vol. 32, pp. 201–271. Elsevier, Amsterdam (2000)

61. Vinciarelli, A., Salamin, H., Pantic, M.: Social signal processing: understanding social interactions through nonverbal behavior analysis (PDF). In: 2009 IEEE Conference on Computer Vision and Pattern Recognition, CVPR 2009, June 2009. https://doi.org/10.1109/CVPRW.2009.5204290

62. Kalma, A.: Gazing in triads: a powerful signal in floor apportionment. Br. J. Soc. Psychol. **31**(1), 21–39 (1992)

63. Levinson, S.C., Torreira, F.: Timing in turn-taking and its implications for processing models of language. Front. Psychol. **6** (2015). ISSN 1664–1078. https://doi.org/10.3389/fpsyg.2015.00731

64. Delaherche, E., Chetouani, M., Mahdhaoui, A., Saint-Georges, C., Viaux, S., Cohen, D.: Interpersonal synchrony: a survey of evaluation methods across disciplines. IEEE Trans. Affect. Comput. **3**(3), 349–365 (2012). ISSN 1949–3045. https://doi.org/10.1109/T-AFFC.2012.12

65. Hospedales, T., Antoniou, A., Micaelli, P., Storkey, A.: Meta-learning in neural networks: a survey. arXiv:2004.05439 [cs, stat], November 2020

66. Garnelo, M., et al.: Neural processes. arXiv:1807.01622 [cs, stat] (2018)

67. Sutskever, I., Vinyals, O., Le, Q.V.: Sequence to sequence learning with neural networks. In: Ghahramani, Z., Welling, M., Cortes, C., Lawrence, N.D., Weinberger, K.Q. (eds.) Advances in Neural Information Processing Systems, vol. 27, pp. 3104–3112. Curran Associates Inc. (2014)

68. Cho, K., et al.: Learning phrase representations using RNN encoder-decoder for statistical machine translation. arXiv:1406.1078 [cs, stat], September 2014

69. Kim, H., et al.: Attentive neural processes. arXiv:1901.05761 [cs, stat], July 2019

70. Singh, G., Yoon, J., Son, Y., Ahn, S.: Sequential neural processes. In: Advances in Neural Information Processing Systems, vol. 32 (2019). https://arxiv.org/abs/1906.10264

71. Yoon, J., Singh, G., Ahn, S.: Robustifying sequential neural processes. In: International Conference on Machine Learning, pp. 10861–10870. PMLR, November 2020

72. Willi, T., Schmidhuber, J.M., Osendorfer, C.: Recurrent neural processes. arXiv:1906.05915 [cs, stat], November 2019

73. Kumar, S.: Spatiotemporal modeling using recurrent neural processes. Master of Science thesis, Carnegie Mellon University, p. 43 (2019)

74. Alameda-Pineda, X., Yan, Y., Ricci, E., Lanz, O., Sebe, N.: Analyzing free-standing conversational groups: a multimodal approach. In: Proceedings of the 23rd ACM International Conference on Multimedia, pp. 5–14. ACM Press (2015). ISBN 978-1-4503-3459-4. https://doi.org/10.1145/2733373.2806238

75. Zhang, L., Hung, H.: On social involvement in mingling scenarios: detecting associates of F-formations in still images. IEEE Trans. Affect. Comput. (2018)

76. Kendall, A., Cipolla, R.: Geometric loss functions for camera pose regression with deep learning. arXiv:1704.00390 [cs], May 2017

77. Vaswani, A., et al.: Attention is all you need. arXiv:1706.03762 [cs], June 2017

78. Ha, D., Eck, D.: A neural representation of sketch drawings. arXiv:1704.03477 [cs, stat], May 2017

79. Bowman, S.R., Vilnis, L., Vinyals, O., Dai, A.M., Jozefowicz, R., Bengio, S.: Generating sentences from a continuous space. arXiv:1511.06349 [cs], May 2016

80. Vazquez, M., Steinfeld, A., Hudson, S.E.: Maintaining awareness of the focus of attention of a conversation: a robot-centric reinforcement learning approach. In: 2016 25th IEEE International Symposium on Robot and Human Interactive Communication (RO-MAN), pp. 36–43. IEEE, New York, August 2016. ISBN 978-1-5090-3929-6. https://doi.org/10.1109/ROMAN.2016.7745088

81. Raman, C., Tan, S., Hung, H.: A modular approach for synchronized wireless multimodal multisensor data acquisition in highly dynamic social settings. arXiv preprint arXiv:2008.03715 (2020)

82. Raman, C., Hung, H.: Towards automatic estimation of conversation floors within f-formations. In: 2019 8th International Conference on Affective Computing and Intelligent Interaction Workshops and Demos (ACIIW), pp. 175–181. IEEE (2019)

83. Le, T.A., Kim, H., Garnelo, M.: Empirical evaluation of neural process objectives. In: NeurIPS workshop on Bayesian Deep Learning, . 71 (2018)

84. Kingma, D.P., Ba, J.: Adam: a method for stochastic optimization. arXiv:1412.6980 [cs], January 2017

85. Paszke, A., et al.: Pytorch: an imperative style, high-performance deep learning library. In: Advances in Neural Information Processing Systems, vol. 32, pp. 8024–8035. Curran Associates Inc. (2019)

86. Falcon, W.A., et al.: Pytorch lightning. GitHub. Note: https://github.com/PyTorchLightning/pytorch-lightning, 3, 2019

87. Rienks, R., Poppe, R., Poel, M.: Speaker prediction based on head orientations. In: Proceedings of the Fourteenth Annual Machine Learning Conference of Belgium and the Netherlands (Benelearn 2005), pp. 73–79 (2005)

88. Farenzena, M., et al.: Social interactions by visual focus of attention in a three-dimensional environment. Expert Syst. 30(2), 115–127 (2013). ISSN 02664720. https://doi.org/10.1111/j.1468-0394.2012.00622.x

89. Ba, S.O., Odobez, J.-M.: Recognizing visual focus of attention from head pose in natural meetings. IEEE Trans. Syst. Man Cybern. Part B (Cybern.) 39(1), 16–33, February 2009. ISSN 1083–4419. https://doi.org/10.1109/TSMCB.2008.927274

# Photo-Realistic 360° Head Avatars
# in the Wild

Stanislaw Szymanowicz[✉], Virginia Estellers, Tadas Baltrušaitis,
and Matthew Johnson

Microsoft, Washington, USA
szymanowiczs@gmail.com,
{virginia.estellers,tadas.baltrusaitis,matjoh}@microsoft.com

**Abstract.** Delivering immersive, 3D experiences for human communication requires a method to obtain 360° photo-realistic avatars of humans. To make these experiences accessible to all, only commodity hardware, like mobile phone cameras, should be necessary to capture the data needed for avatar creation. For avatars to be rendered realistically from any viewpoint, we require training images and camera poses from all angles. However, we cannot rely on there being trackable features in the foreground or background of all images for use in estimating poses, especially from the side or back of the head. To overcome this, we propose a novel landmark detector trained on synthetic data to estimate camera poses from 360° mobile phone videos of a human head for use in a multi-stage optimization process which creates a photo-realistic avatar. We perform validation experiments with synthetic data and showcase our method on 360° avatars trained from mobile phone videos.

**Keywords:** Neural rendering · Photo-realistic avatars · Mobile phone

## 1 Introduction

Immersive interaction scenarios on Mixed Reality devices require rendering human avatars from all angles. To avoid the uncanny valley effect, these avatars must have faces that are photo-realistic. It is likely that in the future virtual spaces will become a ubiquitous part of every-day life, impacting everything from a friendly gathering to obtaining a bank loan. For this reason we believe high-quality, 360° avatars should be affordable and accessible to all: created from images captured by commodity hardware, *e.g.,* from a handheld mobile phone, without restrictions on the surrounding environment (Fig. 1).

Obtaining data to train a 360° photo-realistic avatar 'in the wild' is challenging due to the potential difficulty of camera registration: traditional Structure-from-Motion pipelines rely on reliable feature matches of static objects across different images. Prior work limits the captures to a 120° frontal angle which allows the use of textured planar objects that are amenable to traditional feature detectors and descriptors (*e.g.,* a book, markers, detailed wall decoration). However, in many 360° captures from a mobile phone in unconstrained environments

© The Author(s), under exclusive license to Springer Nature Switzerland AG 2023
L. Karlinsky et al. (Eds.): ECCV 2022 Workshops, LNCS 13803, pp. 660–667, 2023.
https://doi.org/10.1007/978-3-031-25066-8_38

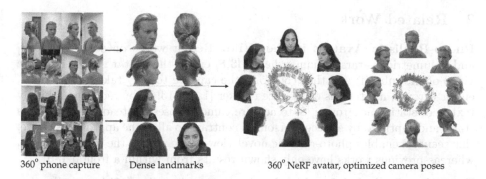

360° phone capture    Dense landmarks    360° NeRF avatar, optimized camera poses

**Fig. 1.** Our system creates photo-realistic 360° avatars from captures from a mobile phone capture and without constraints on the environment. Cameras are registered from full 360° pose variation, and our multi-stage optimization pipeline allows for high quality avatars.

neither the background nor the foreground can be depended upon to provide a source of such matches.

There are several properties of 360° captures in the wild which pose serious challenges to camera registration and avatar model learning. First, the space being captured is likely to either have plain backgrounds (*e.g.,* white walls) and/or portions of the capture in which the background is an open space, leading to defocus blur and the inclusion of extraneous, potentially mobile objects (*e.g.,* pets, cars, other people). Second, in order to obtain the needed details on the face and hair the foreground subject will likely occupy much of the frame. While the face can provide some useful features for camera registration, its non-planar nature combined with changes in appearance due to lighting effects make it less than ideal. Further, while the back of the head can produce many features for tracking the matching can become highly ambiguous due to issues with hair, *i.e.,* specular effects and repeated texture.

To address the challenges of 360° captures we propose a multi-stage pipeline to create 3D photo-realistic avatars from a mobile phone camera video. We propose using head landmarks to estimate the camera pose. However, as most facial landmark detectors are not reliable at oblique or backward-facing angles, we propose using synthetic data to train landmark detectors capable of working in full 360° range. We use the predicted landmarks to provide initialization of the 6DoF camera poses, for a system which jointly optimizes a simplified Neural Radiance Field with the camera poses. Finally, we use the optimized camera poses to train a high quality, photo-realistic NeRF of the subject.

The contributions of our work are three-fold: (1) a reliable system for camera registration which only requires the presence of a human head in each photo, (2) a demonstration of how to leverage synthetic data in a novel manner to obtain a DNN capable of predicting landmark locations from all angles, and (3) a multi-stage optimization pipeline which builds 360° photo-realistic avatars with high-frequency visual details using images obtained from a handheld mobile phone 'in the wild'.

## 2  Related Work

**Photo-Realistic Avatars in the Wild.** Recent works on surface [1,5,13] and volumetric avatars of human heads [3,8] enroll the avatar from a dynamic video of the subject, and they constrain the capture to only take frontal views, either through limitations of the face tracker [1,3,5,13] or by requiring a highly textured, static background [8] to acquire camera poses. Instead, our subject is static and captured by another person. In contrast to all of the approaches above, this scenario enables photo-realistic novel-view renders from the full 360° range, whereas previous works have only shown results from within a frontal 180°.

**Detecting Occluded Landmarks.** Landmark detection [12] is a challenging task even under ±60° yaw angle variation [14]. Existing datasets and approaches are limited to ±90° [2,14], partially due to the difficulty of reliably annotating dense landmarks in such data. Fortunately, recent work on using synthetic data in face understanding [11] has shown evidence of being able to detect landmarks even under severe occlusions. We push this capability to the limit: we aim to detect all landmarks even if completely occluded, *i.e.*, observing the back of the head. Unlike prior work, we use a synthetic dataset with full 360° camera angle variations. The synthetic nature of the dataset allows us to calculate ground truth landmark annotations which could not be reliably labelled on real data.

**Camera Poses.** We estimate initial camera poses from detected face landmarks, similarly to object pose detection from keypoints[?]. We are not constrained by the requirement of highly textured planar scene elements which Structure-from-Motion (SfM) methods impose. We are therefore able to capture our training data in the wild: we perform robust camera registration when a head is observed in the image, while SfM approaches can fail to register all cameras when insufficient matches are available, such as when observing dense dark hair.

Optimizing camera poses jointly with NeRF has been the topic of BARF [6] and NeRF-- [10]. We use the optimization method from BARF, but it only uses one, coarse, NeRF network. Therefore, for maximum visual quality we run a second optimization, starting from the optimized camera poses but initializing the coarse and fine networks from scratch and optimizing them jointly with camera poses. We hence achieve the high-frequency visual detail that the two-stage sampling provides, while optimizing the camera poses to avoid artefacts (Fig. 2).

## 3  Method

We estimate camera pose by solving a Perspective-n-Point (PnP) problem that matches 3D points in the world and their corresponding 2D projections in the images. The 2D image points are dense face landmarks while the 3D world points are the same landmarks annotated on a template face model [11]. We use standard convolutional neural networks (CNN) to estimate landmarks and standard RANSAC PnP solvers to solve for camera pose. Our results are possible because our landmark detector, trained with synthetic data, provides good landmarks for 360° views of a face within ±60° of elevation.

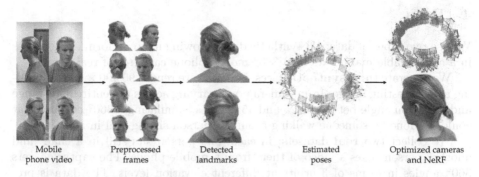

| Mobile phone video | Preprocessed frames | Detected landmarks | Estimated poses | Optimized cameras and NeRF |

**Fig. 2.** We build a 360° Neural Radiance Field of a person from a video captured from a mobile phone. In each selected frame we detect dense landmarks, and we use them to register the cameras from all angles. In the joint optimization process, we refine camera poses jointly with NeRF, obtaining a highly detailed volumetric avatar.

**Landmarks for Pose Estimation.** The main difference between our landmark detector and competing approaches is the number and quality of landmarks, specially for non-frontal views. While a human can consistently label frontal face images with tens of landmarks, it is impossible to annotate an image of the back of the head with hundred of landmarks in a manner consistent in 3D. Using synthetic data allows us to generate images with perfectly consistent labels. We use the framework introduced by Wood *et al.* [11] to render a synthetic training dataset of 100k images that contain extreme head and camera pose variations, covering the 360° range. We use it to train an off-the-shelf ResNet-101 landmark detector model to directly regress 703 landmark positions from an image [12].

The camera poses estimated from landmarks might contain errors because the occluded landmarks are a rough estimate based on the priors learned by the CNN, but they are good enough for our avatar generation system.

**Avatar Generation.** We choose to represent our 3D avatar with a Neural Radiance Field [7] (NeRF): a volumetric representation of the scene because it can generate novel views with a high level of photo-realism and can handle human hair and accessories seamlessly [9]. This is important for in-the-wild captures where no assumptions of subjects appearance are possible.

We use the original representation, architecture and implementation details of NeRF [7]. We optimize the camera poses jointly with the radiance field in a coarse-to-fine manner, as in BARF [6]. While BARF [6] uses only the coarse MLP, we improve on its proposed optimization scheme with a second optimization step to achieve higher fidelity. The steps are as follows: (1) Initialize camera poses from landmarks and optimize them jointly with the coarse MLP following BARF [6]. Discard the coarse MLP. (2) Initialize camera poses from step 1 and optimize them jointly with both coarse and fine MLPs initialized from scratch.

## 4    Data

We use two types of data: (1) synthetic data, allowing us to perform experiments in a controllable manner and (2) 360° mobile phone captures of real people.

We generate three **synthetic** faces. For each, we render 90 400 × 400 px training and 10 testing images from camera poses varying across the entire 360° range and elevation angle between −25 and 45 degrees, simulating a 360° capture that could be done by someone walking around a person sitting still in a chair.

We collect two **real** datasets: in each, one person sits still in a chair, and another person takes a video of them from a mobile phone. The capture covers 360° angles in forms of 3 orbits at different elevation levels. The data is pre-processed for training the avatar by selecting 100 images without motion blur, undistorting them, cropping to a square, segmenting out the background and resizing to 400 × 400 resolution.

## 5    Experiments

**Dense Landmark Detection and Pose Registration.** Figure 3 shows dense landmark detection on real and synthetic images of faces. The results on real images of the back of the head show that a network trained with synthetic data learns correct priors to estimate plausible landmarks from weak cues like ears or the head orientation. With these landmarks, we can register cameras from images of 360° angles around the head (blue frustums in Fig. 3) where traditional SfM pipelines like COLMAP (red frustums in Fig. 3) partially fail because images of the back of the head do not have enough reliable matches.

**Fig. 3.** Detected landmarks are plausible on both synthetic and real faces from all viewpoints and allow to register all cameras (blue frustums). In our sequences SfM pipelines like COLMAP only register front-facing cameras (red frustums). (Color figure online)

**Avatars of Synthetic Faces.** We compare avatar quality as trained from 1) ground truth poses, 2) poses as estimated from dense landmarks, 3) poses optimized jointly with NeRF when initialized from dense landmarks. Figure 4 qualitatively compares novel view renders from different approaches and its associated table provides quantitative metrics w.r.t. ground truth camera poses.

Training a NeRF avatar with noisy poses estimated from landmarks leads to artefacts in the renders, but using them as initialization and optimizing jointly

Landmark + PnP                    Optimized

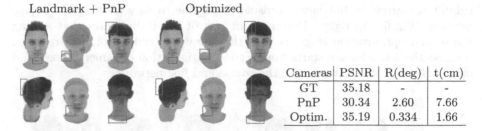

| Cameras | PSNR | R(deg) | t(cm) |
|---------|-------|--------|-------|
| GT | 35.18 | - | - |
| PnP | 30.34 | 2.60 | 7.66 |
| Optim. | 35.19 | 0.334 | 1.66 |

**Fig. 4.** Novel viewpoints rendered from a NeRF trained with camera poses from PnP exhibit artefacts (left). Optimizing (Optim.) poses jointly with NeRF removes the artefacts (middle), recovers the visual quality of NeRF trained with Ground Truth cameras (GT) and the registration error (right). Best viewed zoomed in.

with NeRF removes the artefacts and successfully recovers camera poses. The error in camera poses, leads to a significant drop in PSNR (Fig. 4) of novel-view renders. However, optimizing camera poses jointly with NeRF leads to novel-view quality on par with that obtained using the ground truth cameras.

**Fig. 5.** Novel view renders of 360° avatars trained from images captured from a phone.

**Avatars of Real Faces.** We test our full pipeline on two real captures. Figure 5 shows novel-view renders of avatars trained on 360° capture. The detail in renders is visible even from the back, where the initial camera poses were likely with the most error. We hypothesize that white regions on the torso stem from being cropped by segmentation masks in training data.

Figure 6 shows the comparison of our method to the baselines on real data: (1) using camera poses as estimated from the dense landmarks, (2) using cameras from manually joined models output by off-the-shelf Structure-from-Motion software COLMAP, (3) optimizing a coarse NeRF jointly with camera poses initialized with the estimate from dense landmarks, (4) two-stage optimization process where second stage optimizes both fine and coarse NeRF from scratch with poses initialized at the output of (3).

Error in camera poses result in avatars with artefacts (Fig. 6, left). NeRF optimized jointly with camera poses in the same manner as in BARF does not exhibit the artefacts but lacks in visual detail due to only training the coarse network (Fig. 6, top right). Bottom right row of Fig. 6 illustrates that adding the second optimization stage recovers the fine detail while avoiding artefacts, because the second stage starts from a good initialization of camera poses, and optimizes them jointly with both the coarse and fine network.

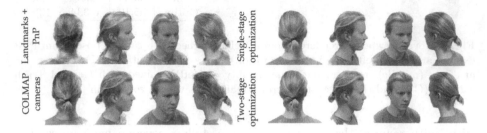

**Fig. 6.** Training a NeRF with cameras with a registration error results in artefacts in renders from novel viewpoints (left). Using BARF [6] removes the artefacts, but lacks the high-frequency details (top right). Adding our second optimization stage leads to an avatar with high-frequency detail but without artefacts (bottom right).

## 6 Conclusions

**Limitations and Future Work.** First, while our current formulation only renders a static avatar, recent work [4] could be used to animate it using geometry-driven deformations. Secondly, we train NeRF on images with background segmented out, and segmentation errors, *e.g.*, on the hair, can lead to incorrectly learnt colour which is then revealed from different viewpoints resulting in artefacts. Future work should aim to refine masks in the optimization process by using the projection of the density field. Finally, we observed that the camera pose estimates for the subject with hair covering the ears resulted in larger camera pose error. While in our experiments we observed that the pose optimization converged to a good solution, future work should (1) investigate the basin of convergence of camera poses and (2) test the method on more subjects.

**Conclusions.** In this paper we demonstrate that we can obtain a 360° avatar from a mobile phone video of a subject gathered 'in the wild'. We train a facial landmark DNN on synthetic data which is sampled from the full 360° range of view angles. We then show that the landmarks predicted by this DNN can be used to initialize the camera poses for the mobile phone image sequence. This is the first stage of an optimization pipeline which we demonstrate can progressively fine-tune the camera poses while training a NeRF that can be used to create photo-realistic, highly detailed images of the subject from novel views, including from the back and side. This can be used as the foundation for a 360° avatar of the kind shown in [4].

# References

1. Cao, C., et al.: Authentic volumetric avatars from a phone scan. ACM Trans. Graph. **41**(4) (2022). https://doi.org/10.1145/3528223.3530143
2. Deng, J., et al.: The menpo benchmark for multi-pose 2D and 3D facial landmark localisation and tracking. Int. J. Comput. Vis. **127**(6), 599–624 (2018). https://doi.org/10.1007/s11263-018-1134-y
3. Gafni, G., Thies, J., Zollhöfer, M., Nießner, M.: Dynamic neural radiance fields for monocular 4d facial avatar reconstruction. In: Proceedings of the IEEE/CVF Conference on Computer Vision and Pattern Recognition (CVPR) (2021)
4. Garbin, S.J., et al.: VolTeMorph: realtime, controllable and generalisable animation of volumetric representations (2022). https://arxiv.org/abs/2208.00949
5. Grassal, P.W., Prinzler, M., Leistner, T., Rother, C., Nießner, M., Thies, J.: Neural head avatars from monocular RGB videos. In: Proceedings of the IEEE/CVF Conference on Computer Vision and Pattern Recognition (CVPR) (2022)
6. Lin, C.H., Ma, W.C., Torralba, A., Lucey, S.: BARF: bundle-adjusting neural radiance fields. In: Proceedings of the IEEE/CVF International Conference on Computer Vision (ICCV) (2021)
7. Mildenhall, B., Srinivasan, P.P., Tancik, M., Barron, J.T., Ramamoorthi, R., Ng, R.: NeRF: representing scenes as neural radiance fields for view synthesis. In: Proceedings of the European Conference on Computer Vision (ECCV) (2020)
8. Park, K., et al.: Nerfies: Deformable neural radiance fields. In: Proceedings of the IEEE International Conference on Computer Vision (ICCV) (2021)
9. Vicini, D., Jakob, W., Kaplanyan, A.: A non-exponential transmittance model for volumetric scene representations. SIGGRAPH **40**(4), 136:1–136:16 (2021). https://doi.org/10.1145/3450626.3459815
10. Wang, Z., Wu, S., Xie, W., Chen, M., Prisacariu, V.A.: NeRF−: Neural radiance fields without known camera parameters. arXiv preprint arXiv:2102.07064 (2021)
11. Wood, E., et al.: Fake it till you make it: Face analysis in the wild using synthetic data alone (2021)
12. Wood, E., et al.: 3D face reconstruction with dense landmarks. arXiv preprint arXiv:2204.02776 (2022)
13. Zheng, Y., Abrevaya, V.F., Bühler, M.C., Chen, X., Black, M.J., Hilliges, O.: I M Avatar: Implicit morphable head avatars from videos. In: 2022 IEEE/CVF Conference on Computer Vision and Pattern Recognition (CVPR) (2022)
14. Zhu, X., Lei, Z., Liu, X., Shi, H., Li, S.Z.: Face alignment across large poses: a 3D solution. In: 2016 IEEE Conference on Computer Vision and Pattern Recognition (CVPR) (2016)

# AvatarGen: A 3D Generative Model for Animatable Human Avatars

Jianfeng Zhang[1][✉], Zihang Jiang[1], Dingdong Yang[2], Hongyi Xu[2], Yichun Shi[2], Guoxian Song[2], Zhongcong Xu[1], Xinchao Wang[1], and Jiashi Feng[2]

[1] National University of Singapore, Singapore, Singapore
zhangjianfeng@u.nus.edu
[2] ByteDance, Beijing, China

**Abstract.** Unsupervised generation of clothed virtual humans with various appearance and animatable poses is important for creating 3D human avatars and other AR/VR applications. Existing methods are either limited to rigid object modeling, or not generative and thus unable to synthesize high-quality virtual humans and animate them. In this work, we propose AvatarGen, the first method that enables not only non-rigid human generation with diverse appearance but also full control over poses and viewpoints, while only requiring 2D images for training. Specifically, it extends the recent 3D GANs to clothed human generation by utilizing a coarse human body model as a proxy to warp the observation space into a standard avatar under a canonical space. To model non-rigid dynamics, it introduces a deformation network to learn pose-dependent deformations in the canonical space. To improve geometry quality of the generated human avatars, it leverages signed distance field as geometric representation, which allows more direct regularization from the body model on the geometry learning. Benefiting from these designs, our method can generate animatable human avatars with high-quality appearance and geometry modeling, significantly outperforming previous 3D GANs. Furthermore, it is competent for many applications, *e.g.*, single-view reconstruction, reanimation, and text-guided synthesis.

## 1 Introduction

Generating diverse and high-quality virtual humans (avatars) with full control over their pose and viewpoint is a fundamental but extremely challenging task. Solving this task will benefit many applications like immersive photography visualization [68], virtual try-on [32], VR/AR [22,61] and image editing [19,67].

Conventional solutions rely on classical graphics modeling and rendering techniques [7,9,12,55] to create avatars. Though offering high-quality, they typically require pre-captured templates, multi-camera systems, controlled studios,

---

J. Zhang and Z. Jiang: Equal contribution.

**Supplementary Information** The online version contains supplementary material available at https://doi.org/10.1007/978-3-031-25066-8_39.

and long-term works of artists. In this work, we aim to make virtual human avatars widely accessible at low cost. To this end, we propose the first 3D-aware avatar generative model that can *generate* 1) high-quality virtual humans with 2) various appearance styles, arbitrary poses and viewpoints, 3) and be trainable from only 2D images, thus largely alleviating the effort to create avatars.

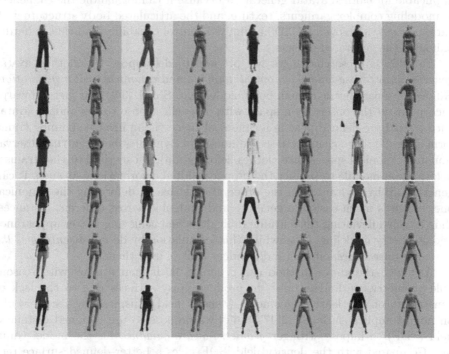

**Fig. 1.** Our AvatarGen model can generate clothed avatars with diverse appearance from arbitrary poses and viewpoints (top), animatable with specific pose signals (bottom). The examples are shown in gif and best viewed with Adobe Reader (click and play).

The 3D-aware generative models have recently seen rapid progress, fueled by introducing implicit neural representation (INR) methods [6,36,37,43] into generative adversarial networks [2,3,16,40,42]. However, these models are limited to relatively simple and rigid objects, such as human faces and cars, and mostly fail to generate clothed human avatars whose appearance is highly sundry because of their articulated poses and great variability of clothing. Besides, they have limited control over the generation process and thus cannot animate the generated objects, *i.e.*, driving the objects to move by following certain instructions. Another line of works leverage INRs [37] to learn articulated human avatars for reconstructing a single subject from one's multi-view images or videos [5,41,46,47,62]. While being able to animate the avatars, these methods are *not generative* and cannot synthesize novel identities and appearances.

Aiming at generative modeling of animatable human avatars, we propose AvatarGen, the first model that can generate *novel* human avatars with full control over their poses and appearances. Our model is built upon EG3D [2], a recent method that can generate high quality 3D-aware human faces via introducing a new tri-plane representation method. However, EG3D is not directly applicable for clothed avatar generation because it cannot handle the challenges in modeling complex garments, texture, and the articulated body structure with various poses. Moreover, EG3D has limited control capability and thus it hardly animates the generated objects.

To address these challenges, we propose to decompose the clothed avatar generation into *pose-guided canonical mapping* and *canonical avatar generation*. Guided by a parametric human body model (*e.g.*, SMPL [34]), our method warps each point in the observation space with a specified pose to a standard avatar with a fixed pre-defined pose in a canonical space via an inverse-skinning transformation [20]. To accommodate the non-rigid dynamics between the observation and canonical spaces (like clothes deformation), our method further trains a deformation module to predict the proper residual deformation. As such, it can generate arbitrary avatars in the observation space by deforming the canonical one, which is much easier to generate and shareable across different instances, thus largely alleviating the learning difficulties and achieving better appearance and geometry modeling. Meanwhile, this formulation by design *disentangles the pose and appearance*, offering independent control over them.

Although the above method can generate 3D human avatars with reasonable geometry, we find it tends to produce noisy surfaces due to the lack of constraints on the learned geometry (density field). Inspired by recent works on neural implicit surface [42,47,58,65], we propose to use a signed distanced field (SDF) to impose stronger *geometry-aware guidance* for the model training. Compared with the density field, SDF gives a better-defined surface representation, which facilitates more direct regularization on learning the avatar geometries. Moreover, the model can leverage the coarse body model from SMPL to infer reasonable signed distance values, which greatly improves quality of the clothed avatar generation and animation. The SDF-based volume rendering techniques [42,58,65] are used to render the low resolution feature maps, which are further decoded to high-resolution images with the StyleGAN generator [2,26].

As shown in Fig. 1, trained from 2D images without using any multi-view or temporal information and 3D geometry annotations, AvatarGen can generate a large variety of clothed human with diverse appearances under arbitrary poses and viewpoints. We evaluate it quantitatively, qualitatively, and through a perceptual study; it strongly outperforms previous state-of-the-art methods. Moreover, we demonstrate it on several applications, like single-view 3D reconstruction and text-guided synthesis.

We make the following contributions. 1) To our best knowledge, AvatarGen is the first model able to generate a large variety of animatable clothed human avatars without requiring multi-view, temporal or 3D annotated data. 2) We propose a human avatar generation pipeline that achieves realistic appearance and geometry modeling, with full control over the pose and appearance.

## 2 Related Works

**Generative 3D-Aware Image Synthesis.** Generative adversarial networks (GANs) [15] have recently achieved photo-realistic image quality for 2D image synthesis [23–26]. Extending these capabilities to 3D settings has started to gain attention. Early methods combine GANs with voxel [38,39,60], mesh [30,56] or point cloud [1,29] representations for 3D-aware image synthesis. Recently, several methods represent 3D objects by learning an implicit neural representation (INR) [2,3,10,16,40,42,53]. Among them, some methods use INR-based model as generator [3,10,53], while some others combine INR generator with 2D decoder for higher-resolution image generation [16,40,64]. Follow-up works like EG3D [2] proposes an efficient tri-plane representation to model 3D objects, StyleSDF [42] replaces density field with SDF for better geometry modeling and Disentangled3D [57] represents objects with a canonical volume along with deformations to disentangle geometry and appearance modeling. However, such methods are typically not easily extended to non-rigid clothed humans due to the complex pose and texture variations. Moreover, they have limited control over the generation process, making the generated objects hardly be animated. Differently, we study the problem of 3D implicit generative modeling of clothed human, allowing free control over the poses and appearances.

**3D Human Reconstruction and Animation.** Traditional human reconstruction methods require complicated hardware that is expensive for daily use, such as depth sensors [7,12,55] or dense camera arrays [9,17]. To alleviate the requirement on the capture device, some methods train networks to reconstruct human models from RGB images with differentiable renderers [14,63]. Recently, neural radiance fields [37] employ the volume rendering to learn density and color fields from dense camera views. Some methods augment neural radiance fields with human shape priors to enable 3D human reconstruction from sparse multi-view data [5,48,54,62]. Follow-up improvements [4,31,46,47,59] are made by combining implicit representation with the SMPL model and exploiting the linear blend skinning techniques to learn animatable 3D human modeling from temporal data. However, these methods are not generative, *i.e.*, they cannot synthesize novel identities and appearances. Instead, we learn fully generative modeling of human avatars from only 2D images.

## 3 Method

Our goal is to build a generative model for diverse clothed 3D human avatar generation with varying appearances in arbitrary poses. The model is trained from 2D images without using multi-view or temporal information and 3D scan annotations. Its framework is summarized in Fig. 2.

### 3.1 Overview

**Problem Formulation.** We aim to train a 3D generative model $G$ for geometry-aware human synthesis. Following EG3D [2], we associate each

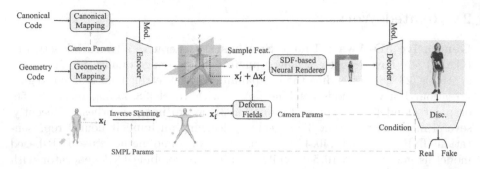

**Fig. 2.** Pipeline of AvatarGen. Taking the canonical code and camera parameters as input, the encoder generates tri-plane based features of a canonical posed human avatar. The geometry code is applied to modulate the deformation field module which deforms the sampled points in the observation space to the canonical space under the guidance of the pose condition (SMPL parameters). The deformed spatial positions are used to sample features on the tri-plane, which are then rendered as low-resolution features and images using the SDF based neural renderer. Finally, the decoder decodes the feature images to high resolution images. The generator with is optimized a camera and pose conditioned discriminator via adversarial training.

training image with a set of camera parameters $\mathbf{c}$ and pose parameters $\mathbf{p}$ (in SMPL format [34]), which are obtained from an off-the-shelf pose estimator [27]. Given a random latent code $\mathbf{z}$ sampled from Gaussian distribution, and a new camera $\mathbf{c}$ and pose $\mathbf{p}$ as conditions, the generator $G$ can synthesize a corresponding human image $I = G(\mathbf{z}|\mathbf{c}, \mathbf{p})$. We optimize $G$ with a discriminator $D$ via adversarial training.

**Framework.** Figure 2 illustrates the framework of our proposed generative model. It takes camera parameters and SMPL parameters (specifying the generated pose) as inputs and generates 3D human avatar and its 2D images accordingly. Our model jointly processes the random canonical code and camera parameters by a canonical mapping module (implemented by an MLP) to generate the intermediate latent code that modulates the convolution kernels of the encoder. This encoder then generates a tri-plane based features [2] corresponding to a canonical pose representation. Regarding the pose control, our model first jointly processes a random geometry code and the input pose condition via an MLP-based geometry mapping module, and outputs latent features. Then, given a spatial point $\mathbf{x}$ in the observation space, the output latent features are processed by a deformation filed module, generating non-rigid residual deformation $\Delta \mathbf{x}'$ over the inverse skinned point $\mathbf{x}'$ in the canonical space. We sample the features from the tri-plane according to the deformed spatial position $\mathbf{x}' + \Delta \mathbf{x}'$, which are then transformed into appearance prediction (i.e., color features) and geometry prediction (i.e., SDF-based features) for volume rendering. We will explain these steps in details in the following sections.

## 3.2    Representations of 3D Avatars

It is important to choose an efficient approach to represent 3D human avatars. The recent EG3D [2] introduces a memory-efficient tri-plane representation for 3D face modeling. It explicitly stores features on three axis-aligned orthogonal planes (called tri-planes), each of which corresponds to a dimension within the 3D space. Thus, the intermediate features of any point of a 3D face can be obtained via simple lookup over the tri-planes, making the feature extraction much more efficient than NeRF that needs to forward all the sampled 3D points through MLPs [37]. Besides, the tri-plane representation effectively decouples the feature generation from volume rendering, and can be directly generated from more efficient CNNs instead of MLPs.

Considering these benefits, we also choose the tri-plane representation for 3D avatar modeling. However, we found directly adopting it for clothed human avatars generation results in poor quality. Due to the much higher degrees of freedom of human bodies than faces, it is very challenging for the naive tri-plane representation model to learn pose-dependent appearance and geometry from only 2D images. We thus propose the following new approaches to address the difficulties.

## 3.3    Generative 3D Human Modeling

There are two main challenges for 3D human generation. The first is how to effectively integrate pose condition into the tri-plane representations, making the generated human pose fully controllable and animatable. One naive way is to combine the pose condition $\mathbf{p}$ with the latent codes $\mathbf{z}$ and $\mathbf{c}$ directly, and feed them to the encoder to generate tri-plane features. However, such naive design cannot achieve high-quality synthesis and animation due to limited pose diversity and insufficient geometry supervision. Besides, the pose and appearance are tightly entangled, making independent control impossible. The second challenge is that learning pose-dependent clothed human appearance and geometry from 2D images only is highly under-constrained, making the model training difficult and generation quality poor.

To tackle these challenges, our AvatarGen decomposes the avatar generation into two steps: *pose-guided canonical mapping* and *canonical avatar generation*. Specifically, AvatarGen uses SMPL model [34] to parameterize the underlying 3D human body. With a pose parameterization, AvatarGen can easily deform a 3D point $\mathbf{x}$ within the observation space with pose $\mathbf{p}_o$ to a canonical pose $\mathbf{p}_c$ (an "X"- pose as shown in Fig. 2) via Linear Blend Skinning [21]. Then, AvatarGen learns to generate the appearance and geometry of human avatar in the canonical space. The canonical space is shared across different instances with a fixed template pose $\mathbf{p}_c$ while its appearance and geometric details can be varied according to the latent code $\mathbf{z}$, leading to generative human modeling.

Such a task factorization scheme facilitates learning of a generative canonical human avatars and effectively helps the model generalize to unseen poses, achieving animatable clothed human avatars generation. Moreover, it by design disentangles pose and appearance information, making independent control over them feasible. We now elaborate on the model design for these two steps.

**Pose-Guided Canonical Mapping.** We define the human 2D image with SMPL pose $\mathbf{p}_o$ as the *observation* space. To relieve learning difficulties, our model attempts to deform the observation space to a *canonical* space with a predefined template pose $\mathbf{p}_c$ that is shared across different identities. The deformation function $T : \mathbb{R}^3 \mapsto \mathbb{R}^3$ thus maps spatial points $\mathbf{x}_i$ sampled in the observation space to $\mathbf{x}'_i$ in the canonical space.

Learning such a deformation function has been proved effective for dynamic scene modeling [44,49]. However, learning to deform in such an implicit manner cannot handle large articulation of humans and thus hardly generalizes to novel poses. To overcome this limitation, we use the SMPL model to explicitly guide the deformation [4,31,46]. SMPL defines a skinned vertex-based human model $(\mathcal{V}, \mathcal{W})$, where $\mathcal{V} = \{\mathbf{v}\} \in \mathbb{R}^{N\times3}$ is the set of $N$ vertices and $\mathcal{W} = \{\mathbf{w}\} \in \mathbb{R}^{N\times K}$ is the set of the skinning weights assigned for the vertex w.r.t. $K$ joints, with $\sum_j w_j = 1, w_j \geq 0$ for every joint.

We use the inverse-skinning (IS) transformation to map the SMPL mesh in the observation space with pose $\mathbf{p}$ into the canonical space [20]:

$$T_{\mathrm{IS}}(\mathbf{v}, \mathbf{w}, \mathbf{p}) = \sum_j w_j \cdot (R_j \mathbf{v} + \mathbf{t}_j), \tag{1}$$

where $R_j$ and $\mathbf{t}_j$ are the rotation and translation at each joint $j$ derived from SMPL with pose $\mathbf{p}$.

Such formulation can be easily extended to any spatial points in the observation space by simply adopting the same transformation from the nearest point on the surface of SMPL mesh. Formally, for each spatial points $\mathbf{x}_i$, we first find its nearest point $\mathbf{v}^*$ on the SMPL mesh surface as $\mathbf{v}^* = \arg\min_{\mathbf{v}\in\mathcal{V}} ||\mathbf{x}_i - \mathbf{v}||_2$. Then, we use the corresponding skinning weights $\mathbf{w}^*$ to deform $\mathbf{x}_i$ to $\mathbf{x}'_i$ in the canonical space as:

$$\mathbf{x}'_i = T_o(\mathbf{x}_i|\mathbf{p}) = T_{\mathrm{IS}}(\mathbf{x}_i, \mathbf{w}^*, \mathbf{p}). \tag{2}$$

Although the SMPL-guided inverse-skinning transformation can help align the rigid skeleton with the template pose, it lacks the ability to model the pose-dependent deformation, like cloth wrinkles. Besides, different identities may have different SMPL shape parameters $\mathbf{b}$, which likely leads to inaccurate transformation. To alleviate these issues, AvatarGen further trains a deformation network to model the residual deformation to complete the fine-grained geometric deformation and to compensate the inaccurate inverse-skinning transformation by

$$\Delta\mathbf{x}'_i = T_\Delta(\mathbf{x}'_i|\mathbf{w}, \mathbf{p}, \mathbf{b}) = \mathrm{MLPs}(\mathrm{Concat}[\mathrm{Embed}(\mathbf{x}'_i), \mathbf{w}, \mathbf{p}, \mathbf{b}]), \tag{3}$$

where $\mathbf{w}$ is the canonical style code mapped from the input latent code $\mathbf{z}$. We concatenate it with the embedded $\mathbf{x}'_i$ and SMPL pose $\mathbf{p}$ and shape $\mathbf{b}$ parameters and feed them to MLPs to yield the residual deformation. The final pose-guided deformation $T_{o\to c}$ from the observation to canonical spaces is formulated as

$$T_{o\to c}(\mathbf{x}_i) = \mathbf{x}'_i + \Delta\mathbf{x}'_i = T_o(\mathbf{x}_i|\mathbf{p}) + T_\Delta(T_o(\mathbf{x}_i|\mathbf{p})|\mathbf{w}, \mathbf{p}, \mathbf{b}). \tag{4}$$

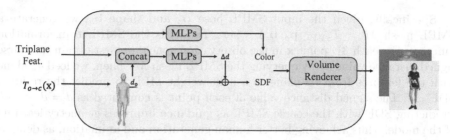

**Fig. 3.** Our proposed geometry-aware human modeling module. It first predicts color and SDF values using the sampled tri-plane features and corresponding point location in the canonical space. Then it feeds them to volume renderer module to generate the raw image and features. The color for each sampled point is directly predicted from the tri-plane feature with MLPs. We use the SMPL model to guide the prediction of SDF and obtain a coarse signed distance value $d_o$ which is concatenated with the input tri-plane feature for predicting the residual distance $\Delta d$. The final SDF is $d_o + \Delta d$. See 3.4 for more details.

**Canonical Avatar Generation.** After deforming 3D points sampled in the observation space to the canonical space, we apply AvatarGen with the tri-plane representation for canonical avatar generation. More concretely, it first generates tri-plane via a StyleGAN generator by taking the latent code $\mathbf{z}$ and camera parameters $\mathbf{c}$ as inputs. Then, for each point deformed via SMPL parameters $\mathbf{p}$ in the canonical space, the model queries tri-plane to obtain the intermediate feature and maps it to color-based feature $c$ and density $\sigma$ for volume rendering. As such, it generates clothed human appearance and geometry in the canonical space with a predefined canonical pose, which alleviates the optimization difficulties and substantially helps our learning of high-quality avatar generation with disentangled pose and appearance control.

### 3.4 Geometry-Aware Human Modeling

To improve geometry modeling of AvatarGen, inspired by recent neural implicit surface works [42,47,58,65], we adopt signed distance field (SDF) instead of density field as our geometry proxy, because it introduces more direct geometry regularization and guidance. To achieve this, our model learns to predict signed distance value rather than density in tri-plane for volume rendering.

**SMPL-Guided Geometry Learning.** Although SDF has a well-defined surface representation and introduces several regularization for geometry learning, how to use it for generative human modeling is still non-trivial due to the complicated body articulation and pose-dependent deformation. We therefore leverage the SMPL model as a guidance for the geometry-aware generation and combine it with a residual SDF network (as shown in Fig. 3), that models the surface details (including hair and clothing) not represented by SMPL.

Specifically, given the input SMPL pose $\mathbf{p}_o$ and shape $\mathbf{b}_o$, we generate a SMPL mesh $M = T_{\text{SMPL}}(\mathbf{p}_o, \mathbf{b}_o)$, where $T_{\text{SMPL}}$ is the SMPL transformation function. For each 3D point $\mathbf{x}$ in the observation space, we first obtain its coarse signed distance value $d_o$ by querying the SMPL mesh $M$. Then, we feed $d_o$ alone with the features from tri-plane to a light-weight MLP to predict the residual SDF $\Delta d$. The signed distance value of each point is computed as $d = d_o + \Delta d$. Predicting SDF with the coarse SMPL as guidance improves geometry learning of the model, thus achieving better human generation and animation, as demonstrated in our experiments. We also introduce a SMPL-guided regularization for SDF learning as elaborated in Sect. 3.5.

**SDF-Based Volume Rendering.** Following [42], we adopt SDF-based volume rendering to obtain the final output images. For any point $x$ on the sampled rays, we first deform it to $\bar{x} = T_{o \rightarrow c}(x)$ by pose-guided canonical mapping. We query feature vector $F(\bar{x})$ for position $\bar{x}$ from the canonical tri-plane and then feed it into two MLP layers to predict the color feature $c = \text{MLP}_c(F(\bar{x}))$ and the signed distance $d = d_o + \Delta d = d_o + \text{MLP}_d(F(\bar{x}), d_o)$. We then convert the signed distance value $d_i$ of each point $\mathbf{x}_i$ alone a ray $r$ to density value $\sigma_i$ as $\sigma_i = \frac{1}{\alpha} \cdot \text{Sigmoid}(\frac{-d_i}{\alpha})$, where $\alpha > 0$ is a learnable parameter that controls the tightness of the density around the surface boundary. By integration along the ray $r$ we can get the corresponding pixel feature as

$$I(r) = \sum_{i=1}^{N} \left( \prod_{j=1}^{i-1} e^{-\sigma_j \cdot \delta_j} \right) \cdot \left( 1 - e^{-\sigma_i \cdot \delta_i} \right) \cdot c_i, \tag{5}$$

where $\delta_i = ||\mathbf{x}_i - \mathbf{x}_{i-1}||$. By aggregating all rays, we can get the entire image feature which is then feed into a StyleGAN decoder [26] to generate the final high-resolutions synthesized image.

## 3.5   Training

We use the non-saturating GAN loss $L_{\text{GAN}}$ [26] with R1 regularization $L_{\text{Reg}}$ [35] to train our model end-to-end. We also adopt the dual-discriminator proposed by EG3D [2]. It feeds both the rendered raw image and the decoded high-resolution image into the discriminator for improving consistency of the generated multi-view images. To obtain better controllability, we feed both SMPL pose parameters $\mathbf{p}$ and camera parameters $\mathbf{c}$ as conditions to the discriminator for adversary training. To regularize the learned SDF, we apply eikonal loss to the sampled points as:

$$L_{\text{Eik}} = \sum_{\mathbf{x}_i} (||\nabla d_i|| - 1)^2, \tag{6}$$

where $\mathbf{x}_i$ and $d_i$ denote the sampled point and predicted signed distance value, respectively. Following [42], we adopt a minimal surface loss to encourage the model to represent human geometry with minimal volume of zero-crossings. It penalizes the SDF value close to zero:

$$L_{\text{Minsurf}} = \sum_{\mathbf{x}_i} \exp(-100 d_i). \tag{7}$$

To make sure the generated surface is consistent with the input SMPL model, we incorporate the SMPL mesh as geometric prior and guide the generated surface to be close to the body surface. Specifically, we sample vertices $\mathbf{v} \in \mathcal{V}$ on the SMPL body surface and then use it as query to deform to canonical space and sample features from the generated tri-plane and minimize the signed distance.

$$L_{\text{SMPL}} = \sum_{\mathbf{v} \in \mathcal{V}} ||\text{MLP}_d(F(T_{o \to c}(\mathbf{v})))||. \tag{8}$$

The overall loss is finally formulated as

$$L_{\text{total}} = L_{\text{GAN}} + \lambda_{\text{Reg}} L_{\text{Reg}} + \lambda_{\text{Eik}} L_{\text{Eik}} + \lambda_{\text{Minsurf}} L_{\text{Minsurf}} + \lambda_{\text{SMPL}} L_{\text{SMPL}}, \tag{9}$$

where $\lambda_*$ are the corresponding loss weights.

## 4 Experiments

**Datasets.** We evaluate methods of 3D-aware clothed human generation on three real-world fashion datasets: MPV [11], DeepFashion [33] and UBCFashion [66]. They contain single clothed people in each image. We align and crop images according to the 2D human body keypoints, following [13], and resize them to $256 \times 256$ resolution. Since we focus on human avatar generation, we use a segmentation model [8] to remove irrelevant backgrounds. We adopt an off-the-shelf pose estimator [27] to obtain approximate camera and SMPL parameters. We filter out images with partial observations and those with poor SMPL estimations, and get nearly 15K, 14K and 31K full-body images for each dataset, respectively. Horizontal-flip augmentation is used during training. We note these datasets are primarily composed of front-view images-few images captured from side or back views. To compensate this, we sample more side- and back-view images to re-balance viewpoint distributions following [2]. We will release our data processing scripts and pre-processed datasets.

### 4.1 Comparisons

**Baselines.** We compare our AvatarGen against four state-of-the-art methods for 3D-aware image synthesis: EG3D [2], StyleSDF [42], StyleNeRF [16] and GIRAFFE-HD [64]. All these methods combine volume renderer with 2D decoder for 3D-aware image synthesis. EG3D and StyleNeRF adopt progressive training to improve performance. StyleSDF uses SDF as geometry representation for regularized geometry modeling.

**Quantitative Evaluations.** Table 1 provides quantitative comparisons between our AvatarGen and the baselines. We measure image quality with Fréchet Inception Distance (FID) [18] between 50k generated images and all of the available real images. We evaluate geometry quality by calculating Mean Squared Error (MSE) against pseudo groundtruth (GT) depth-maps (*Depth*)

**Table 1.** Quantitative evaluation in terms of FID, depth, pose and warp accuracy on three datasets. Our AvatarGen outperforms all the baselines significantly

| | MPV | | | | DeepFashion | | | | UBCFashion | | | |
|---|---|---|---|---|---|---|---|---|---|---|---|---|
| | FID ↓ | Depth ↓ | Pose ↓ | Warp ↓ | FID ↓ | Depth ↓ | Pose ↓ | Warp ↓ | FID ↓ | Depth ↓ | Pose ↓ | Warp ↓ |
| GIRAFFE-HD | 26.3 | 2.12 | .099 | 31.4 | 25.3 | 1.94 | .092 | 34.3 | 27.0 | 2.03 | .094 | 35.2 |
| StyleNeRF | 10.7 | 1.46 | .069 | 26.2 | 20.6 | 1.44 | .067 | 22.8 | 15.9 | 1.43 | .065 | 20.5 |
| StyleSDF | 29.5 | 1.74 | .648 | 19.8 | 41.0 | 1.69 | .613 | 20.4 | 35.9 | 1.76 | .611 | 13.0 |
| EG3D | 18.6 | 1.52 | .077 | 20.3 | 16.2 | 1.70 | .065 | 14.8 | 17.7 | 1.66 | .070 | 23.9 |
| AvatarGen (Ours) | **6.5** | **0.83** | **.050** | **4.7** | **9.6** | **0.86** | **.052** | **6.9** | **8.7** | **0.94** | **.059** | **6.0** |

and poses (*Pose*) that are estimated from the generated images by [27,52]. We also introduce an image warp metric (*Warp*) that warps side-view image with depth map to frontal view and computes MSE against the generated frontal-view image to further evaluate the geometry quality and multi-view consistency of the model. For additional evaluation details, please refer to the appendix. From Table 1, we observe our model outperforms all the baselines w.r.t. all the metrics and datasets. Notably, it outperforms baseline models by significant margins (69.5%, 63.1%, 64.0% in FID) on three datasets. These results clearly demonstrate its superiority in clothed human avatar synthesis. Moreover, it maintains state-of-the-art geometry quality, pose accuracy and multi-view consistency.

**Qualitative Results.** We show a qualitative comparison against baselines in the left of Fig. 4. It can be observed that compared with our method, StyleSDF [42] generate 3D avatar with over-smoothed geometry and poor multi-view consistency. In addition, the noise and holes can be observed around the generated avatar and the geometry details like face and clothes are missing. EG3D [2] struggles to learn 3D human geometry from 2D images and suffers degenerated qualities. Compared with them, our AvatarGen generates 3D avatars with high-quality appearance with better view-consistency and geometric details.

## 4.2   Ablation Studies

We conduct the following ablation studies on the Deepfashion dataset as its samples have diverse poses and appearances.

**Geometry Proxy.** Our AvatarGen uses signed distance field (SDF) as geometry proxy to regularize the geometry learning. To investigate its effectiveness, we also evaluate our model with density field as the proxy. As shown in Table 2a, if replacing SDF with density field, the quality of the generated avatars drops significantly—11.1% increase in FID, 38.6% and 66.8% increases in Depth and Warp metrics. This indicates SDF is important for the model to more precisely represent clothed human geometry. Without it, the model will produce noisy surface, and suffer performance drop.

**SMPL Body SDF Priors.** AvatarGen adopts a SMPL-guided geometry learning scheme, *i.e.*, generating clothed human body SDFs on top of the coarse

**Table 2.** Ablations on deepfashion. In (e), *w/o* denotes without using SMPL SDF prior, *Can.* or *Obs.* denote SDF prior from canonical or observation spaces

(a) The effect of different geometry proxies used in the model.

| Geo. | FID↓ | Depth↓ | Warp↓ |
|---|---|---|---|
| Density | 10.8 | 1.40 | 20.8 |
| SDF | 9.6 | 0.86 | 6.9 |

(b) The effect of different SDF prediction schemes.

| SDF Scheme | FID↓ | Depth↓ | Warp↓ |
|---|---|---|---|
| Raw | 10.8 | 0.94 | 7.9 |
| Residual | 9.6 | 0.86 | 6.9 |

(c) The effect of different SMPL body SDF priors.

| SDF Prior | FID↓ | Depth↓ | Warp↓ |
|---|---|---|---|
| w/o | 14.3 | 1.12 | 8.3 |
| Can. | 10.4 | 0.89 | 7.5 |
| Obs. | 9.6 | 0.86 | 6.9 |

(d) Deformation schemes. IS, RD are inverse skinning and residual deformation.

| Deformation | FID↓ | Depth↓ | Warp↓ |
|---|---|---|---|
| RD | - | - | - |
| IS | 10.7 | 0.93 | 7.7 |
| IS+RD | 9.6 | 0.86 | 6.9 |

(e) Different number of KNN in inverse skinning deformation.

| KNN | FID↓ | Depth↓ | Warp↓ |
|---|---|---|---|
| 1 | 9.6 | 0.86 | 6.9 |
| 2 | 10.4 | 0.90 | 7.4 |
| 3 | 13.1 | 1.08 | 10.2 |
| 4 | 16.3 | 1.14 | 15.3 |

(f) Different number of points sampled each ray.

| Ray Steps | FID↓ | Depth↓ | Warp↓ |
|---|---|---|---|
| 12 | 12.4 | 1.04 | 7.7 |
| 24 | 10.7 | 0.92 | 7.5 |
| 36 | 10.0 | 0.89 | 7.2 |
| 48 | 9.6 | 0.86 | 6.9 |

**SMPL body mesh.** As shown in Table 2c, if removing SMPL body guidance, the performance drops significantly, *i.e.*, 32.9%, 23.2%, 16.9% increase in FID, Depth and Warp losses. This indicates the coarse SMPL body information is important for guiding AvatarGen to better generate clothed human geometry. We also evaluate the performance difference between SMPL body SDFs queried from observation (*Obs.*) or canonical (*Can.*) spaces. We see the model guided by body SDFs queried from observation space obtains better performance as they are more accurate than the ones queried from the canonical space. Moreover, we study the effect of SMPL SDF regularization loss in Eqn. (8). If removing the regularization loss, the performance in all metrics drops (FID: 10.8 *vs.* 9.6, Depth: 0.96 *vs.* 0.86, Warp: 7.8 *vs.* 6.9), verifying the effectiveness of the proposed loss for regularized geometry learning.

**SDF Prediction Schemes.** Table 2b shows the effect of two SDF prediction schemes—predicting raw SDFs directly or SDF residuals on top of the coarse SMPL body SDFs. Compared with predicting the raw SDFs directly, the residual prediction scheme delivers better results, since it alleviates the geometry learning difficulties.

**Deformation Schemes.** Our model uses a pose-guided deformation to transform spatial points from the observation space to the canonical space. We also evaluate other two deformation schemes in Table 2d: 1) residual deformation [44]

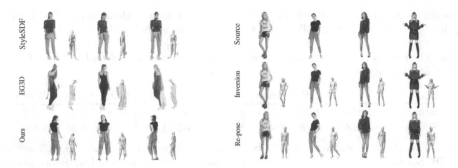

**Fig. 4.** (Left) Qualitative comparison of multi-view rendering and geometry quality against baselines including EG3D [2] and StyleSDF [42]. (Right) Single-view 3D reconstruction and reanimation result of AvatarGen. Given source image, we reconstruct both color and geometry of the human, who is re-posed by taking novel SMPL parameters as input and animated accordingly.

only ($RD$), 2) inverse-skinning deformation [20] only ($IS$). When using RD only, the model training does not converge, indicating that learning deformation implicitly cannot handle large articulation of humans and lead to implausible results. While using IS only, the model achieves a reasonable result (FID: 10.7, Depth: 0.93, Warp: 7.7), verifying the importance of the explicitly pose-guided deformation. Further combining IS and RD (our model) boosts the performance sharply—10.3%, 7.5% and 10.4% decrease in FID, Depth and Warp metrics, respectively. Introducing the residual deformation to collaborate with the posed-guided inverse-skinning transformation indeed better represents non-rigid clothed human body deformation and thus our AvatarGen achieves better appearance and geometry modeling.

**Number of KNN Neighbors in Inverse Skinning Deformation.** For any spatial points, we use Nearest Neighbor to find the corresponding skinning weights for inverse skinning transformation (Eqn. (2)). More nearest neighbors can be used for obtaining skinning weights [31]. Thus, we study how the number of KNN affects model performance in Table 2e. We observe using more KNN neighbors gives worse performance. This is likely caused by 1) noisy skinning weights introduced by using more neighbors for calculation and 2) inaccurate SMPL estimation in data pre-processing step.

**Number of Ray Steps.** Table 2f shows the effect of the number of points sampled per ray for volume rendering. With only 12 sampled points for each ray, AvatarGen already achieves acceptable results, *i.e.*, 12.4, 1.04 and 7.7 in FID, Depth and Warp losses. With more sampling points, the performance monotonically increases, demonstrating the capacity of AvatarGen in 3D-aware avatars generation.

### 4.3   Applications

**Single-View 3D Reconstruction and Re-pose.** The right panel of Fig. 4 shows the application of our learned latent space for single-view 3D reconstruction. Following [2], we use pivotal tuning inversion (PTI) [51] to fit the target images (top) and recover both the appearance and the geometry (middle). With the recovered 3D representation and latent code, we can further use novel SMPL parameters (bottom) to re-pose/animate the human in the source images.

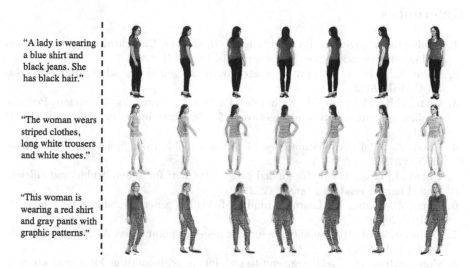

"A lady is wearing a blue shirt and black jeans. She has black hair."

"The woman wears striped clothes, long white trousers and white shoes."

"This woman is wearing a red shirt and gray pants with graphic patterns."

**Fig. 5.** Text-guided (left) synthesis results of AvatarGen with multi-view rendering (right).

**Text-Guided Synthesis.** Recent works [28,45] have shown that one could use a text-image embedding, such as CLIP [50], to guide StyleGAN2 for controlled synthesis. We also visualize text-guided clothed human synthesis in Fig. 5. Specifically, we use StyleCLIP [45] to manipulate a synthesized image with a sequence of text prompts. The optimization based StyleCLIP is used as it is flexible for any input text. From the figure, our AvatarGen is able to synthesis different style human images given different text prompts. This clearly indicates that Avatar-Gen can be an effective tool for text-guided portrait synthesis where detailed descriptions are provided.

## 5   Conclusion

This work introduced the first 3D-aware human avatar generative model, Avatar-Gen. By factorizing the generative process into the canonical avatar generation and deformation stages, AvatarGen can leverage the geometry prior and effective tri-plane representation to address the challenges in animatable human avatar generation. We demonstrated AvatarGen can generate clothed human avatars with arbitrary poses and viewpoints. Besides, it can also generate avatars from multi-modality input conditions, like natural language description and 2D images (for inverting). This work substantially extends the 3D generative models from objects of simple structures (e.g., human faces, rigid objects) to articulated and complex objects. We believe this model will make the creation of human avatars more accessible to ordinary users, assist designers and reduce the manual cost.

# References

1. Achlioptas, P., Diamanti, O., Mitliagkas, I., Guibas, L.: Learning representations and generative models for 3d point clouds. In: ICML (2018)
2. Chan, E.R., et al.: Efficient geometry-aware 3d generative adversarial networks. In: CVPR (2022)
3. Chan, E.R., Monteiro, M., Kellnhofer, P., Wu, J., Wetzstein, G.: pi-gan: Periodic implicit generative adversarial networks for 3d-aware image synthesis. In: CVPR (2021)
4. Chen, J., et al.: Animatable neural radiance fields from monocular rgb videos. arXiv (2021)
5. Chen, M., et al.: Geometry-guided progressive nerf for generalizable and efficient neural human rendering. arXiv (2021)
6. Chen, Z., Zhang, H.: Learning implicit fields for generative shape modeling. In: CVPR (2019)
7. Collet, A., et al.: High-quality streamable free-viewpoint video. ACM Trans. Graph. **34**(4), 1–3 (2015)
8. Contributors, P.: Paddleseg, end-to-end image segmentation kit based on paddlepaddle. https://github.com/PaddlePaddle/PaddleSeg (2019)
9. Debevec, P., Hawkins, T., Tchou, C., Duiker, H.P., Sarokin, W., Sagar, M.: Acquiring the reflectance field of a human face. In: Proceedings of the 27th annual conference on Computer graphics and interactive techniques (2000)
10. Deng, Y., Yang, J., Xiang, J., Tong, X.: Gram: Generative radiance manifolds for 3d-aware image generation. In: CVPR (2022)
11. Dong, H., et al,: Towards multi-pose guided virtual try-on network. In: ICCV (2019)
12. Dou, M.,et al.: Fusion4d: Real-time performance capture of challenging scenes. ACM Trans. Graph. **35**(4),1–13 (2016)
13. Fu, J., et al.: Stylegan-human: A data-centric odyssey of human generation. arXiv (2022)
14. Gomes, T.L., Coutinho, T.M., Azevedo, R., Martins, R., Nascimento, E.R.: Creating and reenacting controllable 3d humans with differentiable rendering. In: WACV (2022)
15. Goodfellow, I., et al.: Generative adversarial nets. In: NeurIPS (2014)
16. Gu, J., Liu, L., Wang, P., Theobalt, C.: Stylenerf: A style-based 3d-aware generator for high-resolution image synthesis. In: CVPR (2022)
17. Guo, K., et al.: The relightables: Volumetric performance capture of humans with realistic relighting. ACM Trans. Graph. **38**(6), 1–9 (2019)
18. Heusel, M., Ramsauer, H., Unterthiner, T., Nessler, B., Hochreiter, S.: Gans trained by a two time-scale update rule converge to a local nash equilibrium. In: NeurIPS (2017)

19. Hong, F., Zhang, M., Pan, L., Cai, Z., Yang, L., Liu, Z.: Avatarclip: Zero-shot text-driven generation and animation of 3d avatars. ACM Trans. Graph. arXiv preprint arXiv:2205.08535 (2022)
20. Huang, Z., Xu, Y., Lassner, C., Li, H., Tung, T.: Arch: Animatable reconstruction of clothed humans. In: CVPR (2020)
21. Jacobson, A., Baran, I., Kavan, L., Popović, J., Sorkine, O.: Fast automatic skinning transformations. ACM Trans. Graph. 31(4), 1–10 (2012)
22. Jiang, B., Hong, Y., Bao, H., Zhang, J.: Selfrecon: Self reconstruction your digital avatar from monocular video. In: CVPR (2022)
23. Karras, T., Aila, T., Laine, S., Lehtinen, J.: Progressive growing of GANs for improved quality, stability, and variation. In: ICCV (2018)
24. Karras, T., et al.: Alias-free generative adversarial networks. In: NeurIPS (2021)
25. Karras, T., Laine, S., Aila, T.: A style-based generator architecture for generative adversarial networks. In: CVPR (2019)
26. Karras, T., Laine, S., Aittala, M., Hellsten, J., Lehtinen, J., Aila, T.: Analyzing and improving the image quality of StyleGAN. In: CVPR (2020)
27. Kolotouros, N., Pavlakos, G., Black, M.J., Daniilidis, K.: Learning to reconstruct 3d human pose and shape via model-fitting in the loop. In: ICCV (2019)
28. Kwon, G., Ye, J.C.: Clipstyler: Image style transfer with a single text condition. In: Proceedings of the IEEE/CVF Conference on Computer Vision and Pattern Recognition, pp. 18062–18071 (2022)
29. Li, R., Li, X., Fu, C.W., Cohen-Or, D., Heng, P.A.: Pu-gan: a point cloud upsampling adversarial network. In: ICCV (2019)
30. Liao, Y., Schwarz, K., Mescheder, L., Geiger, A.: Towards unsupervised learning of generative models for 3D controllable image synthesis. In: CVPR (2020)
31. Liu, L., Habermann, M., Rudnev, V., Sarkar, K., Gu, J., Theobalt, C.: Neural actor: Neural free-view synthesis of human actors with pose control. ACM Trans. Graph. 40(6), 1–16 (2021)
32. Liu, T., et al.: Spatial-aware texture transformer for high-fidelity garment transfer. In: IEEE Transaction on Image Processing (2021)
33. Liu, Z., Luo, P., Qiu, S., Wang, X., Tang, X.: Deepfashion: Powering robust clothes recognition and retrieval with rich annotations. In: CVPR (2016)
34. Loper, M., Mahmood, N., Romero, J., Pons-Moll, G., Black, M.J.: Smpl: A skinned multi-person linear model. ACM Trans.Graph. 34(6), 1–6 (2015)
35. Mescheder, L., Geiger, A., Nowozin, S.: Which training methods for gans do actually converge? In: International conference on machine learning, pp. 3481–3490. PMLR (2018)
36. Mescheder, L., Oechsle, M., Niemeyer, M., Nowozin, S., Geiger, A.: Occupancy networks: Learning 3d reconstruction in function space. In: CVPR (2019)
37. Mildenhall, B., Srinivasan, P.P., Tancik, M., Barron, J.T., Ramamoorthi, R., Ng, R.: Nerf: Representing scenes as neural radiance fields for view synthesis. In: ECCV (2020)
38. Nguyen-Phuoc, T., Li, C., Theis, L., Richardt, C., Yang, Y.L.: HoloGAN: Unsupervised learning of 3D representations from natural images. In: ICCV (2019)
39. Nguyen-Phuoc, T., Richardt, C., Mai, L., Yang, Y.L., Mitra, N.: BlockGAN: Learning 3D object-aware scene representations from unlabelled images. In: NeurIPS (2020)
40. Niemeyer, M., Geiger, A.: Giraffe: Representing scenes as compositional generative neural feature fields. In: CVPR (2021)
41. Noguchi, A., Sun, X., Lin, S., Harada, T.: Neural articulated radiance field. In: ICCV (2021)

42. Or-El, R., Luo, X., Shan, M., Shechtman, E., Park, J.J., Kemelmacher-Shlizerman, I.: Stylesdf: High-resolution 3d-consistent image and geometry generation. In: CVPR (2022)
43. Park, J.J., Florence, P., Straub, J., Newcombe, R., Lovegrove, S.: Deepsdf: Learning continuous signed distance functions for shape representation. In: CVPR (2019)
44. Park, K., et al.: Nerfies: Deformable neural radiance fields. In: ICCV (2021)
45. Patashnik, O., Wu, Z., Shechtman, E., Cohen-Or, D., Lischinski, D.: Styleclip: Text-driven manipulation of stylegan imagery. In: ICCV (2021)
46. Peng, S., et al.: Animatable neural radiance fields for human body modeling. In: ICCV (2021)
47. Peng, S., Zhang, S., Xu, Z., Geng, C., Jiang, B., Bao, H., Zhou, X.: Animatable neural implicit surfaces for creating avatars from videos. arXiv (2022)
48. Peng, S., et al.: Neural body: Implicit neural representations with structured latent codes for novel view synthesis of dynamic humans. In: CVPR (2021)
49. Pumarola, A., Corona, E., Pons-Moll, G., Moreno-Noguer, F.: D-nerf: Neural radiance fields for dynamic scenes. In: CVPR (2021)
50. Radford, A., et al.: Learning transferable visual models from natural language supervision. In: ICML (2021)
51. Roich, D., Mokady, R., Bermano, A.H., Cohen-Or, D.: Pivotal tuning for latent-based editing of real images. ACM Trans. Graph. **42**(1), 1–3 (2021)
52. Saito, S., Simon, T., Saragih, J., Joo, H.: Pifuhd: Multi-level pixel-aligned implicit function for high-resolution 3d human digitization. In: CVPR (2020)
53. Schwarz, K., Liao, Y., Niemeyer, M., Geiger, A.: Graf: Generative radiance fields for 3d-aware image synthesis. In: NeurIPS (2020)
54. Su, S.Y., Yu, F., Zollhöfer, M., Rhodin, H.: A-nerf: Articulated neural radiance fields for learning human shape, appearance, and pose. In: NeurIPS (2021)
55. Su, Z., Xu, L., Zheng, Z., Yu, T., Liu, Y., Fang, L.: Robustfusion: Human volumetric capture with data-driven visual cues using a rgbd camera. In: ECCV (2020)
56. Szabó, A., Meishvili, G., Favaro, P.: Unsupervised generative 3D shape learning from natural images. arXiv (2019)
57. Tewari, A., BR, M., Pan, X., Fried, O., Agrawala, M., Theobalt, C.: Disentangled3d: Learning a 3d generative model with disentangled geometry and appearance from monocular images. In: CVPR (2022)
58. Wang, P., Liu, L., Liu, Y., Theobalt, C., Komura, T., Wang, W.: Neus: Learning neural implicit surfaces by volume rendering for multi-view reconstruction. In: NeurIPS (2021)
59. Weng, C.Y., Curless, B., Srinivasan, P.P., Barron, J.T., Kemelmacher-Shlizerman, I.: HumanNeRF: Free-viewpoint rendering of moving people from monocular video. In: CVPR (2022)
60. Wu, J., Zhang, C., Xue, T., Freeman, W.T., Tenenbaum, J.B.: Learning a probabilistic latent space of object shapes via 3D generative-adversarial modeling. In: NeurIPS (2016)
61. Xiang, D., et al.: Modeling clothing as a separate layer for an animatable human avatar. ACM Trans. Graph. **40**(6), 1–5 (2021)
62. Xu, H., Alldieck, T., Sminchisescu, C.: H-nerf: Neural radiance fields for rendering and temporal reconstruction of humans in motion. In: NeurIPS (2021)
63. Xu, X., Loy, C.C.: 3D human texture estimation from a single image with transformers. In: ICCV (2021)
64. Xue, Y., Li, Y., Singh, K.K., Lee, Y.J.: Giraffe hd: A high-resolution 3d-aware generative model. In: CVPR (2022)

65. Yariv, L., Gu, J., Kasten, Y., Lipman, Y.: Volume rendering of neural implicit surfaces. In: NeurIPS (2021)
66. Zablotskaia, P., Siarohin, A., Zhao, B., Sigal, L.: Dwnet: Dense warp-based network for pose-guided human video generation. In: BMVC (2019)
67. Zhang, J., et al.: Editable free-viewpoint video using a layered neural representation. ACM Trans. on Graph. **40**(4), 1–8 (2021)
68. Zhang, J., et al.: Neuvv: Neural volumetric videos with immersive rendering and editing. ACM Trans. on Graph. arXiv preprint arXiv:2202.06088 (2022)

# INGeo: Accelerating Instant Neural Scene Reconstruction with Noisy Geometry Priors

Chaojian Li[1]([⊠]), Bichen Wu[2], Albert Pumarola[2], Peizhao Zhang[2], Yingyan Lin[1], and Peter Vajda[2]

[1] Rice University, Houston, USA
{cl114,yingyan.lin}@rice.edu
[2] Meta Reality Labs, Irvine, USA
{wbc,apumarola,stzpz,vajdap}@fb.com

**Abstract.** We present a method that accelerates reconstruction of 3D scenes and objects, aiming to enable instant reconstruction on edge devices such as mobile phones and AR/VR headsets. While recent works have accelerated scene reconstruction training to minute/second-level on high-end GPUs, there is still a large gap to the goal of instant training on edge devices which is yet highly desired in many emerging applications such as immersive AR/VR. To this end, this work aims to further accelerate training by leveraging geometry priors of the target scene. Our method proposes strategies to alleviate the noise of the imperfect geometry priors to accelerate the training speed on top of the highly optimized Instant-NGP. On the NeRF Synthetic dataset, our work uses half of the training iterations to reach an average test PSNR of >30.

## 1 Introduction

3D scene reconstruction has become a fundamental technology for many emerging applications such as Augmented Reality (AR), Virtual Reality (VR), and Mixed Reality (MR). In particular, many consumer-facing AR/VR/MR experiences (*e.g.,* virtually teleporting 3D scenes, people, or objects to another user) require the capability to instantly reconstruct scenes on a local headset, which is particularly challenging as it requires both reconstruction quality and real-time speed.

Recently, neural rendering with implicit functions has recently become one of the most active research areas in 3D representation and reconstruction. Notably, Neural Radiance Field (NeRF) [8] and its following works demonstrate that a learnable function can be used to represent a 3D scene as a radiance field, which can then be combined with volume rendering [7] to render novel view images, estimate scene geometry [4,12,17], perform 3D-aware image editing such as relighting [15], and so on. Despite the remarkable improvement in reconstruction

---

C. Li and B. Wu—Equal contribution.
C. Li—Work done while interning at Meta Reality Labs.

© The Author(s), under exclusive license to Springer Nature Switzerland AG 2023
L. Karlinsky et al. (Eds.): ECCV 2022 Workshops, LNCS 13803, pp. 686–694, 2023.
https://doi.org/10.1007/978-3-031-25066-8_40

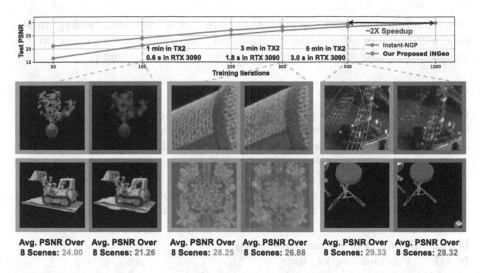

**Fig. 1.** INGeo accelerates neural reconstruction training by ~2× over the current SotA – Instant-NGP [9].

quality, NeRF is extremely computationally expensive. For example, it requires >30 s to render an image and days of training time to reconstruct a scene on even high-end GPUs [8]. Recent works begins to focus on accelerating training of scene reconstruction. Specifically, one recent trend is to replace implicit function-based volume representations with explicit grid-based or point-based ones [1,2, 9,16,18]. The advantage is that unlike implicit representation, where volume features are entangled, grid-based representations are spatially disentangled and thus the gradient updates at one location will not interfere with features that are far away [9]. Using grid-based representation, recent works have been able to accelerate the training of NeRF from days to minutes [1,2] to even seconds [9] on high-end GPUs. However, instant reconstruction on compute-constrained edge devices (mobile phones and AR/VR headsets) is still not possible.

This work sets out to *reduce* the above gap, and boosts training efficiency by leveraging geometry priors to eliminate spatial redundancy during training. To do so, we leverage the intuition that most 3D scenes by nature are inherently sparse as noted by recent works [1,2,5,9,18,19]. Prior works [1,2,6,9] have proposed to leverage a 3D occupancy grid to indicate whether a grid cell contains visible points or not. Such a occupancy grid is gradually updated based on the estimations of the volume density as training progresses. Though this technique has been proven to be effective accelerating training in previous works [1,2,6,9], obtaining a decent occupancy grid during training still takes a nontrivial amount of time, as discussed in Sect. 3.1. Meanwhile, in many scenarios, as also noted by [3,18], geometry of a scene, in the form of depth images, point-clouds, *etc.*, can be obtained *a priori* from many sources, such as RGB-D sensors and depth estimation algorithms (structure-from-motion, multi-view stereo, *etc.*). These geometry priors can be converted to an occupancy grid that can be used in the reconstruction training. Despite that they are sparse and noisy, we hypothesize that with appropriate strategies to mitigate the impact of the noise, we can leverage such noisy geometry priors to further accelerate training.

## 2    Related Works

**Grid-Based Volume Representation:** As discussed in [9], one disadvantage of NeRF's implicit function based scene representation is that spatial features are entangled. In comparison, many recent works [1,2,9,16,19] that successfully accelerated training have switched to explicit grid-based representations, where volume features are spatially disentangled. To reduce the cubic complexity of the grid representation, these works adopted different compression techniques such as tensor factorization, hashing, multi-resolution, sparsifying voxels, and so on.

**Using Geometry Priors in NeRF:** Previous works have also tried to utilize geometry priors to boost the performance of NeRF in terms of training, rendering, and geometric estimation [3,10,12,12,17,18]. Depth-Supervised NeRF [3] utilized sparse depth estimation as an extra supervision in addition to RGB. It observed faster training and better few-shot performance. PointNeRF [18] directly utilizes prior point-cloud to represent a radiance field. It enjoys the benefit of avoiding sampling in empty space and achieved good speedup over NeRF.

To **summarize** the relationship between our work and prior works: our work is built on top

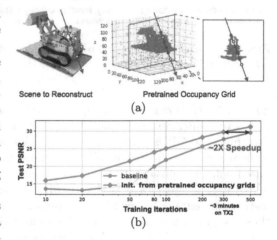

**Fig. 2.** Visualization of the pretrained occupancy grid in (a) suggests the reason why it can accelerate the training process by $\sim 2\times$ in (b) is that the occupancy grid can eliminate spatial redundancy in training.

of the Instant-NGP [9], the current record holder of the fast training, and we further accelerated it. We convert noisy geometry prior to an occupancy grid to reduce spatial redundancy during training and mitigate hash collision of Instant-NGP. This is more straightforward than Depth-Supervised NeRF [3], which use prior depth as training signals.

## 3    Method

### 3.1    Can Pretrained Occupancy Grids Accelerate Training?

To obtain the occupancy grid, previous works [2,9] typically sample points in the 3D volume, compute densities, and update the occupancy grid. In practice, this boot-strapping strategy (iteratively update the occupancy grid and volume representation) usually converges well and can accelerate the training by concentrating training samples around critical regions of the scene. However, at the start of the training, we do not have a reliable occupancy grid, and we have to spend computational budgets to obtain one, which can take a non-trivial amount of time (e.g., it takes about 30 min on TX2 to obtain an occupancy grid with 88 % IoU as compared to the final one on Lego scene [8]). Thus, reducing the training time spent on learning the occupancy grid is critical for instant reconstruction of the scene.

We hypothesize that if an occupancy grid is provided *a priori*, we can further accelerate training. To verify this, we conduct an experiment to re-train an Instant-NGP model but with its occupancy grid initialized from a pretrained model. The visualization of the pretrained occupancy grid can be found in Fig. 2 (a). As shown in Fig. 2 (b), even though we simply inherited the occupancy grid from a pretrained model, while the majority of the model weights (i.e., 89% even if we use floating point numbers to represent the dense occupancy grid) are randomly initialized, we still observed significant training speedup – about 2x faster to reach an average of 30 PSNR on the NeRF synthetic dataset [8].

## 3.2   Initialize Occupancy Grids with Geometry Priors

As noted in [3,18], in many application scenarios, such as on AR/VR headsets, geometry priors of a 3D scene can be obtained from many sources, such as RGB-D cameras and depth-estimation [13,14]. They can be converted to an occupancy grid of the scene. To illustrate this, we use the Lego scene from the NeRF-Synthetic dataset [8] as an example

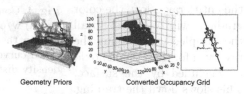

Geometry Priors          Converted Occupancy Grid

**Fig. 3.** Visualization of the point-cloud obtained by COLMAP [13] and the converted occupancy grid.

to obtain a point-cloud using COLMAP [13,14] and convert it to an occupancy grid, as shown in Fig. 3.

We notice that the occupancy grid obtained from point-cloud is similar to the one obtained from the pretrained model. We also notice missing points in the point-cloud and missing regions in the occupancy grid (e.g., bottom board below the bulldozer in Fig. 3), which could have an negative impact on the scene reconstruction.

So the question is how to design strategies to mitigate the impact of the noise and leverage the noisy geometry priors to accelerate training. Our strategy can be described as the following three aspects: density scaling, point-cloud splatting, and updating occupancy grid.

**Density Scaling:** When we first load an initialized occupancy grid to the model, to our surprise, we did not observe obvious speedup over the baseline using a random occupancy grid, as shown in the training speed curve in Fig. 4 (a). After investigation, we found that the reason can be explained as the following: an initialized occupancy grid can ensure we sample and train points with the positive grid cells. However, the initial density values inside the cells are computed by the model, whose weights are randomly initialized. At the beginning of the training, the density prediction of the model is generally low. As an evidence, see the rendered image in Fig. 4 (a), note that the image looks quite transparent. Also note the density along the ray is generally low.

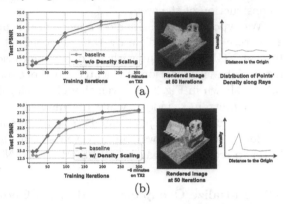

This is problematic, since in a well-trained model, densities around the object surface are generally much higher. This means that the accumulated transmittance will quickly drop to 0 after the ray encounters the surface. As a result, during back-propagation, the gradient of the color prediction will mainly impact the surface points. However, if the density along the entire ray is low, the color gradient will broadcast to all the sampled points along the ray, even for those that should be occluded by the surface. This slows down the training.

**Fig. 4.** Comparing the effectiveness of geometry priors (a) w/o and (b) w/ the proposed **density scaling** on the Lego scene [8]. Compare their training speed curves, rendered images at 50 iterations, and density distribution along a ray.

To fix this, we decide to include an inductive bias to make sure that rays terminate soon after it encounters the object surface that is identified by the occupancy grid. This can be implemented by simply scale up the initial density prediction by a factor. Our experiments show that this made the densities at surface points much larger, as shown in Fig. 4 (b), the rendered images becomes more "solid", and we also observed more significant speedup over the baseline.

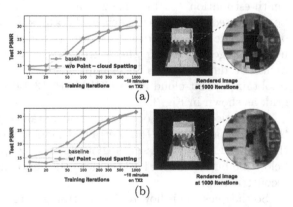

**Fig. 5.** Compare (a) w/o (i.e., only w/ the proposed density scaling) and (b) w/ the proposed **point-cloud splatting** on Lego dataset [8] in terms of the training efficiency over baselines (e.g., training from random initialization) and rendered images.

**Point-Cloud Splatting:** When converting a point-cloud to an occupancy grid, the default option is to set the element of the occupancy tensor to positive if a point is within its corresponding cell. However, point-clouds are typically sparse and incomplete, and the corresponding occupancy grid also contains a lot of missing regions. After adopting the density scaling, despite the significant performance gain at the initial iterations of the training, we notice that the PSNR plateaus very soon. Visualization of the rendered images show missing regions, as a result of incomplete occupancy grid, see the zoom-in in Fig. 5 (a). To address this, one simple strategy is to not to treat each point as infinitesimal, but let it splat to nearby grid cells within a radius. This can help complete small missing regions, at the cost of adding more empty grid cells. We used this strategy to initialize the occupancy grid, and notice significant performance improvement both quantitatively and qualitatively, as shown in Fig. 5 (b), although it still cannot completely remove the mission region.

**Updating Occupancy Grids:** Sparsity and errors in the geometry prior are inevitable and we cannot completely rely on the initial occupancy grid throughout the training. To address this, we adopt a new strategy to combine initialized occupancy grid with continuously updated ones. More specifically, we start with an initialized occupancy grid, but also follow previous works [2,9] to re-evaluate volume density at a certain frequency, and update the occupancy grid. We assume that the initial occupancy grid obtained from geomery priors (depth

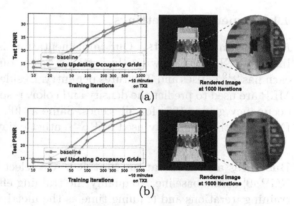

**Fig. 6.** Compare (a) w/o (i.e., only w/ the proposed density scaling and point-cloud splatting) and (b) w/o the proposed **updating occupancy grids** on Lego dataset [8] in terms of the training efficiency over baselines (e.g., training from random initialization) and rendered images.

sensor, depth-estimation) are sparse but accurate (low-recall but high precision). Therefore, we always mark the initially occupied cells as positive, and add new cells whose estimated density are larger than a threshold. This effectively addressed the missing point issue, as shown in Fig. 6.

## 4    Experiments

We run experiments to verify whether noisy geometry priors can further accelerate the training of scene reconstruction on top of the highly optimized Instant-NGP. We first summarize our experimental settings in Sect. 4.1 and report the training speed comparison in Sect. 4.2.

**Table 1.** Comparing our propose INGeo with the SotA efficient training solution, Instant-NGP [9], in terms of the achieved PSNR under given training cost (i.e., training iterations or training time).

| Method | # Iter. | Time on TX2 | Test PSNR | | | | | | | | |
|---|---|---|---|---|---|---|---|---|---|---|---|
| | | | Mic | Ficus | Chair | Hotdog | Materials | Drums | Ship | Lego | Avg |
| Instant-NGP [9] | 10 | 6 s | 15.16 | 14.85 | 12.94 | 14.41 | 15.36 | 13.04 | 17.95 | 13.57 | 14.66 |
| INGeo | | | 22.21 | 20.54 | 15.41 | 15.97 | 18.11 | 15.77 | 18.81 | 14.61 | **17.68 (↑ 3.02)** |
| Instant-NGP [9] | 100 | 1 min | 25.85 | 20.26 | 22.78 | 24.01 | 22.15 | 18.12 | 15.07 | 21.85 | 21.26 |
| INGeo | | | 27.94 | 25.92 | 24.00 | 25.17 | 22.02 | 20.99 | 21.33 | 24.63 | **24.00 (↑ 2.74)** |
| Instant-NGP [9] | 500 | 5 min | 31.72 | 29.17 | 29.84 | 33.01 | 25.92 | 23.37 | 23.97 | 29.60 | 28.32 |
| INGeo | | | 32.47 | 29.55 | 31.18 | 33.24 | 26.04 | 23.71 | 27.27 | 31.16 | **29.33 (↑ 1.01)** |
| Instant-NGP [9] | 1 K | 10 min | 33.03 | 30.27 | 31.60 | 34.28 | 26.77 | 24.27 | 26.06 | 31.70 | 29.75 |
| INGeo | | | 33.66 | 30.41 | 32.72 | 34.41 | 27.02 | 24.49 | 28.67 | 32.76 | **30.52 (↑ 0.77)** |
| Instant-NGP [9] | 30 K | 5 h | 35.68 | 32.34 | 34.83 | 36.99 | 29.46 | 25.85 | 30.25 | 35.60 | 32.63 |
| INGeo | | | 36.17 | 31.54 | 35.04 | 36.98 | 29.15 | 25.82 | 30.95 | 35.47 | **32.64 (↑ 0.01)** |

## 4.1  Experiments Settings

**Models and Datasets.** Our implementation is based on the open-sourced Instant-NGP [9]. Following the settings in [9], our model uses 16 hash tables, each has $2^{19}$ entries and 2 features per entry. Besides, a 1-layer MLP and a 2-layer MLP are used to predict the density and color, respectively. The resolution of the occupancy grid is $128^3$ for all scenes. Same as [9], we benchmark ours proposed INGeo on the commonly used NeRF-Synthetic dataset [8] with 8 scenes.

**Baselines and Evaluation Metrics.** We select the highly optimized Instant-NGP [9] as the baseline. To qualify the training efficiency, we use the number of training iterations and training time as the metrics for training cost, and PSNR on the test set as the metrics for reconstruction quality. The training time is measured on an embedded GPU, NVIDIA Jetson TX2 [11]. We did not include other works as baseline, since Instant-NGP is significantly faster than all the existing methods that we are aware of.

## 4.2  Training Speed Comparison with SotA

As demonstrated in Fig. 1 and Table 1, we observe that (1) to reach an average test PSNR of 30 on the 8 scenes on the NeRF synthetic dataset, our method uses half of the training iterations than Instant-NGP. (2) Under different training budgets (from 6 s to 10 min), our proposed INGeo consistently achieves ↑ 0.77 % ∼ ↑ 3.02 % higher average test PSNR than Instant-NGP. (3) With a much longer training budget (30 K iterations, 5 h on a TX2 GPU), the advantage of our work diminishes, but INGeo does not hurt the performance, and is still slightly better (↑ 0.01 PSNR) than Instant-NGP. Based on the observations above, we conclude that the proposed INGeo makes it possible to achieve instant neural scene reconstruction on edge devices, achieving > 30 test PSNR within 10 min of training on the embedded GPU TX2, and > 24 PSNR within 1 min of training. To check the individual effectiveness of our proposed noise mitigation techniques – density scaling, point-cloud splatting, and updating occupancy grid, we refer readers to Sect. 3.2 and Figs. 4–6.

# 5  Conclusion

In this paper, we present INGeo, a method to accelerate reconstruction of 3D scenes. Our method is built on top of the SotA fast reconstruction method, Instant-NGP, and we utilized geometry priors to further accelerate training. We proposed three strategies to mitigate the negative impact of noise in the geometry prior, and our method is able to accelerate Instant-NGP training by 2 × to reach an average test PSNR of 30 on the NeRF synthetic dataset. We believe this work bring us closer to the target of instant scene reconstruction on edge devices.

# References

1. Yu, A., Fridovich-Keil, S., Tancik, M., Chen, Q., Recht, B., Kanazawa, A.: Plenoxels: Radiance fields without neural networks (2021)
2. Chen, A., Xu, Z., Geiger, A., Yu, J., Su, H.: Tensorf: Tensorial radiance fields. arXiv preprint arXiv:2203.09517 (2022)
3. Deng, K., Liu, A., Zhu, J.Y., Ramanan, D.: Depth-supervised nerf: Fewer views and faster training for free. arXiv preprint arXiv:2107.02791 (2021)
4. Endres, F., Hess, J., Sturm, J., Cremers, D., Burgard, W.: 3-d mapping with an rgb-d camera. IEEE Trans. Rob. 30(1), 177–187 (2013)
5. Kondo, N., Ikeda, Y., Tagliasacchi, A., Matsuo, Y., Ochiai, Y., Gu, S.S.: Vaxnerf: Revisiting the classic for voxel-accelerated neural radiance field. arXiv preprint arXiv:2111.13112 (2021)
6. Liu, L., Gu, J., Zaw Lin, K., Chua, T.S., Theobalt, C.: Neural sparse voxel fields. Adv. Neural. Inf. Process. Syst. 33, 15651–15663 (2020)
7. Max, N.: Optical models for direct volume rendering. IEEE Trans. Visual Comput. Graph. 1(2), 99–108 (1995)
8. Mildenhall, B., Srinivasan, P.P., Tancik, M., Barron, J.T., Ramamoorthi, R., Ng, R.: NeRF: representing scenes as neural radiance fields for view synthesis. In: Vedaldi, A., Bischof, H., Brox, T., Frahm, J.-M. (eds.) ECCV 2020. LNCS, vol. 12346, pp. 405–421. Springer, Cham (2020). https://doi.org/10.1007/978-3-030-58452-8_24
9. Müller, T., Evans, A., Schied, C., Keller, A.: Instant neural graphics primitives with a multiresolution hash encoding. arXiv preprint arXiv:2201.05989 (2022)
10. Neff, T., et al.: Donerf: Towards real-time rendering of compact neural radiance fields using depth oracle networks. In: Computer Graphics Forum, vol. 40, pp. 45–59. Wiley Online Library (2021)
11. NVIDIA Inc.: NVIDIA Jetson TX2 (2021). https://www.nvidia.com/en-us/autonomous-machines/embedded-systems/jetson-tx2/ Accessed Jan 1 2020
12. Roessle, B., Barron, J.T., Mildenhall, B., Srinivasan, P.P., Nießner, M.: Dense depth priors for neural radiance fields from sparse input views. arXiv preprint arXiv:2112.03288 (2021)

13. Schönberger, J.L., Frahm, J.M.: Structure-from-motion revisited. In: Conference on Computer Vision and Pattern Recognition (CVPR) (2016)
14. Schönberger, J.L., Zheng, E., Pollefeys, M., Frahm, J.M.: Pixelwise view selection for unstructured multi-view stereo. In: European Conference on Computer Vision (ECCV) (2016)
15. Srinivasan, P.P., Deng, B., Zhang, X., Tancik, M., Mildenhall, B., Barron, J.T.: Nerv: Neural reflectance and visibility fields for relighting and view synthesis. In: CVPR (2021)
16. Sun, C., Sun, M., Chen, H.T.: Direct voxel grid optimization: Super-fast convergence for radiance fields reconstruction. arXiv preprint arXiv:2111.11215 (2021)
17. Wei, Y., Liu, S., Rao, Y., Zhao, W., Lu, J., Zhou, J.: Nerfingmvs: Guided optimization of neural radiance fields for indoor multi-view stereo. In: ICCV (2021)
18. Xu, Q., Xu, Z., Philip, J., Bi, S., Shu, Z., Sunkavalli, K., Neumann, U.: Point-nerf: Point-based neural radiance fields. arXiv preprint arXiv:2201.08845 (2022)
19. Yu, A., Li, R., Tancik, M., Li, H., Ng, R., Kanazawa, A.: PlenOctrees for real-time rendering of neural radiance fields. In: ICCV (2021)

# Number-Adaptive Prototype Learning for 3D Point Cloud Semantic Segmentation

Yangheng Zhao[1] , Jun Wang[2] , Xiaolong Li[3] , Yue Hu[1] , Ce Zhang[3] ,
Yanfeng Wang[1,4], and Siheng Chen[1,4(✉)]

[1] Shanghai Jiao Tong University, Shanghai, China
{zhaoyangheng-sjtu,18671129361,wangyanfeng,sihengc}@sjtu.edu.cn
[2] University of Maryland, College Park, USA
junwang@umiacs.umd.edu
[3] Virginia Tech, Blacksburg, USA
{lxiaol9,zce}@vt.edu
[4] Shanghai AI Laboratory, Shanghai, China

**Abstract.** 3D point cloud semantic segmentation is one of the fundamental tasks for 3D scene understanding and has been widely used in the metaverse applications. Many recent 3D semantic segmentation methods learn a single prototype (classifier weights) for each semantic class, and classify 3D points according to their nearest prototype. However, learning only one prototype for each class limits the model's ability to describe the high variance patterns within a class. Instead of learning a single prototype for each class, in this paper, we propose to use an adaptive number of prototypes to dynamically describe the different point patterns within a semantic class. With the powerful capability of vision transformer, we design a *Number-Adaptive Prototype Learning (NAPL)* model for point cloud semantic segmentation. To train our NAPL model, we propose a simple yet effective *prototype dropout* training strategy, which enables our model to adaptively produce prototypes for each class. The experimental results on SemanticKITTI dataset demonstrate that our method achieves 2.3% mIoU improvement over the baseline model based on the point-wise classification paradigm.

**Keywords:** Point cloud · Semantic segmentation · Prototype learning

## 1 Introduction

3D scene understanding is critical for numerous applications, including metaverse, digital twins and robotics [3]. As one of the most important tasks for 3D scene understanding, point cloud semantic segmentation provides point-level understanding of the surrounding 3D environment and gets increasing attention.

A popular paradigm for 3D point cloud semantic segmentation follows the point-wise classification, where an encoder-decoder network extracts point-wise features and feeds them into a classifier predicting label, as shown in Fig. 1(a).

L. Karlinsky et al. (Eds.): ECCV 2022 Workshops, LNCS 13803, pp. 695–703, 2023.
https://doi.org/10.1007/978-3-031-25066-8_41

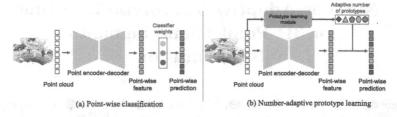

(a) Point-wise classification    (b) Number-adaptive prototype learning

**Fig. 1.** Difference between point-wise classification (PWC) and number-adaptive prototype learning (NAPL). The PWC paradigm [8,9,16] learns a single prototype (classifier weights) for each class, while our proposed NAPL uses a prototype learning module to adaptively produce multiple prototypes for each class.

Following the spirit of prototype learning in image semantic segmentation [15], the point-wise classification model can be viewed as learning one prototype (classifier weights) for each semantic category, and assigning points with the label of the nearest prototype. However, the common single-prototype-per-class design in point-wise classification models limits the model's capacity in the semantic categories with high intra-class variance. More critically, the 3D point cloud data we are interested in is sparse and non-uniform. The issues of distance variation and the occlusion in 3D point cloud can make the geometric characteristics of objects of the same category very different, and this challenge is even more significant in large-scale 3D data. Experiments show that one prototype per class is usually insufficient to describe those patterns with high variations; see Fig. 3.

To better handle the data variance, an intuitive idea is to use more than one prototype for each category. However, we have no prior knowledge about how many prototypes each category needs, and too many prototypes per category may increase the computational costs while also lead to potential overfitting issues. The question is – can we find a smarter way to identify the necessary prototypes and effectively increase existing models' capacity? In this work, we propose to use an adaptive way to set the number of prototypes per semantic category, as shown in Fig. 1(b). We call this paradigm as Number-Adaptive Prototype Learning (NAPL). To instantiate the proposed NAPL model, inspired by the recent work [5,11], we use a transformer decoder to learn adaptive number of prototypes for each category. Unlike previous work [5,11], which is limited by learning one prototype for each semantic category, we design a novel *prototype dropout* training strategy, to enable the model adaptively produce prototypes for each class. The experimental results on SemanticKITTI [1] dataset show that by plugging our design to a common encoder-decoder network, our method achieves a 2.3% mIoU gain than the baseline point-wise classification model.

## 2    Related Work

### 2.1    3D Point Cloud Semantic Segmentation

3D point cloud semantic segmentation has been widely used in metaverse, digital twins, robotics and autonomous driving [4,13]. Based on different representations,

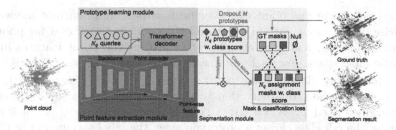

**Fig. 2.** The proposed NAPL consists of three parts: point feature extraction module, prototype learning module, and segmentation module. During training, we randomly drop $M$ prototypes to enable NAPL to adaptively produce prototypes.

existing 3D semantic segmentation methods can be divided into three categories: projection-based, point-based and voxel-based. The projection-based methods, SqueezeSegV3 [14] and RangeNet++ [9], project the 3D point cloud into the 2D plane, and do feature learning and segmentation on the projected 2D image. Alternatively, the point-based methods, PointNet [10], RandLA [8], and KPConv [12], learn point-wise features from the raw point cloud with the specifically designed multi-layer perceptron (MLP) and convolution kernels. The voxel-based methods [6,16], including MinkUNet and Cylinder3D, discretize the space into regular grids, and leverage 3D CNN networks to extract features. In this work, we design our NAPL based on the MinkUNet [6].

## 2.2 Mask Classification for Image Segmentation

In the 2D image segmentation task, inspired by the pioneering work DETR [2], there is a trend to leverage the mask classification paradigm for semantic segmentation. Among those, Segmenter [11] proposes to use a set of learnable queries to predict class masks for semantic segmentation. MaskFormer [5] proposes to unbind the queries from categories, and uses the learnable queries to predict mask embeddings for mask prediction and mask classification. Our model adapts the mask prediction paradigm of MaskFormer from 2D image segmentation to 3D point cloud semantic segmentation. Jointly with the proposed prototype dropout training strategy, our model produces an adaptive number of prototypes for one class, which naturally meets the necessity of 3D semantic segmentation.

## 3 Methodology

In this section, we first revisit the current 3D semantic segmentation paradigm. We then introduce the proposed number-adaptive prototype learning for 3D point cloud semantic segmentation and a novel strategy to train our model.

### 3.1 Overview of 3D Semantic Segmentation Paradigm

Given a frame of 3D point cloud $X \in \mathbb{R}^{N \times 3}$ with $N$ points, the goal of 3D point semantic segmentation is to predict a semantic class label $c \in \{1, 2, \ldots, C\}$

for each point. For the point-wise classification paradigm, current models [8,9] comprise of two main parts: i) an encoder-decoder network $\phi(\cdot)$ for point-wise feature extraction, and ii) a classifier $\psi$ to project the point features into the semantic label space. For each point $x_i \in X$, its feature $f_i = \phi(x_i) \in \mathbb{R}^D$ is fed into $\psi$ for $C$-way classification: $p(c|i) = \exp\left(w_c^\top f_i\right)/\sum_{c'=1}^{C} \exp\left(w_{c'}^\top f_i\right)$, where $p(c|i) \in [0,1]$ is the probability that the $i$th point belongs to the $c$th class, and $\psi$ is parameterized by $W = [w_1, \ldots, w_C] \in \mathbb{R}^{C \times D}$ with $w_c \in \mathbb{R}^D$ a learnable vector for the $c$th class. From a prototype view, the label assignment of point $x_i$ is $\hat{c}_i = \arg\max_c\{w_c^\top f_i\}_{c=1}^{C}$, where $w_c$ can be viewed as a prototype of class $c$.

As mentioned in the introduction, the single prototype per class largely limits the model's ability to describe the high variance pattern within a class.

### 3.2 Number-Adaptive Prototype Learning

Different from the single-prototype-per-class design in the point-wise classification paradigm, we propose to use an adaptive number of prototypes for each class. Our model generally comprises of two parts: i) a point feature extraction module $\phi(\cdot)$ to extract point-wise feature, ii) a prototype learning module $g(\cdot)$ which takes the point cloud as input, and produces $K$ prototype vectors $P = [p_1, p_2, \ldots, p_K] \in \mathbb{R}^{K \times D}$ with the corresponding class label $\{l_k \in \{1, 2, \ldots, C\}\}_{k=1}^{K}$. It is worth noting that the number of class-$c$ prototypes $k_c = |\{p_k|c_k = c\}|$ depends on the input point cloud $X$, so the total prototype number $K = \sum_{c=1}^{C} k_c$ is also input-dependent. Finally, the point $x_i$ is labeled as the class of its nearest prototype, which can be formulated as: $\hat{c}_i = c_{k^*}, k^* = \arg\min_k \text{dist}(p_k, f_i)_{k=1}^{K}$, where $\text{dist}(\cdot, \cdot)$ is the distance between two vectors.

### 3.3 Model Architecture

We now introduce the implementation details of the proposed NAPL. Fig. 2 overviews our model, which consists of 3 modules: point feature extraction module, prototype learning module, and segmentation module.

**Point Feature Extraction Module.** The point feature extraction module (PFEM) takes the raw 3D point cloud as input and extracts point-wise features. It consists of a backbone network to learn compact point features and a decoder network to predict point-wise features $F = [f_1, f_2, \ldots, f_N] \in \mathbb{R}^{N \times D}$, where $N$ denotes the number of points, and $D$ denotes the feature dimension. We pre-train a per-point classification backbone to initialize our model's feature extraction.

**Prototype Learning Module.** The core challenge of the number-adaptive prototype learning paradigm is the implementation of the prototype learning module. Inspired by the recent work [5,11] in image segmentation, we leverage a transformer decoder as our prototype learning module. It takes $N_q$ learnable query vectors and the intermediate point features as input, and then progressively updates query vectors with point features using attention blocks. Finally, it outputs $N_q$ prototype proposals $P = [p_1, p_2, \ldots, p_{N_q}] \in \mathbb{R}^{N_q \times D}$ and the corresponding prototype class score $S = [s_1, s_2, \ldots, s_{N_q}] \in [0,1]^{N_q \times (C+1)}$. The additional class label $C + 1$ denotes that the corresponding prototype does not belong to any semantic class, which enables that an adaptive number $K \leq N_q$ of prototypes are kept for segmentation.

**Segmentation Module.** Given the point-wise feature $F$ and prototype proposals $(P, S)$, we follow the semantic inference procedure of MaskFormer [5] to predict the semantic label for each point. Specifically, the $i$th point's semantic label $\widehat{c}_i$ is obtained by $\widehat{c}_i = \underset{c \in \{1,2,\ldots,C\}}{\arg\max} \sum_{k=1}^{N_q} \mathrm{sigmoid}(f_i^\top p_k) \cdot s_k^c$, where $f_i \in \mathbb{R}^D$ is the $i$th point's feature and $s_k^c$ denotes the $c$ th element of $s_k$.

### 3.4 Prototype Dropout Training Strategy

To enable our model to adaptively produce prototypes for segmentation, we design a simple yet effective training strategy, namely prototype dropout training.

The label assignment in Sect. 3.2 is not differentiable. To facilitate prototype-learning module training, we formulate the assignment between points and prototypes as a set of soft assignment masks $\{m_k = \mathrm{sigmoid}(F \cdot p_k) \in [0,1]^N\}_{k=1}^{N_q}$, and arrange the model prediction as a set of class-mask pairs $z = \{(s_k, m_k)\}_{k=1}^{N_q}$. Ground truth semantic labels are arranged as the same class-mask pairs $z^{\mathrm{gt}} = \{(c_k^{\mathrm{gt}}, m_k^{\mathrm{gt}}) | c_k^{\mathrm{gt}} \in \{1, 2, \ldots, C\}, m_k^{\mathrm{gt}} \in \{0,1\}^N\}_{k=1}^{N_{\mathrm{gt}}}$, where the $i$th element of $m_k^{\mathrm{gt}}$ suggests whether the point $x_i$ is assigned to the prototypes of class $c_k^{\mathrm{gt}}$. However, simply padding the ground truth set with "no object" tokens $\varnothing$ will push $N_{\mathrm{gt}}$ of prototypes to the nearest annotated segments, and the remaining $N_q - N_{\mathrm{gt}}$ prototypes to $\varnothing$, resulting in a degraded solution with one prototype per class.

To encourage adaptive number of prototypes per class, we randomly drop out the class-mask pairs of $M$ prototypes, with the rest denoted as $z'$. We then pad $z^{\mathrm{gt}}$ to the same size as $z'$ and calculate the cross entropy loss and mask loss $\mathcal{L}_{\mathrm{mask}}$ between $z'$ and $z^{\mathrm{gt}}$ under a minimal matching $\sigma(\cdot)$ to jointly optimize the prototype class prediction and the point-prototype assignment:

$$\mathcal{L}(z', z^{\mathrm{gt}}) = \sum_{i=1}^{N_q - M} \left[ -\log s'_{\sigma(i)}\left(c_i^{\mathrm{gt}}\right) + \mathbb{1}_{\{c_i^{\mathrm{gt}} \neq \varnothing\}} \mathcal{L}_{\mathrm{mask}}\left(m_i^{\mathrm{gt}}, m'_{\sigma(i)}\right) \right],$$

where for the padded token in $z^{\mathrm{gt}}$, whose class label $c_i^{\mathrm{gt}} = \varnothing$, we only calculate the cross entropy loss. For simplicity, we use the same $\mathcal{L}_{\mathrm{mask}}$ as DETR [2].

## 4  Experiments

### 4.1  Implementation Details

**Dataset.** SemanticKITTI [1] is a widely used benchmark for 3D semantic segmentation. We follow [8] to use the standard training and validation set splits.

**Table 1.** Quantitative comparison on SemanticKITTI dataset [1]. The proposed NAPL outperforms the recent 3D semantic segmentation methods.

| Method | mIoU | Car | Bicycle | Motorcycle | Truck | Other-vehicle | Person | Bicyclist | Motorcyclist | Road | Parking | Sidewalk | Other-ground | Building | Fence | Vegetation | Trunk | Terrain | Pole | Traffic-sign |
|---|---|---|---|---|---|---|---|---|---|---|---|---|---|---|---|---|---|---|---|---|
| Test set | | | | | | | | | | | | | | | | | | | | |
| PointNet [10] | 14.6 | 46.3 | 1.3 | 0.3 | 0.1 | 0.8 | 0.2 | 0.2 | 0.0 | 61.6 | 15.8 | 35.7 | 1.4 | 41.4 | 12.9 | 31.0 | 4.6 | 17.6 | 2.4 | 3.7 |
| RandLANet [8] | 53.9 | 94.2 | 26.0 | 25.8 | 40.1 | 38.9 | 49.2 | 48.2 | 7.2 | 90.7 | 60.3 | 73.7 | 20.4 | 86.9 | 56.3 | 81.4 | 61.3 | 66.8 | 49.2 | 47.7 |
| KPConv [12] | 58.8 | 96.0 | 30.2 | 42.5 | 33.4 | 44.3 | 61.5 | 61.6 | 11.8 | 88.8 | 61.3 | 72.7 | 31.6 | 90.5 | 64.2 | 84.8 | 69.2 | 69.1 | 56.4 | 47.4 |
| SqueezeSegv3 [14] | 55.9 | 92.5 | 38.7 | 36.5 | 29.6 | 33.0 | 45.6 | 46.2 | 20.1 | 91.7 | 63.4 | 74.8 | 26.4 | 89.0 | 59.4 | 82.0 | 58.7 | 65.4 | 49.6 | 58.9 |
| RangeNet++ [9] | 52.2 | 91.4 | 25.7 | 34.4 | 25.7 | 23.0 | 38.3 | 38.8 | 4.8 | 91.8 | 65.0 | 75.2 | 27.8 | 87.4 | 58.6 | 80.5 | 55.1 | 64.6 | 47.9 | 55.9 |
| SalsaNext [7] | 59.5 | 91.9 | 48.3 | 38.6 | 38.9 | 31.9 | 60.2 | 59.0 | 19.4 | 91.7 | 63.7 | 75.8 | 29.1 | 90.2 | 64.2 | 81.8 | 63.6 | 66.5 | 54.3 | 62.1 |
| Ours (NAPL) | 61.6 | 96.6 | 32.3 | 43.6 | 47.3 | 47.5 | 51.1 | 53.9 | 36.5 | 89.6 | 67.1 | 73.7 | 31.2 | 91.9 | 67.4 | 84.8 | 69.8 | 68.8 | 59.1 | 59.2 |
| Validation set | | | | | | | | | | | | | | | | | | | | |
| PWC | 62.3 | 96.2 | 21.5 | 62.0 | 78.6 | 50.8 | 68.5 | 87.4 | 0.0 | 93.9 | 51.0 | 81.3 | 1.2 | 90.1 | 59.2 | 87.8 | 66.1 | 73.9 | 64.3 | 50.0 |
| Ours (NAPL) | 64.6 | 97.4 | 38.2 | 71.5 | 74.3 | 66.2 | 71.1 | 81.6 | 0.0 | 93.1 | 48.4 | 80.2 | 0.2 | 90.0 | 62.6 | 89.0 | 68.0 | 77.2 | 66.8 | 52.2 |

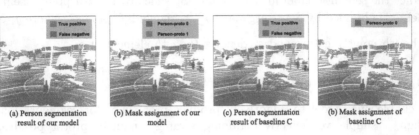

(a) Person segmentation result of our model    (b) Mask assignment of our model    (c) Person segmentation result of baseline C    (b) Mask assignment of baseline C

**Fig. 3.** We visualize a segmentation result of the "person" class. (a) and (c) shows that our model is better than baseline C. (b) and (d) explains the observation: a single prototype cannot cover the high variance intra-class patterns.

**Model Architecture.** Without loss of generality, we use a MinkUNet [6] without classifier as our PFEM, which is a fully convolutional voxel-based model with four stages. The input voxel size is 0.05m. We use the fourth stage feature of PFEM and $N_q = 50$ queries as the input of the prototype learning module.

**Training Details.** We use AdamW optimizer and poly learning rate schedule with an initial learning rate of $10^{-3}$ for transformer and point decoder, and $10^{-4}$ for pre-trained backbone. We set the number of dropout prototypes $M = 10$. Our model is trained with batch size of 16 on 4 RTX 3090 GPUs for 20 epochs.

## 4.2 Results

We use the mean intersection of union (mIoU) [1] as our evaluation metric. The results are reported from both the validation and the test set of SemanticKITTI.

**Quantitative Evaluation.** In Table 1, we compare our number-adaptive prototype learning model with the existing 3D point cloud semantic segmentation models [7–10,12,14] and the baseline point-wise classification model class-by-class. The result shows that our proposed number-adaptive prototype learning paradigm is better than the traditional point-wise classification paradigm. Specifically, in most of the classes where instances have different patterns including person, other-vehicle et al., our model has made significant improvements.

**Table 2.** Model component ablation study on SemanticKITTI val-set.

|       | A     | B     | C     | Full  |
|-------|-------|-------|-------|-------|
| PFEM  | ✓     | ✓     | ✓     | ✓     |
| T     |       | ✓     | ✓     | ✓     |
| PBW   |       |       | ✓     | ✓     |
| PD    |       |       |       | ✓     |
| mIoU  | 62.30 | 48.86 | 63.67 | **64.62** |

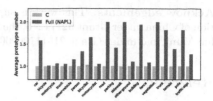

**Fig. 4.** Statistics of the average number of prototypes per class.

**Ablation Study.** In Table 2, we further study the effectiveness of individual components in our model, including the transformer decoder (T), pre-trained backbone weights (PBW), and the prototype dropout training strategy (PD). The results show that: i) directly adding a transformer module to the point cloud segmentation model and training them together greatly harms the performance by 13.44%; ii) using a pre-trained backbone makes the model easy to train and boost the result by 1.37%; and iii) the prototype dropout strategy can further promote the model performance by 0.95%.

**Handling Challenging Cases.** In Fig. 3, we show a person segmentation case from the validation set to discuss the need for multiple prototypes for each class. Fig. 3(a) and (c) show the segmentation results of our model and baseline model C, respectively. The points in green are the true-positive points and the points in red are false-negative. Our model correctly segments all the points in this scene, while C misses the points belonging to the shorter person. Fig. 3(b) and (d) reveal the deeper reason: a single prototype cannot cover all the points of different persons, while two prototypes can describe the different patterns of people, and thus make a better segmentation. The visualization shows the superiority of our proposed number-adaptive prototype learning paradigm.

**Prototype Number Analysis.** Figure 4 presents the average number of prototypes for each class in each frame. With the prototype dropout strategy, our model can adaptively produce prototypes for each class, while the baseline model C using the same model architecture can only produce one prototype for each class. This result shows the effectiveness of our proposed training strategy.

## 5    Conclusions

In this paper, we propose a novel number-adaptive prototype learning paradigm for 3D point cloud semantic segmentation. To realize this, we leverage a transformer decoder in our model to learn prototypes for semantic categories. To enable training, we design a prototype dropout strategy to promote our model to produce number-adaptive prototypes for each class. The experimental results and visualization on SemanticKITTI demonstrate the effectiveness of our design.

**Acknowledgements.** This work is supported by National Natural Science Foundation of China under Grant 62171276, the Science and Technology Commission of Shanghai Municipal under Grant 21511100900 and CALT Grant 2021-01.

# References

1. Behley, J., et al.: SemanticKITTI: a dataset for semantic scene understanding of lidar sequences. In: Proceedings of the IEEE/CVF International Conference on Computer Vision, pp. 9297–9307 (2019)
2. Carion, N., Massa, F., Synnaeve, G., Usunier, N., Kirillov, A., Zagoruyko, S.: End-to-End object detection with transformers. In: Vedaldi, A., Bischof, H., Brox, T., Frahm, J.-M. (eds.) ECCV 2020. LNCS, vol. 12346, pp. 213–229. Springer, Cham (2020). https://doi.org/10.1007/978-3-030-58452-8_13
3. Chen, S., Liu, B., Feng, C., Vallespi-Gonzalez, C., Wellington, C.: 3D point cloud processing and learning for autonomous driving: impacting map creation, localization, and perception. IEEE Signal Process. Maga. **38**(1), 68–86 (2020)
4. Chen, X., Milioto, A., Palazzolo, E., Giguere, P., Behley, J., Stachniss, C.: SuMa++: efficient lidar-based semantic slam. In: 2019 IEEE/RSJ International Conference on Intelligent Robots and Systems (IROS), pp. 4530–4537. IEEE (2019)
5. Cheng, B., Schwing, A., Kirillov, A.: Per-pixel classification is not all you need for semantic segmentation. Adv. Neural Inf. Process. Syst. **34**, 17864–17875 (2021)
6. Choy, C., Gwak, J., Savarese, S.: 4D spatio-temporal convnets: minkowski convolutional neural networks. In: Proceedings of the IEEE/CVF Conference on Computer Vision and Pattern Recognition, pp. 3075–3084 (2019)
7. Cortinhal, T., Tzelepis, G., Erdal Aksoy, E.: SalsaNext: fast, uncertainty-aware semantic segmentation of LiDAR point clouds. In: Bebis, G., et al. (eds.) ISVC 2020. LNCS, vol. 12510, pp. 207–222. Springer, Cham (2020). https://doi.org/10.1007/978-3-030-64559-5_16
8. Hu, Q., et al.: Randla-net: efficient semantic segmentation of large-scale point clouds. In: Proceedings of the IEEE/CVF Conference on Computer Vision and Pattern Recognition, pp. 11108–11117 (2020)
9. Milioto, A., Vizzo, I., Behley, J., Stachniss, C.: Rangenet++: fast and accurate lidar semantic segmentation. In: 2019 IEEE/RSJ International Conference on Intelligent Robots and Systems (IROS), pp. 4213–4220. IEEE (2019)
10. Qi, C.R., Su, H., Mo, K., Guibas, L.J.: Pointnet: deep learning on point sets for 3D classification and segmentation. In: Proceedings of the IEEE Conference on Computer Vision and Pattern Recognition, pp. 652–660 (2017)
11. Strudel, R., Garcia, R., Laptev, I., Schmid, C.: Segmenter: transformer for semantic segmentation. In: Proceedings of the IEEE/CVF International Conference on Computer Vision, pp. 7262–7272 (2021)
12. Thomas, H., Qi, C.R., Deschaud, J.E., Marcotegui, B., Goulette, F., Guibas, L.J.: Kpconv: flexible and deformable convolution for point clouds. In: Proceedings of the IEEE/CVF International Conference on Computer Vision, pp. 6411–6420 (2019)
13. Wu, P., Chen, S., Metaxas, D.N.: Motionnet: joint perception and motion prediction for autonomous driving based on bird's eye view maps. In: Proceedings of the IEEE/CVF Conference on Computer Vision and Pattern Recognition, pp. 11385–11395 (2020)

14. Xu, C., et al.: SqueezeSegV3: spatially-adaptive convolution for efficient point-cloud segmentation. In: Vedaldi, A., Bischof, H., Brox, T., Frahm, J.-M. (eds.) ECCV 2020. LNCS, vol. 12373, pp. 1–19. Springer, Cham (2020). https://doi.org/10.1007/978-3-030-58604-1_1
15. Zhou, T., Wang, W., Konukoglu, E., Van Gool, L.: Rethinking semantic segmentation: a prototype view. In: Proceedings of the IEEE/CVF Conference on Computer Vision and Pattern Recognition, pp. 2582–2593 (2022)
16. Zhu, X., et al.: Cylindrical and asymmetrical 3D convolution networks for lidar segmentation. In: Proceedings of the IEEE/CVF Conference on Computer Vision and Pattern Recognition, pp. 9939–9948 (2021)

# Self-supervised 3D Human Pose Estimation in Static Video via Neural Rendering

Luca Schmidtke[1,2]($\boxtimes$), Benjamin Hou[1], Athanasios Vlontzos[1],
and Bernhard Kainz[1,2]

[1] Imperial College London, London, UK
l.schmidtke@imperial.ac.uk
[2] Friedrich-Alexander-Universität Erlangen-Nürnberg, DE, Erlangen, Germany

**Abstract.** Inferring 3D human pose from 2D images is a challenging and long-standing problem in the field of computer vision with many applications including motion capture, virtual reality, surveillance or gait analysis for sports and medicine. We present preliminary results for a method to estimate 3D pose from 2D video containing a single person and a static background without the need for any manual landmark annotations. We achieve this by formulating a simple yet effective self-supervision task: our model is required to reconstruct a random frame of a video given a frame from another timepoint and a rendered image of a transformed human shape template. Crucially for optimisation, our ray casting based rendering pipeline is fully differentiable, enabling end to end training solely based on the reconstruction task.

**Keywords:** Self-supervised learning · 3D human pose estimation · 3D pose tracking · Motion capture

## 1 Introduction

Inferring 3D properties of our world from 2D images is an intriguing open problem in computer vision, even more so when no direct supervision is provided in the form of labels. Although this problem is inherently ill-posed, humans are able to derive accurate depth estimates, even when their vision is impaired, from motion cues and semantic prior knowledge about the perceived world around them. This is especially true for human pose estimation. Self-supervised learning has proven to be an effective technique to utilise large amounts of unlabelled video and image sources. On a more fundamental note, self-supervised learning is hypothesised to be an essential component in the emergence of intelligence and cognition. Moreover, self-supervised approaches allow for more flexibility in

**Supplementary Information** The online version contains supplementary material available at https://doi.org/10.1007/978-3-031-25066-8_42.

L. Karlinsky et al. (Eds.): ECCV 2022 Workshops, LNCS 13803, pp. 704–713, 2023.
https://doi.org/10.1007/978-3-031-25066-8_42

domains such as the medical sector where labels are often hard to come by. In this paper we focus on self-supervised 3D pose estimation from monocular video, a key element of a wide range of applications including motion capture, visual surveillance or gait analysis.

Inspired by previous work, we model pose as a factor of variation throughout different frames of a video of a single person and a static background. More formally, self-supervision is provided by formulating a conditional image reconstruction task: given a pose input different from the current image, what would that image look like if we condition it on the given pose? Differently from previous work, we choose to represent pose as a 3D template consisting of connected parts which we transform and project to two-dimensional image space, thereby inferring 3D pose from monocular images without explicit supervision.

More specifically, our method builds upon the recent emergence and success of combining deep neural networks with an explicit 3D to 2D image formation process through fully differentiable rendering pipelines. This inverse-graphics approach follows the analysis by synthesis principle of generative models in a broader context: We hope to extract information about the 3D properties of objects in our world by trying to recreate their perceived appearance on 2D images. Popular rendering techniques rely on different representations including meshes and polygons, point clouds or implicit surfaces. In our work we make use of volume rendering with a simple occupancy function or density combined with a texture field that assign an occupancy between $[0, 1]$ and RGB colour value $c \in \mathbb{R}^3$ for every point defined on a regular 3D grid.

## 2  Related Work

**Monocular 3D Human Pose Estimation.** Human pose estimation in general is a long standing problem in computer vision with an associated large body of work and substantial improvements since the advent of deep-learning based approaches. Inferring 3D pose from monocular images however remains a challenging problem tackled by making use of additional cues in the image or video such as motion or multiple views from synchronised cameras or introducing prior knowledge about the hierarchical part based structure of the human body.

**Lifting from 2D to 3D.** Many works break down the problem into first estimating 2D pose and subsequently estimate 3D pose either directly [19], by leveraging self-supervision through transformation and reprojection [15] or a kd-tree to find corresponding pairs of detected 2D pose and stored 3D pose [4].

**Motion Cues From Video.** Videos provide a rich source of additional temporal information that can be exploited to limit the solution space. [2,8,10,16] use recurrent architectures in the form of LSTMs or GRUs to incorporate temporal context while [23] employ temporal convolutions and a reprojection objective.

**Multiple Views.** Other approaches incorporate images from multiple, synchronised cameras to alleviate the ill-posedness of the problem. [22,31] and [24] fuse multiple 2D heatmaps while [26,27] utilize multi-view consistency as a form of additional supervision in the objective function.

**Human Body Prior.** Using non-paremetric belief propagation, [29] estimate the 2D pose of loosely-linked human body parts from image features and use a mixture of experts to estimate a conditional distribution of 3D poses. Many more recent approaches rely on features extracted from convolutional neural networks [14]. Many works such as [7,10,11] make use of SMPL [18], a differentiable generative model that produces a 3D human mesh based on disentangled shape and pose parameters. [32] leverage kinematic constraints to improve their predictions while [12] leverage a forward kinematics formulation in combination with the transformation of a 2D part-based template to formulate self-supervision in form of image reconstruction similar in some ways to our approach.

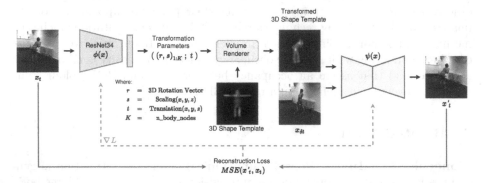

**Fig. 1.** Our method – left to right– An input frame $x_t$ passes through the pose extractor encoder $\phi$ and produces the transformation parameters for each skeletal node of the shape template **T**. The transformed template is then rendered and concatenated with a random frame of the same sequence, $x_{\delta t}$, and is passed into an auto-encoder that's tasked to reconstruct the original frame.

**Human Neural Rendering.** Recently, neural rendering approaches, *ie.* fully differentiable rendering pipelines, have gained a lot of attention. Volume rendering techniques [17,20] have been demonstrated to be powerful tools to infer 3D properties of objects from 2D images when used in combination with neural networks. The end-to-end differentiability offers the intriguing opportunity to directly leverage pixel-wise reconstruction losses as a strong self-supervision signal. This has sparked a number of very recent works estimating human 3D shape and pose via neural radiance fields [13,20,30] or signed-distance function based rendering [6].

# 3   Method

Our approach relies on self-supervision through image reconstruction conditioned on a transformed and rendered shape template. The images are sampled from a video containing a single person moving in front of a static background. More formally, the goal is to reconstruct a number of frames $(\mathbf{x}_{t_1}, \mathbf{x}_{t_n})$ from random time points $t_1, ..., t_n$ in a video with access to *one* frame $\mathbf{x}_{t_k}$, again sampled randomly, and rendered images of transformed templates $\mathbf{T_1}, ..., \mathbf{T_n}$.

Our method can be viewed in two distinct steps; regression of template transformation parameters and image reconstruction, where both steps are parameterized using deep convolutional neural networks. An encoder network $\phi$ regresses rotation, translation and scale parameters from frame $\mathbf{x}_t$ in order to transform each skeletal node of a 3D shape template, $\mathbf{T}$. The generator network $\psi$ takes as input; (a) a frame $\mathbf{x}_{\delta t}$, from a different time instance of the same sequence where the same person assumes a different pose, and (b) a rendered image of the transformed 3D template while being tasked to reconstruct frame $\mathbf{x}_t$.

The encoder $\phi$ consists of a convolutional neural network for feature extraction followed by a number of linear layers and a reshape operation. The generative network $\psi$ resembles a typical convolutional encoder-decoder structure utilised for image translation, where feature maps are subsequently downsampled via strided convolutions and the number of features increases. For the decoder we utilise bilinear upsampling and spatially adaptive instance normalisation (SPADE) [21] to facilitate semantic inpainting of the rendered template image.

## 3.1   Template and Volume Rendering

**Shape Template:** A shape template, $\mathbf{T}$, consists of $K$ Gaussian ellipsoids that are arranged in the shape of a human. Each skeletal node, denoted as $\mathbf{T}_k$, is defined on a regular volumetric grid and represents a single body part. All ellipsoids are parameterized by their mean $\mu_k$ and co-variance $\Sigma_k$. On the volumetric grid wee define two functions: a scalar field $f : \mathbb{R}^3 \to [0, 1]$ that assigns a value to each point $(x, y, z)$ on the grid — in the volume rendering literature it is commonly referred to as the occupancy function, and a vector field $c : \mathbb{R}^3 \to C \subset \mathbb{R}^3$ specifying the RGB-colour for each point, commonly referred to as the colouring function.

**Raycasting and Emission Absorption Function:** We make use of an existing implementation of the raycasting algorithm shipped with the PyTorch3D package [25] to render the template image. Given a camera location $\mathbf{r}_0 \in \mathbb{R}^3$, rays are "emitted" from $\mathbf{r}_0$ that pass through each pixel $\mathbf{u}_i \in \mathbb{R}^3$ lying on a 2D view plane $S$ by sampling uniformly spaced points along each ray starting from the intersecting pixel:

$$\mathbf{p}_j = \mathbf{u}_i + j\delta s, \tag{1}$$

where $j$ is the step and $\delta s$ the step size that depends on the maximum depth and number of points along each ray.

The colour value at each pixel location $\mathbf{u}_i$ is then determined by a weighted sum of all colour values of the points sampled along the ray:

$$\mathbf{c}_i = \sum_{j=0}^{J} w_j \mathbf{c}_j \tag{2}$$

The weights $w_j$ are computed by multiplying the occupancy function $f(x)$ with the transmission function $T(x)$ evaluated at each point $\mathbf{p}$ along the ray:

$$w_j = f(\mathbf{p}_j) \cdot T(\mathbf{p}_j), \tag{3}$$

where $T(\mathbf{x})$ can be interpreted as the probability that a given ray is not terminated, i.e. fully absorbed, at a given point $\mathbf{x}$ and is computed as the cumulative product of the complement of the occupancy function of all $k$ points up until $\mathbf{p}_j$:

$$T(\mathbf{x}) = \prod_{k}(1 - f(\mathbf{x}_k)) \tag{4}$$

Repeating this for all pixels in the view plane results in a 2D projection of our 3D object representing the rendered image $\mathbf{f}_r \in \mathbb{R}^{3 \times h \times w}$.

## 3.2  Pose Regression and Shape Transformation

In order to estimate the skeletal pose of a given frame, we use the encoder network, $\phi : \mathbb{R}^{3 \times h \times w} \to \mathbb{R}^{3K+3}$ based on the ResNet-34 architecture [1].

The encoder maps a color input image of size $h \times w$ to $K$ rotation and scale vectors, $(\mathbf{r}, \mathbf{s})_{1:K} \in \mathbb{R}^3$, and a single global translation vector, $\mathbf{t} \in \mathbb{R}^3$ for the camera. $K$ denotes the number of transformable parts in the template. Here, rotation is parameterised via axis-angle representation, and is subsequently converted to 3D transformation matrices using the Rodrigues' rotation formula. Combined with the scaling parameter for each axis, the resulting matrix defines the affine mapping, excluding the sheer component, for spatial transformation of each skeletal node.

After construction of the 3D transformation matrix, each Gaussian ellipsoid of the template $\mathbf{T}$, with occupancy $f_k(\mathbf{x})$ and colour field $c_k(\mathbf{x})$, gets transformed according to the regressed parameters. Finally, utilising the aforementioned ray-tracing method we render an image based upon our transformed template by summing together all transformed occupancy and colour fields and clipping to a maximum value of 1:

$$\tilde{\mathbf{T}}_k = \Omega_k(\mathbf{T}_k) \tag{5}$$

$$\mathbf{f}_r = \mathcal{R}\left(\sum_k f_k(\mathbf{x}), \sum c_k(\mathbf{x})\right), \tag{6}$$

where $\mathcal{R}$ denotes the rendering operation.

### 3.3 Kinematic Chain

Instead of relying on an additional loss to enforce connectivity between body parts as in [28], we define a kinematic chain along which each body part is reconnected to its parent via a translation after rotation and scale have been applied.

Given a parent and child body part with indices $n$ and $m$ respectively, we define anchor-points $\mathbf{a}_i^n$ and $\mathbf{a}_j^m \in \mathbb{R}^3$ on each part representing the area of overlap in the non-transformed template (see Fig. 2 right). If body part $m$ is being transformed, the position of the anchor point changes: $\tilde{\mathbf{a}}_j^m = H\mathbf{a}_j^m$, where $H \in \mathbb{R}^{3 \times 3}$ specifies a transformation matrix. To ensure continuous connectivity, we apply the transformation for the child body part in an analogous way, it is reconnected with the parent node by applying translation $\mathbf{t} = \tilde{\mathbf{a}}_i^n - \tilde{\mathbf{a}}_i^m$. We first transform the core, and then proceed with all other parts in an iterative fashion along the kinematic chain as depicted in Fig. 2 left.

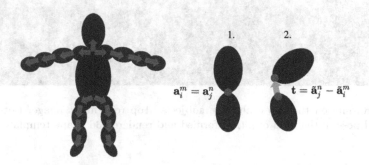

**Fig. 2.** Illustration of the kinematic chain. Red circles denote anchor-points. Following a transformation of the upper part, the translation $\mathbf{t} = \tilde{\mathbf{a}}_j^n - \tilde{\mathbf{a}}_i^m$ is applied to enforce continuity. (Color figure online)

### 3.4 Loss Function

The loss function is a sum of individual components: the reconstruction loss as the pixel wise $l^2$-norm between the decoder output and the original image and a boundary loss of the form

$$\mathcal{L}_{bx}^i = \begin{cases} |\hat{a}_{x,i}|, & \text{if } |\hat{a}_{x,i}| > 1 \\ 0, & \text{otherwise} \end{cases} \qquad \mathcal{L}_{bx} = \sum_i \mathcal{L}_{bx}^i \qquad \mathcal{L}_b = \mathcal{L}_{bx} + \mathcal{L}_{by}, \quad (7)$$

where $\hat{a}_{x,i}$ is the x-component of a projected and transformed anchor point. Note that we normalise image coordinates to $(-1, 1)$.

We also regularise the pose regression via the $l^2$-norm of the rotation vector $\mathbf{r}$ and decay this term linearly to 0 after 500 iterations. Overall, our objective function is:

$$\mathcal{L} = \mathcal{L}_{\text{recon}} + \mathcal{L}_b + \alpha * \sum_K \|\mathbf{r}_k\|_2, \quad \alpha = \min(1 - 0.02 * \text{iter}) \qquad (8)$$

## 4   Experiments

We train and evaluate our model on Human 3.6M [3], a motion-capture dataset
including 11 actors performing various activities while being filmed by four different cameras in a studio setting. Following [5,28], we train on subjects 1, 5, 6,
7, test on 9 and 11 and restrict activities to mostly upright poses, resulting in
roughly 700,000 images for training. We sample video frames in pairs containing the same person in different poses, but with the same background utilising
bounding boxes derived from the masks and utilise the Adam optimizer [9].

**Fig. 3.** Results on the two evaluation subjects. Top row: input image. Bottom row:
predicted pose in the form of a transformed and rendered 3d shape template.

## 5   Results

We restrict our evaluation to qualitative results in Fig. 3. These demonstrate
that the concept of self-supervision through conditional image translation can
be extended to 3D pose estimation. However, there are several issues that still
need to be solved: The model is currently not able to distinguish left and right,
as can be observed in Fig. 3 (fourth image from the right), where the subject is
facing away from the camera, but the template remains in the front-facing configuration. The model also mostly generates limbs facing away from the camera
(third image from the right). We hypothesise that due to depth ambiguity in 2D
and limitations in pose variety due to the restricted sampling the decoder can
perfectly reconstruct the image despite the wrong orientation of the limb.

## 6   Conclusion

We presented preliminary results for a method to estimate human pose in 3d from
monocular images without relying on any landmark labels. Despite issues with
depth ambiguity the qualitative results are encouraging and demonstrate the
feasibility of combining differentiable rendering techniques and self-supervision.
A straightforward improvement would be weak supervision in the form a small

labelled dataset. Replacing the image translation task with a purely generative approach with separate fore- and background similarly to [33] might prove to be very successful in extending the approach to non-static backgrounds as well.

**Acknowledgements.** Supported by EPSRC EP/S013687/1.

# References

1. He, K., Zhang, X., Ren, S., Sun, J.: Deep residual learning for image recognition. In: 2016 IEEE Conference on Computer Vision and Pattern Recognition (CVPR), pp. 770–778 (2016)
2. Hossain, M.R.I., Little, J.J.: Exploiting temporal information for 3D human pose estimation. In: Ferrari, V., Hebert, M., Sminchisescu, C., Weiss, Y. (eds.) ECCV 2018. LNCS, vol. 11214, pp. 69–86. Springer, Cham (2018). https://doi.org/10.1007/978-3-030-01249-6_5
3. Ionescu, C., Papava, D., Olaru, V., Sminchisescu, C.: Human3.6m: large scale datasets and predictive methods for 3d human sensing in natural environments. IEEE Trans. Pattern Anal. Mach. Intell. **36**(7), 1325–1339 (2014)
4. Iqbal, U., Doering, A., Yasin, H., Krüger, B., Weber, A., Gall, J.: A dual-source approach for 3d human pose estimation from a single image. Comput. Vis. Image Underst. **172**, 37–49 (2018)
5. Jakab, T., Gupta, A., Bilen, H., Vedaldi, A.: Self-supervised learning of interpretable keypoints from unlabelled videos. In: 2020 IEEE/CVF Conference on Computer Vision and Pattern Recognition (CVPR), pp. 8784–8794 (2020)
6. Jiang, B., Hong, Y., Bao, H., Zhang, J.: Selfrecon: self reconstruction your digital avatar from monocular video. ArXiv abs/2201.12792 (2022)
7. Kanazawa, A., Black, M.J., Jacobs, D.W., Malik, J.: End-to-end recovery of human shape and pose. In: IEEE Conference on Computer Vision and Pattern Recognition (CVPR), pp. 7122–7131. IEEE Computer Society (2018)
8. Katircioglu, I., Tekin, B., Salzmann, M., Lepetit, V., Fua, P.: Learning latent representations of 3D human pose with deep neural networks. Int. J. Comput. Vis. **126**(12), 1326–1341 (2018)
9. Kingma, D.P., Ba, J.: Adam: a method for stochastic optimization. CoRR abs/1412.6980 (2015)
10. Kocabas, M., Athanasiou, N., Black, M.J.: Vibe: Video inference for human body pose and shape estimation. In: The IEEE Conference on Computer Vision and Pattern Recognition (CVPR) (2020)
11. Kolotouros, N., Pavlakos, G., Black, M.J., Daniilidis, K.: Learning to reconstruct 3d human pose and shape via model-fitting in the loop. In: ICCV (2019)
12. Kundu, J.N., Seth, S., Jampani, V., Rakesh, M., Babu, R.V., Chakraborty, A.: Self-supervised 3d human pose estimation via part guided novel image synthesis. In: 2020 IEEE/CVF Conference on Computer Vision and Pattern Recognition (CVPR), pp. 6151–6161 (2020)
13. Kwon, Y., Kim, D., Ceylan, D., Fuchs, H.: Neural human performer: learning generalizable radiance fields for human performance rendering. ArXiv abs/2109.07448 (2021)
14. LeCun, Y., Bengio, Y.: Convolutional Networks for Images, Speech, and Time Series, pp. 255–258 (1998)

15. Li, Y., Li, K., Jiang, S., Zhang, Z., Huang, C., Xu, R.Y.D.: Geometry-driven self-supervised method for 3d human pose estimation, vol. 34, pp. 11442–11449 (2020). https://ojs.aaai.org/index.php/AAAI/article/view/6808
16. Lin, M., Lin, L., Liang, X., Wang, K., Cheng, H.: Recurrent 3d pose sequence machines. CoRR abs/1707.09695 (2017). http://arxiv.org/abs/1707.09695
17. Lombardi, S., Simon, T., Saragih, J.M., Schwartz, G., Lehrmann, A.M., Sheikh, Y.: Neural volumes. ACM Trans. Graph. (TOG) **38**, 1–14 (2019)
18. Loper, M., Mahmood, N., Romero, J., Pons-Moll, G., Black, M.J.: SMPL: a skinned multi-person linear model. ACM Trans. Graphics (Proc. SIGGRAPH Asia) **34**(6), 248:1–248:16 (2015)
19. Martinez, J., Hossain, R., Romero, J., Little, J.J.: A simple yet effective baseline for 3d human pose estimation. In: Proceedings IEEE International Conference on Computer Vision (ICCV). IEEE, Piscataway (2017)
20. Mildenhall, B., Srinivasan, P.P., Tancik, M., Barron, J.T., Ramamoorthi, R., Ng, R.: NeRF: representing scenes as neural radiance fields for view synthesis. In: Vedaldi, A., Bischof, H., Brox, T., Frahm, J.-M. (eds.) ECCV 2020. LNCS, vol. 12346, pp. 405–421. Springer, Cham (2020). https://doi.org/10.1007/978-3-030-58452-8_24
21. Park, T., Liu, M.Y., Wang, T.C., Zhu, J.Y.: Semantic image synthesis with spatially-adaptive normalization. In: 2019 IEEE/CVF Conference on Computer Vision and Pattern Recognition (CVPR), pp. 2332–2341 (2019)
22. Pavlakos, G., Zhou, X., Derpanis, K.G., Daniilidis, K.: Coarse-to-fine volumetric prediction for single-image 3D human pose. In: 2017 IEEE Conference on Computer Vision and Pattern Recognition (CVPR), pp. 1263–1272 (2017)
23. Pavllo, D., Feichtenhofer, C., Grangier, D., Auli, M.: 3D human pose estimation in video with temporal convolutions and semi-supervised training. In: Conference on Computer Vision and Pattern Recognition (CVPR) (2019)
24. Qiu, H., Wang, C., Wang, J., Wang, N., Zeng, W.: Cross view fusion for 3D human pose estimation. In: 2019 IEEE/CVF International Conference on Computer Vision (ICCV), pp. 4341–4350 (2019)
25. Ravi, N., et al.: Accelerating 3D deep learning with pytorch3d. arXiv:2007.08501 (2020)
26. Rhodin, H., Salzmann, M., Fua, P.V.: Unsupervised geometry-aware representation for 3D human pose estimation. ArXiv abs/1804.01110 (2018)
27. Rhodin, H., et al.: Learning monocular 3D human pose estimation from multi-view images. In: 2018 IEEE/CVF Conference on Computer Vision and Pattern Recognition, pp. 8437–8446 (2018)
28. Schmidtke, L., Vlontzos, A., Ellershaw, S., Lukens, A., Arichi, T., Kainz, B.: Unsupervised human pose estimation through transforming shape templates. In: Proceedings of IEEE Conference on Computer Vision and Pattern Recognition (CVPR) (2021)
29. Sigal, L., Black, M.J.: Predicting 3D people from 2D pictures. In: Perales, F.J., Fisher, R.B. (eds.) Articulated Motion and Deformable Objects, pp. 185–195 (2006)
30. Su, S.Y., Yu, F., Zollhoefer, M., Rhodin, H.: A-nerf: articulated neural radiance fields for learning human shape, appearance, and pose (2021)
31. Tomè, D., Toso, M., de Agapito, L., Russell, C.: Rethinking pose in 3D: multi-stage refinement and recovery for markerless motion capture. In: 2018 International Conference on 3D Vision (3DV), pp. 474–483 (2018)

32. Xu, J., Yu, Z., Ni, B., Yang, J., Yang, X., Zhang, W.: Deep kinematics analysis for monocular 3D human pose estimation. In: Proceedings of the IEEE/CVF Conference on Computer Vision and Pattern Recognition (CVPR) (2020)
33. Yang, Y., Bilen, H., Zou, Q., Cheung, W.Y., Ji, X.W.: Learning foreground-background segmentation from improved layered gans. In: 2022 IEEE/CVF Winter Conference on Applications of Computer Vision (WACV), pp. 366–375 (2022)

# Racial Bias in the Beautyverse: Evaluation of Augmented-Reality Beauty Filters

Piera Riccio$^{(\boxtimes)}$ and Nuria Oliver

ELLIS Alicante, Alicante, Spain
{piera,nuria}@ellisalicante.org

**Abstract.** This short paper proposes a preliminary and yet insightful investigation of racial biases in beauty filters techniques currently used on social media. The obtained results are a call to action for researchers in Computer Vision: such biases risk being replicated and exaggerated in the Metaverse and, as a consequence, they deserve more attention from the community.

**Keywords:** Self-representation · Racial bias · Ethics

## 1 Introduction

The Metaverse may be conceived as the culmination of the digitization of our lives and our society, leveraging key technological advances in fields, such as Augmented, Virtual and Mixed Reality and Artificial Intelligence. From a societal perspective, the Metaverse is considered to be the next stage in the development of current social media platforms [2]. In this regard and similarly to what happens on social media, the broad adoption and use of the Metaverse by potentially billions of users poses significant ethical and societal challenges, including the need to develop an inclusive environment, respecting the diversity of its users [10]. Thus, it is of paramount importance to ensure that the enabling technologies of the Metaverse do not create, replicate or even exacerbate patterns of discrimination and disadvantage towards specific groups of users: fairness and diversity should be at the foundation of its development [25].

In this paper, we focus on diversity in self-representation in the Metaverse. Specifically, we study the existence of implicit racial biases behind the Augmented-reality (AR)-based selfie beautification algorithms that are pervasive in social media platforms. We leverage Computer Vision techniques to perform such a study and argue that current user behaviors observed in today's social media platforms may be analyzed as an anticipation of what will happen in the Metaverse. We refer to the set of these new self-representation aesthetic norms as the *Beautyverse*. Note that existing and under-studied biases in the *Beautyverse* could lead to harmful appearances in avatar representations in the Metaverse [12,17]. Thus, we highlight the importance of coupling the proposal of novel technical contributions for the Metaverse with a comprehensive, multidisciplinary study of their societal implications.

L. Karlinsky et al. (Eds.): ECCV 2022 Workshops, LNCS 13803, pp. 714–721, 2023.
https://doi.org/10.1007/978-3-031-25066-8_43

## 2    Related Work

Self-representation in the digital space is a key factor in online social media platforms that will also shape the social interactions in the Metaverse [9]. In current social media platforms, self-representation is expressed through selfies (photos of the self); in the Metaverse, selfies are translated to avatars, which are common in other types of online environments, such as video games. Behind the self-representation through avatars there is a will to create an ideal version of the self [5,13,16], including both the personality [3] and the appearance [4,14] of the avatars.

Our research focuses on the improvements on the appearance. In recent years, AR-based selfie filters that *beautify* the original faces have become very popular on social media platforms [20]. Previous work has linked these filters to the definition and adoption of new facial aesthetics [24], with significant social and cultural impact, such as an exponential increase in teen plastic surgeries [8] and mental health issues [1]. These filters have been widely criticized for perpetuating racism [15], since the beautifying modifications applied to the original faces include lightening the skin tone, reducing the size of the nose and making the eyes bigger and lighter, which imply that people should look *whiter* to be considered beautiful [6]. In addition, the Eurocentrism of beauty filters is also shown in the *colonization of ethnic features as an aesthetic* [11], accepting certain features only when applied on the faces of white people, and rejecting them in other cases [21]. Moreover, the perpetuation of discriminating and racist beauty ideals in the *Beautyverse* could lead to significant cultural damage and dysmorphia when applied to avatars, given that avatars do not require an underlying physical reality, as selfies do.

In this paper, we aim to shed light on the existing dynamics that create the aesthetic norms and ideals of the *Beautyverse*. We address this challenge from a computational perspective, with the intent of bringing sociological and anthropological research questions to the Computer Vision community.

## 3    The Implicit Racial Bias in AR-Based Beauty Filters

The aim of our research is to leverage Computer Vision techniques to understand key characteristics of the *Beautyverse*, with a special focus on racial biases. We report results of preliminary analyses on the FAIRBEAUTY dataset [19]. This dataset has been designed to enable the study of the implications of AR-based beauty filters on social media. It was built by beautifying the faces of the FAIRFACE dataset [7] via the application of eight AR-based filters available on Instagram. The filters were chosen based on their popularity and were directly applied on the images of the FAIRFACE dataset [7], as illustrated in Fig. 1.

Given the diversity in the FAIRFACE dataset, FAIRBEAUTY enables the study of the differential impact of beautification filters on faces of different ages, genders and races. Previous work has shown that the applied beautification filters homogenize the appearance of the faces, and thus increase the similarity among

**Fig. 1.** Example of the eight different beauty filters applied to the left-most image from the FAIRFACE dataset [7].

individuals [19]. It has also been shown that the level of homogenization does not impact the performance of state-of-the-art face recognition models. In summary, these filters modify the faces so they conform to the *same* beauty standard while preserving the identity of the individuals [19].

In this paper, we study the existence of implicit racial biases in the beautification filters. Specifically, we answer the following research question: do AR-based beautification filters encode a canon of beauty of white people? We describe next the experimental setup to address our research question.

## 3.1  Experimental Setup

In our experiments, we investigate whether beauty filters implicitly make beautified individuals of all races look *whiter*. To tackle this question, we leverage two state-of-the-art race classification algorithms: DeepFace [23] and FairFace [7]. The faces in FAIRFACE are labeled according to seven different racial groups, namely: Black, East Asian, Indian, Latino Hispanic, Middle Eastern, Southeast Asian, and White. In our experiments, we randomly sample a subset of 5,000 faces for each race. We compare the performance of the race prediction algorithms on the face images from FAIRFACE [7] and the corresponding beautified version in FAIRBEAUTY [19].

The first race classification model used in the experiments is DeepFace [23], which is a lightweight face recognition and facial attribute analysis framework. It is available in the deepface Python library and it is based on different face recognition models. The framework is trained to recognize four attributes: age, gender, emotion and race. In this paper, we focus on race recognition with pre-training on the VGGFace2 dataset [18]. The second race classification model is the one released with the publication of the FAIRFACE dataset [7]. In this case, the race predictor is based on ResNet34.

## 3.2   Results

In this section, we present the race classification results on original and beautified faces. Tables 1 and 2 depict the average predicted value of the label *White* (mean and standard deviation) in each racial group by the DeepFace and FairFace algorithms, respectively. The values are averaged over the 5,000 randomly selected faces for each of the 7 races. The right-most column on the tables presents the prediction loss, i.e. the difference in the race classification performance between the original and the beautified datasets.

As shown on the Tables, the predicted value of the label *White* significantly increases in the beautified faces of *all races* when compared to the original, non-beautified images. Moreover, there is a significant loss in the performance of the race classification algorithm when applied to the beautified faces of most races except for the images labeled as *Whites*, whose performance increases in the beautified version of the original faces. In other words, there is a larger probability to classify the beautified faces –independently of their race– as white.

**Table 1.** Race classification results of the DeepFace algorithm, applied to 5,000 images. The first and second columns depict the predicted value of the label *white* in the original FAIRFACE and the beautified FAIRBEAUTY datasets, respectively; the third column contains to the race prediction loss, i.e. the difference in the race classification performance between the original and the beautified datasets.

|  | Original | Beautified | True prediction loss |
|---|---|---|---|
| Race: | | | |
| Black | $4.35 \pm 0.18$ | $7.23 \pm 0.23$ | $-7.36\%$ |
| East Asian | $10.84 \pm 0.27$ | $14.00 \pm 0.29$ | $-7.83\%$ |
| Indian | $10.07 \pm 0.21$ | $15.09 \pm 0.26$ | $-7.73\%$ |
| Latino | $20.71 \pm 0.28$ | $26.85 \pm 0.33$ | $-4.00\%$ |
| Middle Eastern | $27.73 \pm 0.31$ | $35.35 \pm 0.35$ | $-4.05\%$ |
| Southeast Asian | $8.88 \pm 0.23$ | $12.48 \pm 0.26$ | $-11.23\%$ |
| White | $53.46 \pm 0.44$ | $57.89 \pm 0.44$ | $+4.43\%$ |

To deepen the understanding of the results in Tables 1 and 2, we report the confusion matrices obtained with each of the models in Figs. 2 and 3. The left-hand side of the images depicts the confusion matrix on the original faces whereas the right-hand-side depicts the confusion matrix on the beautified version of the faces. Given the differences in behavior between the two race classification algorithms, we discuss the results separately. With respect to DeepFace (Fig. 2), we observe high prediction accuracies on faces labeled as Asian and Black (78.2% and 80.0% respective accuracies), and significantly lower on the rest of racial groups (ranging between 39.5% for Middle Eastern and 69.3% for White). After beautification, the accuracy on faces labeled as *White* increases (74.6%), whereas the accuracies on faces with all other labels significantly decrease, mostly due to a significant increase in the misclassification of the images as White (right-most column on the confusion matrix).

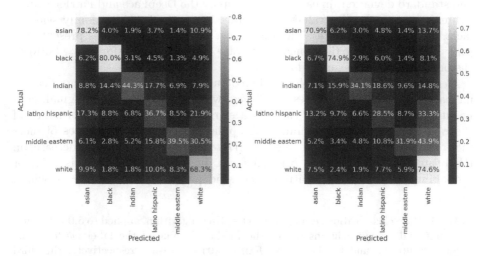

**Fig. 2.** Confusion matrices of the race prediction, obtained through DeepFace [23]. Left-hand side refers to the original images, right-hand side refers to the images after beautification.

Regarding the FairFace race classification algorithm (Fig. 3), we observe in that the highest classification accuracies are obtained for faces labeled as White and Black, with poor performances in the rest of groups (ranging between 15.7% for Southeast Asian and 40.0% for Indian). After beautification, the classification performance decreases in all cases, except for faces labeled as White and Latino-Hispanic, which are better classified. Moreover, the probability of misclassifying faces from all racial groups as White significantly increases (first column in the confusion matrix). In general, the FairFace classifier seems to be biased towards the *White* label, misclassifying faces from all other groups as being White even in the original, non-beautified case. This effect is exacerbated after beautification: the accuracy on the *White* label increases (87.3%), to the detriment of all the other labels except for the Latino-Hispanic label.

Our experiments yield results that, while preliminary, highlight societal and cultural issues that would need deeper investigation. In particular, the *Beautyverse* not only homogenizes the visual aesthetics of faces as reported in [19], but seems to make them conform with a canon of beauty of white people. As social media platforms (and the Metaverse) aim to reach a globalized community of users, it is unacceptable that the technologies that populate these platforms replicate intrinsic and subtle biases that perpetuate historic discrimination and privileges.

**Table 2.** Race classification results of the FairFace algorithm, applied to 5,000 images. The first and second columns depict the predicted value of the label *white* in the original FairFace and the beautified FairBeauty datasets, respectively; the third column contains to the race prediction loss, i.e. the difference in the race classification performance between the original and the beautified datasets.

|  | Original | Beautified | True prediction loss |
|---|---|---|---|
| Race: | | | |
| Black | $2.64 \pm 0.17$ | $5.98 \pm 0.24$ | $-5.99\%$ |
| East Asian | $30.85 \pm 0.33$ | $32.14 \pm 0.33$ | $-1.15\%$ |
| Indian | $41.98 \pm 0.44$ | $45.00 \pm 0.42$ | $-9.25\%$ |
| Latino | $27.86 \pm 0.26$ | $28.97 \pm 0.41$ | $+15.69\%$ |
| Middle Eastern | $28.58 \pm 0.22$ | $32.08 \pm 0.23$ | $-0.5\%$ |
| Southeast Asian | $27.02 \pm 0.20$ | $27.66 \pm 0.20$ | $-1.91\%$ |
| White | $76.37 \pm 0.42$ | $81.44 \pm 0.36$ | $+5.07\%$ |

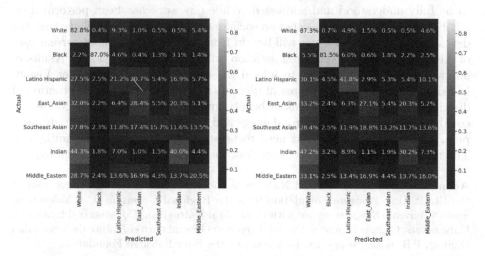

**Fig. 3.** Confusion matrices of the race prediction, obtained through FairFace [7]. Left-hand side refers to the original images, right-hand side refers to the images after beautification.

## 4 Future Work and Conclusion

As an imminent direction of future work, we plan to investigate the impact of beautification filters on higher resolution images. The images in the FairFace dataset only have a resolution of approximately 300 pixels, which severely limits the beautification process. We expect the effect of the beauty filters to be significantly more prominent on higher resolution images with more detailed facial features. Hence, while the results of our experiments could be seen as a worst-case scenario, we believe that it would be important to perform a similar study on higher resolution images, which are also expected in the Metaverse.

Quantitatively evaluating the existence of a racial bias in the *Beautyverse* is certainly an interesting and important endeavor. However, to counteract this issue, we need to shed light on the specific features implemented in these filters that contribute to the *whitening* of the faces. We plan to leverage state-of-the-art explainability frameworks (e.g. [22]) to automatically identify the areas in the images that are responsible for the shift in the classification. Explaining these results could lead to two relevant insights: (1) first, it would shed light on the specific features –associated with canons of beauty of white people– that are embedded in the beautification filters and that are considered desirable on social media; and (2) second, it would enable us to assess to what extent this racial bias is related to the beautification filters and to what extent it is intrinsic to the algorithms that classify face images according to race, as we have observed in the FairFace classifier.

Moreover, we are not aware of any extensive user study to investigate the existence of racial biases in AR-based beauty filters. If we aim to develop an inclusive Metaverse where anyone is welcome, we believe that these issues need to be fully understood and addressed. While our work has been performed on AR-based filters applied to selfies on social media platforms, we believe that the development of the Metaverse will benefit from an understanding of current uses of AR technologies for self-representation on social media. Thus, the results of our work should be valuable to inform the development of Computer Vision-based technologies for self-representation in the Metaverse. The perpetuation of discriminating and racist ideals of beauty applied to avatars in the Metaverse could lead to significant cultural damage and mental health issues (e.g. dysmorphia, anxiety, depression) that need to be studied, understood and mitigated. The research described in this paper contributes to such an understanding.

**Acknowledgements.** P.R. and N.O. are supported by a nominal grant received at the ELLIS Unit Alicante Foundation from the Regional Government of Valencia in Spain (Convenio Singular signed with Generalitat Valenciana, Conselleria d'Innovació, Universitats, Ciència i Societat Digital, Dirección General para el Avance de la Sociedad Digital). P.R. is also supported by a grant by the Banc Sabadell Foundation.

# References

1. Abi-Jaoude, E., Naylor, K.T., Pignatiello, A.: Smartphones, social media use and youth mental health. Can. Med. Assoc. J. **192**(6) (2020). https://doi.org/10.1503/cmaj.190434
2. Anderson, J., Rainie, L.: The metaverse in 2040. Pew Research Center (2022)
3. Bessière, K., Seay, A.F., Kiesler, S.: The ideal elf: identity exploration in world of Warcraft. Cyberpsychol. Behav. **10**(4), 530–535 (2007)
4. Ducheneaut, N., Wen, M.H., Yee, N., Wadley, G.: Body and mind: a study of avatar personalization in three virtual worlds. In: Proceedings of the SIGCHI Conference on Human Factors in Computing Systems, pp. 1151–1160 (2009)
5. Higgins, E.T.: Self-discrepancy: a theory relating self and affect. Psychol. Rev. **94**(3), 319 (1987)
6. Jagota, V.: Why do all the snapchat filters try to make you look white? June 2016

7. Karkkainen, K., Joo, J.: FairFace: face attribute dataset for balanced race, gender, and age for bias measurement and mitigation. In: Proceedings of the IEEE/CVF Winter Conference on Applications of Computer Vision (WACV), pp. 1548–1558, January 2021

8. Khunger, N., Pant, H.: Cosmetic procedures in adolescents: what's safe and what can wait. Indian J. Paediatr. Dermatol. **22**(1), 12–20 (2021). https://doi.org/10.4103/ijpd.IJPD_53_20

9. Kolesnichenko, A., McVeigh-Schultz, J., Isbister, K.: Understanding emerging design practices for avatar systems in the commercial social VR ecology. In: Proceedings of the 2019 on Designing Interactive Systems Conference, pp. 241–252 (2019)

10. Lee, L.H., et al.: All one needs to know about metaverse: a complete survey on technological singularity, virtual ecosystem, and research agenda. arXiv preprint arXiv:2110.05352 (2021)

11. Li, S.: The problems with Instagram's most popular beauty filters, from augmentation to eurocentrism, July 2020

12. Maloney, D.: Mitigating negative effects of immersive virtual avatars on racial bias. In: Proceedings of the 2018 Annual Symposium on Computer-Human Interaction in Play Companion Extended Abstracts, CHI PLAY 2018 Extended Abstracts, pp. 39–43. Association for Computing Machinery, New York (2018). https://doi.org/10.1145/3270316.3270599

13. Manago, A.M., Graham, M.B., Greenfield, P.M., Salimkhan, G.: Self-presentation and gender on myspace. J. Appl. Dev. Psychol. **29**(6), 446–458 (2008)

14. Messinger, P.R., Ge, X., Stroulia, E., Lyons, K., Smirnov, K., Bone, M.: On the relationship between my avatar and myself. J. Virtual Worlds Res. **1**(2) (2008). https://doi.org/10.4101/jvwr.v1i2.352

15. Mulaudzi, S.: Let's be honest: snapchat filters are a little racist, January 2017. https://www.huffingtonpost.co.uk/2017/01/25/snapchat-filters-are-harming-black-womens-self-image_a_21658358/

16. Mummendey, H.D.: Psychologie der Selbstdarstellung (1990)

17. Neely, E.L.: No player is ideal: why video game designers cannot ethically ignore players' real-world identities. ACM SIGCAS Comput. Soc. **47**(3), 98–111 (2017)

18. Parkhi, O.M., Vedaldi, A., Zisserman, A.: Deep face recognition. British Machine Vision Association (2015)

19. Riccio, P., Psomas, B., Galati, F., Escolano, F., Hofmann, T., Oliver, N.M.: OpenFilter: a framework to democratize research access to social media AR filters. In: Thirty-sixth Conference on Neural Information Processing Systems Datasets and Benchmarks Track (2022). https://openreview.net/forum?id=VF9f79cCYdZ

20. Ryan-Mosley, T.: Beauty filters are changing the way young girls see themselves, April 2021. https://www.technologyreview.com/2021/04/02/1021635/beauty-filters-young-girls-augmented-reality-social-media/

21. Ryan-Mosley, T.: How digital beauty filters perpetuate colorism, August 2021

22. Selvaraju, R.R., Das, A., Vedantam, R., Cogswell, M., Parikh, D., Batra, D.: Grad-CAM: why did you say that? arXiv preprint arXiv:1611.07450 (2016)

23. Serengil, S.I., Ozpinar, A.: LightFace: a hybrid deep face recognition framework. In: 2020 Innovations in Intelligent Systems and Applications Conference (ASYU), pp. 23–27. IEEE (2020). https://doi.org/10.1109/ASYU50717.2020.9259802

24. Shein, E.: Filtering for beauty. Commun. ACM **64**(11), 17–19 (2021)

25. Woodruff, A., Fox, S.E., Rousso-Schindler, S., Warshaw, J.: A qualitative exploration of perceptions of algorithmic fairness. In: Proceedings of the 2018 CHI Conference on Human Factors in Computing Systems, CHI 2018, pp. 1–14. Association for Computing Machinery, New York (2018). https://doi.org/10.1145/3173574.3174230

# LWA-HAND: Lightweight Attention Hand for Interacting Hand Reconstruction

Xinhan Di[1]($\boxtimes$) and Pengqian Yu[2]①

[1] BLOO Company, Shanghai, China
xinhan.di@blooxr.com
[2] Singapore, Singapore

**Abstract.** Hand reconstruction has achieved great success in real-time applications such as visual reality and augmented reality while interacting with two-hand reconstruction through efficient transformers is left unexplored. In this paper, we propose a method called lightweight attention hand (LWA-HAND) to reconstruct hands in low flops from a single RGB image. To solve the occlusion and interaction challenges in efficient attention architectures, we introduce three mobile attention modules. The first module is a lightweight feature attention module that extracts both local occlusion representation and global image patch representation in a coarse-to-fine manner. The second module is a cross image and graph bridge module which fuses image context and hand vertex. The third module is a lightweight cross-attention mechanism that uses element-wise operation for cross attention of two hands in linear complexity. The resulting model achieves comparable performance on the Inter-Hand2.6M benchmark in comparison with the state-of-the-art models. Simultaneously, it reduces the flops to $0.47GFlops$ while the state-of-the-art models have heavy computations between $10GFlops$ and $20GFlops$.

**Keywords:** Interacting hand reconstruction · Efficient transformers · InterHand2.6M

## 1 Introduction

Monocular single hand pose and shape recovery has recently witnessed great success owing to deep neural networks [2,9,19,49,52] in industrial applications such as virtual reality (VR), augmented reality(AR), digital shopping, robotics, and digital working. However, interacting hand reconstruction is more difficult and remains unsolved for real-world applications such as hand tracking in Fig. 1. First, it is difficult for networks to align hand poses with image features as feature extractors are confused by severe mutual occlusions and appearance similarity. In addition, interaction between two-hands is hard to be represented during network training. Finally, it is not trivial to design efficient network architectures which can formulate the occlusion and interaction of two hands and meet the requirement of low latency on mobile hardware at the same time.

P. Yu—Independent Researcher.

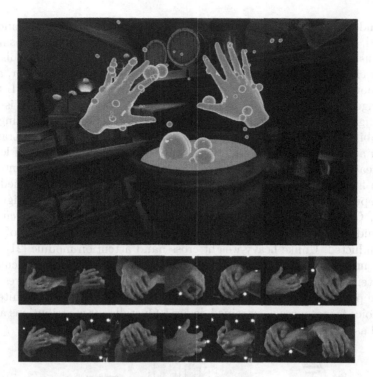

**Fig. 1.** Hand tracking application in visual reality. Here interacting hand reconstruction remains a challenge.

Good results have been produced by the monocular double hands tracking methods based on depth estimation [20,29,30,38,40]. Although these depth-based frameworks are studied for years, the algorithm complexity limit the ubiquitous application of the methods. Recently, a monocular RGB based two-hand reconstruction is proposed through building a tracking dense matching map [41]. However, its tracking procedure is not able to solve fast motion and the prior knowledge between interacting hands is not fully used through deep networks. Since the open of the large scale two-hand dataset InterHand2.6M [27], learning based image reconstruction methods are well explored through the building of different network architectures. Such examples include the 2.5D heatmaps for hand joint positions estimation [8,17,27] and the attention map for the extraction of sparse and dense image features [21,45]. However, these methods consume a lot of computation power and restrict their applications on mobile devices such as VR/AR glass and robots. In contrast, mobile vision transformers [5,25,26,39] has achieved great success in vision tasks and deployment on mobile devices. The lightweight attention based methods are likely to represent the occlusions and interaction of two-hands on mobile devices.

Motivated by the above mentioned challenges, we propose lightweight attention hand (LWA-HAND), a mobile attention and graph based single image reconstruction method. Firstly, two-stream GCN is utilized to regress mesh vertices of each hand in a coarse-to-fine-manner, similar to traditional GCN [9] and Intaghand [21]. However, for the two-hand reconstruction in a low energy consumption, naively using a two-stream network with normal attention modules fails to represent occlusion and interaction of two hands in low latency. Moreover, application of normal feature extraction [11] and extra attention modules [21] for contact image and graph features leads to high flops. To address theses issues, we equip a lightweight vision transformer for feature extraction of interacting hands. This mobile transformer utilizes the data stream of MobileNet [34] and transformers with a two-way bridge. This bridge enables bidirectional fusion of local features of each hand and global features of contexts of the occlusions and interaction between hands. Furthermore, a pyramid cross domain module is applied to fuse image representation and hand representation in a coarse-to-fine and lightweight manner. Global priors of image domain with very few tokens is calculated in a pyramid way and a direct fusion of the prior and representation of hand is then conducted. Unlike heavy-weight cross hand attention module [21], a separate attention mechanism with linearly complexity [26] reduces the calculation of cross attention to encodes interaction context into hand vertex features. Therefore, the proposed lightweight modules makes the whole network architecture a good choice for resource-constrained devices. Overall, our contributions are summarized as:

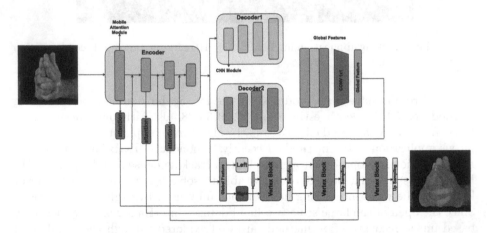

**Fig. 2.** Proposed mobile network architecture. The green block is the mobile attention encoder. The two yellow block are decoders of 2.5D heatmap and 2D segmentation. The stacked bridges represent the attention part (in red), the fusion part (in green) and the vertex attention part (in pink), respectively. The mobile attention encoder, stacked bridges, fusion part and vertex attention part are the proposed modules for the efficient hand reconstruction tasks. (Color figure online)

1. We propose a mobile two-hand reconstruction method based on lightweight attention mechanism named LWA-HAND (illustrated in Fig. 2) and demonstrate the effectiveness of mobile attention for the two-hand reconstruction task.

2. We propose a lightweight feature attention module to extract both local occlusion representation and global image patch representation in a coarse-to-fine manner, producing fusion of these two representation with low latency.

3. We propose a cross image and graph bridge module to fuse image context and hand vertex. It constructs a pyramid bridge to extract global context features of two hands with very few tokens in the attention mechanism, connecting context features directly to the hand vertex domain without extra transforms of calculation.

4. We propose a lightweight cross attention mechanism which uses element-wise operation for cross attention of two hands with linear complexity [26], reducing the flops of attention operation in the representation of interacting hands.

For the construction of interacting hands, our method reduces the calculation to $470M$ flops and achieves comparable results with existing solutions based on heavy-weight networks of 10 times of flops on the InterHand 2.6M benchmark. This demonstrates the ability of the mobile transformers in the construction of interacting hands on resource-constrained devices.

## 2 Related Work

### 2.1 Hand Reconstruction

Hand reconstruction is studied for decades, including single hand reconstruction, two-hand reconstruction, and mesh regression. Some of the existing methods are already applied to virtual reality, robots and remote medical operations. The most related work of hand reconstruction is reviewed below.

**Single Hand Reconstruction.** Hand pose estimation and gesture recognition are well studied [12,42] before deep learning era. Then, estimation of 3D hand skeleton from a single image has achieved great success [3,28,36,53]. Since the popular parametric hand model MANO [31] and a variety of large scale datasets [15,27,35,54] are available, there are various methods to reconstruct both a hand pose and shape [1,4,9,19,22,37,48,52]. Among all of these existing methods, the most recent transformer-based models [21–23] produce the best results, showing the ability of the attention mechanism for the representation of the global relationship between two hand mesh vertices. However, this excellent performance relies on heavy weight attention mechanism which is impractical on mobile devices. Therefore, we propose our methods based on mobile transformers.

**Two-Hand Reconstruction.** Although almost all single-hand reconstruction methods produce good results on the single hand reconstruction task, interacting hands remains a challenge for the hand reconstruction task. First, previous methods simultaneously reconstruct body and hand [6,16,32,43,50,51], each hand is treated separately and hand interaction cases with heavy occlusion could not be handled. A method is based on a multi-view framework to reconstruct high-quality interactive hand motions, however, its hardware setup is costly and the calculation is heavy energy consumption. Other monocular tracking methods based on kinematic formulations are sensitive to fast motion and tracking failure regardless of whether a depth sensor [20,29,30,38,40] or an RGB camera [41] is applied. However, their dense mapping strategy relied on the correspondences between hand vertices and image pixels which is difficult to be calculated. In contrast, deep learning based methods [8,17,27,33,47] are able to reconstruct single frame two-hand interaction through a variety of feature maps. They rely on 2.5D heatmaps to estimate hand joint positions [8,27], extract sparse image features [44], reconstruct each hand respectively and fine-tune later [17,32]. However, the sparse representation of a single hand such as local image features encoded in the 2.5D heatmaps is not able to produce the efficient representation of hand surface occlusions and hands interaction context. Therefore, dense feature representation based on attention and graph [21] are studied to well learn the occlusion and context in the training. Unfortunately, the normal attention mechanism is computationally expensive and is hard to be deployed on mobile devices. In this paper, we propose several designs of lightweight attention modules to reduce the calculation and energy consumption.

## 2.2    Real-Time Hand Reconstruction

Currently, a variety of hand reconstruction methods are applied to mobile devices which require low latency and low energy consumption. In order to drive virtual and augmented reality (VR/AR) experiences on a mobile processor, a variety of hand reconstruction methods are proposed based on mobile network. For example, inverted residuals and linear bottlenecks [34] are used to build base blocks for hand pose, scale and depth prediction [10]. Precise landmark localization of 21 $2.5D$ coordinates inside the detected hand regions via regression is estimated to address CPU inference on the mobile devices [46]. However, these mobile methods based on mobile CNN blocks lack the ability of reconstruct hands with occlusions and interaction. The attention mechanisms and graph is not applied to build representation of local and global context of interacting hands. Therefore, we propose mobile attention and graph modules which can both handle challenging occlusions and interacting context with low flops.

## 3    Formulation

### 3.1    Two-Hand Mesh Representation

Similar with the state-of-the-art models [21,27,45], The same mesh topology of the popular MANO [31] model is adopted for each hand which contains $N = 778$ vertices. To assist the mobile attention mechanism, dense matching encoding for each vertex similar to [21,41] as positional embedding is used. As shown in Fig. 2, our LWA-HAND has a hierarchical architecture that reconstructs hand mesh using a variety of coarse-to-fine blocks with different types such as mobile attention module, domain bridges between image context and hand vertex and pyramid hand vertex decoding modules [21]. To construct the coarse-to-fine mesh topology and enable the building of bridge between image context and hand representation in a coarse-to-fine manner, we leverage the graph coarsening method similar in [21] and build $N\ b = 3$ level submeshes with vertex number $N_0 = 63$, $N_1 = 126$, $N_2 = 252$ and reserve the topological relationship between adjacent levels for upsampling. After the third block, a simple linear layer is employed to upsample the final submesh ($N_2 = 252$) to the full MANO mesh ($N = 778$), producing the final two-hand vertices.

## 3.2    Overview

The proposed system contains three main parts: mobile vision attention encoder-decoder (green block and yellow blocks in Fig. 2), pyramid attention-graph bridges (red, green and pink blocks in Fig. 2), and mobile interacting attention module (vertex representation in Fig. 2). Given a single RGB image, a global feature vector $F_G$ is firstly produced through feeding it to an mobile vision attention encoder. Simultaneously, several bundled feature maps $\{Y_t \in \mathbb{R}^{C_t \times H_t \times W_t}, t = 0, 1, 2\}$, where $t$ indicates that the $t$-th feature level corresponds to the input of the $t$-th bridge in the domain bridges, $H_t \times W_t$ is the resolution of the feature maps which gradually increases, and $C_t$ is the feature channel. The domain bridges take $Y_t, t = 0, 1, 2$ as input and transform them to hidden features $Z_t, t = 0, 1, 2$ through attention operation. Then, the bridges directly fuse global context features $Z_t, t = 0, 1, 2$ with hand vertex features in a coarse-to-fine manner. Note that, at each lever, the stream runs through a vision mobile attention module, a domain bridge, and a graph decoded module. These modules are illustrated in Fig. 2, 3, 4 and 5 and will be discussed in Sect. 4.1, Sect. 4.2 and Sect. 4.3, separately.

# 4    Light Former Graph

Existing work of interacting hands reconstruction [21,27,45] is built with convolutional encoder and attention module with lots of energy consumption. In this paper, we propose a lightweight former graph architecture as shown in Fig. 2 to represent the interaction and occlusion of two-hands. The Light former graph is consisted of three main modules based on lightweight blocks including lightweight feature attention module, pyramid cross image and graph bridge module and lightweight cross hand attention module. They are introduced in the following.

## 4.1    Lightweight Feature Attention Module

Lightweight convolutional neural networks (CNNs) such as MobileNets [13,14,34] efficiently encode local features by stacking depthwise and pointwise convolutions. Mobile vision transformers and its follow-ups [7,24,39] achieve global features

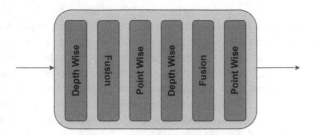

**Fig. 3.** Mobile module with attention.

through tokens. Inspired by the above advantages, a recent mobile architecture [5] is applied to connect local features and global features as shown in Fig. 3.

This module takes an image as the first input $\{\mathbf{X} \in \mathbb{R}^{(H \times W \times 3)}\}$ and applies inverted bottleneck blocks [34] to extract local features. Besides, learnable parameters (tokens) $\mathbf{Z} \in \mathbb{R}^{M \times d}$ are taken as the second input where $M$ and $d$ are the number and dimension of tokens, respectively. Similar to [21], these tokens are randomly initialized, and a small number of tokens ($M < 7$) is applied to represent a global prior of the image. Therefore, the operation of inverted bottleneck blocks and tokens results in much less computational effort.

In order to make fusion of the global and local features in the encoder, the lightweight encoding is computed by:

$$F_{X_i^0} = H(F_{X_{i-1}^1}, Z_i) = \text{Fusion}[\text{CNN}_{\text{Depth-wise}}(F_{X_{i-1}^1}), Z_i], \tag{1}$$

$$F_{X_i^1} = H(F_{X_i^0}) = \text{CNN}_{\text{Point-wise}}(F_{X_i^0}), \tag{2}$$

where $i = 0, 1, 2, 3, ..., N_{\text{stack}}$, $N_{\text{stack}}$ is the number of stacks, $X_i^0, X_i^1, i > 0$ are local features of the two stages in the $i$-th stack, and $X_0 = X_0^0 = X_0^1$ represents the input image. Here $Z_i$ represents the global features in the $i$-th stack, $\text{CNN}_{\text{Depth-wise}}$ represents the depth wise convolution, $\text{CNN}_{\text{Point-wise}}$ represents the point wise convolution, Fusion represents the contact operation. $F_{X_i^0}$ represents the feature of the stage 0 at the $i$-th stack, $F_{X_i^1}$ represents the local feature of the stage 1 at the $i$-th stack, $H(X_{i-1}^1, Z_i)$ represents the convolutional functions with the input $F_{X_{i-1}^1}$ and $Z_i$. The fusion operation directly contact global tokens $Z_i$ with local features $\text{CNN}_{\text{Depth-wise}}(X_{i-1}^1)$ rather than other attention mechanism with heavy calculation.

## 4.2 Pyramid Cross Image and Graph Bridge Module

Communication between CNN and transformer through a bridge is an efficient way to produce fusion of different domain representation [5,25]. Moreover, communication between image and graph is demonstrated as an efficient way to feed context features into the representation of hand vertex. The pyramid cross image and graph bridge module build two types of communication: communication between local features and global tokens and communication between global features and hand vertex. Furthermore, the communication between the

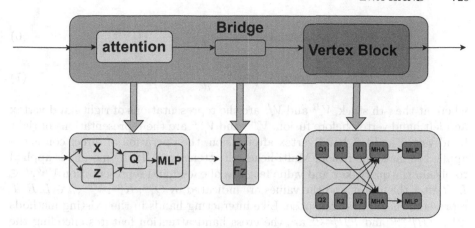

**Fig. 4.** Graph bridge module. The yellow block is the attention module for building an attention map between the local feature $X$ and the global feature $Z$ through an attention matrix $Q$ and MLP module. The fusion module (in green) is the direct contact module for the feature $Fx$ and $Fz$. The pink vertex module is built with cross hand attention module with two $Q$ matrix $Q_1$ and $Q_2$, two key matrix $K_1$ and $K_2$, two value matrix $V_1$ and $V_2$. (Color figure online)

triple domains is built in a coarse-to-fine manner (see, Fig. 4): Firstly, unlike the two-way bridge to connect local and global features [5], the local features and global features are communicated through an one-way bridge. A lightweight cross attention is applied where $(\mathbf{W}^Q, \mathbf{W}^K, \mathbf{W}^V)$ are the matrices of three projections, only $\mathbf{W}^Q$ is used. Specifically, the lightweight cross attention from local features map $\mathbf{X}$ to global tokens $\mathbf{Z}$ is computed by:

$$A_{F_{X_i^1} \to Z_i} = [\text{Attn}(\bar{z}_i \mathbf{W}_i^Q, \bar{x}_i, \bar{x}_i)]_{i=1:h} \mathbf{W}^O, \tag{3}$$

where the local feature $F_{X_i^1}$ and global tokens $Z_i$ are split into $h$ heads as $F_{X_i^1} = [\bar{x}_1, ..., \bar{x}_h]$, $Z_i = [\bar{z}_1, ..., \bar{z}_h]$ for multi-head attention. The split for the $i$-th head is represented as $\bar{z}_i \in \mathbb{R}^{M \times \frac{d}{h}}$, $W_i^Q$ is the query projection matrix for the $i$-th head, and $W_o$ is used to combine multiple heads together.

Secondly, at the $i$-th stack, the global features $Z_i$ are mapped to the domain of graph representation in the bridge, denoted as $M_i$. Here, $M_i$ is computed by:

$$M_i = A_{F_{X_i^1} \to Z_i} F_{X_i^1} \tag{4}$$

where $A_{F_{X_i^1} \to Z_i}$ is the attention matrix calculated previously. Here $F_{X_i^1}$ is the local feature at the $i$-th stack and $Z_i$ is the global feature at the $i$-th stack.

The mapping is calculated in two stages: direct connection stage and cross attention stage. At the first stage, the global feature $M_i$ is directly contacted with hand vertex representation $V_i$ at the $i$-th stage for each hand. At the second stage, a cross attention mechanism is applied between two hands [21] for deep fusion. The two-stage fusion is calculated as the following:

$$V_i^{Ro} = \text{Fusion}(V_i^R, M_i), \quad V_i^{Lo} = \text{Fusion}(V_i^L, M_i), \tag{5}$$

$$FH_i^{R \to L} = \text{softmax}\left(\frac{f(Q_i^{Lo})f(K_i^{Ro})}{d}\right)f(V_i^{Ro}), \tag{6}$$

$$FH_i^{L \to R} = \text{softmax}\left(\frac{f(Q_i^{Ro})f(K_i^{Lo})}{d}\right)f(V_i^{Lo}), \tag{7}$$

where at the $i$-th stack, $V_i^R$ and $V_i^L$ are the representations of right hand vertex and left hand vertex before fusion, $V_i^{Ro}$ and $V_i^{Lo}$ are the representations of right hand vertex and left hand vertex after fusion, the operation of direct contact is applied to make the fusion. Multi-head self-attention (MHSA) module is applied to obtain the query, key and value features of each hand representation $V_i^{ho}, h \in L, R$ after the fusion, and the values are indicated by $Q_i^{ho}, K_i^{ho}, V_i^{ho}, h \in L, R$. $T$ represents the matrix transpose. Like interacting hands in the existing methods [21], $FH_i^{R \to L}$ and $FH_i^{L \to R}$ are the cross hand attention features encoding the correlation between two hands. $d$ is a normalization constant and $f$ represents the function with the three features as input respectively. The cross-hand attention features are merged into the hand vertex features by a pointwise MLP layer $fp$ as

$$FH_i^{La} = fp(FH_i^{Lo} + FH_i^{R \to L}), \tag{8}$$

$$FH_i^{Ra} = fp(FH_i^{Ro} + FH_i^{L \to R}), \tag{9}$$

where $FH_i^{La}$ and $FH_i^{Ra}$ are the output hand vertex features at the $i$-th stack, which act as the input of both hands at the next stack.

### 4.3   Lightweight Cross Hand Attention Module

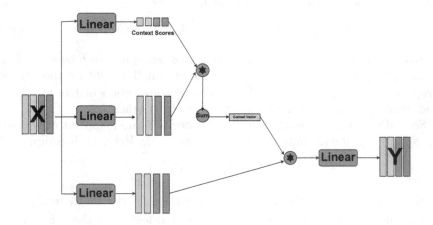

**Fig. 5.** Cross hand attention module.

In order to reduce the calculation of the above mentioned cross attention between two hands, a separable self-attention structure [26] is applied to construct a lightweight cross hand attention module. The input $FH_i^h, h \in L, R$ is processed using three branches: input $I$, key $K$ and value $V$ as shown in Fig. 5.

Firstly, the input branch $I$ is responsible to calculate the value of query features $Q_i^{ho}, h \in L, R$. The input branch $I$ maps each $d$ dimensional token in $FH_i^{ho}, h \in L, R$ to a scalar using a linear layer with weights $W_i^I \in \mathbb{R}^d$. This linear projection is an inner-product operation and computes the distance between latent token $L$ and $FH_i^h$, resulting in a $k$-dimensional vector. A softmax operation is then applied to this $k$ dimensional vector to produce context scores $c_s \in \mathbb{R}^k$ representing the value of query features.

Secondly, the context scores $c_s$ are used to compute a context vector $c_v$. Specifically, the input $FH_i^h$ is linearly projected to a $d$ dimensional space using $K$ branch with weights $W_k \in \mathbb{R}^{d \times d}$ to produce an output $FH_{ik}^h \in \mathbb{R}^{k \times d}$ representing the value of key features. The context vector $c_v \in \mathbb{R}^d$ is then computed as a weighted sum of $FH_{ik}^h$:

$$c_v = \sum_{i=1}^k c_s(i) FH_{ik}^h(i). \tag{10}$$

Finally the contextual information encoded in $c_v$ is shared with all tokens in $FH_i^h$. The input $FH_i^h$ is linearly projected to a $d$ dimensional space using the value branch $V$ with wights $W_v \in \mathbb{R}^{d \times d}$ to produce an output $FH_{iv}^h$ representing the value of the value features.

## 4.4  Loss Function

For training the proposed mobile network and a pair comparison with the state-of-the-art models [21,27,45], vertex loss, regressed joint loss and mesh smooth loss are utilized. It is used in IntagHand [21], other loss items are not applied in order to demonstrate the efficiency of our proposed mobile network architecture.

**Vertex Loss.** Firstly, $L1$ loss is applied to supervise the $3D$ coordinates of hand vertices and MSE loss to supervise the $2D$ projection of vertices:

$$L_v = \sum_{i=1}^N ||\mathbf{V}_{h,i} - \mathbf{V}_{h,i}^{GT}||_1 + ||\Pi(V_{h,i}) - \Pi(V_{h,i}^{GT})||_2^2, \tag{11}$$

where $V_{h,i}$ is $i$-th vertex, $h = L, R$ means left or right hand, and $\Pi$ is the $2D$ projection operation. Vertex loss is applied to each submesh.

**Regressed Joint Loss.** Secondly, joint error loss is applied to supervise the estimation of joints. By multiplying the predefined joint regression matrix $J$, hand joints can be regressed from the predicted hand vertices. The joint error is penalized by the following loss:

$$L_J = \sum_{i=1}^V ||JV_{h,i} - JV_{h,i}^{GT}||_1 + \sum_{i=1}^V ||\Pi(JV_{h,i}) - \Pi(JV_{h,i}^{GT})||_2^2. \tag{12}$$

**Mesh Smooth Loss.** Thirdly, the smooth loss item is applied to supervise the geometric smoothness. In detail, two different smooth losses are applied. First, the normal consistency between the predicted and the ground truth mesh is regularized:

$$L_n = \sum_{f=1}^{F} \sum_{e=1}^{3} ||e_{f,i,h} \cdot n_{f,h}^{GT}||_1, \tag{13}$$

where $f$ is the face index of the hand mesh, $e_{f,i}(i = 1, 2, 3)$ are the three edges of face $f$ and $n_f^{GT}$ is the normal vector of this face calculated from the ground truth mesh. Second, we the $L1$ distance of each edge length between the predicted mesh and the ground truth mesh is minimized:

$$L_e = \sum_{e=1}^{E} ||e_{i,h} - e_{i,h}^{GT}||_1. \tag{14}$$

## 5    Experiment

### 5.1    Experimental Settings

**Implementation Details.** Our network is implemented using PyTorch. The proposed network with efficient modules are trained in an end-to-end manner.

**Training Details.** The Adam optimizer [18] is applied for training on 4 NVIDIA RTX 2080Ti GPUs with the minibatch size for each GPU set as 32. The whole training takes 120 epochs across 1.0 days, with the learning rate decaying to $1 \times 10^{-5}$ at the 50-th epoch from the initial rate $1 \times 10^{-4}$. During training, data augmentations including scaling, rotation, random horizontal flip and color jittering are applied [21].

**Evaluation Metrics.** In fair comparison with other state-of-the-art models, the mean per joint position error (MPJPE) and mean per vertex position error (MPVPE) in millimeters are compared. Besides, we follow Two-Hands [49] to scale the length of the middle metacarpal of each hand to 9.5 cm during training and rescale it back to the ground truth bone length during evaluation. This is performed after root joint alignment of each hand.

### 5.2    Datasets

**InterHand2.6M Dataset [27].** All networks in this paper are trained on Inter-Hand2.6M [27], as it's the only dataset with two-hand mesh annotation. The interacting two-hand (IH) data with both human and machine (H+M) annotated is applied for the two-hand reconstruction task. In detail, $366K$ training samples and $261K$ testing samples from InterHand2.6M are utilized. At preprocessing, we crop out the hand region according to the $2D$ projection of hand vertices and resize it to $256 \times 256$ resolution [21].

## 5.3  Quantitative Results

We first compare our proposed mobile network with the state-of-the-art two-hand reconstruction methods and recent two-hand reconstruction methods. The first model [27] regresses 3D skeletons of two hands directly. The second model [45] predicts the pose and shape parameters of two MANO [31] models. The third model [21] predicts the vertex of interacting-hands with graph and attention modules. For a fair comparison, we run their released source code on the same subset of Inter-Hand2.6M [27]. Comparison results are shown in Table 1. Figure 6 shows the results of our proposed model. It is clearly shown in Table 1 that our method achieved comparable MPJPE and MPVPE as the state-of-the-art models. Furthermore, we reduces the flops to $0.47GFlops$ while the flops of the state-of-the-art models are around $10GFlops$ or higher value.

**Table 1.** Comparison with the state-of-the-art models on performance and flops. The mean per joint position error (MPJPE) and mean per vertex position error (MPVPE) are calculated in millimeters. The flops are calculated in GFlops.

| Model | MPJPE | MPVPE | Total flops | Image part | Pose part |
|---|---|---|---|---|---|
| Inter-hand [27] | 16.00 | – | 19.49 | 5.37 | 14.12 |
| Two-hand-shape [45] | 13.48 | 13.95 | 28.98 | 9.52 | 19.46 |
| Intag-hand [21] | 8.79 | 9.03 | 8.42 | 7.36 | 1.06 |
| Ours | 12.56 | 12.37 | 0.47 | 0.25 | 0.22 |

# 6  Discussion

## 6.1  Conclusion

We present the proposed mobile method to reconstruct two interacting hands from a single RGB image. In this paper, we first introduce a lightweight feature attention module to extract both local occlusion representation and global image patch representation in a coarse-to-fine manner. We next propose a cross image and graph bridge module to fuse image context and hand vertex. Finally, we propose a lightweight cross attention mechanism which uses element-wise operation for cross attention of two hands. Comprehensive experiments demonstrate the comparable performance of our network on InterHand2.6M dataset, and verify the effectiveness and practicability of the proposed model in the real-time applications with low flops.

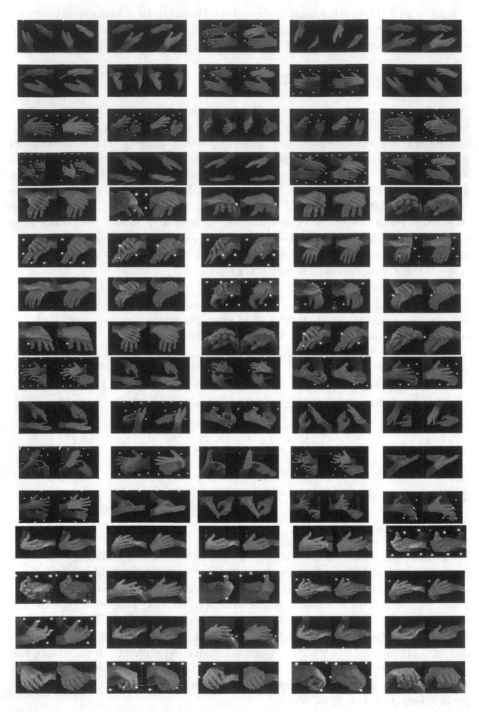

**Fig. 6.** Illustrations for interacting hand reconstruction by our proposed model.

## 6.2   Limitation

The major limitation of our method is the high MPJPE and MPVPE. The proposed mobile modules reduce the flops while increase the error at the same time. We are studying efficient methods to further reduce the error for mobile application of two-hand reconstruction.

# References

1. Baek, S., Kim, K.I., Kim, T.K.: Pushing the envelope for RGB-based dense 3D hand pose estimation via neural rendering. In: Proceedings of the IEEE/CVF Conference on Computer Vision and Pattern Recognition, pp. 1067–1076 (2019)
2. Boukhayma, A., de Bem, R., Torr, P.H.: 3D hand shape and pose from images in the wild. In: Proceedings of the IEEE/CVF Conference on Computer Vision and Pattern Recognition, pp. 10843–10852 (2019)
3. Cai, Y., Ge, L., Cai, J., Yuan, J.: Weakly-supervised 3D hand pose estimation from monocular RGB images. In: Ferrari, V., Hebert, M., Sminchisescu, C., Weiss, Y. (eds.) ECCV 2018. LNCS, vol. 11210, pp. 678–694. Springer, Cham (2018). https://doi.org/10.1007/978-3-030-01231-1_41
4. Chen, X., et al.: Camera-space hand mesh recovery via semantic aggregation and adaptive 2D–1D registration. In: Proceedings of the IEEE/CVF Conference on Computer Vision and Pattern Recognition, pp. 13274–13283 (2021)
5. Chen, Y., et al.: Mobile-former: bridging MobileNet and transformer. In: Proceedings of the IEEE/CVF Conference on Computer Vision and Pattern Recognition, pp. 5270–5279 (2022)
6. Choutas, V., Pavlakos, G., Bolkart, T., Tzionas, D., Black, M.J.: Monocular expressive body regression through body-driven attention. In: Vedaldi, A., Bischof, H., Brox, T., Frahm, J.-M. (eds.) ECCV 2020. LNCS, vol. 12355, pp. 20–40. Springer, Cham (2020). https://doi.org/10.1007/978-3-030-58607-2_2
7. Dong, X., et al.: CSWin transformer: a general vision transformer backbone with cross-shaped windows. In: Proceedings of the IEEE/CVF Conference on Computer Vision and Pattern Recognition, pp. 12124–12134 (2022)
8. Fan, Z., Spurr, A., Kocabas, M., Tang, S., Black, M.J., Hilliges, O.: Learning to disambiguate strongly interacting hands via probabilistic per-pixel part segmentation. In: 2021 International Conference on 3D Vision (3DV), pp. 1–10. IEEE (2021)
9. Ge, L., et al.: 3D hand shape and pose estimation from a single RGB image. In: Proceedings of the IEEE/CVF Conference on Computer Vision and Pattern Recognition, pp. 10833–10842 (2019)
10. Han, S., et al.: MEgATrack: monochrome egocentric articulated hand-tracking for virtual reality. ACM Trans. Graph. (ToG) **39**(4), 87–91 (2020)
11. He, K., Zhang, X., Ren, S., Sun, J.: Deep residual learning for image recognition. In: Proceedings of the IEEE Conference on computer vision and Pattern Recognition, pp. 770–778 (2016)
12. Heap, T., Hogg, D.: Towards 3D hand tracking using a deformable model. In: Proceedings of the Second International Conference on Automatic Face and Gesture Recognition, pp. 140–145. IEEE (1996)
13. Howard, A., et al.: Searching for MobileNetV3. In: Proceedings of the IEEE/CVF International Conference on Computer Vision, pp. 1314–1324 (2019)

14. Howard, A.G., et al.: MobileNets: efficient convolutional neural networks for mobile vision applications. arXiv preprint arXiv:1704.04861 (2017)
15. Joo, H., et al.: Panoptic studio: a massively multiview system for social motion capture. In: Proceedings of the IEEE International Conference on Computer Vision, pp. 3334–3342 (2015)
16. Joo, H., Simon, T., Sheikh, Y.: Total capture: a 3D deformation model for tracking faces, hands, and bodies. In: Proceedings of the IEEE Conference on Computer Vision and Pattern Recognition, pp. 8320–8329 (2018)
17. Kim, D.U., Kim, K.I., Baek, S.: End-to-end detection and pose estimation of two interacting hands. In: Proceedings of the IEEE/CVF International Conference on Computer Vision, pp. 11189–11198 (2021)
18. Kingma, D.P., Ba, J.: Adam: a method for stochastic optimization. arXiv preprint arXiv:1412.6980 (2014)
19. Kulon, D., Guler, R.A., Kokkinos, I., Bronstein, M.M., Zafeiriou, S.: Weakly-supervised mesh-convolutional hand reconstruction in the wild. In: Proceedings of the IEEE/CVF Conference on Computer Vision and Pattern Recognition, pp. 4990–5000 (2020)
20. Kyriazis, N., Argyros, A.: Scalable 3D tracking of multiple interacting objects. In: Proceedings of the IEEE Conference on Computer Vision and Pattern Recognition, pp. 3430–3437 (2014)
21. Li, M., et al.: Interacting attention graph for single image two-hand reconstruction. In: Proceedings of the IEEE/CVF Conference on Computer Vision and Pattern Recognition, pp. 2761–2770 (2022)
22. Lin, K., Wang, L., Liu, Z.: End-to-end human pose and mesh reconstruction with transformers. In: Proceedings of the IEEE/CVF Conference on Computer Vision and Pattern Recognition, pp. 1954–1963 (2021)
23. Lin, K., Wang, L., Liu, Z.: Mesh graphormer. In: Proceedings of the IEEE/CVF International Conference on Computer Vision, pp. 12939–12948 (2021)
24. Liu, Z., et al.: Swin transformer: hierarchical vision transformer using shifted windows. In: Proceedings of the IEEE/CVF International Conference on Computer Vision, pp. 10012–10022 (2021)
25. Mehta, S., Rastegari, M.: MobileViT: light-weight, general-purpose, and mobile-friendly vision transformer. arXiv preprint arXiv:2110.02178 (2021)
26. Mehta, S., Rastegari, M.: Separable self-attention for mobile vision transformers. arXiv preprint arXiv:2206.02680 (2022)
27. Moon, G., Yu, S.-I., Wen, H., Shiratori, T., Lee, K.M.: InterHand2.6M: a dataset and baseline for 3D interacting hand pose estimation from a single RGB image. In: Vedaldi, A., Bischof, H., Brox, T., Frahm, J.-M. (eds.) ECCV 2020. LNCS, vol. 12365, pp. 548–564. Springer, Cham (2020). https://doi.org/10.1007/978-3-030-58565-5_33
28. Mueller, F., et al.: GANerated hands for real-time 3D hand tracking from monocular RGB. In: Proceedings of the IEEE Conference on Computer Vision and Pattern Recognition, pp. 49–59 (2018)
29. Mueller, F., et al.: Real-time pose and shape reconstruction of two interacting hands with a single depth camera. ACM Trans. Graph. (ToG) 38(4), 1–13 (2019)
30. Oikonomidis, I., Kyriazis, N., Argyros, A.A.: Tracking the articulated motion of two strongly interacting hands. In: 2012 IEEE Conference on Computer Vision and Pattern Recognition, pp. 1862–1869. IEEE (2012)
31. Romero, J., Tzionas, D., Black, M.J.: Embodied hands: modeling and capturing hands and bodies together. arXiv preprint arXiv:2201.02610 (2022)

32. Rong, Y., Shiratori, T., Joo, H.: FrankMocap: fast monocular 3D hand and body motion capture by regression and integration. arXiv preprint arXiv:2008.08324 (2020)
33. Rong, Y., Wang, J., Liu, Z., Loy, C.C.: Monocular 3D reconstruction of interacting hands via collision-aware factorized refinements. In: 2021 International Conference on 3D Vision (3DV), pp. 432–441. IEEE (2021)
34. Sandler, M., Howard, A., Zhu, M., Zhmoginov, A., Chen, L.C.: MobileNetV2: inverted residuals and linear bottlenecks. In: Proceedings of the IEEE Conference on Computer Vision and Pattern Recognition, pp. 4510–4520 (2018)
35. Simon, T., Joo, H., Matthews, I., Sheikh, Y.: Hand keypoint detection in single images using multiview bootstrapping. In: Proceedings of the IEEE Conference on Computer Vision and Pattern Recognition, pp. 1145–1153 (2017)
36. Spurr, A., Song, J., Park, S., Hilliges, O.: Cross-modal deep variational hand pose estimation. In: Proceedings of the IEEE Conference on Computer Vision and Pattern Recognition, pp. 89–98 (2018)
37. Tang, X., Wang, T., Fu, C.W.: Towards accurate alignment in real-time 3D hand-mesh reconstruction. In: Proceedings of the IEEE/CVF International Conference on Computer Vision, pp. 11698–11707 (2021)
38. Taylor, J., et al.: Articulated distance fields for ultra-fast tracking of hands interacting. ACM Trans. Graph. (TOG) $36$(6), 1–12 (2017)
39. Touvron, H., Cord, M., Douze, M., Massa, F., Sablayrolles, A., Jégou, H.: Training data-efficient image transformers & distillation through attention. In: International Conference on Machine Learning, pp. 10347–10357. PMLR (2021)
40. Tzionas, D., Ballan, L., Srikantha, A., Aponte, P., Pollefeys, M., Gall, J.: Capturing hands in action using discriminative salient points and physics simulation. Int. J. Comput. Vis. $118$(2), 172–193 (2016)
41. Wang, J., et al.: RGB2Hands: real-time tracking of 3D hand interactions from monocular RGB video. ACM Trans. Graph. (ToG) $39$(6), 1–16 (2020)
42. Wang, Y., et al.: Video-based hand manipulation capture through composite motion control. ACM Trans. Graph. (TOG) $32$(4), 1–14 (2013)
43. Xiang, D., Joo, H., Sheikh, Y.: Monocular total capture: posing face, body, and hands in the wild. In: Proceedings of the IEEE/CVF Conference on Computer Vision and Pattern Recognition, pp. 10965–10974 (2019)
44. Xiao, B., Wu, H., Wei, Y.: Simple baselines for human pose estimation and tracking. In: Ferrari, V., Hebert, M., Sminchisescu, C., Weiss, Y. (eds.) ECCV 2018. LNCS, vol. 11210, pp. 472–487. Springer, Cham (2018). https://doi.org/10.1007/978-3-030-01231-1_29
45. Zhang, B., et al.: Interacting two-hand 3D pose and shape reconstruction from single color image. In: Proceedings of the IEEE/CVF International Conference on Computer Vision, pp. 11354–11363 (2021)
46. Zhang, F., et al.: MediaPipe hands: on-device real-time hand tracking. arXiv preprint arXiv:2006.10214 (2020)
47. Zhang, H., et al.: PyMAF: 3D human pose and shape regression with pyramidal mesh alignment feedback loop. In: Proceedings of the IEEE/CVF International Conference on Computer Vision, pp. 11446–11456 (2021)
48. Zhang, X., et al.: Hand image understanding via deep multi-task learning. In: Proceedings of the IEEE/CVF International Conference on Computer Vision, pp. 11281–11292 (2021)
49. Zhang, X., Li, Q., Mo, H., Zhang, W., Zheng, W.: End-to-end hand mesh recovery from a monocular RGB image. In: Proceedings of the IEEE/CVF International Conference on Computer Vision, pp. 2354–2364 (2019)

50. Zhang, Y., Li, Z., An, L., Li, M., Yu, T., Liu, Y.: Lightweight multi-person total motion capture using sparse multi-view cameras. In: Proceedings of the IEEE/CVF International Conference on Computer Vision, pp. 5560–5569 (2021)
51. Zhou, Y., Habermann, M., Habibie, I., Tewari, A., Theobalt, C., Xu, F.: Monocular real-time full body capture with inter-part correlations. In: Proceedings of the IEEE/CVF Conference on Computer Vision and Pattern Recognition, pp. 4811–4822 (2021)
52. Zhou, Y., Habermann, M., Xu, W., Habibie, I., Theobalt, C., Xu, F.: Monocular real-time hand shape and motion capture using multi-modal data. In: Proceedings of the IEEE/CVF Conference on Computer Vision and Pattern Recognition, pp. 5346–5355 (2020)
53. Zimmermann, C., Brox, T.: Learning to estimate 3D hand pose from single RGB images. In: Proceedings of the IEEE International Conference on Computer Vision, pp. 4903–4911 (2017)
54. Zimmermann, C., Ceylan, D., Yang, J., Russell, B., Argus, M., Brox, T.: Frei-HAND: a dataset for markerless capture of hand pose and shape from single RGB images. In: Proceedings of the IEEE/CVF International Conference on Computer Vision, pp. 813–822 (2019)

# Neural Mesh-Based Graphics

Shubhendu Jena[✉], Franck Multon, and Adnane Boukhayma

Inria, Univ. Rennes, CNRS, IRISA, M2S, Rennes, France
shubhendu.jena@gmail.com

**Abstract.** We revisit NPBG [2], the popular approach to novel view synthesis that introduced the ubiquitous point feature neural rendering paradigm. We are interested in particular in data-efficient learning with fast view synthesis. We achieve this through a view-dependent mesh-based denser point descriptor rasterization, in addition to a foreground/background scene rendering split, and an improved loss. By training solely on a single scene, we outperform NPBG [2], which has been trained on ScanNet [9] and then scene finetuned. We also perform competitively with respect to the state-of-the-art method SVS [42], which has been trained on the full dataset (DTU [1] and Tanks and Temples [22]) and then scene finetuned, in spite of their deeper neural renderer.

## 1 Introduction

Enabling machines to understand and reason about 3D shape and appearance is a long standing goal of computer vision and machine learning, with implications in numerous 3D vision downstream tasks. In this respect, novel view synthesis is a prominent computer vision and graphics problem with rising applications in free viewpoint, virtual reality, image editing and manipulation, as well as being a corner stone of building an efficient metaverse. The introduction of deep learning in the area of novel view synthesis brought higher robustness and generalization in comparison to earlier traditional approaches. While the current trend is learning neural radiance fields (e.g. [29,56,61]), training and rendering such implicit volumetric models still presents computational challenges, despite recent efforts towards alleviating these burdens (e.g. [17,25,49,62]). An appealing alternate learning strategy [2,41,42,53], achieving to date state-of-the-art results on large outdoors unbounded scenes such as the Tanks and Temples dataset [22], consists of using a pre-computed geometric representation of the scene to guide the novel view synthesis process. As contemporary successors to the original depth warping techniques, these methods benefit from a strong geometry prior to constrain the learning problem, and recast its 3D complexity into a simpler 2D neural rendering task, providing concurrently faster feed-forward inference.

Among the latter, NPBG [2] is a popular strategy, being core to several other neural rendering based methods (e.g. [35,37,60,64]). Learnable descriptors

---

**Supplementary Information** The online version contains supplementary material available at https://doi.org/10.1007/978-3-031-25066-8_45.

are appended to the geometry points, and synthesis consists of rasterizing then neural rendering these features. It is practical also as it uses a lightweight and relatively simpler architecture, compared to competing methods (e.g. [41,42]).

Our motivation is seeking a data-efficient, fast, and relatively lightweight novel-view synthesis method. Hence we build on the idea of NPBG [2], and we introduce several improvements allowing it to scale to our aforementioned goals. In particular, and differently from NPBG [2]: We introduce denser rasterized feature maps through a combination of denser point rasterization and face rasterization; We enforce view-dependency explicitly through anisotropic point descriptors; As the foreground and background geometries differ noticeably in quality, we propose to process these two feature domains separately to accommodate independently for their respective properties; Finally, we explore a self-supervised loss promoting photo-realism and generalization outside the training view corpus. The improvement brought by each of these components is showcased in Table 3.

By training simply on a single scene, our method outperforms NPBG [2], even though it has been additionally trained on ScanNet [9] and then further fine-tuned on the same scene, in terms of PSNR, SSIM [58] and LPIPS [66], and both on the Tanks and Temples [22] (Table 1) and DTU [1] (Table 2) datasets. The performance gap is considerably larger on DTU [1]. Our data efficiency is also illustrated in the comparison to state-of-the-art method SVS [42] (Tables 1, 2). We achieve competitive numbers despite their full dataset trainings, their deeper convolutional network based neural renderers, and slower inference. We also recover from some of their common visual artifacts as shown in Fig. 6.

## 2    Related Work

While there is a substantial body of work on the subject of novel view synthesis, we review here work we deemed most relevant to the context of our contribution.

**Novel View Synthesis.** The task of novel view synthesis essentially involves using observed images of a scene and the corresponding camera parameters to render images from novel unobserved viewpoints. This is a long explored problem, with early non deep-learning based methods [7,11,15,24,46,69] using a set of input images to synthesize new views of a scene. However, these methods impose restrictions on the configuration of input views. For example, [24] requires a dense and regularly spaced camera grid while [10] requires that the cameras are located approximately on the surface of a sphere and the object is centered. To deal with these restrictions, unstructured view synthesis methods use a 3D proxy geometry of the scene to guide the view synthesis process [5,23]. With the rise of deep-learning, it has also come to be used extensively for view synthesis, either by blending input images to synthesize the target views [16,54], or by learning implicit neural radiance fields followed by volumetric rendering to generate the target views [29], or even by using a 3D proxy geometry representation of the scene to construct neural scene representations [2,32,41,42,53].

**View Synthesis w/o Geometric Proxies.** Early deep-learning based approaches combine warped or unwarped input images by predicting the corresponding blending weights to compose the target view [8,54]. Thereafter, several approaches came up leveraging different avenues such as predicting camera poses [67], depth maps [19], multi-plane images [12,68], and voxel grids [20]. Recently, implicit neural radiance fields (NeRF) [29] has emerged as a powerful representation for novel view synthesis. It uses MLPs to map spatial points to volume density and view-dependent colors. Hierarchical volumetric rendering is then performed on the predicted point colors to render the target image. Some of the problems associated with NeRF [29] include higher computational cost and time of rendering complexity, the requirement of dense training views, the lack of across-scene generalization, and the need for test-time optimization. A number of works [3,13,18,26,30,40,59,62,63] have tried addressing these limitations. In particular, Spherical Harmonics [51] have been used to speed up inference of Neural Radiance Fields by factorizing the view-dependent appearance [62]. Nex [59] introduced a related idea, where several basis functions such as Hemi-spherical harmonics, Fourier Series, and Jacobi Spherical Harmonics were investigated, and concluded that learnable basis functions offer the best results. Some other methods, like pixelNeRF [63], GRF [55], IBRNet [57] and MVSNeRF [6] proposed to augment NeRFs [29] with 2D and 3D convolutional features collected from the input images. Hence, they offer forward pass prediction models, i.e. test-time optimization free, while introducing generalization across scenes. While these methods are promising, they need to train on full datasets to generalize well, while at the same time evaluating hundreds of 3D query points per ray for inference similar to NeRF [29]. Hence, both training and inference often takes quite long for these methods. We note that besides encoders [6,55,57,63], implicit neural representations can also be conditioned through meta-learning e.g. [33,48,50].

**View Synthesis Using Geometric Proxies.** Different from the work we have discussed so far, several recent methods utilize a geometric reconstruction of the scene to construct neural scene representations and consequently use them to synthesize target views. These geometric proxies can either be meshes [41,42,53] or point clouds [2,27,34,52]. SVS [42] and FVS [41] utilize COLMAP [44,45] to construct a mesh scaffold which is then used to select input views with maximum overlap with the target view. FVS [41] then uses a sequential network based on gated recurrent units (GRUs) to blend the encoded input images. SVS [42] operates on a similar principle except that the geometric mesh scaffold is used for on-surface feature aggregation, which entails processing or aggregating directional feature vectors from encoded input images at each 3D point to produce a new feature vector for a ray that maps this point into the new target view. Deferred neural rendering (DNR) [53] proposes to learn neural textures encoding the point plenoptic function at different surface points alongside a neural rendering convolutional network. It is infeasible for very large meshes due to the requirement of UV-parametrization. On the other hand, NPBG [2] operates on a similar idea by learning neural descriptors of surface elements jointly with the rendering network, but uses point-based geometry representation instead of meshes. Similarly, [27,52] use COLMAP [44,45] to reconstruct a colored point cloud of the scene in question, which is used alongside a neural renderer to

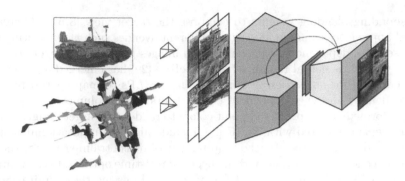

**Fig. 1.** Overview: an automatic split of the scene geometry is used to rasterize foreground/background mesh-borne view-dependent features, through both point based and mesh based rasterizations. A convolutional U-Net [43] maps the feature images into the target image.

render target images. Our method combines both approaches in the sense that it uses both point cloud and mesh representation of the scene as geometric proxies. Like NPBG [2], we also learn neural descriptors of surface elements, but since point clouds of large unbounded scenes are often sparse, using meshes helps in enhancing the richness of the rasterized neural descriptors. Hence, using both point clouds and meshes help us in achieving a balance between accuracy and density of the rasterized scene neural descriptors, which we will explain in detail in the upcoming sections and also through an ablative analysis.

## 3   Method

Given a set of calibrated images of a scene, our goal is to generate novel views from the same scene through deep learning. To this end, we design a forward pass prediction method, expected to generalize to target views including and beyond input view interpolation. Using a geometry proxy of the scene, we set view-dependent learnable descriptors on the vertices, and we split the scene automatically into a dense foreground and a sparser background. Each of these areas are rasterized through PyTorch3D's point based and mesh based rasterizations [39]. A convolutional neural renderer translates and combines the resulting image features into the target color image. Figure 1 illustrates this pipeline. Our method can be trained on a single scene by fitting the point descriptors and learning the neural renderer weights jointly. It can also benefit from multi-scene training through the mutualization of the neural renderer learning. We present in the remaining the different components of our method.

**Fig. 2.** We introduce denser feature images compared to NPBG [2]. Left: NPBG's rasterization. Center: our point based rasterization. Right: our mesh based rasterization. Top row shows feature PCA coefficients. Bottom row shows the resulting rasterization mask (Occupied/unoccupied pixels).

**Preprocessing.** We need to obtain a geometry representing the scene from the training images as a preprocessing stage. Standard structure-from-motion (SfM) and multi-view stereo (MVS) pipelines can be used to achieve this [44, 45]. In this respect, we chose to use the preprocessed data from SVS [42] and FVS [41], and so the preprocessing steps are identical to these methods. The first step involves running structure-from-motion [44,45] on the images to get camera intrinsics $\{K\}$ and camera poses as rotation matrices $\{R\}$, and translations $\{T\}$. The second step involves running multi-view stereo on the posed images, to obtain per-image depth maps, and then fusing these into a point cloud. Finally, Delaunay-based triangulation is applied to the obtained point cloud to get a 3D surface mesh $\mathcal{M}$. These steps are implemented following COLMAP [44,45].

### 3.1 Dual Feature Rasterization

Given a target camera pose $R \in SO(3)$, $T \in \mathbb{R}^3$, the rendering procedure starts with the rasterization of learnable mesh features into the target image domain. The mesh $\mathcal{M}$ consists here of vertices (i.e. points) $\mathcal{P} = \{p_1, p_2, ..., p_N\}$ with neural point descriptors $\mathcal{K} = \{k_1, k_2, ..., k_N\}$, and faces $\mathcal{F} = \{f_1, f_2, ..., f_M\}$.

We noticed initially that having denser feature images improves the performance of the neural rendering. Hence, differently from NPBG [2], we use the PyTorch3D [39] renderer which allows us to obtain denser feature images for point cloud based rasterization. Furthermore, we notice additionally that using PyTorch3D's mesh based rasterization provides even denser feature images (c.f. Fig. 2). Hence, we propose to rasterize the scene geometry features using both modes.

**Point Cloud Based Rasterization.** PyTorch3D [39] requires us setting a radius $r$ in a normalized space which thresholds the distance between the target view pixel positions $\{(u, v)\}$ and the projected 3D points of the scene onto the target view, i.e. $\{\Pi(p) : p \in P\}$. If this distance is below the threshold for a given pixel $(u, v)$, we consider this point to be a candidate for that pixel, and we finally pick the point $p_{u,v}$ with the smallest $z$ coordinate in the camera coordinate frame. This writes:

$$\mathcal{P}_{u,v} = \{p \in \mathcal{P} : ||\Pi(p) - (u,v)||_2 \leq r\}. \tag{1}$$

$$p_{u,v} = \underset{p \in \mathcal{P}_{u,v}}{\arg\min} \, p_z, \quad \text{where} \quad p = (p_x, p_y, p_z). \tag{2}$$

The neural descriptor of the chosen point $p_{u,v}$ is projected onto the corresponding pixel position to construct the rasterized feature image. We set a radius of $r = 0.006$ to achieve a balance between accuracy of the projected point positions and the density of the rasterized feature images. The feature descriptors for each pixel are weighed inversely with respect to the distance of the projected 3D point to the target pixel, which can be expressed as $w_{u,v} = (1 - ||\Pi(p_{u,v}) - (u,v)||_2^2)/r^2$ where $||\Pi(p_{u,v}) - (u,v)||_2 < r$. Finally the point feature image $\{k_{u,v}^{\text{pt}}\}$ can hence be expressed as follows:

$$k_{u,v}^{\text{pt}} = w_{u,v} \times \mathcal{K}(p_{u,v}), \tag{3}$$

where $\mathcal{K}(p_{u,v})$ is the neural descriptor of mesh vertex $p_{u,v}$.

**Mesh Based Rasterization.** On the other hand, the mesh rasterizer in PyTorch3D [39] finds the faces of the mesh intersecting each pixel ray and chooses the face with the nearest point of intersection in terms of the z-coordinate in camera coordinate space, which writes:

$$\mathcal{F}_{u,v} = \{f \in \mathcal{F} : (u,v) \in \Pi(f)\}. \tag{4}$$

$$f_{u,v} = \underset{f \in \mathcal{F}_{u,v}}{\arg\min} \, \hat{f}_z, \quad \text{where} \quad \hat{f} = (\hat{f}_x, \hat{f}_y, \hat{f}_z). \tag{5}$$

$\hat{f}$ represents here the intersection between ray $(u, v)$ and face $f$ in camera coordinate frame. Finally, to find the feature corresponding to each pixel, the barycentric coordinates of the point of intersection $\hat{f}_{u,v}$ of the face $f_{u,v}$ with the corresponding pixel ray are used to interpolate the neural point descriptors over the face. By noting $(p_i, p_j, p_k)$ as the vertices making up face $f_{u,v}$, the mesh feature image $\{k_{u,v}^{\text{mesh}}\}$ can be expressed as follows:

$$k_{u,v}^{\text{mesh}} = \alpha \mathcal{K}(p_i) + \beta \mathcal{K}(p_j) + \gamma \mathcal{K}(p_k), \tag{6}$$

where $(\alpha, \beta, \gamma)$ are the barycentric coordinates of $\hat{f}_{u,v}$ in $f_{u,v}$.

While we expect the point cloud rasterization to be more accurate, the mesh rasterization provides a denser feature image, albeit less accurate and fuzzier due to the dependence on the quality of the geometric triangulation being used. The final feature image combines the best of both worlds as it consists of the mesh and point feature rasterized images concatenated together, i.e. $k_{u,v} = \left[ k_{u,v}^{\text{pt}}, k_{u,v}^{\text{mesh}} \right]$.

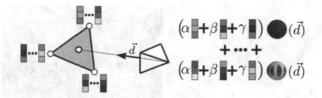

**Fig. 3.** Differently from NPBG [2], we introduce mesh-based view-dependent feature rasterization. We learn per-point Spherical Harmonic coefficients, interpolated via the barycentric coordinates of the ray-face intersection.

## 3.2 Anisotropic Features

Novel view synthesis entails learning scene radiance functions as perceived appearance can vary according to the viewing angle. While the neural point descriptors in NPBG [2] are not view direction dependent per se, the neural rendering convolutional network could in theory model such view-dependency, even without taking the camera parameters of the target view explicitly as input. That is, the spatial disposition and neighborhood of these rasterized descriptors in image domain depends on the target view. However, incorporating view dependency in the geometry descriptors by design is bound to improve this aspect within such novel view synthesis strategy. Hence, we define anisotropic neural point descriptors in this work, and we implement this idea efficiently using Spherical Harmonics (SH). Spherical Harmonics have been long used as a popular low-dimensional representation for spherical functions to model e.g. Lambertian surfaces [4,38] or even glossy surfaces [51]. They have been also recently introduced as a means of alleviating the computational complexity of volumetric rendering of implicit neural radiance fields (NeRFs) [62].

We adapt our point descriptors to auto-decode Spherical Harmonic coefficients rather than point features directly. Reformulating our earlier definition, a descriptor for a point (i.e. vertex) $p$ can be expressed now as the set of coefficients:

$$\mathcal{K}(p) = \left( k_p^{l,m} \right)_{0 \leq l \leq l_{max}}^{-l \leq m \leq l}, \tag{7}$$

where $k^{l,m} \in \mathbb{R}^8$, 8 being the desired final feature dimension. We use 3 SH bands hence $l_{max} = 2$.

Evaluating a view dependent point feature consists of linearly combining the Spherical Harmonic basis functions $\Phi_l^m : [0, 2\pi]^2 \to \mathbb{R}$ evaluated at the viewing angle corresponding to a viewing direction $\vec{d}$. The rasterized point feature at pixel $(u, v)$, as introduced in Eq. 3, can thus be finally expressed as:

$$k_{u,v}^{\mathrm{pt}} = w_{u,v} \times \sum_l \sum_m k_{p_{u,v}}^{l,m} \Phi_l^m(\vec{d}). \tag{8}$$

**Fig. 4.** COLMAP [44,45] geometry for scene "Truck" of Tanks and Temples [22]. Left: point cloud. Right: mesh.

We recall that in concordance with definitions in the previous Sect. 3.1, $p_{u,v}$ is the point-rasterized vertex at location $(u, v)$.

Similarly, the rasterized mesh feature at pixel $(u, v)$, as introduced in Eq. 6, can be expressed as:

$$k_{u,v}^{\text{mesh}} = \sum_l \sum_m (\alpha k_{p_i}^{l,m} + \beta k_{p_j}^{l,m} + \gamma k_{p_k}^{l,m}) \Phi_l^m(\vec{d}). \tag{9}$$

Here again, $(p_i, p_j, p_k)$ is the triangle rasterized at pixel $(u, v)$, and $\alpha$, $\beta$, $\gamma$ are the barycentric coordinates of the ray intersection with that triangle. Figure 3 illustrates the former equation.

We note the view direction $\vec{d}$ for a pixel $(u, v)$ can be expressed as a function of the target camera parameters as follows:

$$\vec{d} = RK^{-1} \begin{pmatrix} u \\ v \\ 1 \end{pmatrix} + T. \tag{10}$$

The rasterized view-dependent feature images ($k_{u,v} = \left[k_{u,v}^{\text{pt}}, k_{u,v}^{\text{mesh}}\right]$) are subsequently fed to a U-Net [43] based convolutional renderer to obtain the final novel rendered images. We will be detailing upon this neural renderer next.

### 3.3  Split Neural Rendering

Upon observing the geometry obtained from running COLMAP [44,45], especially on large unbounded scenes, such as the scenes in the Tanks and Temples dataset [22], the reconstruction is considerably more dense, detailed and precise for the main central area of interest where most cameras are pointing, as can be seen in Fig. 4. The remaining of the scene geometry is sparse and less accurate. Hence, we argue that feature images rasterized from these foreground and background areas lie in two relatively separate domains, as the former is richer and more accurate than the latter.

As such, we propose to split the proxy geometry into a foreground and background sub-meshes (c.f. Fig. 1). The split is automatically performed following NeRF++ [65]. Essentially, the center of the foreground sphere is approximated

**Fig. 5.** Left: foreground rasterization. Right: background rasterization. Top row: mesh based rasterization. Bottom row: point based rasterization. We visualize feature PCA coefficients.

as the average camera position $(\bar{T})$, and its radius is set as 1.1 times the distance from this center to the furthest camera: $r_{fg} = \max(||T - \bar{T}||_2)$. We separately rasterize the features from both areas (c.f. Fig. 5), and we process these foreground and background feature images with two separate yet identical encoders, focusing each on processing feature images from their respective domains. While there are 2 separate encoders, the features share a single decoder with skip connections coming in from both encoders. Other than these aspects, our neural renderer follows the multi-scale architecture in NPBG [2].

### 3.4 Hybrid Training Objective

We experiment with a mix of supervised and self-supervised losses for training our method: $L = L_{\text{rec}} + L_{\text{GAN}}$. The loss is used to perform gradient descent on the parameters $\theta$ of the neural renderer $f_\theta$ and the scene point descriptors $\mathcal{K}$ jointly.

**Supervised Loss.** The supervised loss follows the perceptual reconstruction loss in NPBG [2], where we urge the network to reproduce the available groundtruth images $(I_{gt})$ in feature space, i.e. :

$$L_{\text{rec}} = \sum_l ||\Psi_l(f_\theta(\{k_{u,v}\})) - \Psi_l(I_{gt})||_1. \tag{11}$$

$f_\theta(\{k_{u,v}\})$ is the output image from our network using rasterized features $\{k_{u,v}\}$. $\Psi_l$ represents the $l^{\text{th}}$ feature map from a pretrained VGG19 network [47], where $l \in \{\text{'conv1\_2', 'conv2\_2', 'conv3\_2', 'conv4\_2', 'conv5\_2'}\}$.

**Unsupervised Loss.** We introduce a GAN [14] loss to encourage the photo-realism of the generated images and also improve generalization outside the training camera view-points. Specifically, with our entire model so far being the generator, we use a discriminator based on the DCGAN [36] model. Besides sampling from the training cameras, we additionally sample artificial views following RegNeRF [31]. We note that for these augmented views we do not have a target ground-truth image, and hence only the GAN loss is back-propagated. To obtain the sample space of camera matrices, we assume that all cameras roughly focus on a central scene point $\bar{T}$, as described in Sect. 3.3. The camera positions $\tilde{T}$ are sampled using a spherical coordinate system as follows: $\tilde{T} = \bar{T} + \tilde{r}[\sin\theta\sin\phi, \cos\theta, \sin\theta\cos\phi]$, where $\tilde{r}$ is sampled uniformly in $[0.6r_{fg}, r_{fg}]$, $r_{fg}$ being the foreground radius as defined in Sect. 3.3. $\phi$ is sampled uniformly in $[0, 2\pi]$. $\theta$ is sampled uniformly around the mean training camera elevation within $\pm 1.5$ its standard deviation. For a given camera position, the camera rotation is defined using the camera "look-at" function, using target point $\bar{T}$ and the up axis of the scene.

## 4    Results

**Datasets.** To demonstrate the effectiveness of our approach for novel view synthesis in the context of large unbounded scenes, we choose the Tanks and Temples dataset [22]. It consists of images in Full HD resolution captured by a high-end video camera. COLMAP [44,45] is used to reconstruct the initial dense meshes/point clouds as well as to obtain the camera extrinsics and intrinsics for each scene. For quantitative evaluation, we follow the protocol in FVS [41] and SVS [42]. 17 of the 21 scenes are used for training and for the rest of the scenes, i.e., "Truck", "Train", "M60", and "Playground", the images are divided into a fine-tuning set and a test set. We also compare our method to prior approaches on the DTU [1] dataset, which consists of over 100 tabletop scenes, with the camera poses being identical for all scenes. DTU [1] scenes are captured with a regular camera layout, which includes either 49 or 64 images with a resolution of $1200 \times 1600$ and their corresponding camera poses, taken from an octant of a sphere. We use scenes 65, 106 and 118 for fine-tuning and testing purposes and the others are used for training. For scenes 65, 106 and 118, of the total number of images and their corresponding camera poses, 10 are selected as the testing set for novel view synthesis and the rest are used for fine-tuning.

**Training Procedures.** We perform two types of training. In full training, we follow FVS [41] and SVS [42] and jointly optimize the point neural descriptors of the training scenes along with the neural renderer. To test on a scene, we use the pretrained neural renderer and fine-tune it while learning the point descriptors of that scene, using the fine-tuning split. In single scene training, we perform the latter without pretraining the neural renderer.

For single scene trainings, we train our network for 100 epochs regardless of the scene. For full training, we pretrain our neural renderer for 15 epochs on all training scenes followed by fine-tuning the entire network for 100 epochs as before on specific scenes. For all our experiments, we use a batch size of 1 with

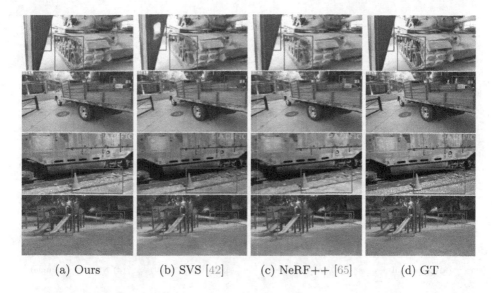

(a) Ours          (b) SVS [42]          (c) NeRF++ [65]          (d) GT

**Fig. 6.** Qualitative comparison on Tanks and Temples [22].

Adam [21] optimizer, and learning rates of $10^{-1}$ and $10^{-4}$ for the point neural descriptors and the neural renderer respectively.

**Table 1.** Quantitative comparison on Tanks and Temples [22]. Deeper shades represent better performance.

| Methods | Truck | | | M60 | | | Playground | | | Train | | |
|---|---|---|---|---|---|---|---|---|---|---|---|---|
| | PSNR ↑ | SSIM ↑ | LPIPS ↓ | PSNR ↑ | SSIM ↑ | LPIPS ↓ | PSNR ↑ | SSIM ↑ | LPIPS ↓ | PSNR ↑ | SSIM ↑ | LPIPS ↓ |
| LLFF [28] | 10.78 | 0.454 | 60.62 | 8.98 | 0.431 | 71.76 | 14.40 | 0.578 | 53.93 | 9.15 | 0.384 | 67.40 |
| EVS [8] | 14.22 | 0.527 | 43.52 | 7.41 | 0.354 | 75.71 | 14.72 | 0.568 | 46.85 | 10.54 | 0.378 | 67.62 |
| NeRF [29] | 20.85 | 0.738 | 50.74 | 16.86 | 0.701 | 60.89 | 21.55 | 0.759 | 52.19 | 16.64 | 0.627 | 64.64 |
| NeRF++ [65] | 22.77 | 0.814 | 30.04 | 18.49 | 0.747 | 43.06 | 22.93 | 0.806 | 38.70 | 17.77 | 0.681 | 47.75 |
| FVS [41] | 22.93 | 0.873 | 13.06 | 16.83 | 0.783 | 30.70 | 22.28 | 0.846 | 19.47 | 18.09 | 0.773 | 24.74 |
| SVS [42] | 23.86 | 0.895 | 9.34 | 19.97 | 0.833 | 20.45 | 23.72 | 0.884 | 14.22 | 18.69 | 0.820 | 15.73 |
| NPBG [2] | 21.88 | 0.877 | 15.04 | 12.35 | 0.716 | 35.57 | 23.03 | 0.876 | 16.65 | 18.08 | 0.801 | 25.48 |
| Ours (Single) | 23.88 | 0.883 | 17.41 | 19.34 | 0.810 | 24.13 | 23.38 | 0.865 | 23.34 | 17.35 | 0.788 | 23.66 |
| Ours (Full) | 24.03 | 0.888 | 16.84 | 19.54 | 0.815 | 23.15 | 23.59 | 0.870 | 22.72 | 17.78 | 0.799 | 24.17 |

**Metrics.** We report our performance for view synthesis, in line with previous seminal work, using three image fidelity metrics, namely Peak signal-to-noise ratio (PSNR), structural similarity (SSIM) [58] and learned perceptual image patch similarity (LPIPS) [66].

**Quantitative Comparison.** For our method, we show both single scene training (Ours (Single)) and full dataset training (Our (Full)) results. We relay the performances of methods [2,8,28,29,41,42,65] as reported in SVS [42].

Table 1 shows a quantitative comparison of our method with the recent state-of-the-art on Tanks and Temples [22]. Most methods underperform on these challenging large unbounded scenes apart from NPBG [2], FVS [41] and SVS [42].

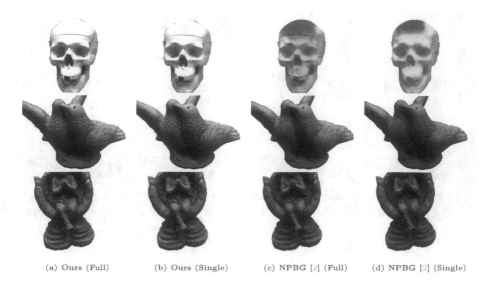

(a) Ours (Full)      (b) Ours (Single)      (c) NPBG [2] (Full)      (d) NPBG [2] (Single)

**Fig. 7.** Qualitative comparison on DTU [1].

**Table 2.** Quantitative comparison on DTU [1]. Left/Right: view interpolation/extrapolation. Deeper shades represent better performance.

| Methods | 65 | | | 106 | | | 118 | | |
|---|---|---|---|---|---|---|---|---|---|
| | PSNR ↑ | SSIM ↑ | LPIPS ↓ | PSNR ↑ | SSIM ↑ | LPIPS ↓ | PSNR ↑ | SSIM ↑ | LPIPS ↓ |
| LLFF [28] | 22.48/22.07 | 0.935/0.921 | 9.38/12.71 | 24.10/24.63 | 0.900/0.886 | 13.26/13.57 | 28.99/27.42 | 0.928/0.922 | 9.69/10.99 |
| EVS [8] | 23.26/14.43 | 0.942/0.848 | 7.94/22.11 | 20.21/11.15 | 0.902/0.743 | 14.91/29.57 | 23.35/12.06 | 0.928/0.793 | 10.84/25.01 |
| NeRF [29] | 32.00/28.12 | 0.984/0.963 | 3.04/8.54 | 34.45/30.66 | 0.975/0.957 | 7.02/10.14 | 37.36/31.66 | 0.985/0.967 | 4.18/6.92 |
| FVS [41] | 30.44/25.32 | 0.984/0.961 | 2.56/7.17 | 32.96/27.56 | 0.979/0.950 | 2.96/6.57 | 35.64/29.54 | 0.985/0.963 | 1.95/6.31 |
| SVS [42] | 32.13/26.82 | 0.986/0.964 | 1.70/5.61 | 34.30/30.64 | 0.983/0.965 | 1.93/3.69 | 37.27/31.44 | 0.988/0.967 | 1.30/4.26 |
| NPBG [2] | 16.74/15.44 | 0.889/0.873 | 14.30/19.45 | 19.62/20.26 | 0.847/0.842 | 18.90/21.13 | 23.81/24.14 | 0.867/0.879 | 15.22/16.88 |
| Ours (Single) | 26.78/20.85 | 0.957/0.925 | 9.64/12.91 | 29.98/25.40 | 0.931/0.909 | 12.75/13.70 | 31.43/26.52 | 0.946/0.931 | 11.73/11.13 |
| Ours (Full) | 28.98/22.90 | 0.970/0.943 | 7.15/11.13 | 30.67/25.75 | 0.939/0.917 | 12.10/13.10 | 32.39/27.97 | 0.956/0.941 | 10.62/10.07 |

Although they could achieve promising results with only single scene training, methods based on volumetric neural rendering (NeRF [29] and NeRF++ [65]) are famously computationally expensive to train and render. Even by training on a single scene only, our method outperforms NPBG [2] in almost all scenes. We note that the neural renderer of NPBG [2] here was pretrained on Scan-Net dataset [9]. Our method produces competitive results with respect to the best performing methods on this benchmark, i.e. FVS [41] and SVS [42]. It is interesting to observe in particular that with merely single scene training, our method outperforms the state-of-the-art SVS on scene "Truck", while coming as a close second in almost all other scenes in PSNR and SSIM [58]. FVS [41] and SVS [42] perform exceedingly well in all metrics but require training on the entire dataset, with a considerably larger training time than our single scene training. In particular, their lower LPIPS values could be attributed to the use of a considerably deeper neural renderer than ours, consisting of 9 consecutive U-Nets [43]. Ours is a much lighter single U-Net [43].

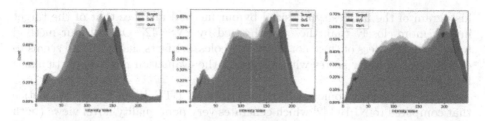

**Fig. 8.** Red, Blue and Green histograms for images in the "Truck" scene of Tanks and Temples [22] for our method, SVS [42] and the groundtruth. (Color figure online)

Table 2 reports quantitative comparison on the DTU [1] dataset. We use the view interpolation and extrapolation setting, as adopted by SVS [42] and FVS [41] for a fair comparison with other methods. The setting comprises of 6 central cameras to evaluate view interpolation and 4 corner cameras to evaluate view extrapolation. As has been observed from the experiments conducted in SVS [42], LLFF [28] and EVS [8] perform decently on DTU [1], while NeRF [29], FVS [41] and SVS [42] excel on it. NPBG [2], on the other hand, performs very poorly due to lack of data per scene (approximately 39 images) making the point feature autodecoding less efficient. Our method, despite using point feature autodecoding equally, gains considerably on NPBG [2] and performs relatively close to FVS [41]. Again, as observed for the Tanks and Temples [22] dataset, our method is able to rapidly train for a single scene and perform relatively well, and close to methods which have been trained and finetuned on the entire dataset such as FVS [41] and SVS [42], which makes it very practical in the sense that it is able to achieve reasonable results with limited training time and data.

**Qualitative Comparison.** Qualitative results on Tanks and Temples [22] of some of the competitive contemporary methods are summarized in Fig. 6. In general, we notice that SVS [42] excels in the synthesis quality of the novel views. This is also reflected in the LPIPS [66] metric, where SVS [42] performs better than the competition. NeRF++ [65] on the other hand tends to underperform and produces blurs and artifacts, particularly if we look at the results of the "M60" and "Train" scenes.

Although SVS [42] performs well overall, especially in background regions, it tends to display an unnatural "smoothing effect" on the image in certain regions, such as the rocks indicated in the "Train" scene or in the track of the tank in the "M60" scene. We suspect that this is due to their neural renderer which contains 9 consecutive U-Nets [43], which might cause the view-dependent feature tensor to be oversmoothed. Furthermore, SVS [42] results sometimes contain major artifacts like holes, missing structures or transparent parts in certain regions, such as the ones indicated in the "M60", "Truck" and "Playground" scenes respectively in Fig. 6. Another potential issue that we observed was an apparent color shift in some of the results of SVS [42]. To investigate this, we plot separate color histograms of the synthesized outputs of SVS [42], our method and the target images for the "Truck" scene in Fig. 8. We notice in this figure that the histogram of the images synthesized by our method matches that of the target images more closely than those synthesized by SVS [42]. Overall, our method tends to display less of the color shifts and holes/artifacts that SVS [42] seems to exhibit despite lower scores with respect to the evaluation metrics, particularly for LPIPS [66].

Next, we present some qualitative results on DTU [1] in Fig. 7. We notice that compared to NPBG [2] which generates very poor quality novel views (both for single-scene finetuning and full training) presumably due to lack of adequate training data, our method offers way better visual results which we believe is due to a denser input feature image fed to the neural renderer, on account of point based rasterization with bigger radius and the mesh based rasterization. In fact, our method, with just single-scene finetuning performs better than NPBG [2] even when fully trained, which is also supported by the quantitative results presented in Table 2.

**Table 3.** Quantitative ablation on Tanks and Temples [22]. Best/second best performances are emboldened/underlined respectively.

| Methods | Truck | | | M60 | | | Playground | | | Train | | |
|---|---|---|---|---|---|---|---|---|---|---|---|---|
| | PSNR ↑ | SSIM ↑ | LPIPS ↓ | PSNR ↑ | SSIM ↑ | LPIPS ↓ | PSNR ↑ | SSIM ↑ | LPIPS ↓ | PSNR ↑ | SSIM ↑ | LPIPS ↓ |
| Original NPBG | 21.55 | 0.807 | 27.58 | 17.48 | 0.757 | 33.39 | 22.26 | 0.808 | 28.79 | 16.08 | 0.697 | 34.58 |
| Bigger radius | 22.26 | 0.832 | 23.55 | 18.76 | 0.784 | 28.73 | 22.61 | 0.821 | 26.62 | 16.15 | 0.717 | 30.89 |
| With mesh | 22.70 | 0.850 | 20.76 | **19.95** | _0.811_ | 24.26 | 22.78 | 0.838 | 25.22 | 15.75 | 0.730 | 29.01 |
| Dir. features | 23.40 | 0.872 | 18.57 | 19.26 | 0.810 | _24.05_ | 23.27 | 0.857 | 24.42 | 17.13 | 0.782 | 24.79 |
| Split scene | _23.64_ | _0.880_ | _17.99_ | 19.03 | **0.812** | **24.04** | _23.32_ | _0.865_ | **23.11** | **17.37** | _0.787_ | _24.13_ |
| Ours (Single) | **23.88** | **0.883** | **17.41** | _19.34_ | 0.810 | 24.13 | **23.38** | **0.865** | _23.34_ | _17.35_ | **0.788** | **23.66** |

**Ablation Studies.** In this section, we conduct an ablative analysis to justify the choice of our final architecture. We ablate in the single-scene training scenario using all testing scenes of Tanks and Temples [22]. We progressively add components to our baseline architecture (i.e. NPBG [2]) until we reach our final model to demonstrate their individual contributions to our performance. The results are summarized in Table 3. "Original NPBG" is NPBG [2] in the single-scene setting. "Bigger radius" is the "Original NPBG" with a bigger rasterization radius as discussed in Sect. 3.1, as opposed to NPBG [2] which has a rasterization radius of half a pixel. "With mesh" includes the mesh rasterized and interpolated feature image mentioned in Sect. 3.1. These two previous components lead to denser feature images. "Directional features" incorporates view dependency in the geometry descriptors using Spherical Harmonics (SH) as discussed in Sect. 3.2, while "Split scene" splits the proxy geometry into foreground and background, rasterizes and encodes each features separately. Our final model uses an additional GAN [14] loss during training as described in Sect. 3.4. Overall throughout all scenes, the numbers witness the consistent improvement brought by the various components.

# 5  Conclusion

We improved in this work on the Neural Point Based Graphics (i.e. NPBG [2]) model for novel view synthesis, by providing a new data-efficient version that can achieve superior results by training solely on a single scene. SVS [42] still produces Superior LPIPS [66] synthesis performance. As future work, we will investigate and improve on this aspect of our method, while attempting to maintain memory and compute efficiency.

# References

1. Aanæs, H., Jensen, R.R., Vogiatzis, G., Tola, E., Dahl, A.B.: Large-scale data for multiple-view stereopsis. Int. J. Comput. Vis. **120**(2), 153–168 (2016)
2. Aliev, K.-A., Sevastopolsky, A., Kolos, M., Ulyanov, D., Lempitsky, V.: Neural point-based graphics. In: Vedaldi, A., Bischof, H., Brox, T., Frahm, J.-M. (eds.) ECCV 2020. LNCS, vol. 12367, pp. 696–712. Springer, Cham (2020). https://doi.org/10.1007/978-3-030-58542-6_42
3. Barron, J.T., Mildenhall, B., Tancik, M., Hedman, P., Martin-Brualla, R., Srinivasan, P.P.: Mip-NeRF: a multiscale representation for anti-aliasing neural radiance fields. In: Proceedings of the IEEE/CVF International Conference on Computer Vision, pp. 5855–5864 (2021)
4. Basri, R., Jacobs, D.W.: Lambertian reflectance and linear subspaces. IEEE Trans. Pattern Anal. Mach. Intell. **25**(2), 218–233 (2003)
5. Buehler, C., Bosse, M., McMillan, L., Gortler, S., Cohen, M.: Unstructured lumigraph rendering. In: Proceedings of the 28th Annual Conference on Computer Graphics and Interactive Techniques, pp. 425–432 (2001)
6. Chen, A., et al.: MVSNeRF: fast generalizable radiance field reconstruction from multi-view stereo. In: Proceedings of the IEEE/CVF International Conference on Computer Vision, pp. 14124–14133 (2021)
7. Chen, S.E., Williams, L.: View interpolation for image synthesis. In: Proceedings of the 20th Annual Conference on Computer Graphics and Interactive Techniques, pp. 279–288 (1993)
8. Choi, I., Gallo, O., Troccoli, A., Kim, M.H., Kautz, J.: Extreme view synthesis. In: Proceedings of the IEEE/CVF International Conference on Computer Vision, pp. 7781–7790 (2019)
9. Dai, A., Chang, A.X., Savva, M., Halber, M., Funkhouser, T., Nießner, M.: ScanNet: richly-annotated 3D reconstructions of indoor scenes. In: Proceedings of the IEEE Conference on Computer Vision and Pattern Recognition, pp. 5828–5839 (2017)
10. Davis, A., Levoy, M., Durand, F.: Unstructured light fields. In: Computer Graphics Forum, vol. 31, pp. 305–314. Wiley Online Library (2012)
11. Debevec, P.E., Taylor, C.J., Malik, J.: Modeling and rendering architecture from photographs: a hybrid geometry-and image-based approach. In: Proceedings of the 23rd Annual Conference on Computer Graphics and Interactive Techniques, pp. 11–20 (1996)
12. Flynn, J., Neulander, I., Philbin, J., Snavely, N.: Deepstereo: learning to predict new views from the world's imagery. In: Proceedings of the IEEE Conference on Computer Vision and Pattern Recognition, pp. 5515–5524 (2016)

13. Garbin, S.J., Kowalski, M., Johnson, M., Shotton, J., Valentin, J.: FastNeRF: high-fidelity neural rendering at 200fps. In: Proceedings of the IEEE/CVF International Conference on Computer Vision, pp. 14346–14355 (2021)
14. Goodfellow, I., et al.: Generative adversarial nets. In: Advances in Neural Information Processing Systems, vol. 27 (2014)
15. Gortler, S.J., Grzeszczuk, R., Szeliski, R., Cohen, M.F.: The lumigraph. In: Proceedings of the 23rd Annual Conference on Computer Graphics and Interactive Techniques, pp. 43–54 (1996)
16. Hedman, P., Philip, J., Price, T., Frahm, J.M., Drettakis, G., Brostow, G.: Deep blending for free-viewpoint image-based rendering. ACM Trans. Graph. (TOG) **37**(6), 1–15 (2018)
17. Hedman, P., Srinivasan, P.P., Mildenhall, B., Barron, J.T., Debevec, P.: Baking neural radiance fields for real-time view synthesis. In: Proceedings of the IEEE/CVF International Conference on Computer Vision, pp. 5875–5884 (2021)
18. Jain, A., Tancik, M., Abbeel, P.: Putting nerf on a diet: semantically consistent few-shot view synthesis. In: Proceedings of the IEEE/CVF International Conference on Computer Vision, pp. 5885–5894 (2021)
19. Kalantari, N.K., Wang, T.C., Ramamoorthi, R.: Learning-based view synthesis for light field cameras. ACM Trans. Graph. (TOG) **35**(6), 1–10 (2016)
20. Kar, A., Häne, C., Malik, J.: Learning a multi-view stereo machine. In: Advances in Neural Information Processing Systems, vol. 30 (2017)
21. Kingma, D.P., Ba, J.: Adam: a method for stochastic optimization. arXiv preprint arXiv:1412.6980 (2014)
22. Knapitsch, A., Park, J., Zhou, Q.Y., Koltun, V.: Tanks and temples: benchmarking large-scale scene reconstruction. ACM Trans. Graph. (ToG) **36**(4), 1–13 (2017)
23. Kopf, J., Cohen, M.F., Szeliski, R.: First-person hyper-lapse videos. ACM Trans. Graph. (TOG) **33**(4), 1–10 (2014)
24. Levoy, M., Hanrahan, P.: Light field rendering. In: Proceedings of the 23rd Annual Conference on Computer Graphics and Interactive Techniques, pp. 31–42 (1996)
25. Li, Q., Multon, F., Boukhayma, A.: Learning generalizable light field networks from few images. arXiv preprint arXiv:2207.11757 (2022)
26. Liu, L., Gu, J., Zaw Lin, K., Chua, T.S., Theobalt, C.: Neural sparse voxel fields. In: Advances in Neural Information Processing Systems, vol. 33, pp. 15651–15663 (2020)
27. Meshry, M., et al.: Neural rerendering in the wild. In: Proceedings of the IEEE/CVF Conference on Computer Vision and Pattern Recognition, pp. 6878–6887 (2019)
28. Mildenhall, B., et al.: Local light field fusion: practical view synthesis with prescriptive sampling guidelines. ACM Trans. Graph. (TOG) **38**(4), 1–14 (2019)
29. Mildenhall, B., Srinivasan, P.P., Tancik, M., Barron, J.T., Ramamoorthi, R., Ng, R.: NeRF: representing scenes as neural radiance fields for view synthesis. In: Vedaldi, A., Bischof, H., Brox, T., Frahm, J.-M. (eds.) ECCV 2020. LNCS, vol. 12346, pp. 405–421. Springer, Cham (2020). https://doi.org/10.1007/978-3-030-58452-8_24
30. Niemeyer, M., Barron, J.T., Mildenhall, B., Sajjadi, M.S., Geiger, A., Radwan, N.: RegNeRF: regularizing neural radiance fields for view synthesis from sparse inputs. arXiv preprint arXiv:2112.00724 (2021)
31. Niemeyer, M., Barron, J.T., Mildenhall, B., Sajjadi, M.S., Geiger, A., Radwan, N.: RegNeRF: regularizing neural radiance fields for view synthesis from sparse inputs. In: Proceedings of the IEEE/CVF Conference on Computer Vision and Pattern Recognition, pp. 5480–5490 (2022)

32. Niemeyer, M., Mescheder, L., Oechsle, M., Geiger, A.: Differentiable volumetric rendering: learning implicit 3D representations without 3D supervision. In: Proceedings of the IEEE/CVF Conference on Computer Vision and Pattern Recognition, pp. 3504–3515 (2020)

33. Ouasfi, A., Boukhayma, A.: Few'zero level set'-shot learning of shape signed distance functions in feature space. arXiv preprint arXiv:2207.04161 (2022)

34. Pittaluga, F., Koppal, S.J., Kang, S.B., Sinha, S.N.: Revealing scenes by inverting structure from motion reconstructions. In: Proceedings of the IEEE/CVF Conference on Computer Vision and Pattern Recognition, pp. 145–154 (2019)

35. Prokudin, S., Black, M.J., Romero, J.: SMPLpix: neural avatars from 3D human models. In: Proceedings of the IEEE/CVF Winter Conference on Applications of Computer Vision, pp. 1810–1819 (2021)

36. Radford, A., Metz, L., Chintala, S.: Unsupervised representation learning with deep convolutional generative adversarial networks. arXiv preprint arXiv:1511.06434 (2015)

37. Raj, A., Tanke, J., Hays, J., Vo, M., Stoll, C., Lassner, C.: ANR: articulated neural rendering for virtual avatars. In: Proceedings of the IEEE/CVF Conference on Computer Vision and Pattern Recognition (CVPR), pp. 3722–3731, June 2021

38. Ramamoorthi, R., Hanrahan, P.: On the relationship between radiance and irradiance: determining the illumination from images of a convex lambertian object. JOSA A **18**(10), 2448–2459 (2001)

39. Ravi, N., et al.: Accelerating 3D deep learning with PyTorch3D. arXiv preprint arXiv:2007.08501 (2020)

40. Reiser, C., Peng, S., Liao, Y., Geiger, A.: KiloNeRF: speeding up neural radiance fields with thousands of tiny MLPs. In: Proceedings of the IEEE/CVF International Conference on Computer Vision, pp. 14335–14345 (2021)

41. Riegler, G., Koltun, V.: Free view synthesis. In: Vedaldi, A., Bischof, H., Brox, T., Frahm, J.-M. (eds.) ECCV 2020. LNCS, vol. 12364, pp. 623–640. Springer, Cham (2020). https://doi.org/10.1007/978-3-030-58529-7_37

42. Riegler, G., Koltun, V.: Stable view synthesis. In: Proceedings of the IEEE/CVF Conference on Computer Vision and Pattern Recognition, pp. 12216–12225 (2021)

43. Ronneberger, O., Fischer, P., Brox, T.: U-Net: convolutional networks for biomedical image segmentation. In: Navab, N., Hornegger, J., Wells, W.M., Frangi, A.F. (eds.) MICCAI 2015. LNCS, vol. 9351, pp. 234–241. Springer, Cham (2015). https://doi.org/10.1007/978-3-319-24574-4_28

44. Schonberger, J.L., Frahm, J.M.: Structure-from-motion revisited. In: Proceedings of the IEEE Conference on Computer Vision and Pattern Recognition, pp. 4104–4113 (2016)

45. Schönberger, J.L., Zheng, E., Frahm, J.-M., Pollefeys, M.: Pixelwise view selection for unstructured multi-view stereo. In: Leibe, B., Matas, J., Sebe, N., Welling, M. (eds.) ECCV 2016. LNCS, vol. 9907, pp. 501–518. Springer, Cham (2016). https://doi.org/10.1007/978-3-319-46487-9_31

46. Seitz, S.M., Dyer, C.R.: View morphing. In: Proceedings of the 23rd Annual Conference on Computer Graphics and Interactive Techniques, pp. 21–30 (1996)

47. Simonyan, K., Zisserman, A.: Very deep convolutional networks for large-scale image recognition. arXiv preprint arXiv:1409.1556 (2014)

48. Sitzmann, V., Chan, E.R., Tucker, R., Snavely, N., Wetzstein, G.: MetaSDF: metalearning signed distance functions. In: NeurIPS (2020)

49. Sitzmann, V., Rezchikov, S., Freeman, B., Tenenbaum, J., Durand, F.: Light field networks: neural scene representations with single-evaluation rendering. In: Advances in Neural Information Processing Systems, vol. 34, pp. 19313–19325 (2021)
50. Sitzmann, V., Zollhöfer, M., Wetzstein, G.: Scene representation networks: Continuous 3D-structure-aware neural scene representations. In: Advances in Neural Information Processing Systems, vol. 32 (2019)
51. Sloan, P.P., Kautz, J., Snyder, J.: Precomputed radiance transfer for real-time rendering in dynamic, low-frequency lighting environments. In: Proceedings of the 29th Annual Conference on Computer Graphics and Interactive Techniques, pp. 527–536 (2002)
52. Song, Z., Chen, W., Campbell, D., Li, H.: Deep novel view synthesis from colored 3D point clouds. In: Vedaldi, A., Bischof, H., Brox, T., Frahm, J.-M. (eds.) ECCV 2020. LNCS, vol. 12369, pp. 1–17. Springer, Cham (2020). https://doi.org/10.1007/978-3-030-58586-0_1
53. Thies, J., Zollhöfer, M., Nießner, M.: Deferred neural rendering: image synthesis using neural textures. ACM Trans. Graph. (TOG) $38(4)$, 1–12 (2019)
54. Thies, J., Zollhöfer, M., Theobalt, C., Stamminger, M., Nießner, M.: IGNOR: image-guided neural object rendering. arXiv preprint arXiv:1811.10720 (2018)
55. Trevithick, A., Yang, B.: GRF: learning a general radiance field for 3D representation and rendering. In: Proceedings of the IEEE/CVF International Conference on Computer Vision, pp. 15182–15192 (2021)
56. Wang, P., Liu, L., Liu, Y., Theobalt, C., Komura, T., Wang, W.: NeuS: learning neural implicit surfaces by volume rendering for multi-view reconstruction. arXiv preprint arXiv:2106.10689 (2021)
57. Wang, Q., et al.: IBRNet: learning multi-view image-based rendering. In: Proceedings of the IEEE/CVF Conference on Computer Vision and Pattern Recognition, pp. 4690–4699 (2021)
58. Wang, Z., Bovik, A.C., Sheikh, H.R., Simoncelli, E.P.: Image quality assessment: from error visibility to structural similarity. IEEE Trans. Image Process. $13(4)$, 600–612 (2004)
59. Wizadwongsa, S., Phongthawee, P., Yenphraphai, J., Suwajanakorn, S.: NeX: real-time view synthesis with neural basis expansion. In: Proceedings of the IEEE/CVF Conference on Computer Vision and Pattern Recognition, pp. 8534–8543 (2021)
60. Wu, M., Wang, Y., Hu, Q., Yu, J.: Multi-view neural human rendering. In: Proceedings of the IEEE/CVF Conference on Computer Vision and Pattern Recognition, pp. 1682–1691 (2020)
61. Yariv, L., Gu, J., Kasten, Y., Lipman, Y.: Volume rendering of neural implicit surfaces. In: Advances in Neural Information Processing Systems, vol. 34, pp. 4805–4815 (2021)
62. Yu, A., Li, R., Tancik, M., Li, H., Ng, R., Kanazawa, A.: Plenoctrees for real-time rendering of neural radiance fields. In: Proceedings of the IEEE/CVF International Conference on Computer Vision, pp. 5752–5761 (2021)
63. Yu, A., Ye, V., Tancik, M., Kanazawa, A.: pixelNeRF: neural radiance fields from one or few images. In: Proceedings of the IEEE/CVF Conference on Computer Vision and Pattern Recognition, pp. 4578–4587 (2021)
64. Zakharkin, I., Mazur, K., Grigorev, A., Lempitsky, V.: Point-based modeling of human clothing. In: Proceedings of the IEEE/CVF International Conference on Computer Vision, pp. 14718–14727 (2021)
65. Zhang, K., Riegler, G., Snavely, N., Koltun, V.: NeRF++: analyzing and improving neural radiance fields. arXiv preprint arXiv:2010.07492 (2020)

66. Zhang, R., Isola, P., Efros, A.A., Shechtman, E., Wang, O.: The unreasonable effectiveness of deep features as a perceptual metric. In: Proceedings of the IEEE Conference on Computer Vision and Pattern Recognition, pp. 586–595 (2018)
67. Zhou, T., Brown, M., Snavely, N., Lowe, D.G.: Unsupervised learning of depth and ego-motion from video. In: Proceedings of the IEEE Conference on Computer Vision and Pattern Recognition, pp. 1851–1858 (2017)
68. Zhou, T., Tucker, R., Flynn, J., Fyffe, G., Snavely, N.: Stereo magnification: learning view synthesis using multiplane images. arXiv preprint arXiv:1805.09817 (2018)
69. Zitnick, C.L., Kang, S.B., Uyttendaele, M., Winder, S., Szeliski, R.: High-quality video view interpolation using a layered representation. ACM Trans. Graph. (TOG) **23**(3), 600–608 (2004)

# One-Shot Learning for Human Affordance Detection

Abel Pacheco-Ortega[1]([✉])(iD) and Walterio Mayol-Cuervas[1,2](iD)

[1] University of Bristol, Bristol, UK
{abel.pachecoortega,walterio.mayol-cuevas}@bristol.ac.uk
[2] Amazon, Seattle, WA 98109, USA

**Abstract.** The diversity of action possibilities offered by an environment, a.k.a affordances, cannot be addressed in a scalable manner simply from object categories or semantics, which are limitless. To this end, we present a one-shot learning approach that trains on one or a handful of human-scene interaction samples. Then, given a previously unseen scene, we can predict human affordances and generate the associated articulated 3D bodies. Our experiments show that our approach generates physically plausible interactions that are perceived as more natural in 60–70% of the comparisons with other methods.

**Keywords:** Scene understanding · Affordances detection · Human interactions · Visual perception · Affordances

## 1  Introduction

Coined by James J. Gibson in [3], affordances refer to the action possibilities offered by the environment to an agent. He claimed that living beings perceive their environment in terms of such affordances.

An artificial agent with object, semantics and human affordances detection capabilities would be able to identify elements, their relations and the locations in the environment that support the execution of actions like stand-able, walk-able, place-able, and sit-able. This enhanced scene understanding is helpful in the Metaverse, where virtual agents should execute

**Fig. 1.** Trained in a one-shot manner, our approach detects human affordances and hallucinates the associated human bodies interacting with the environment in a natural and physically plausible way

actions or where scenes must be populated by humans performing a given set of interactions.

We present a direct representation of human affordances that extracts a meaningful geometrical description through analysing proximity zones and clearance space between interacting entities in human-environment configurations.

L. Karlinsky et al. (Eds.): ECCV 2022 Workshops, LNCS 13803, pp. 758–766, 2023.
https://doi.org/10.1007/978-3-031-25066-8_46

Our approach can determine locations in the environment that support them and generate natural and physically plausible 3D representations (see Fig. 1). We compare our method with state-of-the-art intensively trained methods.

## 2    Related Work

Popular interpretations of the concept of affordances refer to them as *action possibilities* or *opportunities of interaction* for an agent/animal that are perceived directly from the shape and form of the environment/object.

The affordances detection from RGB images was explored by Gupta et al. [4] with a voxelised geometric estimator. Lately, data-intensive approaches were used by Fouhey et al. [2] with a detector trained with labels on RGB frames from the NYUv2 dataset [13] and by Luddecke et al. [9] with a residual neural network trained with a lookup table between affordances and objects parts on the ADE20K dataset [20].

Other approaches go further by synthesising the detected human-environment interaction. The representation of such interactions has been showcased with human skeletons in [7,8,15]; nevertheless, their representativeness cannot be reliably evaluated because contacts, collisions, and the naturalness of human poses are not entirely characterised.

Closer to us, efforts with a more complex interaction representation over 3D scenes have been explored. Ruiz and Mayol [12] developed a geometric interaction descriptor for non-articulated, rigid object shapes with good generalisation in detecting physically feasible interaction configurations. Using the SMPL-X human body model [10], Zhang et al. [18] developed a context-aware human body generator that learnt the distribution of 3D human poses conditioned on the depth and semantics of the scene from recordings in the PROX dataset [5]. In a follow-up effort, Zhang et al. [17] developed a purely geometric approach to model human-scene interactions by explicitly encoding the proximity between the body and the environment, thus only requiring a mesh as input. Lately, Hassan et al. in [6] learnt the distribution of contact zones in human body poses and used them to find environment locations that better support them.

Our main difference from [5,6,17,18] is that ours is not a data-driven approach; ours does not require the use of most, if not all, of a labelled dataset, e.g. around 100K image frames in PROX [5]. Just one if not a few examples of interactions are necessary to train our detector, as in [12], but we extend the descriptor to consider the clearance space of the interactions and their uses and optimise with the SMPL-X human model after positive detection.

## 3    Method

### 3.1    A Spatial Descriptor for Spatial Interactions

Inspired by recently developed methods that have revisited geometric features such as the bisector surface for scene-object indexing [19] and affordance detection [12], our affordance descriptor (see Fig. 2) expands on the Interaction Bisector Surface (IBS) [19], an approximation of the well-known Bisector Surface (BS)

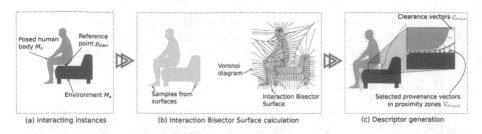

(a) Interacting instances    (b) Interaction Bisector Surface calculation    (c) Descriptor generation

**Fig. 2.** Illustrative 2D representation of our training pipeline. (a) Given a posed human body $M_h$ interacting with an environment $M_e$ on a reference point $p_{train}$, (b) we extract the Interaction Bisector Surface from the Voronoi diagram of sample points on $M_h$ and $M_e$, then (c) we use the IBS to characterise the proximity zones and the surrounding space with provenance and clearance vectors

[11]. Given two surfaces $S_1, S_2 \in \mathbb{R}^3$, the BS is the set of sphere centres that touch both surfaces at one point each.

Our one-shot training process requires 3-tuples $(M_h, M_e, p_{train})$, where $M_h$ is a posed human body mesh, $M_e$ is an environment mesh and $p_{train}$ is a reference point on $M_e$ where the interaction is supported.

Let $P_h$ and $P_e$ denote the sampling points on $M_h$ and $M_e$, respectively; their IBS $\mathcal{I}$ is defined as:

$$\mathcal{I} = \Big\{ p \mid \min_{p'_h \in P_h} \|p - p'\| = \min_{p'_e \in P_e} \|p - p'\| \Big\} \tag{1}$$

We operate the Voronoi diagram $\mathcal{D}$ generated with $P_h$ and $P_e$ to produce $\mathcal{I}$. By construction, every ridge in $\mathcal{D}$ is equidistant to the couple of points that define it. Then, $\mathcal{I}$ is composed of ridges in $\mathcal{D}$ generated because of points from both $P_h$ and $P_e$. An IBS can reach infinity, but we limit $\mathcal{I}$ by clipping it with the bounding sphere of $M_h$ augmented $ibs_{rf}$ times in its radius. A low sampling rate degenerates on an IBS that pierces the boundaries of $M_h$ or $M_e$. A higher density of samples is critical in those zones where the proximity between the interacting meshes is small. We use three stages to populate $P_h$ and $P_e$: 1) We generate Poisson disk sample sets [16] of $ibs_{ini}$ points on each $M_e$ and $M_h$. 2) *Counterpart sampling* strategy. We append to $P_e$ the closest points on $M_e$ to elements in $P_h$, and equally, we integrate into $P_h$ the closest point on $M_h$ to samples in $P_e$. We executed the *counterpart sampling* strategy $ibs_{cs}$ times. 3) *Collision point sampling* strategy. We calculate a preliminary IBS and test it for collisions with $M_h$ and $M_e$; if they exist, we add as samples the points where collisions occur as well as their counterpart points. We perform the *collision point sampling* strategy until we get an IBS that does not pierce $M_h$ nor $M_e$.

To capture the regions of interaction proximity on our enhanced IBS, as mentioned above, we use the notion of provenance vectors [12]. The *provenance vectors* of an interaction start from any point on $\mathcal{I}$ and finish at the nearest point on $M_e$. Formally:

$$V_p = \{(a, \vec{v}) \mid a \in \mathcal{I}, \ \vec{v} = \arg\min_{e \in M_e} \|e - a\| - a\} \tag{2}$$

where $a$ is the starting point of the delta vector $\vec{v}$ to the nearest point on $M_e$. *Provenance vectors* inform about the direction and distance of the interaction; the smaller the vector, the more noteworthy is for the description of the interaction. Let $V_p' \subset V_p$ the subset of *provenance vectors* that ends at any point in $P_e$; we perform a weighted randomised sampling selection of elements from $V_p'$ with the weight allocation as follows:

$$w_i = 1 - \frac{|\vec{v}_i| - |\vec{v}_{min}|}{|\vec{v}_{max}| - |\vec{v}_{min}|}, \ i = 1, \ 2, \ \ldots, \ |P_e| \tag{3}$$

where $|\vec{v}_{max}|$ and $|\vec{v}_{min}|$ are the norms of the biggest and smallest vectors in $V_p'$ respectively. The selected *provenance vectors* $\mathcal{V}_{train}$ integrate into our affordance descriptor with an adjustment to normalise their positions with the defined reference point $p_{train}$:

$$\mathcal{V}_{train} = \{(a_i', \vec{v}_i) \mid a_i' = a_i - p_{train}, \ i = 1, \ 2, \ \ldots, \ num_{pv}\} \tag{4}$$

where $num_{pv}$ is the number of samples from $V_p'$ to integrate into our descriptor.

However, the *provenance vectors* are, on their own, insufficient to capture the whole nature of the interaction on highly articulated objects such as the human body. We expand this concept by taking a more comprehensive description that includes a set of vectors to define the surrounded space necessary for the interaction. Given $S_H$ an evenly sampled set of $num_{cv}$ points on $M_h$, the *clearance vectors* that integrate to our descriptor $\mathcal{C}_{train}$ are defined as follows:

$$\mathcal{C}_{train} = \{(s_j', \vec{c}_j) \mid s_j' = s_j - p_{train}, \ s_j \in S_H, \ \vec{c}_j = \psi(s_j, \hat{n}_j, \mathcal{I})\} \tag{5}$$

$$\psi(s_j', \hat{n}_j, \mathcal{I}) = \begin{cases} d_{max} \cdot \hat{n}_j & \text{if } \varphi(s_j, \hat{n}_j, \mathcal{I}) > d_{max} \\ \varphi(s_j, \hat{n}_j, \mathcal{I}) \cdot \hat{n}_j & \text{otherwise} \end{cases} \tag{6}$$

where $p_{train}$ is the defined reference point, $\hat{n}_i$ is the unit surface normal vector on sample $s_j$, $d_{max}$ is the maximum norm of any $\vec{c}_j$, and $\varphi(s_j, \hat{n}_j, \mathcal{I})$ is the distance travelled by a ray with origin $s_j$ and direction $\hat{n}_i$ until collision with $\mathcal{I}$.

Formally, our affordances descriptor is defined as:

$$f : (M_h, M_e, p_{train}) \longrightarrow (\mathcal{V}_{train}, \mathcal{C}_{train}, \hat{n}_{train}) \tag{7}$$

where $\hat{n}_{train}$ is the unit surface normal vector of $M_e$ at $p_{train}$.

## 3.2  Human Affordances Detection

Let $\mathcal{A} = (\mathcal{V}_{train}, \mathcal{C}_{train}, \hat{n}_{train})$ an affordance descriptor, we define its rigid transformations as:

$$\Omega(\mathcal{A}, \phi, \tau) = (\mathcal{V}_{\phi\tau}^A, \ \mathcal{C}_{\phi\tau}^A, \ \hat{n}_{train}) \tag{8}$$

$$\mathcal{V}_{\phi\tau}^A = \{(a_i'', \vec{v}_i) \mid a_i'' = R_\phi \cdot a_i' + \tau, \ (a_i', \vec{v}_i) \in \mathcal{V}_{train}\}$$

$$\mathcal{C}_{\phi\tau}^A = \{(s_i'', \vec{c}_i) \mid b_i'' = R_\phi \cdot s_i' + \tau, \ (s_i', \vec{c}_i) \in \mathcal{C}_{train}\}$$

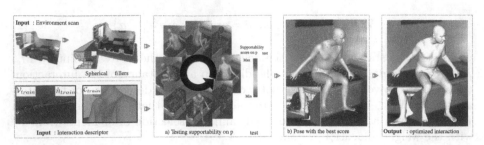

**Fig. 3.** We determine supportability of interaction on a given point by (a) measuring compatibility of surface normal, as well as provenance and clearance vector over different rotated configurations. (b) After a positive detection, the body pose is optimised to generate a natural and physically plausible interaction

where $\tau \in \mathbb{R}^3$ is the translation vector, $\phi$ is the rotation around $z$ defined by the rotation matrix $R_\phi$.

We determine that a test location $p_{test}$ on an environment $M_{test}$ with a unit surface normal vector $\hat{n}_{test}$ supports a trained interaction $\mathcal{A}$ if the angle difference between $\hat{n}_{train}$ and $\hat{n}_{test}$ is less than a threshold $\rho_{\vec{n}}$, and its translated descriptor at $p_{test}$ has a good alignment of provenance vectors and a gated number of clearance vector that collide with $M_{test}$ in any of the $n_\phi$ different $\phi$ values used during the test.

After corroborating the match between train and test normal vectors, we transform the interaction descriptor $\mathcal{A}$ with $\tau = p_{test}$ and $n_\phi$ different $\phi$ values within $[0, 2\pi]$. For each calculated 3-tuple $(\mathcal{V}_{\phi\tau}^A, \mathcal{C}_{\phi\tau}^A, \hat{n}_{train})$, we generate a set of rays $R_{pv}$ defined as follows:

$$R_{pv} = \left\{ (a_i'', \hat{\nu}_i) \mid \hat{\nu}_i = \frac{\vec{vi}}{\|\vec{vi}\|}, \ (a_i'', \vec{v}_i) \in \mathcal{V}_{\phi\tau}^A \right\} \tag{9}$$

where $a_i''$ is the starting point, and $\hat{\nu}_i \in \mathbb{R}^3$ is the direction of each ray. Then we extend each ray in $R_{pv}$ by $\epsilon_i^{pv}$ until collision with $M_{test}$ as

$$(a'' + \epsilon_i^{pv} \cdot \hat{\nu}_i) \in M_{test}, \quad i = 1, 2, \ldots, num_{pv} \tag{10}$$

and compare with the magnitude of each correspondent provenance vector in $\mathcal{V}_{\phi\tau}^A$. When any element in $R_{pv}$ extends beyond a predetermined limit $max_{long}$, the collision with the environment is classified as non-colliding. We calculate the alignment score $\kappa$ as a sum of the differences between the extended rays and the trained *provenance vectors* with

$$\kappa = \sum_{\forall i | \epsilon_i^{pv} \leq max_{long}} |\epsilon_i^{pv} - \vec{v}_i| \tag{11}$$

The higher the $\kappa$ value, the less supportability of the interaction on $p_{test}$. We experimentally determine interaction-wise thresholds for the sum of differences

**Fig. 4.** Action planning as a further step. Left: 3 affordances evaluated in an environment. Right: scores are used to plan concatenated action milestones

$max_\kappa$ and the number of missing ray collisions $max_{missings}$ that allow us to score the affordance capabilities on $p_{test}$.

*Clearance vectors* are meant to fast detect collision configurations by the calculation of ray-mesh intersections. Similarly to *provenance vectors*, we generate a set of rays $R_{cv}$ with origins and directions determined by $\mathcal{C}^A_{\phi\tau}$. We extend the rays in $R_{cv}$ until collision with the environment and calculate its extension $\epsilon^{cv}_j$. Extended rays with $\epsilon^{cv}_j \leq \|\vec{c}_j\|$ are considered as possible collisions. In practise, we also track an interaction-wise threshold to refuse supportability due to collisions $max_{collisions}$. A sparse distribution of clearance vectors on noisy meshes results in collisions not detected by *clearance vectors*. To improve, we enhance scenes with a set of *spherical fillers* that pad the scene (see Fig. 3).

Every human-environment interaction trained from the PROX dataset [5] has an associated SMPL-X characterisation that we use to optimise the human pose with previously determined body contact regions, the AdvOptim loss function presented in [17] and the SDF values of the scene.

## 4   Experiments

We evaluate the physical plausibility and the perception of the naturalness of the human-environment interactions generated. Our baselines are the approaches presented in PLACE [17] and POSA [6].

PROX [5] is a dataset with 12 scanned indoor environments and 20 recordings with data of subjects interacting within them. We divide PROX into train and test sets following the setup in [17]. To generate our descriptors, we get data from 23 manually selected frames with subjects sitting, standing, reaching, lying, and walking. We also test on 7 rooms from MP3D [1] and 5 rooms of Replica [14].

We generate the IBS surface $\mathcal{I}$ with an initial sampling set of $ibs_{ini} = 400$ points on each surface, with the *counterpart sampling* strategy executed $ibs_{cs} = 4$ times and a cropping factor of $ibs_{rf} = 1.2$. Our descriptors are made up of $num_{pv} = 512$ *provenance vectors* and $num_{cv} = 256$ *clearance vectors* extended up to $d_{max} = 5[cm]$. In testing, we use a normals angle difference threshold of $\rho_{\vec{n}} = \pi/3$, check for supportability on $n_\phi = 8$ different directions and extend *provenance vectors* up to $max_{long} = 1.2$ times the sphere radius used for cropping $\mathcal{I}$ during training.

**Physical Plausibility Test.** We use the non-collision and contact scores as in [17], but include an additional cost metric that indicates the collision depth

**Table 1.** Physical plausibility. Non collision, contact and collision depth scores ($\uparrow$: benefit, $\downarrow$: cost) before and after optimization. Best results boldface

| Model | Optimizer | Non collision$^\uparrow$ | | | Contact$^\uparrow$ | | | Collision depth$^\downarrow$ | | |
|---|---|---|---|---|---|---|---|---|---|---|
| | | PROX | MP3D | Replica | PROX | MP3D | Replica | PROX | MP3D | Replica |
| PLACE | w/o | 0.9207 | 0.9625 | 0.9554 | 0.9125 | 0.5116 | 0.8115 | 1.6285 | 0.8958 | 1.2031 |
| PLACE | SimOptim | 0.9253 | 0.9628 | 0.9562 | 0.9263 | 0.5910 | 0.8571 | 1.8169 | 1.0960 | 1.5485 |
| PLACE | AdvOptim | 0.9665 | 0.9798 | 0.9659 | 0.9725 | 0.5810 | 0.9931 | 1.6327 | 1.1346 | 1.6145 |
| POSA (contact) | w/o | **0.9820** | 0.9792 | 0.9814 | 0.9396 | 0.9526 | 0.9888 | 1.1252 | 1.5416 | 2.0620 |
| POSA (contact) | optimized | 0.9753 | 0.9725 | 0.9765 | **0.9927** | **0.9988** | **0.9963** | 1.5343 | 2.0063 | 2.4518 |
| Ours | w/o | 0.9615 | **0.9853** | **0.9931** | 0.5654 | 0.3287 | 0.4860 | **0.1648** | **0.1326** | **0.2096** |
| Ours | AdvOptim | 0.9816 | **0.9853** | 0.9883 | 0.9363 | 0.6213 | 0.8682 | 0.6330 | 0.8716 | 0.8615 |

between the generated body and the scene. We generate 1300 interacting bodies per model in each scene and report the averages of the scores in Table 1. In all datasets, bodies generated with our optimised model present high non-collision as well as low contact and collision-depth scores.

**Perception of Naturalness Test.** Every scene in our datasets is used equally in the random selection of 162 test locations. We use the optimised version of the models to generate human-environment interactions at test locations and evaluate their perceived naturalness on Amazon Mechanical Turk. Each MTurk performs 11 randomly selected assessments, including two control questions, by observing interactions with dynamic views. Three different MTurks evaluate every item. In a side-by-side evaluation, we simultaneously present outputs from two different models. Answers to "Which example is more natural?" show that our human-environment configurations are preferred on 60.7% and 72.6% of the comparisons with PLACE and POSA, respectively. In an individual evaluation, where every interaction generated is assessed with the question "The human is interacting very naturally with the scene. What is your opinion?" using a 5-point Likert scale (from 1 for "strongly disagree" to 5 for "strongly agree"), the mean and standard deviations of the evaluations are: PLACE $3.23 \pm 1.35$, POSA $2.79 \pm 1.18$, and ours $3.39 \pm 1.25$.

## 5    Conclusion

Our approach generalises well to detect interactions and generate natural and physically plausible body-scene configurations. Understanding a scene in terms of action possibilities is a desirable capability for autonomous agents performing in the Metaverse (see Fig. 4).

**Acknowledgments.** Abel Pacheco-Ortega thanks the Mexican Council for Science and Technology (CONACYT) for the scholarship provided for his studies with the scholarship number 709908. Walterio Mayol-Cuevas thanks the visual egocentric research activity partially funded by UK EPSRC EP/N013964/1.

# References

1. Chang, A., et al.: Matterport3D: learning from RGB-D data in indoor environments. In: International Conference on 3D Vision (3DV) (2017)
2. Fouhey, D.F., Wang, X., Gupta, A.: In defense of the direct perception of affordances. arXiv preprint arXiv:1505.01085 (2015). https://doi.org/10.1002/eji.201445290
3. Gibson, J.J.: The theory of affordances. In: Perceiving, Acting and Knowing. Toward and Ecological Psychology. Lawrence Eribaum Associates (1977)
4. Gupta, A., Satkin, S., Efros, A.A., Hebert, M.: From 3D scene geometry to human workspace. In: 2011 IEEE Conference on Computer Vision and Pattern Recognition (CVPR), pp. 1961–1968, June 2011. IEEE. https://doi.org/10.1109/CVPR.2011.5995448. http://ieeexplore.ieee.org/document/5995448/
5. Hassan, M., Choutas, V., Tzionas, D., Black, M.J.: Resolving 3D human pose ambiguities with 3D scene constraints. In: Proceedings of the IEEE/CVF International Conference on Computer Vision, pp. 2282–2292 (2019)
6. Hassan, M., Ghosh, P., Tesch, J., Tzionas, D., Black, M.J.: Populating 3D scenes by learning human-scene interaction. In: Proceedings of the IEEE/CVF Conference on Computer Vision and Pattern Recognition, pp. 14708–14718 (2021)
7. Jiang, Y., Koppula, H.S., Saxena, A.: Modeling 3D environments through hidden human context. IEEE Trans. Pattern Anal. Mach. Intell. **38**, 2040–2053 (2016). https://doi.org/10.1109/TPAMI.2015.2501811
8. Li, X., Liu, S., Kim, K., Wang, X., Yang, M.H., Kautz, J.: Putting humans in a scene: learning affordance in 3D indoor environments. In: Proceedings of the IEEE Conference on Computer Vision and Pattern Recognition, pp. 12368–12376 (2019)
9. Luddecke, T., Worgotter, F.: Learning to segment affordances. In: The IEEE International Conference on Computer Vision (ICCV) Workshops, pp. 769–776. IEEE, October 2017. https://doi.org/10.1109/ICCVW.2017.96. http://ieeexplore.ieee.org/document/8265305/
10. Pavlakos, G., et al.: Expressive body capture: 3D hands, face, and body from a single image. In: Proceedings of the IEEE/CVF Conference on Computer Vision and Pattern Recognition, pp. 10975–10985 (2019)
11. Peternell, M.: Geometric properties of bisector surfaces. Graph. Models **62**(3), 202–236 (2000). https://doi.org/10.1006/gmod.1999.0521
12. Ruiz, E., Mayol-Cuevas, W.: Geometric affordance perception: leveraging deep 3D saliency with the interaction tensor. Front. Neurorobot. **14**, 45 (2020)
13. Silberman, N., Hoiem, D., Kohli, P., Fergus, R.: Indoor segmentation and support inference from RGBD images. In: Fitzgibbon, A., Lazebnik, S., Perona, P., Sato, Y., Schmid, C. (eds.) ECCV 2012. LNCS, vol. 7576, pp. 746–760. Springer, Heidelberg (2012). https://doi.org/10.1007/978-3-642-33715-4_54
14. Straub, J., et al.: The replica dataset: a digital replica of indoor spaces. arXiv preprint arXiv:1906.05797 (2019)
15. Wang, X., Girdhar, R., Gupta, A.: Binge watching: scaling affordance learning from sitcoms. In: Proceedings of the IEEE Conference on Computer Vision and Pattern Recognition, pp. 2596–2605 (2017)
16. Yuksel, C.: Sample elimination for generating poisson disk sample sets. In: Computer Graphics Forum, vol. 34, pp. 25–32 (2015). https://doi.org/10.1111/cgf.12538
17. Zhang, S., Zhang, Y., Ma, Q., Black, M.J., Tang, S.: PLACE: proximity learning of articulation and contact in 3D environments. In: 8th International Conference on 3D Vision (3DV 2020) (virtual) (2020)

18. Zhang, Y., Hassan, M., Neumann, H., Black, M.J., Tang, S.: Generating 3D people in scenes without people. In: The IEEE/CVF Conference on Computer Vision and Pattern Recognition (CVPR) (2020). https://github.com/yz-cnsdqz/PSI-release/
19. Zhao, X., Wang, H., Komura, T.: Indexing 3D scenes using the interaction bisector surface. ACM Trans. Graph. **33**(3), 1–14 (2014). https://doi.org/10.1145/2574860. http://dl.acm.org/citation.cfm?doid=2631978.2574860
20. Zhou, B., Zhao, H., Puig, X., Fidler, S., Barriuso, A., Torralba, A.: Scene parsing through ADE20K dataset. In: Proceedings of the IEEE Conference on Computer Vision and Pattern Recognition (2017)

# Author Index

Printed in the United States
by Baker & Taylor Publisher Services